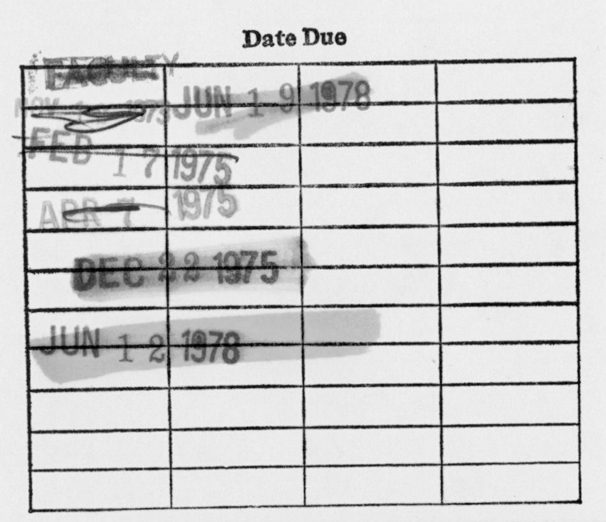
Date Due
JUN 1 9 1978
FEB 1 7 1975
APR 7 1975
DEC 2 2 1975
JUN 1 2 1978

NEW MEXICO
EASTERN
UNIVERSITY
E
N
M
U
LIBRARY

RESPONSES OF PLANTS TO ENVIRONMENTAL STRESSES

PHYSIOLOGICAL ECOLOGY

A Series of Monographs, Texts, and Treatises

EDITED BY

T. T. KOZLOWSKI

University of Wisconsin
Madison, Wisconsin

T. T. KOZLOWSKI. Growth and Development of Trees, Volume I – 1971; Volume II – 1971

DANIEL HILLEL. Soil and Water: Physical Principles and Processes, 1971

J. LEVITT. Responses of Plants to Environmental Stresses, 1972

In Preparation

T. T. KOZLOWSKI. Seed Biology (in three volumes)

YOAV WAISEL. The Biology of Halophytes

V. B. YOUNGNER AND C. M. MCKELL. The Biology and Utilization of Grasses

RESPONSES OF PLANTS TO ENVIRONMENTAL STRESSES

J. LEVITT

Division of Biological Sciences
University of Missouri
Columbia, Missouri

1972

ACADEMIC PRESS New York and London

ACADEMIC PRESS, INC.
111 Fifth Avenue, New York, New York 10003

United Kingdom Edition published by
ACADEMIC PRESS, INC. (LONDON) LTD.
24/28 Oval Road, London NW1 7DD

LIBRARY OF CONGRESS CATALOG CARD NUMBER: 74-190729

PRINTED IN THE UNITED STATES OF AMERICA

CONTENTS

Chapter 5. Low-Temperature Stresses—The Freezing Process

Chapter 6. Freezing Injury

Chapter 7. Freezing Resistance—Types, Measurement, and Changes

Chapter 8. Factors Related to Freezing Tolerance

Chapter 9. Theories of Freezing Injury and Resistance

Chapter 10. Molecular Basis of Freezing Injury and Tolerance

Chapter 11. **High-Temperature or Heat Stress**

Chapter 12. **Heat Resistance**

WATER STRESS

Chapter 13. **Water Deficit (or Drought) Stress**

Chapter 14. **Drought Avoidance**

Chapter 15. **Drought Tolerance**

Chapter 16. The Measurement of Drought Resistance

RADIATION STRESSES

Chapter 17. Radiation Stress—Visible and Ultraviolet Radiation

Chapter 18. Ionizing Radiations

SALTS AND OTHER STRESSES

Chapter 19. Salt and Ion Stresses

Chapter 20. Miscellaneous Stresses

INTERRELATIONS

Chapter 21. **Comparative Stress Responses**

PREFACE

For many years, bits of information have been accumulating on the effects of stresses on plants. I have long felt the need to integrate these in an attempt to discover the basic principles. This need has now become more urgent due to the increasing importance of stress injuries, largely as a result of man's activities. Previously known stresses are becoming more important, and new ones are constantly arising. The practical aim is, therefore, to learn how to control the stresses, or to decrease the injuries they produce.

But the practical goal, though sufficient in itself, is not the sole reason for investigating environmental stresses. It has been said that to understand the normal cell we must study the abnormal cell. To paraphrase this statement, if we wish to understand life we must also study death. The causes of death as a result of exposure to environmental stresses are, therefore, of fundamental importance to all biology, and, for that matter, to all human activities since these are all impossible without life. An understanding of the nature of environmental stresses and of the plant's responses to them may, therefore, help to answer the age-old question: What is life?

It is, therefore, essential that we understand how stresses produce their injurious effects and how living organisms defend themselves against stresses. Why then confine our attention to plants? The simplest answer, of course, is my ignorance. But there is also another reason. The plant has succeeded in developing defenses against stresses that the animal (with few exceptions) has not developed, for instance, against freezing and drought. These also happen to be the stresses that have been most intensively studied. As a result, the research on animals has been mainly confined to responses of quite a different kind. At this stage, therefore, the resistance of plants to environmental stresses is a field in itself. This does not mean that investigations of other organisms can be completely ignored. Some of the most important aids to our understanding of the effects of stresses on plants have come from investigations of animal cells and microorganisms. Such information must, of course, be included.

I have covered four stresses in previous publications: "Frost Killing and Hardiness of Plants" (1941, Burgess, Minneapolis), "The Hardiness of Plants" (1956, Academic Press, New York), Frost, drought, and heat resistance (1958, *Protoplasmatologia* **6**), and Winter hardiness in plants [1966, *in* "Cryobiology" (H. T. Meryman, ed.), Academic Press, New York]. The first two are now out of print, and all are out-of-date. This monograph will include essentially all the environmental stresses which have been intensively investigated (with the exception of mineral deficiencies, which comprise too broad and involved a field to be incorporated with other stresses) and will attempt to bring the information on the above four stresses up-to-date. An attempt will then be made to analyze the possibilities of developing unified concepts of stress injury and resistance. The aim of this synthesis is, therefore, a comprehensive, unified, and molecular point of view. Descriptive aspects of the plant's responses have been largely excluded. For a diagnostic approach to the problem, the reader is referred to Threshow (1970, "Environment and Plant Response," McGraw-Hill, New York).

Only too often in the history of science, parallel investigations by different investigators have led to parallel but different systems of nomenclature. This has occurred in the field of stress research. Any attempt to integrate the results of such parallel investigations requires the adoption of a single, exactly defined terminology. In the case of stresses, this terminology should be applicable to all organisms, plant as well as animal. I have, therefore, attempted to introduce such a uniform terminology in this monograph. The earlier term "frost" has, for instance, been discarded in favor of "freezing," which is now used more generally by cryobiologists. Similarly, the term "tolerance" is adopted in place of the older "hardiness." It is my hope that such adoptions will clarify rather than confuse the concepts.

Unfortunately, the information explosion has prevented an all-inclusive integration. I tender my apologies to all investigators whose important contributions have not been included.

J. Levitt

STRESS CONCEPTS

CHAPTER 1

STRESS AND STRAIN TERMINOLOGY

The responses of plants to the severities of their environment have occupied the attention of man long before the beginnings of the science of biology (Levitt, 1941). To the farmer, plants that survive in these environments are "hardy," those that do not are "tender." The scientist, however, requires a more quantitative terminology. Therefore, in recent years, biologists have adopted the term *stress* for any environmental factor potentially unfavorable to living organisms, and *stress resistance* for the ability of the plant to survive the unfavorable factor. Unfortunately, although stress has been exactly defined in mechanics, no such exact terminology has been developed in biology. Since the lack of an exact terminology in science commonly leads to a lack of exact concepts, an attempt will first be made to apply the definitions of mechanics to biology. It must be recognized at the outset, however, that the mechanical and the biological stresses are not completely identical, and that, therefore, the terminology can be transferred only up to a point.

A. Physical Stress and Strain

According to Newton's laws of motion, a force is always accompanied by a counter force (Duff, 1937). If a body A exerts a force on body B, then body B must also exert a counter force on A. The two forces are called action and reaction and are parts of an inseparable whole, known as a stress. When subjected to a stress, a body is in a state of strain. The external force produces internal forces between contiguous parts of the body leading to a change in size or shape. The magnitude of the stress is the force per unit area. The magnitude of the strain is the change in dimension (e.g., length or volume) of the body.

Up to a point, which is specific for each body, a strain may be completely reversible. Such reversible strains are said to be *elastic*. Beyond this point, the strain will be only partially reversible, and the irreversible part is called the *permanent set* (Fig. 1.1). The permanent set is also called a *plastic strain*. The elastic strain produced in a specific body as a result of a specific stress will always be the same, and the strain is proportional to the stress. Therefore,

$$\text{stress/strain} = M$$

The constant M is known as the modulus of elasticity of the body, which differs for different bodies: the greater the modulus, the more elastic the body. The more elastic, the body, the greater is its resistance to defor-

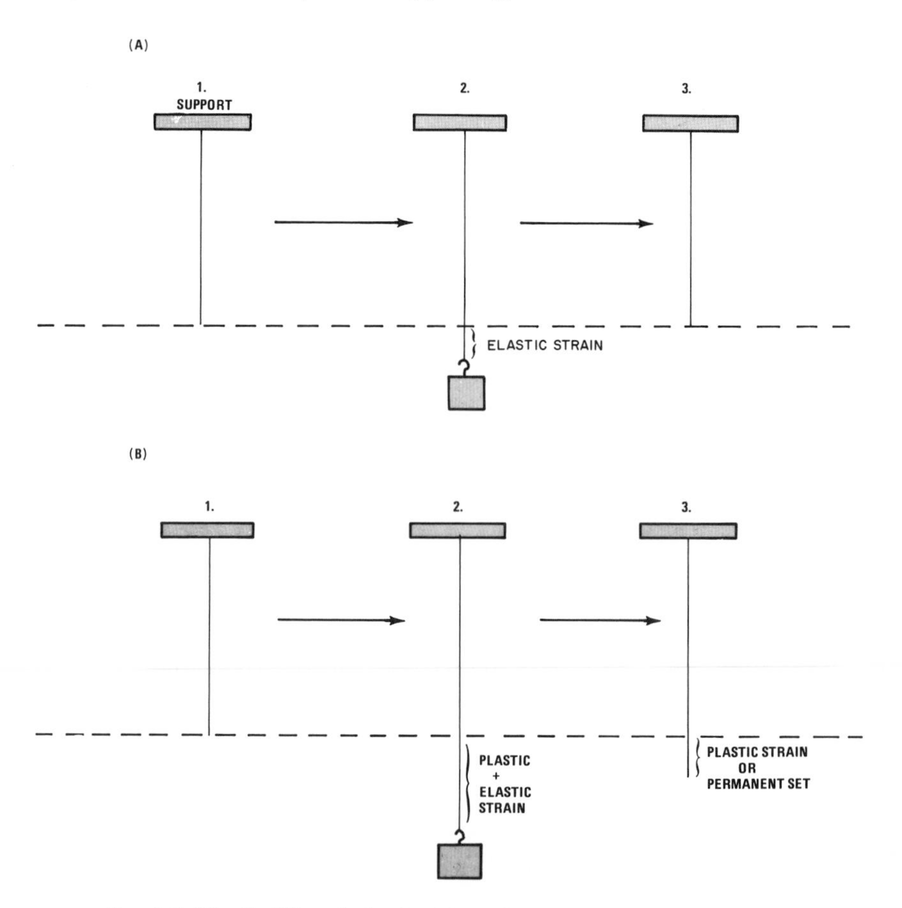

Fig. 1.1. Elastic (A) and plastic (B) strains in a simple physical system.

mation (i.e., the larger the stress required to produce a unit strain). It should be noted that elasticity is *not* the same as elastic extensibility, which is a measure of the maximum possible elastic (i.e., reversible) strain. Unlike elastic strains, plastic strains are not constant for specific stresses, since they may increase with the time the body is subjected to the stress, and may eventually lead to rupture of the body. There is, therefore, no modulus of plasticity.

B. Biological Stress and Strain

Biological stresses differ from mechanical stresses in two main ways. First, since the plant is able to erect barriers between its living matter and the environmental stress, the stress must be measured not in units of force but in units of energy. Second, the term stress in biology always has a connotation of possible injury—i.e., of irreversible or plastic strain. A biological stress may, therefore, be defined as any environmental factor capable of inducing a potentially injurious strain in living organisms. Since the biological stress is not necessarily a force, the biological strain is also not necessarily a change in dimension. The living organism may, however, show a physical strain or change (e.g., cessation of cytoplasmic streaming) or a chemical strain (a shift in metabolism). If either strain is sufficiently severe, the organism may suffer a permanent set, i.e., injury or death. Like the physical body, a specific organism will undergo a specific strain when subjected to a specific stress. It will, therefore, have its own modulus of elasticity, or resistance to physical or chemical change. By analogy with disease resistance, the term "elastic resistance" is more in agreement with biological terminology than modulus of elasticity. In biological systems, unlike physical systems, "plastic resistance" is more commonly measured than elastic resistance. Since plastic strains may be dependent on the time exposed to the stress, the time factor must be measured whenever the plastic resistance of biological systems is determined. The above stress terminology for the two systems is compared in Table 1.1.

The stress resistance of biological organisms is, therefore, of two main types. *Elastic resistance* is a measure of the organism's ability to prevent reversible or elastic strains (physical or chemical changes) when exposed to a specific environmental stress. *Plastic resistance* is a measure of its ability to prevent irreversible or plastic strains and, therefore, injurious physical or chemical changes.

One advantage of a precise biological terminology based on an analogy

TABLE 1.1 Stress Terminology

Term	Physical sense	Biological sense
Stress	A force acting on a body (F/A = dynes/cm^2 or bars)	An external factor acting on an organism (e.g., bars of water stress)
Strain	A change in dimension produced by a stress	Any physical or chemical change produced by a stress
Elastic strain	A reversible change in dimension	A reversible physical or chemical change
Plastic strain	An irreversible change in dimension	An irreversible physical or chemical change
Modulus of elasticity (or elastic resistance)	Stress/elastic strain	Intensity of external factor/amount of reversible physical or chemical change
Modulus of plasticity (or plastic resistance)	Not measured	Intensity of external factor producing a standard irreversible physical or chemical change[a]

[a] The organism must be exposed to the stress for a standard time.

with mechanics now becomes apparent. The term resistance to environmental stresses has, until now, been used only for plastic resistance. The concept of an elastic resistance has not been clearly recognized. There is, therefore, a whole new field in stress physiology waiting to be investigated—a determination of the comparative elastic resistances of different organisms and an attempt to discover the mechanisms involved. As an example, when a corn plant is cooled from 30° to 5°C, its growth comes to a complete stop. Wheat, on the other hand, continues to grow, though at a slower rate. In both cases, when returned to the normal growing temperature, normal growth is resumed. The strain is, therefore, reversible, i.e., elastic. Why does the corn plant suffer a greater elastic strain than the wheat plant when cooled? Or, using resistance terminology, what is the cause of the greater elastic resistance of wheat than corn when cooled?

Another advantage is that the importance of the time factor becomes obvious in the case of plastic strains. The plastic stretch of a wire may be just as dependent on the time exposed to the stress as on the stress itself. Similarly, injury to an organism is just as dependent on the time exposed to a high-temperature stress as on the high temperature used. On the other hand, this is not completely true of freezing stresses, as will be seen below.

There are two pronounced differences, however, between the responses of a nonliving body and of a living organism to stress.

1. Plastic strains in biological systems may be reparable. As in the case of the physical systems, the plastic strain will increase with the stress, producing more and more injury; the plastic strain is irreversible only in the spontaneous (thermodynamic) sense. The plant may be able to repair the strain by an active expenditure of metabolic energy. As the stress increases, the plastic strain also increases until the "rupture" point, when the strain is irreversible both thermodynamically and by metabolic repair, and the plant is killed. It is obvious, then, that stress resistance has two main components: (a) The innate internal properties (or "forces") of the plant which oppose (i.e., resist) the production of a strain by a specific stress. (b) The repair system which reverses the strain. Only the first of these is analogous to the modulus of elasticity in physical systems.

2. Living organisms are adaptable. They are, therefore, capable of changing gradually in such a way as to decrease or prevent a strain when subjected to a stress. Both the elastic and plastic resistances of a plant to a specific stress may, therefore, increase (or decrease). This adaptation may be either stable, having arisen by evolution over a large number of generations, or unstable, depending on the developmental stage of the plant and the environmental factors to which it has been exposed. The unstable adaptation must, of course, also have arisen by evolution, but the hereditary potential is wide enough to permit large changes during the growth and development of the organism.

This adaptation is important both in the case of elastic and plastic strains. Plastic strains are by definition injurious. Therefore the adaptation leading to increased plastic resistance will obviously prevent injury by a stress which injures the unadapted organism. This kind of adaptation has been called "resistance adaptation" by Precht *et al.* (1955), since the adaptation implies a resistance to injury. Injury due to elastic strain would seem to be precluded, by analogy with nonliving systems. Although elastic strains are reversible by removal of the stress and therefore, by definition, are noninjurious, it must be realized that if they are maintained for a long enough period, they may lead to injury and even death. This may simply be due to the inability of the organism to compete with others that undergo less elastic strain when subjected to the same stress (e.g., mesophiles versus psychrophiles at low temperatures). The elastic strain may also eventually injure the plant even in the absence of competition, due to a disturbance of the metabolic balance. Thus, a low-temperature stress may simply decrease the rates of all metabolic processes reversibly, but not all may be decreased to the same degree. Therefore, if the stress is maintained for a long enough period, the strain may conceivably lead to an accumulation of toxic intermediates or to a deficiency of essential intermediates. In either case, a long enough exposure to the stress may injure or kill the organism. An adapted organism, on the other hand, may live, grow, complete its life cycle, and regenerate in the presence of the stress. This kind of adaptation has been called "capacity adaptation" by Precht (1967). Resistance adaptation may not permit growth and may merely prevent the plastic strain and therefore the injury until the stress is removed or decreased to the level permitting growth and development

Nevertheless, both adaptations involve a *resistance* to the effects of a stress; on the one hand, a resistance to elastic strain, and, on the other, a resistance to plastic strain. It seems better, then, to include them both under the heading of stress resistance, using the terms elastic adaptation and plastic adaptation for Precht's "capacity adaptation" and "resistance adaptation," respectively. A plant showing elastic adaptation would develop elastic resistance, and one with plastic adaptation, plastic resistance (Diag. 1.1).

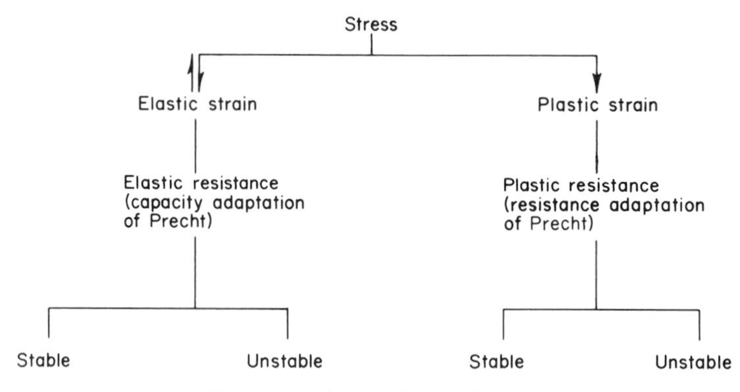

DIAG. 1.1. Adaptation to Stress.

In the case of animals, elastic adaptation has been intensively investigated and plastic adaptation has been largely ignored (Precht, 1967). In plants, the reverse has been the case. The subject of stress resistance in plants will, therefore, be confined essentially to plastic resistance. Nevertheless, there is a dependence of plastic resistance on elastic strain. This is illustrated in Fig. 1.2, which relates the two types of strain to the stress and the time factor. The vertical graph, to the left, illustrates the instantaneous response. With increase in stress, there is an increase in elastic (reversible) strain up to the yield point, the strain being proportional to the stress. Beyond the yield point, a plastic (irreversible) strain occurs, and the strain increases more rapidly than the stress. The horizontal graph with the two vertical panels illustrates the time-dependent response.

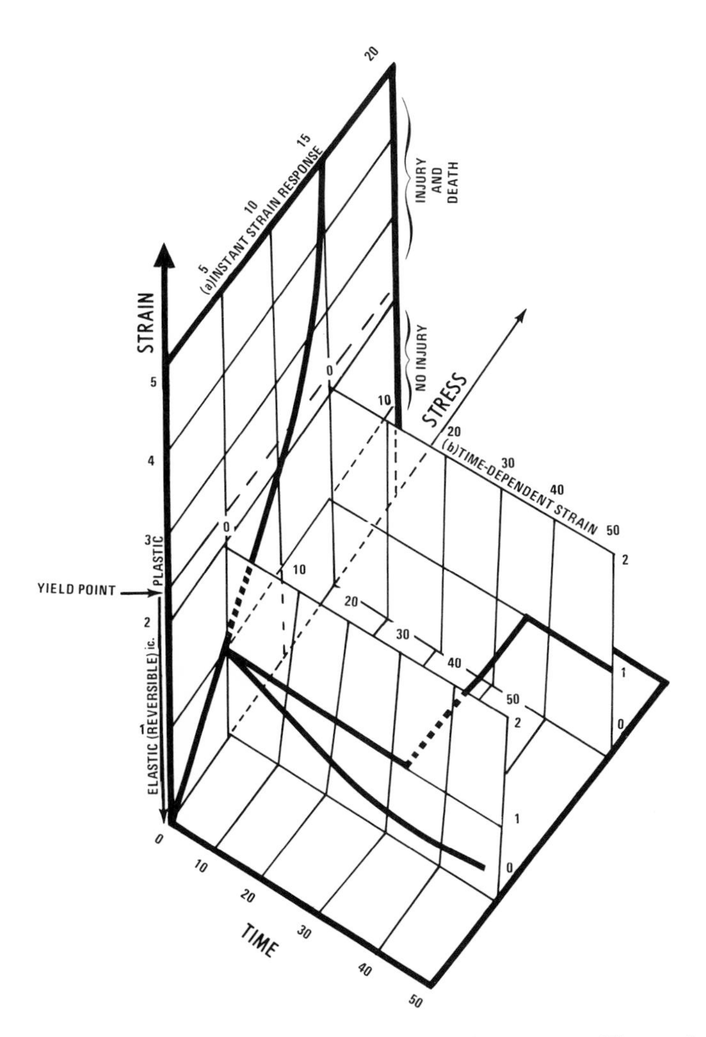

FIG. 1.2. Adaptation of plants to stress. For explanation see text. Time scale may be in seconds, minutes, hours, or days, depending on the particular stress.

If a small stress, producing only an elastic strain is maintained for some time, two kinds of adaptation are possible (first vertical panel). (1) The strain may decrease with time to a constant low value, leading to elastic (Precht's "capacity") adaptation, or (2) the strain may remain constant. In the latter case, even though the specific strain measured is constant, secondary changes may be induced in the plant, leading to a plastic (Precht's "resistance") adaptation. As a result, if the stress is now increased to a point that produced a plastic strain in the unadapted plant, no plastic strain now occurs because of plastic adaptation (second vertical panel).

This time-dependent response of the plant to an elastic strain occurs only in certain plants. There are actually three possible responses. (1) The elastic strain may remain constant and fully reversible, without leading to other changes. (2) The elastic strain may be converted to an indirect plastic (injurious) strain (see Chapter 2). (3) The elastic strain may lead to secondary changes which induce either elastic or plastic adaptation (Fig. 1.2). The distinction between the "tender" and "hardy" plants of the practical man now becomes clear. Those plants in which the first or second time responses occur are, therefore, unable to adapt and belong to the "tender" group. Those in which the third effect occurs become adapted in the process and belong to the "hardy" group.

CHAPTER 2

THE NATURE OF STRESS INJURY AND RESISTANCE

A. Stress Injury

Although it is not possible to eliminate all of the stresses to which a plant may be exposed, it is possible to modify them or the strains that they are capable of producing. It is, therefore, essential that we understand how the stresses produce their injurious effects, and how some living organisms succeed in surviving stresses that injure others.

When a stress acts on a plant, it may produce an injury in different ways (Diag. 2.1).

1. It may induce a direct plastic strain which produces the injury. This may be called *direct stress injury* and may be recognized by the speed of its appearance. In such cases, the plant may be killed by very brief exposures to the stress (seconds or minutes). An example is the rapid freezing strain produced by a sudden low-temperature stress. If the protoplasm

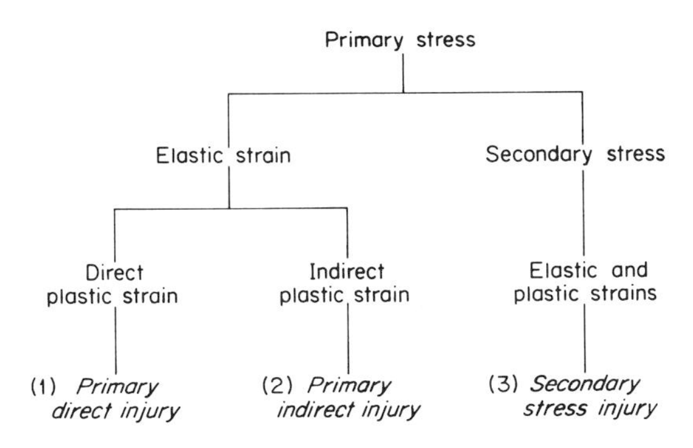

DIAG. 2.1. Kinds of stress injury.

freezes, the ice crystals may lacerate the plasma membrane, thereby producing instant loss of semipermeability and death of the cell.

2. The stress may produce an elastic strain which is reversible and, therefore, not injurious of itself. If maintained for a long enough time, this elastic (reversible) strain may give rise to an indirect plastic (irreversible) strain which results in injury or death of the plant. This may be called *indirect stress injury*. Indirect injury may be recognized by the long exposure (hours or days) to the stress before injury is produced. An example is a chilling stress, which exposes the plant to a low temperature too high to induce freezing. The strains may be mainly elastic—the slowdown of all the physical and chemical processes in the plant—and, therefore, not injurious of themselves. The slowdown may not be uniform for all processes and may produce a disturbance in the cell's metabolism, leading to a deficiency of a metabolic intermediate, or production of a toxic substance.

3. A stress may injure a plant, not by the strain it produces, but by giving rise to a second stress. A high temperature, for instance, may not be injurious of itself, but may produce a water deficit which may injure the plant. This may be called *secondary stress injury*. Since the secondary stress may require some time to develop, secondary stress injury may also require relatively long exposures to the primary stress. The secondary stress, in its turn, may also produce a direct or an indirect injury. Furthermore, it is conceivable that it may give rise to a tertiary stress, etc. The injury (whether direct or indirect) is, therefore, primary if caused by a primary stress, and secondary if caused by a secondary stress, etc. However, when any one stress is considered, only three kinds of injury will be discussed: (1) primary direct stress injury, (2) primary indirect stress injury, and (3) secondary stress injury (a tertiary stress will be indicated in one case). The direct and indirect injuries of the secondary stress will be dealt with in the chapters that discuss it as a primary stress.

B. Stress Resistance

1. Kinds of Resistance

The stress resistance of a plant may be defined as the stress necessary to produce a specific strain. The stress that is just sufficient to produce an elastic (i.e., reversible) cessation of growth, for instance, may be measured. In general, however, only the resistance to the plastic strain is measured. Some standard plastic strain (or injury) must be chosen, for instance, zero strain, or the stress that is just insufficient to induce a plastic strain. This is called the sublethal point, and is synonymous with the yield point in physical systems. The most commonly chosen standard strain, however, is the 50% killing point. Stress resistance then becomes the stress that is just sufficient to produce 50% killing. This value may be referred to as the K_{50} (the killing point) or the LD_{50} (the 50% lethal dose).

Environmental stresses are of two main types—biotic and physicochemical ones. The former belongs to the field of pathology and ecology and will not be considered in this monograph. Among the physicochemical stresses, to which a plant may be exposed, some that have been unimportant until now may become more important due to man's explorations of outer space. At least eight are known to give rise to resistance adaptations (Diag. 2.2). The degree of resistance that can be developed will, of course, vary with the organism, but the kind of resistance may also vary. Two basic types are theoretically possible:

(a) Stress avoidance is stress resistance by avoiding thermodynamic equilibrium with the stress. The plant with avoidance is able to exclude the stress, either partially or completely, either by means of a physical

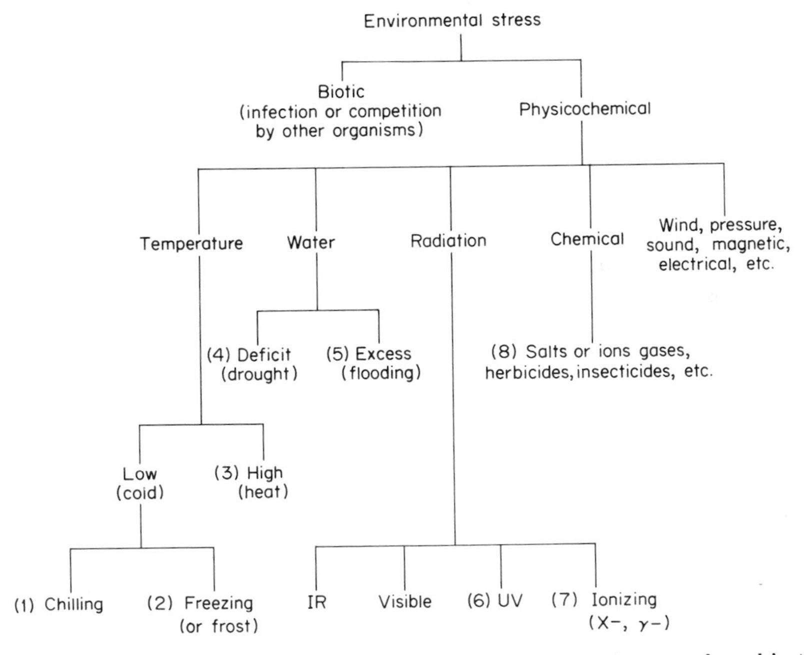

Diag. 2.2. Kinds of environmental stresses to which an organism may be subjected. Resistance against each of the numbered stresses is known.

barrier which insulates its living cells from the stress, or by a steady state exclusion of the stress (a chemical or metabolic barrier). By avoiding the stress, it also avoids the strain.

(b) Stress tolerance is stress resistance by an ability to come to thermodynamic equilibrium with the stress without suffering injury. The plant with stress tolerance is able to prevent, decrease, or repair the *strain* induced by the stress. Though tolerating the stress, it may either avoid or tolerate the strain. Stress resistance, therefore, includes the following components illustrated in Diag. 2.3 below:

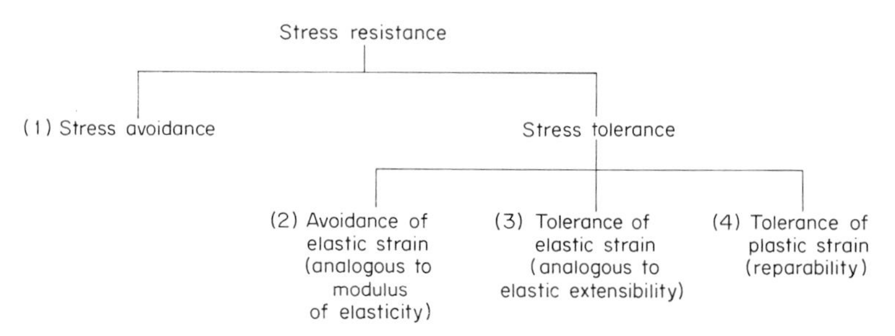

It should be pointed out that the term tolerance is sometimes used in the literature for tolerance of an external stress without determining whether or not the stress is tolerated internally. This is particularly true of the usage of the term "salt tolerance." In this monograph, all such uncertain cases will be called resistance, indicating that either avoidance or tolerance is involved. The meaning of the two main kinds of resistance becomes clear when they are considered in relation to specific stresses (Table 2.1).

TABLE 2.1 Twofold Nature of Stress Resistance

Stress	Condition of resistant plant cells exposed to the stress and surviving due to	
	Avoidance	Tolerance
(1) Low (chilling) temperatures	Warm	Cold
(2) Low (freezing) temperatures	Unfrozen	Frozen
(3) High temperatures	Cool	Hot
(4) Drought	High water potential	Low water potential
(5) Radiation	Low absorption	High absorption
(6) Salt (high conc.)	Low salt conc.	High salt conc.
(7) Flooding (O_2 def.)	High O_2 content	Low O_2 content

Practical men have used the term "hardiness" as a synonym for resistance to a stress, particularly in the case of freezing, drought, and heat resistance. The concept of hardiness is synonymous with toughness, and usually implies tolerance and not avoidance. The most common case is plants that harden (or become resistant) to freezing temperatures during the normal exposure to low temperatures in the fall. This hardening is an increase in tolerance. In at least some cases, however, hardening has been shown to involve an increase in avoidance. Soybeans that are allowed to wilt for some days develop a cuticle less permeable to water than before wilting (Clark and Levitt, 1956). They are therefore able to recover from wilting without a decrease in the environmental water stress. Consequently, they increase their drought avoidance but they show no increase in drought tolerance. If hardening is taken to be strictly an increase in tolerance, such an increase in avoidance may be called a "pseudohardening."

2. Resistance to Individual Stresses

These are the theoretical possibilities, but each kind of stress must be examined individually in order to find out whether or not the plant has succeeded in developing the two kinds of resistance. The following brief survey will be discussed more fully in later chapters.

a. Low Temperature (or Cold)

Table 2.2 illustrates how closely the plant temperature agrees with that of its environment at low temperatures even below the freezing point. The inability of the plant to control its own temperature is further shown by the radiation cooling of plants below the air temperature at night and by the radiation heating of the cambium on the sunny side of trees to as high as 30°C above the shady side in winter. Plants are thus *poikilotherms*—they

TABLE 2.2 Temperatures (°C) of Tree Trunks (Elm and Red Fir) and of the Surrounding Air[a]

Air	Tree
−13 to −15	−12 to −14
−2.0	−1.0
−2.5	−0.5
−15.2	−14

[a] See Levitt, 1956.

tend to assume the temperature of their environment. This means, by definition, that they cannot develop low-temperature avoidance. Exceptions are the fleshy inflorescences of the Araceae and so-called snow plants that develop temperatures above that of their environment due to rapid growth and the consequent rapid release of respiratory heat. However, this occurs only at environmental temperatures not low enough to induce injury.

The inescapable conclusion is that existing *low-temperature resistance can only be tolerance.*

b. High Temperature

The temperatures of leaves at high air temperatures in the full sun have long been a matter of controversy. The question has been whether or not the radiant heating by the sun can be counteracted by the transpirational cooling. At moderately high temperatures (up to 40°C) it now appears established, both from experimental evidence and theory, that the leaf temperature cannot be more than a very few degrees below the air temperature. At very high temperatures (40°–50°C) this is not true, and leaf temperatures may be considerably (as much as 15°C) below the air temperature, in this way exhibiting pronounced high-temperature avoidance. The leaves of most plants, however, survive in spite of temperatures above that of the air, and they must therefore possess high-temperature tolerance.

c. Water Deficit

Aquatic plants and lower land plants are *poikilohydric* (Walter, 1955)—they rapidly come to equilibrium with the water in their environment. In some cases, they may be hydrated in the morning and air dry in the afternoon. These poikilohydric plants must therefore be drought tolerant. Higher land plants, on the other hand, may be considered *homoiohydric*, since they normally remain turgid and therefore near 100% relative humidity, although daily exposed to the much lower relative humidity of the surrounding air. They, therefore, possess drought avoidance.

d. Radiation

Ionizing radiations are highly penetrating and highly absorbed by the plant. Plants that survive such radiations must possess tolerance. Ultraviolet radiations may, however, be largely reflected, transmitted, or absorbed by the plant surface. Survival of these radiations may therefore be due to avoidance.

e. SALTS

Halophytes are plants that grow in high salt soils. It has long been known that these plants contain high cell sap concentrations due to absorption of large quantities of salts. Osmotic values as high as 200 atm have been recorded. However, in at least some cases, these values are too high due to a redissolving of surface-excreted salts by the extracted juice. Even when care is taken to wash off this surface-excreted salt, high values have been found (100–130 atm) and these plants must therefore possess salt tolerance. In the case of varietal differences in mildly resistant plants (e.g., barley), it has been found that the more resistant variety excludes the salt better than the less resistant variety and therefore owes its resistance to avoidance.

The general conclusion from all the above examples is that in the case of nearly all stresses that have been investigated plants have succeeded in developing both tolerance and avoidance. Since some plants develop the one more than the other, the following are the three general groups into which resistant plants can be classified:

1. Tolerance but no avoidance (the tolerant nonavoiders).
2. Tolerance and avoidance (the tolerant avoiders).
3. Avoidance but no tolerance (the intolerant avoiders).

Tolerance seems to be the more primitive adaptation, and avoidance more advanced (Table 2.3). This is reasonable, since tolerance involves an equilibrium state, and avoidance requires development by the plant of a mechanism to avoid equilibrium and to replace it by the steady state. Avoidance is also a more efficient adaptation; by avoiding the stress, the plant avoids both the elastic and the plastic strain. It is, therefore, able not only to survive when exposed to the stress, but also to metabolize, develop, and complete its life cycle. Tolerance (assuming plastic, but no elastic resistance), on the other hand, merely permits the plant to survive

TABLE 2.3 KINDS OF RESISTANCE FOUND IN LOWER VERSUS HIGHER PLANTS

Stress	Resistance in	
	Lower plants	Higher plants
Freezing	Solely tolerance	Solely tolerance
Heat	Solely tolerance	Mainly tolerance
Drought	Solely tolerance	mainly avoidance

until such time that the stress is removed and the plant can recommence its normal metabolism, growth, and development.

On the same basis, drought resistance seems to be the most advanced of the three types of resistance. This is also reasonable, since freezing and heat resistance no doubt had to be developed by plants before the evolution of species capable of completing their entire life cycle on land.

C. Kinds of Stress Tolerance

It is obvious that the mechanism of resistance in the case of any one stress will depend on whether the plant owes its resistance to avoidance or tolerance; the avoidance mechanism must be completely different from the tolerance mechanism. Even if only tolerance is investigated, we must be prepared for more than one kind of resistance. If both direct and indirect injury are produced by a single stress, there will be at least two tolerance mechanisms. Each kind of stress must, therefore, be investigated for the kinds of injury produced, before any attempt is made to determine the mechanism of resistance. The stress tolerance mechanism will further depend on whether it is due to strain avoidance or strain tolerance; since these two resistance mechanisms may conceivably be applied at three levels: the stress, the elastic strain, and the plastic strain. The six possible combinations are listed in Table 2.4. It must be emphasized that in each case, avoidance and tolerance are used in a quantitative sense.

The first resistance type, stress avoidance, permits the plant to avoid both the elastic and plastic strains. The second resistance type, stress tolerance, is possible only by development of one or other (or both) of the third or fourth types—avoidance or tolerance of elastic strain. The fourth type, in its turn, is synonymous with the fifth type, for tolerance of an elastic strain is, by definition, avoidance of a plastic strain. Intolerance of an elastic strain occurs at the point of permanent set, when the elastic

TABLE 2.4 The Six Possible Resistance Combinations

Factor	Resistance	
	Avoidance	Tolerance
Stress	(1) Stress avoidance	(2) Stress tolerance
Elastic strain	(3) Avoidance of elastic strain	(4) Tolerance of elastic strain
Plastic strain	(5) Avoidance of plastic strain	(6) Tolerance of plastic strain

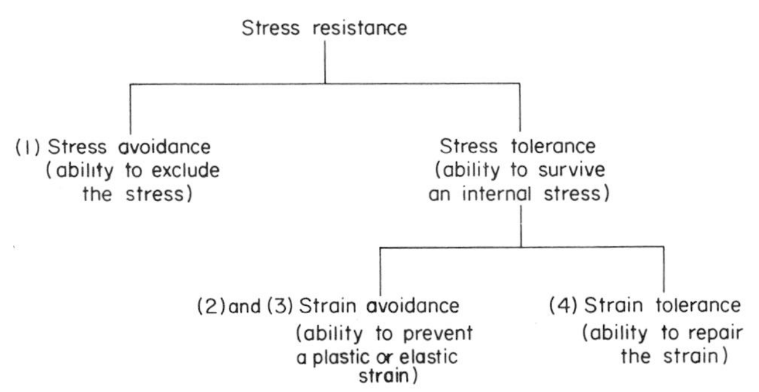

DIAG. 2.4. The four possible different stress resistance mechanisms.

strain becomes an irreversible plastic strain, which is, therefore, not avoided. At first sight, the sixth resistance type appears impossible, since a plastic strain is irreversible and, therefore, cannot be tolerated. The living organism, however, may be capable of repairing, and therefore reversing a thermodynamically irreversible plastic strain. The above six resistance mechanisms, therefore, reduce to four (Diag. 2.4). Only two of these have their counterparts in physical systems. The second mechanism—avoidance of elastic strain—would be proportional to the modulus of elasticity (stress/strain), which is really a measure of the stress that must be imposed in order to produce unit elastic strain. Obviously, the greater the stress that must be employed to produce this unit strain, the greater the elastic strain avoidance. The third mechanism—tolerance of elastic strain—is analogous to elastic extensibility. The greater the elastic extensibility of a body, the more it can be extended without suffering a plastic strain and therefore the greater its tolerance of elastic strain or avoidance of plastic strain.

The significance of Precht's terminology also becomes clearer, for his "capacity adaptation" is simply a measure of the organisms avoidance of elastic strain, and his "resistance adaptation" a measure of its tolerance of elastic strain.

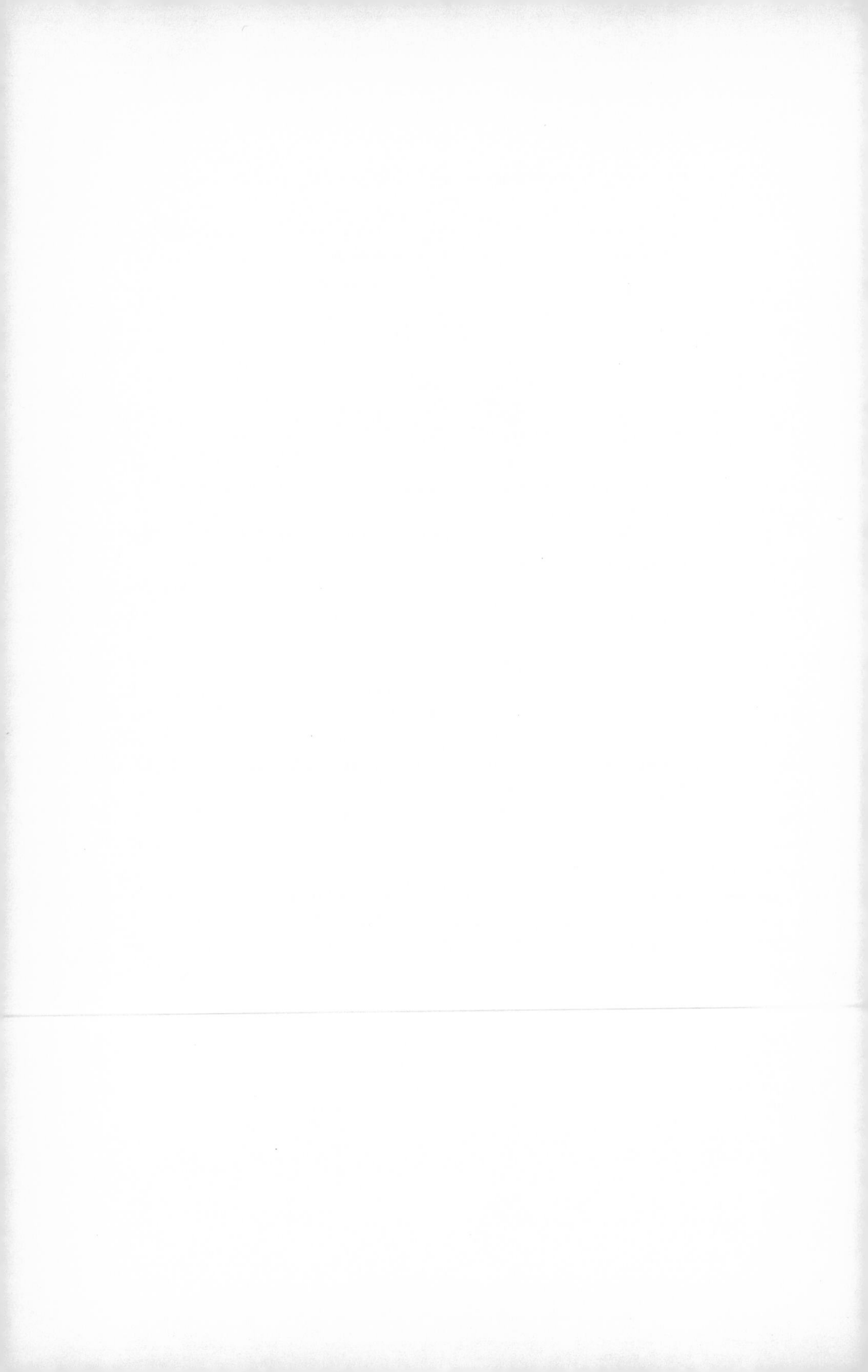

TEMPERATURE STRESSES

CHAPTER 3

LOW-TEMPERATURE STRESS—LIMITS OF TOLERANCE

The effects of low temperatures on living organisms have long interested biologists, both from practical and theoretical points of view. Only in recent years, however, have the practical applications been seriously exploited, with the preservation of food, blood, semen, cultures, and tissues. The resulting explosive increase in research reached a climax in 1958 with the introduction of a new name for the field—*cryobiology*, the biology of freezing temperatures (Parkes, 1964). In 1963, the Society for Cryobiology was organized. A monograph on Cryobiology (Meryman, 1966a) appeared, and now the plant life of snow and ice has been treated separately in yet another monograph on Cryovegetation (Kol, 1968). Long before the development of a separate science, biologists attempted to discover whether or not there is a theoretical limit to the low-temperature stress that living cells can survive. This question is now important from a practical as well as a theoretical point of view.

A. Dehydrated Protoplasm

Low-temperature resistance, as mentioned previously (Chapter 2), is tolerance in nearly all cases. In order to understand low-temperature *resistance* it is, therefore, first necessary to answer the question of whether there is a limit to the low-temperature *tolerance* of plants. This problem was approached long ago by a number of investigators. Since they were searching for maximum tolerance, only the most resistant of plant parts were tested—either lower plants, or higher plants in the dry and dormant states (seeds, spores, pollen grains). These resistant cells and tissues were able to survive the lowest temperatures to which they

TABLE 3.1 SURVIVAL OF VERY LOW TEMPERATURES BY PLANTS.[a]

Plant or plant part	Temperature	Exposure time	Reference
Seeds	−100°C	4 days	de Candolle, 1895
Seeds	−190°C	110 hr	Brown and Escombe, 1897
Seeds	−250°C	6 hr	Thiselton-Dyer, 1899
Bacteria and yeast	−190°C	6 months	Macfayden, 1900
Seeds	−190°C	130 hr	Becquerel, 1907
Fungi and algae	−190°C	13 hr	Kärcher, 1931
Seeds	−190°C	60 days	Lipman and Lewis, 1934
Mosses (protonema)	−190°C	50 hr	Lipman, 1936a
Seeds and spores	1—4°K	44 hr	Lipman, 1936b
Fungus mycelium and bacteria	−190°C	48 hr	Lipman, 1937
Spores and pollen grains	−273°C (within a few thousandths of a degree)	2 hr	Becquerel, 1954

[a] All were able to grow after the low temperature exposure. (See also Luyet and Gehenio, 1938.) (−190°C means liquid air was used, −250°C, liquid hydrogen.)

were exposed, even down to a fraction of a degree above absolute zero (Table 3.1). Since the higher plant tissues were "dry" (i.e., air dry), the small amount of moisture present in them was "bound" and, therefore, was not converted to ice at even the lowest temperatures used. These results, therefore, simply showed that dehydrated cells which are unable to freeze at low temperatures can survive the lowest temperatures without suffering any injury.

B. Hydrated Protoplasm

Some lower plant (Table 3.1) and animal (Smith *et al.*, 1951) cells in the normally moist condition can also survive exposure to extremely low temperatures without any special precautions. Yet when the above, fully tolerant dry seeds are allowed to imbibe water, they are killed by very slight freezes (Table 3.2). Similarly, when the water content of pollen grains was high enough (about 36%) to permit X-ray detection of ice crystals at −25°C, the pollen no longer survived the low temperatures (Ching and Slabaugh, 1966). It has long been known, of course, that the normally hydrated protoplasm of most plants (except those that overwinter

TABLE 3.2 Survival of Low Temperatures by Dry and Hydrated Plants

Plant	Low temperature treatment	Survival		Reference
		Dry	Hydrated	
Seeds	$-25°$ to $-40°C$ for 15 hr	All germinated	None germinated	Göppert, 1830
Seeds	$-190°C$ for 130 hr	All germinated	None germinated	Becquerel, 1907
Corn kernels	32°–28°F	25% Moisture all germinated	75–85% Moisture none germinated	Kiesselbach and Ratcliff, 1918[a]
Yeast	$-113°C$	Unaffected	Vacuolate cells all killed; young (nonvacuolate) uninjured	Schumacher, 1875
Ranunculus tubers	$-190°C$ for 18 days	9% Moisture all survived	30–50% Moisture all killed	Becquerel, 1932
Wheat seed	$-190°C$ for 2 min	10.6% H_2O all germinated	25.1% H_2O none germinated	Lockett and Luyet, 1951
Alfalfa seed	$-20°C$ for 1 day	90% Germinated	None germinated	Tysdal and Pieters, 1934

[a] Similar results were obtained by Jensen, 1925; Steinbauer, 1926; Stuckey and Curtis, 1938; and McRostie, 1939.

in cold climates) is normally killed by even slight freezing. What, then, is the limit of low-temperature tolerance in the case of the normally hydrated protoplasm of plants?

Luyet (1937) has shown that even in the normally hydrated state, plant cells that are killed by slight freezes, can nevertheless survive immersion in liquid nitrogen if both the cooling and the rewarming rates are ultrarapid (10,000–100,000°C/sec). He achieved these extremely rapid rates by plunging thin strips of tissue (e.g., one cell thick) directly from room temperature into the liquid nitrogen, followed by direct transfer to a warm (25°–30°C) aqueous solution. In this way, rates of cooling and warming as high as 150,000°C/sec have been obtained (Luyet, 1951). He first considered this a vitrification process, the rapidity of the cooling maintaining the water in the supercooled, noncrystalline or vitrified (i.e., glassy) state. This conclusion was based on the absence of double refraction (which is characteristic of crystals) when viewed under the microscope.

According to the definition suggested recently (1968) by a committee appointed by the National Research Council, "a glass or vitreous substance is a solid giving a typical, amorphous phase X-ray pattern and capable of exhibiting the glass transition" (Angell and Sare, 1970). The glass transition, in turn, is "that phenomenon in which a vitreous phase exhibits with changing temperature a more or less sudden change in the derivative thermodynamic properties, such as heat capacity and expansion coefficient, from crystal- to liquidlike values. The temperature of the transition is called the glass transition temperature." It has also been called the vitrification point. The true vitrification point for water has recently been calculated to be 162 ± 1°K or −111°C (Miller, 1969). On the basis of this definition, true vitrification was probably not achieved in the above experiments; later investigations, using X-rays, revealed the presence of submicroscopic crystals (Table 3.3), at least in some cases of rapid freezing such as used by Luyet. Nevertheless, whether true vitrification is achieved, or a pseudovitrification with crystals too small to be detected microscopically, this method may permit a strip of onion epidermis to survive liquid air without injury, though the cells are otherwise killed by a very moderate freeze (e.g., −5°C). Algae (twenty-three strains of five genera) have also survived freezing in liquid nitrogen (Hwang and Horneland, 1965).

Even if the hydrated cells or tissues survive such ultrarapid freezing and thawing without injury, it does not follow that this absence of injury is independent of the length of time exposed to the low temperature. Meryman (1966a) has demonstrated crystal growth at temperatures as low as −130°C. At −80°C, for instance, they grew from 200 to 10,000 Å in

TABLE 3.3 SUMMARY OF RESULTS RELATING STRUCTURES OBTAINED BY THE DEPOSITION OF WATER VAPOR ON A SUBSTRATE TO LOW TEMPERATURES[a]

Experimental method	Temperature ranges (°C) −180 −160 −140 −120 −100 −80 −60			Reference
X-ray diffraction	Amorphous	Semicrystalline	Hexagonal	Burton and Oliver, 1935
Calorimetric	Amorphous	Crystalline		Staronka, 1939
X-ray diffraction	Small crystals	Intermediate range not investigated	Hexagonal	Vegard and Hillesund, 1942
Electron diffraction	Small crystals	Cubic	Hexagonal	König, 1942
Calorimetric	Vitreous	Crystalline		Pryde and Jones, 1952
Electron diffraction	Crystal growth poor	Cubic	Hexagonal	Honjo *et al.*, 1956
Calorimetric	Amorphous	Crystalline		Ghormley, 1956
Calorimetric	Vitreous	Crystalline		De Nordwall and Staveley, 1956
Electron diffraction	Amorphous or small crystals	Cubic	Hexagonal	Blackman and Lisgarten, 1957

[a] The terms used are those of the respective authors. (From Blackman and Lisgarten, 1958; see Merryman, 1966b.)

TABLE 3.4 Ice Form and Temperature[a]

Temperature (°C)	Ice form	
0	Temp. drop ↓	Temp. rise
−80	Hexagonal	Hexagonal ↑
−100	│	Cubic + hexagonal ↑
−140	↓	Cubic (diamond) ↑
−192	Hexagonal	Amorphous

[a] Adapted from Shikama, 1963.

8 min, becoming microscopically visible. Another possible source of injury is crystal shape which depends on the temperature and the order of cooling and warming (Table 3.4). Consequently, if the ultrarapid cooling method is successful, maintaining the crystals below microscopic size, the protection would be only temporary unless the cells are stored at temperatures well below −130°C.

Even for short freezes, the method does not work with all living cells. Even when it does succeed in keeping the cells viable, if they contain 70–80% water, the tissue must be less than 0.1 mm thick. Thicker pieces of tissue do not permit the extremely rapid temperature drop (and rise) needed to prevent the growth of crystals large enough to be injurious within the cells. If, however, the water content of the tissue is lower than 70–80%, thicker pieces may be frozen rapidly enough to prevent injury (Luyet, 1951). In the case of wheat grains frozen in liquid nitrogen (−190°C) from seconds to minutes, and then maintained at −120°C, ice can be detected by X-ray analysis if the moisture content is above 33% (Radzievsky and Shekhtman, 1955). Similar results were obtained by Sun (1958). Small pea seedlings (7–12 mm long), excised from their cotyledons, survive exposure to liquid nitrogen if they are first dried to a moisture content of 27–40%. In the case of larger seedlings, with low water content, only the stem tip survived. This relationship between size and the protective effect of low water content may perhaps explain the failure of other investigators to obtain survival of liquid nitrogen (e.g., Genevès, 1955). Nevertheless, even when the above precautions are taken, some animal cells are killed by too rapid cooling, whereas a slower cooling (e.g., cooled to −79°C in 5 min) permits survival (Smith *et al.*, 1951). The speed of cooling (and warming) is particularly important within a temperature

TABLE 3.5 The Critical Low Temperature Zone for Injurious Crystallization of Ice

Living material	Temp. zone (°C)	Reference
General	Freezing point to −30	Luyet, 1940
Red blood cells	−4 to −40	Lovelock, 1953
Pasteurella tularensis	−30 to −45	Mazur *et al.*, 1957
Gill pieces of oyster	−40 to −50	Asahina, 1958
Sucrose solution	Freezing point to −32	Rey, 1961
Denaturation of catalase	−12 to −75	Shikama, 1963
Denaturation of myosin	−20 to −72	Shikama, 1963

range which may be called the danger zone. Published ranges for this danger zone vary somewhat with both the investigator and the organism used (Table 3.5). In general, it is between the freezing point of the material and about −30° to −40°C. This zone presumably includes the temperature range in which the size of the immediately formed crystals is large enough to damage the cell. Some plants seem able to survive indefinitely even when frozen within this danger zone. Thus, of 291 strains of molds stored for 5 years in a freezer at −17° to −21°C, only 15 were not viable (Carmichael, 1962). This method was successful even in the case of some fungi which do not survive freeze-drying. On the other hand, freeze-drying, which usually involves a slower freezing than used by Luyet, permits the survival of pollen, particularly if stored at −25°C (Snope and Ellison, 1963; Layne, 1963). Similarly, three species of blue-green algae showed no decline in viability when lyophilized (freeze-dried), even after storage at 25°C for 5 years (Holm-Hansen, 1967). In the case of four species of green algae, the same treatment did result in a significant decline in viability.

There are two other methods of passing through the danger zone without injury, in addition to the ultrarapid cooling method. If the cells are first treated with protective substances such as glycerol and dimethylsulfoxide (DMSO), they may be cooled slowly (e.g. $\frac{1}{4}$–1°C/sec) without being killed by liquid nitrogen. Such substances are called cryoprotectants if they prevent injury during the freezing. This method works well for animal cells (although not for all; Sherman, 1962) and for tissue cultures of some plants, but has not proved successful for normal plant tissues. Thus, cultured cells of flax and *Haplopappus gracilis* survived −50°C for up to a month if frozen in a medium containing 10% DMSO, cooled at a rate of 5–10°/min, and thawed quickly in warm water at 40°C (Quatrano, 1968). The method failed with *Eucalyptus camaldulensis*. Eight fungal cultures failed to survive freezing in the presence of either glycerol or DMSO

(Hwang and Howells, 1968). In the case of cells permeable to glycerol (epidermis of *Campanula* species), 40–50% survive rapid freezing in liquid nitrogen followed by rapid thawing, if they have been allowed to imbibe 15–20% glycerol (Holzl and Bancher, 1968). This method is lethal if cells are stored for some time at −70°C (Richter, 1968a). When the glycerol concentration is increased to over 70%, the solutions solidify in the vitreous state and the cells are uninjured. These cells, however, are unusually indifferent to glycerol, surviving even a transfer to 100% glycerol without apparent injury when returned to mixed solutions (Richter, 1968b). The cells are, however, injured by ethylene glycol and DMSO. Some protective action by glycerol, ethylene glycol, etc., has been reported for sections of collards (Samygin and Matveeva, 1967). Glycerol was also somewhat protective to winter wheat (Trunova, 1968), though high concentrations were injurious.

A third method has proved successful for partially resistant plant cells (Sakai, 1958). If a hardened plant twig is cooled slowly down to a temperature which is noninjurious (e.g., −15° to −30°C) and allowed to come to freezing equilibrium at this temperature, it may then be plunged into liquid nitrogen without injury. This method has been confirmed by Tumanov *et al.* (1959) and by Krasavtsev (1961), and the temperature survived was extended to that of liquid hydrogen. In the case of some extremely hardy plants, such as leaves of *Pinus strobus* in midwinter, survival in liquid nitrogen has been obtained simply by cooling at a rate of 3°C/hr (Parker, 1959b, 1960). Extremely hardy cells from winter twigs of mulberry trees can even survive rapid immersion in liquid nitrogen from room temperature and subsequent rewarming in water at 35°C (Sakai, 1968a). Sakai succeeded in using this method, in combination with cryoprotectants in even less tolerant cells, which could not survive freezing below −10°C. These cells were able to survive immersion in liquid nitrogen and subsequent rapid rewarming provided that they were previously treated with an isotonic or slightly hypertonic glucose solution, and were blotted to remove the excess solution before immersion in liquid nitrogen.

It is difficult to generalize from all the above results. At least in the case of a large number of both resistant and nonresistant plants, it can be concluded that if the danger zone (0° to −40°C approx.) can be passed without injury, either by an extreme speed of cooling or by a more gradual cooling in the presence of protective substances, or in their absence in the case of hardy cells, their protoplasm is able to survive the lowest temperatures without injury. In the case of plants that have not responded to such methods, perhaps more sophisticated methods will eventually lead to success.

It must be emphasized, however, that the above successful survival of low temperatures by protoplasm in the hydrated state, does not prove that the hydrated protoplasm possesses plastic resistance to all forms of injury produced by low-temperature stress. The special methods used are successful only if they prevent the occurrence of a specific strain which is normally produced at the low temperatures, and which normally kills the hydrated protoplasm of unadapted plants. The adapted plant has developed its own methods of achieving survival of the low-temperature stress (see Chapters 7 and 8) in the absence of the above artificial protective measures.

CHAPTER 4

CHILLING INJURY AND RESISTANCE

A. Chilling Stress

Although it has been recognized for centuries, the chilling stress is still very difficult to define quantitatively. As early as 1778, Bierkander reported of some eight species that were killed at 1° to 2°C above the freezing point (Molisch, 1896). Göppert (1830) obtained similar results. Of 56 species of tropical plants tested by Hardy in 1844, some 25 were killed at 1° to 5°C (see Molisch, 1896). Molisch (1897) suggested that low-temperature damage in the absence of freezing should be called chilling injury (Erkältung) as opposed to freezing injury (Erfrieren). On the basis of the above results, a chilling temperature can be defined as any temperature that is cool enough to produce injury but not cool enough to freeze the plant. Therefore, in the case of an elastic strain, the theoretical definition of a chilling stress would be the number of degrees (or kelvins) that the environmental temperature is below optimum for the plant activity being measured (e.g., growth). In practice, this definition would be unworkable, since the optimum temperature for any plant process is not necessarily a constant, but may vary with other conditions. In most cases, plants do not suffer chilling injury until the temperature drops below 10°C. This may, therefore, be taken as the arbitrary zero point, and the plastic chilling stress may be defined as the number of degrees below 10°C. Unfortunately, rice (when in flower) and sugar cane may suffer chilling injury at 15°C (Adir, 1968; Tsunoda *et al.*, 1968), so that this definition would not apply to them. Similarly, the germination of cacao seeds is decreased at temperatures below 14°C (Boroughs and Hunter, 1963). Furthermore, exceptional plants (e.g., some fungi) may be killed at freezing temperatures in the absence of freezing—i.e., in the undercooled state (Lindner, 1915; Onoda,

1937). The term chilling temperature must, therefore, refer to different temperature ranges for different plants and must be defined for a specific group of plants. Even in the case of plants for which the 10°C threshold does apply, the strain does not necessarily parallel the stress, for in some cases injury is greater at a higher than at a lower chilling temperature (see below).

B. Chilling Injury

Although chilling injury is primarily observed in plants from tropical or subtropical climates, certain cells of plants from temperate climates may be injured. At a foliage temperature of 0°–3°C, chilling produces sterility in wheat, when the pollen is in the stage of the first nuclear division (Toda, 1962). Any plant exposed to a chilling temperature may be expected to undergo an elastic (reversible) strain, due to the slowdown of all chemical reactions, and the much smaller slowdown of physical processes. Chilling temperatures may also induce a plastic (irreversible) strain, and the plant may suffer injury or death. All three kinds of stress injury may occur: (1) direct, (2) indirect, and (3) secondary stress injury.

1. Direct Injury

Seible (1939) divides plants that suffer chilling injury into two types according to the speed of the reaction. The first type (e.g., *Episcia, Achimenes, Gloxinia*) show injured spots after hours or at the latest after a day, due to death of the protoplasm and infiltration of the intercellular spaces. The second type (e.g., *Tradescantia, Solanum, Coleus*) are more resistant. They remain perfectly normal and turgid for a full day but become soft and wilted only after chilling 5–6 days. Many fruits would fit into Seible's second type. Some apples, for instance, are injured by prolonged storage at temperatures below 36° to 40°C; but the fruit of tropical plants is much more sensitive—bananas may be injured by a few hours at temperatures below 55°F (Pentzer and Heinze, 1954). Seible's first type is presumably direct injury, the second type indirect injury. However, there is no sharp line between the two. Many crop plants native to warm climates are intermediate between Seible's two types, showing injury after 24–48 hr at 0.5° to 5°C (Sellschop and Salmon, 1928). Nevertheless, at least some cases of injury appear to occur too rapidly to be explainable by metabolic disturbances, since chemical reactions occur slowly at chilling temperatures (Table 4.1). This is particularly true of Möbius' observations of injury due

TABLE 4.1 Rapid Chilling Injury

Plant	Plant temp.	Air temp.	Exposure time	Injury	Reference
Begonia metallica		−5°C	1–2 min	Older leaves died	Möbius, 1907
	6°C approx.	−10	1–2 min	Leaves fell off	Möbius, 1907
Tradescantia zebrina	6°C approx.	−10	1–2 min	Wilted	Möbius, 1907
Fittonia argyroneura	6°C approx.	−10	1–2 min	Wilted	Möbius, 1907
Callisia repens	2½°C approx.	−10	1½ min	Wilted	Möbius, 1907
Aspergillus niger	−2°C		2 hr	Killed	Bartetzko, 1909
Sea algae	1°–2°C		12–72 hr	Killed	Biebl, 1939
	−2°C		12 hr (no ice formation)	Killed	Biebl, 1939
Coffee leaves	3°C		6 hr	Leaf discoloration	Franco, 1958

to exposure for 1–2 min. In such cases, the injury due to sudden chilling occurs so rapidly it may be called a *cold shock*. Cold shock is obviously a direct chilling injury, since it occurs too rapidly for either indirect stress effects to appear, or for development of a secondary stress. It has been explained by a sudden increase in permeability. Cohn described a *pseudoplasmolysis* in *Spirogyra* cells suddenly exposed to 0°C (see Molisch, 1897). Greeley (1901) and Livingston (1903) observed the same phenomenon. This was explained by an increase in permeability resulting in leakage of cell solutes. The postulated leakage has, indeed, been confirmed by Lieberman *et al.* (1958). Chilled sweet potatoes showed five times as much leakage as the controls, almost all of it being K^+. In this case, somewhat longer chilling periods were required than for cold shock. Direct injury, therefore, may perhaps occur as a result of several hours' chilling, as well as when chilling abruptly for a few minutes. In the case of *Coleus*, a clear increase in conductivity of the petioles occurred after 5–14 hr at 0.5° to 4.0°C (Katz and Reinhold, 1965). It continued to rise almost linearly for several hours, attaining a constant value after 20 hr. The damage at this time is apparently irreversible since the conductivity continued to rise if the petioles were transferred to room temperature after chilling for 22 hr.

These measured increases in cell permeability as a result of somewhat longer chilling periods, seem to differ from the increase proposed by early investigators to explain chilling shock; the shock produced a rapid, spontaneous pseudoplasmolysis. In contrast, the epidermal cells of the coleus petioles plasmolyzed more slowly or not at all when chilled (Katz and Reinhold, 1965). However, the difference may be more apparent than real. The slower plasmolysis occurred in a hypertonic solution, and due to the increase in permeability, the plasmolyte may have penetrated the cell, slowing down or preventing plasmolysis.

All these results point to a damage to the cell membrane as a result of chilling. Lyons *et al.* (1964) suggest that the chilling injury is due to the relatively inflexible mitochondria in chilling sensitive plants (tomato fruit, sweet potato root, bean and corn seedlings), compared to their greater ability to swell in the chilling resistant plants (cauliflower buds, turnip roots, pea seedlings). They explain this difference by the higher content of unsaturated fatty acids in the resistant plants. Lyons and Asmundson (1965), therefore, attempt to explain chilling sensitivity by a solidification of the lipids of the plasma membrane. Using artificial systems, they showed that the freezing points of mixtures of palmitic and linoleic, or palmitic and linolenic acids (the predominant fatty acids in plants) decrease slowly as the unsaturated fatty acid is increased to 60 mole%. Beyond this percentage, the freezing point is depressed more markedly by each addition.

The differences are pronounced at percentages that approximate the composition in plant membrane lipids. Casas *et al.* (1965) also suggest that membrane damage in cotyledonary tissues of cacao is a direct result of cold treatment. Restoration of viability following heat treatment could be due to a reversal of such a change. The sensitivity to low temperatures of a maize mutant (M11) also seems to be associated with membrane sensitivity (Millerd *et al.*, 1969). When grown in the light at 15°C, its plastids contain little pigment, they are deficient in ribosomes, their ultrastructure is abnormal, and the membrane shows an extreme sensitivity to light.

In all the above cases, the sole criterion of direct injury is the speed of the chilling effect. By definition, however, direct injury is due to a stress that produces a direct plastic strain. Unfortunately, due to our incomplete understanding of the nature and properties of the cell membranes, it is not clear whether or not the above described rapid chilling injury is, in fact, due to a direct plastic strain. On the basis of the above hypothesis of lipid solidification, the injury may be due either to a direct plastic strain or to an indirect plastic strain resulting from an elastic strain followed by a secondary plastic strain. Thus, if chilling solidifies the membrane lipids, it may be assumed that this change in state would be accompanied by contraction, leading to "cracks" in the membrane. Below 4°C these "cracks" would be accentuated due to the expansion of the water (and, therefore, of the whole cell within the membrane), at least in the case of the plasmalemma. If rewarmed quickly, before a lethal loss of cell contents, and if reliquefaction of the lipids reversed the loss of semipermeability, this would be an elastic and noninjurious strain. Injury due to longer chilling periods would then be due to an irreversible loss of cell contents. If, however, solidification of the membrane lipids involves a simultaneous irreversible loss in semipermeability, then the chill-induced strain would be plastic, producing direct injury. Distinction between these two alternatives is obviously difficult, and perhaps merely a matter of semantics. It is, therefore, not surprising that the same mechanism has been suggested for indirect injury (see below).

2. Indirect Injury

Slow chilling injury is much more common than cold shock, and may require days or weeks of exposure to the temperature stress before producing injury (Table 4.2). Many suggestions have been made as to the cause of injury. An increase in permeability has been found, as in the case of direct injury, although the increase was only to three times the original value, after 4 weeks at 0°C, in the case of mature-green, chilling-sensitive

TABLE 4.2 CHILLING INJURY IN PLANTS FROM WARM CLIMATES[a]

Species	Time for first injury to appear	Time for complete killing (days)
Episcia bicolor Hook.	18 hr	5
Sciadocalyx warcewitzii Regel	24 hr	5
Eranthemum tricolor Nichols	48 hr	4–5
Eranthemum couperi Hook.	3–5 days	10
Boehmeria argentea Linden	8 days	20
Iresine acuminata	11 days	19
Uhdea bipinnatifida Kunth	15 days	16
Eranthemum nervosum R.Br.	20 days	30–35

[a] Exposed continuously to 1.4°–3.7°C in diffuse light and covered to prevent transpiration. From 28 species listed by Molisch (1897).

tomato fruit (Lewis and Workman, 1964). Chilling-resistant cabbage showed no increase. Since the change occurs so slowly, it is not necessarily due to a true increase in cell permeability. Ion absorption is an active process dependent on respiratory energy. If this source of energy is cut off, either by an uncoupling of ATP synthesis, or by a decrease in the aerobic oxidative processes of respiration, some of the actively absorbed ions could leak out. This would appear as an increase in permeability. It, therefore, appears possible that the slower, indirect injury is due to a gradual metabolic upset. Several kinds of metabolic disturbances have been described.

a. STARVATION

In the case of plants from temperate climates, a low temperature above freezing produces only an elastic (reversible) strain and the plant is normally uninjured. A similar elastic strain was early suggested as a possible cause of injury in chilling sensitive plants from tropical or subtropical climates. At chilling temperatures, respiration rate may exceed the rate of photosynthesis, and this may conceivably lead eventually to starvation (Molisch, 1896). However, all observed cases of chilling injury appear to occur long before the reserves are used up. There is, therefore, no direct experimental confirmation of this hypothesis.

Nevertheless, this concept must not be discarded too hurriedly. It has long been known that photosynthesis has a very high activation energy at low temperatures and, therefore, decreases more rapidly than respiration. This is a property of the carbon fixation cycle and has been explained by a mechanism for bleeding off intermediates in the cycle at low temperatures (Selwyn, 1966). As a result, chilling-sensitive plants may be below the compensation point at chilling temperatures, in which case carbohydrates will be broken down more rapidly than they are synthesized. If this condition is maintained long enough, starvation may conceivably occur, if not throughout the plant, at least in localized regions. Starvation of nonphotosynthesizing plant parts may, for instance, result from the inhibition of translocation by chilling temperatures (Geiger, 1969). The inhibition may be due to an effect on the sink or on the translocatory path. Recovery of translocation occurs rapidly in sugar beet and in a northern ecotype of Canada thistle, and very slowly in bean and in a southern ecotype of the thistle. In the case of the tropical grass, *Digitaria decumbens*, the high starch content accumulated during daylight disappears at night at normal temperatures but remains in the leaves when exposed to 10°C (Hilliard and West, 1970). This inhibition of starch translocation out of the chloroplasts appears to account for the decreased photosynthesis and growth even at day temperatures of 30°C. In this case, root starvation is conceivable.

b. Protein Breakdown

Protein breakdown at low temperatures without an equally rapid resynthesis has been suggested as a cause of injury, either due to a deficiency of proteins or a toxicity by the products of hydrolysis (amino acids, NH_3). Wilhelm (1935) produced evidence of such hydrolysis in the case of beans and tomato plants exposed to low-temperature stress, but Seible (1939) pointed out that an even greater hydrolysis occurs in control (unchilled) plants kept in the dark (Table 4.3). Minamikawa *et al.* (1961) failed to detect any hydrolysis of proteins in the mitochondria of chilled sweet potatoes (10 days at 0°C) although their respiratory activity declined. They concluded that proteins are not likely to be degraded during chilling injury. Razaev (1965), however, did observe proteolysis in chilling-sensitive plants,

TABLE 4.3 Protein Breakdown in Tomato Plants Exposed to Chilling Temperatures[a]

Leaf no.	Number days treated	Ratio of protein N: soluble N		
		Light controls	Coldroom plants	Dark controls
1	2	15.9	9.7	8.2
2	4	19.8	9.2	6.7
3	7	16.9	8.2	4.6
4	11	—	13.3	2.0

[a] From Seible, 1939.

but not in others resistant to chilling injury (corn and wheat, respectively).

In the case of microorganisms, the relatively high minimum temperature for the growth of mesophiles is commonly explained by a derangement of biochemical mechanisms that regulate the synthesis and activity of enzymes, although inability to transport solutes across membranes has also been suggested (Rose, 1968). According to Ingraham (1969), the regulated enzymes of bacteria may become so sensitive to feedback effectors that they cannot function at low temperatures. In other cases, the component proteins of essential aggregates lose their ability to form aggregates. There are also some that cannot synthesize functional ribosomes. This effect has been established in the case of *E. coli* (Friedman *et al.*, 1969). The cells of this organism are unable to divide at temperatures below 8°C. Ribosomal subunits accumulate at these temperatures, due to a block in the formation of the initiation complex, and, therefore, of protein synthesis. The exact step at which the cold-induced block occurred was not identified. Preliminary experiments indicated that the same block occurred in chick embryos.

c. Respiratory Upset

The respiration rate of cucumber fruit increased to a plateau at the chilling temperature (41°F) then decreased (Eaks and Morris, 1956). The increase occurred at the onset of injury, the decrease on death. A change in oxygen concentration (1–100%) had little effect on the respiration rate at 41°F, although it had the usual effect at 59°F. Conversely, CO_2 increased the injury at 41°F but had no effect at 59°F. Similar results were obtained with sweet potato (Lewis and Morris, 1956). It was uninjured at 59°F, injured at 50°F, and its respiration accelerated. The same result was obtained with the cotyledons of cacao seed (Ibanez, 1964). Chilled cotyledons showed a higher initial respiration than normal, but after 6 hr it was below normal. Eight minutes at 6°C reduced seed viability; 10 min at 4°C inhibited it completely. The speed of this chilling injury indicates that it is probably a direct injury, for instance, an increase in permeability of membranes (see Lyons and Raison below). If this permitted more rapid access of enzyme to substrate, the respiratory rate could conceivably increase. In the case of cotton (Amin, 1969), the order of events appears to be the opposite of that in all the above cases: respiration was severely reduced below 15°C, but exposures of 12 hr were followed by a subsequent increase in the respiration rate of the leaves and roots at 25°C.

Low-temperature injury in the rose variety Baccara leads to malformed flowers called "Bullhead" and black colored petals (Halevy and Zieslin, 1968). Three changes are actually involved. (1) Flower malformation was produced experimentally by chilling the plants to 5°C for 17 nights at the early stage of flower development. The blackening was due to (2) an increase in anthocyanin content to as much as two to five times that of the normal flowers, and (3) an accumulation of oxidation products of polyphenols. The lower the temperature, the darker the color of the petals. The black flowers contained an active polyphenolase system, but no activity could be detected in the normal flowers.

Effects of chilling on cytoplasmic streaming may possibly be related to these respiratory changes, since streaming depends on utilization of respiratory energy. In all the chilling-sensitive plants tested (tomato, watermelon, honeydew, tobacco, sweet potato), streaming was just perceptible in the trichomes or ceased completely after 1–2 min at 10°C (Lewis, 1956). In the chilling-resistant plants, on the other hand (radish, carrot, filaree—*Erodium cicutarium*), streaming continued even at 0°–2.5°C.

Just what phase of the aerobic process is inhibited by the chilling has not yet been demonstrated, although some attempts have been made to measure individual steps in the process. Lewis and Workman (1964) exposed mature green tomato fruit to 0°C for 4 weeks. Within 12 days, the tomato lost two-thirds of its capacity to esterify phosphate at 20°C, and one-half of its original capacity at 0°C. In contrast, chilling-resistant cabbage leaves showed a steady rise in capacity to esterify phosphate during a 5-week exposure to the same chilling temperature. In the case of sweet potatoes, however, the chilling injury affects the oxidative activity as much as the phosphorylating capacity, and the P/O ratio remains constant (Minamikawa *et al.*, 1961). The respiratory activity increased slightly during the first 10 days at 0°C, then declined sharply. They concluded that the declining respiratory rate was limited by the oxidative system and not by the concentration of ADP and ATP.

A possible cause of the respiratory rise in fruit associated with chilling injury is an increased production of ethylene, which is known to stimulate respiration. When citrus fruit were chilled by a temperature regime of 20°Day/5°Night, the ethylene in the fruit increased as much as 20–25 times (Cooper *et al.*, 1969). In the case of avocado, two varieties showed an increase and were injured, a third showed no increase and was uninjured.

Lyons and Raison (1970a) explain the effect of chilling on respiration by the same mechanism as used above for "direct" injury. Arrhenius plots of the respiration rates of mitochondria from chilling-sensitive plants (tomato and cucumber fruit, sweet potato root) showed a linear drop from 25°C down to 9°–12°C. From this point down to 1.5°C the slope increased. Chilling-resistant plants (cauliflower buds, potato tubers, beet roots), on the other hand, showed a linear decrease over the entire temperature range. They explain this by a phase change as a result of a physical effect of temperature on some membrane component such as lipids.

d. Toxins

Plank (see Smith, 1954) has explained injury by the accumulation of a cell toxin due to disturbances in the normal balance of biochemical processes. Injury would then depend on whether the rate of accumulation of the toxin exceeds the rate of its dispersal. In plums, the evidence is in agreement with this concept (Smith, 1954). Thus, injury due to 21 days' storage at 31°F was prevented if the fruit were warmed to 65°F for 1 to 2 days at about the seventeenth day. Furthermore, the rate of injury was more rapid at 40° than at 34°F. Similar results have been obtained by others (Pentzer and Heinze, 1954).

A possible connection between the toxin concept and the above described changes in respiration rate has been provided by earlier work by Banga (1936). He showed that in green tomatoes and bananas, the ratio of CO_2 evolution in nitrogen (anaerobic respiration) to that in air (aerobic respiration) was 0.7 at 13°C and 1.3–1.4 at 5°–6°C. Therefore, the chilling temperature has a greater inhibiting effect on the aerobic portion of respiration, for it decreases aerobic respiration twice as much as it decreases anaerobic respiration. After some time at the chilling temperature, aerobic respiration might be so greatly inhibited that anaerobic respiration would take over. This explanation is supported by the above finding that oxygen concentrations of 1–100% failed to affect respiration at chilling temperatures. More direct support has been produced by Murata (1969), who investigated chilling injury in banana fruit at 4° to 6°C. Acetaldehyde and ethanol contents increased in the chilled fruits after transfer to 20°C. There was an accumulation of α-keto acids in the peel and of browning substances (polyphenols) around the vascular tissue. Similarly, lima beans are injured when allowed to imbibe water at 5°–15°C. This chilling injury can be prevented if the beans are first allowed to absorb water vapor up to a content of 20% (Pollock, 1969). The liquid water would interfere with oxygen uptake, the water vapor would not.

It has long been known that continued anaerobic respiration kills most higher plants within 24–48 hr at room temperature. At chilling temperatures, it would undoubtedly take longer. It is, therefore, possible that the above described increases in respiration rate at chilling temperatures may actually be due to an enhancement of anaerobic respiration relative to aerobic respiration (the increase being due to a Pasteur effect), leading to the accumulation of toxic products and, therefore, injury. This injury might, then, result in the secondary decrease in respiration rate.

Other toxic products may also conceivably arise at the chilling temperatures. Cucumber leaves in a moist chamber at 0°C showed injury within 5 hr when exposed to light (Kislyuk, 1964). The injury was proportional to the irradiation, and the leaves did not photosynthesize. Photoperoxides were formed which oxidized the substrate.

e. Biochemical Lesions

This source of injury has been investigated much more thoroughly in animals than in plants (R. A. Peters, 1963). A biochemical lesion is an abnormality in the metabolism of an organism leading to a deficiency of an essential intermediate metabolite. Chilling-sensitive bacteria have provided particularly clear cases of biochemical lesions. A cold-sensitive mutant of *E. coli* ceases growth at about 20°C, versus 10°C in the more normal strain (Ingraham, 1969). When the mutant was supplied with histidine, the growth curves of the two strains were identical. It has been suggested (Ketellapper and Bonner, 1961) that chilling injury in higher plants is also due to a biochemical lesion. Attempts were, therefore, made to prevent the injury by supplying the organism with a number of possible intermediates that might be deficient. Some partial success was claimed in preliminary reports, but no clear-cut results have as yet appeared. Other investigators, however, have produced evidence which may be interpreted in this way. Podin (1966) was able to show that a specific reaction (transformation of lutein to violaxanthin) does not occur below 10°C in the chilling-sensitive banana, but continues down to 0°C in the chilling-resistant *Bergenia*. Similarly, chilled tobacco plants exhibited symptoms of nitrogen deficiency and there was a four-to-five fold increase in chlorogenic acid, and lesser increases in other substances (Koeppe *et al.*, 1970). The chilling temperature may have shifted the plant's metabolism from the normal pathway to an abnormal one, leading to a deficiency in some intermediates of nitrogen metabolism. A parallel explanation has been proposed for cottonseed killed by exposure to 5°C for 12 hr during hydration (Christiansen, 1968). Even a 30-min exposure may induce subsequent root abnormalities. Sensitivity to chilling persists during the initial 2–4 hr of hydration. If, however, the seeds are imbibed for 4 hr at 31°C and then dried, they are immune to chilling injury. Christiansen suggests that an irreversible event occurs during early hydration at normal germinating temperatures, but that it is blocked or disrupted by chilling. Chilling injury in the case of lima beans is apparently due, at least partly, to leaching of compounds absorbing at 264 mμ (Pollock, 1969), and, therefore, to a deficiency. The injury occurs if they are exposed to 5°–15°C during the initial stages of imbibition. It is avoided if the axes are just allowed to absorb water vapor; but longer exposures to chilling temperatures injure even these high moisture axes.

Unfortunately, such indirect evidence may frequently be interpreted in more than one way. A recent mutant of *Zea mays* has an extremely high chilling temperature (Millerd and McWilliam, 1968). It is able to produce chlorophyll only at temperatures above 17°C. When photooxidation is minimized, chlorophyll accumulates and seedlings can photosynthesize efficiently at a low temperature. This result points to the interrelationships between the different metabolic upsets and the difficulty in drawing a sharp line between them. Thus the photooxidation may be thought of as producing a toxic product (see above) or as preventing the synthesis of an essential component (chlorophyll) and, therefore, producing a biochemical lesion. It may also damage membranes.

3. Secondary Stress Injury

That desiccation may occur at low temperatures was first indicated by Sachs (1864; see Molisch, 1896). He found that plants such as tobacco and cucumber begin to wilt if their roots are cooled to temperatures just above zero. This may lead, eventually, to death by desiccation. Molisch (1896) confirmed these results. He also covered some plants with bell jars containing large pieces of wet filter paper before cooling them, and immersed the leaves of others in snow and ice water. In these ways he was able to expose the plants to temperatures just above freezing in the absence of transpiration. Even under these conditions chilling injury occurred (Table 4.2). It is also possible to prevent the secondary stress injury at chilling temperatures by severing the shoot from the root system and standing it in water. Sugar cane wilts if the root temperature drops to 15°C.

4. Mechanism of Chilling Injury

As in the case of the primary chilling injury, the secondary desiccation stress can also be explained by a permeability change. Desiccation of the plant can be due to one of two changes: a decreased water absorption or an increased water loss. Since there is no primary chilling injury, there can be no membrane damage. On the other hand, permeability of the cell to water is markedly affected by temperature, and Q_{10} values have been reported varying from low values characteristic of physical processes to high values typical of chemical processes. In those cases when only the roots were chilled, the permeability of the root cells was lowered without any change in permeability of the leaf cells. Water absorption would, therefore, be retarded although water loss was unchanged, and the result would be desiccation of the leaves. By analogy, if the whole plant is chilled and the leaves wilt, this must be due to a greater chilling-induced decrease in root than in leaf cell permeability to water.

So many changes have been found to accompany chilling, that there seems at first sight to be no single basic cause of the injury. Nevertheless, all the recorded observations can be grouped into two classes—permeability changes and metabolic changes. It is, therefore, conceivable that one of these may be the primary effect of chilling, and the other may result from this primary effect. In the case of the most rapid kinds of "direct" injury, permeability changes have been reported, although the injury is too rapid to be due to metabolic changes at the chilling temperatures. In these cases, at least, membrane damage must be the primary effect of chilling. Since there is no sharp line separating the rapid, direct injury from the slower, indirect injury, the differences between the two are more likely to be quantitative than qualitative. It is, therefore, reasonable to explain all the different kinds of chilling injury on the basis of a primary permeability change (Fig. 4.1).

(1) Direct injury or cold shock can be explained by a sudden, chilling-induced increase in permeability of the plasmalemma (the outer protoplast membrane), leading to loss of cell contents and ultimate death of the cell.

(2) Indirect injury. (a) Starvation injury or any injury due to a disturbance of photosynthesis may be due to an injurious change in chloroplast permeability. Respiratory disturbances could be due to similar injurious changes in mitochondrial permeability. (b) A small increase in mitochondrial permeability would explain the increase in respiratory rate at the onset of injury, for this would facilitate contact between substrate and enzyme. Even if the permeability change is severe enough to inhibit the mitochondrial system, anaerobic respiration would take over, and due to the Pasteur effect, this would increase the rate of CO_2 evolution. (c) Toxic products would then arise in the anaerobic process. (d) Similarly, biochemical lesions could occur, due to a deficiency of essential intermediates of aerobic respiration. (e) Protein synthesis could be inhibited due to a deficiency of these intermediates, e.g., ATP. Other associated changes could be similarly explained.

It, therefore, appears possible that all three types of chilling injury—direct, indirect, and secondary stress—may be caused by a single effect: a change in cell permeability. As suggested by Lyons and co-workers (see

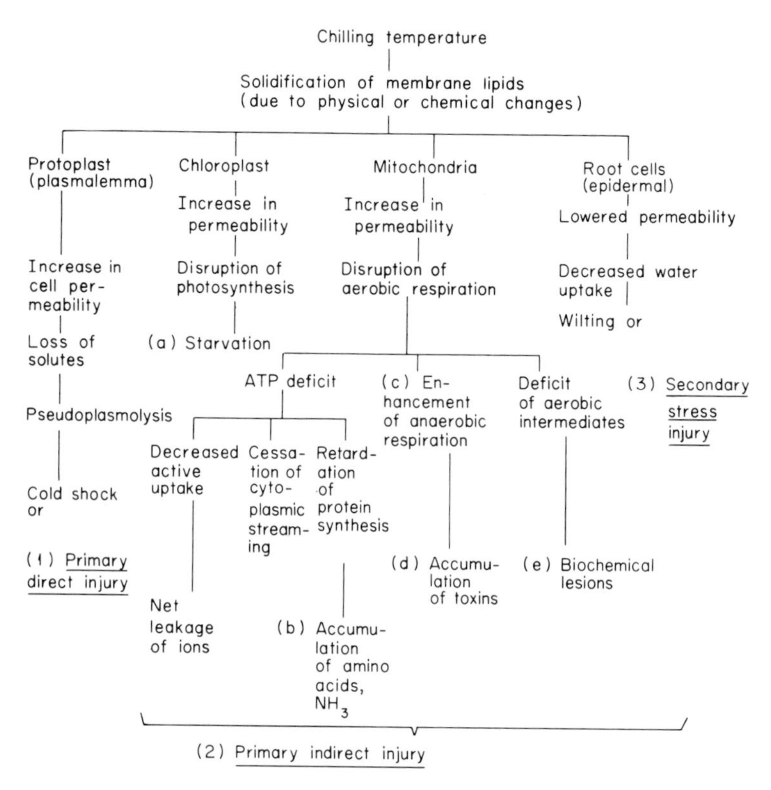

FIG. 4.1. Proposed dependence of all the kinds of chilling injury on a primary change in membrane lipids.

above) this may be due to a change in state of the membrane lipids, from liquid to solid. The cause of this change in state may be simply physical, due to the high melting point of the lipids, or chemical, due to their peroxidation.

C. Chilling Resistance

All plants from temperate climates and all psychrophiles among microorganisms routinely survive exposure to chilling temperatures and, therefore, are fully chilling resistant. Plants from tropical and semitropical climates, on the other hand, show varying degrees of resistance. Attempts have, therefore, been made to develop methods of measuring chilling resistance quantitatively, e.g., in rice (Adir, 1968; Tsunoda *et al.*, 1968). However, the usual method is simply to expose the plants to an arbitrary chilling temperature for an arbitrary time and to observe injury.

Since plants are poikilotherms, chilling resistance must be due to tolerance. Early evidence indicated that plants which are susceptible to chilling injury may harden in the field, becoming more resistant and therefore capable of surviving chilling temperatures (Sellschop and Salmon, 1928). Similarly, greenhouse plants normally susceptible to chilling injury when grown at 17°C become hardy if grown at 12°C for 2 months. Wheaton and Morris (1967) hardened tomato seedlings by as little as 3 hr at 12.5°C, though maximum protection was achieved in 48 hr. Sweet potato roots, however, failed to harden during this treatment, although effects on the respiratory behavior were observed. Pea seedlings, which are injured by exposure for 3 hr at −3°C (without freezing) may be hardened to survive this treatment by three daily exposures to 5°C for 3 hr (Kuraishi *et al.*, 1968). Kushmrenko and Morozova (1963) succeeded in hardening cucumber seedlings against chilling injury at 3°–12°C, by cooling them to 22°C for 18 hr a day during 4 days. Solov'ev and Nezgovorov (1968) hardened cucumber leaves at 10°C, against chilling injury at 3°–5°C.

The Russian investigators have reported increases in chilling resistance of corn plants as a result of a variety of treatments. Al'tergot and Bukhol'tsev (1967) hardened corn seedlings by alternating a gradual drop in temperature from the optimum to zero, with a gradual rise back to the optimum. Treatment of the seed of corn or cotton with tetramethylthiuram disulfide (TMTD) before planting, considerably increased the survival in cold and wet weather (e.g., 7°–10°C, Radchenko *et al.*, 1964; Nezgovorov and Solov'ev, 1965). This may, of course, be a protection against microorganisms in the soil, rather than against true chilling injury. Other substances used in sprays (e.g., 2,4-D and KCl + NH_4NO_3 + boric acid) have been reported to protect cucumber leaves against chilling injury (Solomonovskii and Pomazova, 1967). Attempts have been made to relate resistance to a number of factors—respiration rate and conversion of stored carbohydrates (Mishustina, 1967), the concentration of amino acids and proteins (Petrova, 1967), fat content, and metabolic rate on sprouting (Beletskaya, 1967).

Chilling resistance of citrus may also be affected by nutrient treatment, a deficiency decreasing resistance (Del Rivero, 1966). Buckwheat shows an increased resistance to both chilling and cold shock (−3°C for 5 min) when potassium supply is increased, but an increase in calcium increases resistance to the former and decreases resistance to the latter (Korovin and Frolov, 1968).

Artifically induced chilling resistance has been reported by application of the chemical substances picolinic acid and Dexon to cotton plants (Amin, 1969). The treated plants recovered better after chilling exposures to 15°C. Since these substances are respiratory inhibitors they are thought to prevent the respiratory disturbances normally induced by the chilling, due to a protection of specific systems by the inhibitors.

Practically no attempts have been made to identify the changes that occur during hardening. Kuraishi *et al.* (1968), however, found an increase in the ratio of NADPH:NADP. This would seem to indicate that the chilling treatment tends to lower the reduction potential of the living cells and the hardening treatment counteracts this by developing a greater reduction potential. Cotton seedlings chilled at 5°C show a continuing decrease in ATP concentration (Stewart and Guinn, 1969). If returned to optimum temperature after 1 day, the initial ATP concentration was restored, but not after 2 days of chilling. The decrease in ATP on chilling was prevented by hardening for 2 days at 15°C immediately before chilling. The ATP level of hardened was higher than that of unhardened seedlings, increasing more in the roots than in the leaves. When the hardened plants were exposed to chilling (5°C), the ATP level increased in the leaves and decreased in the roots, the increase leveling off after 2 days of chilling.

D. Mechanism of Chilling Resistance

Lyons *et al.* (1964) found a higher content of polyunsaturated fatty acids in the mitochondria of chilling resistant- than in those of chilling-sensitive plants. Similarly, when fed acetate-2-^{14}C at 10°–40°C for 5 hr, seeds of castor bean, sunflower, and flax showed an increased formation of unsaturated fatty acids at the low temperatures (Harris and James, 1969). One of the major factors was an increase in availability of oxygen, which is the rate-limiting factor for desaturation.

The chemical mechanisms involved in such changes of unsaturation are not known. According to Christopherson (1969), the formation of lipid peroxides from the unsaturated fatty acids of membranes is probably highly injurious. He showed that glutathione peroxidase can catalyze the reduction of the hydroperoxides of all the polyunsaturated fatty acids that occur in the subcellular membranes of rat liver. It is, therefore, possible that the membrane lipids are normally protected against such peroxidation by the high reducing power maintained in actively metabolizing cells. When, however, the cells are cooled to chilling temperatures, their metabolism slows down and their reducing power decreases. The membrane lipids may then perhaps form peroxides, and lose their property of semipermeability. In the case of plants hardened to chilling, an increase in the ratio of NADPH:NADP occurs (Kuraishi *et al.*, 1968), which may prevent the lipid peroxidation, for instance, by scavenging the oxygen.

On the basis of these results, and of the concept of chilling injury described above, chilling-resistant plants may be divided into three groups. (1) Permanently chilling-resistant plants would always possess membrane lipids with a low solidification temperature and, therefore, a high degree of unsaturation. Plants of this group survive chilling without any hardening treatment. (2) Chilling-sensitive plants that suffer direct or rapid chilling injury would possess membrane lipids with a high solidification temperature due to a high degree of saturation. Chilling these plants would quickly solidify their membrane lipids, leading to direct (rapid) injury (cold shock). A hardening treatment would induce chilling resistance in these plants only if it increased the degree of unsaturation of the membrane lipids. (3) Chilling-sensitive plants that suffer indirect (slow) chilling injury would possess membrane lipids with a low solidification temperature and, therefore, with a high degree of unsaturation. They would, therefore, survive brief chilling without injury. During chilling, however, peroxidation of the membrane lipids would occur, resulting in solidification and, therefore, injury. A hardening treatment would induce chilling resistance in these plants by preventing peroxidation of the membrane lipids at chilling temperatures, perhaps by maintaining a high enough concentration of reducing substances (e.g., NADPH) to scavenge the oxygen.

The two kinds of hardening would, therefore, be basically different. In the one case (group 2) dehydrogenation or oxidation of the membrane lipids would have to occur, in the other (group 3) peroxidation of the membrane lipids would have to be prevented or reversed. It is, of course, possible that plants of group 2 would have to resynthesize their membrane lipids at the hardening temperature, in order to obtain them in the unsaturated state. In this case, a high concentration of ATP would also have to be maintained. It is, therefore, conceivable that their resistance also depends on specific protein (enzyme) properties.

CHAPTER 5

LOW-TEMPERATURE STRESSES—THE FREEZING PROCESS

A. The Freezing Stress

A chilling stress commonly includes only temperatures above the freezing point of water and is, therefore, incapable of producing a change in the state of the plant's water. A freezing stress, on the other hand, occurs only if the external temperature is below the freezing point of water, and may be defined as the freezing potential of the low-temperature stress. If the plant does freeze this may, of course, induce secondary strains in other components of the plant. In terms of the primary strain (ice formation), a freezing stress can only be measured by the freezing potential of the plant's environment. Since the concentration of a plant's cell sap varies within a wide range, the freezing potential of any one temperature will differ for different plants. For any one plant, however, the lower the temperature, the greater the amount of ice formed at equilibrium until all the freezable water has crystallized. The simplest measure of freezing stress is, therefore, the number of degrees (or kelvins) the environmental temperature is below the freezing point of pure water at atmospheric pressure (i.e., $-T$°C where T is the temperature in °C). The freezing strain will, therefore, always increase with the freezing stress until a point is reached where no more water can freeze (see below). Instead of the stress being proportional to the strain, as in the case of a physical system, the increase in the freezing strain decreases logarithmically with each equal additional freezing stress.

B. Observations of Frozen and Thawed Tissues

The freezing of plants has long been a controversial subject for scientists from many disciplines. The frost splitting of trees, which occurs with a "crack like that of a gun," led to the belief that the plant tissues expand on freezing and may ultimately rupture because of this expansion (Bobart, 1684; Chomel, 1710; Du Hamel and de Buffon, 1740; Strömer, 1749; Thouin, 1806; Hermbstädt, 1808). Schübler (1827) pointed out that this splitting occurs only in thick trees (1½–2 ft in diameter) and not at all in thin ones (a few inches in diameter), although the temperature drops much lower in the latter. He further states that when a tree is split in this way, no significant injury occurs; on the other hand, it is the youngest twigs that are injured first, although they never suffer such splits. Nevertheless, even the limpness of thawed herbaceous plants was thought to be due to cell rupture (Senebier, 1800; Thouin, 1806; Hermbstädt, 1808). Plants that survive freezing were believed able to prevent such expansion or even to contract due to the presence of oils that shrink on freezing (Bobart, 1684; Du Hamel and de Buffon, 1740; Strömer, 1749; Reum, 1835).

This "rupture theory" has been accepted even relatively recently (Goodale, 1885, Kerner von Marilaun, 1894; West and Edlefsen, 1917; Goetz and Goetz, 1938; Bugaevsky, 1939a), perhaps because it is based on the sound physical fact that water expands on freezing. Since the plant consists mostly of water, the assumption that it, too, expands on freezing was a logical one. Others besides Schübler began to doubt the theory. Du Petit-Thouars (1817) found it difficult to believe that some plants could survive such expansion, yet he knew that many survive freezing. Soon after, Göppert (1830) microscopically examined literally thousands of plants in search of cell rupture. In agreement with previous observations, he found that the juice could be easily squeezed out of freeze-injured leaves, but he was unable to find any torn cell walls, although the cells were somewhat collapsed. Others confirmed his observations (Morren, 1838; Lindley,

TABLE 5.1 Volume Changes on Freezing of Plant Tissues[a]

Plant part	Percentage change in volume	Reference
Leaves	−25	Hoffmann, 1857
Petioles and midribs	−1 to −3½ (length)	Sachs, 1860
Beet root and pumpkin fruit	0	Sachs, 1860
Bark of twigs	−13.5	Wiegand, 1906a
Wood of twigs	−2.5	Wiegand, 1906a
Yeast cells	−10	Molisch, 1897
Spirogyra cells	Diameter reduced to ⅓	Molisch, 1897
Cladophora cells	Diameter reduced by 20%	Molisch, 1897

[a] See also Müller-Thurgau, 1880.

1842; Schacht, 1857; Martens, 1872; see Prillieux, 1872; Schumacher, 1875). Nägeli (1861) pointed out, in fact, that cell walls can stretch much more than the small amount that would occur even if all the cell's water froze. He also showed that freeze-killed spirogyra cells will collapse if transferred to glycerin after thawing. This could not happen if there were any tears in the wall. The final death blow to the rupture theory was the discovery that tissues actually contract, instead of expanding, on freezing (Table 5.1). Even freeze splitting of trees can be explained by an asymmetrical contraction (Caspary, 1857).

The explanation of plant contraction during freezing required direct microscopic examination of frozen tissues. This showed that the rupture theory was unsound at the outset, for instead of the ice forming inside the cell, as had been tacitly assumed, it normally occurs outside it—in the intercellular spaces or on the surface of the tissues (Table 5.2). The ice crystals grow to masses larger than the cells (Fig. 5.1), and the cells contract and even collapse (Figs. 5.1 and 5.2). This contraction may be so severe that, in the case of the epidermal cells with colored vacuoles, the opposite sides of the cell can be seen to come in contact with each other (Iljin, 1933b, 1934; Siminovitch and Scarth, 1938). Since the cells are firmly connected with each other, this results in a contraction of the tissues and of the organ as a whole; at the same time, the air from the intercellular spaces is squeezed out (Lindley, 1842; Wiegand, 1906b). The ice crystals may be confined to certain regions, forming such large masses (as much as

TABLE 5.2 Observations of Extracellular Ice Formation in the Plant

Plant	Reference	Plant	Reference
Daphne, hydrangea, iris, fritillaria	du Petit-Thouars, 1817	Twigs of *Acer negundo*	Dalmer, 1895
Many	Göppert, 1830	Algae, agave, aloe, beet	Molisch, 1897
Exotic plants	Caspary, 1854	Algae, potatoes, beets	Wiegand, 1906a
Potatoes, beets	Schacht, 1857	Tree twigs, buds	Wiegand, 1906b
Pumpkin slices	Sachs, 1860	Cabbage, fungi	Schander and Schaffnit, 1919
Iris germanica, etc.	Prillieux, 1869	Fucaceae	Kylin, 1917
Nitella syncarpa	Kunisch, 1880	*Buxus sempervirens*	Steiner, 1933
Roots and tubers	Müller-Thurgau, 1880	Peach buds	Dorsey, 1934
		Cortical cells of trees	Siminovitch and Scarth, 1938

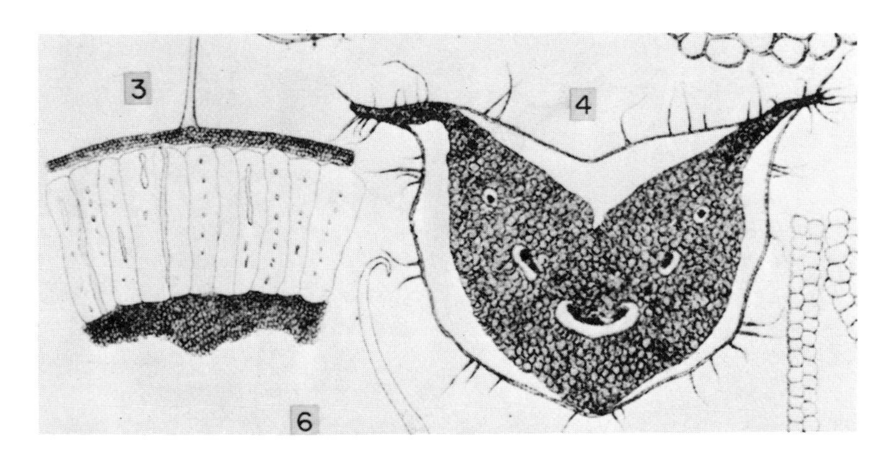

FIG. 5.1. *Left:* large ice masses formed between the much smaller cells. *Right:* contraction of tissue caused by ice masses formed beneath the epidermis. From Prillieux, 1869.

1000 times the size of a cell; Müller-Thurgau, 1886) that tissues are pushed apart (Caspary, 1854; Schacht, 1857; Fig. 5.1). Even moss leaves that have no intercellular spaces form ice crystals between the walls which split them apart and grow at the expense of water from the cells (Modlibowska and Rogers, 1955). On thawing, the tissues are limp and have a water-soaked appearance (Prillieux, 1869); when the ice in the intercellular spaces is converted to water, the spaces are essentially free of air, and the cells are flaccid due to the loss of water to the ice masses. This led earlier workers to believe that cell rupture had occurred. If the tissues are uninjured, the intercellular water is soon reabsorbed by the cells. As they regain their turgor, air enters the spaces, and the water-soaked appearance is quickly lost. If thawing is gradual enough, the water may be reabsorbed by the cells as soon as it forms, and no infiltration with water occurs. Injured cells are unable to reabsorb the water (Wiegand, 1906b).

The freeze-killed cells characteristically show "frost plasmolysis"—a contraction of the dead protoplast, leaving a large space between it and its cell wall (Fig. 5.2). This phenomenon was mentioned by several of the early investigators and studied carefully by Buhlert (1906). The observations of "frost plasmolysis" were always made on dead cells after they were thawed. The term is, therefore, an unfortunate one, since it has led some to believe that this "plasmolysis" is supposed to occur while the cell is frozen (Becquerel, 1949). Actually, it is due to the cell contraction (cell wall as well as protoplast) during extracellular ice formation and the

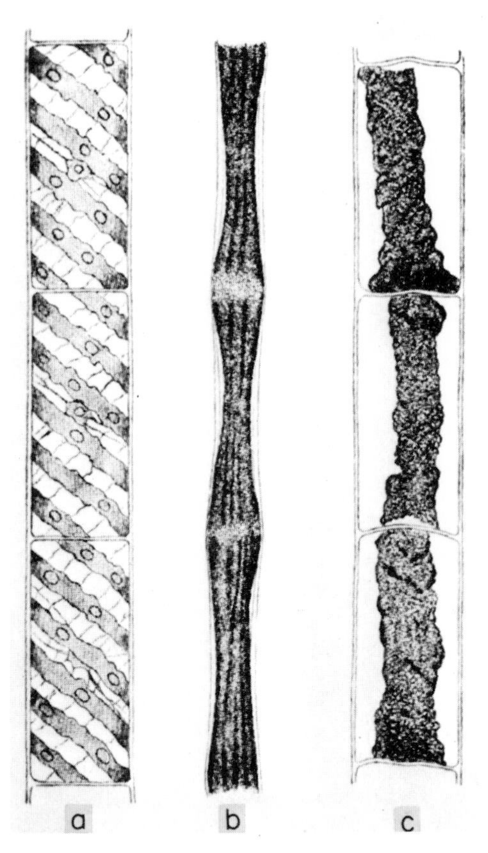

FIG. 5.2. Spirogyra (× 300): (a) Normal, unfrozen; (b) frozen extracellularly, showing cell collapse without any ice inside the cell; (c) thawed, showing "frost plasmolysis." From Molisch, 1897.

inability of the dead protoplast to reabsorb the water formed in the intercellular spaces on thawing of the extracellular ice. As a result, the cell wall expands back to nearly its original shape, while the dead protoplast remains contracted, giving a false appearance of plasmolysis. This would seem to indicate that the injury had occurred during the freezing process and the cells were nonfunctional by the time the ice began to thaw.

Although ice is normally formed extracellularly, many observations of intracellular ice formation have been made in the laboratory (Table 5.3; Fig. 5.3). The freezing occurs in sudden flashes in one cell at a time, the crystals appearing to be both within and outside the vacuole (e.g., in onion epidermis; Chambers and Hale, 1932). When freezing is relatively slow, the ice may form between the wall and the tonoplast (Asahina, 1956). This may be the same phenomenon as the "thin hull of ice inside the wall" previously described by Schander and Schaffnit (1919). Onoda (1937) explains this exceptional type of ice formation as follows. The protoplasm of the cells is strongly dehydrated by ice growth in adjacent cells. As a result, it separates from the wall. The space formed becomes filled with ice, resulting in pseudoplasmolytic freezing. This description, however, is not in accord with the cell collapse observed by Iljin and others. Ice formation between the wall and the protoplasm in *Nitella* had previously been described by Cohn and David (1871) and in *Conferva fracta* by Göppert (1883). In cortical cells of trees, with thick protoplasm layers, Siminovitch and Scarth (1938) saw the ice form first at one end of the protoplast, and from there spread around each side of the vacuole to the other end. Only when the freezing of the protoplasm was complete did the vacuole freeze. In one case, in fact, the protoplasm froze but the vacuole remained unfrozen. Stuckey and Curtis (1938) also state that the cytoplasm appears to freeze before the vacuole. The flashlike, cell-by-cell freezing that occurs when ice forms intracellularly has been strikingly recorded in more recent years by the moving picture camera (Luyet, Modlibowska).

In nearly all cases (Table 5.3), the intracellular ice formation was brought about by rapid freezing, e.g., by supercooling well below the

TABLE 5.3 Observations of Intracellular Ice Formation in the Plant

Plant	Temperature at which frozen	Reference
Apple		Morren, 1838
Pumpkin	−12 to −20°R	Sachs, 1860
Nitella syncarpa	−3 to −4°C	Cohn and David, 1871
Roots and tubers	−10°C	Müller-Thurgau, 1880
Conferva fracta		Göppert, 1883
Codium bursa	−11°C	Molisch, 1897
Tradescantia crassula (staminal hairs)	−6.5°C	Molisch, 1897
Agave, aloe, beet	Rapid cooling	Molisch, 1897
Tradescantia discolor and *T. guianensis*	Slow freezing	Molisch, 1897
Tradescantia and *Vallisneria spiralis*	−4 to −5°C	Schaffnit, 1910
Cabbage	−5°C	Schander and Schaffnit, 1919
Fungi	Rapid freezing	Schander and Schaffnit, 1919
Cabbage, tree cortex	−12°C	Siminovitch and Scarth, 1938

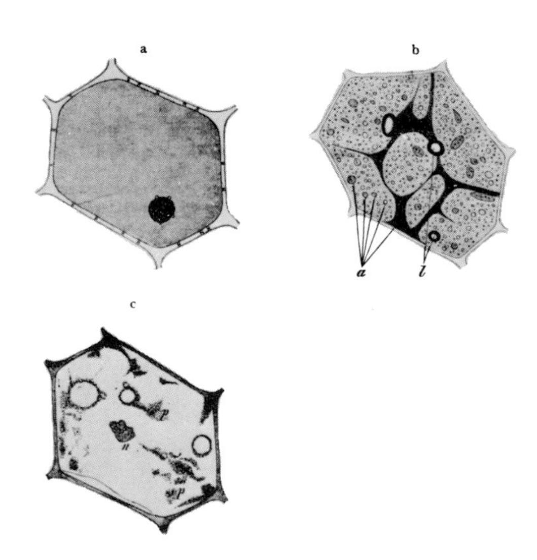

FIG. 5.3. Epidermal cell of *Tradescantia discolor* (× 300). (a) Normal cell filled with red cell sap; (b) frozen intracellularly. Most of cell sap converted to ice, leaving concentrated, dark red unfrozen sap at (a). Gas bubbles at l. (c) Thawed, dead cell. Wall and nucleus stained red, coagulated protoplasm at p. From Molisch, 1897.

freezing point of the tissue and then inducing or waiting for ice formation. Not only was freezing more rapid than normally occurs in nature, but, to facilitate observations, sections of tissues were used. That intracellular ice formation is not due to sectioning was shown by Siminovitch and Scarth (1938), who observed it in whole potted plants cooled from 0° to −10°C in one-half hour. This resulted in death of the plant. Mazur (1963, 1970) has examined the kinetics of water loss from cells at subzero temperatures and concluded that slow cooling decreases the probability of intracellular freezing.

The fact that ice can form intracellularly in sections of tissue frozen in the laboratory leads to the suspicion that this kind of ice formation may also occur in nature, if the freezing is sufficiently rapid. It is possible, for instance, that the so-called "sunscald" may be due to this. On a cold winter day, the sun shining on one side of a tree may raise its temperature so high (25°–30°C above the shady side) that thawing occurs. When the sun suddenly disappears behind a cloud (or another object) the temperature of the thawed tissue drops rapidly, and refreezing may perhaps occur so suddenly as to result in intracellular freezing and death. This explanation has never been confirmed experimentally, and there are other equally valid explanations of sunscald (see Godman, 1959). Nevertheless, some observations are in agreement with it. According to Weiser (1970), evergreen foliage, which is not injured at −87°C during slow freezing in winter, is killed at −10°C when thawed tissues are frozen rapidly (8°–10°C/min). Foliage on the southwest side of these plants were actually found to cool at these rates on sunny winter days in Minnesota, when the sun moved behind an obstruction. He, therefore, concluded that intracellular freezing is the cause of sunscald. An exceptional kind of intracellular freezing with death of the cells may occur in apple flowers in spring (Modlibowska, 1968). During the first phase of freezing, ice forms extracellularly, lifting the skin of the receptacle. This may result in some mechanical damage which heals rapidly on return to favorable conditions. During the second phase of freezing, ice forms intracellularly in the placenta, styles, and ovules, causing death.

A carefully applied electrometric method (Olien, 1961) seems capable of distinguishing between intracellular and extracellular freezing, by enabling calculation of the relative content of liquid extracellular water. According to these results, leaves of hardened barley show only extracellular freezing and are eventually injured by it. Leaves of unhardened barley show the reverse pattern indicating a sudden loss of solutes before extracellular freezing can occur, perhaps due to intracellular freezing. A simpler method is by following the change in permeability of the tissues to gases (Le Saint, 1957), which can only be explained by the replacement of the intercellular gases by ice crystals. The process of ice formation in sections of tissue under laboratory conditions has been thoroughly described and illustrated by Asahina (1956).

Artificial freezing has produced another phenomenon not normally occurring in nature. Many fruits and vegetables, when frozen for purposes of preservation, have their cell walls ruptured (Woodroof, 1938; Mohr and Stein, 1969). The freezing process in this case is, of course, initiated at a much lower external temperature than occurs in nature, and is rapid enough to permit intracellular freezing. Furthermore, the cells in most fruit and some vegetables are very large, thin-walled, and have a very high water content. They are, therefore, more likely to show wall rupture than the smaller, drier, thicker walled cells of overwintering plants. Nevertheless, if frozen slowly, even artificially frozen tissues may show the kind of extracellular freezing observed in nature. In the case of fruits and vegetables frozen artificially, the size of ice crystals may be increased as much as 500 times by slowing down the rate of freezing (Woodroof, 1938). The largest crystals may be up to 1000 times the size of the cells in asparagus or spinach. Such large crystals result in crushing and distortion of the cells, just as in nature (see Fig. 5.1).

C. The Cause of Extracellular Freezing

It has long been a question as to why ice does form extracellularly under normal conditions. On the basis of the available evidence, the process may be suggested to occur as follows. It has been shown that ice normally crystallizes first in the large vessels both in the case of leaves (Asahina, 1956) and even in defoliated mulberry trees (Kitaura, 1967). Freezing proceeds along the vessels from a few nucleation points and reaches all parts of the shoots at a relatively high velocity (about 34 cm/min/°C), proportional to the supercooling (also called undercooling or subcooling). Freezing in the vessels is to be expected since their large diameter does not favor undercooling, and their dilute sap probably has the highest freezing point of any of the plant's water. Once ice forms in the vessels, it will spread throughout the plant body. Each living cell is surrounded by a highly lipid plasma membrane. It has been shown that ice crystals are barred by this membrane from inoculating the cell contents (Chambers and Hale, 1932). Consequently, the ice from the vessels spreads only throughout the part of the plant external to the living cells, i.e., the intercellular spaces. The crystals will form here at the expense of the water vapor in the air and of the surface film of water on the cell walls.

Freezing is far less likely to extend into the cell walls (except, of course, into the pits in the walls), because their water-containing microcapillaries are less than 0.1μ in diameter (since they cannot be resolved by the optical microscope). The vapor pressure of water in capillaries can be calculated from the equation (Moor, 1960):

$$p = p_0 e^{-c/r}$$

where p = vapor pressure of water in capillaries; p_0 = normal vapor pressure of pure water; $c = 10^{-7}$ cm; r = radius of capillary. Such calculations reveal that the vapor pressure of the water in the microcapillaries of the cell wall will be lowered by more than 1%. This will, of course, lower the freezing point slightly. More important, undercooling of water is greatly increased in capillaries.

At the instant before ice formation in the intercellular spaces, the cell contents must be practically at temperature and vapor pressure equilibrium with the immediately adjacent wall and intercellular spaces. When intercellular ice begins to form (as a result of seeding from the vessels), the vapor pressure in the intercellular spaces drops sharply (Fig. 5.4). Due to the capillary size of the cells, undercooling of their contents will be favored, and due to protection by the lipid plasma membrane, seeding with the extracellular ice crystals will be prevented. As soon as the tissue

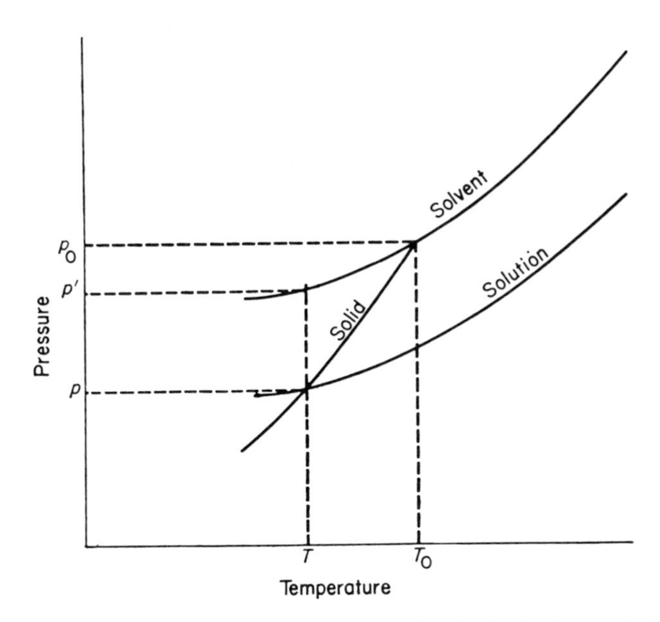

FIG. 5.4. Accelerated decrease in vapor pressure of water in the frozen state with drop in temperature. At temperatures below the freezing point, both pure solvent and solution in the supercooled state have higher vapor pressures than that of the solid (ice) at the same temperature. p_0 = vapor pressure of pure water at the freezing point; p = lower vapor pressure of solution at its freezing point T; p' = higher vapor pressure of supercooled water at the same temperature T. (From Getman and Daniels 1937.)

temperature drops below the freezing point of the cell contents to the smallest degree, their vapor pressure will be higher than that of the intercellular ice at the same temperature (Fig. 5.4). Consequently, water will diffuse through the semipermeable lipid plasma membrane to the intercellular ice. In this way, these ice crystals on the external surface of the cell wall will grow at their bases, and the cell itself will contract due to the water loss. If the temperature continues to drop at a gradual enough rate, this diffusion of cell water to the external ice loci will continue, and will steadily increase the cell sap concentration. If temperature equilibrium is then attained, vapor pressure equilibrium will also soon occur. However, when the cell contents are at the same temperature and vapor pressure as the intercellular ice, they must be exactly at their freezing point (Fig. 5.4). Consequently, the plant can be kept frozen indefinitely at this constant temperature without any possibility of ice forming in the remaining cell contents. At this temperature equilibrium, the tissues will contain strongly contracted living cells with an unfrozen cell sap sufficiently concentrated to have a freezing point equal to that of their prevailing temperature. The major part of the tissue will now consist of intercellular spaces filled with ice. There will be little air in these spaces, since during the temperature drop the solid matter of the tissues contracts at the same time as the freezing water expands into the intercellular spaces. This squeezes and pushes the intercellular air out of the tissues. Such a simultaneous contraction of the structural components of the plant and expansion of the contained water on freezing has long ago been given as the explanation for the winter splitting of tree trunks that sometimes occurs with a noise like the crack of a gun (see above).

It must be realized that the rate of this diffusion of water to ice loci outside the cells is limited by the permeability of the lipid plasma membrane surrounding the living cell. Although this membrane is highly permeable to water (relative to solutes), it significantly slows up water movement, so that this is not as rapid as the movement of water through the aqueous cell wall. Therefore, if the temperature drop is rapid enough, this diffusion to the extracellular ice cannot occur with sufficient speed to increase the concentration of the cell contents as rapidly as the temperature drops. The rapidly cooled cell may, therefore, eventually reach a temperature sufficiently below its freezing point to induce spontaneous ice formation intracellularly.

The above description of "slow" extracellular freezing assumes a rapid enough temperature drop to permit a spread of the ice front throughout the plant body, so that the ice crystals contact all or nearly all the cells. This kind of freezing is more likely to occur when relatively rapid, such as in artificial "slow" freezes. Under natural conditions, the freeze may be gradual enough to prevent the ice front from spreading throughout the plant body. It may then be confined to specific regions, growing slowly at the expense of water diffusing to it from relatively distant unfrozen regions (see Fig. 5.1).

If hoarfrost forms on the leaves, as commonly occurs in fall and spring (Kitaura, 1967), this may inoculate the internal tissues rather than spontaneous ice formation in the vessels, particularly in the case of tissues with narrow vessels or tracheids; but the net result would be the same since the plasma membrane would again block seeding of the cell contents. Ice nuclei may, of course, form without inoculation from the outside. Kaku and Salt (1967) suggest that such ice nucleation takes place at sites associated with the cell walls and not on nucleators suspended in the water.

It must be realized that even the slow freezing of sections under the microscope differs, in some respects, from the freezing of a normal whole plant. If the section is immersed in an aqueous medium this will, of course, freeze first, and the subsequent series of events will be very different from that described above (Genevès, 1955).

The ability of the plasma membrane to act as a barrier against inoculation of the cell contents by extracellular ice crystals, is undoubtedly due to its lipid nature. Even nonliving membranes that are freely permeable to liquid water may be completely impermeable to growing ice crystals (Lusena and Cook, 1953). According to Mazur's (1963) calculations, if ice forms at a sufficiently low temperature due to rapid cooling, the ice crystals may then be small enough to penetrate the plasma membrane and therefore to induce intracellular freezing.

D. Eutectic Points

It is important to know whether more and more ice continues to form in the plant as the temperature drops, or whether a point is soon reached at which all the water suddenly freezes. The latter conception was at one time believed true, since it is known that many solutions have a "eutectic (or cryohydric) point" at which all the solute crystallizes and, therefore, the remaining water in the absence of solute also solidifies. The importance that such a eutectic point would have is illustrated by Asahina's (1962a) experiments with sea urchin eggs. When frozen in KNO_3 solution with a eutectic point at −2.9°C, the cells were killed at this temperature, although they normally survived much more severe freezing. Such a eutectic point, however, is clear-cut only in pure solutions. A mixture such as Earle's salt solution containing glucose and glycerine has no real eutectic point but a eutectic zone in which a slow and progressive melting occurs as the temperature of the deep frozen solution rises, becoming marked from −38°C upward (Rey, 1961). Some solutes, such as sugars, tend to prevent the crystallization of others; and since there are many solutes in the cell sap and sugars among them, the plant cannot be expected to have a specific eutectic point.

Actual observation of the freezing process in plant tissues has indeed corroborated this expectation. In many cases, a "double freezing point" has been found and the two points were, at first, interpreted as the freezing point and the eutectic point, respectively. The second freezing point was readily eliminated by either killing the tissues or blotting the surface of the cut tissues (Luyet and Gehenio, 1937); it was, in any case, at too high a temperature (−2° to −3°C) for the eutectic points of any of the substances known to be present in large quantities in the plant.

These results, however, still fail to show whether or not there is any temperature at which all the freezable water in the plant can be said to have crystallized. Attempts have therefore been made to measure the ice formed in plants at different temperatures using three main methods: (1) the dilatometer method which measures the ice by its expansion on freezing (Fig. 5.5); (2) the calorimeter method which measures the ice formed by the heat of crystallization released (Fig. 5.5), and (3) the ice flotation method which measures the change in specific gravity on freezing (Scholander *et al.*, 1953). The second is the most commonly used since it does not require the use of possibly toxic liquids, and it can be applied at

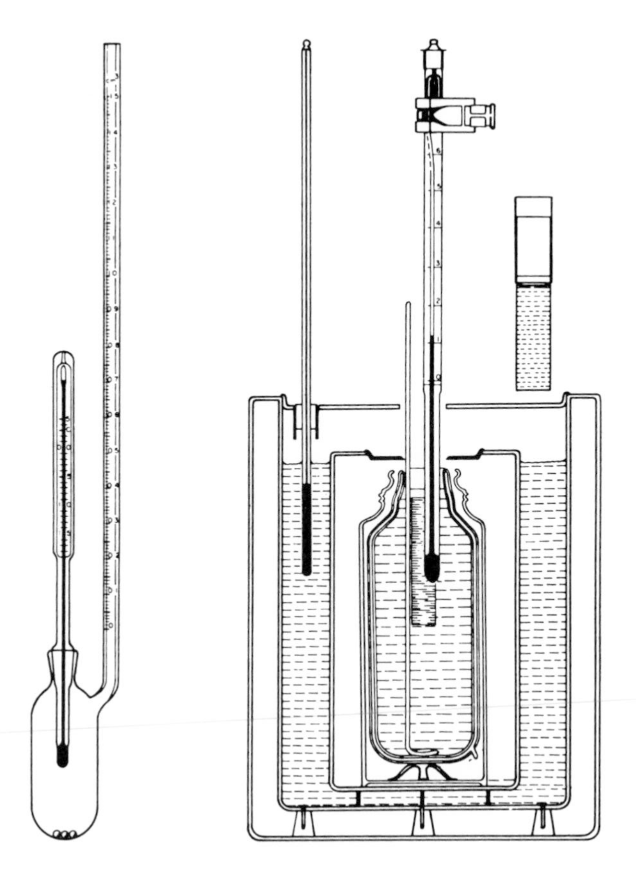

FIG. 5.5. *Left:* Dilatometer vessel used to measure ice formation in plant tissues by measuring the expansion due to freezing when submerged in oil. From Lebedincev, 1930. *Right:* Calorimeter for measuring ice formation in tissues by the heat of fusion. (From Levitt, 1939.)

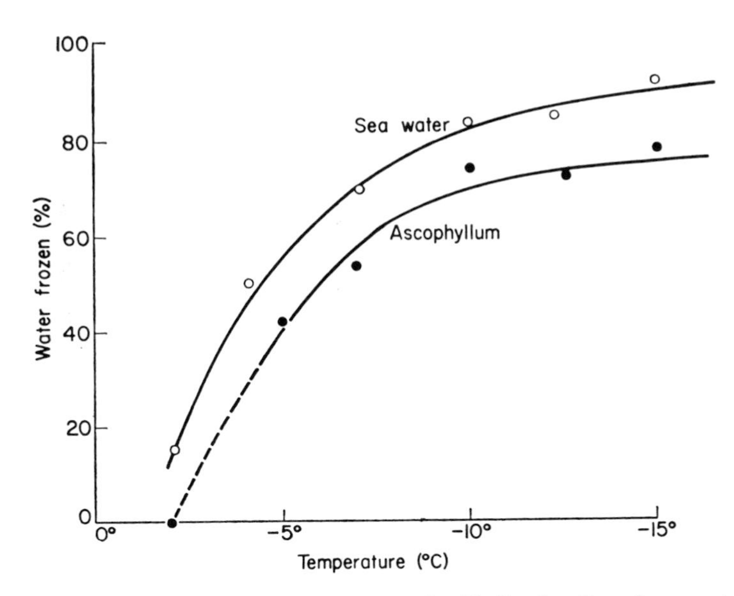

Fig. 5.6. The freezing of algal water compared with the fraction of sea water that freezes at the same temperature. (From Kanwisher, 1957.)

lower temperatures than the first method, i.e., at temperatures at which the expansion is too small to be easily measured. The third method can be used only for air-free tissues. By use of these methods, measurements of ice formation in the plant have repeatedly shown that crystallization continues down to at least −30°C (Scholander *et al.*, 1953), or even down to −72°C in yeast (Wood and Rosenberg, 1957). The freezing curve parallels that of a pure solution (Fig. 5.6) although a smaller fraction of the water freezes due to binding by the tissue colloids. There are, of course, difficulties in the use of these methods, and some assumptions must be made in calculating the amount of ice formed (see Wood and Rosenberg, 1957). Krasavtsev (1968), for instance, has used the calorimetric method to measure ice formation down to temperatures as low as −60°C, but the calorimetric method requires accurate calculations of heat exchange due to processes other than the heat of fusion of ice (Johansson, 1970). Since heat capacity and other values are not accurately known for temperatures as low as −60°C, the calculations of the amounts of ice formed at these low temperatures are unreliable. That the results are essentially correct for moderate freezing temperatures is attested to by similar results obtained by use of an indirect dehydration-melting point method. This method of estimating ice formation eliminates the assumptions adopted in the above methods although replacing them by others (Salt, 1955).

E. The Double Freezing Point

Although no eutectic point can be detected for plant tissues, a freezing point can be readily determined. The plant's temperature nearly always drops below this point before freezing occurs (Fig. 5.7). There is, therefore, also an undercooling point (Table 5.4), which may vary much more than the freezing point. It may even be altered by mechanical shocks (Luyet and Hodapp, 1938). The freezing point itself is not constant. It is lower in living than in dead tissues (Table 5.5), and varies somewhat with the rate of temperature drop (Maximov, 1914; Walter and Weismann, 1935; Luyet and Gehenio, 1937). However, if the cooling rate is slow enough, the freezing point of the living tissue is identical with that of dead tissue (Luyet and Galos, 1940). In most cases the true freezing point of the living tissue also coincides with the freezing point of the expressed juice, though in some cases it may be 0.5°C below it (Marshall, 1961). As mentioned above, living tissues may show two freezing points. Although this double freezing point cannot be due to a eutectic point because both freezing points are too high, it is still in need of an explanation. It occurs in living but not

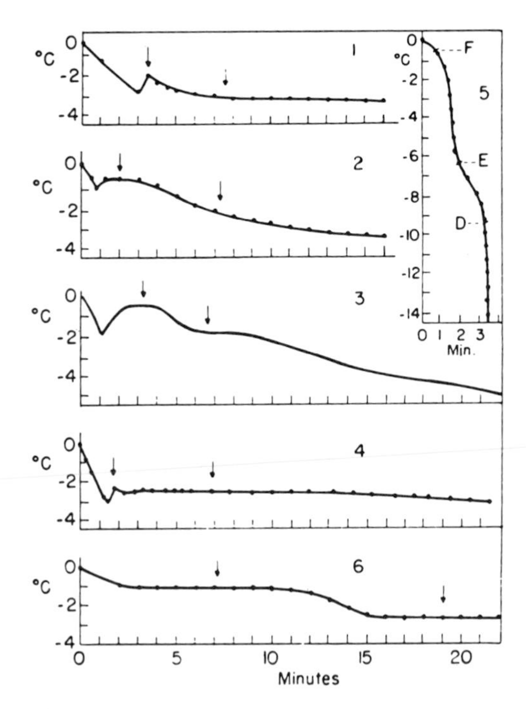

FIG. 5.7. The double freezing point as obtained by several investigators. (From Luyet and Gehenio, 1937.)

TABLE 5.4 UNDERCOOLING AND FREEZING POINTS OF SOME TISSUES (°C)

Plant	Undercooling point	Freezing point	Reference
Potatoes	−6.1	−0.98	Müller-Thurgau, 1880
Aspergillus niger			
in 1% dextrose	−6	−0.29	Bartetzko, 1909
in 20% dextrose	−8	−2.89	Bartetzko, 1909
in 50% dextrose	−14	−8.9	Bartetzko, 1909
Red beet root	−6.83	−1.88	Maximov, 1914
Tussilago farfara	−2.70	−1.83	Maximov, 1914
Potato tuber	−1.60 to −2.42	−0.85 to −1.57	Maximov, 1914

in dead tissue (Luyet and Gehenio, 1937). Furthermore, it can be suppressed in living tissue by a preliminary surface drying, accentuated by soaking in water. Luyet and Gehenio suggest that the first freezing point is due to freezing of the extracellular water, while the second is due to freezing of intracellular water.

Other explanations have also been proposed—that the first freezing point is due to freezing of the damaged surface layer of tissue and the second due to freezing of the inner tissues (Aoki, 1950). Levitt (1958) has proposed a turgor theory to explain it. Ice forms first at the cut surface of the piece of tissue. Since it is so small the whole piece of tissue is essentially in vapor pressure equilibrium; and since there is no barrier to the propagation of the ice throughout the intercellular spaces, the ice spreads forming a continuous thin layer on the cell surfaces. If the tissue has been rinsed with water before freezing, it is in the turgid state at the instant of freezing and, therefore, has a higher water potential than that of its expressed sap.

TABLE 5.5 THE FREEZING POINTS OF LIVING AND DEAD TISSUES (°C)

Plant	Living	Dead	Reference
Phaseolus leaves	−1.1	−0.5	Müller-Thurgau, 1880
Cypripedium leaves	−2.0	−0.5	Müller-Thurgau, 1880
Red beet roots	−2.15 to −2.55	−1.25	Maximov, 1914
Tussilago farfara	−2.03 to −3.35	−0.97	Maximov, 1914
Helleborus viridus	−4.07 to −5.88	−2.07 to −2.15	Maximov, 1914
Potato tubers	−1.0	−0.5	Müller-Thurgau, 1880
	−1.22	−0.63	Maximov, 1914
	−0.87 to −2.00	−0.69 to 0.80	Walter and Weismann, 1935
	−1.2 to −2.0	−0.5 to −0.75	Luyet and Gehenio, 1937

The freezing point of the cells is therefore *higher than the freezing point of the cell sap at atmospheric pressure*. This results in a high first freezing point somewhere between 0°C and the freezing point of the cell sap depending on the balance between the temperature gradient and the amount of water able to freeze in the tissues. This initial, relatively rapid, ice formation at the cell surfaces removes enough water from the living cells to cause loss of turgor. The result is a *marked* lowering of the freezing point due to the double effect of turgor loss and a slight increase in cell sap concentration. From this point down, much larger quantities of ice can form. As an example of this, a fully turgid piece of tissue with a freezing point lowering (of the cell sap) of −2°C would have about 10% of its water frozen between 0° and −2°C and 45% frozen between −2° and −4°C. This would give the appearance of two freezing points. The double freezing point would, therefore, occur whether freezing is intracellular or extracellular, although showing up more sharply in the former case due to greater undercooling. It should be pointed out that if Aoki's explanation is correct, the double freezing point is an artifact that occurs only in cut tissue. If the turgor theory is correct, it can occur also in normal, turgid, undamaged tissue, though perhaps depending on the rate of temperature drop. Kaku and Salt (1967) have recently reinvestigated this question. However, since the double freezing point is apparently dependent on artificial conditions of freezing, there may be more than one explanation of the phenomenon, depending on just what these artificial conditions are. This plurality of explanations appears to be borne out in the case of stem sections of *Cornus stolonifera*. The number of freezing points varied with the season (McLeester *et al.*, 1968). During summer and winter, there were two distinct points, in early autumn three distinct points, and in spring one prominent first freezing point which tended to mask the second point.

Weiser (1970) calls these "exotherms," recognizing the fact that they are not true freezing points. His third exotherm occurred at the death of the plant. It can, therefore, be logically explained as follows. Due to the resistance to flow of water through the plasma membrane, as the tissue is cooled progressively, the cell contents will always be supercooled a little below its freezing point. At the moment of death, the plasma membrane becomes freely permeable, and there is a surge of water through it to the extracellular ice loci. This sudden freezing of a larger amount of water than before death, and the consequent release of a large amount of heat of fusion will result in a flattening or even a slight rise in the freezing curve. On the basis of this explanation, the third exotherm should occur only when cooling is relatively rapid, e.g., when cooled from 0° to −36°C in about 20 min. (Weiser, 1970).

CHAPTER 6

FREEZING INJURY

A. Occurrence

In contrast to chilling injury, which occurs in tropical and subtropical plants, but not in plants from temperate zones, freezing (frost or cryo-) injury may occur in all plants. It is, therefore, far more prevalent than chilling injury and has been more intensively studied. Many descriptions of it are found in both the old and the new literature (see Levitt, 1956). Those plants that are subject to chilling injury (as well as some that are not) are usually killed by the first touch of frost (Molisch, 1897). On the other hand, some that are native to cold climates may be frozen solid at the lowest temperatures without injury (Scholander *et al.*, 1953). Between these two extremes all gradations occur (Table 6.1). Even for a single plant the range of freeze-killing temperatures may be large, depending on its physiological state (Table 6.1). Does the freezing stress always produce injury in the same way, or is there more than one kind of injury?

B. Primary Direct Freezing Injury

By definition, direct injury due to the freezing process, can occur only as a result of intracellular (or more correctly intraprotoplasmal) freezing. Extracellular freezing does not permit direct contact between the ice and the protoplasm and therefore cannot produce direct freezing injury. Intra-protoplasmal freezing, on the other hand, always kills the cells, if the ice crystals are large enough to be detected microscopically (see Chapters 3 and 5). The crystals, presumably, damage the protoplasmic structure,

TABLE 6.1 KILLING TEMPERATURES FOR PLANTS IN THE FROZEN STATE[a]

Species	Killing temperature (°C) when frozen		Reference
	Unhardened	Hardened	
Potato tuber	−1.53		Maximov, 1914
Red beet root	−2.15		Maximov, 1914
Wheat		−12–15	Tumanov and Borodin, 1930
Cabbage	−2.1	−5.6	Levitt, 1939
Vaccinium vitis idea	−2.5	−22	Ulmer, 1937
Erica carnea	−3–4	−18–19	Ulmer, 1937
Sempervivum glaucum	−3.0	−25	Kessler, 1935
Rhododendron ferrugineum	−4	−28	Ulmer, 1937
Globularia nudicaulis	−4	−19	Ulmer, 1937
Globuluria cordifolia	−4	−18–19	Ulmer, 1937
Saxifraga caesia	−4	−29–30	Ulmer, 1937
Homogyne alpina	−4	−18	Ulmer, 1937
Saxifraga aizoon	−4	−18–19	Ulmer, 1937
Hedera helix	−4.5	−18.5	Kessler, 1935
Rhododendron hirsutum	−5	−28–29	Ulmer, 1937
Saxifraga cordifolia	−5	−19	Kessler, 1935
Carex firma	−5–6	−29–30	Ulmer, 1937
Pinus mugo	−6	−40–41	Ulmer, 1937
Empetrum nigrum	−6	−29	Ulmer, 1937
Juniperus nana	−6–8	−26	Ulmer, 1937
Pinus cembra	−9	−38	Ulmer, 1937
Pinus cembra	−10	−40	Pisek, 1950

[a] From Levitt, 1956.

perhaps by lacerating the membranes and destroying their semipermeability. Intracellular freezing injury has rarely been observed in nature, as for instance the injury to apple blossoms (Chapter 5). As mentioned above (Chapter 5) it may possibly explain the sunscald injury on the sunny side of trees. The recent explanation by Olien *et al.* (1968) for the winterkilling of barley in Michigan and in the northeastern United States, apparently also involves a direct freezing injury. The damage most commonly occurs during cold weather following a midwinter thaw. They propose that it may depend on the type of ice structures which develop in the lower crown tissue. Since ice structure can be important only if the ice penetrates the protoplasm, this would certainly be a direct freezing injury.

C. The Time Factor as Evidence of the Kind of Injury

The best evidence of primary direct freezing injury is, therefore, the observation of intracellular freezing. Such observations are difficult and are rarely, if ever, made in frozen plants. The next best criterion is the speed of the process, since direct injury due to any stress is rapid, compared to indirect injury by the same stress, and since rapid freezing has been shown to favor intracellular freezing (Chapter 5).

1. *Rates of Freezing and Thawing*

Normally, it is not the rates of freezing and thawing that are measured, but merely the rates of cooling and warming. The actual rate of freezing will depend not only on the cooling rate, but also on the degree of supercooling, the properties of the tissue, etc. In order to eliminate one of these complications (supercooling), rate of freezing has been defined as the ratio of the temperature difference to the time difference between the beginning of the freezing process and the time of the final temperature (Rottenberg, 1968). Even if used as a relative measure, this gives only the average value and may fail to detect an extremely rapid initial rate following supercooling. The actual rate of freezing has rarely been measured. In the lemon, a subcooling to −5.2°C induced a rate of spread of freezing of 15 cm/min (Lucas, 1954). In a qualitative way, however, rate of cooling may be an indirect measure of the relative rate of freezing in a specific plant part or tissue frozen under identical conditions, and rate of warming an indirect measure of the relative rate of thawing. The quantitative ranges for "slow" and "rapid" cooling are several orders below the range for protective ultrarapid cooling (Table 6.2). As a general rule slow cooling in a freezing chamber means no more than 1°–2°C/hr. In nature, the rate is usually less, but there are exceptions. Biel *et al.* (1955) inserted thermocouples into the

TABLE 6.2 Approximate Ranges for Different Speeds of Cooling

Rate of cooling	Sections of tissue	Whole plant
Slow	1–2°C/min or less	1–2°C/hr or less
Rapid	5–20°C/min	5–20°C/hr
Ultrarapid for relatively dry tissues	100°C/sec or more	
Ultrarapid for fully moist tissues	10,000°C/sec or more	

stolons of ladino clover and found maximum cooling rates of 10°F/hr. Even these rates were fast enough to produce rapid cooling injury. Sprague (1955), observed more injury in ladino clover and alfalfa when cooled from 36°–10° or 5°F at a rate of 7° to 10°F/hr than at slower rates. Rates of cooling of 5°C/hr have frequently been found to produce rapid cooling injury (see Levitt, 1958). This rate is incapable of producing rapid freezing injury in the case of sections (Levitt, 1957a), since it is slow enough to permit the small amount of ice formation to occur outside the cells. When freezing cells under the microscope, 1°–2°C/min may be considered slow freezing since, if the tissue is inoculated with ice, it will probably still freeze extracellularly at these rates of cooling. On the other hand, a drop of only 1/3°C/min is rapid enough to induce intracellular freezing in whole plants that are not inoculated (Siminovitch and Scarth, 1938).

In general, then, sections must be cooled about 60 times as rapidly as whole plants in order to induce rapid freezing injury. This can be explained if the following two assumptions are made: (1) rapid freezing injury is due to intracellular freezing (Mazur and Schmidt, 1968), and (2) whether or not intracellular freezing occurs depends on the rate of diffusion of water from the cell interior as well as the rate of temperature drop. Some evidence for both these assumptions has already been presented and more is given below. In the case of sections one living cell thick, the total distance that water must diffuse from the cell interior to the ice is one-half the cell width, since ice crystals occur on both surfaces of the section. In the plant as a whole, ice forms first in and around the xylem vessels. Since some living cells in a leaf may be as much as 1 mm from the nearest vessel, the maximum distance that the water has to diffuse at the moment of freezing would be 50 times that in the sections, if the cells are 40 μ wide, and this would slow up the diffusion of the water from the cell by a factor of 50. This factor is therefore of the right order to account for the fact that the freezing rate required to produce injury in sections is 60 times more rapid than that for whole plants.

If we take into account this difference between sections and whole plants, we must conclude that in both cases, rapid cooling increases the freezing injury and, therefore, raises the killing temperature (Levitt, 1956). In some cases, rate of cooling has no effect on the freezing injury. Warming rates also may or may not affect freezing injury, although increased injury due to rapid warming has been found less frequently than increased injury due to rapid cooling (Levitt, 1956). Many of the negative results are readily understood. A tender plant is killed by the first touch of frost no matter how slowly it is cooled or warmed. Similarly, a plant already killed by a severe freeze cannot be brought back to life by slow warming.

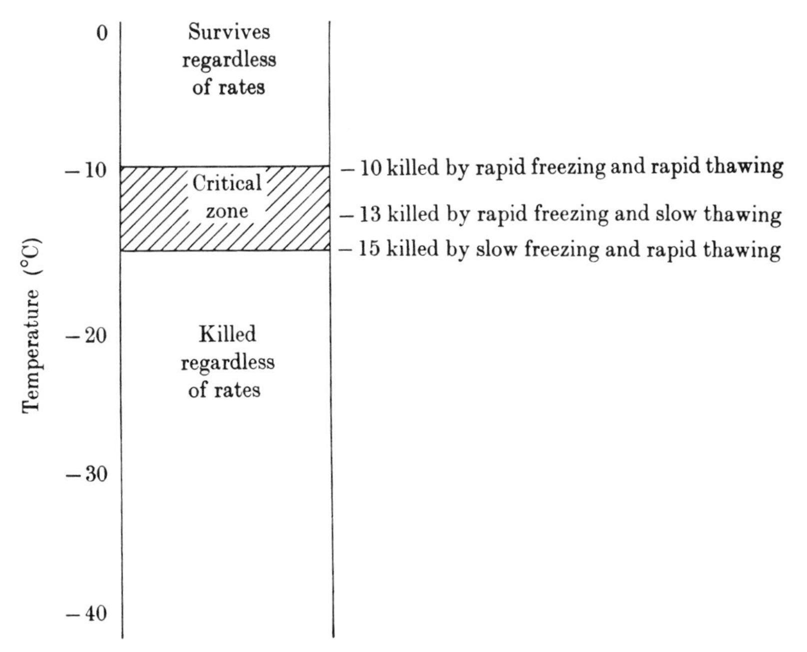

FIG. 6.1. Effect of rates of freezing and thawing on a plant of moderate hardiness. (From Levitt, 1966b.)

All these results with different rates of freezing and thawing indicate that there is a critical freezing zone for any one hardy plant (Fig. 6.1). Above and below this zone, rates of cooling and warming have no effect. In the upper temperature zone no injury occurs regardless of the rates of cooling and warming. In the lowest zone, the plant is killed no matter how slowly it is cooled or warmed. In the critical zone, however, both rates are important. Death occurs at the top of the zone when both cooling and warming are rapid, at the middle or bottom when one is rapid, the other slow, below it when both are slow.

2. Length of Time Frozen

The earlier quantitative evidence on the injurious effect of the length of time the plant is kept frozen is not very conclusive. For relatively short periods of time (2–24 hr) the injury appears to be independent of the time, once equilibrium has been attained (Levitt, 1956). For longer periods of time (1–30 days) injury appears to increase with the length of time frozen. Unfortunately, however, the freezing temperatures are usually not kept sufficiently constant, and it is conceivable that, at least in some cases, the injury is due to repeated partial thawing and refreezing (see below). Sakai (1956a) has clearly shown the relationship between length of time frozen and amount of injury. The time at −10°C survived by mulberry and poplar increased with increase in hardiness, reaching 30 days for mulberry and 90 for poplar. Freezing at −5°C was fully survived 360 days by poplar. Browning due to injury occurred after 60, 30, and 10 days at −10°, −20°, and −30°C, respectively (Table 6.3). In the case of sea urchin eggs, the freezing injury increases with the time frozen at −10° or −20°C, during freezes of 1–72 hr (Asahina, 1967). Here too, however, there was little change for the first 6 hr, and sometimes for the first 24 hr.

3. Repeated Freezing and Thawing

As in the case of rates of cooling and thawing, repeated freezing and thawing can be expected to increase the injury only within a critical temperature zone that will be specific for the plant and for its physiological state. In one case, for instance, when the freezing and thawing was repeated seven times, the killing temperature was raised from −21°C for a single freeze to −13°C (Winkler, 1913). Some grass varieties seem to be more sensitive to repeated freezing than others (Thomas and Lazenby, 1968).

4. Postthawing Treatment

It has long been known that death may not occur until hours or even days after thawing, and that the injury may sometimes be reversible. It is, therefore, standard procedure to keep the thawed plant cool (0°–5°C) for about 24 hr after thawing. If sections of tissue are transferred to $CaCl_2$ solutions after thawing, they show less injury than if transferred to H_2O (Levitt, 1957a).

TABLE 6.3 The Time (Days) Survived by Twigs Frozen at Different Temperatures

	Temperature frozen and stored at			
Tree	−5°C	−10°C	−20°C	−30°C
Mulberry	180	30	10	½
Poplar	360	90	50	20

[a] From Sakai, 1956a.

These effects of the time factor demonstrate that in some cases freezing injury occurs rapidly and is, therefore, presumably direct injury; in other cases, it occurs too slowly for direct injury. These conclusions may be examined more logically on the basis of the moment of injury.

D. The Moment of Freezing Injury

1. Injury during Freezing

When freezing injury occurs during rapid cooling to temperatures that fail to injure when slowly cooled (see above), this is indirect evidence that the injury must occur during the freezing process. More direct evidence has been obtained in a few cases. Some plants undergo a color change or develop a specific odor when killed by any method. Some of these plants show these changes when frozen, and, therefore, must be dead while in the frozen state (Molisch, 1897). Furthermore, these changes occur essentially immediately on freezing. Unfortunately, it was not determined whether this color or odor characteristic of dead cells was accompanied by extracellular or intracellular freezing. Cells of spruce and cherry, however, showed a yellow fluorescence when killed by either flash (1°/min) or slow freezing, during which intracellular and extracellular freezing were observed, respectively (Krasavtsev, 1962). According to Le Saint (1966), when freezing is fatal the evolution of CO_2 ceases, leading to the conclusion that irreversible changes during the freezing cause death.

2. Injury while Frozen

Since injury may increase with time after 24 hr at the same freezing temperature (see above), there must be a second moment of injury—while frozen but after freezing equilibrium has been attained. This, therefore, cannot be a direct freezing injury.

3. Injury during Thawing

The fact that the injury may be increased by increasing the speed of thawing is explainable only if some injury may also occur during the thawing process. Yoshida and Sakai (1968) suggest that the important factor may be the rate of rehydration of the cells at temperatures near the freezing point of the tissues. Leaves frozen at −15°C were killed by thawing in water at 20°C, but survived if first warmed for 2–3 min at −2° to −5°C before transfer to the water at 20°C. Whatever the explanation, if injury occurs during thawing, it cannot be due to direct freezing injury.

4. Injury after Thawing

The effect of the temperature after thawing and of the medium in which the cell is immersed, indicates that injury can either progress or be repaired after thawing. Further evidence can be obtained by observing the cells immediately on thawing. They may still be alive even though the plant will be dead several days later. Any injury developed after thawing cannot be direct freezing injury. Even if the difference is due to repair, it is not likely to be due to direct freezing injury, for there is no evidence that intracellular freezing injury can be repaired by controlling the thawing or postthawing conditions.

From the above considerations of the time factor and of the four moments of injury, it must be concluded that besides the primary direct freezing injury, there are one or more other kinds of injury.

E. Primary Indirect Freezing Injury

Since intracellular freezing, with few exceptions, always causes instant death, it appears unlikely that it can ever lead to indirect injury. However, what of the exceptional cells that have been reported to survive intracellular freezing? If survival is due to "vitrification" or to the formation of intracellular crystals too small to be observed microscopically, the cell apparently remains alive as long as this condition is maintained (see above). If, however, the crystals are allowed to grow large enough, exactly the same kind of injury occurs as when these larger crystals are formed at the instant of freezing—primary direct freezing injury. Recent observations of tumor cells (Asahina *et al.*, 1970) have revealed that they may remain in the translucent state, with intracellular crystals too small to interfere with ordinary transmitted light even at temperatures above −30°C. It, therefore, appears possible that the exceptional cases of survival of intracellular freezing that have been reported in the literature are all of this type. We must, therefore, conclude that when intracellular freezing injures the cell, it is always due to direct injury, and that there is no evidence for the existence of indirect injury due to the primary freezing process.

F. Secondary Freezing Injury

There are three kinds of freezing injury due to secondary stresses induced by the primary freezing stress. (1) Freeze-smothering is due to a gas stress imposed by the ice barrier to gas diffusion. (2) Freeze-desiccation is due to a freeze-induced water stress, leading to a net evaporation from the plant. (3) Freeze-dehydration is due to a freeze-induced water stress, leading to exosmosis of water from the cells to ice centers within the plant.

1. Freeze-Smothering

Plants injured due to an ice covering are said to suffer smothering injury. If the implications of the term are correct, it must involve a secondary stress, since it indicates a disruption of normal respiration due to a gas stress rather than a primary stress effect of the ice. There is considerable evidence in favor of this interpretation. Thus killing may occur at higher temperatures if the plants are covered by an ice sheet for a considerable period of time, than in the case of control plants frozen in air. In the case of alfalfa, injury was manifest only after 7–10 days and it could not be ascribed to any direct mechanical effect of the ice (Sprague and Graber, 1940). After 20 days under the ice sheet, mortality was high. Sprague and Graber (1940) attributed this to the accumulation of toxic products of aerobic and anaerobic respiration, since ice inhibits the diffusion of CO_2 and other respiratory products.

In favor of this explanation, plants immersed in water at 1°C were injured at approximately the same rate and to the same degree as those frozen in ice at −4°C. When CO_2 was bubbled through the water, survival was more rapidly reduced and injuries were more intense. However, CO_2 was only one of several gases that appeared toxic. Later tests (Sprague and Graber, 1943) showed that even at the highest concentrations of CO_2 used, a longer storage time was needed to cause injury than when the plants were frozen in blocks of ice. This led them to conclude that the external concentration of CO_2 was not directly toxic, but that respiratory compounds accumulated internally until they reached a toxic concentration. Thus, injury similar to that produced by storage in blocks of ice could be induced by removal of both CO_2 and O_2.

Not all plants show the same sensitivity to encasement in ice. When frozen in this way at a temperature that is noninjurious when frozen in air (−3°C) ladino clover died within 12–14 days and white clover survived 4 weeks or more (Smith, 1949). A relationship to CO_2 concentration was again indicated, since this was consistently higher for the ladino clover. In general, Smith (1952) found that the survival of ice encasement among legumes was approximately in the same order as their winter hardiness. He suggests that this may be due to a higher level of metabolic activity in the less hardy legumes. This is in agreement with Bula and Smith (1954), who found that the rate of loss of carbohydrates in legumes during winter dormancy was inversely proportional to hardiness.

2. *Freeze-Desiccation*

Under natural conditions, winter injury due to desiccation has frequently been reported or inferred from the lower transpiration rates of the more northerly distributed species (Bates, 1923; Iwanoff, 1924; Walter, 1929; Thren, 1934; Michaelis, 1934; Rouschal, 1939). It has even been suggested that the higher concentrations of the cell sap in alpine plants than in individuals of the same species from the plains indicates a better ability to remove water from the soil (Senn, 1922). Direct determinations have failed to produce any evidence of winter injury due to freeze-desiccation in very cold climates (Hildreth, 1926). Twigs do not dry out appreciably at temperatures below 41°F, and above this temperature the rate of water transfer to the twigs is quite adequate to prevent injury from the transpirational loss (Wilner, 1952). Although desiccation does not seem to be a cause of winter injury in the climates of Minnesota (Hildreth) or of Canada (Wilner), it may possibly be a factor in other climates characterized by a loss of water from thawed leaves at a time when translocation of water to them is impossible due to freezing of other parts of the plant. A possible example is the commonly observed death of evergreens in regions above the snow line when exposed to the wind for a few days without snow cover (Pisek, 1962). Resistance to this kind of winter injury may depend on the transpiration rate (Larcher, 1957). In eastern Hokkaido (Japan), minimum temperatures reached −30°C and soil temperatures at 10 cm depth remained below zero for 3.5 months even on the southern slopes (Sakai, 1970a). Temperatures of stems and leaves of young conifers in winter rose to about 17° and 9°C at midday on the southern and northern slopes, respectively. They remained unfrozen for 6 and 2 hr, respectively, during daytime. Under these conditions, the conifers on the southern slopes were intensely desiccated toward the end of February. Damage was observed as browning of the stem bark. Artificial desiccation induced a similar browning. Fir, spruce, and arborvitae which were damaged by this kind of desiccation were able to survive freezing at −50°C (fir) or even −120°C (spruce and arborvitae).

3. Freeze-Dehydration

If the ice forms extracellularly (see Chapter 5), the cell is dehydrated and, therefore, collapses as a whole in exactly the same manner as if the water is removed by evaporation. Therefore, extracellular freezing is due to a secondary water stress, leading to a dehydration strain. It may be suggested that, due to the much lower temperature, a freeze-dehydration may not injure, even though equal in intensity to an evaporative dehydration that injures at higher temperatures. What, then, is the evidence of injury due to freeze-dehydration?

a. Freezing and Drought Tolerance

There is ample evidence of a correlation between freezing tolerance and drought tolerance, which will be documented in Chapter 21.

b. The Four Moments of Injury

In the absence of ice coverings and of long periods of freeze-desiccation, there are only two kinds of freezing injury. Since intracellular freezing injures instantly, the other three moments of injury can occur only as a result of extracellular freezing. On the other hand, the existence of intracellular freezing injury during the first moment does not eliminate the possibility of extracellular freezing injury during the same first moment. The fact that a single slow cooling (which normally prevents intracellular freezing) can injure the plant no matter how briefly it is held in the equilibrium frozen state and no matter how gradually it is thawed, or how carefully it is protected against postthawing injury, is strong evidence for the existence of extracellular freezing injury during the first moment. Evidence of this kind of injury has been produced for winter wheat (Salcheva and Samygin, 1963). It, therefore, seems safe to conclude that the secondary, freeze-dehydration injury due to extracellular freezing, may injure during any one of the four moments of injury: (1) during freezing, (2) while frozen at equilibrium, (3) during thawing, and (4) after thawing.

c. Frost Plasmolysis

Intracellular freezing does not remove the water from the cell and, therefore, does not lead to cell collapse as it does in the case of cell desiccation. This difference provides a method of distinguishing between primary direct freezing injury (due to intracellular freezing), and secondary freeze-dehydration injury (due to extracellular freezing). Plants killed by freezing were long ago observed to show "frost plasmolysis" (see Chapter 5). It is difficult to explain this phenomenon except by the death of the extracellularly frozen and, therefore, contracted cell. On thawing, the dead protoplast is freely permeable and, therefore, unable to reabsorb the thaw water osmotically. The cell wall, on the other hand, is elastic and, therefore, snaps back to nearly its original position as water enters between it and the dead, contracted protoplast. The result is a "frost plasmolysis" of the dead cell.

d. Ice Formation

Direct observation of plant tissues in nature has repeatedly revealed that the ice formation is extracellular and has failed to detect intracellular freezing. It can, of course, be suggested that as long as ice formation is extracellular no injury results, but as soon as a small amount of intracellular freezing occurs (perhaps too small to be detected) the cell is killed. There is neither evidence of this amount of intracellular freezing, nor evidence that its occurrence would be fatal. On the contrary, all observations of killing due to intracellular freezing have been confined to large amounts of intracellular ice formation. More important, however, are the many records of injury in the presence of extracellular freezing and in the complete absence of intracellular freezing. Müller-Thurgau (1886) long ago succeeded in removing the pieces of ice from frozen potato tubers and showed that they were many times the size of the adjacent cells which were soft and gave no signs of being frozen and yet were dead. In agreement with these results, Terumoto (1960a) recently observed large ice masses between the cells of table beets frozen at −4° to −5°C. They occurred mostly concentrically in the vascular bundle ring region. The specific conductivity and betacyanin content of this ice increased from the inner to the outer side of the ice masses. This indicated a gradual killing of the living cells external to the ice ring during the growth of the extracellular ice mass. Such a result would have been impossible if ice had formed within the cells, for this would have brought to a stop the growth of the extracellular ice masses; and if the extracellular freezing has not injured the cells, this ice would be uniformly free of electrolyte and betacyanin. Asahina (1954) has also reported injury to potato sprout tissue when frozen extracellularly at −3°C for even a short time. Similarly, a marine alga (*Enteromorpha intestinalis*) is killed only when frozen below −25°C, and the "frost plasmolysis" of the cells indicates that this was due to extracellular freezing (Terumoto, 1961).

Many observers of artificially frozen cells, although agreeing that intracellular freezing is nearly always fatal and that extracellular freezing may be harmless, have nevertheless observed fatal injury to cells clearly frozen only extracellularly (e.g., Molisch, 1897; Iljin, 1933b; Siminovitch and Scarth, 1938; Modlibowska and Rogers, 1955; Asahina, 1956; Terumoto, 1967). Thus when ice forms within cells having colored sap, the colorless crystals are visible between the sap that becomes darker and darker, but when ice forms extracellularly, the cells become colorless at the middle due to collapse. The latter kind of freezing can occur with or without death (Siminovitch and Scarth, 1938). True frost plasmolysis, which involves ice formation between the protoplast and the cell wall, but not within the protoplast, can also occur with or without injury (Asahina, 1956). Even in the case of insects, extracellular freezing may result in injury. Salt (1962) showed this by the method of freeze-substitution. Certain large cells (e.g., of the fat body) did freeze internally, but the remaining smaller cells of the larvae showed no sign of intracellular freezing, whether they survived the freezing or were killed by it.

e. Intracellular Freezing

The point at which intracellular freezing occurs (even in a single kind of cell of a single plant) is not an exact one, but varies markedly even though the conditions are maintained as nearly constant as possible. Therefore, intracellular freezing is incapable of explaining the relatively exact frost-killing points (Chapter 7) found by so many investigators for a single variety of plant grown and hardened in a standard way. Due to this, and to the above mentioned correlation between freezing and drought tolerance, it appears obvious that extracellular freezing is the general cause of frost killing of higher plants in nature.

There may, of course, be exceptions. It has already been mentioned that intracellular freezing may conceivably be the cause of injury in certain specialized cases, e.g., "sunscald." Nevertheless, on the basis of all the above evidence, it must be concluded that in higher plants, the major cause of freezing injury is the secondary stress, producing freeze-dehydration by extracellular freezing. In microorganisms, such as yeast, the major cause is apparently the primary freezing stress, producing direct injury by intracellular freezing (Mazur and Schmidt, 1968).

f. Cell Changes due to Freezing Injury

Besides "frost plasmolysis," a number of other changes have been observed in cells killed by freezing. Thus, the chloroplasts of wheat leaves injured by extracellular freezing, have been described as severely swollen and cloudy green immediately after thawing (Salcheva and Samygin, 1963). It is unfortunately difficult to evaluate some of these observations since (1) it is usually not determined whether they are due to intracellular or extracellular freezing, and (2) the specificity of the changes is usually not investigated, e.g., by comparing them with changes in cells killed by other methods.

CHAPTER 7

FREEZING RESISTANCE—TYPES, MEASUREMENT, AND CHANGES

A. Possible Types of Resistance

It was indicated in Chapter 2 that low-temperature avoidance does not occur in plants because they are poikilotherms. It does not necessarily follow that the same simplification can be applied to freezing resistance. The following Diagram (7.1) presents the possibilities:

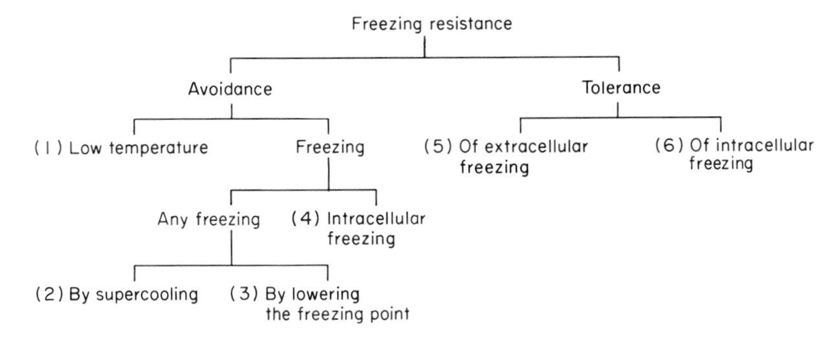

Before attempting to investigate and explain freezing resistance, it is obviously essential to know which of the six possible types actually exist.

1. Avoidance of Low Temperature

The plant's temperature usually closely follows that of its environment, even in the case of bulky, insulated tree trunks, although, of course, lagging behind it during periods of rise or fall (Levitt, 1956). The ineffectiveness of plant structures as insulators against low temperature is readily shown by artificial cooling. Thus, in spite of the presence of surrounding scales, the growing point within the large buds of chestnut (*Aesculus hippocastanum*) cools practically as rapidly as the surrounding atmosphere when the latter's temperature drops from +20° to −20°C in 30 min (Fig. 7.1). Similar results have been obtained by Pisek (1958). In the case of leaves and buds, their temperature may actually drop below that of the air on a cold night: 2.5°C in the case of buds, 3°–5°C in the case of tomato leaves when the sky is clear (Jenny, 1953; Shaw, 1954). Even pine needles may show this drop (Table 7.1). In full sun, on the other hand, plant temperatures may rise several degrees above that of the air. An extreme example is the marked difference (as much as 30°C) between the cambial temperature of a tree in the sun and that in the shade in midwinter (Harvey, 1923; Eggert, 1944, Sakai, 1966). Both these differences are simply due to radiation from and to the plant, respectively. Unlike warm-blooded animals, plants are poikilotherms, i.e., they are unable to maintain a constant tissue temperature different from that of their environment. Some tissues, e.g., fleshy inflorescences of Araceae, have long been known to raise their temperatures well above that of their environment due to the heat produced by respiration, although this occurs only at temperatures well above freezing. Similarly, some early spring plants have been said to melt the surrounding snow. In neither case is a constant tissue temperature produced.

Yet this respiratory production of heat has led to the question of the existence of low-temperature avoidance in plants. Respiratory heat, of course, results from the utilization of reserves. In cold climates, loss of reserves would be more damaging to the overwintering plant than to the animal, since the deciduous plant cannot replenish its reserves during the winter or even during the early spring, and since only those deciduous and evergreen plants that still possess their reserves essentially intact at win-

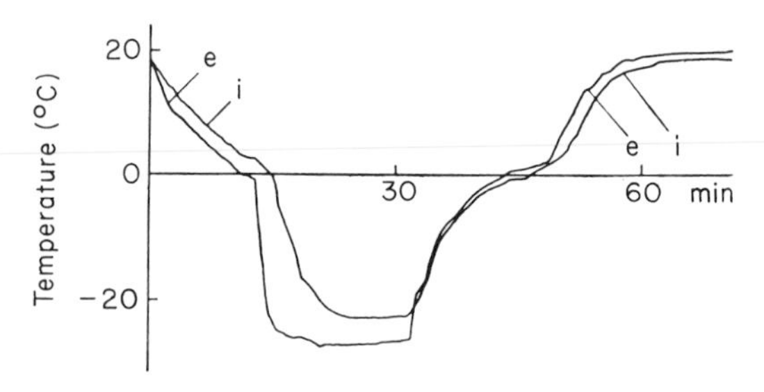

Fig. 7.1. Internal temperature of a chestnut bud (i) compared with the external temperature (e), during a 30-min drop in the latter from +20°C to below −20°C. From Genevès, 1957.

TABLE 7.1 RADIATION COOLING OF *Pinus cembra* NEEDLES BELOW AIR TEMPERATURE (°C) AT 07.00 HOUR[a]

Date (in November)	Air temperature	Degree of cloudiness (increasing from 1–10)	Cooling of needles below air temperature
16	−14.6	1	−6
17	−11.0	9	−2
18	−13.5	10	−1.5
19	−14.4	0	−6.5
20	−10.6	0	−6.5
21	−10.8	5	−4
22	− 9.5	4	−4.5
23	− 8.8	0	−6.5
24	− 9.1	10	−1.5
25	−14.0	10	−1.5
26	−18.5	2	−5.5
27	−13.5	0	−6.5

[a] From Tranquillni, 1958.

ter's end can produce the rapid spring growth on which their survival depends. Most important, because of the much larger specific surface of plants than of animals, it would require a much more rapid utilization of reserves in order for the plant to maintain its temperature above that of its environment than in the case of the animal. There is no solution to this problem for the plant as a whole, since the decrease in specific surface needed to reduce the loss of heat (and therefore also the utilization of reserves) would also reduce the rate of accumulation of reserves by decreasing the photosynthetic surface. Adaptation has, in fact, proceeded in the opposite direction; for plants with the best developed low-temperature resistance may possess the lowest respiratory rates (see below), and therefore the lowest heat production at low temperatures.

It is, therefore, useless to look for plants that might have a well developed low-temperature avoidance due to the heat produced by respiration. Other types of low-temperature avoidance may, however, occur to a greater or lesser degree. The cambial cells of tree trunks (and any other living cells of the trunk) are somewhat protected from the extremes of atmospheric temperatures by the insulating effect of the bark; yet this protection may be completely inadequate (see below). Although possessing no true avoidance with respect to their own environments, roots and plants that pass the winter under a snow cover are able to avoid the lower temperatures of the above ground atmosphere. It has, in fact, been shown

(Torssell, 1959) that a rape variety with less tolerance may survive better than another with greater tolerance because of its greater ability to remain covered by snow, i.e., its greater low-temperature avoidance. One special case of temporary low-temperature avoidance occurs when freezing commences. The heat released by the crystallization of water may be sufficient to keep plant temperatures above that of the air for a short time, but this must be of very limited value to the plant. That it plays no part in freezing resistance follows from the commonly found inverse relationship between water content and freezing resistance (see Chapter 8).

2. Avoidance of Freezing by Supercooling

It is now known that many insects survive winter without freezing, i.e., in the supercooled state (Salt, 1950; Asahina *et al.*, 1954). Their freezing resistance is, therefore, due to avoidance. It might be reasonable to seek the same phenomenon among plants. It was, at one time, actually believed that only those plants that avoid ice formation in their tissues survive, and that a frozen plant is invariably a dead plant (see Levitt, 1956). It was even suggested that when an axe bounces off a frozen tree trunk in winter, the trunk actually freezes at the instant the axe touches it! It did not seem to disconcert the proponents of this theory that ice formation throughout a tree trunk takes considerable time, and that this tree (that they could not deny had ice in it), nevertheless remained alive. Continuous recording of the temperature changes during cooling have now clearly demonstrated that the sap of trees is, indeed, frozen in winter (Lybeck, 1959). Actual measurements of the ice formed in the tissues indicated that all the sap was frozen and only the water bound to the wood was unfrozen. Even the trunks of large elm trees (86 cm in diameter) are frozen during winter in northern Japan (Sakai, 1966).

Direct observation in the open under the microscope (Wiegand, 1906b), revealed the presence of ice during winter in the twigs of trees. It could, of course, again be objected (ignoring the appearance and feel of the leaves before sectioning) that they were actually unfrozen until cut by the razor blade. However, the very fact that the ice crystals were always found to occur in the intercellular spaces proved that it had been present before sectioning; for it is now known (see Chapters 5 and 6) that if the ice had formed suddenly at the touch of the razor blade, it would occur intracellularly and not extracellularly. In the case of evergreen plants, even without touching them it can be seen that their leaves are frozen at temperatures below freezing, for not only do they become darker in color but they move stiffly in the wind.

TABLE 7.2 Observations of Supercooling of Plants in Nature

Plant	Plant part	Supercooling temperature	Reference
Eight out of 27 species	Buds	−18°C	Wiegand, 1906a
Four out of 27 species	Buds	−26.5°C	Wiegand, 1906a
Plum	Flower buds	−21°F	Dorsey and Strausbaugh, 1923
Pyrola	Leaves	−32°C	Lewis and Tuttle, 1920
Caragana		−21°C	Novikov, 1928
Hedera helix		−20°C	Iljin, 1934
Pine and fir	Needles	−21°F	Clements, 1938
Grain	Roots	−11°C	Zacharowa, 1926
Olive	Leaves	−10°C	Larcher, 1959
Azalea	Flower buds	−43°C	Weiser, 1970

A few cases of pronounced supercooling have been reported in nature (Table 7.2). It is the only resistance mechanism possessed by olive leaves, for they are invariably killed by freezing (Larcher, 1959). Potato leaves survive −6°C due to supercooling, but are frozen and killed at −8°C (Li and Weiser, 1969b). According to Weiser (1970), supercooling is the survival mechanism utilized by flower buds of woody plants, in general, and they are always killed when they freeze. This is, apparently, not true of vegetative buds of woody plants, since Wiegand (1906a) observed numerous crystals at −26.5°C in living buds that showed no ice at −18°C. Their low supercooling points were believed due to their low water contents (20–30%) and their small cell size. Even under artificial conditions, supercooling is commonly observed (Lucas, 1954), and in order to induce freezing under the microscope without excessive supercooling, it is usually necessary to inoculate sections with an ice crystal. Orange fruit with a freezing point of about −2°C may supercool as much as 3°C below this point before freezing (Hendershott, 1962b). In spite of the above cases of pronounced supercooling, in most plants it is confined to a very few degrees and its duration is brief. Thus, thousands of plant parts (leaves, roots, stems, flowers, fruits) were all found to freeze readily in a room at −3°C (Whiteman, 1957).

It is easy to understand why supercooling is so rare among higher plants although common in insects. The ability of the water to remain supercooled varies inversely with the diameter of the capillary in which it occurs. The water is, therefore, much more likely to supercool in the case of the smaller insect cells than in the larger cells of higher plants. Even if some of the plant's cells are small enough to favor supercooling, this may not help since ice formation commonly begins in the large, well filled vessels and from here spreads throughout the plant (Asahina, 1956). Furthermore, insects can keep very still and be protected from winds, whereas the leaves and branches of trees are readily moved by any breeze; such movement is one of the best ways to induce ice formation in supercooled water. They may also apparently be inoculated by hoarfrost on leaves (Asahina, 1956). In some cases, in fact, this may be a requirement for freezing, at least during brief freezes. Thus undercooling may be maintained (at $-4°C$) for some time in etiolated pea shoots even if the intercellular spaces are filled with water, provided that the shoots are paraffin coated and, therefore, protected from external inoculation (Le Saint, 1956). Green shoots, however, are unstable in the undercooled state and freeze spontaneously (Le Saint, 1958). Finally, even in the case of insects, undercooling is not permanent (Salt, 1950), and freezing eventually occurs at irregular intervals over long periods of time. Since a sudden freezing after a severe undercooling is far more likely to be fatal than the gradual freezing that occurs when there is no marked undercooling (see below), it is obvious that a marked undercooling in plants can do more harm than good and, therefore, is not likely to be selected as a survival factor. It is more likely to be useful during moderate freezes of short duration. Yet even during the brief night frost of spring, flowers of fruit trees are commonly unable to remain supercooled, and it is necessary to confer artificial avoidance by spraying them with water (Rogers *et al.*, 1954). Such artificial avoidance is itself dangerous, for if the heat released on freezing of the spray water is insufficient to keep the tissue temperature above its freezing point, the cells will freeze even more readily (due to inoculation with the ice crystals) and be killed. Thus, it appears safe to conclude that undercooling plays no role in the freezing resistance of most higher plants.

3. *Avoidance of Freezing by Lowering the Freezing Point*

A sufficiently low vapor pressure due to low water content, coupled with high content of water-binding substances and high cell sap concentration, can prevent ice formation by lowering the freezing point. In some larvae, a concentration of glycerol as high as 5 molal has been found during hibernation in the fall, and is lost again in spring (Salt, 1958). The freezing point of these larvae is thus depressed to as low as $-17.5°C$. This extraordinarily high concentration has never been found in normally hydrated parts of higher plants, although, of course, an even greater avoidance arises during the drying out of seeds (and some buds; see above) previous to overwintering. The highest cell sap concentration ever found in normally turgid vegetative plants is equivalent to an osmotic pressure of about 200 atm (and the dependability of these values is doubtful). In nonhalophytes this value seldom exceeds or even equals 50 atm. The latter value would yield a freezing point no lower than about −4°C.

4. Avoidance of Intracellular Freezing

An avoidance of primary direct freezing injury is demonstrated by resistance to intracellular freezing. The first direct evidence of this kind of freezing resistance was obtained by cooling sections under the microscope end inoculating them with ice at different temperatures (Siminovitch and Scarth, 1938). Intracellular freezing occurred at slower rates of freezing (i.e., higher inoculating temperatures) in unhardened than in hardened plants (Table 7.3). In confirmation of these results, the cells of hardened wheat leaves showed intracellular freezing only if cooled at a rate of more than 1°C/3 min, and the meristematic cells of tillering nodes at a rate of more than 1°C/5 min (Salcheva and Samygin, 1963). Intracellular freezing occurred at slower rates of cooling in the unhardened plants, although even in these, the rates required were more rapid than those normally occurring in nature. Aronsson and Eliasson (1970) have successfully used survival of very rapid freezing (by cooling from +4° to −22°C in 10 min or about 3°/min) to measure the seasonal changes in hardiness of pine needles. All these results demonstrate that avoidance of intracellular freezing increases with freezing resistance.

TABLE 7.3 Avoidance of Intracellular Freezing in Resistant (Hardened) and Nonresistant (Unhardened) Cells[a]

Temp. at inoculation (°C)	Average percentage of cells frozen intracellularly			
	Catalpa		*Cornus*	
	Unhardened	Hardened	Unhardened	Hardened
−2 to −3	75	15	25	0
−3 to −4	90	30	15	0
−4 to −5	100	55	35	0
−5 to −6			90	10

[a] Reproduced by permission of the National Research Council of Canada from Siminovitch and Scarth (1938). *Can. J. Res. C16*, 467–481.

5. *Freezing Tolerance*

Of the above four possible types of avoidance, it can, therefore, be concluded that with the exception of a few cases of complete freezing avoidance, only the avoidance of intracellular freezing is characteristic of freezing-resistant plants, in general. Avoidance of *intra*cellular freezing is a survival mechanism only if *extra*cellular freezing is tolerated. Therefore, this type of resistance would be useless in the absence of freezing tolerance; and we must conclude that all freezing resistance of plants is primarily freezing tolerance.

Since primary direct freezing injury (due to intraprotoplasmal freezing) is always fatal, and since there is no primary indirect freezing injury, the only possible freezing tolerance is of the secondary, freeze-dehydration stress. Freezing tolerance, in other words, is tolerance of extracellular freezing. Several methods have been developed for measuring it.

B. Measurement of Freezing Tolerance

1. *Field Survival*

The ability of plants to survive the severities of winter has long been known as *winter hardiness* and has, therefore, been measured by field survival. The surviving plants were classified as hardy, while those that did not survive, as tender. Gradations between these two extremes, such as "moderately hardy" or "semihardy" were later introduced and developed into a semiquantitative rating (cf. Table 8.10), on the basis of percentage field survival of a "test winter"—a winter severe enough to kill the most tender and to damage those of intermediate hardiness in graded degrees. Test winters occur on an average of once every 10 years and, therefore, the need for a quicker and more accurate method of measurement soon became apparent.

2. *Artificial Freezing*

Freezing chambers in which the hardiness of plants could be tested quickly and quantitatively were first introduced by Harvey (1918). By freezing a series of apple varieties at a previously determined temperature, it was possible to obtain a graded series of injuries, depending on the resistance or hardiness of the variety (Table 7.4). This method was developed to a high degree by the Swedish investigators (Åkerman, 1927), who gave

TABLE 7.4 COMPARISON OF HARDINESS RATING FROM ARTIFICIAL FREEZING TESTS (IN ORDER OF DECREASING HARDINESS) AND FROM FIELD EXPERIENCE[a]

Order of apple varieties from artificial freezing test	Field experience (Horticultural Society rating, 1 = maximum hardiness)
1 Charlamoff	2
2 Hibernal	1
3 Duchess (Oldenburg)	1
4 Patten	1
5 Wealthy	2
6 Windsor Chief	3
7 McIntosh	4
8 Fameuse	4
9 University	4
10 Northwestern	4
11 Wolf River	5
12 Jonathan	4
13 Delicious	4
14 King David	4
15 Black Ben[b]	6
16 Paragon[b]	6
17 Lansingburg[b]	6

[a] Adapted from Hildreth, 1926.
[b] Not recommended by the Horticultural Society.

a numerical rating for hardiness from 1 to 10 (Table 8.3). The usefulness of this procedure was soon recognized and it was applied particularly as an aid to the breeding and selection of hardy wheat varieties of high quality. Similar numerical ratings are still used for the selection of hardy plants of many kinds (Manis and Knight, 1967). For this purpose, the varieties must be compared when in their hardened state (see below), since some wheat varieties may change their hardiness rank relative to the others as they develop (Roberts and Grant, 1968).

In the several decades since its introduction by Harvey, artificial freezing tests have been used for many different kinds of plants (Emmert and Howlett, 1953; Coffman, 1955; Rachie and Schmid, 1955; Johansson *et al.*, 1955; Amirshahi and Patterson, 1956). This method has even been used in the field, in the case of 2-year-old grapefruit trees (Cooper *et al.*, 1954), as well as for winter turnips, rape, and rye (Johansson and Torssell, 1956). In the latter case, temperatures as low as −28°C were obtained over an area of a square meter. Greatest sensitivity to freezing was found at the flowering state, e.g., rye was injured at 0° to −2°C. Liquid nitrogen has recently been used as the refrigerant for controlled freezing of dormant twigs (Weaver and Jackson, 1969).

Even the highly quantitative numerical rating of the Swedish investigators is only relative, and does not permit a direct comparison with the varieties of another species that may require a different freezing temperature for such a rating. As in all scientific measurements, the ultimate goal must be an absolute rather than a relative measure of the quantity. This is especially necessary for the solution of theoretical problems, such as the quantitative relationships between freezing and drought resistance (see Chapter 21).

The need for an absolute measurement is most simply met by determining the "frost-killing point," the freezing temperature required to kill 50% of the plant (Fig. 7.2). The temperature determined has sometimes been either the "ultimate frost-killing point" resulting in 100% killing or the "incipient frost-killing point" that just begins to cause injury (Pisek, 1958). The above curve illustrates that the most readily measured point is the 50% killing point which has now become standard.

The frost-killing point must be determined under standard conditions. Although no single uniform standardization has been adopted by all investigators, the following five steps have been proposed as basic requirements (Levitt, 1956):

1. The plants must be inoculated to ensure freezing.

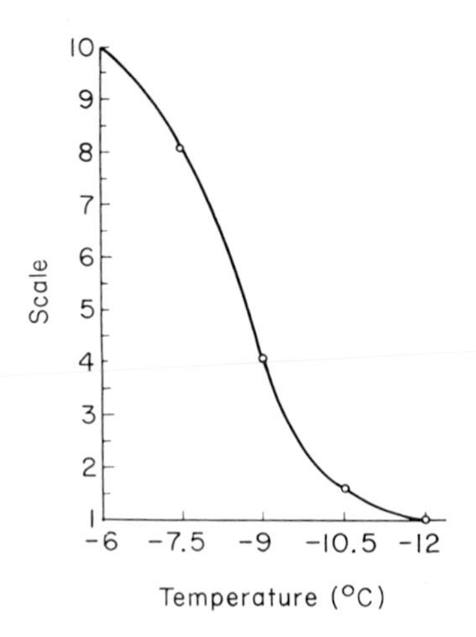

Fig. 7.2. Determination of 50% killing point, in this case, −9°C. Adapted from Johansson *et al.*, 1955.

2. Cooling must be at a standard rate.
3. A single freeze must be used for a standard length of time.
4. Thawing must be at a standard rate of warming.
5. Postthawing conditions must be standardized

In order to achieve these basic requirements, the following procedure has been used. A series of shoots or leaves are taken from a number of plants of a genetically pure strain and grown under identical conditions. These are suspended in a freezing chamber at 0°C and after reaching temperature equilibrium the cut surfaces are sprinkled with snow to prevent appreciable supercooling. The temperature is then dropped at the rate of 2°C/hr. At each of a series of temperatures ranging above and below the suspected killing temperature, several shoots or leaves are quickly transferred to a room at a temperature just above freezing, where they are allowed to thaw and recover for 24 hr. They are then transferred to room temperature where their cut bases are stood in water. The percentage injury is estimated as soon as it becomes clear-cut (2–7 days depending on the species). The selected temperature is that which produces 50% killing—the frost-killing point.

Artificial freezing tests have usually been found to give excellent agreement with winter survival in the field (Table 7.4). Yet some differences between the two may be expected since some of the field conditions are not duplicated by the freezing test. Torssell (1959) has shown that a less frost-hardy variety may have a greater "field hardiness," using the latter term for both low-temperature avoidance and freezing tolerance. This is to be expected since the freezing test measures only freezing tolerance, and ignores the ability of the plant to avoid low-temperature injury by, for instance, remaining under the snow. Another complication may arise in those cases where freezing injury occurs mainly in the spring (Till, 1956). The order of hardiness may then conceivably be different from that in midwinter or after artificial hardening. In apples, only the fall measurements of hardiness may agree with field experience in some climates (Emmert and Howlett, 1953). In some cases, the length of the rest period or the ability to reharden after temporary losses of hardiness during midwinter warm spells may be the deciding factors (Brierley and Landon, 1946).

In summary, if the artificial freezing tests are intended to give a complete picture of the plant's freezing tolerance, measurements must be made at frequent intervals throughout the year. According to Scheumann (1968), there are at least four important aspects of this annual freezing curve. (1) The plant's readiness for hardening is important for tolerance of early frosts. (2) The extent of its hardening potential is important for winter hardiness. (3) The stability of its hardiness is important during periods of widely fluctuating temperatures. (4) The time of bud bursting and flowering is important for tolerance of late frosts. As pointed out by Larcher (1968) a complete picture of the plant's hardiness further requires measurements on different parts of the plant, since injury to these parts may range from 0 to 100%.

Even the method of using the freezing test may conceivably introduce differences between artificial and field testing. If estimates of injury are based on percentage killing of the plant's foliage (e.g., in winter annuals), this may show little relationship to subsequent yield, since a plant with all or nearly all its foliage killed may still yield well. Yield is, in these cases, more closely related to the percentage of plants completely killed. It must also be remembered that measurements of freezing resistance by the artificial freezing method determine the resistance of the plant under standard conditions of slow freezing and slow thawing with optimum conditions for extracellular freezing and postthawing recovery. It has never really been adequately determined whether or not the relative ratings of varieties would be altered by changing these conditions.

A departure from this standardized procedure has been introduced by Aronsson and Eliasson (1970). Shoots of 1 to 2-year-old seedlings of Scots pine (*Pinus sylvestris*) were first stored at 4°C, then plunged into a deep-freeze at −12°, −22°, −32°, or −44°C. The temperature of the needles dropped to the deep-freeze temperature in about 10 min, a cooling rate of about 1.5–5°C/min. Although the lowest temperatures were too severe, at −22°C they obtained the same kind of seasonal curve as is normally obtained by the slow-freezing method. Yet this method presumably measures avoidance of primary direct freezing injury, i.e., avoidance of intracellular freezing.

Artificial freezing tests have generally given such good agreement with field experience (and have even revealed errors in the latter: Hildreth, 1926) that it must be concluded that the freezing tolerance, which is measured in these tests, is by far the major factor in the overwintering of plants. For most purposes, the frost-killing point is all that is needed. For some theoretical considerations, it has been proposed to define hardiness (and therefore tolerance) as the numerical difference between the frost-killing point and the freezing point (Levitt, 1956):

$$H = (T_{k50} - T_{\Delta})$$

where H = frost hardiness; T_{k50} = frost-killing point for 50% killing; T_{Δ} = freezing point. The advantage of this quantity over the frost-killing point is that it yields a value of zero for plants killed at the first touch of frost, and increasingly positive values for those with lower (negative) frost-killing points.

3. *Methods of Estimating Freezing Injury*

One weakness in the above methods is that the estimate of injury is subjective. This can be overcome by the electrical conductivity method originated by Dexter (1932) and now used regularly by many investigators (e.g., Wilner, 1955, 1960). Since cell injury results in increased permeability to the electrolytes in the cell sap, the greater the injury the greater the conductivity of the extract. The conductivity value of the tissues or an extract of them is, therefore, a direct measure of the injury produced by freezing. Other substances besides inorganic electrolytes will, of course, also diffuse out of the injured cells. The release of amino acids and other ninhydrin reacting substances was used by Siminovitch *et al.* (1962) as a sensitive measure of freezing injury.

An interesting variant of the conductivity method has been described by Greenham and Daday (1957, 1960). The ratio of conductance of a high frequency to that of a low frequency current drops from 10 in the living plant to 1 when a tissue is killed. With this simple method, a fine electrode can be inserted directly into the leaf and injury can be determined repeatedly in different parts of the leaf at different times after thawing. In most of the earlier investigations, the conductivity method was used only to measure the relative injuries suffered by a series of varieties frozen under identical conditions. The results are far more useful when the freezing temperature is determined that produces the conductivity corresponding to 50% killing (e.g., Aronsson and Eliasson, 1970). One difference between the subjective method of estimating injury and the conductivity method is that the former determines the injury several days after freezing and the latter immediately or a few hours afterward. The two values may perhaps differ somewhat due to postthawing injury or recovery. Lapins (1962), for instance, was able to estimate recovery more reliably after a 3-week forcing period in a growth chamber than by the electrical conductivity method.

Estimations of injury may also be made immediately after freezing by vital staining of sections with neutral red and observation of the percentage of living (stained) cells. Simple plasmolysis may also be used either alone or in combination with vital staining. The reduction of colorless triphenyltetrazolium chloride (TTC) to the red form on the cut surface of tissues has also been used successfully (Parker, 1953, 1958; Torssell and Hellstrom, 1955; Larcher and Eggarter, 1960; Larcher, 1969). A 25% decrease in

TTC absorbance following the freeze treatment correlated closely with visual survival checks in the case of bermudagrass, and served as a good index of viability (Ahring and Irving, 1969). McLeester *et al.* (1969) have made use of the multiple freezing point of plant tissues as a test of viability in evaluating frost hardiness. In preference to such indirect methods, Tjurina (1968) recommends grafting of chilled shoots of trees onto seedling roots, protecting the lower parts of the shoots against the effects of the low temperature.

For many years, investigators have searched for an indirect "measuring stick" to evaluate hardiness without having to freeze the plant, e.g., sugar content, bound water (see below). To this day, the search continues, and sometimes the methods work beautifully for one series of plants but not for another, e.g., sugar content is directly proportional to hardiness in a series of twelve wheat varieties, but not in the case of other varieties or other grains (see Chapter 8). Frequently, methods that seem promising at first, fail in later tests. Thus, attempts to evaluate the hardiness of alfalfa varieties by the extent of germination in solutions of graded osmotic pressures gave variable results and, in contrast to earlier reports, proved unreliable (Heinrichs, 1959). An indirect method that has worked satisfactorily so far, but has not been tried by many investigators, is the deplasmolysis method (Scarth and Levitt, 1937; Siminovitch and Levitt, 1941; Siminovitch and Briggs, 1953a; Sakai, 1955b, 1956a). It must be cautioned, however, that this method works only if plasmolysis is maintained for a sufficiently long period (e.g., some hours). It is not precise enough for small varietal differences. Impedance has recently been correlated with hardiness of forty-three peach-bearing trees (Weaver *et al.*, 1968) and with increases in apparent freezing resistance of alfalfa (Hayden *et al.*, 1969). However, scion diameter was also correlated with hardiness in the same forty-three peach trees, and impedance was not correlated with the hardiness of nonbearing trees (Weaver *et al.*, 1968). Similarly, impedance readings did not separate the winter hardy from the tender cultivars of red raspberry (Craig *et al.*, 1970), although they did indicate when the rest period began in the fall and the resumption of growth in the spring.

In the case of 87 wheat varieties and 15 barley varieties a correlation coefficient of 0.83 to 0.85 has been obtained between SH content of the homogenate and winter hardiness (Schmuetz *et al.*, 1961, Schmuetz, 1962, 1969). The SH measurements, however, had to be made under rigidly standardized conditions (see Chapter 8). An equally good correlation has been obtained between ascorbic acid content and winter hardiness in the case of several wheat varieties (Schmuetz, 1969; see Chapter 8).

In spite of the promising results described above, past experience indicates that no one indirect "measuring stock" can be trusted as a measure of relative freezing resistance in all plants. Sakharova and Yakupov (1969) have, in fact, recommended measuring a long list of variables. In the final analysis, however, direct freezing tests are essential for fully reliable measurements of freezing resistance, whether it is due to tolerance or avoidance.

C. Changes in Freezing Tolerance

1. Seasonal Changes

Some plants are killed by the first touch of frost. For instance, a leaf temperature of −1.5°C for 10 to 15 min produces enough freezing to kill banana leaves (Shmueli, 1960). Others survive the lowest winter temperatures without injury, and all gradations exist between these extremes. Yet the frost-killing point, although determined in a standard way, is not a constant even for a genetically pure strain, but varies markedly with the stage of development and with several environmental factors that can alter resistance. In the case of even the most resistant species, the actual tolerance varies from a minimum of practically none in the new, spring growth to the maximum value of midwinter (Fig. 7.3). This increase in freezing tolerance is called frost hardening, cold hardening, cold acclimation, etc. Due to this seasonal change, even conifers that are among the most hardy of plants may be severely injured by light summer frost (Pomerleau and Ray, 1957). Very tender species (e.g., those from tropical climates), on the other hand, never develop any frost resistance no matter what the stage of growth or time of the year. Under artificial conditions, tropical species of willows are exceptions to this rule (Sakai, 1970b). When grown in Sapporo and hardened in the same manner as the northern species, they developed as high a degree of tolerance, surviving −50°C or even liquid nitrogen. Nevertheless, in the vast majority of cases, when hardiness is at its maximum, it also varies markedly from species to species (Table 6.1) and from variety to variety (Tables 7.3 and 8.10). It does not follow that a species or variety with the maximum tolerance at its full development necessarily is also the most tolerant of a series of species or varieties at all times of the year. The rates of hardening may vary independently of the maximum attained. A hardy apple variety, for instance, shows far greater hardiness in the fall than the less hardy variety, but the difference between the two is slight in midwinter (Fig. 7.3). Some plants, in fact, are "resistance stable" and their hardiness fails to change much with external conditions (Larcher, 1954). These may, therefore, be the most freezing tolerant of their com-

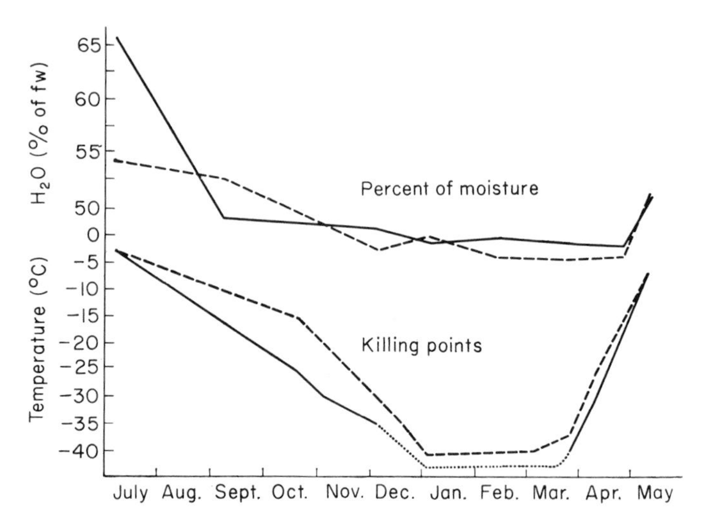

FIG. 7.3. Seasonal variations in frost-killing points and in moisture contents of Duchess (solid line) and Jonathan (broken line) apple twigs. (From Hildreth, 1926.)

munity in summer (e.g., a killing point of −12° to −16°C in *Citrus trifoliata*) and the least tolerant in winter. The most extreme examples of resistance-stable plants are to be found among lower plants. *Porphyridium cruentum* shows normal optimum growth above 27°C, but no damage results from rhythmic alternation of periods of frost at −22°C in the dark with periods of 27°C in the light (Rieth, 1966). Whole taxonomic groups may be resistance stable. Thus, the freezing tolerance of arctic mosses and liverworts was not found to differ from that of tropical mosses (Biebl, 1967b). Many algae, however, are resistance labile (Schölm, 1968). In the case of higher plants of temperate climates, the autumn rise in freezing tolerance is a universal phenomenon and has been reported in innumerable species and varieties tested by investigators throughout the world. The same is true of the spring drop in freezing tolerance. The more developed the bud, the sooner it loses its resistance in spring (Mair, 1968). Yet it must not be concluded that development and freezing tolerance in higher plants are indissolubly related. The genes for reaction to cold in the coleoptile stage of wheat are not linked to the genes of development (Goujon *et al.*, 1968).

The pronounced seasonal change in freezing tolerance is even found in overwintering insects. Although the insect is usually resistant only in the pupal state, the adult carabid beetle has been found to tolerate −35°C during winter (Miller, 1969). In summer, the beetle is killed if frozen at

−6.6°C. Even in winter, however, relatively slow cooling is required to avoid damage.

Many marine algae are highly tolerant of freezing, but show no seasonal change in tolerance (Terumoto, 1967). Due to their marine environment, they are exposed to relatively small seasonal temperature changes. Land plants, on the other hand, are exposed to marked seasonal temperature changes and these parallel the seasonal changes in freezing tolerance (Fig. 7.4).

2. Environmental Factors

A decrease in freezing tolerance occurs during winter when the plants are exposed to warm weather for 2 weeks (Göppert, 1830) or even a few hours (mosses, Irmscher, 1912; evergreen trees, Pisek, 1950) and an increase in freezing tolerance, when they are exposed to low temperatures (Haberlandt, 1875; Schaffnit, 1910; Irmscher, 1912; Chandler, 1913; Gassner and Grimme, 1913). It is now standard procedure to "harden off" plants by exposing them for a week or two to temperatures a few degrees above the freezing point. The "threshold temperature" above which hardening does not occur is usually 5°–10°C (Harvey, 1922). The precise threshold temperature for hardening or dehardening is difficult to determine, and varies with the species and probably the variety. Some hardening of cab-

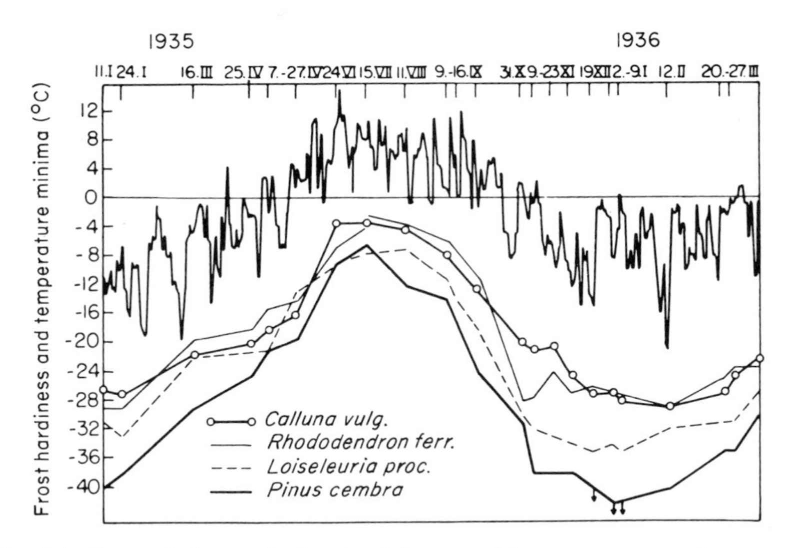

FIG. 7.4. Seasonal changes in freezing tolerance of four evergreen species compared with the daily temperature minima. (From Ulmer, 1937.)

bage may occur at 12°C, but none occurs at 18°C (Le Saint, 1966). Similar results have been obtained for *Mentha viridis* (Codaccioni, 1968) in the case of artificial tissue cultures. These survived −7°C for 10 hr after growth at 12°C, but not after growth at 20°C; hardening occurred only if the cultures were supplied with 2% glucose. In the case of some young trees, any temperature above 13°C led to a loss in freezing tolerance (Sakai, 1967). Nevertheless, poplars held at 15°C increased in tolerance from a killing temperature of −2° to −30°C over a 2-month period (Sakai and Yoshida, 1968a). The more hardy of eighteen citrus types were able to harden at higher temperatures (70° F D/50°F N or 60°F D/40°F N) than the less hardy types (Young, 1969b).

A temperature of 0° to 5°C will induce greater hardening than 5°–10°C. Once the maximum hardening possible at this temperature has been attained, a second stage increase may still occur at a temperature just below 0°C: −4°C in the case of barley at all stages and in sprouting wheat (Dantuma and Andrews, 1960), −2.5° to −5°C in the case of mulberry (Sakai, 1956b). Tumanov and his co-workers in the USSR have long emphasized the "second stage" hardening that may occur at temperatures well below the freezing point. He lists three periods in the preparation of plants for hibernation (Tumanov, 1969): (1) the onset of dormancy, (2) the first stage of hardening at about 0°C, and (3) the second stage of hardening during a gradual lowering of the temperature below 0°C. The freezing tolerance of cherry and apple twigs was markedly increased by prolonged (5–20 day) exposure to −5° and subsequently −10°C (Krasavtsev, 1969). Further hardening occurred during 1 day at −20° and −30°C. Birch and poplar twigs did not require this preliminary gradual freezing treatment. They survived −60°C if cooled gradually and −30° (in the case of birch) and −50°C (in the case of poplar) if cooled rapidly.

These "periods" or "stages" are not necessarily qualitatively different. At least in the case of cabbage, the differences are purely quantitative (H. Kohn, unpublished). When hardened for successive 2-week-periods, hardening at 5°C (day and night) led to a drop in the killing temperature which then remained constant. Lowering the temperature to 5°C D/0°N lowered the killing point further, again reaching a plateau. A third hardening at 0°D/0°N further lowered the killing point to a third plateau. Finally, after exposure to −3°C, a fourth lowering occurred, to a killing temperature of −20°C. This demonstrates that one can obtain as many "stages" of hardening as desired, simply by means of a graded series of hardening treatments.

Some results have indicated that alternating warm and cold temperatures are at least as effective as constant low temperatures (Harvey, 1918,

1930; Tumanov, 1931; Tysdal, 1933; Angelo *et al.*, 1939). Others have failed to obtain as hardy plants with alternating temperatures (Peltier and Kiesselbach, 1934; Suneson and Peltier, 1934; Day and Peace, 1937). Suneson and Peltier (1938) seem to have resolved these differences, for they obtained maximum resistance by exposure to alternating temperatures during November and December, followed by sustained low temperature for 3 weeks.

The prevailing temperature under natural conditions also markedly affects the freezing tolerance. The above ground parts of the plant may have 11.5°C greater freezing tolerance than the below ground parts (Till, 1956). Even among the above ground parts, the most exposed and therefore colder parts of the plant are more freezing tolerant than parts covered by snow (Brierley and Landon, 1954). Among the roots, the deeper ones are killed by freezing temperatures that fail to injure the shallower ones (Smirnova, 1959). Similarly, a colder winter has frequently been observed to result in greater freezing tolerance (Kohn, 1959), and midwinter thaws to result in a partial loss of freezing tolerance. Finally, the cortical cells on the south (warmer) side of trees in northern Japan are less tolerant of freezing than those on the north (cooler) side (Sakai, 1966). In general, freezing tolerance has been found to fluctuate throughout the winter, increasing as the temperature drops, decreasing as it rises (e.g., Proebsting, 1959).

Low temperature by itself is incapable of inducing hardening, at least in the case of winter annuals, biennials, and seedlings of perennials. The hardening of plants when alternating low and high temperatures are used has been shown to occur only if light is supplied during the high-temperature period (Dexter, 1933; Tysdal, 1933). Even continuous low temperatures in the absence of light are incapable of inducing hardiness in winter annuals (Tumanov, 1931; Dexter, 1933; Pfeiffer, 1933; Constantinescu, 1933; Andersson, 1944). Many have also found a reduction in hardiness as a result of darkening the plant (Lidforss, 1907; Weimer, 1929; Angelo *et al.*, 1939). No hardening of cabbage seedlings occurred in the dark at +4°C or in the light at +18°C (Le Saint, 1966), but normal hardening occurred when they were exposed to both low temperature and light. A threshold illumination of about 1000 fc was required for grains (see above) and also for young conifers (McGuire and Flint, 1962; Scheumann and Börtitz, 1965). Douglas fir seedlings failed to harden in the dark at 2.5°C even after several weeks (van den Driessche, 1969b), but they did harden at low light intensities (40 or 100 fc). An exceptional slight hardening in the dark occurred after growth in high-intensity light with a 16-hr photoperiod. Light enhances the rate of hardening of *Hedera helix*, but it is not essential for the hardening process (Steponkus and Lanphear, 1968a). The second stage of hardening, on exposure to temperatures below 0°C, may occur in the dark (Tumanov and Trunova, 1963; Kohn and Levitt, 1965). Even at this stage, however, light is necessary in the case of conifers (Scheumann and Börtitz, 1965). In the case of cabbage, 2000 lux was sufficient for hardening to a tolerance of −7°C, and some hardening was obtained even at 1250 lux (Le Saint, 1966). Much greater hardening of cabbage (survival of −20°C) was obtained when illuminated with 1000 fc, and cooled to a series of successively lower hardening temperatures (Kohn and Levitt, 1965).

The need for light is apparently due to a need for photosynthesis, since if the leaves are chlorotic, the plants are unable to harden even when exposed to light, although when allowed to become green by spraying with ferrous sulfate, they harden normally (Rosa, 1921). Similarly, if exposed to CO_2-free air, hardening does not occur even in the light (Dexter, 1933). Plants with abundant organic reserves are the exceptions, since they harden markedly at 0°C even in the dark (Dexter, 1933). This is not true of cabbage seedlings kept in the dark at the hardening temperature (Le Saint, 1966). Nevertheless, once hardened in the light, they maintained their hardiness in the dark for at least 2 weeks. Furthermore, if the part of the shoot that is capable of hardening is kept in the dark, and the remaining leaves are illuminated, the darkened part will harden due to translocation from the illuminated part. Since only the younger leaves are capable of hardening, this method works if the upper leaves are darkened and the lower ones illuminated, but if this is reversed, the darkened (older) leaves do not harden. Steponkus and Lanphear (1967a), in agreement with Le Saint, found that light results in the production of a promoter of hardiness in *Hedera helix*, which could be translocated to a darkened receptor. Labeling with ^{14}C indicated that the translocatable promoter was sucrose.

Even in the case of roots, which are commonly much less resistant than the shoots, light may be important. Thus the roots of apple trees in winter may be killed by −15° to −18°C, although the shoots survive until frozen at −40°C (Tumanov and Khvalin, 1967). Improvement of the hardening conditions increased root hardiness very little. If, however, they were kept in the air and light for 3–5 months before hardening, they hardened to the same degree as the shoots.

Temperature and light are thus the two main environmental factors controlling the development of freezing tolerance in plants. Artificial hardening by control of these two factors is, in fact, capable of producing a degree of hardiness equal to that developed under natural conditions, e.g., in the cases of cabbage (Kohn and Levitt, 1965) and *Hedera helix* (Steponkus and Lanphear, 1967b). This is true even of extremely hardy plants such as dogwood, which is capable of surviving −100°F or lower after either natural or artificial hardening (Van Huystee *et al.*, 1967), but to achieve this extreme degree of hardening under artificial conditions, it was necessary also to control the photoperiod (see below).

Other factors must, of course, be optimum in order for the plant to harden maximally on exposure to hardening levels of temperature and light. Many reports (see Levitt, 1956) indicate that full hardening is not obtained in the presence of excess nitrogen or of insufficient potassium, phosphorus, or even calcium. Contradictory results have been obtained in the case of the latter deficiencies (Levitt, 1956). Furthermore, cabbage seedlings that have grown excessively due to an excess of nitrogen may attain essentially the same degree of hardiness as normal plants, provided that they are exposed to optimum light and temperature regimes for hardening (see Kohn and Levitt, 1966). A deficiency of water may also induce some freezing tolerance (see Levitt, 1956). This will be discussed in connection with drought resistance.

3. *Relationship to Growth and Development*

Low temperature and adequate light intensity are not always able to induce the hardening of potentially hardy plants. Newly formed buds of evergreens, for instance, fail to harden at low temperatures and with normal light, even though they may survive −30°C during the subsequent winter (Winkler, 1913). They also lose their ability to harden when they begin to develop into shoots during spring (West and Edlefsen, 1917, 1921; Roberts, 1922; Knowlton and Dorsey, 1927; Field, 1939; Geslin, 1939). Similarly, the changes in tolerance may not follow the temperature changes at certain times of the year (Ulmer, 1937). This is clearly shown by the effects of exposure to low and high temperatures for 1 day on the hardiness of the plant at different times of the year (Pisek, 1953).

Even if kept constantly at the hardening temperature, the plant does not retain its maximum hardiness indefinitely. Sprouting winter wheat, for instance, reaches its maximum hardening at 1.5°C in the dark after about 5 weeks, and hardiness decreases rapidly between the seventh and eleventh weeks (Andrews *et al*, 1960). Although this drop in hardiness may sometimes be due to loss of reserves (Jung and Smith, 1960), it may also occur without an appreciable loss, e.g., the earlier spring development in the less tolerant species or varieties may be indicated by an earlier appearance of starch (Sergeev *et al*, 1959). The metabolic changes that occur in preparation for spring growth apparently lead to loss of freezing tolerance even though the plants are exposed to optimum hardening temperature and light. The mere cessation of growth and development in the fall may itself confer some hardiness without the aid of hardening temperatures (e.g., at 20°–30°C; Larcher, 1954; Sakai, 1955a). In many cases the plant actually enters a nongrowing "rest period," and freezing tolerance has been frequently related to the depth or length of this period (Levitt, 1956). Thus, the reduction in freezing tolerance of plants brought indoors during winter occurs only if they are no longer in their rest period (Lidforss, 1907; Meyer, 1932; Kessler, 1935). This has been confirmed in the case of *Acer negundo* and *Viburnum plicatum*. When in the naturally hardened state, their dormant condition retarded the loss of tolerance on exposure to 70°F (Irving and Lanphear, 1967b).

This correlation between rest period and freezing tolerance does not occur in all plants (Pojarkova, 1924). Some plants may be dormant although possessing no freezing tolerance (Clements, 1938), and others may survive the winter without a rest period (Walter, 1949). In the case of prevernal, remoral ephemerals, the dormant stages were the least resistant to cold (Goryshina and Kovaleva, 1967). Even some hardy woody plants (*Acer negundo, Viburnum plicatum tomentosum*) develop tolerance independently of bud dormancy (Irving and Lanphear, 1967b). In the case of very hardy plants, such as dogwood, the high degree of freezing tolerance may be maintained in winter long after its rest period is over (van Huystee *et al.*, 1967). In many cases, the importance of the rest period is believed due to prevention of growth and the accompanying loss of freezing tolerance during winter warm spells (Brierley and Landon, 1946). In opposition to this concept, raspberry canes may deharden at high temperatures more rapidly when in the resting than when in the nonresting state (Weiser, 1970). Nevertheless, these and other observations have led many investigators to point to growth per se, rather than stage of development as the factor that prevents hardening. In the case of winter annuals, hardiness is inversely related to rate of growth in the fall (Buhlert, 1906; Schaffnit, 1910; Hedlund, 1917; Klages, 1926; Worzella, 1932; Mark, 1936; Kolomycev, 1936; Vassiliev, 1939). In the case of British and North African varieties of tall fescue, there was a clear, inverse relationship between the ability of the varieties to survive low temperature and their ability to grow rapidly during winter (Robson and Jewiss, 1968). It is generally found, in fact, that if plants are growing rapidly, they cannot be frost hardened (Rivera and Corneli, 1931; Dexter *et al.*, 1932), whereas treatments that retard growth increase hardening (Chandler, 1913; Harvey, 1918; Rosa, 1921; Collison and Harlan, 1934; Shmelev, 1935; Kessler and Ruhland, 1938), although exceptions may occur (Kuksa, 1939). An increase in tolerance can frequently be obtained by withholding water from the plant to a sufficient degree to induce some wilting and stunting of growth. Conversely, as mentioned above, a marked decrease in tolerance often results from heavy nitrogen fertilization leading to a rapid, succulent growth, and, as in the case of the normal spring growth, the plants lose their ability to harden normally on exposure to hardening temperatures in the light. Similarly, growth-promoting substances sometimes decrease freezing tolerance, whereas growth-inhibiting substances sometimes increase freezing tolerance (see Chapter 8); but there are exceptions (Tumanov and Trunova, 1958).

The relationship of freezing tolerance to growth and development is clearly demonstrated by controlling the photoperiod. Hardening is improved by short days, both in the case of woody plants (Moschkov, 1935; Bogdanov, 1935) and in herbaceous plants (Dexter, 1933; Timofejeva, 1935; Saprygina, 1935; Saltykovskij and Saprygina, 1935; Sestakov, 1936; Rudorf, 1938; Suneson and Peltier, 1938; Tysdal, 1933; Frischenschlager, 1937; Smith, 1942; Ahring and Irving, 1969). In fact, the annual curve for freezing tolerance is as clearly correlated with the change in photoperiod as with the temperature (Fig. 7.5). In at least some cases, the normal autumn hardening can be prevented if the plants have previously been induced to continue their growth by maintaining them in a long photo-

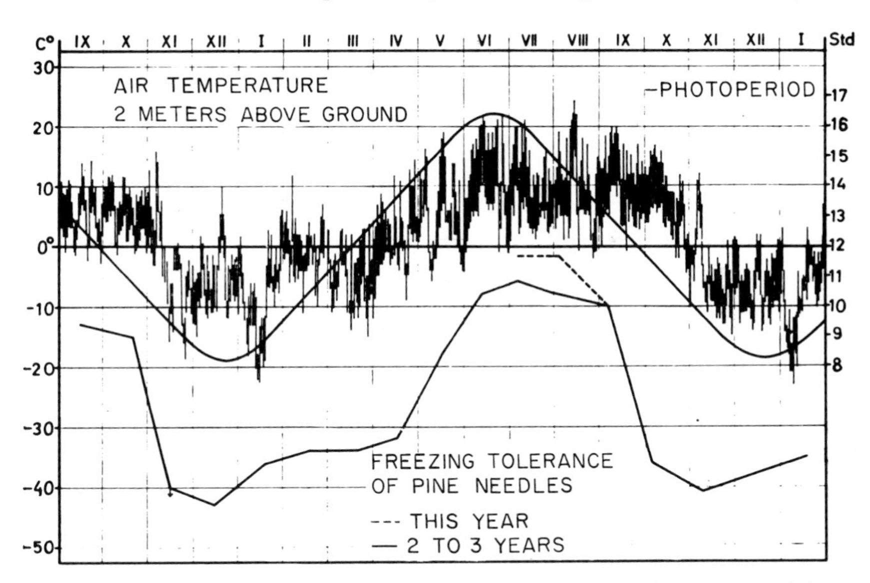

FIG. 7.5. Seasonal changes in air temperature (daily extremes), photoperiod (upper curve), and freezing tolerance (lower curve) of pine needles from September 1965 to January 1967. (From Schwarz, 1968.)

period. This can be seen in some modern cities where the shoots of introduced trees and shrubs close to street lights may be winter-killed, whereas the other parts of the plants or others like them are uninjured due to their safe distance from weak lights (Kramer, 1937). In the case of arctic plants, Biebl (1967c) concluded that the fall increase in freezing tolerance is dependent primarily on the shortening of the day length and only secondarily on the decrease in temperature. When different species or varieties are compared, a direct correlation frequently exists between length of the critical photoperiod and freezing tolerance (e.g., in wheat varieties) (Rimpau, 1958), and an inverse correlation between the growth effect of a long photoperiod and hardiness (Schmalz, 1957). Adequate hardening may sometimes occur, however, in spite of long photoperiods (Kneen and Blish, 1941). Cabbage seedlings harden as well when grown before and during hardening at any photoperiod from 8 to 24 hr, provided that optimum temperature and light regimes are used for hardening (Kohn and Levitt, 1965). In the case of Douglas fir seedlings, day length, temperature, and light were all important for the development of freezing tolerance, but only temperature affected the loss of tolerance (van den Driessche 1969b). Long days retarded the development of tolerance in autumn, but not later.

Vernalization of winter annuals (by storage of the imbibed seeds at 0°–5°C for 30–60 days) also stimulates growth and decreases hardiness (Schmalz, 1958), though exceptions occur. Since the low temperature used to vernalize plants does induce some hardening, the vernalized may be more hardy than the nonvernalized plants if these have not been subjected to hardening temperatures (Saltykovskij and Saprygina, 1935; Timofejeva, 1935; Vetuhova, 1936). If both are exposed to hardening temperatures, the nonvernalized always become more tolerant of freezing than the vernalized plants (Vetuhova, 1936, 1938, 1939). In many cases, those varieties that require the longest cold treatment for vernalization are the most hardy; there are many exceptions (Hayes and Aamodt, 1927; Martin, 1932; Quisenberry and Bayles, 1939; Saltykovskij and Saprygina, 1935; Straib, 1946). Furthermore, the longer the vernalization period, the greater the effect of an increased photoperiod in lowering the freezing tolerance (Rimpau, 1958).

The following evidence, therefore, indicates that freezing tolerance is inversely related to growth and development.

1. The rapid spring growth is essentially unable to harden.
2. Preparation for spring growth is accompanied by a loss of freezing tolerance, even at hardening temperatures.
3. The fall cessation of growth is accompanied by an increase in freezing tolerance.

4. The relative growth rate of winter annuals in the fall is inversely related to their relative hardiness.

5. Artificial stimulation of growth by excess nitrogen fertilization, by long days, by vernalization, or by growth regulators (see below) is accompanied by a loss of tolerance or of ability to harden. Artificial retardation of growth by wilting or by growth inhibitors is accompanied by an increase in freezing tolerance.

This evidence has led to the conclusion that freezing tolerance (T_F) is inversely related to growth:

$$T_F \propto 1/\text{growth}$$

A more careful examination of the results, however, reveals that even in the above cases, there are many exceptions. Furthermore, a distinction must be made between growth *per se* and developmental stage. At the same time as growth occurs and tolerance decreases due to photoperiodic inductions, the length of the embryonic spike increases (Sestakov and Smirnova, 1936; Sestakov and Sergeev, 1937), and the plant passes through the "second stage of development." In nearly all the above cases, a developmental change accompanied the change in freezing tolerance, e.g., the rapid spring growth includes reproductive development, the varieties of winter annuals that grow more rapidly in the fall also are less dependent on vernalization for their development. Conversely, the smallest effects on freezing tolerance are produced by those treatments that affect growth but have little effect on development (wilting, nitrogen fertilization, certain growth regulators, etc.)

It, therefore, appears likely that the true inverse relation is between freezing tolerance and developmental stage, rather than growth, so that

$$T_F \propto 1/\text{development}$$

Evidence in favor of this interpretation has recently been produced (Cox and Levitt, 1969). When the growth rate of individual cabbage leaves was measured, the freezing tolerance achieved, when exposed to hardening temperature and light, was *directly* proportional to growth rate, whether the latter was measured at the moment of transfer to hardening conditions or during the hardening period (Fig. 7.6). Cabbage is a biennial and does not undergo reproductive development until transferred back to warm temperatures after exposure to a low (vernalizing) temperature. The logical conclusion is that hardiness is inversely related to development, and not to growth *per se*, and even development may be directly related (see below).

Many attempts have been made to relate the age of a plant to its freezing tolerance (see Levitt, 1956). It is impossible to separate age from de-

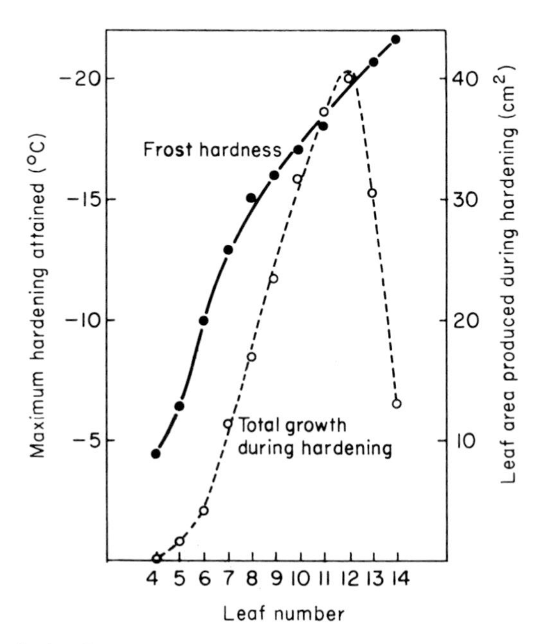

FIG. 7.6. Correlation between maximum hardening achieved (●——●) and total growth (– – –) produced during hardening period in leaves 4–12 of cabbage plants. (From Cox and Levitt, 1969.)

velopmental stage, and the age factor may simply be considered as another method of determining the developmental state. In the case of *Quercus ilex,* for instance, the greatest increase in hardening occurs within the first 5 years (Larcher, 1969), but the full capacity for freezing tolerance is not attained until the tree enters its reproductive phase. The roots, on the other hand, do not change their tolerance pattern with age.

As mentioned above, however, there is evidence that development and freezing tolerance are not indissolubly related. Siminovitch *et al.* (1967b) have pointed to "seasonal rhythms" in the tree as being more important than temperature *per se.* Schwarz (1968) believes that it is these rhythms, rather than the developmental stage, that control freezing tolerance. Thus, when he maintained *Pinus cembra* at a constant temperature (15°C) in a growth chamber throughout the year, tolerance varied in exactly the same way as in others kept in the open (Fig. 7.7). In the second year, however, the annual amplitude of freezing tolerance decreased at the constant temperature. Similarly, when the photoperiod was kept constant, the annual fluctuations in freezing tolerance were much smaller (Fig. 7.7). Schwarz, therefore, concluded that the annual change in freezing tolerance depends on three factors: temperature, photoperiod, and the internal rhythm of the plant.

D. The Nature of Freezing Tolerance

As indicated above, the only freezing tolerance developed by the plant is tolerance of the secondary freezing stress—the water stress induced by freezing. From the basic concepts of stress tolerance (Chapter 2) this can conceivably be of two kinds (Diag. 7.2):

Tolerance of secondary freezing stress
(= tolerance of freeze-induced water stress)

(1) Avoidance of dehydration strain (2) Tolerance of dehydration strain

It is of fundamental importance to examine the evidence for these two types of freezing tolerance.

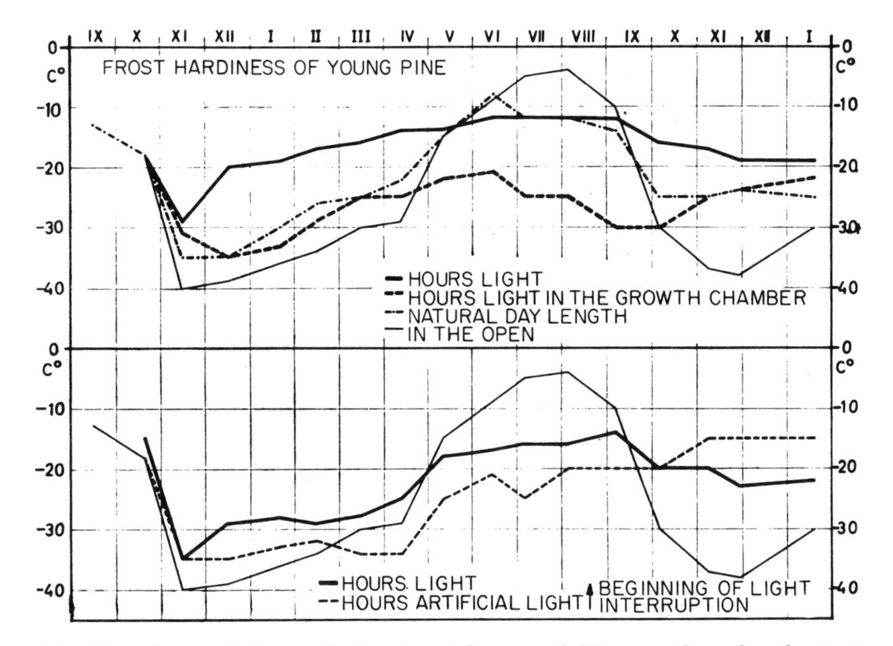

FIG. 7.7. Seasonal changes in freezing tolerance of *Pinus cembra* when kept at a constant temperature (15°C) in a growth chamber, and when left under natural conditions (thin, continuous line). (From Schwarz, 1968.)

1. Avoidance of Dehydration Strain

As shown above, the plant soon comes to thermodynamic equilibrium with the secondary freezing stress; it is, therefore, unable to decrease the freeze-induced water stress. It can, however, decrease the amount of ice formed in its tissues, and therefore, the dehydration strain. Stress tolerance of this kind is proportional to the fraction of water kept in the unfrozen state at any one freezing temperature; and plants possessing such stress tolerance would require a lower freezing temperature to produce an injurious dehydration strain. Assuming that the plant sap acts as an ideal solution, this kind of tolerance would be proportional to the freezing point lowering of the plant sap. Many plants do, indeed, show a direct relationship between freezing point lowering and freezing tolerance (Chapter 8), indicating that at least part of their tolerance is due to an increased avoidance of dehydration strain.

Since the cell sap may not act as an ideal solution, more direct evidence is necessary. The amount of water unfrozen at freezing temperatures has been measured calorimetrically by several investigators. In the case of two molluscs differing in freezing tolerance, the difference was completely accounted for by avoidance of dehydration strain, both being killed when 64% of their water was frozen (Table 7.5). Human red blood cells, with a smaller freezing tolerance, were also killed at the same percentage dehydration, though muscle cells required a greater degree of freezing dehydration (Table 7.5). In the case of two wheat varieties (Johansson and Krull, 1970), each increase in freezing tolerance during hardening was accompanied by a calorimetrically measurable decrease in percentage freeze-dehydration at −9.0°C. These calorimetrically measured values all agreed

TABLE 7.5 Percentage Dehydration in a Number of Animal Cells at Their Freeze-Killing Temperatures

Species or tissue	Freeze-killing temperature (°C)	Percentage water frozen at killing temperature	Reference
Muscle	−2	78	Moran, 1929
Molluscs			(see Salt, 1955)
Venus mercenaria	−6	64	Williams and Meryman, 1970
Mytilus edulis	−10	64	Williams and Meryman, 1970
Human red cells	−3	64	Meryman, 1970

TABLE 7.6 PERCENTAGE WATER FROZEN IN A HARDY (SAMMETSVETE) AND A TENDER (CAPELLE DESPREZ) WHEAT VARIETY AT DIFFERENT DEGREES OF HARDENING[a]

Days hardened	Freezing point depression (°C)	Frost-killing temperature (°C)	Percentage of water frozen at −9.0°C: Observed calorimetrically, Intact plants	Observed calorimetrically, Expressed sap	Observed calorimetrically, Residue	Calculated cryoscopically, Intact plants
Sammetsvete						
0	0.97	− 6.5	92.0	93.6	85.5	91.8
10	1.38	− 9.0	82.7	84.4	79.1	83.0
20	1.64	−11.5	79.2	81.2	75.6	78.2
Capelle Desprez						
0	0.78	above − 6.0	93.4	96.2	84.3	93.2
10	1.15	− 7.5	87.4	89.7	81.9	87.4
20	1.42	−10.0	82.7	84.0	80.4	82.7

[a] From Johansson and Krull, 1970.

perfectly with the cryoscopically calculated values (from the freezing point lowerings) throughout the hardening period (Table 7.6). The hardening achieved was, unfortunately, too small to reveal the difference in hardiness between the two varieties. The hardier variety, under field conditions, is capable of developing two to three times the maximum freezing tolerance obtained artificially by Johansson and Krull. It is interesting to note that the observed degrees of dehydration were identical in the two varieties, when they possessed a freezing tolerance equal to the temperature at which the measurements were made (82.7% when frost-killing temperatures were −9.0° and −10.0°C, respectively).

Following up these results, Johansson (1970) investigated eleven varieties of three species calorimetrically: wheat (7), rye (2), and turnip rape (2). In all cases, increased tolerance was accompanied by an ability to decrease the freeze-dehydration at any one temperature; the percentage of water frozen at the lower killing temperature of the more tolerant (hardened) plants was about the same as the percentage at the higher killing temperature of the less tolerant (unhardened) plants (Table 7.7).

It should be possible, from these measurements, to determine whether or not there is a specific freeze-dehydration at which all cells are killed.

TABLE 7.7 Relationship of Freezing Tolerance to Freezing Dehydration[a]

Plant	Variety	Days hardened	Frost-killing temperature	Percentage of water frozen at frost-killing temperature
Wheat	Dalavårvete	0	− 5.5	87.0
		13	− 9.0	87.2
	Norre	0	− 6.5	86.7
		12	−10.5	87.3
		25	−12.5	86.6
	Odin	0	− 7.5	85.5
		11	−10.5	87.2
		18	−12.0	87.6
		36	−13.5	84.4
	Skandia IIIB	0	− 7.5	85.9
		11	− 9.0	87.9
		21	−10.0	87.3
	Sammetsvete	0	− 8.5	85.6
		11	−12.0	87.0
		20	−12.0	87.1
		37	−13.5	84.0
	Starke	0	− 9.0	85.5
		12	−11.0	87.1
		19	−12.0	88.0
	Virtus	14	−11.5	86.3
Rye	Kungs II	8	− 7.5	86.3
		12	−12.5	86.9
		14	−13.0	87.3
Turnip rape	Storrybs	0	− 5.5	(88.8)
		13	− 9.5	87.4
	Rapids II	0	− 6.5	87.4
		13	− 8.0	(89.0)
		32	−10.5	86.7
		38	−11.0	84.5
Wheat	Capelle Desprez	0	− 7.5	87.8
		23	−11.0	86.3
	Eka Nowa	0	− 8.0	87.9
		14	−13.0	88.0
		35	−13.5	85.9
	Sammetsvete	0	− 8.5	87.6
		13	−11.5	87.8
	Starke	0	− 9.0	86.5
		20	−12.0	86.2

TABLE 7.7 (Continued)

Plant	Variety	Days hardened	Frost-killing temperature	Percentage of water frozen at frost-killing temperature
	Mironowskaja 808	0	−10.0	86.7
		14	−15.0	87.0
		34	−16.0	86.3
Rye	Kungs II	0	−10.5	85.8
		20	−12.5	86.9
	Ensi	0	− 9.5	88.0
		22	−16.0	87.5

[a] From Johansson, 1970.

The above values for eleven varieties of three plant species all fall between 84–89%. These values are considerably higher than for the four kinds of animal cells given above (Table 7.5). This difference is undoubtedly at least partly due to the large vacuoles in plant cells. The percentage dehydration of the protoplasm must be much less, since it consists largely of hydrophilic colloids as opposed to the essentially pure solution of the vacuole. Johansson and Krull (1970) expressed the sap (mainly vacuolar) and measured the water frozen in the sap and residue separately. Since the residue consists essentially of vacuole-free material, it should give an average value for cell wall and protoplasm. The percentage dehydration of the residue at the frost-killing temperature was 79 and 80.5% for the two wheat varieties, respectively. This value is close to the one for muscle, but considerably larger than for the other animal cells.

2. Tolerance of Dehydration Strain

In the above investigations by Johansson, the increase in stress tolerance appears to be fully accounted for by an increased avoidance of freeze-dehydration. In some cases, in fact, there seems to be an overshoot, and the more stress-tolerant plants cannot survive as high a freeze-dehydration at their lower killing temperatures as survived by the less stress-tolerant plants at their higher killing temperatures (Table 7.7). This relationship, however, may not necessarily hold true for all plants. Johansson and Krull (1970) showed that calculations from cryoscopic values of the amount of water frozen at the frost-killing point give values identical with the calorimetrically measured ones (Table 7.6). Since many cryoscopic values are available in the literature, it is, therefore possible to find out how widespread this type of tolerance is.

Calculations from values obtained by Tumanov and Borodin (1930; see Levitt, 1956) for eight hardy wheat varieties not used by Johansson, yield considerably higher freeze-dehydrations (90–92%) at the frost-killing temperatures, than those obtained by Johansson (84–89%). In fact, when a large number of wheat varieties from all over the world are compared, there is relatively little relation between freezing point lowering and varietal hardiness (W. Schmuetz, personal communication). In the case of some of the hardiest plants, this lack of a correlation is even more pronounced. Needles of *Pinus cembra* do show a steady lowering of their freezing point during hardening, but the change is so small that the calculated percentage dehydration at the frost-killing point rises steadily from a low of 84.5%, in the least tolerant state, to a high of 95.7%, in the most tolerant state (Table 7.8). An objection may be made that these calculated values for pine are not comparable to the measured values in wheat. A comparison of the two, however, reveals that the calculated percentage freeze-dehydration at the frost-killing temperatures of pine that fell within the range of Johansson's wheats (−10° to −15°C) are 84.5 and 89%, in perfect agreement with Johansson's measured values for wheat (84–89%). In the case of wheat itself, the calorimetrically observed and the cryoscopically calculated values agreed perfectly throughout the range tested, including as high as 93.4% (Table 7.6). There is no escaping the conclusion, therefore, that in the case of pine needles, the increase in freezing tolerance during hardening is mainly due to an increase in tolerance of the freeze-induced dehydration strain.

TABLE 7.8 Seasonal Changes in Current Needles of *Pinus cembra*[a]

Date	Percentage water	Freezing point lowering	Frost-killing temperature	Calculated percentage water frozen at frost-killing temperature	Calculated gram water unfrozen/gram dry matter
9/23/42	59	1.55	−10	84.5	0.22
10/8/42	56	1.64	−15	89.0	0.14
11/27/42	57	1.78	−35	94.9	0.06
1/11/43	57	1.73	−40	95.7	0.06
3/28/43	55	1.69	−30	94.4	0.07
5/3/43	50	1.96	−13	85.0	0.15

[a] Adapted from Pisek, 1950.

TABLE 7.9 NATURE OF INCREASE IN FREEZING TOLERANCE OF CABBAGE DURING HARDENING[a]

Hardening	Freezing point lowering (°C)	Frost-killing temperature	Percentage water frozen at frost killing temperature: Measured (calorimetrically)	Percentage water frozen at frost killing temperature: Calculated (cryoscopically)
Unhardened	0.81	−2.1	60	61.5
Hardened	1.257	−5.6	75	77.5

[a] From Levitt, 1939.

That hardening may involve both factors, even in the case of herbaceous plants, was shown by direct calorimetric measurements of ice formation in cabbage plants (Levitt, 1939). Although the hardening process involved a large increase in avoidance of dehydration strain (a 50% increase in freezing point lowering) there was also a very definite increase in tolerance of the freeze-induced dehydration strain, from 60% in the unhardened to 75% in the hardened (Table 7.9). The cryoscopically calculated values agreed almost perfectly with the calorimetrically measured values (Table 7.9). These measurements also agree almost perfectly with Johansson's; he found 62% of the water frozen in unhardened wheat leaves at −2.65°C, compared to the above 60% in the cabbage at −2.1°C. Results from a third laboratory are also in agreement with these two values. Williams and Meryman (1970) concluded that a removal of 65% of the osmotically active water is the limit tolerated by grana isolated from unhardened spinach chloroplasts. It is, of course, doubtful whether chloroplast grana can be expected to show the same tolerance of dehydration as in the whole cell, or even the whole protoplasm. Furthermore, their tolerance when isolated may differ from their normal tolerance when protected within the protoplasm. Nevertheless, the agreement encourages the acceptance of the above results as valid.

3. *Interactions between Avoidance and Tolerance of Dehydration Strain*

Since all these results from different laboratories agree with each other quantitatively whenever comparisons can be made, it must be concluded that all are equally trustworthy. An increase in tolerance of the freezing stress is, therefore, due to an increase in (1) avoidance of the dehydration strain, or (2) tolerance of the dehydration strain, or both. It is of funda-

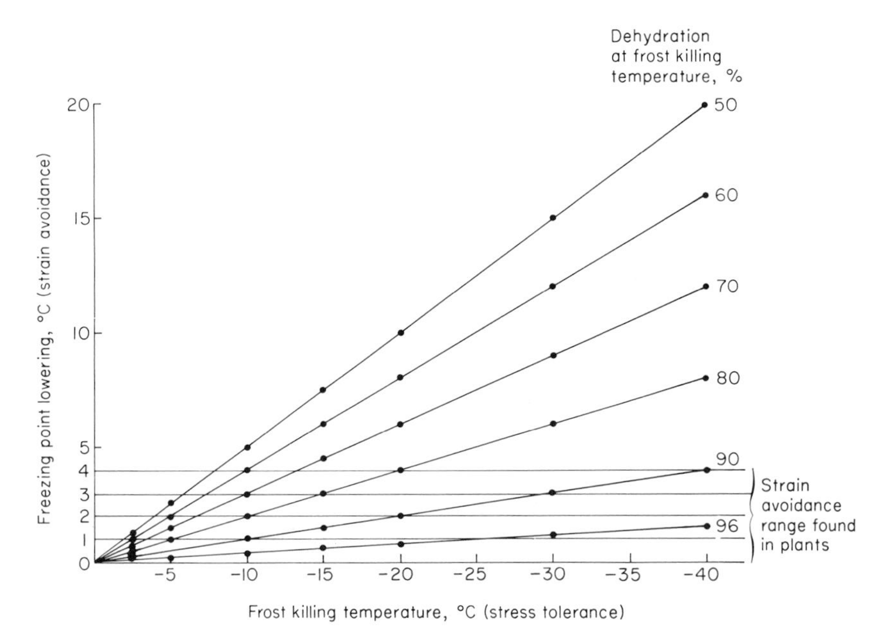

Fig. 7.8. Dependence of freezing stress tolerance (abscissa) on dehydration strain avoidance (freezing-point lowering of plant sap—ordinate) and dehydration strain tolerance (percentage dehydration at frost-killing temperature).

mental importance to understand the quantitative interrelationships between these two components of freezing tolerance.

On the basis of the above conclusions, if the tolerance of the dehydration strain is known, the stress tolerance can be calculated from the freezing point lowering. The values for a series of dehydration strains are given in Fig. 7.8 for a range of stress tolerance from 0° to −40°C. Since the lowest freezing points recorded in the literature for the hardiest plants are about −4.0°C, the possibilities open to the plant range between this value and 0°C. There are several obvious conclusions (Table 7.10).

1. Plants with freezing point lowerings of 0.5°C (osmotic potential of −6 atm) will be killed by the "first touch of frost" (i.e., not lower than −1.6°C) unless their tolerance of the freeze-induced dehydration strain is higher than 70% This explains the extreme tenderness of all freshwater aquatic plants, succulents, and many other plants.

2. Plants with freezing point lowerings of 2°C will not be killed by any temperature above −5.5°C unless their tolerance of freeze-induced dehydration is less than 70%. This explains the ability of even the completely unhardened trees to survive slight freezes in summer.

TABLE 7.10 STRESS TOLERANCE OF PLANTS WITH SPECIFIC FREEZING-POINT LOWERINGS AND SPECIFIC TOLERANCES OF FREEZE-INDUCED DEHYDRATION STRAIN

	Frost-killing temperatures (°C) at dehydration tolerances of 50–96%					
Freezing-point lowering (°C)	50	60	70	80	90	96
4	−8	−10	−13.5	−20	−40	
3	−6	− 7.5	−10	−15	−30	−70 approx.
2	−3.5	− 4	− 5.5	− 8.5	−16.5	−42.5
1	−2	− 2.5	− 3.5	− 5	−10	−25
0.5	−1	− 1.3	− 1.6	− 2.5	− 5	−13

3. Plants with a tolerance of freeze-induced dehydration in the neighborhood of 80%, can increase their stress tolerance from as low as −2.5°C to as high as −20°C, by simply increasing their freezing point lowering from 0.5° to 4.0°C. This simple method of hardening has apparently been adopted by several varieties of grains and other winter annuals.

4. Plants with a tolerance of freeze-induced dehydration that does not exceed 50%, cannot harden markedly by increasing their freezing point lowering. This is why plants such as potato tubers, which undergo a marked increase in freezing point lowering on exposure to hardening temperatures, nevertheless, fail to harden appreciably.

5. In order to tolerate the severe freezing stresses of northern climates plants must develop at least a 90% tolerance of freeze-induced dehydration.

In all the above cases, tolerance of freeze-induced dehydration is the primary factor in freezing tolerance. When the tolerance of freeze-induced dehydration is low, no amount of avoidance of freeze-induced dehydration can help the plant. When the tolerance of freeze-induced dehydration is high, the plant can harden markedly by a relatively small increase in avoidance of freeze-induced dehydration, which is, therefore, a secondary factor in freezing tolerance. Nevertheless, both factors are required for marked freezing tolerance.

CHAPTER 8

FACTORS RELATED TO FREEZING TOLERANCE

A. Morphological and Anatomical Factors

In view of the above inverse relationships between freezing tolerance and development, the morphological characteristics associated with dormancy or with retarded development might reasonably be sought in hardy plants. Conversely, the characteristics of advanced development might be expected in tender plants. These relationships have, indeed, often been reported. Thus, when varieties of a species differing in hardiness are compared, it is frequently found that hardiness varies inversely with the size characteristics of the plant, e.g., height, leaf length, internode length, and especially cell size (Levitt, 1956). Hardy varieties of flax, for instance, form a rosette during the winter, in contrast to the upright growth of the sensitive lines (Omran *et al.*, 1968). This relationship can be expected to hold true only if the measured characters are the result of development during the hardening period, for instance, in the case of winter annuals. It is not surprising, therefore, that species of native herbaceous plants that complete their growth before the hardening period may show the opposite relationship, e.g., the taller the plant the greater the freezing tolerance (Till, 1956). This is presumably due to the selective effect of the lower temperatures to which the taller plants are exposed. It has also frequently been reported that hardiness is directly related to the degree of "ripening" of the twigs (e.g., as judged by cork formation; Sakai, 1955a). These relationships do not by any means always hold true; nor can they be expected to. When, for instance, a plant at the right stage of development is exposed to low temperature and adequate light intensity, it undergoes a rapid increase in hardiness yet it does not shrink in size, nor do its cells become smaller. In fact, no gross morphological or anatomical changes occur during this hardening period, nor do they occur during the dehardening period that takes place before growth commences.

On the other hand, changes in form and arrangement of the protoplasmic constituents may conceivably occur during hardening. It has long been known, for instance, that chloroplasts frequently accumulate in a belt around a cell in the hardened state instead of being uniformly distributed throughout its surface as in the growing nonhardy state. Here again, however, a marked increase in hardiness may occur without any change in chloroplast distribution. According to Heber (1959a), chloroplasts of wheat increase in size during hardening due to an accumulation of low molecular weight sugars and water-soluble proteins (see below). It must be concluded from all these observations that when a relationship between hardiness and morphological or anatomical characteristics occurs this is indirect, due to the accompanying physiological factors. This is true even of development itself, for the growth of tender plants can be brought to a stop and they can even have a well-defined rest period without developing any hardiness (see above). These facts have long been known and led to many investigations of the individual physiological or physicochemical characters of the plant in relation to hardiness.

B. Physiological Factors

Unfortunately, the total number of factors involved in freezing tolerance is unknown. Genetic investigations have so far succeeded only in pointing to a multifactor relationship. In the case of *Brassica oleracea*, for instance, the genetic evidence suggested that two dominant, epistatic genes conditioned freezing tolerance (Bouwkamp, 1969). An essentially unlimited number of factors has been investigated. The factor most commonly chosen, is the quantity of a specific substance (i.e., its accumulation) in relation to freezing tolerance.

1. Accumulation of Substances

a. Total Solutes

A vast number of measurements have revealed that, in general, cell sap concentration increases with freezing tolerance. This parallel has been found (1) during hardening (Fig. 8.1), (2) when species or varieties differing in hardiness are compared (see Levitt, 1956), and (3) when solutes are fed artificially to potentially tolerant plants (Fig. 8.2). Even simple refrac-

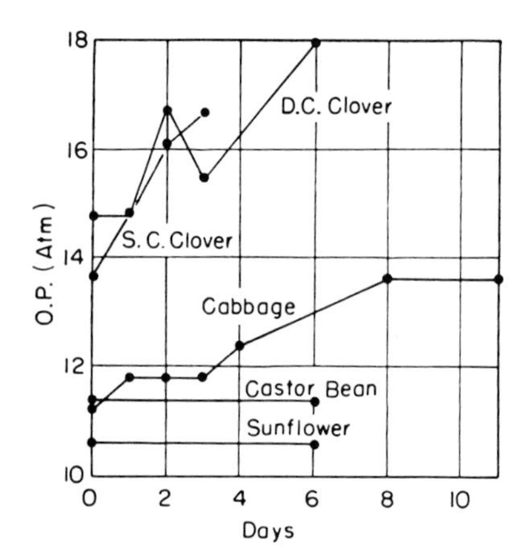

FIG. 8.1. Increase in osmotic concentration of hardy (cabbage and clover) leaves during hardening at 5°C, and absence of any increase in tender (castor bean, sunflower) leaves. From Levitt and Scarth, 1936. (Reproduced by permission of the National Research Council of Canada from the *Can. J. Res.* C14.)

tometer measurements of total soluble dry matter showed a close correlation with frost resistance in kale (Thompson and Taylor, 1968). This method had earlier been used successfully by other investigators (see Levitt, 1941). In a few cases of supposed varietal differences (Magistad and Truog, 1925; Civinskij, 1934), freezing tolerance is not involved, since the plants are killed by very slight freezes and the so-called "hardier" variety is simply better able to avoid freezing by virtue of a lower freezing point.

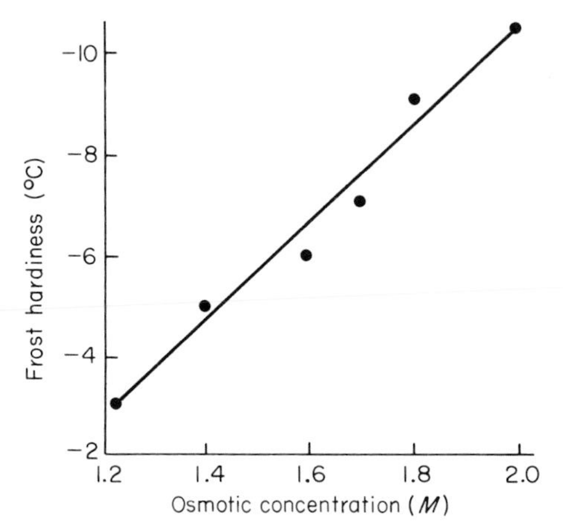

FIG. 8.2. Increase in freezing tolerance (ordinate) of gardenia leaves with increase in osmotic concentration due to sugar (glucose) feeding. (From Sakai, 1962.)

Such tender plants are usually incapable of undergoing any increase in solutes on exposure to hardening temperatures (Rein, 1908; Pantanelli, 1918; Rosa, 1921; Meindl, 1934; Schlösser, 1936) unlike frost-hardy plants (Fig. 8.1). Potato tubers are exceptions, for although unable to harden, they show a marked increase in solutes at hardening temperatures (Müller-Thurgau, 1882; Apelt, 1907). Negative results have also been obtained, primarily when comparing varieties, both the more hardy and the less hardy varieties, showing the same increase in solute concentration with hardening (see Levitt, 1956). Some of the negative results may be due to a passive loss of water (Grahle, 1933; Pisek *et al.*, 1935). This error is possible only when freezing-point determinations are used. The loss of water can, however, be compensated for by measuring the water content and using the product of this times the freezing-point lowering (Tranquillini, 1958). Yet even this corrected value, as well as values obtained by the plasmolytic method, may fail to show a correlation with varietal tolerance (see Levitt, 1956). This proves that there are some real exceptions. Furthermore, the very investigators who have succeeded in demonstrating the most striking correlations when some varieties are compared, have completely failed in other cases. Wheat varieties, for instance, sometimes give excellent results; barley varieties usually show no correlation between freezing tolerance and solute concentration; and some of the hardiest evergreens (e.g., *Picea*; see Levitt, 1956) show relatively little change in cell sap concentration although their tolerance changes seasonally from one extreme to the other.

Some attempts have been made to find out whether the relationship between cell sap concentration and freezing tolerance is a direct one. It is

TABLE 8.1 FROST KILLING OF *Aspergillus niger* GROWN IN DEXTROSE SOLUTIONS OF DIFFERENT CONCENTRATIONS[a]

Conc. of dextrose (%)	Frost-killing point (°C)	Isotonic $NaNO_3$ (%)	Freezing point of $NaNO_3$ (°C)
1	−2	9	−3.3
10	−4	16	−5.6
20	−9	24	−8.2
30	−14	30.5	−10.2
40	−22	36.5	−11.9
50	living at −26	38.5	−12.5

[a] From Bartetzko, 1909.

easy to show that increasing the concentration of solutes in the nutrient medium markedly increases tolerance (Table 8.1; Maximov, 1908; Chandler, 1913). This certainly produces other changes besides the increase in cell sap concentration (e.g., in cell size and total growth; Todd and Levitt, 1951). When rapidly penetrating substances (e.g., glycerine, urea, ethyl alcohol) are used, such secondary changes are presumably avoided and a small, but definite lowering, of the frost-killing point occurs (Table 8.2; Åkerman, 1927; Iljin, 1935a). In the case of succulents, however, there was little or no increase in freezing tolerance (Kessler, 1935). Even when considerable quantities of sugar are absorbed from a solution, there is only a slight increase in the freezing tolerance of cabbage leaves (Dexter, 1935). On the other hand, it is possible to lower freezing tolerance very markedly by chloroform treatment without affecting the cell sap concentration (Kessler, 1935; Ulmer, 1937).

It seems obvious from these results that the increase in cell sap concentration may sometimes be a factor, but is unable by itself to account for all the increase in freezing tolerance that normally accompanies it. Furthermore, plants with osmotic concentrations as high as 30 atm may be killed by light frosts (Walter, 1931). Consequently, the specific solutes involved have been examined in the hope of clearing up these discrepancies.

b. Sugars

Although the above plasmolytic and freezing-point measurements merely determine total osmotic concentration, there is ample evidence that

TABLE 8.2 Effect of Solute Uptake on Frost Hardiness[a]

Plant	Treatment	Δ of solution used (°C)	Δ of sap after solute uptake (°C)	% Killing
Cabbage	None		0.780	80 at −4°C
	KCl	0.775	1.145	40
	Glycerine	2.82	1.780	20
	NH_4Cl	0.360	0.950	100
Cowpeas	Sucrose	1.570	1.230	0.0 at −3°C
	Glucose	1.740	1.250	49.9
	Glycerine	1.575	1.160	0.0
	KCl	0.730	1.130	66.6
	NH_4Cl	0.725	1.140	41.6
	Water	0.00	0.870	66.6

[a] From Chandler, 1913.

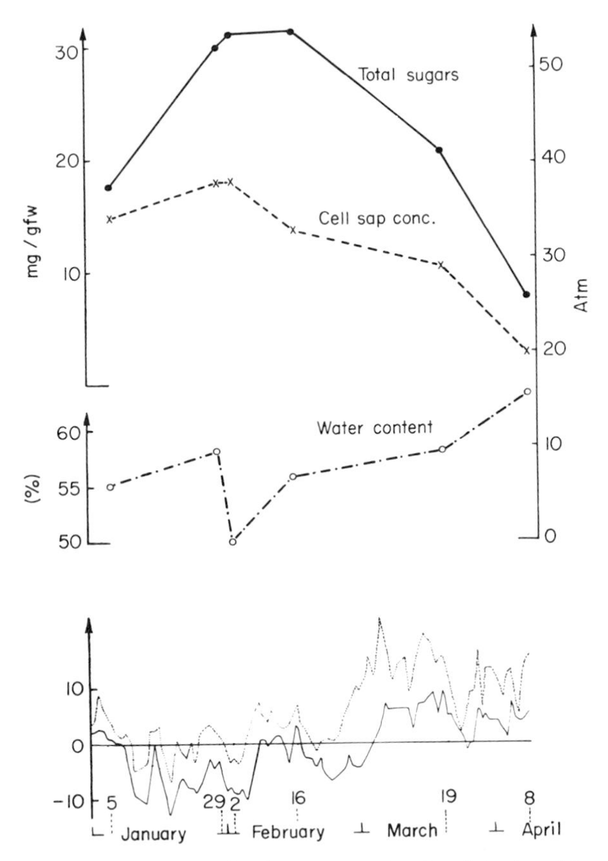

FIG. 8.3. Parallel changes in total cell sap concentration (atm.) and in sugar content (mg/gfw) of sycamore during seasonal changes in freezing tolerance. (From Le Saint and Catesson, 1966.)

the major changes are due to changes in concentration of sugars. Thus, when both measurements are made, the seasonal curves run parallel (Fig. 8.3). These and many other measurements made over the years, all clearly show that sugars normally increase in the fall as plants harden, and decrease in the spring as they deharden (Levitt, 1956). Such changes in sugar content occur in both woody and herbaceous perennials as well as in winter annuals. Even when different tissues of the same woody plant are compared during late fall and winter, the sugar content of the tissue is commonly proportional to its freezing tolerance (Levitt, 1956). In the case of tea plants, for instance, both decrease from the outer cortex to the inner cortex

TABLE 8.3 Relationship between Sugar Content and Frost Hardiness in Wheat[a]

Variety	Relative sugar content	Relative frost hardiness (I = highest)
Sammet	100	I
Svea II	87	II
Thule II	67	IV
Standard	66	IV
Sol II	65	IV
Pansar II	48	V
Extra Squarehead II	44	VI
Danish small wheat	41	VII
Wilhelmina	39	VIII
Perl summer wheat	29	IX
Halland summer wheat	22	X

[a] From Åkerman, 1927.

and xylem, and from here to the least tolerant pith (Sugiyama and Simura, 1968a). Sugar content has even been used as a measure of relative varietal hardiness, e.g., in wheat (Table 8.3). Thermophilic cereals, on the other hand, fail to accumulate oligosaccharides at hardening temperatures (Babenko and Gevorkyan, 1967). Many other wheat varieties also fail to show this parallel; when different species of cereals are compared, the relationship of sugar content to freezing tolerance may or may not become evident (Levitt, 1956). Even hardy species of halophytes with high salt contents accumulate considerable quantities of sugars in winter, at the expense of a part of their salt content (Kappen, 1969a). Yet the most resistant of the three species showed the smallest sugar increase. Unlike the hardy glycophytes, the osmotic concentration of the sugars in these halophytes was, in all cases, lower than that of the salts.

The relationship between sugar content and freezing tolerance may become more pronounced from late fall to late winter, judging by the inverse relationship between tolerance and respiration rate in grains (Newton *et al.*, 1931), apples (Fig. 8.4), legumes (Bula and Smith, 1954), etc. In the spring, tolerance markedly decreases even at temperatures that permit the development of maximum tolerance in the late fall (e.g., +2° to −10°C), and the sugars are simultaneously converted to starch (Levitt, 1956). In midsummer, exposure to these same hardening temperatures fails to induce a rise in freezing tolerance or an increase in sugar content (e.g., in *Tilia*; see Levitt, 1956).

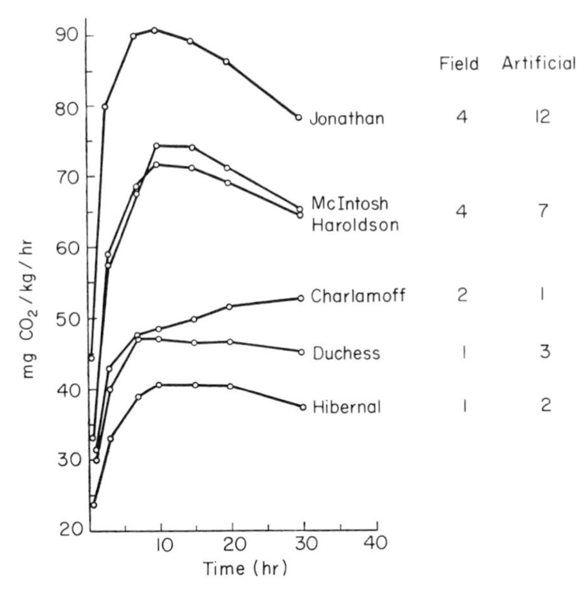

FIG. 8.4. Inverse relation between CO_2 output at +6°C and hardiness (see Table 7.4) of apple twigs. (From De Long *et al.*, 1930.)

In some plants (e.g., mulberry; Sakai, 1961) any treatment that increases the sugar content increases hardiness and any treatment that decreases the sugar content lowers hardiness. A similar intimate relationship between sugar content and freezing tolerance was observed in black locust at all times of the year (Yoshida and Sakai, 1967). No other factor showed this close relationship. This was not true, however, of poplar. When held at 15°C for 2 months, its freezing tolerance increased from a killing temperature of −2° to −30°C, without an appreciable increase in sugar (Sakai and Yoshida, 1968a). Conversely, cells that do not increase in freezing tolerance, markedly increased in sugar content. Other investigators have found similar exceptions. Thus, although sugars appear to play a role in the hardening of *Hedera helix*, there was a lack of parallelism between the two (Steponkus and Lanphear, 1968b). In the case of chestnut trees, a positive correlation between hardiness and sugar content occurred during winter, but not from April to October (Sawano, 1965). In the case of wheat, more sucrose accumulated in the roots of the more resistant than in those of the less resistant variety, but there was no difference between the stems (Musich, 1968). In fact, the total sugar content was less in the stems of the more resistant variety. Some plants with the highest sugar contents possess little or no hardiness (e.g., potatoes kept at low temperatures and sugar cane). Conversely, some of the most hardy species and varieties have lower contents than less hardy ones; mediterranean evergreens, for instance, show little or no increase during the hardening period (Pisek, 1950; Larcher, 1954). This is also true of another evergreen—*Juniperus chinensis* (Pellett and White, 1969). Finally, some marine algae are highly tolerant of freezing in spite of very low sugar contents, and their freezing tolerance is actually lowered by sugars (Terumoto, 1962).

That sugars can truly increase hardiness was shown by early sugar feeding experiments (Chandler, 1913). Those plants fed sugars were able to survive freezes that killed the controls (Table 8.2); but the differences were small in this as well as in later experiments by other investigators. Some failed to obtain any differences, and in only one case were large differences reported (Fig. 8.2).

These discrepancies can all be explained on the basis of the concepts of freezing tolerance described previously (Chapter 7). If the effect of the sugar increase is purely osmotic, leading to an increase in avoidance of freeze-dehydration, a hardened plant with a higher tolerance of freeze-dehydration will show a greater increase in freezing tolerance per unit sugar added than will an unhardened plant with a lower tolerance of freeze-dehydration. Thus, for values of 80 and 50%, respectively, the hardened plant will have its killing temperature lowered a full 5°C (from −5° to −10°C) by the same increase in sugar content as causes only a 2°C increase (from −1° to −3°C) in the case of the unhardened plant (Table 8.4). Furthermore, this difference is an underestimate, since as the sugar concentration increases, the effect of an equal addition on the freezing point lowering increases. Thus, the less hardy the plant, the smaller (and, therefore, the

TABLE 8.4 Relative Effect on Freezing Tolerance of an Equal Addition of Sugar to the Cell Sap of a Hardened and an Unhardened Plant

	Unhardened		Hardened	
	Original	After addition of 9 gm glucose per 100 ml	Original	After addition of 9 gm glucose per 100 ml
Tolerance of freeze-dehydration (%)	50	50	80	80
Freezing-point lowering (°C)	0.5	1.5	1.0	2.0
Freeze-killing temperature (°C)	− 1	− 3	− 5	−10

TABLE 8.5 Effect of Infiltration with Organic Substances on the Hardiness of Plants

Species	Substance	Treatment	Osmotic potential (atm)	Frost-killing temp. (°C) Determined	Frost-killing temp. (°C) Calculated	Reference
Cabbage	Glycerol	Control	−20.25	− 7		
		Infiltrated twice	−27.0	−11	−10	Levitt, 1957b
		Control	−25.2	−12.5		
		Infiltrated once	−32.3	−15.5	−15.5	Levitt, 1957b
Saxifrage		Control	−21.9	−16		
		Infiltrated	−28.7	−20	−20.5	Levitt, 1957b
Cabbage	Dextrose	Control	−16.1	− 7.5		
		Infiltrated	−20.0	− 9.5	− 9	Levitt, 1959a
	Fructose	Control	−15.0	− 7.0		
		Infiltrated	−18.6	− 9.5	− 9	Levitt, 1959a

more difficult to detect) the increase in hardiness that results from a specific increase in sugar content. In the case of plants already in the hardened condition, it has, in fact, been possible to increase the hardiness significantly by infiltrating with sugars and other solutes, and the actual increase observed was always the expected amount on the basis of the above calculations (Table 8.5). Similarly, Perkins and Andrews (1960) obtained greater protection in the hardier wheat variety than in the less hardy one when both absorbed sugar. Further evidence that the protective effect by the infiltrated sugar was simply osmotic in nature, follows from the results of Sakai (1960b). A variety of sugars, as well as polyhydric alcohols, acetamide, and urea were capable of increasing hardiness when taken up in the transpiration stream, although inorganic salts were ineffective. The order of effectiveness of 0.7 *M* solutions for woody plants is shown in the following scheme:

Ethylene glycol, Glycerol, Sucrose 〉 Glucose, Xylose, Acetamide 〉 Raffinose 〉 Mannitol 〉

Urea 〉 Glycocoll, Ethanol 〉 Water control 〉 Balanced salt solution, KNO_3

Amino acids were ineffective and so was the sugar galactose (Perkins and Andrews, 1960). *Hedera helix*, on the other hand, increased significantly in freezing tolerance when fed sucrose solutions, but not with equimolar solutions of glucose, galactose, or mannitol (Steponkus and Lanphear, 1968b). Concentrations ranging from 5 to 500 m*M* were equally effective, inducing maximum tolerance in excised leaves. These results do not agree with those of Sakai. In none of the above cases was the amount of sugar that entered the cells determined. Differences in both permeability and active absorption may account for some of the differences in effectiveness of different substances, as well as the discrepancies found between investigators.

In a later investigation (Sakai, 1962), the osmotic increases were determined and penetration was found to be an important factor. Unlike the infiltration experiments (Levitt, 1957b, 1959a) these results indicate greater increases in hardiness of gardenia leaves by sugar uptake than can be accounted for by the increase in cell sap concentration; for instance, controls with a concentration of 1.2 *M* were killed by −3°C, but after sugar uptake the cell sap concentration was 2.0 *M* and the killing point, −11°C. In these experiments, not only is the amount that enters into the cells unknown and nonuniform (more is probably taken up by the cells adjacent to the bundles), but the amount that does enter may then be metabolized. Infiltration of leaves avoids these difficulties as the exact amount entering the cells is known, and since the whole process can be conducted within 12–24 hr at close to 0°C, metabolism and temperature-induced changes in hardiness are reduced to a minimum.

The importance of metabolism of the sugars in long-term absorption experiments is clearly demonstrated by Tumanov and Trunova (1957, 1963). They increased the freezing tolerance of wheat seedlings by keeping their roots in 12% sucrose solutions for 14 days in the dark. The seedlings were originally somewhat hardy (killing temperature, −10°C), but they became much hardier (killing temperature, −28.5°C) during this period. It was necessary, however, to maintain them at a hardening temperature (+2°C) during the feeding period, in order to obtain this large increase in tolerance. Freezing was gradual (3 days at −4°, 1 day at −7°, 1 day at −10°, and 1 day at −13°C) before the final temperature lowering. The same degree of hardening was obtained by 6 hr of feeding per day as from 24 hr. The optimal sugar concentration for this treatment was 20%. Galactose was ineffective; lactose and maltose were less effective than sucrose; glucose and rhamnose were nearly equally effective; raffinose and fructose were as effective as sucrose. According to Trunova (1963), lactose penetrated the cell but remained unchanged. The total sugars did not increase, but decreased due to utilization in respiration and growth. In the earlier results, however, the sugar content of the tillering nodes increased from 24.7 to 55.1% of the dry weight. Pentoses gave no protection.

If the sugar feeding leads to an increase in sugar content and cell sap concentration, the sugars must accumulate in the vacuole. If, however, the sugars are metabolized, they must enter the protoplasm. Sugiyama and Simura (1968b) attempted to decide this question by immersing tea shoots into 0.1 and 0.3 *M* ^{14}C-labeled sucrose solutions at 14°–18°C for 2 days. This led to increases in freezing tolerance, in osmotic concentration, and in content of total sugars. Radioautographs indicated that the ^{14}C was found only in the thin cytoplasmic layer adjacent to the cell wall. Nevertheless, the osmotic concentration of the plant sap increased 25–30% and the total sugar content by about 30–35%. Since the thin cytoplasmic layers appeared to account for only about one-tenth the cell volume, if the original sugar concentration in the protoplasm and vacuole were about the same, this would require an increase in the cytoplasmic sugar content of about three times that in the vacuole, and a corresponding increase in its volume. Since the radioautographs do not indicate any increase in thickness of the cytoplasmic layer, it is difficult to accept their conclusions.

A specific effect of ribose, not obtainable with glucose, has been reported (Jeremias, 1956). The methods used were not valid (Levitt, 1958) and more recent work from the same laboratory has disproved the conclusions of this work (see below).

Just which sugars increase normally during hardening was not clear from the early investigations. In the case of certain trees and vines it is sucrose (Siminovitch and Briggs, 1954; Sakai, 1957; Steponkus and Lanphear, 1967a, 1968b). Some results with grains indicate hexoses (Heber, 1958a), while others indicate sucrose (Hylmö, 1943; Johansson *et al.*, 1955). In the later stages of hardening, wheat accumulates raffinose and perhaps another high molecular weight sugar (Gunar and Sileva, 1954; Johansson *et al.*, 1955). With the use of modern, chromatographic methods, Parker (1959a) found that raffinose and stachyose increased markedly in the bark and leaves of six conifers during fall. Sucrose and sometimes glucose and fructose also increased. Broad-leaved, deciduous trees did not show an increase in raffinose and stachyose although broad-leaved evergreens did. In *Pinus strobus* (Parker, 1959b), the raffinose content (from 0 to 1.5% of fresh weight) followed hardiness changes both in fall and spring, although sucrose, glucose, and fructose failed to show any such close correlation with hardiness. On the other hand (Sakai, 1960a), the proportions of the different sugars were found to vary in eighteen species of woody plants, but no specific sugar was consistently correlated with hardiness. In six species, sucrose, glucose, and fructose showed marked increases during hardening, but raffinose and stachyose did not. In one species (mulberry), the latter two sugars also increased, but less so than the above sugars. Results with some other plants are summarized in Table 8.6.

TABLE 8.6 Sugars Found to Increase in Plants in Parallel with Freezing Tolerance

Plant	Glucose	Fructose	Sucrose	Melibiose	Raffinose	Stachyose	Reference
Conifers (6)	+ (Sometimes)	+ (Sometimes)	+		+	+	Parker, 1959a
Halophytes	+	+	+		+		Kappen, 1969a
Apple	+		+		+		Benko and Pillar, 1965
Arctic and alpine plants					+ (Mainly)		Billings and Mooney 1968
Sycamore (phloem and cambium)	+	+	+	+	+		Le Saint and Catesson, 1966
Cabbage	+	+	+	+	+		Le Saint, 1966
Tea			+				Sugiyama and Simura, 1968a
Poplar and willow	+						Sakai, 1962
Grapefruit	+	+	+				Young, 1969a

The kind of carbohydrate accumulated is, of course, dependent on the normal carbohydrate metabolism of the particular plant. Cold storage of chicory and dandelion roots caused a breakdown of inulin and of high molecular weight oligosaccharides (polymers of more than ten subunits) to oligosaccharides of a lower degree of polymerization (Rutherford and Weston, 1968). In some grass leaves, the major accumulation at low temperatures is due to fructosans (Smith, 1968a). Even alcohols related to carbohydrates may be accumulated. In some insects, for instance, glycerol increases during hardening and this acts as an antifreeze or at least favors undercooling (see above). Salt (1961) has, in fact, suggested that perhaps all freezing-tolerant insects and possibly other invertebrates contain glycerol or some equally protective substances, although exceptions containing glycerol may not be resistant (Salt, 1957). Both kinds of exceptions occur. Thus, in spite of its relatively large glycerol content (2–3%), a carpenter ant was found to be killed by freezing at −10°C, whereas the larvae of two species of butterfly survived freezing at −15°C without any glycerol or other polyhydric alcohol (Takehara and Asahina, 1960b). Even when glycerol was effective in a caterpillar, considerable freezing tolerance developed before glycerol formation, and maximum tolerance was attained with 2% glycerol, although it continued to accumulate beyond this point (Takehara and Asahina, 1961). Furthermore, injection of glycerol to the amount of 3% of the body weight failed to enhance their freezing tolerance. Analyses of plants (Sakai, 1961) failed to reveal any glycerol or other polyhydric alcohol in appreciable quantity in most of the woody plants tested. In a few species (gardenia, apple, mountain ash, pomegranate) polyhydric alcohols (mannitol, sorbitol, glycerol) amount to about 40% of the total sugar content and may therefore play some role in hardiness (Sakai, 1960b). The accumulation of glycerol during the hardening of insects and its disappearance during dehardening is apparently due to glycogen $\rightleftharpoons$ glycerol conversions (Takehara and Asahina, 1960a). The lack of glycogen in higher plants may therefore perhaps explain the absence of this metabolic adaptation to low temperature.

What can be concluded from this vast mass of data relating or not relating sugars to freezing tolerance? The many records of a correlation with freezing tolerance strongly indicate that, at least in some plants, sugars must play some role in the mechanism of tolerance. Although the correlations under natural conditions may be due to some other relationship common to the two, feeding experiments are more conclusive. They indicate that sugars may increase freezing tolerance in two ways. (1) The osmotic effect—by accumulating in the vacuole, sugars can decrease the amount of ice formed, and therefore increase the avoidance of freeze-induced dehydration. On this basis, sugar is a secondary factor, and cannot induce tolerance of freeze-induced dehydration. This would explain the existence of tender plants with high sugar contents (e.g., sugarcane). Only those plants possessing a marked tolerance of freeze-induced dehydration would be able to show a detectable increase in freezing tolerance due to an increase in sugar content. (2) The metabolic effect—sugars, as such, have little or no effect on freezing tolerance, but by being metabolized in the protoplasm at low, hardening temperatures, they produce unknown protective changes. Some of these changes may conceivably increase the tolerance of freeze-induced dehydration. There are several lines of evidence in favor of this explanation. Very hardy plants can survive much lower temperatures than their sugar content can explain, on the basis of avoidance of freeze-induced dehydration. The fact that the hardiest plants commonly accumulate their sugars in the fall before the main increase in freezing tolerance can be explained by either effect.

c. Water (Total and Bound)

Water content is frequently inversely related to hardiness (Fig. 7.3, also see Levitt, 1956). The practical man, in fact, has long related frost hardiness to "maturity" or "ripening" of the tissues, which he usually considers inversely related to water content. This may be partly due to displacement of the water in the cell during the accumulation of sugars. There are, again, many exceptions. Some succulents (e.g., species of *Saxifraga* and *Sempervivum*) with water contents as high as 90–95% nevertheless develop a high degree of hardiness; among three species of halophytes, water content was lower in winter than in summer in two of the species, but there was no difference in the third, most tolerant species (Kappen, 1969a). Similarly, the water content of sycamore twigs fluctuates during winter in a manner that does not parallel freezing tolerance (Le Saint and Catesson, 1966). On the other hand, the percentage moisture in both the root and top of *Juniperus chinensis* decreased during the fall, the rate paralleling the increase in freezing tolerance (Pellett and White, 1969).

It was suspected early that only one component of the total water content, the so-called "bound water" is important. This water was presumed to be held so tightly that it could not freeze. Excellent correlations were, at first, obtained between this quantity and frost hardiness in grains. These are the same plants that had previously shown an equally good correlation between sugars and hardiness. It was soon realized that sugars themselves bind water, and therefore determinations of bound water on the whole plant may reveal no more than sugar analyses or, more simply, determinations of cell sap concentrations. The reason for this is that the methods used in these early investigations were not valid. Determinations were made at much too high temperatures, and therefore measured the water that was prevented from freezing due to the high concentrations of sugars but that was easily frozen at somewhat lower temperatures. Recent attempts have avoided the major part of this error by using much lower temperatures. The bound water values are, therefore, also much lower—33% (0.5 gm/gm dry matter) in wheat grain at −120°C (Radzievsky and Shekhtman, 1955) and 9% of the total water (0.2 gm/gdm) in yeast at −72°C (Wood and Rosenberg, 1957). Recent NMR measurements have also yielded values of 0.3 to 0.5 gm of H_2O/gram protein at −35°C (Kuntz *et al.*, 1969). Nucleic acids, however, proved to be three to five times as hydrated as proteins. As Kuntz *et al.* point out, the very fact that the water signals can be observed by high-resolution NMR suggests that the bound water is not "icelike" in any literal sense, although it is clearly less mobile than liquid water at the same temperature. The quantity bound must, of course, depend on the vapor pressure of the surrounding water. In the case of egg albumin, at a relative humidity of 92% at 25°C, the value is 0.3 gm of H_2O/gram of protein, but as the relative humidity approaches 100%, the bound water exceeds 0.5 gm H_2O/gm (Bull and Breese, 1968). The protein hydration is proportional to the sum of the polar residues minus the amides, which apparently inhibit binding of water. About 6 moles H_2O are complexed per mole of polar residue.

Conclusions of a relationship between freezing tolerance and bound water have frequently been based on the most indirect evidence that actually has nothing to do with bound water. For instance, a tacit assumption has been made by some workers that bound water is a mysterious fraction of the plant's water that is unable to freeze in the living plant, but that can freeze as soon as the plant dies (Hatakeyama, 1957, 1960, 1961). This conclusion is based on the difference between the freezing points of living and dead tissues. It has already been shown (Chapter 5) that this difference can be increased or decreased by simply altering the method of freezing. All these results have been explained by the simple principles of the freezing point method and therefore have nothing to do with bound water (Levitt, 1966b). Thus, using pear tissue held at fairly low temperatures, Marshall (1961) was able to obtain a freezing point 2°F below the freezing point of the juice. In most cases, however, the true freezing point of living pear tissue, when determined correctly, was shown to coincide with the freezing point of the expressed juice. In the exceptional cases, no more than 1°F difference was obtained. In spite of these newer results, hardening is still being explained by an increase in the ratio of bound to free water (Pellett and White, 1969).

The earlier concept of the role of bound water was as a means of protecting by reducing the amount of water frozen. This explanation was also proved incorrect, at least in some cases. Measurements of the amount of ice formed in cabbage showed that the hardened plant survived the freezing of a larger fraction (75%) of its water than did the same plant in the unhardened state (60%; Table 7.9). The role of the bound water might be specifically in the protoplasm rather than in the nonliving vacuole and this might be overlooked since the latter occupies most of the cell. In an attempt to measure the water bound by the protoplasm itself, determinations were later made on the isolated chloroplasts, and again showed a relationship between bound water and hardiness. Although little or no sugars could have remained in the washed chloroplasts, starch was definitely present, and it, too, was shown to be capable of binding water. Even chloroplasts and grana free of starch were, however, found to show this relationship (Levitt, 1959b). It is difficult to evaluate the significance of such results. Measurements are made over long periods of time, and it is therefore possible that the differences are an artifact, due to a greater sensitivity of the chloroplasts and grana from the unhardened plants. In favor of this objection is the difficulty in obtaining consistently uniform results, and the artifactual results obtained in the case of SH measurements (see below).

One of the problems in the above kinds of measurements is the increase in proteins that frequently parallels hardening. If, for instance, the protein content of the chloroplasts is higher in the hardened state (see below), an increase in bound water might simply be due to an increase in the proportion of these water-binding proteins without any increase per unit of protein. It is, therefore, essential to measure the water bound by the proteins themselves. This has been done indirectly by measuring a series of properties of the soluble proteins (Brown *et al.*, 1970). The proteins were extracted from the root tissue of a hardy and a nonhardy alfalfa variety sampled in the field in mid-October, mid-November, and mid-January, when large differences in freezing tolerance occur. Measurements were made of the partial specific volume of the protein and water of the extracts, the specific heat capacities, the temperature of spontaneous nucleation, and the expansion on freezing. Absorption isotherms of the lyophilized protein powders were also determined. Essentially no differences were found between the two varieties, indicating that the hydration properties of the soluble proteins were similar.

d. Amino Acids

Many earlier investigators measured the amino acid content of plants in an attempt to find a correlation with hardiness. The results were not convincing (Levitt, 1956). In some cases (Wilding *et al.*, 1960a, b), amino acid content may increase with hardiness (e.g., red clover), in others, it may fail to show any relationship (e.g., alfalfa varieties); and where the relationship exists, it seems to reflect a general increase in storage of organic nitrogen during the fall, either due to the nitrogen-storing amide asparagine or the accompanying even greater increase in nonamino acid nitrogen. Similar results have been obtained by Romanova (1967), Ostaplyuk (1968), Protsensko and Rubanyuk (1967), and Paulsen (1968). It has been suggested (Heber, 1958b), however, that an increase in amino acids and peptides may occur during fall in the field, due to a partial, reversible injury as a result of slight freezing. In the case of two varieties of winter wheat, the amino acid composition was the same and did not change with hardening (Toman and Mitchell, 1968). In the phloem and cambium of 2- to 3-year-old branches of sycamore, the total amino acids reach a maximum at the end of January when hardiness is presumably maximal, but a second, larger maximum occurs in April when growth is revived and hardiness is largely lost (Le Saint and Catesson, 1966). The changes, therefore, were explained by dependence, during winter, on the temperature and during spring, on the physiological activity of the plant.

Although total amino acid content has not been consistently related to freezing tolerance, the specific amino acid proline has been reported by several investigators to accumulate at hardening temperatures (Heber, 1958b; Le Saint, 1958, 1960, 1966; Markowski *et al.*, 1962; Ostaplyuk, 1967; Protsenko and Rubanyuk, 1967), and to decrease on loss of hardiness or to increase more in more hardy plants. A winter-hardy rose variety had a higher proline content than a nonhardy variety (Ziganigirov, 1968). In the case of forage plants, arginine and alanine were chief among the several amino acids that increased more in the hardy than in the nonhardy variety (Smith, 1968b).

In cabbage, the content of free proline parallels the freezing tolerance of the different organs in hardened and unhardened plants (Le Saint, 1969a), with a maximum in the hardened terminal shoot. It is the only amino acid that accumulates on hardening, increasing from 2–4% of the total amino acid content of the whole unhardened cabbage plant to 60% in the hardened plant. The correlation between proline and hardiness was, in fact, better than between sugars and hardiness; the basal leaves which have little freezing tolerance may be rich in sugars but have little or no proline. Conversely, the terminal part of the shoot which is richest in proline, is not particularly rich in sugar. When hardiness is transmitted from illuminated (older) leaves to darkened (younger) leaves during exposure to hardening temperature (4°C), proline accumulates in the darkened leaves [Le Saint (-Quervel), 1969b]. Conversely, when hardiness is not transmitted from illuminated (younger) leaves to darkened (older) leaves during exposure to hardening temperature (4°C), proline fails to accumulate in the darkened leaves. Le Saint, therefore, concludes that the chemical effector responsible for the transmission of hardening is either proline itself or a substance closely related to its metabolism. It was even possible to induce some hardening by allowing cabbage shoots to absorb proline from a solution (5 gm/liter) at a nonhardening temperature (18°C; Le Saint, 1966). This was accompanied by the same increase in sugars as occurs on exposure to hardening temperatures.

The relationship between proline and freezing tolerance was not as clear-cut in the case of the phloem and cambium of 2- to 3-year-old branches of sycamore (Le Saint and Catesson, 1966). Proline showed the same two maxima as the total amino acid content, the spring maximum being again about double the winter maximum. The authors suggest that the first maximum is related to the maximum winter cold which preceded it and the spring maximum is due to a regeneration of growth. Somewhat similar results have been obtained for buds of white spruce (Durzan, 1968a, b). Free arginine declined in the fall, as proline and the amides contributed more to the seasonal levels of soluble nitrogen. The high levels of proline extended into the early spring, so that the expanding buds contained a high proportion of proline. At this time, proline levels showed a diurnal periodicity, being maximal at sunset and sunrise (Durzan, 1969). Since these daily changes were extremely large (from a minimum of 50 μg N/gfw to a maximum of 700 μg N/gram fresh weight), proline can hardly contribute to freezing tolerance by its mere presence, for these changes occurred at the end of May when hardening does not occur. Labeled carbon (14) from arginine and citrulline and to a lesser extent from γ-guanidinobutyric acid fed to the plant was recovered in the protein fraction mainly as arginine, glutamic acid, and proline.

In the case of wheat, if seeds are germinated for 2 days at 24°C and then incubated at 2° for 3 weeks, there is a manifold increase in free amino acids, with outstandingly high concentration of proline (Scheffer and Lorenz, 1968). In germinating spring wheat, proline does not increase as much as

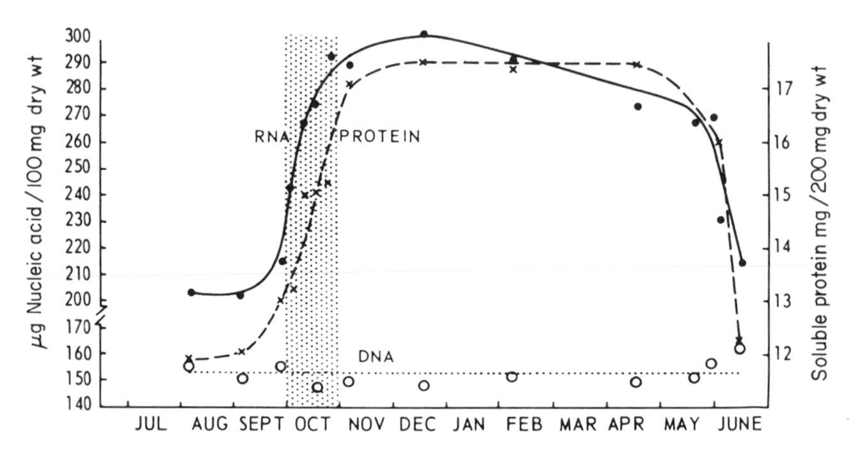

FIG. 8.5. Seasonal changes in water-soluble protein, RNA, and DNA in bark of locust trees. Shaded area is the period of maximum rise in freezing tolerance. (From Siminovitch *et al.*, 1967.)

in winter wheat varieties. Young leaves of winter wheat varieties grown at 24°C contain relatively high concentrations of amino acids, except for proline. In similar plants of the spring wheat variety the proline concentration is rather high. Even in the cotyledons of the peanut (a nonhardy plant) during germination a marked synthesis of proline occurs (Mazelis and Fowden, 1969). They extracted a soluble enzyme system which converts ornithine into proline.

e. Proteins

A striking parallel between soluble protein content and freezing tolerance (Fig. 8.5) has been clearly demonstrated in the cortical cells of trees both during hardening and dehardening (Siminovitch and Briggs, 1949, 1953a, b, 1954; Siminovitch *et al.*, 1967a; Li and Weiser, 1967). This increase in soluble proteins may actually precede the increase in hardiness. After its development in the fall, however, a greater effect is obtained from artificial hardening (Sakai, 1958). A possible factor in the rise of soluble proteins is the increase during hardening in the ability to incorporate glycine into water-soluble proteins (Fig. 8.6). The increase in soluble proteins may sometimes fail to parallel either the main autumn increase in hardiness or the spring decrease or even an artificial increase in mulberry, according to Sakai (1957, 1962). Similarly, the spring swelling of grape buds, which is accompanied by a loss of freezing tolerance, involves an intense accumulation of protein (Cheban, 1968). Finally, under artificial hardening conditions, the increase may not occur; Siminovitch *et al.* (1968) succeeded in hardening bark tissue of locust to a considerable degree (killing temperature −45°C) under conditions preventing an increase in soluble proteins (Table 8.7). The same results have been obtained with poplars kept at 15°C during October and November (Sakai and Yoshida, 1968a). The trees increased in freezing tolerance (the killing point dropping from −2° to −30°C) without any concomitant increase in soluble proteins. On the other hand, when hardened at a more normal temperature of 0°C, during the same two months, the hardening was more extreme (killing point −70°C) and soluble proteins did increase with hardening. Similarly, the normally hardened bark tissue of the locust tree did show a doubling of soluble protein, and survived the temperature of liquid nitrogen (Siminovitch *et al.*, 1968).

Similar results have been obtained with evergreen, broad-leaved woody plants. Although the water-soluble protein increased gradually in the leaves of *Hedera helix*, from summer to winter, freezing tolerance increased steeply in November (Parker, 1962). In spring, the water-soluble protein continued at a high level, while freezing tolerance declined markedly. In

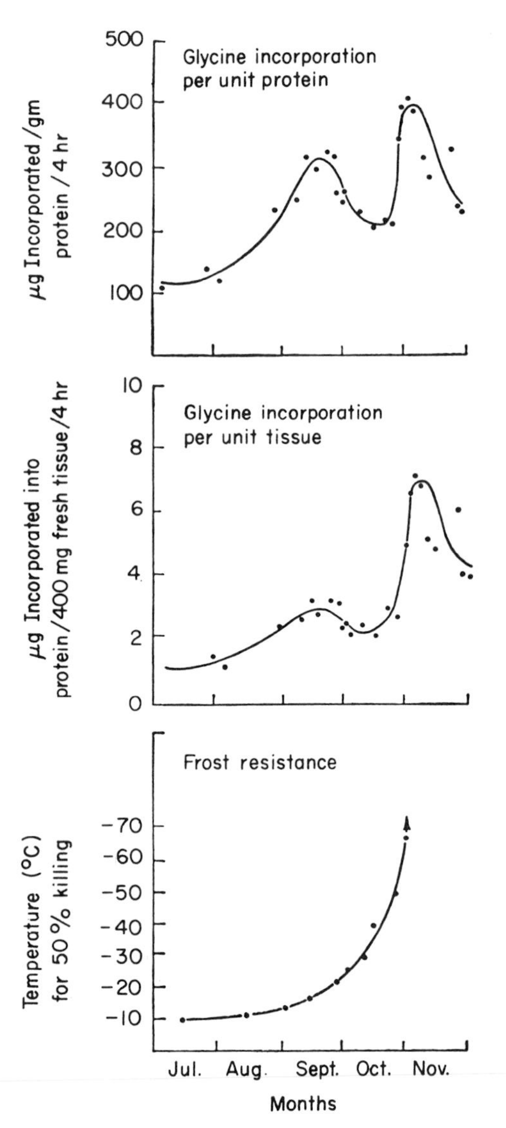

FIG. 8.6. Increase in rate of protein synthesis (measured by rate of incorporation of glycine-^{14}C) during the period of rapid increase in freezing tolerance (frost-resistance). (From Siminovitch, 1963.)

the case of tea leaves, the relationship is more consistent. Soluble protein increased with an increase in freezing tolerance whether this was due to natural or artificial hardening (Sugiyama and Simura, 1967a). The more resistant varieties showed the higher contents of soluble protein. The

TABLE 8.7 CHANGES IN "STARVED" BARK SEGMENTS, ENCIRCLED ON TREES IN AUGUST[a]

Time	Killing temperature (°C)	Content of: Sugars (% dry wt)	Soluble proteins (mg/100 mg/dw)	RNA mg/100 mg/dw	Total lipid (mg/gdw)	Lipid phosphorus (μg/gdw)	Leucine-1-^{14}C incorporated (cpm/400 mg fw/4 hr)
August	−15	7.2	3.7	218	29.3	113	5,650
November (encircled)	−45	2.1	3.9	221	28.3	187	6,150
November control	Not killed by liquid nitrogen	9.1	7.5	312	32.0	210	15,025

[a] From Siminovitch *et al.*, 1968.

soluble protein of the chloroplasts increased from about 45% in September to 55–60% of the total in December to January (Sugiyama and Simura, 1967b). This was the period during which freezing tolerance increased. The content remained constant until April, then decreased with the decrease in freezing tolerance. In *Juniperus chinensis*, total nitrogen showed little relationship to seasonal changes in freezing tolerance (Pellett and White, 1969). Similarly, water-soluble protein did not change materially in the leaves of grapefruit during hardening (Young, 1969a).

Increases in soluble protein have been found during hardening of grains (Johansson *et al.*, 1955). Cytoplasmic tests indicated an increase in both cytoplasmic and basic nuclear proteins (Chian and Wu, 1965). Heber (1959a) found that the protein content of chloroplasts increased during hardening of wheat. In agreement with these results, the ratio of chloroplast to total soluble protein was higher in the more resistant than in the less resistant tea varieties (Sugiyama and Simura, 1967a). Pauli and Mitchell (1960) at first failed to detect any increase in soluble proteins (as percentage of total nitrogen) on hardening of wheat plants, although free amino acids and amides did increase. Later results (Zech and Pauli, 1960) revealed a relationship between soluble protein and hardiness in all three wheat varieties tested, but none with other nitrogen fractions. More recent measurements (Harper and Paulsen, 1967) fail to support this relationship. The crown tissue of field grown "Pawnee" wheat showed a decrease in water-soluble protein with hardening and an increase with dehardening. Similarly, there was no relationship between the hardening of two varieties of winter wheat and the amount of protein eluted from the crown tissue with 0.05, 0.1, or 0.2 *M* NaCl (Toman and Mitchell, 1968). In the case of forage legumes, there was an increase during the fall in total nitrogen, water-soluble nonprotein nitrogen, water-insoluble nitrogen, and water-soluble protein nitrogen/gfw. The water-soluble protein nitrogen was the only fraction consistently higher in both the roots and crowns of the more resistant alfalfa than in the less resistant red clover (Jung and Smith, 1962). The proteins in the microsomes were more closely associated with hardiness than those in the nucleus, the mitochondria, or the cytoplasm (Shih *et al.*, 1967). Furthermore, a significant correlation was found between soluble protein content and hardiness of ten alfalfa varieties, although later investigations failed to confirm this relationship (Smith, 1968). In contrast to the many positive results with legumes, in the case of cabbage, hardening induced a decrease in total nitrogen, protein nitrogen, as well as soluble nitrogen per unit dry matter (Le Saint, 1966). The ratio of protein nitrogen: total nitrogen did not change and no difference was found among the soluble proteins. In our own laboratory, hardening induced a marked increase in soluble proteins if the whole shoot was analyzed after long-continued hardening (Morton, 1969). When individual, rapidly hardened leaves were compared, no increase could be detected (J. Dear, unpublished).

It is now clear that these discrepancies are partly due to the method used for extracting the proteins which may have led, in some cases, to an artificial difference. If lyophilized powders are used, a greater aggregation of the soluble proteins may conceivably occur during the freeze-drying of the less tolerant tissues. If distilled water is the extractant, a higher pH of the hardened tissue (which has been observed in some cases; see Levitt, 1956) may permit extraction of a larger fraction in the soluble state. Thus, in contrast to the above negative results of Le Saint and of Dear, according to Kacperska-Palacz *et al.* (1969), the soluble protein content of hardened cabbage was about 3.2 times as high as in the nonhardened plants, even though only the fourth pair of leaf blades were used in all cases. The leaves were homogenized in distilled water instead of a buffer solution (15 gm/55 ml H_2O), and homogenization was continued for 10 min, instead of the 30 sec, more commonly employed by other investigators. Kacperska-Palacz *et al.* suggest that the difference may be an artifact due to pH changes occurring during hardening. Another possible cause is the greater sensitivity of proteins of nonhardened plants to homogenization (see Schmuetz below), which would result in greater precipitation, during the lengthy (10 min) homogenization period used by them. Similar results have now been obtained in the case of legumes. Faw (1970, personal communication) working in Jung's laboratory, found that, in general, more protein was extracted from a given sample as the pH of the extracting solution increased. Varietal differences were not significant when three of nine buffer solutions were used for protein extraction. The total increase, when it was observed, was the result of general increases in many of the fourteen electrophoretically distinguishable regions. The amount determined densitometrically in ten of these regions was highly correlated with hardiness. It must be cautioned, however, that correlations of this kind can be obtained in the absence of hardening. Proteins accumulate in potato tubers stored at low temperatures, even though no freezing tolerance is developed (Levitt, 1954). Both soluble proteins and 4S RNA may accumulate in potato foliage at hardening temperatures although no true tolerance is developed (Li and Weiser, 1969b). In *Mimulus*, soluble protein increased progressively from a growing temperature of 30°C to growing temperatures of 20° and 10°C (Björkman *et al.*, 1970). These temperatures are not low enough to induce hardening even in hardy plants.

From all these results, the most logical conclusion seems to be that the first stage of hardening (down to a killing temperature of about −30°C) may occur without a detectable increase in soluble protein, but that for freezing tolerance in excess of this, an increase in soluble proteins may be necessary. In some plants, of course, such as alfalfa, it is conceivable that the order may be the reverse of this. It is obvious, however, that the relationship will not be categorically established without a quantitative investigation of the individual proteins. Some attempts have been made to obtain this more specific information.

Indirect evidence of a change in a specific protein was obtained by Soviet investigators and confirmed by Roberts (1967). The energy of activation of invertase from leaves of winter wheat (Kharkov) when grown at 6°C (resulting in a killing temperature of −20°C) was lower than that of the same enzyme obtained from leaves grown at 20°C (killing temperature −10°C). The enzyme from the less hardy (lowest killing temperature −10°C) spring wheat (Rescue) had the identical energy of activation under the above two growing conditions. One explanation suggested by Roberts was that the invertase consists of a number of isozymes with different temperature coefficients. A change in the proportions of these isozymes could then account for the above results.

Coleman *et al.* (1966) investigated the electrophoretic and immunological properties of the soluble proteins of alfalfa roots. A zone of highly charged and/or low molecular weight protein components was more prevalent in the hardened than in the unhardened material. Disc electrophoresis failed to distinguish differences in pattern between varieties differing in hardiness (Gerloff *et al.*, 1967). Certain proteins did increase during hardening, but this occurred in both hardy and nonhardy varieties. Morton (1969) separated some thirty proteins from cabbage leaves by disc electrophoresis but found them all present in both the hardened and unhardened plants. Other investigators have been more successful in detecting specific protein changes.

Of three clones of dianthus, the two winter hardy ones showed a gradual synthesis of two to four new peroxidase isoenzymes from August through November, the tender one showed only a relatively weak initiation of one isoenzyme. The formation of the new isoenzymes preceded the hardening period by several weeks (McCown *et al.*, 1969a). Apple and arborvitae also showed changes in specific protein bands which were separated by disc electrophoresis (Craker *et al.*, 1969). When apple dehardened (from a killing point of −50° to −10°C), the total protein nitrogen remained at about the same value (0.67–0.85% of the dry matter). Of the twenty-seven bands that could be distinguished, only six were observed at all sampling dates. There were four bands present only when the plants were hardy. Most of the other changes were related to bud swell. Arborvitae was hardened during a 60-day period from a killing point of −10° to −50°C. During the first 36 days, the twenty-two bands were very constant, although hardening occurred to a killing temperature of −24°C. During the final 21 days of hardening (from a killing point of −24° to −50°C), four previously present bands disappeared, two appeared and later disappeared, and three appeared and remained. Bermuda grass cultivars, which increased in freezing tolerance by a maximum of 6.0°C showed a loss of density in two protein bands during a 15-day hardening treatment (Davis and Gilbert, 1970). During over-wintering, four bands appeared near the origin. Indoleacetic acid oxidase increased ten times in activity in wheat seedlings held at 2°C for 4 days (Bolduc *et al.*, 1970). This change was inhibited by puromycin or 6-methyl-purine.

It is apparent, from the above results, that specific protein changes do occur during the hardening period. This does not, of course, prove a cause and effect relationship. Some of the changes, in fact, appeared to be related to other plant properties, e.g., bud swell. Similarly, in some cases, no changes in specific proteins could be detected during moderate hardening (to −15° in cabbage and to −24°C in arborvitae). This agrees with the above results of Siminovitch *et al.* (1968) and Sakai and Yoshida (1968a) that the increase in soluble proteins does not accompany the increase in hardening down to −30° to −45°C, but does accompany the hardening in excess of this.

f. Nucleic Acids and Simpler Nucleotides

Increases in RNA accompany the increase in proteins during the fall hardening of locust trees (Fig. 8.5). Similar changes were found in dogwood (Li and Weiser, 1967). In neither case did DNA increase. Cytoplasmic tests on wheat plants revealed an increase in RNA during late fall and early winter (Chian and Wu, 1965). In the case of alfalfa, however, the differences between two varieties differing in hardiness, appeared first in the DNA content during fall hardening, then in RNA content, and finally in water-soluble, TCA precipitable protein content and freezing tolerance (Jung *et al.*, 1967a). The discrepant rise in DNA content of alfalfa may conceivably be due to the presence of meristems (crown buds) in the root samples and their absence from the bark samples used by the above investigators. In 1-year-old apple twigs, RNA began to increase 1 week prior to the rapid increase in freezing tolerance (Li and Weiser 1969a). The sRNA increased 38% in 1 week and the light and heavy rRNA increased 41% in 2 weeks, just prior to and during the stage of rapid hardening. DNA showed a possible slight increase during hardening, but during slow dehardening, RNA decreased while DNA increased. The most dramatic decrease during the 3-week dehardening was in the heavy rRNA (4.0%).

Cytochemical methods have corroborated these nucleic acid changes (Chuvashina, 1962). When radial sections of the bark next to the cambium of 1-year-old apple shoots were stained, hardy varieties showed methylophilic nuclei as early as August, with the maximum between December and February. In the nonhardy varieties, the nuclei were pyroninophilic even in winter. Similar results were obtained with cherry and apple by Sergeeva (1968): histochemical tests showed a correlation between RNA content in winter and the winter hardiness of the variety.

There is some evidence of accumulation of the simpler nucleotides during hardening. According to Trunova (1968), inorganic phosphate is lower and acid-soluble organic phosphate (mainly sugar-phosphate esters) is higher in frost-hardened winter wheat than in the unhardened state. The hardened plants contain more high-energy phosphorus, indicating a high degree of coupling between oxidation and phosphorylation at low positive temperatures. A similar correlation was obtained by Sergeeva (1968) and by Borzakivska and Motruk (1969). In confirmation of such a relationship, infiltration with DNP during the first hardening stage decreased the nucleotide content 25–28% and lowered the freezing tolerance (Trunova, 1969). On the other hand, the nucleotide content greatly decreased during the second hardening phase at −5°C, due presumably to utilization in the hardening process.

Similarly, the application of phosphate fertilizers increased the content of high-energy phosphorus in the tillering nodes of winter wheat and winter rye and raised their freezing tolerance (Kolosha and Reshetnikova, 1967). Lowering the temperature from 0° to −20°C lowered the high-energy phosphorus to a greater degree in the less resistant species. Soaking the plants for 12 days in a 10% sucrose solution caused a rise in high-energy phosphorus, and an increase in reducing sugars as well as in freezing tolerance. In opposition to these results, the content of organic acid-soluble phosphorus decreases markedly in grape buds during the fall transition to the rest period, and increases after the rest period (Cheban, 1968).

As in the case of the proteins, it is again a question whether or not the inverse correlation between nucleic acid content and temperature indicates a role in the hardening process. In the case of four perennial grasses, as the day temperature was increased from 18.3° to 43.8°C, the RNA concentration decreased in a nearly linear manner (Baker and Jung, 1970b). This is the same kind of a correlation as found during hardening, yet it can have no relationship to freezing tolerance since no hardening occurs within this temperature range.

g. Lipids

A positive correlation between lipid content and hardiness was often noted in the early literature (see Levitt, 1941). Some of these early results have long been known to be due to erroneous methods of identifying lipids, and these errors were still being made in relatively recent times (see Pieniazek and Wisniewska, 1954). Modern quantitative methods, however, have succeeded in establishing an increase in lipid content during hardening of locust trees (Siminovitch *et al.*, 1968, Siminovitch *et al.*, 1967a,b) and nearly a doubling of the fatty acid content of alfalfa roots (Gerloff *et al.*, 1966). Similar results have been obtained with a less hardy species (*Citrus sp.*), although only in the hardier varieties (Kuiper, 1969). It has also long been known (e.g., Malhotra, see Levitt, 1941), that low temperature increases the degree of unsaturation of the fatty acids (as measured by the iodine number). This has been corroborated in the case of alfalfa (Gerloff *et al.*, 1966), which accumulated polyunsaturated fatty acids (linoleic and linolenic) during hardening.

In the case of locust trees, Siminovitch's group found that the lipid increase during the first stage of hardening was confined to the phospholipids. The fundamental nature of this increase was revealed by isolating bark segments in August (Siminovitch *et al.*, 1968). These "starved" segments showed none of the capacity to synthesize soluble proteins or RNA or to increase in cytoplasmic substance. Yet they tolerated freezing down to

$-45°C$ in November or December. They did show a slight increase in leucine-^{14}C incorporation, but the largest increase was in phospholipid (Table 8.7). A similar result was obtained in the case of citrus (Kuiper, 1969). The two hardy varieties (Dancy tangerine and Satsuma tangerine) had a smaller amount of neutral lipids and a larger amount of phospholipids than the two tender varieties (Marsh grapefruit and Eureka lemon). The glycolipids showed no significant difference.

Yoshida (1969b) also corroborated the increase in phospholipids during hardening of black locust. Phosphatidylethanolamine, phosphatidylcholine, and an unidentified component increased markedly from autumn to winter, then decreased toward spring. Phosphatidylglycerol accumulated in late autumn and seemed to be transformed into the other three substances under conditions of both artificial and natural hardening. In opposition to the other investigators, Yoshida (1969a) was able to detect some change in two of the seven glycolipids found in the bark tissue: the monogalactosyl diglyceride and the steryl glycoside. The most characteristic change was the seasonal deacylation of the esterified form of steryl glycoside from autumn to winter, and its acylation from early spring to summer. All these results were taken to suggest changes in the membrane components with variations in freezing tolerance.

The results of all these investigators agree that phospholipids (at least certain ones) increase during hardening, and that the degree of unsaturation of the fatty acids, in general, increases with hardening. Again, as in the case of the proteins and nucleic acids, the significance of these correlations must be questioned. When alfalfa plants were grown at 15°, 20°, or 30°C for 4 to 5 weeks, Kuiper (1970) obtained large increases at the lower temperatures in those leaf lipid fractions containing mainly polyunsaturated fatty acids (monogalactose diglyceride and phosphatidylglycine—86% unsaturated) as well as in two fractions with somewhat smaller, though still large, degrees of unsaturation (51–63%: digalactose diglyceride and phosphatidylethanolamine). Those leaf lipids with the lowest unsaturation (33–59%: phosphatidylglycerol, sulfolipid, and phosphatidylinositol) decreased at 15°C. Therefore, the average degree of unsaturation, and the content of at least one phospholipid was higher at the lowest temperature, in agreement with earlier results. Furthermore, larger changes occurred in the leaves of the hardier variety (Vernal) than in those of the less hardy variety (Caliverde). Unfortunately, no measurement of freezing tolerance was made. However, 15° and 20°C are unquestionably too high temperatures to induce hardening. Perhaps this indicates a relation to chilling rather than to freezing tolerance, since 15°C does permit hardening to chilling.

h. Miscellaneous Substances

Many substances have at one time or another been reported to show a correlation with hardiness, e.g., pentosans, anthocyanins, tannins (see Levitt, 1956). In *Hedera helix*, for instance, anthocyanin (as well as total sugar) content of the leaves paralleled hardiness from August to May (Parker, 1962). In each case, however, negative results are at least as common (see Levitt, 1956).

Ascorbic acid content has been reported to increase in many plants during September when prolonged frost sets in (Shmatok, 1958). The content of wheat seedlings was much higher when germination (in the dark) occurred at a hardening temperature (1.5°C) than when at nonhardening temperatures (Andrews and Roberts, 1961; Waisel *et al.*, 1962). The hardier winter varieties developed higher quantities than the less hardy varieties, but only at hardening temperatures. After 6 weeks both hardiness and ascorbic acid decreased. Winter hardy oat varieties (Bronco and Mustang) contained more ascorbic acid than did winter tender varieties (Alamo and Frazier); similarly lines selected by ascorbic acid content proved more hardy in field tests (Futrell *et al.*, 1962). According to Polishchak *et al.* (1968), bound ascorbic acid is converted to free ascorbic acid in the bark of the frost-resistant black walnut on transfer from the cold room to the laboratory. The less resistant Persian walnut showed the opposite change—a rise in bound ascorbic acid. Schmuetz (1969) has established a close relationship between content of reduced ascorbic acid and varietal hardiness of wheat and barley. A fall and winter decrease in bios and auxin has been reported for various citrus and peach trees (Sulakadze, 1961), the greater decrease occurring in the more freezing-tolerant plants. In *Acer negundo*, nonhardening long days induced the greatest activity of gibberellinlike substances, hardening short days or long days with 5° nights induced the highest levels of abscisic acid or a similar substance (Irving, 1969a). Since the long days with 5°C nights are the most capable of developing hardiness, the hardening process appeared to be more closely related to a build up of abscisic acid rather than to a reduction of gibberellin levels.

2. Metabolic Rates

From the above results, we must conclude that hardening is normally accompanied by an accumulation of one or more substances synthesized by the plant—sugars, amino acids, proteins, nucleic acids, lipids, and perhaps several others. This conclusion is supported by the visible increase in the amount of protoplasm per cell (Fig. 8.7). The mere fact that sub-

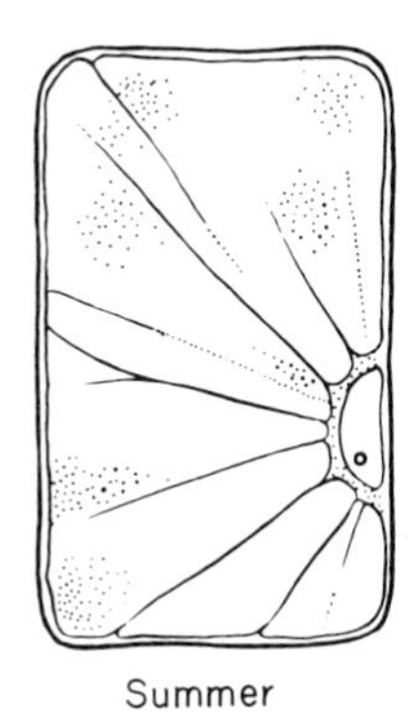

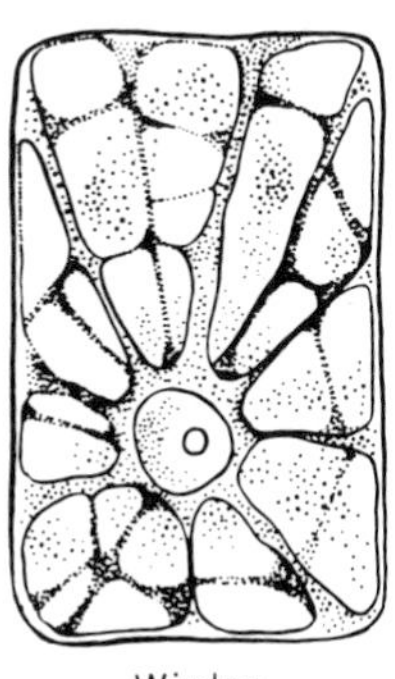

FIG. 8.7. Diagrammatic representation of living cells of locust bark, showing higher content of protoplasm during winter. (From Siminovitch *et al.*, 1967b.)

stances accumulate in the fall is no proof, of course, that they play a role in freezing tolerance. They may simply serve as reserve substances to be used in the spring burst of growth. In favor of this explanation, is the pronounced (although not full) hardening described above, that can occur without the accumulation of proteins, nucleic acids, and lipids, and the absence of sugar accumulation in many cases of full hardening. It is still conceivable, however, that the accumulation is necessary for full hardening. Whether or not the substances play a role in freezing tolerance, the mere fact that they accumulate proves that a fundamental change in the metabolism of the plant normally occurs during hardening.

How can we account for this accumulation of substances during hardening? The accumulation of any substance synthesized by the plant is the net result of an excess of synthesis over breakdown:

$$A = S - B$$

where A = amount accumulated; S = amount synthesized; B = amount broken down.

Therefore, the increased content of any substance under hardening conditions over the content under nonhardening conditions must be due to an increased net synthesis:

$$(S - B)_H > (S - B)_{NH}$$

where H = hardening conditions; NH = nonhardening conditions. Since hardening temperatures are lower than nonhardening temperatures, both the rates of synthesis and of breakdown may be expected to decrease. The greater accumulation of substances in a hardy plant at hardening temperatures than in a tender plant, must then be due to either (1) a smaller decrease in synthesis, or (2) a larger decrease in breakdown:

$$(S_{NH} - S_H)_{HP} < (S_{NH} - S_H)_{TP} \qquad (1)$$

or

$$(B_{NH} - B_H)_{HP} > (B_{NH} - B_H)_{TP} \qquad (2)$$

where HP = hardy plant; TP = tender plant.

Reaction (2) consists of two components: (a) breakdown reactions in the fundamental metabolism of the plant (the metabolism required to keep a cell alive), and (b) breakdown reactions in the additional metabolism that supports the growth and development of the plant (cell division, enlargement, and differentiation). Therefore, (2)—the greater decrease in the breakdown reactions of hardy than of tender plants at hardening temperatures—may be due to a greater decrease either in its fundamental metabolism (reaction 2a) or in the additional metabolism of growth and development (reaction 2b):

$$(B_{FNH} - B_{FH})_{HP} > (B_{FNH} - B_{FH})_{TP} \qquad (2a)$$

$$(B_{GNH} - B_{GH})_{HP} > (B_{GNH} - B_{GH})_{TP} \qquad (2b)$$

where F = fundamental metabolism; G = additional metabolism for growth and development.

Which of these three theoretical possibilities (1), (2a), or (2b) is (or are) actually responsible for the accumulation of substances by hardy plants during the hardening process?

Hardening normally begins in the short days of fall (Chapter 7), which according to the available evidence (see above) induces a decrease in growth stimulators (gibberellins and auxins) and an increase in growth inhibitors (abscisic acid). As a result, growth comes to a complete stop in the case of woody plants which enter a period of dormancy. The growth process is apparently uncoupled from metabolism. The fundamental metabolism continues, and the net synthetic product, which would have been used for growth, accumulates. In the case of winter annuals, the hardiest of which do not attain the extreme hardiness of woody plants, no true dormancy occurs, and the seedlings continue to grow slowly. In many cases, the less hardy varieties and species do not have their growth sufficiently curtailed by the short days (due, presumably, to an insufficient change in growth regulators) and due to this excess growth in the fall cannot accumulate the synthesized substances to as high a degree as the more hardy varieties and species.

In both woody perennials and winter annuals we must, therefore, conclude that the accumulation of substances during the fall is at least partly due to (2b)—a cessation of (woody plants) or a decrease in (winter annuals) the breakdown reactions associated with growth. This effect is so large, that the accumulation will occur even if the breakdown reactions of the fundamental metabolism of the plant (reaction 2a) have not been greatly slowed down, and even if the synthesizing reactions (reaction 1) are markedly slowed down at hardening temperatures. The accumulation of substances, therefore, does not require an inverse relationship between residual (fundamental) breakdown reactions of respiration at hardening temperatures and freezing tolerance; neither does it require a direct relationship between rate of photosynthesis at hardening temperatures and freezing tolerance.

Actual measurements support this conclusion, since they sometimes indicate one relationship, sometimes the reverse. Earlier results indicate an inverse relationship between freezing tolerance and respiration rate in grains (Newton *et al.*, 1931), apples (Fig. 8.4), legumes (Bula and Smith, 1954), etc. Although these measurements have been made at above freezing temperatures, the relationship may hold true even at very low temperatures, for respiration continues at measurable rates at temperatures as low as $-30°C$ (Scholander *et al.*, 1953; Kanwisher, 1957), although the Q_{10} rose precipitously to 20–50, below 0°C (Scholander *et al.*, 1953). Some later results agree with this inverse relationship (e.g., in pears, Zotochnika, 1962), but many indicate the opposite—a direct relationship between freezing tolerance and respiration rate. For instance, when measured at high temperatures (30°C), the more resistant alfalfa variety showed the greater increase in respiration as a result of hardening (Swanson and Adams, 1959). In the case of fruit and nut trees (Yasmykova and Tolmachov, 1967) the more freezing resistant varieties had higher respiration rates. The less freezing resistant varieties had respiratory quotients which rose above 1 during January to February, showing a disturbance in the respiratory process, and a sharp decrease in activity of the enzymes involved in oxygen absorption.

In the case of wheat varieties, there is a close relationship between absorption of micronutrients (^{65}Zn and ^{60}Co) at low temperatures and resistance to freezing (Giosan *et al.*, 1962). The resistant varieties, in fact, were able to maintain a level of absorption at 5° to 7°C close to that at higher temperatures. It is difficult to explain these results except by an increased metabolic rate leading to increased active absorption. Similarly, Kenefick and Swanson (1963) found a higher oxidative rate in mitochondria isolated from winter barley previously exposed to 2°C for 4 weeks, than from plants kept at 16°C. In the case of spruce and Douglas fir, the respiratory rate is so high in the hardened plants that the CO_2 balance may be negative

(Weise and Polster, 1962). Measurements of photosynthesis agree with this result (see below).

The relationship of photosynthesis to freezing tolerance is even more complicated than in the case of respiration. On the one hand, many investigators have shown a clear dependence of freezing tolerance on photosynthesis, e.g., by withholding light or CO_2, hardening was prevented (see above). Some hardening may occur in the dark if starch has accumulated, but it is slight (Dexter, 1933). Even in the case of deciduous trees, the cortical cells of their twigs contain ample chlorophyll that colors the cytoplasm layer an intense green in fall and winter. Furthermore, the chlorophyll content of the shoots has been found to increase from October to March, followed by a decrease (Smol'skaya, 1964). Similarly, the bark of hardier trees has a higher chlorophyll content during winter than that of less hardy trees (Borzakivs'ka, 1965). In agreement with all these results, the alga *Chlorella pyrenoidosa* photosyntehsizes at essentially the same rate at temperatures down to 7° as at 20°C (Steemann-Nielsen and Jorgenson, 1968). This was attributed to an increase of all the enzymes at the low temperature, for the protein content per cell at 7°C is double the amount at 20°C (Jorgensen, 1968).

On the other hand, many investigators have shown that photosynthesis decreases with hardening, from a maximum during the early stages of hardening to a minimum at maximum hardening— in wheat (Anderssen, 1944). Both respiratory and assimilation rates decreased during hardening in the fall in wheat, barley, spinach, lamb's lettuce, and *Picea excelsa* (Zeller, 1951). Rising temperatures in winter increased both, but to a lower level than the same temperatures in spring, and more so in the winter annuals (which had no winter rest period) than in *Picea excelsa* (which did). Assimilation was detected at $-2°$ to $-3°C$ and respiration at $-6°$ to $-7°C$. Pines that began to harden in September, showed a drop in assimilation rate (calculated for 20°C) by the end of October to half that in May at the same light intensity (Tranquillini, 1957). By the beginning of November it dropped to one-tenth its original value. The respiration rate also decreased. On warming, respiration rose more rapidly than photosynthesis, so that CO_2 was released in sunlight. Therefore, the CO_2 balance was negative even above 0°C, and the greater the warming, the greater the CO_2 loss. These results have since been confirmed for conifers by many other investigators, the rate of photosynthesis decreasing rapidly in the fall and reaching a value of zero in late fall or early winter (for white, stone, and bristlecone pines: Shiroya *et al.*, 1966; Bamberg *et al.*, 1967; Schulz *et al.*, 1967; for silver fir: Pisek and Kemnitzer, 1967; for pine, spruce, and juniper: Ungerson and Scherdin, 1965). If the winter is mild enough, photosynthesis may remain active in silver fir (Pisek and Kemnitzer, 1967). However, the longer the trees remain below −5°C, the longer it takes for photosynthesis to recommence. Broad-leaved evergreens show the same changes as found in the conifers, according to Steinhubel and Halas (1969). They measured the photosynthetic increase in dry matter by leaf disks at 20°C and 0.03% CO_2. The disks from plants exposed to normal fall and winter temperatures (below 0°C and down to −20°C) showed a steady decrease in photosynthesis from October, reaching a value of zero in December or January.

There is, of course, the possibility of injury to the photosynthetic apparatus. The net photosynthesis of fir and maple leaves, however, was inhibited by freezing temperatures far above those causing tissue injury (Bauer *et al.*, 1969). In fact, in the case of fir (*Abies alba*), as soon as the needles were shown to form ice (by thermocouple measurements), photosynthesis (measured at 20°C) was inactivated. The depression of photosynthesis following thawing was proportional to the previous cooling. After 2 days of freezing at −8°C, CO_2 was evolved in the light for 2–3 days following thawing. Yet at least 80% of the assimilation tissue was intact, and the leaves finally regained their full photosynthetic capacity. In opposition to photosynthesis, the cooling leads to a reversible rise in respiration (measured at 20°C). The extent of the rise is less in hardened than in unhardened plants—in the case of firs, in the open, it was only half of that in greenhouse plants cooled to the same degree (250 and 500%, respectively, of the original rate). The increase in respiration reaches a peak some 5 hr after thawing, if frozen at −6°C, some 9–12 hr after thawing when the freezing was more severe. The decrease in photosynthesis and the increase in respiration may be partly due to injury, but they occur even at temperatures that do not cause injury.

It should be pointed out that it is *apparent* photosynthesis which is zero in winter. *Actual* photosynthesis is, on the average, positive in winter (Ungerson and Scherdin, 1965). The plant may even adapt so as to continue photosynthesis in the dehydrated (frozen) state. Possible indirect evidence of such adaptation has been obtained in the case of *Pinus contorta* (Morris and Tranquillini, 1969). In winter and spring (December–April) photosynthesis decreased less with decreasing osmotic potential of the root medium than it did in summer.

Other plants, besides the conifers, are capable of photosynthesizing in the frozen and, therefore, dehydrated state. Unlike the conifers, some of them may show net as well as actual photosynthesis. Lichens, that survive −75°C for several days, detectably photosynthesize at −24°C (Lange, 1962b, 1965a). The temperature optimum for photosynthesis was 10°C, and it ceased above 20°C. At −5°C, in spite of ice formation, the rate was still half the optimum. Most of the species ceased CO_2 uptake at −7° to −13°C. Tropical lichens showed little CO_2 fixation below 0°C and their optimum temperature for photosynthesis was 20°C. However, the temperature minimum did not always parallel the habitat of the lichen. By means of ^{14}C labeling, two lichens were shown to assimilate carbon at −11°C at rates 1/5 to 1/9 the rates at +15°C (Lange and Metzner, 1965). Even among the higher plants, *Ranunculus glacialis* and *Geum reptens* (which grow at altitudes as high as 3100 m in the Alps) had compensation points below 0°C. A net CO_2 uptake occurred at −3° to −5°C (Moser, 1969).

The winter deficit of conifers may be serious in the subarctic. In *Pinus sylvestris*, although positive photosynthesis occurs only above −4°C, respiration rate is measurable down to −18°C (Ungerson and Scherdin, 1967). As a result, the daily respiration loss amounted to 186 mg CO_2/100 gm dry needles. The first positive apparent photosynthesis in several conifers did not occur until April 12–17 (Ungerson and Scherdin, 1965). In bristlecone pine, a winter loss of 140 mg CO_2/gm dry weight required 117 hr of photosynthesis during summertime at peak rates to compensate for the loss (Schulz *et al.*, 1967).

Two logical conclusions emerge from all the above evidence. (1) The fall accumulation of substances in hardy plants is due to a cessation or a marked retardation of growth and development. (2) The main function of this accumulation of substances during the fall is twofold: (a) to support the winter metabolism of the plant in the absence of appreciable synthesis, and (b) to support the spring burst of growth. It probably has little effect *per se* on freezing tolerance. This was shown by preventing the accumulation without preventing the major increase in freezing tolerance, by the large winter loss in reserves without an appreciable loss in freezing tolerance, by the existence of hardy plants that do not become truly dormant, and by tender plants that do. This conclusion does not, of course, deny the possibility that a specific substance may have to be synthesized, or one substance may have to be converted into another before freezing tolerance can increase. In the case of cabbage, for instance, the increase in freezing tolerance is proportional to the growth and, therefore, is not accompanied by a general accumulation of substances (e.g., of soluble proteins). Nevertheless a conversion of starch to sugar does occur.

In order to understand the role of metabolism in freezing tolerance, it is therefore necessary to examine specific metabolic processes rather than overall synthesis and breakdown. Of all the substances that accumulate during hardening, the carbohydrates have been most intensively investigated (see above). The accumulation of sugars during hardening is brought about in two ways.

(1) In the case of winter annuals, photosynthesis continues in excess of respiration at hardening temperatures (Andersson, 1944), and the slow growth rate fails to use up the excess photosynthetic products. This is undoubtedly because the hardening temperature may permit a surprisingly high rate of photosynthesis, e.g., in mimulus at 0°C the rate of light-saturated photosynthesis ranged from 11–20% of the maximum rate at higher temperatures (Milner and Hiesey, 1964). Furthermore, this accumulation of excess photosynthetic products during the hardening period is greater in hardier varieties of wheat than in less hardy varieties (Andersson, 1944). Similarly, the rate of ^{14}C incorporation into sucrose was 50% higher in snow plants when in the hardened state than when in the unhardened state (Zhuravlev and Popova, 1968). There was, in fact, little change in rate of photosynthesis from 0° to 10°C. If this sugar accumulation during hardening is indeed an essential component of the hardening process, this would explain why light is essential for hardening to occur. It would also explain the transfer of hardening in the light to the shaded parts of the plant, and the fact that this occurs only if the shaded parts can act as "sinks" to which sugars can be translocated. The hardening that occurs in the dark, due to sugar feeding at hardening temperatures is similarly explainable.

(2) In the case of deciduous perennials, growth slows down and finally ceases before leaf fall. The carbohydrates therefore, accumulate long before the hardening period, mainly in the form of starch. The sugar increase during hardening is, therefore, due to a starch → sugar conversion. The cause of this hydrolysis at low, hardening temperatures has been sought for many years. Müller-Thurgau (1882) ascribed the sugar accumulation in potato tubers to a marked reduction in the respiration rate at the low temperature, without as marked a drop in activity of the hydrolytic enzymes. The sugar accumulation is greater at 0°C than at 3°C even though the respiration rate is minimal at 3°C and rises at 0°C (Hopkins, 1924; Schander *et al.*, 1931—see Snell, 1932; Wright, 1932). Another complication is the stimulating effect of the sugar increase on respiration rate (Snell, 1932). In any case, the respiratory utilization of sugars could not possibly account for the decreased sugar content at high temperatures, for at this rate, all the reserves of the tuber would be used up in a very short time and, as shown above, the sugar loss is actually due to starch formation. Potato tubers, of course, do not develop freezing tolerance.

Among some of the earlier investigators, the most widely held concept is a shift in the starch ⇌ sugar equilibrium caused by the temperature drop (Overton, 1899; Czapek, 1901; Rosa, 1921; Fuchs, 1935; Algera, 1936). According to the laws of thermodynamics, since hydrolysis of starch is an exothermic process the concentration of sugars must increase at low temperatures. Direct evidence of this is the fact that when leaves are floated on 2–5% sugar solutions, rich starch accumulation occurs at 16° to 18°C, but little or none occurs at 0° to 2°C (Czapek, 1901). The sugar must be increased to 7% before some starch forms at 0°C. Unfortunately, the sugar formation directly due to the temperature drop would be relatively slight (Algera, 1936). Furthermore, if this were the complete explanation, all the plants (whether frost-tender or frost-hardy), would show the same sugar accumulation at low temperatures, and it would happen at all times of the year. Neither of these predictions agrees with the facts.

Several other suggestions have been made without any evidence to back them: (1) a weakening of the (starch) synthetic mechanism (Michel-Durand, 1919; Doyle and Clinch, 1927); (2) a dependence of starch synthesis on the temperature, and of starch solution on the time of the year (Weber, 1909); (3) an increased permeability of the starch "membrane" to the enzyme (Coville, 1920); (4) a pH control (Mitra, 1921; but this could not be confirmed by Hopkins, 1924); and (5) a hormone control (Lewis and Tuttle, 1923).

One unanswered question is why does the starch → sugar conversion goes to completion in bark but not in wood cells? This may perhaps be related to the less active metabolism of the latter. It may also be a response to the more severe and rapid temperature changes to which the externally located bark cells are exposed. Fluctuating temperatures might conceivably be important in bringing about starch hydrolysis. There is direct evidence against this explanation, at least in the case of herbaceous plants (potato tubers, cabbage leaves, etc.). The starch → sugar conversion occurs as readily at constant low temperatures as at fluctuating ones. A more likely suggestion is a relationship between starch hydrolysis and oxidation-reduction potential; where oxygen is deficient (e.g., in roots, submerged plants, internal tissues of rhizomes, or woody tissues of twigs), the hydrolysis occurs only partially or not at all.

Evergreen perennials do not seem to fit clearly into either of the above patterns. Douglas fir seedlings resemble the winter annuals. They, too, show a constant rate of photosynthesis in spite of a drop in temperature, provided that they are at a suitable developmental stage (Brix, 1969). When, in the stage of leaf production, they show a pronounced effect of temperature on photosynthesis (as measured by dry matter production) between 13° and 18°C. During bud dormancy, on the other hand, no significant change was detected even between 7° and 24°C. The net assimilation rates were, however, lower during bud dormancy, due supposedly to the absence of "sinks" for use of photosynthetic products. Douglas fir are only moderately hardy and these results were obtained with young seedlings. The most hardy conifers, as shown above, undergo a sharp drop in photosynthesis, to zero net photosynthesis in midwinter at the stage of maximum freezing tolerance. The accumulation of substances in them, therefore, follows the pattern of the deciduous perennials.

The most logical explanation of the starch → sugar hydrolysis is an activation of the hydrolytic enzymes; but early attempts to prove this ended in failure (Müller-Thurgau, 1882). On the contrary, tulip bulbs kept in the cold formed sugars less rapidly on transfer to warm temperatures than did those kept at warm temperatures (Algera, 1936). Many other investigations have led to negative or, at best, confusing results (Table 8.8). In spite of these early failures, more recent investigators cannot avoid the conclusion that some kind of enzyme activation must be involved (Ewart *et al.*, 1953). Some results indicate that such a change does occur in potato tubers (Arreguin and Bonner, 1949). Ewart *et al.* (1953) found both amylase and phosphorylase in the bark of black locust, and they suggest that a differential sensitivity to natural sulfhydryl reagents may control the seasonal changes. Later evidence indicated that phosphorylase was not involved in the starch synthesis (Ewart *et al.*, 1953). More recent results (Terumoto, 1957b) have revealed distinct increases in both phosphorylase and phosphatase activity during the hardening of table beet and a loss, during dehardening, at least in phosphorylase activity. These changes occurred in the leaf and not in the root.

Other enzymes besides those of carbohydrate metabolism have been investigated, again without any general, consistent pattern emerging (Table 8.8). Low temperature may, of course, lead to increased enzyme activity, e.g., in the case of three enzymes found in wheat seeds (Fleischmann, 1959). Correlations have been found between freezing tolerance and catalase activity (Table 8.8). Recently interest seems to have centered on peroxidase, which was first investigated in this connection by Doyle and Clinch (1927). A number of investigations have obtained an inverse relationship between temperature and peroxidase activity. Peroxidase was more active at 2° than at 25° and 37°C in the bulbils and tubers of *Ficaria verna* (Augusten, 1963). Two types of peroxidase were found in wheat embryos. One predominated in embryos germinating at 20°–24°C and the other in embryos germinating at 0°–4°C (Antoniani and Lanzani, 1963). Aging pea stem sections for 34 hr results in a quantitative increase in existing peroxidases (Highkin, 1967). The increase was greater at the lower (22°C) than at the higher (34°C) temperature. The isozyme pattern also depended on the aging temperature. Even tobacco plant cells grown in suspension secrete a peroxidase isozyme at 13°C which is absent at 25° or 35°C (De Jong *et al.*, 1968). In all these cases the low temperature stimulation of peroxidase

TABLE 8.8 Relationship between Enzymatic Activity and Frost Hardiness

Enzyme	Plant	Observation	Reference
		(*a*) *Of carbohydrate metabolism*	
Dextrinase			
Amylase	Conifer leaves	Less during winter	Doyle and Clinch, 1927
Maltase			
Emulsin, invertase	Conifer leaves	Constant	Doyle and Clinch, 1927
Diastase	Alfalfa	More sugar formed by enzyme from hardy plants	Tysdal, 1934
Carbohydrases	Wheat	Greater activity in less hardy	Bereznickaja and Oveckin, 1936a, b
Starch active	Clover	More hydrolysis by enzyme from non-hardy	Greathouse and Stuart, 1937
Invertase		More hydrolytic activity in nonhardy at high temp.	Sisakjan and Rubin, 1939
Invertase	Grains	More synthesis in hardened	Morosov, 1939
		(*b*) *Of fat metabolism*	
Lipase	Conifer leaves	Always negative	Doyle and Clinch, 1927
		(*c*) *Of protein metabolism*	
Proteolytic enzymes	Winter cereals	Related to hardiness	Kling, 1931—see Tysdal, 1934
Proteolytic enzymes	Wheat	Greater activity in less hardy	Bereznickaja and Oveckin, 1936a, b
		(*d*) *Oxidation reduction*	
Peroxidase	Conifer leaves	Related to starch	Doyle and Clinch, 1927
Catalase	Wheat	Direct correlation with hardiness	Newton and Brown, 1931
Catalase	*Pinus sylvestris*	Direct correlation with hardiness	Langlet, 1934
Catalase	Citrus	No relation	Ivanov, 1939

activity seems unrelated to freezing resistance. Gerloff *et al.* (1967) detected increases in both peroxidase and catalase activity (per unit protein or dry weight) during the hardening of alfalfa root. Although the quantities did not clearly differentiate between varieties, peroxidase increased sooner in the hardier varieties.

Avundzhyan *et al.* (1967) compared the enzyme activities in the bleeding sap from four different grape varieties differing in frost hardiness. The proteolytic and peroxidase activities were more stable and the amylase activity higher in one frost-resistant variety than in two nonresistant varieties. The Wisconsin group has investigated peroxidase in many plants in relation to freezing tolerance. They obtained evidence of a relationship between peroxidase activity and hardiness in overwintering carnation plants (Hall *et al.*, 1969). When separated by disc electrophoresis, the peroxidase bands from field-hardened plants showed an increase in number and density in three species (*Sedum*, *Mitchella*, and *Salix* spp.) as compared with those from unhardened greenhouse plants (McCown *et al.*, 1969b,c). In *Cornus*, however, there was a decrease. In *Salix fragilis*, they also failed to observe a relationship, for the peroxidase activity which appeared during hardening, remained after dehardening (Hall *et al.*, 1969). Roberts (1969a) also found a greater intensity of staining of the fastest moving peroxidase isozyme in the leaves of wheat plants grown at 6°C than in similar leaves from plants grown at 20°C. This increase in peroxidase was apparently not associated with freezing tolerance, since there was no difference between hardy and sensitive varieties. According to Polishchak *et al.* (1968), the reverse change occurs in the bark of walnut trees. On transfer from −1.8° to −15°C, the oxidoreductases (ascorbic acid oxidase, polyphenol oxidase, and peroxidase) all decreased. On return to 20°C, peroxidase activity rose again. The other enzymes failed to show this change.

Pogosyan and Sklyarova (1968) suggest that the respiratory system in the stem of grapevines is altered during the fall transition to winter dormancy, the flavine system replacing the cytochrome oxidase system as the terminal oxidase. In winter wheat, nitrate reductase was initially at low activity, increasing slightly during maximum hardening, but increasing rapidly during dehardening (Harper and Paulsen, 1967).

These results all lead to the conclusion that some enzymes increase in activity during the fall hardening, but that this may have little or nothing to do with freezing tolerance.

This is not surprising, particularly in the case of peroxidase, since so many factors may lead to changes in its activity: it may be either inhibited or promoted by indoleacetic acid (IAA) and inhibited by kinetin (Lavee and Galston, 1968), and the IAA may induce one of eight isoperoxidases and repress another (Stuber and Levings, 1969). Nevertheless, there does seem to be a general increase in peroxidase activity at low temperatures—even at low temperatures above the freeze hardening range. Possibly, this is a hardening to chilling rather than freezing. The reducing system that normally protects the membrane lipids against injurious peroxidation (see Chapter 5) may be unable to do so at chilling temperatures. An increase in peroxidase activity would destroy the peroxides rapidly enough to prevent this kind of chilling injury.

Roberts (1969b) suggests "the substitution at hardening temperatures of a modified form of a protein for the form of the functionally identical protein present at higher temperatures," i.e., the formation of isozymes during hardening. There are four lines of evidence against this concept:

(1) The synthesis of a new isozyme is conceivable during relatively long hardening periods, but it is inconceivable during the rapid hardening (an increase in tolerance of 6°–8° within 24 hr) at a temperature that markedly slows down metabolism (0°–5°C).

(2) Conformational changes could and do occur at such low temperatures, but the existence of conformational isozymes is still a subject of controversy (Vesell, 1968), and the evidence to date (Levitt and Dear, 1970) has failed to detect an average conformational change of the proteins during hardening.

(3) Roberts carefully documents the existence of isozymic differences at low temperatures, but they are all associated with chilling resistance. Indeed, they are best documented in organisms that have no freezing tolerance. In the case of animals, for instance, as the temperature is lowered, changes in the velocity constants appear to activate certain isoenzymes which are inactive at the higher temperatures. These isoenzymes are important in the low temperature acclimation of animals (Somero and Hochachka, 1969). These animals are unable to tolerate freezing. Since the development of isoenzymes in animals at low (hardening) temperatures fails to induce freezing tolerance, it seems hardly likely that it will in plants.

(4) The above attempts to discover isozymes associated with freezing tolerance have, indeed, discovered isozymes (e.g., of peroxidase) but have failed to demonstrate a relationship to freezing tolerance.

It is now obvious that if the factors that control freezing tolerance are to be discovered, it is necessary to eliminate all the changes associated with the buildup of reserves for later growth processes, as well as senescent changes, and any others not directly related to freezing tolerance. This goal can be best achieved by a maximum rate of hardening. The aim should be to harden so rapidly that none of these other changes can occur.

3. Protoplasmic Properties

On exposure to hardening temperatures, hardy plants that increase in freezing tolerance also increase in permeability to polar substances; tender plants do not (Fig. 8.8). Later investigations (Granhall, 1943) of a series of wheat varieties differing in hardiness failed to show any difference in this factor. This is to be expected since the differences in hardiness had earlier been accounted for by sugar content (see above). The wheat varieties may, therefore, differ in freezing tolerance due to avoidance of freeze-dehydration, and not due to tolerance of freeze-dehydration, or to avoidance of intracellular freezing. In the mulberry tree (Sakai, 1955b), permeability to water rises with hardiness from October to January, but the maximum is then retained even up to May, although freezing tolerance drops markedly. That the difference in permeability exists during freezing and thawing was shown by Siminovitch and Scarth (1938), for the hardy cells both lost water and gained it more rapidly than the nonhardy cells. Similar results have been recently reported by Asahina (1956). He also showed (1962b) that 1 M urea increases the permeability to water of unfertilized arbacia eggs, enabling them to survive rapid freezing by preventing intracellular ice formation. Similarly, anaerobically grown cells of *E. coli* have a lower permeability to water than the aerobically grown cells, and are more susceptible to intracellular freezing (Nei *et al.*, 1967). Akabane

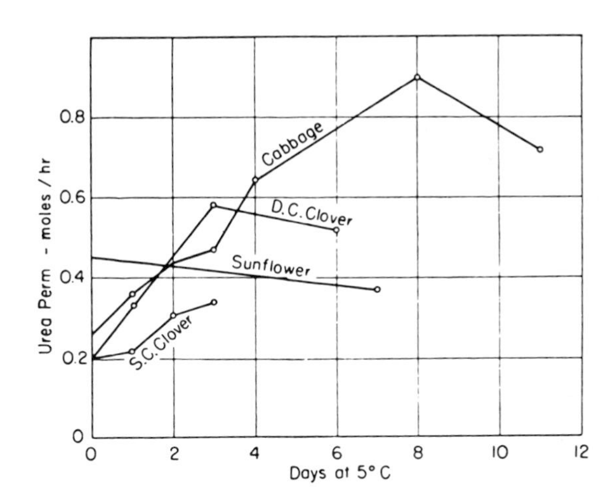

Fig. 8.8. Increase in permeability to urea of hardy (cabbage and clover leaf) cells during hardening at 5°C and absence of any increase in tender (sunflower leaf) cells. From Levitt and Scarth, 1936. (Reproduced by permission of the National Research Council of Canada from the *Can. J. Res.* C14, 297.)

(see Kuiper, 1969) observed that permeability was directly related to the freezing tolerance of the pistil cells of apple flowers. Unlike the other factors discussed in this chapter, protoplasmic permeability is an avoidance factor—avoidance of intracellular freezing. There is no evidence that it is a factor in freezing tolerance.

Many measurements have been made of the effect of hardening on protoplasmic viscosity. This is, unfortunately, a complex property, since protoplasm is not a simple liquid. Protoplasmic "viscosity" is, therefore, actually structural or nonNewtonian viscosity. It has also been given the more general name "consistency." Two diametrically opposite results have been obtained. Some investigators (Kessler and Ruhland, 1938, 1942; Granhall, 1943; Johansson *et al.*, 1955) reported a direct relationship between this protoplasmic consistency and hardiness, others (Levitt and Scarth, 1936; Scarth and Levitt, 1937; Siminovitch and Levitt, 1941; Levitt and Siminovitch, 1940) an inverse relationship. Parker (1958) reports a gel to sol conversion in the vacuole of *Pinus* cells during hardening in the fall but the protoplasm may, of course, behave differently. The differences are at least partly due to errors inherent in the methods used. A common method of measuring protoplasmic consistency is to determine the rate of displacement of the included chloroplasts when subjected to centrifugation. The plastids of nonhardy plants frequently contain large starch grains and are, therefore, heavier and more easily displaced than those of the hardy plant, since the hardening process leads to hydrolysis of starch to sugar and the accumulation of the latter in the vacuole. The difference is enhanced by

TABLE 8.9 Evidence of Stiffening due to Dehydration of the Protoplasm or Its Outermost Layer in Unhardened but not in Hardened Cells[a]

Cell property	Unhardened state	Hardened state
1. Plasmolysis shape	Concave	Convex
2. Cytoplasmic strands	Stiffen and rupture on plasmolysis	Remain fluid on plasmolysis
3. Shape of oil drop injected into cytoplasm of plasmolyzed cell	Flattened on deplasmolysis	Convex on deplasmolysis
4. Microdissection tests	Stiff	Fluid
5. Deplasmolysis after prolonged plasmolysis	Rupture of ectoplasm	Normal deplasmolysis
6. Pseudoplasmolysis on thawing	Common	Rare

[a] Numbers 1–5, see Levitt, 1956; number 6, see Asahina.

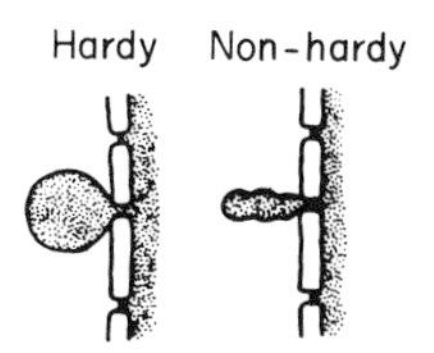

FIG. 8.9. Shape of cytoplasm when forced through a punctured pit in a hardened (left) and an unhardened (right) cell of catalpa in a balanced solution of NaCl + $CaCl_2$ (9:1) with an osmotic potential of −50 atm. (From Scarth, 1941.)

the larger size of the chloroplasts in the hardy state (Heber, 1959a). These differences would indicate a spurious increase in viscosity on hardening. The more direct methods of microdissection, though not quantitative, failed to reveal any significant difference between the two. However, a small difference could exist without being detected by microdissection. Furthermore, due to the higher vacuole concentration in the hardy cell, its protoplasm would be slightly more dehydrated than that of the unhardy cell when both are unplasmolyzed. Consequently, it is reasonable to expect a slight increase in protoplasmic viscosity on hardening simply due to this reduced water content. The pronounced difference between the two is seen when the cells are dehydrated due to plasmolysis or freezing, and this difference is in the opposite direction. The dehydration leads to a marked stiffening of the protoplasm (or at least its outermost layer) in the unhardened cell, but not in the hardened one. This is clearly shown in many ways (Table 8.9 and Figs. 8.9, 8.10, and 8.11).

This protoplasmic stiffening is so closely related to hardiness that it can

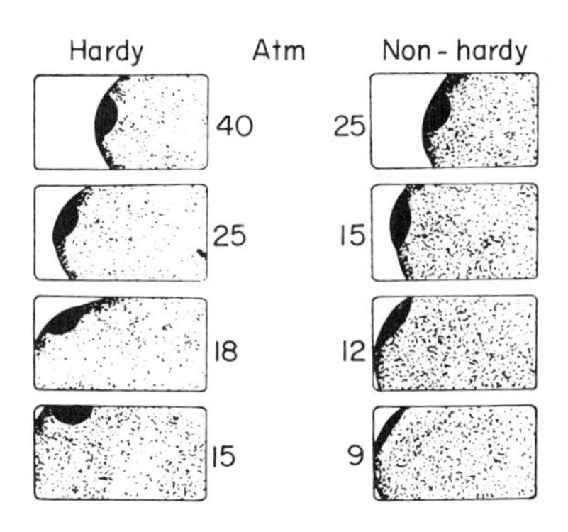

FIG. 8.10. Shape assumed by oil drops injected into cytoplasm of hardened (left) and unhardened (right) cortical cells of cornus during progressive deplasmolysis. Osmotic values (atm) of solutions in equilibrium with cells given by the numbers. (From Scarth, 1941.)

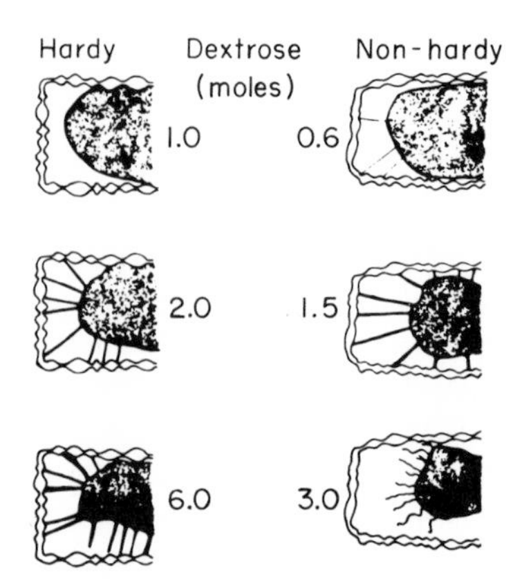

FIG. 8.11. Rupture of cytoplasmic strands during progressive plasmolysis of unhardened cortical cells of hydrangea (right), absence of rupture of hardened cells (left). Solutions in equilibrium with cells from top to bottom: Right (unhardened): 0.6, 1.5, 3.0 *M* dextrose; left (hardened): 1.0, 2.0, 6.0 *M* dextrose. (From Scarth, 1941.)

be used quantitatively as a measure of hardiness by determining relative resistance to plasmolysis and deplasmolysis injury (Scarth and Levitt, 1937; Siminovitch and Briggs, 1953a). Sakai (1957) corroborated this fact but obtained relatively small differences that could be at least partly accounted for by the higher cell sap concentration of the hardy cells. Such results are obtained only if the plasmolysis time is brief. After prolonged plasmolysis, the hardy cells can be deplasmolyzed successfully from the same degree of plasmolysis as is sufficient to kill nonhardy cells (Table 8.10), and therefore this factor cannot be accounted for by any differences

TABLE 8.10 COMPARISON OF DEPLASMOLYSIS IN HARDENED AND UNHARDENED CABBAGE CELLS AFTER PLASMOLYSIS IN TWICE ISOTONIC $CaCl_2$; DEPLASMOLYZED IN DISTILLED WATER[a]

Condition of plant	Osmotic value (M $CaCl_2$)	Percentage of surviving cells: In epidermis and chlorenchyma	Percentage of surviving cells: In pith
Nonhardened	0.16	Trace	0
5-Day hardened	0.23	Most	0
Nonhardened	0.17	Few	0
10-Day hardened	0.25	All	Many

[a] From Scarth and Levitt, 1937.

in cell sap concentration. Terumoto (1967) found that marine algae having a high resistance to plasmolysis were also highly tolerant of freezing.

These protoplasmal differences may be interpreted as evidence of a greater hydrophily of protoplasm in the hardened than in the unhardened state. Siminovitch (1963), however, has concluded that all these protoplasmal changes during hardening, together with the increase in soluble proteins which he showed was due to synthesis during hardening, are part of a general augmentation of protoplasm during the hardening period. He later (1967) gave visual evidence that the augmentation, indeed, does occur. Furthermore, as already mentioned, the greater mobility (and, therefore, presumably hydration) of the protoplasm in the hardened state is not demonstrable in the normally hydrated protoplasm, but arises as a result of the dehydration process. In other words, dehydration by any method so changes the protoplasmic proteins as to reduce their hydrophily to a greater degree the less hardy the plant. This may explain the many contradictory results in the literature when attempts have been made to relate "bound water" to hardiness.

Scarth and Levitt's (1937) discovery that many hardy cells show convex plasmolysis and nonhardy cells, concave plasmolysis, has been confirmed by the Russian workers (e.g., Genkel and Oknina, 1954). They have interpreted this as due to the absorption of plasmodesmata into the protoplasm during the winter rest, the protoplasm losing contact with the cell wall. Although their evidence of plasmodesmal differences was based on observations of fixed and stained cells and is, therefore, unreliable as an indication of the condition in the living cell, it has been partially confirmed by observations of living cells with the aid of phase-contrast microscopy (Pieniazek and Wisniewska, 1954). They call this phenomenon "protoplasm isolation" (Genkel and Zhivukhnia, 1959). They have published numerous papers on the subject, culminating recently (Akad. Nauk., 1968) in a symposium on the "depth of the resting state as judged by the separation of the protoplasm" in the case of winter wheat, native overwintering plants, strawberries, raspberries, pears, citrus fruit, apples, and maple.

It is impossible to accept the above interpretation of Genkel and Oknina; for the photomicrographs of Siminovitch and Levitt (1941) clearly demonstrate the existence of as many protoplasmal strands connecting the plasmolyzed protoplasts of hardened cells to their walls as in the case of the unhardened cells. Some of the strands are clearly attached to opposite sides of pits in the walls and, therefore, represent plasmodesmal connections. The observations of Genkel and his co-workers are, therefore, undoubtedly due to an artifact, and do not apply to the normal, living cells. This interpretation has been supported by a thorough investigation in another Soviet laboratory (Alexandrov and Shukhtina, 1964). A microscopic observation of cells (mounted in silicone oil) of numerous herbaceous and woody overwintering plants was carried out in December, January, and February under natural conditions outdoors as well as after thawing in the laboratory. In all the cases studied, the protoplasm was closely attached to the cell walls. There was no evidence of Genkel's "protoplasm isolation." They, therefore, conclude that the theory of the separation of protoplasm from the cell wall as a means of cell protection against the injurious effects of cold is erroneous.

4. Sulfhydryl (SH) groups

The sulfhydryl group occurs in plant protoplasm mainly as a component of proteins, an essential amino acid (cysteine or CSH), and a peptide (glutathione or GSH). From time to time, this chemical group has been suggested as a factor in freezing tolerance. Probably the first suggestion was by Ivanov (1939), who found an inverse relationship between glutathione content and winter hardiness of citrus varieties. He explained this by the commonly found inverse relationship between growth and hardiness, and the equally common direct relationship between growth and GSH content. Ewart *et al.* (1953) point to qualitative observations by bio-

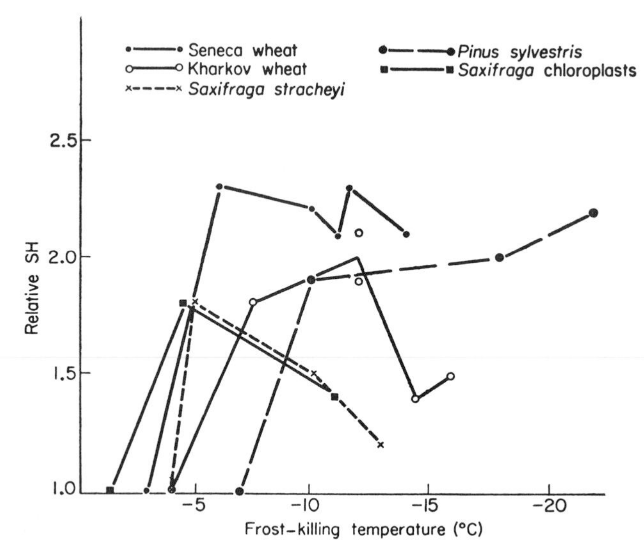

Fig. 8.12. Change in SH content of homogenates on hardening. (Adapted from Levitt *et al.*, 1961.)

TABLE 8.11 FROST HARDINESS AND SH CONTENT IN WHEAT VARIETIES[a]

Variety	Hardiness from field experience	Frost-killing temp. (°C)	SH content of supernatant (μmole/gfw leaves)
Anna Migliori	Very hardy	−15	0.84
Carsten VIII		−14	0.93
Eroica II		−12.5	0.84
Criewener 192	Hardy	−12.5	0.81
Derenburger Silber		−12.5	0.77
Austro Bankut		−12.5	0.73
General v. Stocken		−12.5	0.65
Pfeuffers Schernauer	Moderately hardy	−12.5	0.52
Heine VII		−11	0.58
Panter		−11	0.49
Etoile de Choisy	Slightly hardy	−12.5	0.46

[a] From Schmuetz, *et al.*, 1961.

chemists, which suggest that amylase activity is dependent on the integrity of the SH groups within the enzyme molecule. They, therefore, investigated the amylase system of the living bark of the black locust tree. It proved to be inhibited by some SH reagents but not by others. The inhibition, when it occurred, was readily reversed by cysteine. Similar results were obtained with phosphorylase. They, therefore, suggested that differential sensitivity of enzymes to natural SH reagents may be involved in the regulation of metabolic processes of plants and, particularly, in controlling the seasonal changes in composition of the protoplasm.

These first two investigations, though pointing to a possible relation between SH groups and freezing tolerance, appear to contradict each other. The first indicates an inverse relationship between the two, the second a direct one. This apparent contradiction arose also in later investigations. Measurements made on homogenates or on their supernatant solutions obtained from the leaves of four species of hardy plants revealed an increase in SH content after a hardening period of a few days that resulted in increased freezing tolerance (Fig. 8.12). In the case of some of these plants, this relationship was reversed during later stages of hardening (also see below). A close correlation between SH content and freezing tolerance was obtained when fifteen wheat varieties differing in hardiness were compared (Table 8.11). This relationship held whether SH was determined by the argentometric, amperometric method of analysis or by the completely different colorimetric nitroprusside method (Schmuetz, 1962). In the case of cabbage, the SH changes were detected even in the free chloroplasts, and therefore were undoubtedly due to the proteins (Fig. 8.12). Further evidence was the inability to account for the increase by an increase in GSH, since this was oxidized too rapidly under the experimental conditions used (Levitt *et al.*, 1962). The lower SH content of the unhardened plants was, at least in some cases, associated with a higher SS (disulfide) content and, therefore, due to oxidation (Levitt, 1962).

In the case of vernalizing wheat in the hardened state, the SH content of the proteins was found to be higher relative to SS content (Kohn *et al.*, 1963). In at least one of the wheats, the higher proportion of protein SH to protein SS in the hardened state was an artifact due to a more rapid oxidation of SH to SS in the unhardened material during preparation of the samples for analysis. Similarly, in the case of *Saxifraga*, homogenization in air lowered the SH content, and the degree of this lowering was inversely related to freezing tolerance, due presumably to a more rapid oxidation to SS in the unhardened material (Levitt, 1962).

A thorough investigation by Schmuetz (1969) has now firmly established both the correlation between SH measurements and freezing tolerance, and the artifactual nature of this correlation. He analyzed a total of eighty-seven wheat varieties, including all degrees of winter hardiness to be found in the varieties from western Europe, as well as extremely resistant ones from the U.S.S.R., Finland, and the United States. The SH measurements obtained from three replicates gave a correlation coefficient of 0.85 with the winter-hardiness of the varieties. The wheat seed were germinated in the dark at 20°C in order to eliminate the development of color (which might interfere with the colorimetric determinations) and 6-day-old seedlings were used. Even though the seedlings were not subjected to a hardening treatment, they possessed considerable freezing tolerance (killing temperature −12° to −15°C) and their relative freezing tolerance agreed with the known winter hardiness of each variety.

In contrast to these results, the seedlings of several barley varieties differing in hardiness all had the same SH contents. After 20 days' hardening at 2°C, however, they showed the same close correlation between SH content and winter hardiness as in the case of the wheat varieties—a coefficient of 0.83 for the eighteen varieties tested. These results agree with practical experience; for winter barley are not as hardy as wheat and are more dependent on a hardening treatment for the attainment of their freezing tolerance.

In order to obtain these high correlations between SH content and freezing tolerance, Schmuetz showed that a rigid standardization of the procedure was necessary. The seedlings had to be homogenized and centrifuged rapidly enough to determine the SH content colorimetrically 7 min after the homogenization. In spite of this speed, an oxidation of the SH groups occurred, which was more or less intense, according to the variety. An exact time plan, therefore, had to be followed. The original SH contents in hardy and nonhardy varieties were the same, and the differences at the time of measurement were due to different degrees of oxidation of the SH groups during preparation of the sample. The oxidation was slow in hardy varieties and more rapid in the less hardy varieties.

The (reduced) ascorbic acid content was also correlated with hardiness in wheat and barley. Furthermore, this correlation was obtained whether the ascorbic acid was measured immediately after preparing the plant sap or some time later. Again, the correlation depended on the different rates of oxidation of ascorbic acid to dehydroascorbic acid in the time between homogenization and titration. This was proved by adding a known amount of ascorbic acid to each homogenate (Table 8.12). It is, thus, not the total ascorbic acid (oxidized + reduced) content that is related to winter hardiness in the cereals tested, but that portion, which at the moment of testing, remains in the reduced form. In the case of vernalizing wheat, on the contrary, the total ascorbic (+ dehydroascorbic) acid content does increase (Waisel *et al.*, 1962). Nevertheless, even in this case, there is a relationship between oxidation state and temperature. The temperature optimum of the ascorbic acid oxidizing system shifts during vernalization to lower temperatures, and is 5°C at the 50% vernalization level (Dévay, 1965). Since as-

TABLE 8.12 Relationship of (Reduced) Ascorbic Acid Content to Hardiness of Wheat Varieties (Arranged in Order of Decreasing Hardiness)[a]

	Ascorbic acid content		
Wheat variety	Original	After adding the same amount of ascorbic acid to each	Increase
Minhardi	49	91	42
Carsten VIII	44	83	39
Strubes General v. Stocken	39	72	33
Graf Toerring II	33	56	23
Heines Peko	29	51	22

[a] From Schmuetz, 1969.

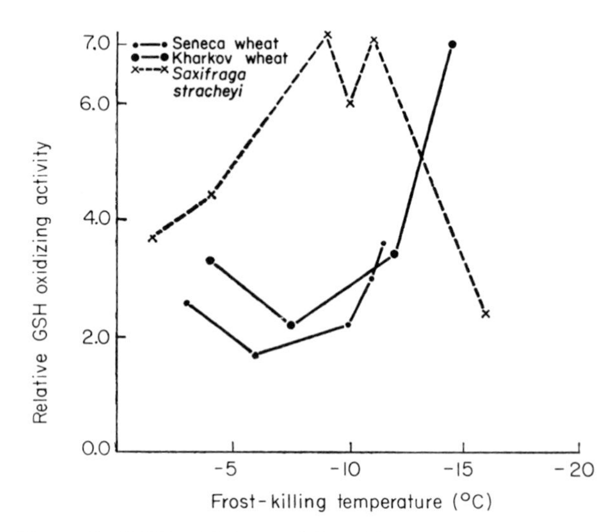

FIG. 8.13. Change in GSH oxidizing activity of homogenates on hardening. (Adapted from Levitt *et al.*, 1962.)

corbic acid is a cofactor for the enzyme GSH reductase, Dévay concludes that new enzymes are synthesized at the low temperature, which lead to an increase in the tissue content of ascorbic acid. This cofactor in its turn regulates the SH content. Dévay's interpretation agrees with earlier determinations of an increase in GSH-oxidizing activity during hardening (Fig. 8.13), although this depended on the hardening stage and the species.

In the case of cabbage seedlings hardened over long periods of time, the artifactual rise in SH was observed only during the early stages of hardening, and was followed by a very marked and steady drop in SH content which paralleled the steady decrease in the killing temperature (Fig. 8.14). In some cases, even the early artifactual rise was not observed, due undoubtedly to homogenization and titration under nitrogen (Kohn and Levitt, 1966). It was at first concluded that this decrease in SH content may be a significant factor in the hardening process, since it paralleled freezing tolerance. Thus, the early artifactual increase in SH with hardening could indicate a protective prevention of oxidation of the SH group, while the later steady decrease in total SH could indicate a decrease in the number of SH groups available for oxidation. Later measurements, however (J. Dear, unpublished), revealed a similar drop in SH content in control plants kept at nonhardening temperatures. The controls showed no appreciable increase in freezing tolerance. The decrease in SH content is, therefore, at least partly, and perhaps solely, an ageing effect.

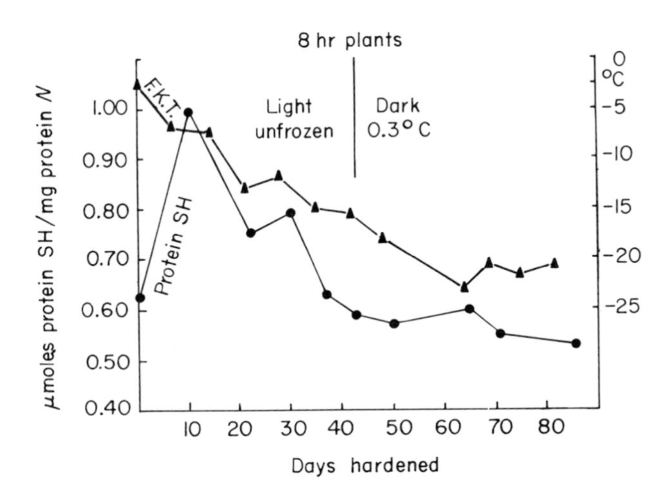

FIG. 8.14. Changes in protein SH content per unit nitrogen and in frost killing temperature (FKT) of short day (8 hr) cabbage plants during hardening. (From Kohn and Levitt 1966.)

It can, therefore, be concluded that freezing tolerance involves an increase in resistance toward oxidation of both the SH groups of the proteins and of ascorbic acid. There is no valid evidence that the decrease in SH observed during later stages of hardening is anything other than an ageing effect.

C. Resistance Induced by Applied Substances

Infiltration with sugars and other neutral substances has been shown (see above) to confer a small osmotically accountable increase in freezing tolerance in the case of plants that already possess some hardiness. A much greater effect of these same substances was first reported by Maximov (1912). Cells were able to survive as low as −30°C if frozen in solutions of nonpenetrating sugars or other nontoxic solutes above their eutectic points (Table 8.13). These results were confirmed by Åkerman (1927) who further showed that the survival depended on plasmolysis of the cells. Iljin (1933a, 1934) was able to extend these experiments and reported survival at extreme low temperatures simply by transferring the frozen, nonhardy cells to the sugar solutions before thawing. This protective effect was also obtained with salts; and in the case of the anion the effectiveness varied with the cation as follows: Ca > Mg > Sr > Ba, Li, Na, K (Terumoto, 1959).

TABLE 8.13 FROST KILLING OF RED CABBAGE WHEN FROZEN IN GLUCOSE SOLUTIONS OF DIFFERENT CONCENTRATIONS[a]

Freezing temperature (°C)	Percentage alive in glucose solutions of given molarities						
	0.0 (H_2O)	0.06	0.13	0.25	0.50	1.0	2.0
− 5.2	50	100	100	100	100	100	100
− 7.8	0	Few	25	100	100	100	100
−11.1	0	0	Few	50	100	100	100
−17.3	0	0	0	Few	25	100	100
−22.0	0	0	0	0	Few	100	100
−32.0	0	0	0	0	0	Few	50

[a] From Maximov, 1912.

The difference between the anions was minor, and oxalate and citrate actually decreased hardiness; Ca^{2+} seemed to prevent intracellular freezing. Sakai (1961) has even reported survival in liquid nitrogen with the use of sugar solutions, but salt solutions were much less effective.

The weakness of all these results is in the use of plasmolysis as the criterion of survival. It has long been known that injured cells may frequently show "tonoplast plasmolysis" even though all the protoplasm except the tonoplast is dead. Iljin admitted that the plasmolyzed cells could not be deplasmolyzed successfully, and this fact was corroborated by others (see Levitt, 1956). Furthermore, as was mentioned earlier, cells that prove to be dead several days after freezing and thawing, may plasmolyze normally immediately after thawing. It is, therefore, difficult to say just how effective these nonpenetrating solutions are.

In the case of some animal cells, it is possible to prevent injury by freezing in solutions of glycerol or DMSO (see Chapter 3). Unlike the above results, which are due to nonpenetrating, plasmolyzing solutions, these results are obtainable only if the substance penetrates the cell. Red blood cells and spermatozoa, for instance, can survive freezing below −40°C in glycerol (Lovelock, 1953). The same substance, as seen above (Tables 8.2 and 8.5), increases the freezing tolerance of higher plant cells by only a few degrees. Recent results (Terumoto, 1960b) with a green alga (*Aegagropilla santeri*) resemble those obtained with animal cells. Substances that penetrated the cells (ethylene glycol and propylene glycol) permitted freezing at −35°C for 2 hr without injury. Nonpenetrating substances (including the most effective one for animals— glycerol) failed to protect against freezing injury. Ethylene glycol protects sections of hardened mulberry twigs from slow warming injury after cooling in liquid nitrogen (Sakai, 1961).

Heber (1967) showed that sugar could protect spinach chloroplasts against injury by a rapid freeze at −20°C. These results were confirmed by Sugiyama and Simura (1967b) for tea chloroplasts. Those suspended in 0.1 to 0.3 *M* sugar solutions were not injured by freezing at −20°C, as judged by their Hill reaction activity. The protection occurred whether glucose, fructose, sucrose, xylose, or raffinose were used. Sucrose showed about 20% better protection. They also showed that when the plant was frozen, damage to the chloroplasts was proportional to freezing injury. Nevertheless, when the leaves were infiltrated with sugar solutions, this failed to protect either the leaves or the chloroplasts from injury on freezing at −20°C. Chloroplast function can be preserved in the frozen state for at least 6 weeks at −20°C in the presence of 1% bovine serum albumin and 10% glycerol (Wasserman and Fleischer, 1968).

Attempts have been made to increase freezing tolerance by applying growth regulators to whole plants, e.g., by spraying. These attempts were not at first successful (Oknina and Markovich, 1951; Moretti, 1953). Kessler and Ruhland (1942) were able to decrease tolerance by inducing growth with heteroauxin and ethylene. Tumanov and Trunova (1958) found that auxin (both free and bound) decreases on hardening. On the other hand, the ability of coleoptiles to harden was appreciably reduced if a high (growth inhibiting) concentration (200 mg/liter) of indoleacetic acid was added to a 12% solution of sucrose. Another exception was the suppression of growth by maleic hydrazide, which failed to induce hardening. Consequently, they concluded that growth rate cannot always be regarded as a reliable inverse index of hardiness. Lona *et al.* (1956; Lona, 1962), however, have reported an increased survival of −3° to −4°C when the applied substances inhibited growth and a decreased survival when they increased growth. In contrast to his results, a fall spray of GA (gibberellic acid 200 ppm), applied in 2 consecutive years to peach trees, allowed the foliage to remain green for 15–20 days longer than the controls, and it delayed the full bloom of the trees, but did not increase the freezing resistance of the flowers (Marlangeon, 1969). In spite of some negative results (see above), maleic hydrazide seems to give the most consistent protection. Even when it does, however, the growth following recovery may be abnormal (Hendershott, 1962a) and the increased tolerance is small (Le Saint, 1966). Young winter wheat seedlings have shown an increase in freezing tolerance following root feeding with ascorbic acid (Andrews and Roberts, 1961). Since such high concentrations were usually required (0.25–

TABLE 8.14 Attempts to Increase Frost Resistance by Means of Applied Growth Regulators

Substance	Conc. (ppm) unless otherwise stated	Plant	Freezing resistance	Reference
Dalapan	10	Litchi trees	Increased	Gaskin, 1959
Maleic hydrazide	500	Litchi trees	Increased	Gaskin, 1959
	1000	Grapefruit and orange trees	Increased	Stewart and Leonard, 1960
	1000–5000	Mulberry trees	Increased	Sakai, 1957
	1000–2000	Citrus trees	Increased	Hendershott, 1962a
Several		Grapefruit	More or less effective	Cooper *et al.*, 1955; Cooper, 1959
Several		Bartlett pears	Increased set	Griggs *et al.*, 1956
Several		Bartlett pears	No effect	Rogers, 1954
Leaf extract (*Bergenia crassifolia*)		Stock and cabbage	Increased	Konovalov, 1955
Amo and phosphon D		Several spp.	Increase	Lona, 1962
Cycocel (CCC)		Wheat	Possible increase	Toman and Mitchell, 1968
B-9		Wheat	Possible increase	Toman and Mitchell, 1968
CCC	200	Tomato	Increase	Kentzer, 1967
2-Alkenylsuccinic acids and other surface-active chemicals	10^{-3} to 10^{-4} M	Strawberry flowers, beans, etc.	Slight increase	Kuiper, 1967
DMSO	5%	Grapevine	No increase	Marlangeon, 1967
Decenylsuccinc acid	1000	Grapevine	No increase	
Alar	2000	Grapevine	No increase	
CCC	2000	Grapevine	No increase	
Phosphon 2000	2000	Grapevine	No increase	
Gibberellin	1000	*Acer negundo*	Lowered	Irving and Lanphear, 1968; Irving, 1969b
B9	3000	*Acer negundo*	Increased	Irving, 1969b

TABLE 8.14 (Continued)

Substance	Conc. (ppm) unless otherwise stated	Plant	Freezing resistance	Reference
Amo 1618	1000	*Acer negundo*	Increased	Irving and Lanphear 1968; Irving, 1969b
Dormin (abscisic acid)		*Acer negundo*	Increased	Irving and Lanphear 1968; Irving, 1969a
Proline	5000	Cabbage	Increased	Le Saint, 1966
Dalapan (in dark)	4–8	Sugar beet	Improved	Corns and Schwerdtfeger, 1954
Trichloropropionic acid (in dark)	4–8	Sugar beet	Improved	Corns and Schwerdtfeger, 1954
Dalapan, TCA, Trichloropropionic acid		Clover, alfalfa, beans	Ineffective	Corns and Schwerdtfeger, 1954
GA	200	Peach trees	Ineffective	Marlangeon, 1969
CCC	2	(fall spray)	Ineffective	
Alar	2		Ineffective	
TD-692	2		Ineffective	
CCC	1000	*Acer negundo*	Not increased under LD and no hardening	Irving, 1969a
B-9	3000		Increased under SD	
Amo	1000		Treatment of hardened failed to retard loss of tolerance	
GA			Accelerated loss of hardiness	
KGA	50–150	Peach	Retardation of floral development reduced freezing damage during bloom	Stembridge and Larue, 1969

TABLE 8.14 (Continued)

Substance	Conc. (ppm) unless otherwise stated	Plant	Freezing resistance	Reference
ABA	20	Apple seedlings	Increased tolerance	Holubowicz and Boe, 1969
GA	100	"	No change	"
CCC (applied in autumn)	1–16 kg/ha	Winter rape	Increased tolerance	Chrominski *et al.*, 1969
CCC	2000	Cabbage	Increased hardening	Kacperska-Palacz *et, al.* 1969
B-9	4000			
Decenylsuccinic acid	10^{-3} and 10^{-4} M	Barley and winter wheat	No increase in non-hardening environment	Green *et al.*, 1970
CCC		Wheat	Improves winter hardiness but not frost tolerance due to increasing depth in soil	Kretschmer and Beyer, 1970

1.0 *M*), the possibility of a purely osmotic effect cannot be eliminated. Freezing tolerance of alfalfa has been reported to increase following foliar applications of uracil, thiouracil, or guanine (Jung, 1962). At the same time, protein and nucleic acid content as well as tissue pH were higher than in the controls (Jung *et al.*, 1967b). In some cases, the treated plants were less hardy than the controls, and these showed the reverse relationship with respect to proteins, nucleic acids, and pH.

Le Saint (1966) did not succeed in hardening cabbage seedlings by supplying them with either sugar or proline in the dark. When the seedlings were stood in solutions of proline (5 gm./liter) in the light at 18°C (a nonhardening temperature) for 2 weeks (with five changes), some increase in tolerance resulted, although it was not as good as by hardening in the light at 4°C. The plants that were fed the proline accumulated the same sugars and in the same amount as the hardened plants. Some of the many reports in the literature are listed in Table 8.14.

Kuiper (1967) has reported increases in freezing resistance due to application of a completely different group of substances which also act as antitranspirants, supposedly by leading to stomatal closure. He used completely tender plants which normally are killed by the first touch of frost (e.g., tomato, beans, strawberry flowers). Since the temperatures used were close to the freezing point (usually −3°C or higher) and freezing periods were brief (1 hr in his early experiments) it may involve freezing avoidance rather than tolerance. An increase in permeability due to the applied substance led him to suggest that this was the resistance factor. If so, this would be a case of avoidance of intracellular freezing. However, the concentration of decenylsuccinic acid (10^{-3} *M*) used to produce the protection was shown to increase the permeability of bean roots by killing them (Newman and Kramer, 1966). Kuiper (1969) applied this concentration of spray to strawberry flowers and 2 hours later exposed them to −6°C for 2 hr. The untreated flowers were all killed, while most of the treated ones survived and set fruit. Hilborn (1966; see Kuiper) confirmed these results but the protection did not last longer than 8 hr. According to Kuiper, the protection seems to last longer for flowers of plum and cherry. Some of the differences, however, are very slight, e.g., 8% survival in the controls, 10–40% in the treated. Other investigators have failed to obtain an increased tolerance (e.g., Marlangeon 1969). On the contrary, Green *et al.* (1970) not only failed to obtain an increase in freezing tolerance of winter wheat treated with decenylsuccinic acid (DSA) in a nonhardening environment, they actually obtained a negative effect. Treated plants were killed by −5.5° (±0.5°), untreated plants by −6.5°C (±0.5°). They also confirmed Newman and Kramer—the increased permeability following DSA treatment was due to injury.

CHAPTER 9

THEORIES OF FREEZING INJURY AND RESISTANCE

Since there are two main kinds of freezing injury, (1) primary direct, due to intracellular freezing and (2) secondary, freeze-induced dehydration due to extracellular freezing, all theories can be classified according to which of these they attempt to explain. The oldest theory, the caloric theory (see Levitt, 1956), actually attempted to explain a kind of injury that does not exist—injury due to loss of heat. According to this concept, resistance was due to release of enough heat to prevent freezing. Since plants do not possess this implied low-temperature avoidance (see Chapter 7), the caloric theory is only of historical interest.

A. Primary Direct Freezing Injury

1. The Rupture Theory

This was the first theory to attempt an explanation of this kind of injury. It ascribed the injury to cell rupture due to expansion on freezing. This theory was disposed of by showing that, in nature, cell rupture due to freezing never occurs (see Chapter 6), and that the tissues normally contract rather than expand. Since air is squeezed out of leaves during freezing, a small pressure does develop; this pressure originates due to a contraction external to the living cells, and is symmetrical. It therefore cannot cause cell rupture. Nevertheless, under artificial conditions intracellular freezing can be induced, and if the ice crystals are large enough to be detected by the optical microscope, is always fatal (Chapter 6). It has, therefore, been frequently proposed that this intracellular freezing does occur in nature and is the cause of freezing injury. Cell rupture does not, however, occur except under very special artificial conditions (see Chapter 6).

2. *The Intracellular Freezing Theory*

In the case of certain microorganisms that are uninjured by dehydration even to the point of air drying, intracellular freezing would undoubtedly be the only possible source of injury. Mazur *et al.* (1957) and Mazur (1961) have concluded that this is true for yeast and other microorganisms which are unaffected by freezing above −10°C but are killed in a critical temperature range of −10° to −30°C. In the case of some (e.g., *Pasteurella*) the range is −30° to −45°C. Thus, rapid cooling to −30°C results in 99.99% killing of yeast cells; slow cooling to the same temperature results in 50% killing (Mazur, 1960). Freeze substitution of the cells by cold ethanol in order to retain the cell shape that occurred in the frozen state, revealed that the rapidly frozen cells retained their original shape and that the slowly frozen were considerably smaller and more flattened. Even in the case of higher plants and animals, at least under experimental conditions, intracellular freezing is the main cause of killing due to rapid freezing (Asahina, 1961). Such evidence has frequently led to the suggestion that all freezing resistance is due to the avoidance of intracellular freezing (e.g., Stuckey and Curtis, 1938). Scarth (1936) was the first to suggest that the higher permeability of hardy cells to water would favor this avoidance, and later to produce evidence that hardy cells do actually show such an avoidance of intracellular freezing (Siminovitch and Scarth, 1938). Both of these observations have since been amply confirmed by other investigators, especially at the Japanese Low Temperature Station at Hokkaido (e.g., Asahina, 1956). Even in the case of animal cells, it has been found (Asahina, 1961) that sea urchin eggs are much more resistant to rapid freezing injury when fertilized than when unfertilized and this is correlated with a four times higher water permeability. This rapid injury resulted from intracellular ice formation. Both were equally resistant to slow freezing injury.

It might be objected that the permeability increase during hardening would actually favor seeding of the cell interior due to penetration of the larger aqueous pores in the lipid plasma membrane by the external ice crystals. Chambers and Hale (1932) first showed that this membrane is an essentially impermeable barrier to such seeding. This has since been clearly demonstrated by the true frost plasmolysis that occurs on ice formation between the protoplast and the cell wall under certain artificial conditions (Asahina, 1956; Modlibowska and Rogers, 1955). The ice mass grows inward from the cell wall by withdrawing water from the unfrozen protoplast which contracts and retreats before the advancing ice. In spite of this constant contact of the protoplast surface with ice, no seeding of the protoplast interior occurs. Mazur (1961) has concluded from theoretical considerations that only at very low temperatures (e.g., −30° to −40°C) is the radius of curvature of the ice crystal small enough to suggest its penetration of pores the size of those in the plasma membrane.

It was even possible (Asahina, 1962b) to increase the permeability of unfertilized egg cells by means of urea, and this enables them to avoid intracellular freezing injury. The Japanese workers (Asahina, 1956; Terumoto, 1957a, 1959; Sakai, 1958), therefore, suggest that not only natural hardening, but even the prevention of freezing injury by protective solutions may be due to the prevention of intracellular freezing. Unlike urea, however, most protective solutes do not increase cell permeability and, therefore, their effects cannot be explained in this way. The evidence indicates that they protect plant cells by inducing plasmolysis (see Chapter 8). The permeability to water is actually decreased by plasmolysis (Levitt *et al.*, 1936), as would be the protoplast area if plasmolysis is convex. If plasmolysis is concave the area would be essentially the same as that of the collapsed cell in the absence of protective solution and therefore would not alter the rate of exosmosis. There is, however, one reason for suggesting that the protective solutions could speed up the exosmosis of water and, therefore, prevent intracellular freezing. The protective solutions decrease the diffusion distance by eliminating the cell wall between the water diffusing from the cell and the ice loci outside the cell walls. However, it seems likely that the unstirred layers external to the plasmalemma and the viscosity of the protective solution would soon counteract this effect.

Olien (1965) has proposed a new concept on the basis of electrical measurements made during the freezing of leaves. He concluded that high molecular weight polysaccharides may act as natural inhibitors of ice formation in the crowns of wheat plants, leading to the formation of slush-like ice instead of a more injurious kind of ice. This implies that the injury is due to a direct effect of the ice crystals. It has already been emphasized, however, that a direct injury by ice is possible only when the ice forms intracellularly (see above). Therefore, this might be considered another intracellular freezing theory.

Although primary direct freezing injury is, by definition, due to intracellular freezing, this does not explain the injury. It is usually assumed to be a mechanical disruption of the protoplasmic structure. Even when the intracellular freezing is conducted in such a way as to produce small (but microscopically detectable) crystals and to maintain the protoplast intact, the killing is just as complete as when the protoplasm is disrupted by larger intracellular crystals (e.g., in amoeba; Smith *et al.*, 1951). Similarly, injury is not due to a purely physical compression of the protoplasm between the frozen vacuole and cell wall, since intraprotoplasmal freezing has been observed to result in death even when the vacuole does not freeze (Siminovitch and Scarth, 1938).

It must, therefore, be recognized that the so-called intracellular freezing theory is a theory only in the sense that it has been used to explain all freezing injury. Since it does not explain how intracellular freezing induces primary direct freezing injury, and since the above attempt to explain all freezing injury by intracellular freezing has been shown to be invalid (Chapter 6), it must be abandoned as a theory. On the contrary, in the case of higher plants, freezing injury under natural conditions is, in general, due to the secondary, freeze-induced dehydration (see Chapter 6), and the only potentially valid theories are those that attempt to explain this kind of injury.

B. Secondary, Freeze-Induced Dehydration Injury

1. Protein Precipitation Theory

It was first suggested by Lidforss (1896) that sugars may protect the protoplasmic proteins against coagulation during freezing. In support of this suggestion, Gorke (1906) froze expressed sap from cereal leaves and other plants at temperatures from $-4°$ to $-40°C$, and observed protein precipitation in the thawed juice. The more resistant the plant to freezing injury, the lower was the temperature required to precipitate the proteins. Adding sugar to the sap before freezing prevented, or at least decreased, the precipitation during subsequent freezing. Gorke, therefore, explained freezing injury as a precipitation of the protoplasmic proteins due to the increased concentration of the cell salts that occurs on freezing, and freezing resistance as a protection of the proteins by the sugars that accumulate on hardening. Gorke's protective effect by sugars was corroborated in the case of grains but not in the case of many other plants (e.g., fruit trees; see Levitt, 1956). Furthermore, as mentioned above, there are many plants that cannot become resistant in spite of high sugar content, so his explanation of freezing resistance is inadequate. It is also incapable of explaining even a small protective effect by plasmolysis in sugar solutions and the existence of thawing as well as freezing injury. Attempts to show a lower salt content in hardy than in nonhardy cells, have failed (see Levitt, 1956). Terumoto (1962), in fact, has shown that lake balls are among the most resistant of algae (they survived −20°C for 26 hr) in spite of their extraordinarily high salt concentration—equivalent to 0.85 *M* NaCl. In the case of marine algae, there was no apparent relationship between the lethal salt concentration at 20°C and the concentration that killed the cells when frozen at −15°C (Terumoto, 1967). Salt injury occurred only in monovalent salt solutions which induced irregularities in the protoplast surface. In divalent salt solutions there was little or no injury and the plasmolyzed cells had a smooth surface. In the case of collards, the hardened plants showed a greater tolerance than the unhardened plants in both NaCl and $CaCl_2$ solutions during either plasmolysis or freezing (Samygin and Matveeva, 1969). This was taken to support Gorke's theory, but, as shown above, hardened plants develop this plasmolysis resistance to nonelectrolytes as well as to salts. Kappen (1969a) has shown that, in spite of their high salt content, the leaves of three halophytic species were able to tolerate freezing at as low as −16° to −20°C in February, with not more than 10% injury, although all were killed in summer by −4° to −7°C. Finally, it has even been possible to increase freezing resistance by inducing the uptake of salts (Sakai, 1961).

All these more recent results add support to the earlier rejection of the salt-precipitation theory by many investigators (see Levitt, 1956, 1958). Nevertheless, support for the theory has come from several laboratories as an explanation of freezing injury both in animal and plant cells. Lovelock (1953) showed that when red blood cells were frozen at different temperatures in 0.16 *M* NaCl containing glycerol in different concentrations, the killing temperature varied inversely with the glycerol concentration. Yet in each case hemolysis occurred at 0.8 *M* NaCl. Meryman (1968) has concluded that injury from extracellular ice in animal cells results from the concentration not just of salts but of any extracellular, nonpenetrating solute. That absolute electrolyte concentration is not the cause of injury was shown by exposing the cells to high concentrations of a penetrating electrolyte—NH_4Cl. In opposition to the nonpenetrating NaCl, which injured at 0.8 *M*, the red blood cells survived 4 *M* NH_4Cl. Even some of Lovelock's own evidence does not support a salt-precipitation theory (Levitt, 1958). Nevertheless, Farrant and Woolgar (1970) have recently supported Lovelock's concept, and interest in the theory continues to appear in the current literature.

Lovelock's concept, however, depends on salt concentrations outside the cell, and, therefore, cannot apply to plant cells, which are surrounded by air and other cells but not by an external solution. At the same time as he was investigating red blood cells (the 1950's), however, support was coming from Ullrich's laboratory, again based on work with grains, as in Gorke's pioneering research (Ullrich and Heber, 1957, 1958, 1961; Heber, 1958a,b, 1959a,b). These investigators thoroughly confirmed the protection by sugars against freezing precipitation of proteins *in vitro* (i.e., in plant juice). They emphasized that in order to be effective *in vivo*, the sugars would have to be in the protoplasm rather than in the vacuole (Heber, 1958a) and suggested that this might explain the lack of correlation between sugars and hardiness in some plants. It was, in fact, possible to show an increased sugar content of the chloroplasts on hardening (Heber, 1959a).

This renewed interest in Gorke's work was perhaps stimulated by Tonzig's (1941) proposal of a formation of mucoprotein complexes between the protective sugar and the proteins. Experimental support for Tonzig's proposal was produced by Jeremias (1956), although earlier results had been negative (Levitt, 1954). The positive results of Jeremias have since been disproved by experiments from the same laboratory. The lack of complex formation was shown by dialysis of the plant juice to which sugar had been added. This enabled complete removal of the protective effect (Ullrich and Heber, 1958). Also, more reliable analyses (Heber, 1959b) failed to reveal any relationship between glycoprotein content and freezing resistance, even when sucrose concentration was closely related to hardiness. These investigators then returned to an explanation similar to that given originally by Gortner (1938)—a protection by the formation of uncombined sugar coats around the protein molecules in place of the water coats that are removed by freezing (Ullrich and Heber, 1957, 1958; Heber, 1958a). They suggest a possible formation of H bridges between the OH groups of the carbohydrates and the polar groups of the protein. This interpretation does not agree with direct determinations of hydrogen bond formation between sugars and proteins. They may be formed quite regularly with pentoses and sometimes with hexoses, but not with disaccharides (Giles and McKay, 1962), indicating that only sugar molecules below a limiting size can penetrate the dissolved protein aggregates and form a hydrogen bond complex. Yet the disaccharides and even larger sugars are correlated with freezing resistance. All of the sugars tried by Ullrich and Heber were equally effective on a molar basis, and the maximum effect was attained at about 0.6 *M*. On the other hand, the results are difficult to explain on a simple osmotic basis on account of the pronounced effect of low concentrations and the lack of any increase at concentrations higher than 0.6 *M*.

Heber and his co-workers now appear to have dealt the death-blow to all their earlier confirmation of Gorke's results. They have demonstrated that none of the soluble enzymes tested lose their activities when the plant is killed by freezing (see below). The "protection" of these soluble proteins by sugars in both Gorke's and their experiments is, therefore, an artifact. At least part of the effect of the sugar is to protect the membrane system of the broken chloroplasts. If these membrane systems are allowed to rupture they precipitate much more easily (Uribe and Jagendorf, 1968).

Nevertheless, Heber and Santarius (1964) did succeed in protecting an insoluble particulate enzyme (ATPase) against inactivation by freezing, by an addition of sugar before the freeze (see below). They again attempt to explain the protection by H bonding. Again, as they admit, disaccharides are just as effective protectants as monosaccharides, even though disaccharides do not form H-bonded complexes, "presumably because a firmly bound water environment (of the sugar molecule) prevents interaction with protein." In order to make their theory plausible, they therefore conclude that "on freezing, part of this environment is frozen out, and interaction may now become possible." However, a simple fact proves that the "water environment" is not frozen out; for it is the retention of this "water environment" that results in the much lower freezing point of a molar sucrose solution than that of a 1 *M* hexose solution. On the other hand, their evidence that disaccharides are even more effective than monosaccharides in low molar concentrations is in agreement with an osmotic protection due to prevention of chloroplast or lamellar rupture. Later results indicate that sucrose is far more effective than monosaccharides (Heber, 1967), so Heber seems to have eliminated both covalently bound sugars in glycoproteins and H bonding of sugars to proteins, as factors in protein protection. Heber (1968) also prepared chloroplast membranes from hardy spinach which were not inactivated by freezing, yet there was not enough sugar present to protect them. Therefore, the natural tolerance cannot be due to any kind of protein–sugar combination.

Santarius (1969), has confirmed Heber's results with chloroplasts, showing that only photophosphorylation and electron transport were affected by freezing and that a number of soluble enzymes were not inactivated. He also obtained the same results with desiccation as with freezing. He suggests that during dehydration by either freezing or desiccation, the concentration of electrolytes in the remaining solution is responsible for inactivation of the chloroplast membranes.

Gorke's theory, therefore, seems to be half retained by Heber's group. Dehydration on freezing is believed to cause a sufficient increase in concentration of salts to inactivate a specific system; but protection of this system in tolerant plants is no longer explained by the accumulation of sugars. Instead, the protective effect is now believed due to a protein.

Heber (1968, 1970) has now succeeded in isolating from hardened spinach leaves first two small proteins (mol. wt. 17,000 and 10,000), then a third, which are not found in nonhardened leaves. They are heat stable and stable against acidification. They are more than twenty to fifty times as effective as sucrose (on a weight basis) in protecting ATPase activity of chloroplast vesicles against inactivation by freezing. Other protein fractions were scarcely or not at all protective. Such results must be interpreted with caution. Both bovine serum albumin and various SH reagents greatly enhance phosphorylation in both red kidney beans and spinach chloroplasts, although having little effect on photoreduction (Howes and Stern, 1969). These substances also protect against inhibition by atabrin, but not against the rapid loss in activity at 0°C. Furthermore, Pullman and Monroy (1963) had previously obtained similar results with mitochondrial ATPase. Both the ATPase and the coupling factor (F_1) in the mitochondria lost their activities on freezing at −55°C (or even at 4°C). In the presence of another mitochondrial protein of low molecular weight, the above two components survive 4 days at −55°C without loss of coupling activity. In spite of this protective effect, the animal tissues from which the mitochondria were obtained (beef heart) are completely intolerant of freezing. It cannot, therefore, be concluded that Heber's proteins are responsible for the freezing tolerance of the spinach leaves.

Similar results have been obtained with the ATPase of *Bacillus megaterium* (Ishida and Mizushima, 1969). This is a membrane-bound enzyme whose activity depends on this binding. When solubilized by dialysis and mild alkaline treatment, its activity is cold labile. On recombination with the alkali-treated membrane in the presence of Ca^{2+} or Mg^{2+} (i.e., by an ionic bond), it is protected against cold inactivation. The general protective effect of proteins against loss in enzymatic activity of other proteins has, in fact, been shown by others. Shikama (1963) found that 0.01% gelatin was much more effective in preventing inactivation on freezing, than were glycerol, glucose, or sucrose.

A second revival of Gorke's concept and, in particular, a support for Tonzig's mucoprotein hypothesis, has come from two completely independent laboratories. When frozen at −20°C, the soluble chloroplast protein of tea leaves was precipitated (Sugiyama and Simura, 1967b). The amount precipitated was inversely proportional to the freezing tolerance of the variety, and decreased seasonally in the more resistant varieties from 20% in September to 5–10% in December–February. They also observed an increase in the sugar content of the soluble chloroplast protein of tea leaves from about 12% in September to 20–25% in winter (November–January). The increase paralleled the freezing resistance of the variety.

However, the sugar content of the insouble chloroplast protein was only about 1% and failed to increase. On the basis of Heber's recent results, this would seem to eliminate glycoproteins as a factor in freezing tolerance, since his results point to the insoluble proteins as the ones involved in freezing injury. Furthermore, Sugiyama and Simura (1966) also found their glycoprotein in rice leaves, yet rice is not tolerant of freezing.

Recent support of the Gorke-Tonzig concept has resulted from a newer method of detecting mucoproteins. These results have led Steponkus and Lanphear (1966) and Steponkus (1968) to explain the hardening of *Hedera helix* on the basis of a two-phase process. The first phase occurs in the light and results in an accumulation of sucrose. The second phase proceeds in the dark and is the rate-limiting step. Four hours of the dark reaction are required for every hour of exposure to the light condition. The dark reaction may correspond to Tumanov's second stage of hardening. According to Steponkus (1969a,b) an alteration of protein structure resulting in an increased affinity for sugars corresponds to the second or dark phase. He suggests two types of protein–sugar interactions during cold hardening of *Hedera helix,* leading to (1) covalently linked protein–sugar complexes or glycoproteins (i.e., mucoproteins), and (2) less stable protein–sugar associations via H bonding, as suggested by Heber, and much earlier by Gortner (see above). In opposition to the results of Sugiyama and Simura with tea, both the soluble and the particulate fractions from hardened tissue of *Hedera helix* showed a higher sugar binding capacity than the corresponding protein fractions from unhardened tissue. Furthermore, sucrose showed a higher affinity for the protein fractions than did glucose.

These results from Steponkus' laboratory are directly opposed to those of Heber's group and others (see above). It is again impossible to explain the greatest "protection" of the proteins by the very sugar (sucrose) that shows the lowest ability to H bond or combine covalently with proteins. Furthermore, it is possible that the greater "binding" of the sugar by the proteins from hardened tissue is simply an adsorption phenomenon due to a higher specific surface of the proteins, since no evidence of a covalent or a H bonding has been produced. The fact that the ^{14}C-labeled sugars became attached to proteins which were already supposedly bonded naturally to the sugars in the plant eliminates the possibility of covalent bonding, for this would prevent exchange with the labeled sugars. Until such objections are met, it must still be concluded that the mass of evidence opposes Gorke's protein precipitation theory, and Tonzig's mucoprotein modification of it. Furthermore, no explanation is attempted of the precise protein change leading to cell injury. Finally, no explanation is given for the lack of protection of animal cells (and some algal cells) by sugars against freezing injury. This has been demonstrated even more forcefully by the positive results in the case of animals. Glycoproteins have recently been found to protect fish against freezing injury (De Vries, 1970). The glycoproteins were unequivocally proved to be present, and it was also proved that the protection they confer is due to freezing avoidance and that the fish are killed as soon as they freeze. Thus, just as in the case of rice, glycoproteins, when present, do not confer freezing tolerance. It, therefore, must be concluded that freezing tolerance is independent of glycoproteins.

Although no mention is made of the salt-precipitation theory, attempts have been made recently to revive the bound water aspect of the theory. Shikama (1963) assumes that water adjacent to the proteins forms cubical "icebergs" on the nonpolar regions, thus stabilizing their structure when frozen. Above −75°C, he suggests that they may be converted to the hexagonal form, leading to breakdown of intramolecular H bonds. At these higher freezing temperatures, however, protective substances might stabilize the iceberg structure.

A comparison of the amount of unfrozen water in living and killed plants, led Tumanov *et al.* (1969) to conclude that the freezing tolerance of winter wheat depends, to a great extent, on the water-retaining power of the cells. They believe that this is due not only to osmotic forces, but also to the "vital" state of the protoplast.

Weiser (1970) has apparently combined the above two concepts in what he labels his "vital water hypothesis." Resistance is supposedly due to the uncovering of hydrophobic groups of proteins at low temperature, leading to the formation of "icebergs" around them, which supposedly binds the water more firmly and, therefore, prevents freezing. This is a proposal of protein denaturation as a protective mechanism against freezing injury; the apolar groups are normally within the native (folded) protein, and only by unfolding (denaturing) can a large proportion come to the surface of the molecule. It is this denatured state that is stabilized by the structured water around these apolar groups (Brandts, 1967). Since a cell with denatured proteins is a dead cell, this cannot possibly be a protective mechanism.

Weiser (1970) defines "vital water" as water which is intimately associated with protoplasmic constituents and necessary for life. According to his concept, this water is released at the instant of frost-killing and freezes, accounting for his third exotherm. This concept has been indisputably put to rest by Johansson and Krull (1970). The amount of water unfrozen (per gram dry matter) was found to be exactly the same in dead as in living wheat plants when frozen either at −2.65° or −9.0°C (Table 9.1). Yet the living plants were uninjured by the freezing in all cases at −2.65°C, and in

TABLE 9.1 Comparison of Unfrozen Water (gm/gdm) in Living Intact Plants, and in Plants Killed by Chloroform Determined Calorimetrically (Winter Wheat Variety Sammetsvete)[a]

Days hardened	Frost-killing temp. (°C)	Unfrozen water (g/gdm)			
		at −2.65°C		at −9.0°C	
		living	dead	living	dead
0	− 6.5	2.88	2.84	0.60	0.62
10	− 9.0	2.56	2.53	0.81	0.80
20	−11.5	2.27	2.31	0.72	0.74

[a] From Johansson and Krull, 1970.

the hardiest state at −9.0°C. Furthermore, the amount of water left unfrozen at −9.0°C (0.6–0.8 gm/gdm) is only slightly above the value for amount bound by proteins (0.5 gm/gdm; see Chapter 8). These measurements are undoubtedly the most accurate to be found in the literature, due to the painstaking measurements of the constants needed for calculation of ice formation from calorimetric measurements.

2. *Iljin's Theory of Mechanical Stress*

On the basis of numerous experiments and observations, Iljin (1933a) proposed his mechanical stress theory of freezing injury and resistance. He was struck by the pronounced cell collapse that occurs on extracellular ice formation, and concluded that the protoplasm of such cells must be subjected to a mechanical stress. If thawing was rapid enough, he observed the cell wall to snap back to its original position, sometimes tearing part of the protoplasmic surface. Due to its lower permeability, the protoplast reabsorbed water less rapidly than the cell wall, and, therefore, showed a temporary "pseudoplasmolysis" during the thawing.

Iljin later went so far as to conclude that the injury occurred only during the thawing, mainly because of his "success" in preventing injury by simply thawing in strongly plasmolyzing solutions. As mentioned above, this was not a true success since he was unable to deplasmolyze these cells successfully. Many phenomena can, however, be explained by Iljin's theory and it was, therefore, widely accepted at least as a working hypothesis. Any factor that reduced the mechanical stress during freezing and thawing would, for instance, confer hardiness. The factors associated with hardiness and the role of avoidance of freeze-dehydration can thus be logically explained by Iljin's theory. The smaller the cell size, the greater the specific surface and, therefore, the less the volume strain per unit surface at any one degree of cell contraction. The greater the sugar concentration, the smaller the water loss and therefore the less the cell contraction and the consequent mechanical stress at any one freezing temperature. An increase in bound water would have the same effect. It would also explain why such factors are of secondary value and therefore not always correlated with hardiness. The primary factor in freezing tolerance would have to be tolerance of freeze-dehydration—the possession of a kind of protoplasm that can be subjected to the mechanical stress without suffering a sufficient strain to produce injury. Siminovitch (Levitt and Siminovitch, 1940) was, in fact, able to show that protoplasm from hardened plants can survive a greater pressure than protoplasm from unhardened plants. The greater stiffening of the unhardened protoplasm on dehydration (see Chapter 8) would ensure its rupture when subjected to a smaller mechanical stress.

In the case of animal cells, Meryman (1970) concludes that freezing injury is related to the removal of a critical proportion of the total cell water, and the associated reduction in cell size beyond a critical volume. This seems to support Iljin's theory. Yet, on the basis of Iljin's theory, animal cells should not suffer freezing injury, since they do not possess the stiff, cellulose cell walls responsible for the mechanical stress during cell collapse. Indeed, Asahina (1962a) observed that unfertilized sea urchin eggs become flattened on extracellular freezing, while fertilized eggs contract without loss of their spherical shape. According to Iljin's theory the former would be expected to suffer more injury; but the two kinds of eggs are equally resistant to extracellular freezing injury. Unfortunately, the strongest evidence that the injury is due to the mechanical stress was the prevention of such injury by the use of protective solutions that plasmolyzed the cells and therefore completely eliminated the mechanical stress on the protoplasm by the cell wall. As mentioned above, Iljin greatly overestimated the protective effect of such treatment by accepting plasmolysis without deplasmolysis as evidence of the complete absence of injury. Results (Terumoto, 1962) with algae (known as lake balls) have shown that when frozen in salt solutions the cells plasmolyze, yet injury may or may not occur, depending on the salt and temperature used. The very existence of plasmolysis and deplasmolysis injury and its correlation with freezing injury (see Chapter 8) opposes Iljin's concept. If injury is dependent on the mechanical stress due to cell collapse, there should be no such thing as plasmolysis and deplasmolysis injury, since the cell wall is separated from the protoplasm and unable to subject it to a mechanical stress.

In other cases, however, observations support Iljin's theory. Winter wheat cells, frozen extracellularly, were severely compressed (Salcheva and Samygin, 1963), leading to death during freezing, and not during thawing. The unhardened cells reached a maximum compression at −7° to −8°C, with several pinched portions. The hardened cells showed maximum compression at −12° to −13°C, but the cells were square with concave walls. The unhardened cells filled out incompletely on thawing and were injured; the hardened smoothed out completely and were uninjured.

Although the above observations lend qualitative support to Iljin's theory, a true test of the theory must be quantitative. According to his concept, the two components of freezing tolerance, which have already been identified as avoidance of dehydration strain and tolerance of dehydration strain (Chapter 7), are really an avoidance of cell contraction and a tolerance of cell contraction, respectively. Unfortunately, measurement of the second component yields the opposite of the expected result. On the basis of cryoscopic measurements in a series of wheat varieties, the calculated tolerance of cell contraction on freezing is *inversely* related to the measured freezing tolerance (Table 9.2). Johansson and Krull (1970) and

TABLE 9.2 Relationship of Cell Contraction on Freezing to Freezing Tolerance[a]

	Frost-killing temperature	Calculated degree of cell contraction at frost-killing temperature
Wheat varieties		
Minhardi	−15	4.9
Lutescens	−15	5.2
Hostianum	−14	5.1
Erythrospermum	−14	5.7
Durable	−14	5.7
Ukrainka	−13	5.1
Zemke	−13	5.9
Schroder	−12	6.2
Pine leaves		
Sept. 23	−10	2.4
Oct. 8	−15	2.5
Nov. 27	−35	2.9
Jan. 1	−40	3.0
Mar. 28	−30	2.8
May 3	−13	2.1

[a] From Levitt, 1956. Adapted from Tumanov and Borodin (1930), and from Pisek (1950).

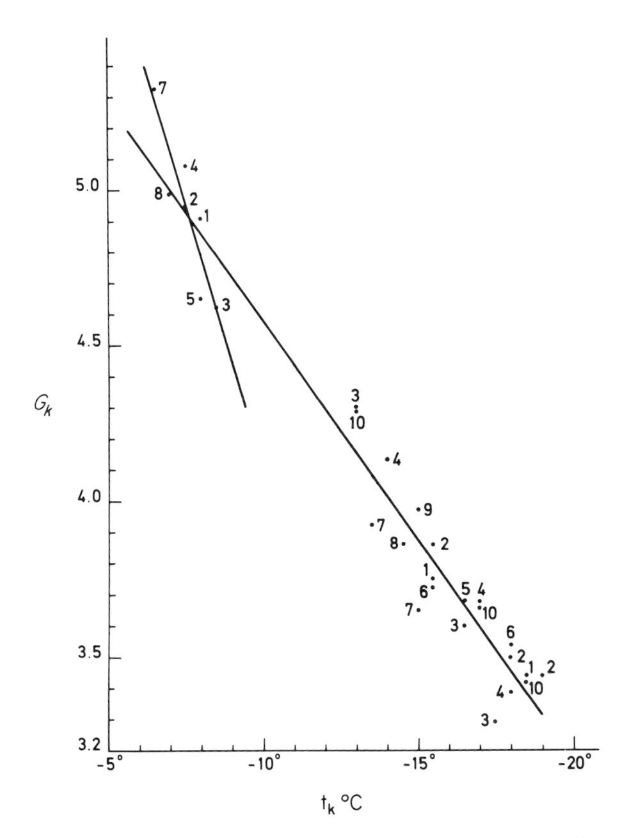

FIG. 9.1. Degrees of cell contraction (G_k) in ten different (numbered) rye varieties at the frost-killing temperatures of the plants (t_k°C). (From Johansson, 1970.)

Johansson (1970) obtained a similar inverse relationship during the hardening of wheat and rye varieties, on the basis of direct calorimetric measurements of ice formation (Fig. 9.1). Freezing tolerance in these grains, therefore, cannot be due to tolerance of cell contraction. It was, however, directly proportional to avoidance of cell contraction.

A similar result was earlier obtained with catalpa cells (Scarth and Levitt, 1937). Calculations revealed that 75% of the cell volume was converted to ice at -6°C when unhardened, whereas due to the low water content, the hardened cells could never attain this amount of cell contraction, even if all their water were frozen. This would indicate that the hardened cells owed their freezing tolerance to avoidance of cell contraction on freezing. Even the very hardy pine had only half the tolerance of cell contraction possessed by the far less hardy wheats (3.0 at -40°C versus

5.0–6.0 at −12° to −15°C, respectively; Table 9.2). This was again due to the much smaller water content of the pine needles (50–60% versus 85–90% of the fresh weight in wheat leaves), leading to avoidance of cell contraction on freezing. In the pine, however, there was a very small but steady rise in tolerance of cell contraction, paralleling the seasonal increase in freezing tolerance (Table 9.2).

All these data can be explained on the basis of Iljin's theory; they indicate that tolerance of cell contraction is either a negligible or a nonexistent factor in freezing tolerance. This does not agree with the many lines of evidence demonstrating a direct relationship between freezing tolerance and tolerance of protoplasmic dehydration (Chapters 7 and 8). These all point to the reverse conclusion—that avoidance of freeze-dehydration (and, therefore, of cell contraction) is the secondary factor and that tolerance of freeze-dehydration is the primary factor (Chapter 7).

Further evidence of this conclusion was obtained in the case of catalpa (Scarth and Levitt, 1937). Although the seasonal hardening led to an enormous increase in avoidance of cell contraction (see above), dehardening occurred in the laboratory without appreciable loss of this avoidance. This can only mean that the loss of freezing tolerance during dehardening was due to a loss of tolerance of freeze-dehydration, without which the marked avoidance of cell contraction was useless.

The explanation of these discrepancies is provided by the above described results with wheat and rye (Table 9.2 and Fig. 9.1). Since there is a straight-line inverse relationship between the freezing tolerances of the ten rye varieties and the degree of cell contraction at the freeze-killing point, it must be concluded that the differences in tolerance between these varieties are solely due to avoidance of freeze-induced cell contraction. This conclusion is in agreement with Åkerman's (1927) earlier direct relation between varietal freezing tolerance of his wheat varieties and sugar content. If, however, the sugars accumulated even to a minor degree in the protoplasm, this would increase not only the avoidance, but also the protoplasmic tolerance of cell contraction, since it would decrease the freeze-dehydration of the protoplasm at any one freezing temperature, and, therefore, the "stiffening" of the protoplasm on dehydration. Since no such increase in tolerance of freeze-dehydration is associated with freezing tolerance of the rye varieties, this indicates that all the sugars are accumulated in the vacuole.

Other evidence leads to the same conclusion. The freeze-dehydration calculated from the freezing points of the rye varieties is, in each case, identical with the value measured calorimetrically. This means that the cell is behaving as an ideal osmotic system. The vacuole may resemble such a system, since it consists of a true solution, but the protoplasm is a colloidal system, and, therefore, cannot be expected to approach an ideal osmotic system. Even the vacuole sap would not be expected to act as an ideal osmotic system when largely freeze-dehydrated. As Johansson points out, this ideal behavior can be explained by two mutually compensating factors. The vacuole solutes consist of a mixture of mainly salts and sugars (chiefly sucrose). The freezing-point lowering per mole decreases with the increase in concentration of salts (due to the decreasing dissociation constant) and increases with increase in concentration of sucrose (due to the binding of water). Consequently, as the vacuole is freeze-dehydrated, the net result is that of an ideal osmotic system.

The tolerance of freeze-dehydration does not merely remain constant in these rye varieties. It actually *decreases* with increase in freezing tolerance. On the other hand, it must be emphasized that this decrease occurs only if measured at the lower freeze-killing temperatures of the more tolerant plants. At this lower temperature, the dehydrated protoplasm would be "stiffer" than at the higher temperature even if at the same degree of dehydration; but it must also be more dehydrated, since the above evidence indicates that essentially *all* the increase in sugar content is in the vacuole. The actual protoplasmic strain produced by the mechanical stress (due to cell contraction) must, of course, depend not only on the severity of the stress but also on the protoplasmic resistance to the stress. It is, therefore, this protoplasmic resistance, and not the cellular freeze-dehydration that comprises the second component of freezing tolerance (Fig. 9.2). On this basis, it should be possible to calculate the relative value of the strain produced at the killing point, if the protoplasmic resistance, as well as the degree of cell contraction is known. Unfortunately, the quantitative effects of temperature and dehydration on the protoplasmic resistance are unknown. On the basis of Johansson's results, it can be assumed empirically, that for every 1°C lowering of the freezing point, the protoplasmic resistance to the mechanical stress is decreased by 0.125 of a unit of the degree of cell contraction. Using this quantity as a correction factor, all his values for the ten rye varieties when corrected for a temperature of -5°C, give a resistance (as measured by the cell contraction at the killing point) of 5.1–5.3. If this same correction factor applies to the wheats, then the variety with the lowest freezing tolerance among the eight investigated by Tumanov and Borodin (Schroder) has a 15–20% greater tolerance of freezing dehydration than the variety with the highest freezing tolerance—Minhardi. If these two were crossed, it might, therefore be possible to select a more hardy wheat from the progeny than either parent.

In the case of the pine leaves, on the other hand, the tolerance of freeze-dehydration during January would be not merely 25% greater than in

September, as indicated by the respective cell contractions at the killing temperatures (3.0 and 2.4, respectively), but nearly 200% greater (6.75 versus 2.4). It is, of course, not likely that the same correction factor can be applied to the protoplasm of such different plants, but the corrected value is undoubtedly closer to the true value than is the uncorrected one.

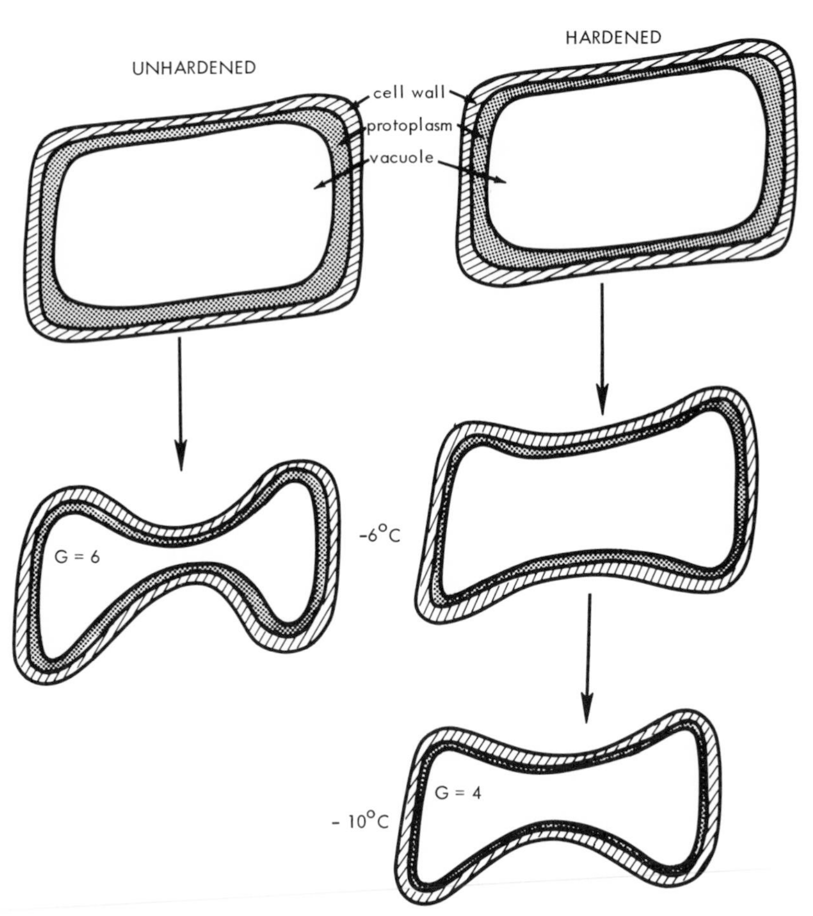

FIG. 9.2. Diagrammatic representation of the relationship of freezing tolerance in Johansson's wheat and rye plants to cell contraction and protoplasmic dehydration. Unhardened cell killed at −6°C, when the cell volume is decreased to one-sixth of the original volume, and the protoplasm is dehydrated to one-half of its original volume. Hardened cell killed at −10°C, although the cell volume is decreased only to one-fourth its original volume. This is because at the lower temperature, the protoplasm is now dehydrated to one-third of its original volume, making it more "brittle" and therefore injured by a smaller mechanical stress due to the smaller degree of cell contraction.

Even in the case of animal cells, freezing tolerance has now been shown to consist of two components. Preliminary results by Meryman's group seemed to indicate that only avoidance of freeze-dehydration was involved, and that injury occurred at about 65% freeze-dehydration (Table 7.5). Following these results, Meryman (1967) compared red blood cells with different levels of artificially induced avoidance of freeze-dehydration. The cells were first equilibrated with a series of concentrations of glycerol (from 0.5 to 2.5 *M*), before gradually freezing them. The higher the concentration of glycerol, the lower was the freezing temperature required to induce 10% hemolysis (ranging from just below −5° to −40°C). Freezing tolerance, therefore, increased due to the increase in avoidance of freeze-dehydration. This was not the only factor, for killing did not occur at the same percentage freeze-dehydration. On the contrary, the greater the freezing tolerance, the larger was the fraction of water left unfrozen at the frost-killing temperature. Again, as in the case of the plant cells, it must be assumed that the protoplasm "stiffness" increased with the drop in frost-killing temperature, and therefore it became more susceptible to injury due to the mechanical stress induced by cell contraction.

It must be concluded, therefore, that there are two basic factors in freeze-dehydration injury: (1) the mechanical stress, which increases with the degree of cell contraction, and (2) the resistance of the protoplasm to this stress (its "modulus of elasticity") which is inversely related to protoplasmic dehydration but which decreases with the drop in temperature. There are thus, basically, two elastic strains due to freeze-dehydration: (1) the protoplasmic strain (or protoplasmic dehydration), and (2) the strain due to the mechanical stress induced by cell contraction. It is the interaction between these two elastic (reversible) strains that produces the plastic (irreversible) injurious strain. The possible nature of the latter will not be considered at this point.

Freezing tolerance, in turn, must depend on two components. (1) Avoidance of freeze-dehydration of the cell as a whole, and therefore of cell contraction. This is a vacuolar property due to the accumulation of solutes. (2) Tolerance of freeze-dehydration. This is a protoplasmic property due to undetermined factors. In the case of the above grains, only the first of these two components increased during hardening. In the case of pine, other grain varieties, cabbage, and many other plants, both components increased during hardening.

This modification of Iljin's concept explains all the above cases that are unexplainable by his original, unmodified theory. Animal cells are killed due to insufficient tolerance of protoplasmic freeze-dehydration. Even the moderate mechanical stress, due to cell contraction in the absence of a rigid cell wall, is sufficient to injure its dehydrated protoplasm. The same is true of plasmolyzed plant cells, when plasmolyzed sufficiently severely to induce a critical dehydration of the protoplasm. Even cells with a low water content and a high cell sap concentration may be killed by a slight freeze, provided that it exceeds the protoplasmic tolerance of freeze-dehydration, even though the cell contraction may be slight.

3. *Other Theories*

The Soviet investigators have adopted Iljin's theory of freezing injury, and have repeatedly attempted to explain freezing resistance on its basis. According to Tumanov's (1967) "new hypothesis" (which is really an extension of Iljin's theory), when the plants prepare for winter, the cell contents change from a sol to a gel, and this renders the cell stable against mechanical deformations and dehydration. A slow drop in freezing point of the solution in the gel lattice protects the cells from ice formation. The transformation also makes the cells more inert chemically. The protective substances act not only as antifreezes but also as plasticizers of the gel. In the most freezing-resistant plants he also assumes fundamental changes in the plasmalemma. Apparently, the solidification is supposed to occur only in the vacuole, whereas the cytoplasm becomes "softer." On the other hand, the gel lattice is supposed to form in the cells as a result of synthesis of water-soluble proteins, which presumably accumulate in the protoplasm. More recently, Tumanov (1969) suggested three kinds of resistance: (1) intracellular freezing, (2) dehydration, and (3) resistance to mechanical deformation. This concept is in general agreement with the above conclusions.

Ewart *et al.* (1953) have proposed that the conversion of starch to sugar during hardening is simply a mechanism of removing potentially injurious starch grains, which could damage the protoplasm physically during the dehydration induced by extracellular freezing. In opposition to this hypothesis, Codaccioni and Le Saint (1966) found that the presence of starch in considerable quantity was no obstacle to freezing tolerance in the case of the unhardened layer of cells surrounding the vascular bundle. These cells possessed a natural freezing tolerance. Similarly, plants grown in continuous light had abundant starch in all the parenchyma cells even when hardened.

Many investigators are now turning to the cell membranes as the seat of freezing injury (see Smith, 1968b; Sakai and Yoshida, 1968a; Modlibowska, 1968; Heber, 1968; Siminovitch *et al.*, 1968). Livne (1968) observed a marked increase in turbidity of soybean phospholipids as a result of freezing and thawing in the presence of 50 m*M* NaCl. Sucrose (0.2 *M*) protected the suspension: the higher the concentration of NaCl, the higher the concentration of sucrose required to retain transparency of the suspension. He, therefore, concludes that these results support a mechanism of freezing injury due to a change in permeability following dehydration. Meryman (1970) has now produced evidence that the salt-concentration injury to blood cells demonstrated by Lovelock is apparently a membrane injury. He concludes that there is a minimum cell volume to which red blood cells can shrink, at which point membrane leakage occurs. The membrane concept will be discussed further in Chapter 10.

CHAPTER 10

MOLECULAR BASIS OF FREEZING INJURY AND TOLERANCE

A. Evidence for a Molecular Basis

On the basis of the above analysis of the available information (Chapters 6–9), the effects of freezing on the plant may be summarized as follows:

(a) Freezing injury consists of two main types: (1) primary direct injury due to intracellular freezing, and (2) secondary freeze-dehydration injury due to extracellular freezing. The former is usually ascribed to a direct physical effect of the intracellular ice on the protoplasm. The latter is most logically explained by an interaction between (i) the severity of the mechanical stress, which decreases with to the degree of cell (mainly vacuolar) contraction, and (ii) the degree of protoplasmic resistance to the mechanical stress, which decreases with the depth of the freezing temperature to which it is exposed, because of the increase in the freeze-induced, protoplasmic dehydration.

(b) Freezing resistance is solely due to avoidance of intracellular freezing in the case of primary direct injury. Freezing tolerance exists only of the secondary freeze-dehydration, and consists of two components: (i) avoidance of freeze-dehydration, and (ii) tolerance of freeze-dehydration.

Both types of stresses are physical, and both lead to physical strains. The primary freezing stress produces a change in state of the protoplasmic water from liquid to solid; the secondary freeze-induced water stress produces a diffusion of protoplasmic water to the external ice loci. It might, therefore, be concluded that the injury is also a physical process, and not due to a biochemical, molecular change. The above two physical strains are elastic and completely reversible. They cannot of themselves be the plastic (irreversible) injurious strain. They must, therefore, produce a secondary plastic strain in the protoplasm which is the immediate cause of the injury. What is the nature of this injurious strain?

1. Primary, Direct Freezing Injury

Protoplasm is so highly structured that it is difficult to conceive of intra-protoplasmal ice formation without some disruption of the organelles, membranes, or simply the groundplasm. If large enough, the expanded crystals must separate the protoplasmic components between which they are found. If this separation disrupts an essential structure, damage must occur. Which components of the cell are likely to be damaged in this way? According to Lyscov and Moshkovsky (1969), DNA degradation may occur by cryolysis in the course of multiple freezing and thawing of solutions. The absence of denaturation was demonstrated by UV spectroscopy, electron microscopy, etc. On the other hand, a decrease in molecular weight was revealed by viscosity and sedimentation measurements. Polydispersity of the DNA rose sharply after cryolysis. The cryolytic degradation was especially effective when a shallow solution was frozen rapidly, producing thin cracks in the ice. Lyscov (1969) suggests that breaks occur in the DNA molecules and that they are situated across the surface of these cracks.

The application of these observations to the living cell would merely be a modern version of the discredited *rupture theory* of freezing injury. Such ice cracks are not likely to occur at the surface of protoplasm which is supported by firm, unruptured cell walls, except in the case of very watery, very thin-walled cells frozen very rapidly (e.g., in fruit and some vegetables; See Chapter 6). Furthermore, it is not likely that rupture of DNA molecules, even if it did occur, could be the cause of the instant death and immediate loss of the semipermeability characteristic of protoplasm frozen intracellularly. On the other hand, ice formation within the protoplasm cannot be fully uniform. Consequently, injury due to intracellular freezing has long been explained by the formation of ice crystals in the protoplasm which protrude into the plasma membrane leaving permeable holes upon thawing. This would explain the need for both rapid freezing (inducing a crystal size too small to protrude through the membrane) and rapid thawing (preventing the growth of these small crystals to a sufficient size for membrane damage) for the survival of intracellularly frozen cells. Some evidence does exist of such hypothetical membrane damage. When chloroplast grana were frozen in liquid nitrogen and thawed, centrifuging precipitated a heavy fraction which was much more severely damaged than the light fraction (Uribe and Jagendorf, 1968). Since the light fragments consisted of swollen grana and the heavy fraction did not swell, it was concluded that some of the thylakoid membranes were rendered highly permeable by freezing and were precipitated as the heavy fraction. Similarly, when *Chlorella* cells were frozen in liquid nitrogen and thawed, they showed increased oxygen consumption for 55 min (Lane and Stiller, 1970). When freezing and thawing were repeated ten times in succession, no oxygen was evolved. This result is best explained by membrane damage during the first freeze, followed later by disruption of metabolism.

Primary direct freezing injury may, therefore, be explained by a purely physical strain resulting in loss of cell semipermeability. This explanation, however, seems unlikely in the case of deep-frozen cells, e.g., in liquid nitrogen. The intraprotoplasmal crystals are formed so rapidly that they are not likely to "grow" into the membranes. On the other hand, the crystals are undoubtedly of a size larger than the individual protein molecules. Therefore, small packets of protein molecules must be dehydrated suddenly and pressed together by the ice crystals that separate them from other protein packets. The close approach of the dehydrated molecules within a packet, together with the pressure exerted on them, would favor protein aggregation. It has even been suggested that the ice crystals act as catalysts (see below). That such aggregation does, indeed, occur has been shown by Luyet and Grell (1936). They observed that freezing split the protoplasm into flocculi easily precipitated by centrifugation. It is, therefore, conceivable that primary direct injury may involve a molecular, biochemical change following the primary physical freezing strain. If the aggregation involved the membrane lipoproteins, this would again result in loss of cell semipermeability.

2. Secondary Freeze-Dehydration Injury

Any molecular changes responsible for primary direct freezing injury may be produced by way of sudden physical changes—physical rupture of a molecule, physical compression of dehydrated molecules, physical perforation of a membrane by crystals, etc. In the case of the secondary freeze-dehydration, on the other hand, the freezing process is gradual, and does not give rise to the above sudden physical changes, except in the case of Iljin's sudden thawing injury which tears the protoplasm from the cell wall. It has already been shown that this is not normally the cause of freezing injury. Any molecular changes must, therefore, be produced more directly by chemical rather than physical action. It has long been known, however, that the temperature coefficient (Q_{10}) of chemical reactions is much larger than that of physical processes. Therefore, the lowering of the plant's temperature to its frost-killing point would be expected to slow down all chemical reactions so much more than it slows down the physical process of diffusion, that freeze-dehydration should occur much more rapidly than any accompanying chemical reactions. Since freezing injury can be essentially instantaneous, this would seem to exclude chemical reactions as a possible cause of the injury.

In agreement with this conclusion, when the lichen *Parmelia physodes* was exposed for 48 hr to an atmosphere containing $^{14}CO_2$, it was not possible to detect labeling of the photosynthetic pigments at −12°C, although photosynthesis still occurred (Godnev *et al.*, 1966). The ^{14}C was incorporated into chlorophyll *a* and *b* only down to −2°C, and into carotene down to −5°C. In some cases, in fact, there is a secondary low temperature-induced deceleration of chemical reactions in excess of that predicted from the Arrhenius equation. Enzymes in frozen, aqueous systems are, in general, less active than in the liquid solutions at the same temperature. This was true of the decrease in invertase activity with a drop in temperature between 12° and −22°C (Lund *et al.*, 1969).

It is not surprising, therefore, that none of the above theories attempts to explain freezing injury on a chemical basis. Even the salt-precipitation theory is basically thought of as inducing injury due to an excessive dehydration of the proteins, leading to a physical aggregation or adsorption.

New information, however, encourages a reexamination of the chemical possibilities. Many (mainly nonenzymatic, bimolecular) reactions are now known to show an anomalous acceleration as a result of freezing (Table 10.1), by as much as 5–1000 times (Grant and Alburn, 1967; Grant, 1969) in spite of the low temperature. Even in the case of enzymatically controlled reactions, there are some exceptions to the above described deceleration. Thus, small increases were observed in frozen systems relative to otherwise identical supercooled systems (e.g., at −4°C) in the rate of hydrolysis of sucrose by invertase (Tong and Pincock 1969). Even nonenzymatic reactions do, however, occur in frozen cells, e.g., the dechlorination of DDT in frozen avian blood stored at −20°C for 9 weeks, supposedly due to a redox reaction (Echobichon and Saschenbrecker, 1967).

More than one factor may be involved in such anomalous accelerations. In the case of enzymatically controlled reactions, the decrease in catalytic velocity with lowering of temperature (expected on thermodynamic grounds) may be partially and sometimes fully offset by an increase in enzyme–substrate affinity (Somero and Hochachka, 1969); but this cannot explain an actual net acceleration. Four other explanations have been proposed (Lund *et al.*, 1969), two of which have received the most attention. According to one interpretation (Grant, 1966) it is due to the ice crystals acting as catalysts. This could explain the greater injury due to intracellular than to extracellular freezing, since only in the former do the ice crystals contact the reactants within the protoplasm.

The second-order kinetics, with up to 1000 times more rapid rates than

TABLE 10.1 Comparative Rates of Nonenzymic Reactions in Frozen and Unfrozen Aqueous Systems[a,b]

Type of reaction	Substrate	Catalyst	Rate of reaction[c,d]	
			Unfrozen	Frozen
Spontaneous hydrolysis	Acetic anhydride	None	30X (+5°C)	X (−10°C)
	β-Propiolactone	None	Considerable	O (5 hr at −10°)
Acid-catalyzed hydrolysis	Acetic anhydride	HCl	X (5°)	3–27 X (−10°)
Base-catalyzed hydrolysis	Acetic anhydride	Acetate	X (5°)	2.7 X (−10°)
Imidazole-catalyzed hydrolysis	β-Propiolactone	Imidazole	X (5°)	7X (−10°)
	Penicillin G (pH 7.7)	Imidazole	X (0°)	18X (−8°); 16X (−18°); 5X (−28°); 1X (−78°)
Acid-catalyzed chemical dehydration	5-Hydro-6-hydroxy-deoxyuridine	HCl	X (30°) O (22°)	12X (−10°) Rapid (−20°)
Oxidation	Ascorbic acid (pH 5.5)	None	X (1°)	3X (−11°)
	Ascorbic acid (pH 5.5)	$CuCl_2$	X (1°)	3.5X (−11°)
Reduction	Potassium ferricyanide	KCN	Stable above freezing	Complete conversion to ferrocyanide in: 7 hr (−12°); 2 hr (−25°); 107 sec (−78°)
Catalyzed decomposition of peroxide	Hydrogen peroxide (pH 7.2)	$FeCl_3$	X (1°)	13–28X (−11°)
	Hydrogen peroxide (pH 7.2)	$CuCl_2$	Stable (1°)	Quite rapid (−11°)
Hydroxy-aminolysis	Amino acid methyl esters (pH 7.2–7.7)	None	X (1°)	1.7–5.5X (−18°)
	Amides (pH 7.0)	Buffer	X (0°)	1.3–7X (−18°)

[a] Substrate concentration ranged from 0.0001 to 0.02 *M*.
[b] From Fennema, 1966.
[c] X = the lower rate of the two.
[d] O = undetectable.

in supercooled liquid solutions, led Pincock and Kiovsky (1966) to conclude that it is a concentration effect. In the case of freeze-dehydration injury, only the latter explanation can apply, since ice forms extracellularly, and the dehydrated cell contents have no contact with the crystals. On the basis of Pincock and Kiovsky's interpretation, at least some of the chemical reactions that occur in the protoplasm may be expected to accelerate as a result of freeze-dehydration. Obviously, not all the reactions will be equally affected, and the net result may be some completely new reactions. Which of the protoplasmic substances are most likely to be involved?

There are three major groups of organic substances essential to the life of protoplasm, and conceivably injuriously altered by freezing: nucleic acids, lipids, and proteins. Unfortunately, little is known about the effects of freezing on the first two of these chemical groups. Shikama (1965) has shown that freezing and thawing do not break down deoxyribonucleic acid (DNA). Similarly, no denaturation of DNA on freezing could be detected by Lyscov (1969). Presumably, RNA would also be unaffected, since its molecular structure is so similar to that of DNA. Although the nucleic acids of freezing tolerant and intolerant plants have not been compared with respect to stability on freezing, Koffler *et al.* (1957) failed to find any difference in stabilities of either RNA or DNA obtained from thermophiles and mesophiles. Finally, it does not seem likely that any change in nucleic acid structure could cause the immediate, complete loss of semipermeability that accompanies freezing death, since nucleic acids are not components of the plasma membrane. Lipids, on the other hand, are essential components of the plasma membrane, and a change in them could easily lead to a rapid loss in semipermeability. Unfortunately, there is very little information as to the effect of freezing on lipids; but since they are not hydrophilic, they contain no water which could be frozen *in situ* or removed to extracellular ice loci. Furthermore, even ice crystals formed outside a lipid layer do not penetrate the layer, as mentioned previously (Chapter 6).

Many proteins, on the other hand, are denatured, aggregated, or dissociated into monomeric units on freezing, and, in some cases, even by cooling in the absence of freezing. Proof of subunit dissociation and reassociation on freezing was produced by Markert (1965). The enzyme L-amino acid oxidase (from snake venom) is inactivated between $-5°$ and $-60°C$, with a maximum at $-20°C$ (Curtis *et al.*, 1968). When slices of mouse liver were frozen slowly (1°C/min) and cooled down to $-25°C$, about half of the activity of succinate cytochrome *c* reductase was lost during one freeze–thaw cycle, independently of the temperature between 0° and $-16°C$, or of the rate of freezing and thawing (Fishbein and Stowell, 1968). The damage occurred during ice crystallization, particularly during the final stages. It was suggested that the damage was probably dependent on supramolecular organization of the enzyme complex, since most monomolecular enzymes are not damaged by freezing.

On the basis of these and many other results, it is safe to assume that the most likely molecular effects of freezing are on the proteins. Many investigators have, therefore, postulated that freezing injury is due to protein changes induced by the freezing process (see Chapter 9). Until recently, no attempt has been made to suggest what the specific chemical changes are, other than something involving denaturation. The only molecular concept proposed to explain freezing injury on the basis of specific chemical changes induced by freezing is the SH hypothesis.

B. The SH Hypothesis

1. The Concept

On the basis of a series of investigations (see Chapter 8), the sulfhydryl-disulfide ($SH \rightleftharpoons SS$) or more simply, the SH hypothesis of freezing injury and resistance was proposed (Levitt, 1962). Injury was assumed to result from the following series of three steps:

(a) Protein molecules approach each other as the protoplasm is progressively freeze-dehydrated.

(b) When they are sufficiently close, chemical combination occurs between the S atoms of adjacent protein molecules, (1) by oxidation of 2 SH groups, or (2) by $SH \rightleftharpoons SS$ interchange, as follows:

$$2\ RSH + \tfrac{1}{2}\ O_2 \rightleftharpoons RSSR + H_2O \qquad (1)$$

$$\begin{matrix} R_1\text{—}S \\ |\quad\ \ | \\ R_2\text{—}S \end{matrix} + R_3SH \rightleftharpoons HSR_1R_2SSR_3 \qquad (2)$$

(c) The protein molecules undergo a conformational change due to this aggregation leading to injury or death. The conformational change could occur during the freeze-dehydration or during the thawing (Fig. 10.1). Freezing tolerance would be due to prevention of intermolecular SS bonding.

2. Supporting Evidence

The $SH \rightleftharpoons SS$ hypothesis was based on the following evidence. (i) An increase in protein SS was found to accompany freezing injury; no such

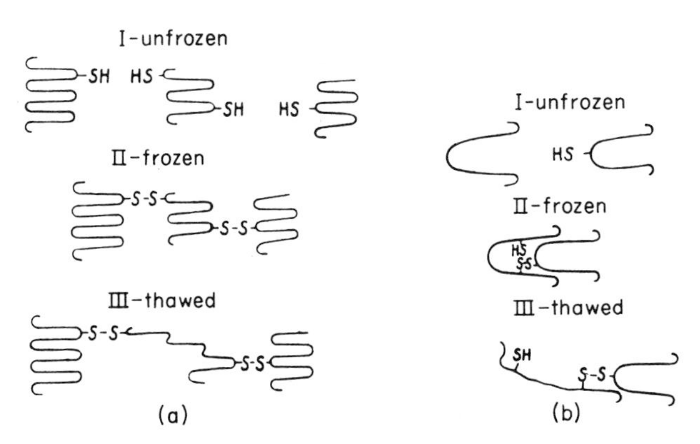

FIG. 10.1. Postulated mechanisms of protein unfolding due to intermolecular SS formation during freezing (from Levitt, 1962).

increase occurred when the freezing failed to injure (Fig. 10.2). More recent measurements have revealed that these early SS values are probably exaggerated, since they involved unfolding of the molecule and, therefore, included the previously masked SH groups. The relative relationship has, however, been confirmed qualitatively by labeling the proteins with ^{14}C-*p*-chloromercuribenzoate and separating them by disc electrophoresis (Morton, 1969). This evidence indicates that only a small fraction of the SH

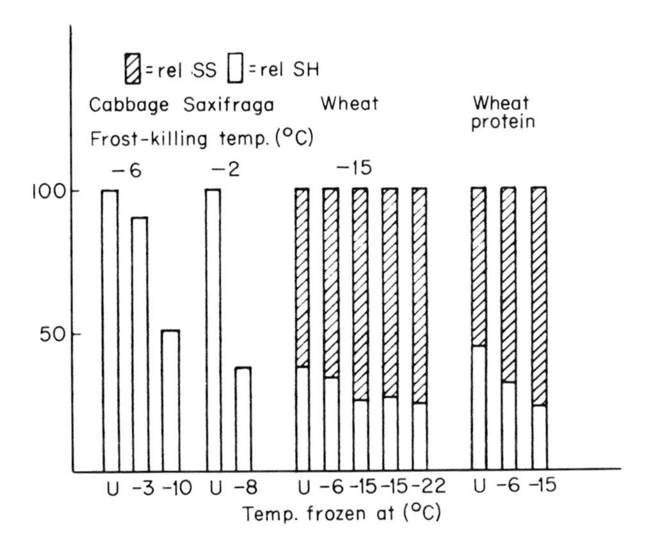

FIG. 10.2. Increase in protein SS as a result of freezing injury (from Levitt, 1967c). U, Unfrozen; ▨ = rel. SS; □ = rel. SH.

groups of the soluble proteins is converted to intermolecular SS bonds on freezing. Reliable measurements on the insoluble proteins have not yet been obtained. (ii) The SH content of the homogenate of a plant (and of the proteins in it) was proportional to freezing tolerance, and this was due to a greater resistance of the tolerant plants to SH oxidation during homogenization (Chapter 8). (iii) An increase in activity of the GSH oxidizing system also paralleled the increase in freezing tolerance, due at least partly to an increase in ascorbic acid. This system could conceivably protect the protein SH groups by scavenging free oxygen, H_2O_2, or other oxidizing agents. For example, glutathione peroxidase can compete with catalase for the common substrate H_2O_2 at low concentrations under physiological conditions (Flohe and Brand, 1969). This could be an important protective device since H_2O_2 is capable of oxidizing protein SH groups to intermolecular SS bonds (Khan *et al.*, 1968). This does not mean, of course, that the GSSG form would persist in the plant. On the contrary, the GSH $\rightleftharpoons$ GSSG system would be effective only so long as respiratory or photosynthetic hydrogens are continuously available to reduce the GSSG as soon as it is formed.

3. Explanation of the Known Facts

The usefulness of an hypothesis can be measured by its ability to explain the known facts and its predictability. The SH hypothesis has been shown to fulfill these two criteria (Levitt, 1967a).

a. The Four Moments of Injury

Injury at the moment of freezing would be due to protein aggregation and unfolding (Fig. 10.1). Injury while maintained in the frozen state could increase due to slow, small rearrangements of the protein molecules, leading to more contacts between SH groups of adjacent molecules over long periods of time. It is also conceivable that injury at the moment of freezing is due to rapid SH $\rightleftharpoons$ SS interchanges; the slower injury while frozen is due to the slower oxidations of SH $\rightleftharpoons$ SS (Fig. 10.1). Injury during the moment of thawing is also illustrated in Fig. 10.1. Injury during the postthawing moment may be due to the increased rate of reaction between the SH groups of adjacent molecules at the higher postthawing temperature and before the molecules are pushed apart by the reimbibition of water. This also explains the well-known increase in injury with increase in postthawing temperature.

b. The Danger Zone

The ability of freezing-intolerant cells to survive freezing below the danger zone (0° to −30° or −40°C), provided that they can be safely carried through this zone, would be due to the cessation or markedly decreased rate of the chemical reactions (e.g., the oxidation of SH ⇌ SS) at such low temperatures. Peroxidation of guaiacol has been detected down to −43° to −45°C in the absence of free water (Siegel *et al.*, 1969). This agrees well with the approximate lower limit of the danger zone. Ultrarapid freezing or freezing in the presence of a cryoprotectant would permit passage through this zone quickly enough, or while keeping the protein molecules far enough apart to prevent intermolecular SS formation.

c. Salt Injury

The relationship of freezing injury to external salt concentration that has been suggested for red blood cells may be due to the increased rate of reaction of SH groups in the presence of salts which has been observed in β-amylase (Warren and Cheatum, 1966).

d. Sucrose Protection

Several investigators (Tumanov and Trunova, 1957; Heber and Santarius, 1964; Steponkus and Lanphear, 1967b) have reported a greater effectiveness of sucrose than of other solutes against freezing injury. Although it is no more effective than other solutes in preventing intermolecular SS formation in Thiogel (see below) during freezing, it is far more effective in preventing the much slower process in unfrozen Thiogel (Andrews and Levitt, 1967).

e. Inverse Relationship between Growth and Tolerance

It has long been known that rapid growth is associated with high SH content. More recent investigations of plant tissue have confirmed this fact (Szalai, 1959; Spragg *et al.*, 1962; Pilet and Dubois, 1968; Koblitz *et al.*, 1967; Klein and Edsall, 1966). High SH values are also associated with cancerous tissues (Wiman, 1964). The larger the number of SH groups per unit protein, the more difficult it must be to prevent their oxidation to intermolecular SS bonds on freezing. A distinction must be made here between (1) a high SH content at temperatures supporting active growth, and (2) an ability to retain all or nearly all the SH groups in the reduced state *during freezing*. The latter property of plant cells will be discussed below.

f. THE DECREASE IN FREEZING TOLERANCE DUE TO NO_3 FERTILIZATION

The chloroplast enzyme system that mediates the reduction of NO_3 involves NADPH as a cofactor (Wessels, 1966). The reduction of the absorbed NO_3 might, therefore, favor the oxidation of protein SH groups on freezing, by competing with the NADPH-dependent GSH system that protects the protein SH groups.

The ability of the SH hypothesis to explain the mechanism of freezing tolerance will be discussed below.

4. Tests of the Hypothesis with Model Systems

The hypothesis must first be tested with pure proteins, in order to determine whether or not freeze-dehydration actually does induce intermolecular SS bonding of protein molecules. Many pertinent results are already available in the literature on proteins. Table 10.2 lists a number of enzymes inactivated by freezing. They are all SH proteins. Among these, lipoyl dehydrogenase is a particularly clear case. In the reduced (SH) form it is

TABLE 10.2 SH-CONTAINING ENZYMES INACTIVATED BY FREEZING[a]

Enzyme	Molecular weight	SH (groups/ molecule)	SS (groups/ molecule)	Protectant against cryoin activation
1. Lactic dehydrogenase	170,000	14		Glutathione, mercapto-ethanol
2. Triosephosphate dehydrogenase	100,000	11–12	0	Mercaptoethanol
3. Glutamic dehydro-genase	1,000,000	90–120	0	Mercaptoethanol
4. Lipoyl dehydrogenase	100,000	8–12	2	Oxidation: SH → SS
5. Catalase	248,000	(Altered by SH regents)		
6. Myosin	594,000	45		
7. 17β-Hydroxy steroid dehydrogenase		(Inactivated by SH reagents)		
8. Succinate dehydro-genase	200,000	(Inactivated by SH reagents)		
9. Phosphoglucomutase	74,000	7.5		

[a] From Levitt, 1966a.

inactivated by freezing; in the oxidized (SS) form it is not inactivated by freezing (Massey *et al.*, 1962). Urease, on the other hand, is a SH protein, yet is extracted from hydrated Jack bean seeds after freezing at −20°C. Successful extraction of the active enzyme requires large quantities of mercaptoethanol (Sehgal and Naylor, 1966), which would protect or regenerate the SH groups. Phosphorylase, however, is another SH protein and can be prepared from freeze-dried potato juice without the addition of a SH protectant (Baum and Gilbert, 1953). Similarly, Ullrich and Heber (1961) were unable to detect any loss in activity of five SH proteins of wheat leaves, as a result of freezing. Many of the most stable proteins, on the other hand, contain SS but no SH groups (Table 12.8). Cytochrome oxidase is a disulfide enzyme and is prepared in the active form by extraction of freeze-dried material (Cooperstein, 1963). It and two other SS proteins (RNase and BSA) can be lyophilized without loss of activity. Myokinase, an SH enzyme, can also be lyophilized without loss of activity, but this has been explained by the presence of stabilizers (Noda, 1958). Ullrich and Heber's negative results may be explained in the same way, since they did not purify the proteins, but froze them in the presence of the rest of the leaf material. A relationship between ATPase inactivation on freezing and SH content has been found in the case of purified myofibrillar proteins of chicken pectoralis (Khan *et al.*, 1968). When frozen at −30°C, the proteins showed a loss of SH content, solubility, ATPase activity, and water-binding capacity. Blocking the SH groups with PCMB (p-chloromercuribenzoate) also inactivated the ATPase, but failed to alter its solubility or water-binding capacity. Oxidation of the SH groups to SS by H_2O_2 had the same effect as freezing. The logical explanation is that freezing induced an aggregation of the proteins by intermolecular SS formation, since ATPase is an SH enzyme (Kuokol *et al.*, 1967), requiring thiol for activity (Heber, 1967). It has, of course, long been known that the extraction of active enzymes is often dependent on the addition of thiols to the extraction medium (Anderson and Rowan, 1967). Obviously, any tying up of the SH group associated with the active site (e.g., by intermolecular SS formation) would certainly inactivate the ATPase. In the case of three dehydrogenases and one transferase, it has even been possible to prevent their inactivation during freezing, by the addition of a small amount of thiol (Table 10.2).

It is questionable, however, whether any of the above experiments involves a true freeze-dehydration, as it occurs in the case of living cells. Protein solutions frozen in test tubes may undergo freeze-dehydration if frozen at high temperatures (e.g., −5°C). The ice then forms first adjacent to the colder tube wall, and as more and more ice crystallizes externally, the solution gradually becomes more and more concentrated in the warmer center of the test tube (Goodin and Levitt, 1970). The freezing of protein solutions, however, is usually performed at much lower temperatures, and the ice crystals probably form rapidly, trapping the protein molecules between them. In the experiments described above with the myofibrillar proteins, which present the most clear-cut evidence of intermolecular SS formation on freezing, freeze-dehydration may have occurred as a secondary process; for although frozen at −30°C, they were subsequently stored at −5°C for 10 weeks, allowing the larger ice crystals to grow at the expense of the smaller crystals, and presumably leaving pockets of concentrated protein between the large crystals.

A true test of the hypothesis obviously requires a model protein system that can be freeze-dehydrated in the same way as the protoplasm of extracellularly frozen cells. This can be achieved with any protein that forms an aqueous gel, by inoculating the slightly supercooled gel surface with an ice crystal. If the temperature is maintained slightly below the freezing point, the ice forms slowly externally to the gel. When this ice sheet is removed, the remaining unfrozen gel is correspondingly thinner due to the freeze-dehydration. Gelatin itself contains no SH groups; but a thiolated form of gelatin, called Thiogel is available. Unlike ordinary gelatin, the melting point of a gel made with Thiogel rises with intermolecular SS formation. The bond formation can, therefore, be followed simply by measuring the

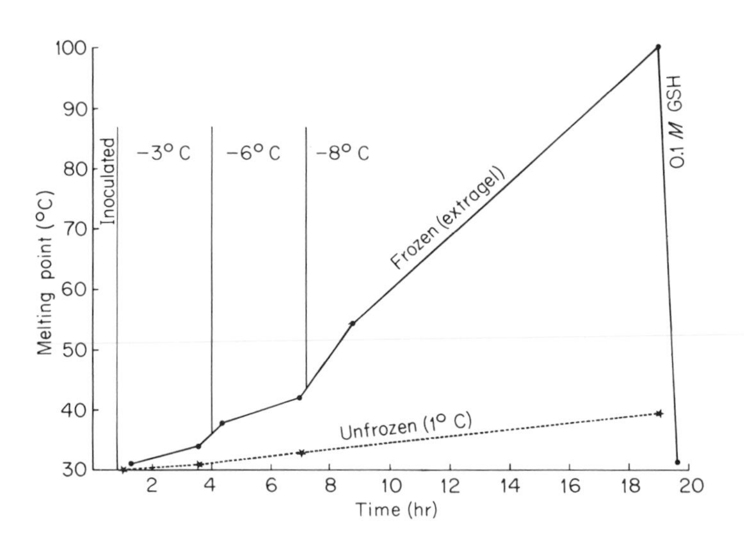

FIG. 10.3. Acceleration of intermolecular SS formation (shown by rise in melting point) in Thiogel on freezing (from Levitt, 1965).

TABLE 10.3 THIOGEL SQUARES STIRRED CONTINUOUSLY IN 0.1 *M* PHOSPHATE BUFFER (pH 7.3) AND IN SIMILAR BUFFER CONTAINING 36% POLYVINYLPYRROLIDONE[a]

	0.1 *M* Buffer		0.1 *M* Buffer containing 36% PVP	
Time (hr)	Thiogel dimensions (mm)	Melting point (°C)	Thiogel dimensions (mm)	Melting point (°C)
0		29		29
1	11 × 11	30	8 × 8	100
4½	10 × 12	34		
7½		39		
12½		90–100		
20	½ hr in 0.1 M GSH			31

[a] From Levitt, 1965.

melting point of the gel. When the gel was freeze-dehydrated (i.e., frozen extracellularly) its melting point rose much more rapidly than when kept unfrozen at 1°C (Fig. 10.3). That this rise in melting point was, indeed, due to intermolecular SS bonding was proved by adding GSH, which reduced the SS groups and returned the melting point to its original value (Fig. 10.3). The hypothesis, therefore, holds for the model system—freeze-dehydration does accelerate intermolecular SS formation. That the dehydration, rather than the freezing *per se*, is the initiating factor was established by dehydrating the Thiogel osmotically at 1°C. The intermolecular SS bonding was accelerated even more than by freeze-dehydration (Table 10.3).

The model system can also be used as a test of the tolerance mechanism. Freezing tolerance has two components: avoidance of freeze-dehydration and tolerance of freeze-dehydration. The first of these is easily tested with the model system. The addition of solutes to the gel confers avoidance by decreasing the degree of freeze-dehydration. It also retarded the intermolecular SS bonding on freeze-dehydration (Fig. 10.4).

Thiogel, however, is a denatured protein and, therefore, cannot strictly test the SH hypothesis, which proposes that denaturation follows intermolecular SS bonding. A native protein, bovine serum albumin (BSA) was, therefore, investigated. Although it does not form a gel, it can be freeze-dehydrated in a dialyzing sac by inoculating the external surface of the sac (Goodin and Levitt, 1970) or more simply by inoculating a solution in a

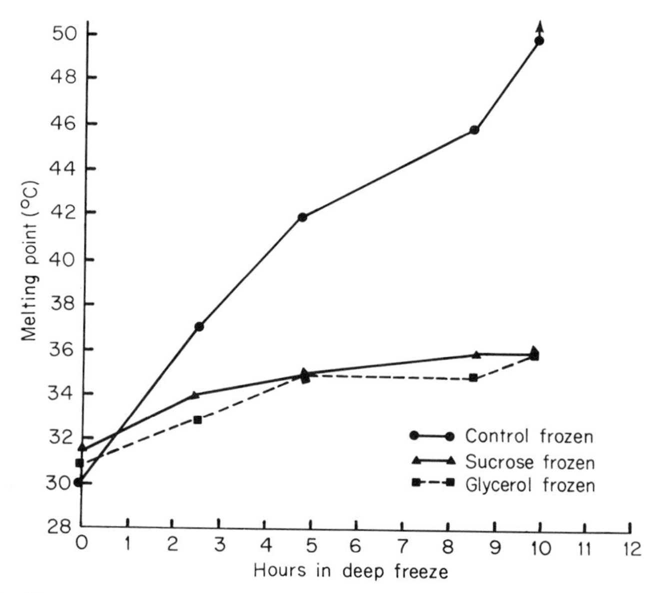

FIG. 10.4. Protection of Thiogel by sucrose and glycerol against intermolecular SS formation (measured by rise in melting point) on freezing. (From Andrews and Levitt, 1967.) ●——●, Control frozen; ▲——▲, sucrose frozen; ■——■, glycerol frozen.

test tube at a temperature just below its freezing point, allowing the freezing to occur from outside inward. Native BSA contains 17 SS groups per molecule and no SH groups. Freeze-dehydration of this native protein failed to induce aggregation detectable by precipitation on centrifuging. When, however, the protein was denatured in 6 *M* urea, and the SS groups were reduced by addition of thiol, the subsequently purified protein aggregated (precipitated) completely on freeze-dehydration. Measurements revealed the formation of SS groups and the aggregate was completely resolubilized in the presence of a thiol and urea, although not by urea alone.

If the SH groups of the denatured and reduced BSA were tied up by combination with a SH reagent (N-ethylmaleimide or NEM), freeze-dehydration in some cases failed to precipitate the protein (Goodin and Levitt, 1970). In other cases, precipitation did occur, particularly if the protein was warmed before rehydration. This was apparently due to intermolecular apolar bonding, since it was reversed by 6 *M* urea. This may, perhaps, explain postthawing injury in plants warmed rapidly after thawing. However, the NEM added seventeen extra hydrophobic groups per protein molecule, and, therefore, these results are not necessarily typical of an unaltered protein.

The results with BSA support the SH hypothesis; for the native SS protein did not aggregate on freeze-dehydration, but the SH protein did. This agrees with the above results with lipoyl dehydrogenase (Massey *et al.*, 1962).

The two model systems clearly establish one part of the SH hypothesis. Freeze-dehydration of a SH protein does induce intermolecular SS bonding. In both cases, however, this involved a denatured protein. Unfortunately, no test has yet been made on a native SH protein. Brandts (1967), however, has shown both by a detailed thermodynamic analysis and by direct measurements that proteins denature (unfold) reversibly at low temperatures in the absence of freezing (e.g., at 0° to 10°C), due to a weakening of the hydrophobic bonds responsible for the tertiary, folded structure. This reversible denaturation may then be followed by aggregation, which converts it to an irreversible form:

$$N \underset{\text{normal T}}{\overset{\text{low T}}{\rightleftharpoons}} D \xrightarrow{\text{low T}} A$$

where N = native protein; D = denatured protein; A = aggregated protein.

In the case of cold labile enzymes, (e.g., ovalbumin), this denaturation is quickly followed by aggregation and, therefore, becomes irreversible. In the case of lysozyme, it appears that only part of it becomes irreversible; if stored in the cold (4°C for 18 hr) then warmed (26° for 30 min) before use, this did not affect its ability to catalyze the lysis of cells of *Micrococcus lysodeikticus,* or the hydrolysis of chitin oligosaccharides (Zehavi, 1969). The cold treatment did, however, markedly lower the transfer function of the enzyme (of chitin oligosaccharides to a glucose acceptor). This was interpreted as a localized conformational change. Results with other enzymes are also explainable by Brandts' scheme. Dimerization of the subunits of D-amino acid oxidase proceeds more favorably above 14° than below 12°C (Henn and Ackers, 1969). Presumably the bonding is hydrophobic in nature. Similarly, inactivation of carboxydismutase by urea was more marked at 0° than at 25°C (Akazawa and Sugiyama, 1969). The enzyme is split reversibly into large and small subunits. Reversible inactivation of pyruvate carboxylase from chicken liver mitochondria occurs on exposure to low temperature (Irias *et al.*, 1969). This is apparently due to dissociation of a tetrameric form of the enzyme into four protomers. On longer exposure to low temperatures, however, further changes apparently occur in the protomers, which prevent reactivation by rewarming, although association to tetramers or aggregates may still occur. Similar effects can be produced by urea at room temperature. Acetyl-CoA affords partial or complete protection against the inactivation and dissociation by cold or urea. It is also protected by ATP and several related substances, as well as many other substances, e.g., inorganic phosphate, methanol, polyols, high concentrations of KCl, and oxaloacetate. A Mg-activated, cold-labile ATPase from beef heart mitochondria is normally bound to a component which prevents cold inactivation (Selwyn and Chappell, 1962).

From all these results, it must be concluded that the cold-induced, reversible denaturation described by Brandts occurs in many proteins, and that it may, indeed, be converted to an irreversible aggregation. On the other hand, it is also apparent that the denaturation and subsequent aggregation may be prevented by the presence of protective substances.

A further observation with a model protein may help to explain the effect of the mechanical stress due to cell contraction. When the fibrous protein keratin is subjected to tension, the mechanical stress greatly facilitates the formation of new SS bonds between previously existing ones (Feughelmann, 1966). In fact, the temperature at which this exchange reaction occurs is lowered from 100° to 20°C by the tension. Subjection of the plant cell to a similar mechanical stress due to freeze-dehydration may be expected to produce the same effect.

On the basis of all these experiments with model proteins, the original SH hypothesis must be modified. Injury would result from the following chain of events:

(a) Low temperature denatures proteins reversibly, unmasking reactive SH groups.

(b) Freeze-dehydration: (i) removes vacuole water, producing cell contraction which applies stress on the protoplasmic proteins, activating their SS bonds, and (ii) removes protoplasmic water, decreasing the distance between the reversibly denatured protein molecules.

(c) Intermolecular bonding due to SH oxidation or SH $\rightleftharpoons$ SS or SS $\rightleftharpoons$ SS interchange, aggregates the proteins irreversibly, killing the cell.

It must be emphasized that the SH hypothesis does not exclude the possibility that bonds other than SS bridges may induce aggregation. H bonds, however, are not likely to be involved due to their strong bonding to water molecules. Hydrophobic bonds are not likely to occur, because they are weakened by the low temperature, and still further by freezing, since this removes the water which is the real cause of hydrophobic bonding by repelling the hydrophobic groups. However, the results of Goodin and Levitt (1970) suggest that, on thawing, hydrophobic groups may lead to aggregation of the proteins before reimbibition of the thaw-water. When embedded in a hydrophobic environment, electrostatic and H bonds are stronger (Koshland and Kirtley, 1966) and, therefore, may help to strengthen these weak bonds. The strongest of all are the SS bonds, and these may confer an irreversibility to otherwise reversible bonding.

5. Tests with Living Cells

One part of the SH hypothesis is now firmly established. Denatured SH proteins are induced to form intermolecular SS bonds by freeze-dehydration. It has also been shown that SS groups form in cells injured by dehydration. It remains to be proved that this is the cause and not the result of freezing injury. Furthermore, perhaps, the plant proteins do not denature reversibly before freezing, as they do in the case of Brandts' model proteins. He has, in fact, demonstrated that some proteins denature at so low a temperature that their solutions must be maintained in the unfrozen state by the addition of an antifreeze, in order to induce reversible denaturation (Brandts *et al.*, 1970). This is undoubtedly why many enzymes are neither denatured nor inactivated by freezing. On the contrary, many of them are preserved over long periods of time in the frozen state. Furthermore, even enzymes that are inactivated by freezing in the pure state will not necessarily be inactivated *in vivo*. It is, therefore, necessary to determine whether any plant proteins are, indeed, inactivated when plants are killed by freezing.

In spite of their earlier evidence supporting Gorke's "frost-precipitation" of proteins, Ullrich and Heber (1961) found that five soluble enzymes which were not sedimented by centrifuging, were neither coagulated nor inactivated by freezing. Even some of the particulate enzymes showed no change in activity after freezing. McCown *et al.* (1969b) were also unable to detect any decrease in enzyme activity in carnation tissue killed by freezing. It is, of course, conceivable that the injury may be due to inactivation of only one or a few key enzymes. In the case of seeds, some proteins have been precipitated by freezing and have been called cryoproteins (Ghetie and Buzila, 1964a). These were apparently concentrated in the cotyledons and were bound to 2–10% lipid (Ghetie and Buzila, 1964b). They were, perhaps, storage proteins and therefore presumably not present in other leaves, and not active as enzymes. Enzymes may also be cryoinactivated. Mitochondrial MDH (malic dehydrogenase) is unstable when frozen (Nakanish *et al.*, 1969) and pectin esterase from both coleus and bean was denatured by freezing and thawing (Lamotte *et al.*, 1969). In the case of mouse liver, the succinate cytochrome *c* reductase complex was particularly sensitive to and suitable for evaluating freeze–thaw injury (Fishbein and Stowell, 1968). Hanafusa (1967) detected marked decreases in enzymatic activity and conformational parameters of fibrous proteins (myosin A and B; H- and L-meromyosin) as a result of freezing (at −30°, −79°, or −196°C), but no change in the conformational parameters of globular proteins (G actin and catalase) in spite of the decrease in enzymatic activity. He concludes that freeze-thawing causes a partial unfolding of the helical structure in the fibrous proteins (25% in myosin A), but not in the globular proteins.

Nevertheless, all the work with plants indicates that in spite of *in vitro* inactivation of some soluble enzymes by freezing, *in vivo* freezing fails to inactivate them. It is, of course, conceivable that a key enzyme may yet be found in plants that is inactivated by *in vivo* freezing. In view of the mass of negative evidence this seems unlikely. Furthermore, enzyme inactivation can injure a plant only by a disturbance in its metabolism. Such a disturbance would occur very slowly at the low temperature inducing freezing injury. These negative results (especially Hanafusa's with globular proteins in contrast to his positive results with fibrous proteins) can be explained by Gaff's (1966) evidence that soluble proteins fail to increase in SS content when the plant is killed by desiccation, but that the insoluble proteins show a significant increase. If the intermolecular bonding of these insoluble proteins is the cause, rather than the result of injury, it should be possible to prevent injury by preventing the intermolecular SS bonding. If SS formation is the result of injury, this method should fail to prevent injury.

Attempts to prevent injury in this way, were first made by adding GSH to sections of tissue before freezing (Levitt and Hasman, 1964). No change in freezing tolerance occurred. Unfortunately, it was not known whether or not the GSH entered the cells. Furthermore, even if it did enter, it is known that some applied thiols may be immediately oxidized on entering the living cell, others may not (Eldjarn, 1965). It was, therefore, decided to use a thiol that (1) is known to penetrate the cell readily, and (2) is not a normal constituent of cells and, therefore, is not rapidly oxidized by the cell's enzymes. This second attempt succeeded in altering the freezing tolerance of the cells, but in the unexpected direction. Sections of unhardened tissue are protected against freezing injury by freezing in sugar solutions (see Chapter 6). When treated with the thiol mercaptoethanol, this protection is removed (Krull, 1966). Similarly, SH reagents which tie up SH groups, actually lowered the freezing tolerance of cabbage cells (Levitt, 1969a). The first of these negative results cannot be fully understood until the nature of the sugar protection is known (see below). They are explainable, however, on the basis of known SH reactions. Too low a concentration of small-molecule thiols to keep the protein SH groups reduced may serve to trigger SH $\rightleftharpoons$ SS interchange reactions, thus increasing the very process against which it is being used to protect the plant. Too high a concentration may split essential intramolecular SS bonds. Similarly, at least one of the SH reagents that lowered freezing tolerance (PCMB), may induce a conformational change in the protein (Sugiyama and Akazawa, 1967). Furthermore, the SH group has been shown to participate in the hydrophobic bonding system of proteins (Heitmann, 1968). Any reagents which combine with an SH group will break this bond and may, therefore, induce unfolding. Thus, although the above prediction from the SH hypothesis is not supported by the tests to date, the negative results may conceivably be due to uncontrolled complications.

Some positive results have, however, been obtained. A natural increase in cell SH, due perhaps to GSH accumulation, occurs on fertilization of sea urchin eggs. This is accompanied by a rise in freezing tolerance (Asahina and Tanno, 1963). Similarly, a small increase in freezing tolerance of rat uterus tissue has been obtained by adding thiol (Wirth *et al.*, 1970). Two kinds of negative results have been obtained by Heber and Santarius (1964) using free, broken spinach chloroplasts, and determining freezing injury by the ability of these chloroplast fragments to synthesize ATP in the light (photophosphorylation). (1) Neither cysteine nor glutathione was able to protect the chloroplasts against inactivation on freezing. (2) No oxidation of SH to SS groups could be detected. These results are at first sight surprising, since the ATPase activity of chloroplasts is inhibited by PCMB or NEM (Sabnis *et al.*, 1970), and Heber (1968) himself had to use mercaptoethanol (ME) to obtain the enzyme in the active state. Unfortunately, they failed to measure the SS content, and merely showed that the SH content was unaltered by freezing. This does not rule out the possibility that SS formation occurred, and that this decrease in SH was compensated for by an unmasking during the denaturation of SH groups previously masked in the native protein. Their own later evidence (Santarius and Heber, 1967) indicates that this, indeed, did happen. They obtained a sharp increase in SH groups (from 47 to 107) when isolated chloroplasts were dehydrated, losing 98–99% of their water content. They interpret this as evidence of denaturation of the proteins, freeing previously masked SH groups. It is also possible, of course, for intermolecular SS bonds to arise by SH $\rightleftharpoons$ SS interchange without any change in proportion of total SH:SS. More recent results (Heber, 1968), however, demonstrate that these tests of the effect of freezing on ATPase activity have no bearing on the SH hypothesis, but are simply dependent on the osmotic properties of these chloroplast vesicles under artificial conditions. Finally, primary direct freezing injury rather than freeze-dehydration injury was probably involved.

The lack of success, so far, in attempts to prevent injury due to intermolecular SS formation by application of thiols, must be balanced against the complications in the use of this method. It is, unfortunately, not possible to generalize as to the effect of thiols on the plant's enzymes. Thiols are known, of course, to activate many enzymes; but they may also inactivate others (Nagatsu *et al.*, 1967), or prevent their reactivation (Kim and Paik, 1968). There are also cases where intermolecular SS bond formation plays a definite positive role, e.g., in the sexual agglutination reaction of yeast (Taylor *et al.*, 1968), and this role would be blocked by thiols. Similarly, fructose diphosphatase is actually activated by disulfides: oxidized mercaptoethanol (ME), ethyl disulfide, and an aromatic disulfide (Pontremoli *et al.*, 1967). It is not even possible to generalize for any specific case as to the effect of SH reagents, since some may be effective and others may not. Thus, the germination of lettuce seed is stimulated by arsenate, and this is enhanced by BAL (dimercaptopropanol) but not by any other thiols (McDonough, 1967). Similarly, an acetaldehyde dehydrogenase from germinating seeds in stabilized by GSH but is inhibited by CSH (cysteine) and ME (Oppenheim and Castelfranco, 1967).

It must, therefore, be concluded that SH reagents are double-edged weapons which may act to kill or cure. It is therefore difficult to predict the effect in the case of any one reagent acting on any one enzyme. It must be even more difficult to predict in the case of living cells. Consequently, a few positive results should carry more weight than the several negative results. In the meanwhile, however, the weak link in the SH hypothesis remains the absence of conclusive evidence that intermolecular SS bonding is the cause, rather than the result of freeze-dehydration injury. Until evidence on this point is forthcoming, the strongest argument for the hypothesis is its ability to explain the known facts (see above and below).

C. Molecular Aspects of Membrane Damage

On the basis of the above evidence, it must be concluded that soluble enzymes, in general, remain active in frost-killed plants. This fact is not surprising, since color, odor, and other enzyme-dependent changes have long been known to occur in frost-killed plants either during freezing or after thawing. There is, in fact, no reason to expect their inactivation; for the relatively slow metabolic changes at freezing temperatures that would result from inactivation of an enzyme do not seem able to explain the instantaneous nature of freezing injury. Molisch (1897) long ago observed the symptoms of injury in the still-frozen plant. Similarly, when a twig is freeze-dehydrated rapidly (cooling it down to −36°C in about one-half hour, Weiser, 1970), there is an exotherm at the frost-killing temperature showing that injury occurred while the plant was frozen. How can this apparently instantaneous injury be explained?

1. Evidence that Freezing Injury is due to Membrane Damage

Although the instantaneous nature of freezing injury cannot be explained by inactivation of a soluble enzyme, it can be explained by membrane damage. The first sign of freezing injury in a thawed plant is the absence of semipermeability. This becomes dramatically evident from the absence of turgor due to the inability of the flaccid cells to reabsorb the thaw-water in the intercellular spaces. Of course, cell death by any method is accompanied by a loss of semipermeability. It may, therefore, be the result, rather than the cause of injury. There are, however, several lines of evidence that point to the semipermeable cell membrane as the locus of freezing injury.

(a) Maximov (1912) protected cells against freezing injury by use of nonpenetrating solutes. Since these protective solutes do not penetrate the membrane, and are immediately effective, Maximov concluded that they must exert their effect on the surface of the protoplasm, and that freezing injury must be due to plasma membrane injury.

(b) The solute–efflux method of evaluating freezing injury is a technique for measuring the freezing injury by the amount of electrolyte which diffuses out of the thawed tissue. Surprisingly, it is capable of distinguishing between plants which suffer no injury at the freezing temperature used; a plant which must be frozen at −15°C in order to injure it, will show a greater loss of electrolyte after freezing at −8°C, than will a plant which must be frozen at −20°C in order to produce a similar effect. The plants apparently do suffer some membrane damage at −8°C, but it is slight enough to be completely repaired on thawing. This result can be explained only if the loss of cell permeability precedes (rather than follows) cell death. Similar evidence is the ability of many leaves to recover their full turgor, even though remaining infiltrated and flaccid for some hours after thawing.

(c) The increase in cell permeability which accompanies hardening must be due to a change in the plasma membrane, since this structure controls the permeability of the protoplast; hardening must involve a change in that part of the cell which is specifically injured by freeze-dehydration.

(d) The prevention of ectoplasmic stiffening on dehydration of the protoplasm, is characteristic of the hardened state. The ectoplast, or outer layer of the protoplast, includes the plasma membrane.

(e) As shown above, it has now become clear from Heber's (1968) more recent results, that the freezing damage to chloroplast fragments is primarily a membrane destruction, and only secondarily ATPase inactivation, because of the dependence of ATPase activity on membrane integrity. Similarly, electron micrographs of freeze-substituted and freeze-etched tomato fruit indicated that membranes are the most sensitive cell components to freezing, the tonoplast being more sensitive than the plasmalemma (Mohr and Stein, 1969). Unfortunately, in both of these cases, primary direct injury rather than freeze-dehydration injury must have been involved. Thus, in opposition to freeze-dehydration, rapid freezing and thawing were less damaging to Heber's chloroplast fragments than slow freezing and thawing.

(f) As mentioned above, when twigs are frozen, presumably extracellularly, but relatively rapidly, there is an exotherm at the instant of killing (Weiser, 1970). The explanation in best agreement with the known facts (see Chapter 9) is as follows. Due to the relatively rapid cooling, water cannot diffuse through the semipermeable membrane sufficiently rapidly for the cell contents to be in near equilibrium with its temperature. The cell contents are, therefore, markedly supercooled. At the instant of loss of semipermeability, the barrier is removed, and the supercooled cell water quickly freezes, releasing sufficient heat of fusion to flatten the curves, or even to produce a slight rise in temperature. This is possible only if the loss of semipermeability occurs either before, or simultaneously with cell death. If the loss in semipermeability occurs appreciably after cell death, the exotherm would occur at a temperature significantly below the frost-killing temperature.

2. Evidence that Membrane Damage is due to Protein Changes

Since the cell membranes consist of proteins and lipids, it must be determined which of these substances is responsible for the membrane damage. Although the first two of the above lines of evidence may be explained by changes in either of these membrane components, only a change in the protein component seems capable of explaining the increase in permeability, since this involves polar substances which presumably enter through the aqueous, protein pores in the lipid membrane. Similarly, the stiffening of the cytoplasm can only be due to the proteins, since solutions of proteins have the colloidal property of structural viscosity, changing progressively from a highly fluid to a fully rigid state on dehydration. The lipids themselves are not hydrated and will behave as solids below their melting point and as liquids above it.

This conclusion, together with the above evidence that intermolecular SS formation on dehydration occurs in the insoluble (presumably membrane) proteins, encourages an attempt to explain freeze-induced membrane injury by the SH hypothesis. Evidence indicates that membrane proteins must be particularly susceptible to intermolecular SS bonding. (a) Although the plasma membrane proteins of plants have not been investigated, those of animals and microorganisms have. The proteins of erythrocyte membranes are higher in SH content than the model system Thiogel (Table 10.4) and therefore may be expected to form intermolecular SS bonds even more readily than Thiogel on freezing. Mitochondrial structural (i.e., membrane) proteins have extraordinarily high SH contents. By use of radioactive maleimides, which localize the reactive SH groups in the mitochondrial membrane, radioactivity was found in the basic, structural proteins and not significantly in the oligomycin insensitive ATPase protein (Zimmer, 1970). A protein involved in active transport of lactose was located in the membrane of *E. coli*, and has a SH group essential for transport (Yariv *et al.*, 1969). Two mercaptans (2,3-dimercaptopropanol and 2-mercaptoethanol) at low concentrations (10–100 μM) inhibit and at higher concentrations (above 1 mM) activate the membrane enzyme

TABLE 10.4 Total Potential SH Groups (SH + 2SS) in Membrane Proteins[a]

Protein	SH groups/ 10,000 mol. wt.	Reference (see Levitt and Dear, 1970)
Thiogel	0.6–0.8	
Chloroplast structural		
Chlorella pyrenoidosa	1.2	Weber, 1962
Allium porrum	0.3	Weber, 1962
Antirrhinum majus	0.32	Weber, 1962
Spinacea oleracea	0.44	Weber, 1962
	1.2	Criddle, 1966
Beta vulgaris	0.5–0.8	Bailey *et al.*, 1966
Mitochondrial structural		
Neurospora	2.0–2.5	Woodward and Munkres, 1966
Yeast	5.0	Woodward and Munkres, 1966
Beef heart	4.6	Criddle *et al.*, 1962
Beef heart	2.5	Woodward and Munkres, 1966
Beef heart	1.0	Lenaz *et al.*, 1968
Erythrocyte membrane	1.0	Morgan and Hanahan, 1966
Erythrocyte membrane	1.0	Mazia and Ruby, 1968
Liver membrane (Eigen)	0.0	Neville, 1969

[a] From Levitt and Dear, 1970.

$(Na^+ + K^+)$-ATPase (Bader *et al.*, 1970). Further evidence of SH groups in the plasma membrane is their existence on the surface of intact Ehrlich ascites tumor cells, human blood platelets, and lymphocytes (Mehrishi and Grassetti, 1969). Similarly, six water-soluble, globular proteins were found on the surface of *Paramecium aurelia*, and were remarkably high in cystine (more than 10%; Reisner *et al.*, 1969). In the case of the only plant membranes investigated—the chloroplast structural proteins—the published results range from values below those for Thiogel to values well above them (Table 10.4). (b) Tests with SH reagents have localized the SH groups in membranes and demonstrated their importance in cell permeability. In the membranes of red blood cells, 7% of the SH groups are readily reactive with *N*-ethylmaleimide, up to 25% with chlormerodrin, and the remaining 75% only with $HgCl_2$ (Rothstein and Weed, 1963). Both the passive and active uptake of substances by red blood cells are affected by SH reagents (Webb, 1966). Epidermal cells of red cabbage were less permeable to urea after treatment at low temperatures (0°–3°C) with four SH reagents and more permeable after treatment with one (Table 10.5). Both changes were reversed by treatment with a thiol (ME), or after longer periods in the absence of thiol or other SH reagents. Leaf disks of *Saccharum officinarum*,

TABLE 10.5 Relative Permeability of Red Cabbage Epidermal Cells to Urea after Treatment at 1°–3°C with the Given SH Reagent[a,b]

SH reagent	Concentration (molar)	Time in solution (hr)	No. of trials	Relative permeability
β-mercaptoethanol (ME)	10^{-2}	1–24	11	0.97 ± 0.15
Dithiothreitol (DTT)	10^{-2}	5–40	4	0.60 ± 0.17
Iodoacetamide (IA)	10^{-4}	7–40	11	0.66 ± 0.10
Methylmercuric iodide (MMI)	10^{-5}	9–24	8	0.59 ± 0.15
N-Ethylmaleimide (NEM)	10^{-5}	12–18	3	0.50 ± 0.10
p-Chloromercuribenzoate (PCMB)	10^{-4}	9–40	11	3.7 ± 1.5

[a] Urea concentration 0.45–0.75 *M*. Treatment solution: 0.1 *M* 9:1 NaCl + $CaCl_2$ with or without (control) the given partial concentration of SH reagent.

$$\text{Relative permeability} = \frac{\text{(time for deplasmolysis in urea after treatment in control solution)}}{\text{(time for deplasmolysis in urea after treatment for same time in solution with SH reagent)}}$$

[b] From Levitt, 1971.

which were first allowed to accumulate large quantities of sugar, secreted about one-third of it into the external medium when treated with 10^{-4} *M* of the SH reagent iodoacetate (Schoolar and Edelman, 1970). No other respiratory inhibitors induced secretion, indicating a selective effect of IA on the cell membranes. Similar results were obtained with strawberry leaves (Younis, 1969). Phenylmercuric acetate poisons many enzymatic systems by reactions with SH and SS groups. It initially increases membrane permeability of plant cells, but later causes denaturation of the proteins (Waisel *et al.*, 1969). The deleterious effect of Cu^{2+} on *Chlorella pyrenoidosa* is due to binding to the cytoplasmic membrane (Steeman-Nielsen *et al.*, 1969). As mentioned above, Cu^{2+} can combine with SH groups. Similarly, contractile elements in the ectoplast (the outer protoplasmic layer) of *Mougeotia* are inhibited by PCMB, and the inhibition is reversed completely by cysteine (Schönbohm, 1969). (c) In agreement with the results obtained with red blood cells, the insoluble proteins (which include the membrane proteins) of cabbage leaves contain both SH and SS groups, unlike the soluble proteins which contain only SH groups (J. Dear, unpublished). The membrane proteins may, therefore, be able to form intermolecular SS bonds without SH oxidation, by SH $\rightleftharpoons$ SS interchange. (d) Membrane lipoproteins contain more hydrophobic groups than do the soluble lipoproteins, which in turn have more than soluble (nonlipo) proteins (Hatch and Bruce, 1968). Since the low temperature-induced unfolding of proteins is due to a weakening of the hydrophobic bonds, the membrane proteins will undergo greater denaturation at low temperatures, and will therefore be more readily aggregated on freezing. These membrane lipoproteins have a marked tendency to self-aggregate after removal of the lipids (Hatch and Bruce, 1968). Since both low temperature and dehydration weaken hydrophobic bonds, if these lipoproteins are linked by hydrophobic bonds, they may become separated from the lipids on freezing, leading to irreversible protein aggregation, loss of semipermeability, and death.

All these facts favor the conclusion that freezing injury is initiated in the plasma membrane, and that the change occurs in the membrane proteins due to intermolecular SS formation, causing coaggregation of proteins. How can this freezing injury arise?

3. Membrane-Hole Hypothesis

All the above evidence can be explained if freezing injury is due to the formation of permeable, nonlipid "holes" in the lipid layer of the plasma membrane. The series of events is, therefore, postulated to be as follows

(Fig. 10.5). (a) Extracellular freezing produces cell contraction and, therefore, a tension on the surface (see Chapter 9). (b) When the contraction and consequently tension become sufficiently severe, a break in the protoplast surface will occur. This break must be located in the thin, bimolecular lipid layer; due to its high degree of apolarity, its cohesive force must be

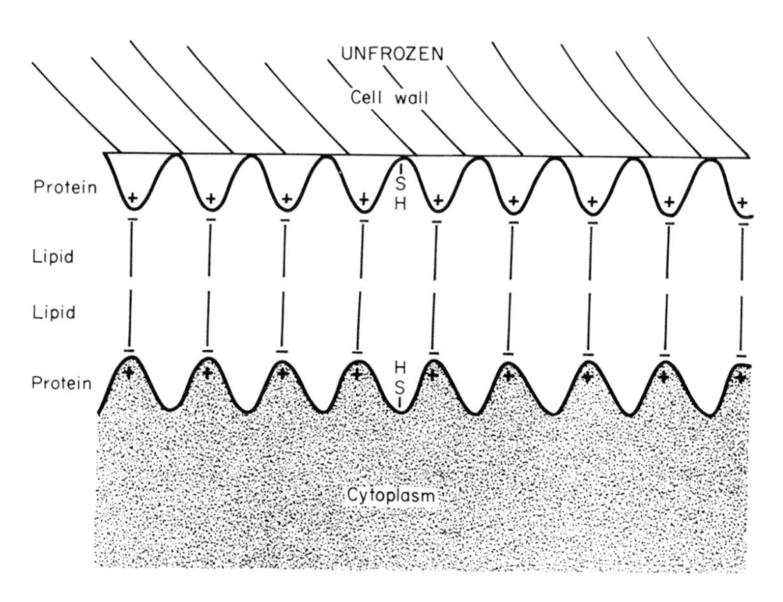

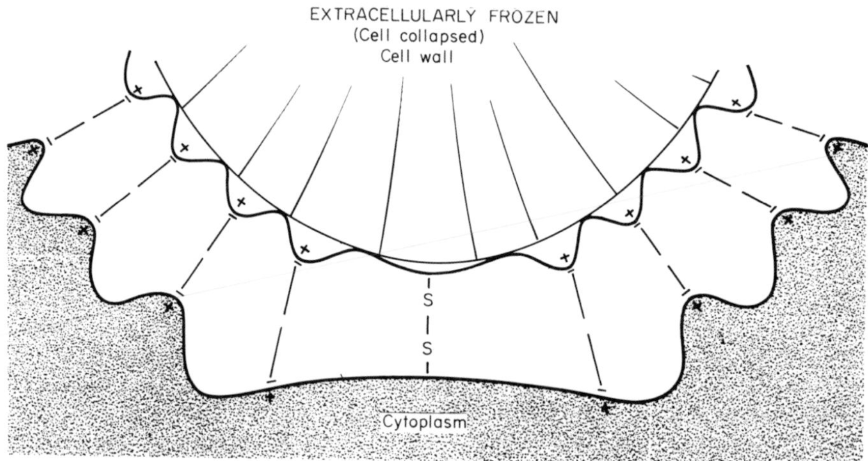

FIG. 10.5. Diagrammatic representation of possible mechanism of membrane protein aggregation due to cell collapse on freezing dehydration. Top: normal; bottom: cell collapse.

TABLE 10.6 Proposed Mechanism of Hardening[a]

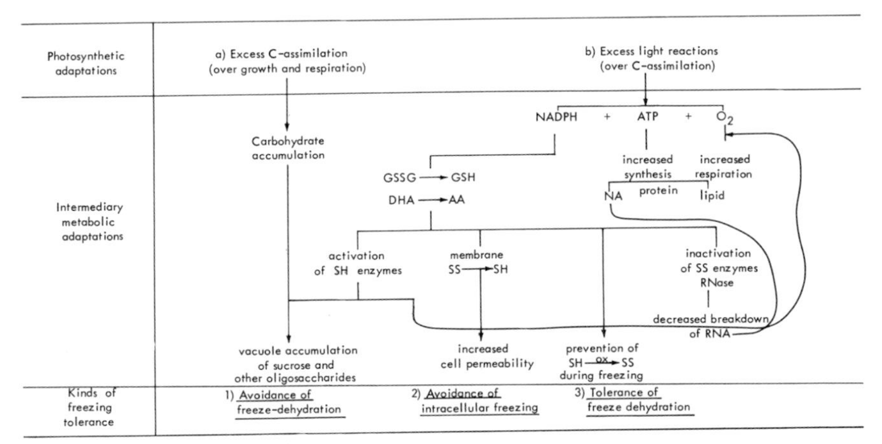

[a] GSSG, oxidized glutathione; GSH, reduced glutathione; DHA, dehydroascorbic acid; AA, oxidized ascorbic acid; NA, nucleic acids.

much smaller than that of the adjacent, much more polar membrane proteins. Direct evidence in support of this conclusion has been obtained by freeze-etching. When membranes are split while in the deep-frozen state, the fracture occurs between the lipid layers (Meyer and Winkelmann, 1969), indicating that this is the region of lowest resistance. (c) As a result of the break in the lipid layer, the two layers of membrane proteins on opposite sides of the lipid layer must come in contact with each other. (d) If this contact results in covalent bonding between the protein layers the hole becomes irreversible, leading to instant efflux of cell contents on thawing, and therefore to death. If covalent bonding does not occur between the protein layers, the lipid layer may, perhaps, become continuous again, on thawing, and no injury would occur.

This concept, of course, depends on the validity of the Davson-Danielli concept of membrane structure. A similar hole, however, could form in the Benson-type monomolecular lipoprotein layer. The low temperature must weaken the hydrophobic bonds linking the lipids with the proteins. On subjection to mechanical stress due to cell contraction, the protein molecules now separate from the lipids. They also become freeze-dehydrated and are able to form intermolecular SS bonds between the groups previously protected by bonding to the lipids. Thus, an aggregate of SS bonded protein molecules would form a permeable "hole" in place of the previously semipermeable lipoprotein complex. The proposed membrane role in both primary and secondary freezing injury is summarized in Table 10.6.

This membrane-hole hypothesis fits well into the above modified SH hypothesis. It is simply necessary to replace "protein" by "membrane protein" and to alter step (c) as follows: Due to the mechanical stress which separates the lipid layer, and the weakening of the hydrophobic bonds linking the lipids to the proteins, membrane proteins aggregate, forming permeable "holes" in the otherwise semipermeable lipoprotein membrane, rendering it freely permeable to both water and solutes.

There is some evidence that SS bridges may also form between soluble proteins and adjacent membrane proteins. Such mixed aggregates are apparently responsible for the loss of semipermeability and the consequent injury in abnormal red blood cells containing Heinz bodies (Jacob *et al.*, 1968). Labeling of SH groups with ^{14}C-*p*-chloromercuribenzoate has led to evidence of a similar mixed aggregation of proteins as a result of freezing injury to cabbage leaves (Morton, 1969). Direct evidence of SS formation in cell membranes as a result of radiation injury has been obtained in the case of isolated erythrocyte ghosts, i.e., the membranes (Sutherland and Pihl, 1968; see Chapter 18). Similar evidence in the case of drought injury has been mentioned above.

The membrane-hole hypothesis explains the known facts about freezing injury which were not adequately explained by the original SH hypothesis.

1. Maximov's method can now be seen to protect, because instead of the membrane being stretched on freezing, it contracts by plasmolysis in the hypertonic solution. No holes in the lipid layer can, therefore, form and the cells survive uninjured. The hypothesis can also explain a paradoxical aspect of Maximov's method. Plasmolysis by itself (without freezing), may lead, in time, to a stiffening of the protoplasm surface, and a consequent rupture on deplasmolysis. How then can a method which injures the cell, protect it against freezing injury? Unfortunately, it is not known how long Maximov's method continues to protect. Maximov froze for only 4–5 hr. If protoplast stiffening in the plasmolyzed cell requires 6 hr at room temperature (25°C), the same degree of freeze-dehydration and freeze-plasmolysis of a cell frozen in isotonic sucrose would require probably sixteen times (2.5^3) or about 90 hr at -5°C. Presumably, therefore, the protective effect of Maximov's method would be lost after freezing for 4 days under the above conditions. On thawing, the plasmolyzed protoplasts would presumably rupture after partial deplasmolysis.

2. The above mentioned protoplasmic stiffening during plasmolysis, which in time leads to deplasmolysis injury, must also be due to aggregation of the osmotically dehydrated membrane proteins; the stiffening occurs at the protoplast surface. Due to the absence of any mechanical stress, however, the aggregation is much less likely to occur. On the other hand, due to plasmolysis, a lateral contraction of the membrane occurs, bringing

the membrane proteins closer together, and eventually leading to intermolecular SS bonding.

3. The increased freezing injury induced by SH reagents opposed the original SH hypothesis since they would tie up the SH groups and prevent the injurious intermolecular SS formation. Most of these reagents have now been shown to decrease permeability (Levitt, 1971). This may, therefore, lead to intracellular freezing and primary direct freezing injury.

D. The Mechanism of Hardening

Metabolism, in general, must continue at hardening temperatures (although at a decreased rate), in order for hardening to occur. Certain specific changes must be included, and if they are not, no other metabolic changes can induce hardening. Potato tubers, for instance, can be maintained essentially indefinitely at 0° to 5°C without suffering injury. Two metabolic changes occur at this temperature that are characteristic of the hardening process: a conversion of starch to sugar and a synthesis of proteins from amino acids (Levitt, 1954). Nevertheless, these changes fail to increase the freezing resistance of the tubers, and they are killed by the slightest freeze. On the basis of the membrane-hole hypothesis, hardening must involve protection of the membrane against freezing injury. The changes leading to this protection must occur in plants that harden, but not in plants, like the potato, that do not harden. What then are these essential changes?

1. *Changes Expected from Membrane-Hole Hypothesis*

a. Avoidance of Freeze-Dehydration

This must, of course, confer freezing tolerance by decreasing the cell contraction and therefore the mechanical stress at any one freezing stress. Hardy plants develop this kind of freezing tolerance by an accumulation of sugars in the cell vacuole. If this is the sole location of the accumulated sugar, it would not alter the tolerance of protoplasmic freeze-dehydration, for at any one freezing temperature, the protoplasm must suffer the same degree of freeze-dehydration, regardless of the solute content in the vacuole. This freezing tolerance due to avoidance of freeze-dehydration would be effective regardless of the hypothesis of freezing injury. The advantage of the membrane-hole hypothesis is its ability to predict the changes that would lead to increased tolerance of freeze-dehydration. These predictions can then be tested experimentally.

b. Tolerance of Freeze-Dehydration

Any change that opposes hole formation in the membrane, or that favors its repair, would be expected to increase freezing tolerance.

1. The lipid layer may change. All lipids are apolar and therefore their molecules attract each other with a very small force. It has, however, been known for some time that the lipids are more fluid, due to greater unsaturation, when synthesized by the plant at low temperatures (see Chapter 8). This fluidity may, perhaps, play a role by permitting the membrane to coalesce across the holes on thawing. The lipids cannot, therefore, be expected to prevent hole formation but they may conceivably repair it on thawing.

Siminovitch *et al.* (1968), however, have suggested another way by means of which lipids could conceivably increase hardiness. They found that total lipids increased during fall hardening by only 20–40%, but that phospholipids increased more than 100%. The increase in the polar phospholipids was, therefore, at least partially at the expense of the nonpolar or neutral lipids. This increase corresponded to the increase in membrane content of the cells since the organelles and the cytoplasm as a whole also showed a 100% increase. On the other hand, starved cells increased markedly in freezing tolerance and in phospholipid content without any increase in organelles or cytoplasm. On the basis of modern concepts of membrane structure, the lipid content of a membrane is fixed (e.g., as a bimolecular leaflet), and, therefore, no increase per unit membrane area would be expected. The only possible way in which an increase in membrane lipids could occur, as Siminovitch *et al.* suggest, is by folding of the protoplast surface. If this occurred, it would, of course, be expected to prevent the tension on the protoplast surface and, therefore, the injury during freezing. This mechanism, however, requires proof of the existence of such folds in the plasma membrane. Electron micrography of plant cells has not, as yet, detected any such folding. Furthermore, this would not be an exclusively lipid mechanism, since the membrane proteins would also have to increase and fold. This is also true of the suggested repair mechanism. Lipids could coalesce across the postulated "holes" only if these holes had not been made irreversible during freezing by intermolecular bonding of the proteins. It, therefore, seems impossible to have an exclusively lipid mechanism of freezing tolerance on the basis of the membrane "hole" hypothesis, and another possible change must always be involved.

2. A change in the properties of the protoplasm may prevent intermolecular bonding between membrane proteins during extracellular freezing. This would be the primary tolerance factor, without which neither the avoidance of freeze-dehydration nor the lipid changes would be able to

confer appreciable freezing tolerance. The question arises as to whether the changes that occur during hardening lead to the development of the above three predicted factors: (1) a change in the protoplasm that prevents intermolecular SS bonding in the membranes, (2) an increase in fluidity of the lipids, and (3) an increase in osmotically effective substances.

2. Metabolic Control of the Hardening Process

The first effect of the low temperature is kinetic—a slowdown of all the metabolic processes. This retardation will not be uniform for all reactions, because of differences in the temperature (or Arrhenius) coefficients. Some reactions may even be stopped altogether, due to lack of substrate (Selwyn, 1966). The net result may be a marked change in the relative quantities of the different substances present in the plant. Thus, when young plants are raised at normal growing temperatures (e.g., 25°C) nearly all of the photosynthetic products are used up in the production of new growth, either as the raw materials or in the energy-supplying respiratory process. A small amount accumulates as sugars or starch. When the temperature is dropped to 5°C growth nearly stops and photosynthesis continues at a decreased rate. In hardy plants, the decrease in utilization of photosynthetic products due to the nearly complete growth stoppage is greater than the decrease in rate of photosynthesis. As a result, the net accumulation of these products is greater than at 25°C, and the concentration of sugars in the plant increases (Fig. 10.6).

Although the net accumulation of carbohydrates at 5°C is greater than at 25°C, the actual rate of photosynthesis definitely and markedly decreases. This is to be expected, since the process is measured by the rate of CO_2 assimilation to carbohydrates, consisting of a series of ordinary dark chemical reactions which have relatively high temperature coefficients. Furthermore, since the carbohydrates are primarily in the soluble form in hardened plants, their accumulation may be sufficient to inhibit carbon assimilation. The "light reactions" of photosynthesis are not usually measured, and since they are more closely related to the true photochemical reactions of photosynthesis, they occur much more rapidly and they must, therefore, have low temperature coefficients. In other words, they must take place at nearly the same rate at 5° as at 25°C. As direct evidence of this, photophosphorylation has been found to occur in spinach at a good rate not only at 0° but even at −10°C (Hall, 1963), whereas oxidative phosphorylation stopped completely at −2°C. At 25°C, the ATP and NADPH produced photosynthetically are undoubtedly nearly all used up by the rapid CO_2 assimilation. At 5°C, due to the markedly decreased CO_2

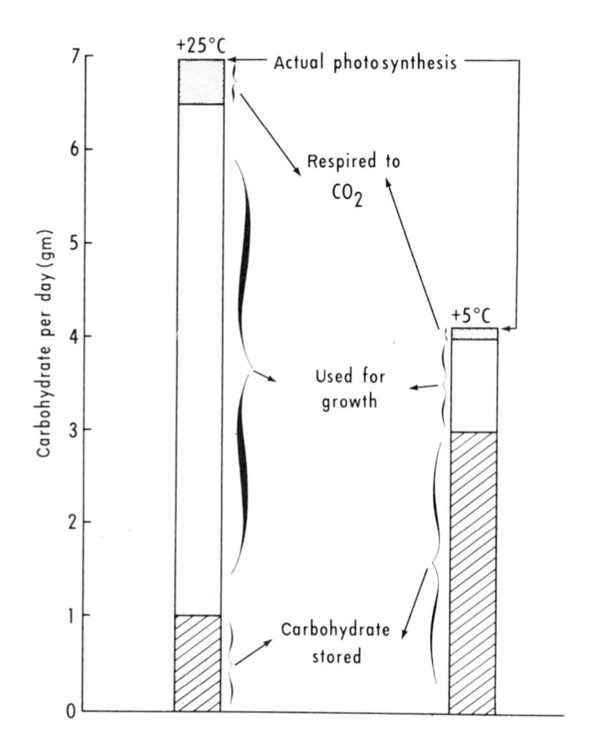

FIG. 10.6. Schematic representation of changes in photosynthesis, growth, and carbohydrate storage due to low (hardening) temperatures. Relative photosynthesis and respiration values based on Andersson (1941). (From Levitt, 1967b.)

assimilation, only a small part is used up photosynthetically, and the major part is therefore available for other metabolic processes (Fig. 10.7).

Even at normal temperatures, a brief red irradiation caused an immediate rise of NADPH level and an immediate drop of $NADP^+$ level in the coleoptilar node of etiolated *Avena* seedlings (Fujii and Kondo, 1969) According to Arnon (1969), if $NADP^+$ turnover ceases and NADPH accumulates, only cyclic photophosphorylation can operate, supporting protein synthesis which requires only the ATP. The accumulated carbohydrate, ATP, and NADPH may, therefore, support an increased synthesis of RNA, proteins, and phospholipids at the low, hardening temperatures. In the case of winter perennials, the accumulation of these substances is even greater because growth inhibitors accumulate in late summer and early fall. As a result, all growth stops and the plant becomes dormant. Dormancy is a state of growth inactivity, but not a state of metabolic inactivity. Dormant seeds of *Avena fatua*, for instance, are capable of synthesizing protein at a rate comparable to that of nondormant seeds (Chen and Varner, 1970). Due to this continued active metabolism in winter perennials, without any utilization in growth, substances accumulate in the fall to an even greater degree than in the case of the winter annuals which do continue to grow, although at a decreased rate. The winter perennials, therefore, attain an even higher degree of freezing tolerance than do the winter annuals.

A basic result of the kinetic changes at hardening temperatures is, therefore, an accumulation of reducing power, produced photosynthetically (by photoelectron transport) but not used up in carbon assimilation. Although this prediction is based on the above theoretical considerations, it is supported experimentally. (1) Kuraishi *et al.* (1968) demonstrated an increased ratio of NADPH/NADP at hardening temperatures. (2) Ascorbic acid in the reduced form has been shown to accumulate during hardening (see Chapter 8). (3) An artificial difference in SH content, between hardy and less hardy tissues, arises due to more rapid oxidation in the less hardy tissues during the first few minutes after homogenization (Schmuetz, 1969; Chapter 8). This difference in oxidation rate between the hardy and less

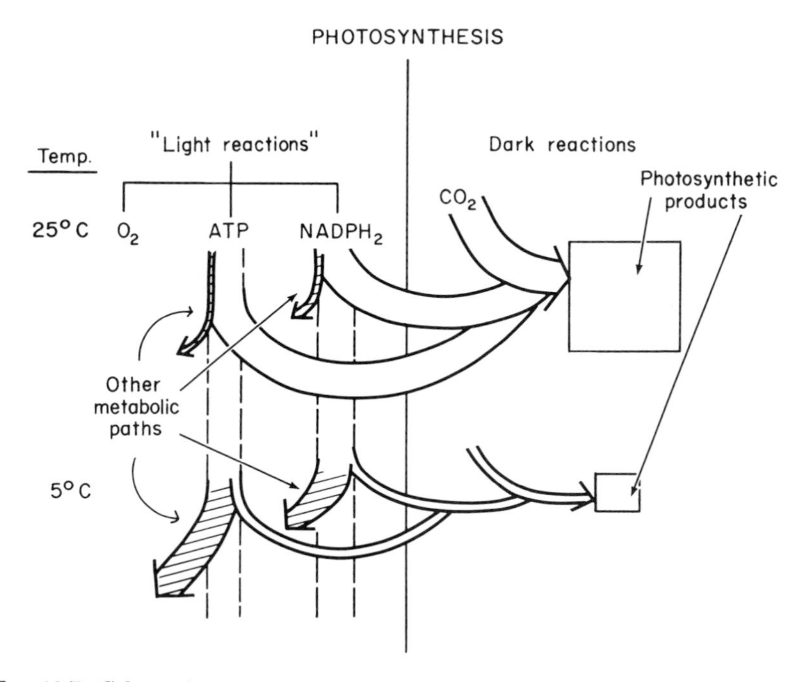

FIG. 10.7. Schematic representation of diversion of ATP and NADPH to nonphotosynthetic pathways at 5°C. (From Levitt, 1967b.)

hardy plants cannot be detected later at room temperature. Benson *et al.* (1949) long ago showed that the photosynthetic reducing potential is available for carbon assimilation for 10 min after the light is turned off, but that after longer periods in the dark, photosynthetic carbon reduction cannot occur. (4) The importance of oxidation as a factor in freezing injury is supported by some experiments with cucumber seedlings. When grown in air, they were killed by a single night at $-2°C$. When grown in 2% O_2 + 98% argon they were not killed until the temperature dropped 5°–8°C below this value (Siegel *et al.*, 1969). An atmosphere of nitrogen and CO_2 was similarly able to protect plants of *Haworthia* at temperatures of $-30°C$.

In the case of photosynthesizing leaves, the ultimate in reducing power would be expected in those most hardy evergreen plants whose net carbon assimilation decreases to zero during winter (see Chapter 8); photosynthesis produces both reducing (NADPH) and oxidizing (O_2) substances. In these plants, therefore, the NADPH accumulates due to the decreased carbon assimilation but the released oxygen is immediately reduced to water due to an equal rate of respiration. The direct relationship between respiration rate and freezing tolerance (see Chapter 8) is, therefore, readily understood.

The internal xylem cells do not contain chlorophyll and are less freezing tolerant than the cortical cells. Presumably even they undergo an increase in reduction potential because they are separated from the air by the surrounding chlorophyll-containing cortical cells. There are, however, nonphotosynthesizing tissues not surrounded by chlorophyll-containing cells, that are fully freezing-resistant, e.g., buds and the cambium. These are meristematic and such tissues have high reduction intensities (Van Fleet, 1954). It must also be remembered that some buds possess freezing avoidance (Chapter 7) and, therefore, may not have as high a freezing tolerance as the photosynthesizing tissues. It is a fact, however, that nonmeristematic, chlorophyll-free tissues such as in the mature root, are far less freezing tolerant.

How is this reduction capacity maintained (a) at very low temperatures when presumably photosynthesis ceases and (b) at night? Many investigators have succeeded in measuring photosynthesis in hardy plants at temperatures below freezing. The extreme has been measured in very hardy lichens below $-20°C$ (Lange, 1965a). If this has any survival value it cannot be due to the negligible accumulation of carbohydrates at these low temperatures, but could be due to the maintenance of the high reduction capacity. It must be realized, of course, that oxidation and reduction reactions always take place simultaneously. The hardy plant must permit the harmless oxidation of carbohydrate reserves, while preventing the harmful oxidation of lipids and protein SH. For example, the Russian investigators (Tumanov and Trunova, 1963) have recently developed methods of hardening wheat plants artificially in complete darkness and therefore, of course, in the absence of photosynthesis. They do this by feeding sugars to the seedlings via their roots, at low temperatures and over about a 2-week-period. This is done only after the plants have begun to harden, and therefore presumably have already undergone the first protoplasmic changes. Furthermore they have found that not all sugars work. Only those that are metabolized by the plant give good results. It is, perhaps, the oxidation of these sugars that prevents the oxidation of lipids or protein SH.

The accumulation of NADPH is apparently insufficient by itself to induce hardening, for it confers chilling tolerance, on peas, but no freezing tolerance (see above). It is presumably necessary to have a complete reduction series, including reduced ascorbic acid, the GSH oxidation-reduction system, etc. This may, perhaps, explain the decrease in freezing tolerance induced by SH reagents. The first effect of these may be to inactivate the reducing system by combining with GSH or some other component.

What role can this increase in reducing power play in the hardening process? On the basis of the SH hypothesis, this could lead to a reduction of SS to SH in some proteins. Experimental evidence of such a change has been obtained by Asahi (1964). Spinach chloroplasts, indeed, reduce protein disulfides in the light, but not in the dark. NADPH was not, however, able to reduce the protein in the dark. Asahi concluded from these and other results, that reduction was due to the photosynthetic electron transport system, but did not involve NADPH. This does, of course, support the general concept of hardening described above, although suggesting the replacement of NADPH by some other reducing substance formed photosynthetically. In the case of the photosynthesizing bacterium, *Chromatium*, direct evidence has been produced of this dependence of protein SH on photosynthesis (Hudock *et al.*, 1965). Triosephosphate dehydrogenase prepared from cells grown in the light had 4.2 SH groups per mole enzyme, whereas the same enzyme extracted from nonphotosynthesizing cells grown on organic medium in the dark had only 2.4 SH groups. During the early hardening of plants, an increase in protein SH has, indeed, been repeatedly observed. Although this is an oxidation artifact, in at least some cases (see Chapter 8), it is still conceivable that a real change in this direction may occur in the living cell, at least in the case of certain enzymes. This would explain the so-far unexplained hydrolysis of starch → sugar at hardening temperatures; β-amylase can be reversibly inactivated by SS interchange and its activity can be depressed by the sulfhydryl reagents iodoacetate and *N*-ethyl maleimide (Spradlin and Thoma, 1970). Similarly, sucrose cleavage by potato sucrose synthetase is activated four times by mercaptoethanol, sucrose synthesis is only slightly activated (Pressey, 1969).

Spradlin *et al.* (1969) suggest that the SH groups of β-amylase are involved in the *in vivo* regulation of enzyme activity. Taka amylase *A* is also a SH protein, containing one masked SH and four SS groups (Seon, 1967). Unfortunately, the specific enzymes involved in the conversion of starch to sugar during hardening have not been identified. Other results, however, agree with the above conclusion of Spradlin *et al.* Starch-filled leaves of plants which have been subjected to low dosages of naturally occurring photochemical oxidants (ozone or peroxyacetyl nitrate) hydrolyze their starch more slowly either in the dark or in the light (Hanson and Stewart, 1970). Similarly, an SS compound, tetramethylthiuram disulfide increases the size of starch grains in potato (Feldman *et al.*, 1968).

On the basis of this concept, sugar accumulation during hardening will be a good measure of freezing tolerance even if it plays no role in the mechanism of tolerance, provided that it is due to enzyme activation by the increase in reducing power. Conversely, sugar accumulation would be completely unrelated to freezing tolerance if due to new synthesis of enzyme. In agreement with this prediction, gibberellic acid is known to increase α-amylase synthesis, leading to starch hydrolysis and germination of barley seed. This increase in sugar during germination is accompanied by a loss of freezing tolerance. This explanation of sugar accumulation on hardening, therefore, predicts the well-known fact that sugars are often correlated with freezing tolerance, often completely unrelated to it. Similarly, the rise in RNA and in subsequent protein synthesis may be due to the reduction of the SS groups of RNase, which would inactivate the enzyme and prevent the breakdown of RNA. This could shift the balance to increased net synthesis of RNA. It must be realized, however, that the ability of the native enzyme to undergo reduction of intramolecular SS bonds, varies with the enzyme. Thus the SS bonds of BSA were much more readily cleaved by ME than were those of RNase (Bewley and Li, 1969). On the other hand, evidence of this mechanism has been produced by use of the thiol oxidizing agent diamide to convert GSH to GSSG within the cells of *E. coli* (Zehavi-Willner *et al.*, 1970). This resulted in inhibition of RNA synthesis. An inverse relationship between RNA content and RNase activity has, indeed, been found in the root of *Lens* (Pilet and Braun, 1970).

A change in SH content of the membrane would explain the increase in cell permeability which is known to occur on hardening. It has long been known that both Cu^{2+} and Hg^{2+} inhibit the entry of glycerol into red blood cells. Since the amount of Hg^{2+} required to alter the membrane properties is of the same order of magnitude as the estimated SH content, Webb (1966) suggests that the change in permeability is due to reaction of the Hg with SH groups (forming S-Hg-S bridges) in and around the membrane pores, thus impeding the passage of substances across the membrane. A similar effect could be produced by SS bridges between adjacent membrane proteins. Consequently, a conversion of SS to SH groups in the membrane proteins could conceivably account for the increase in permeability. In agreement with this conclusion, reagents that combine with SH groups lower the cell permeability (Levitt, 1971), except in the case of PCMB which is known to unfold proteins and, therefore, may enlarge the memgrane "pores."

The final possible role of the increase in reducing power would be to prevent the oxidation of SH groups in the membrane proteins to intermolecular SS bonds. This role would be more directly involved in freezing tolerance.

A relationship to the lipid metabolism is also conceivable, although little is known about the effect of hardening on the lipids. An increase in unsaturation has been recorded (see Chapter 8). This condition may, perhaps arise in conjunction with the photoelectron transport reactions at hardening temperatures. The increase in reducing power may prevent the formation of lipid peroxides from these unsaturated fatty acids, a reaction which is probably highly injurious to membranes (Christophersen, 1969). The danger of peroxidation is demonstrated by the strong acceleration of H_2O_2 decomposition and ascorbic acid oxidation in frozen solutions in the presence of Cu and Fe salts (Grant, 1969). The net result of the increase in reducing power would, on this basis, be to maintain the membrane lipids in a state of maximum fluidity and, therefore, with maximum ability to repair reversible "holes" in the membrane.

The metabolic changes during hardening, are, therefore, capable of explaining the development of the three changes expected on the basis of the membrane-hole hypothesis: (1) An accumulation of soluble carbohydrates due to the smaller decrease in photosynthesis than in growth and respiration, as well as to the activation of carbohydrases. (2) A production (and/or protection) of the SH groups of the membrane proteins by the accumulation of reducing power, due to the smaller decrease in the rate of the light reactions than of the carbon-assimilating reactions, and to a high respiratory rate. (3) An increase in fluidity of the membrane lipids due to unsaturation and protection against peroxidation of the fatty acid chains, and to an increase in phospholipids.

The actual steps in the hardening process may, therefore, be as shown in Table 10.6. What then distinguishes hardy from tender plants enabling these changes to take place in the former but not in the latter? On the basis of the above concept of freezing injury, we may define tender plants as those with proteins which undergo a sufficient degree of reversible denaturation (N $\rightleftharpoons$ D) at hardening temperatures (e.g., 5°C) to inactivate reversibly at least some of the enzymes required for the above metabolic processes that lead to hardening. Hardy plants would, then, possess enzymes which remain in the native state at these hardening temperatures. This would permit the hardy plants to continue their normal metabolism at the low (hardening) temperature, which would lead to a gradual increase in freezing tolerance. In agreement with this assumption, the Q_{10} for the change in rate of growth of cabbage is 2.0–2.3 (Cox and Levitt, 1969), which is characteristic of simple chemical reactions. These results agree with the hypothesis that the enzymes required for growth of cabbage retain the activity expected of the fully native enzyme. Conversely, the well-known higher minimum temperature for growth of tender plants would indicate reversible inactivation of their enzymes at hardening temperatures which are below their minimum for growth.

What can account for this apparent difference in enzyme response to low temperature? If the enzymes of the tender plants possessed a higher content of hydrophobic bonds, they would unfold (denature) more readily at low temperature than would the enzymes of the hardy plant. According to Ingraham (1969), in the case of bacteria, the proteins usually show minor changes due to the weakening of the hydrophobic bonds at low temperatures. In most cases this has little or no impact on their ability to function at low temperature, although, in certain cases, slight conformational changes can lead to inactivation. In the case of higher plants, however, indirect evidence is in agreement with this concept. Hydrophobic bonds become stronger with rise in temperature. Therefore, if the enzymes from tender plants have more hydrophobic bonds, they should require a *higher* temperature for *heat* denaturation than would the enzymes from the hardy plants. This expectation is fulfilled, judging from the higher maximum temperatures for growth in the tender plants (Levitt, 1969b). Furthermore, preliminary amino acid analyses of the leaf proteins of tender and hardy plants supports the above hypothesis. The proteins of the two tender species had a significantly higher hydrophobicity than those of the two hardy species (Butera, 1970). This difference in hydrophobicity of the proteins, together with an interrelated difference in lipid fluidity, can account for all the differences between tender and hardy plants (Table 10.7).

Why, then should hardy plants differ in the degree of hardening which they can undergo at hardening temperatures? This may also depend on the

TABLE 10.7 Proposed Differences between Tender and Hardy (Capable of Hardening) Species[a]

	TENDER		HARDY	
Protoplasmic substances	proteins	lipids	proteins	lipids
Hydrophobicity	high		low	
Melting point		high		low
Properties at hardening temperatures	$N \rightleftharpoons D$		N	
Enzymes	some inactivated reversibly		remain fully active	
Hardening changes	none or only some		see Table 10.6	
Membranes	(extracelluar freezing) ↓ D → A (membrane) ↓ loss of semi-permeability ↓ freezing injury	(0–5°C) solidification or oxidation ↓ loss of semipermeability ↓ chilling injury	(freezing) N ↘ retain normal permeability	(below 5°C) remain liquid ↙ retain normal permeability

[a] N, native; D, denatured; A, aggregated.

degree of hydrophobicity of their proteins. Hardening is a progressive process, which increases only up to a certain point with the length of time at a single hardening temperature. Maximum hardening requires exposure for successive periods of time to 5°, 0°, and −3°C. The lower this hardening temperature, however, the greater the degree of denaturation ($N \rightleftharpoons D$) and, therefore, of inactivation of the enzyme. Consequently, slightly hardy plants may perhaps be able to maintain their enzymes in a native state and therefore to harden only at 5°C, moderately hardy plants may continue to harden at 0°C, and very hardy plants may continue even at −3°C. Thus, white clover was the hardiest of three varieties tested and it continued to accumulate sugars at a later date in the fall (and, therefore, at a lower temperature) than did the others (Smith, 1968b). Ladino, the least hardy, stopped its accumulation at the earliest date (and therefore at the highest temperature). Presumably, the degree of hardening achieved at each of these temperatures would depend on the degree of reversible denaturation and inactivation of the enzymes, and therefore on their hydrophobicity.

In summary, only part of the SH hypothesis has so far been proved, and it still remains to be shown that intermolecular SS bonding is the cause and not the effect of freeze-dehydration injury. Nevertheless, the hypothesis is able to explain all the known facts of freeze-dehydration injury and freezing tolerance, and it is the only one capable of doing this. Finally, it fulfills the role of a working hypothesis by suggesting many experiments that will surely yield information of fundamental importance to the problem of the response of the plant to the freezing stress.

CHAPTER 11

HIGH-TEMPERATURE OR HEAT STRESS

A. Quantitative Evaluation of Stress

As in the case of chilling stress, a high-temperature stress is difficult to evaluate on an absolute basis. It is, nevertheless, possible to classify organisms on the basis of their response to the stress (Table 11.1). (1) Psychrophiles (lovers of cold) grow and develop in a temperature range that includes chilling temperatures (0°–20°C). Any temperature above 15° to 20°C may be a heat stress for them. The term has been used mainly for bacteria and fungi. Algae belonging to this group may actually grow on snow, and, therefore, have been called cryobionts or cryosestonic algae (Hindak and Komarek, 1968). (2) Mesophiles (lovers of middle tempera-

TABLE 11.1 Classification of Organisms According to the Temperature Range Inducing Heat Stress

Classification of organisms	Threshold of high temperature stress (°C)	Organisms included
a. Normally hydrated for growth		
Psychrophiles	15–20	Algae, bacteria, fungi
Mesophiles	35–45	Aquatic and shade higher plants lichens, and mosses
Moderate thermophiles	45–65	Higher land plants, some cryptogams
Extreme thermophiles	65–100	Blue-green algae, fungi, bacteria
b. Air dry cells or tissues	70–140	Pollen grains, seeds, spores, lichens, and mosses

TABLE 11.2 HEAT KILLING TEMPERATURE FOR DIFFERENT PLANTS AND PLANT PARTS

Plant	Heat-killing temperature (°C)	Exposure time	Reference
(a) Lower plants			
Cryptograms	42–47.5	15–30 min	de Vries, 1870
Ulothrix	24		Klebs, 1896 (see Belehradek, 1935)
Mastigocladus	52		Lowenstein, 1903
Blue-green algae	70–75	Few hr	Bünning and Herdtle, 1946
Thermoidium sulfureum	53		Miehe, 1907
Thermophilic fungi	55–62		Noack, 1920
Hydrurus foetidus	16–20	Few hr	Molisch, 1926
Sea algae	27–42	12 hr	Biebl, 1939
Ceramium tenuissimum	38	8.5 min	Ayres, 1916
Gymnodinium Pascheri	18	10 min	Diskus, 1958
(b) Higher plants			
Herbaceous plants			
Nicotiana rustica			
Cucurbita pepo			
Zea mays	49–51	10 min	Sachs, 1864
Mimosa pudica			
Tropaeolum majus			
Brassica napus			
Aquatics	45–46	10 min	Sachs, 1864
Nineteen species	47–47.5	15–30 min	de Vries, 1870
Citrus aurantium	50.5	15–30 min	de Vries, 1870
Opuntia	>65		Huber, 1932
Shoots of iris	55		Rouschal, 1938b
Sempervivum arachnoideum	57–61		Huber, 1935
Succulents	>55	1–2 hr	Huber, 1935
Succulents	53–54	10 hr	Huber, 1935
Potato leaves	42.5	1 hr	Lundegårdh, 1949
Trees			
Pine and spruce seedlings	54–55	5 min	Münch, 1914
Cortical cells of trees	57–59	30 min	Lorenz, 1939
Seeds			
Barley grains (soaked 1 hr)	65	6–8 min	Goodspeed, 1911
Medicago seeds	120	30 min	Schneider-Orelli, 1910
Wheat grains (9% H_2O)	90.8	8 min	Groves, 1917
Wheat (soaked for 24 hr)	60	45–75 sec	Porodko, 1926b
Trifolium pratense seeds	70	Short time	Buchinger, 1929

TABLE 11.2 (Continued)

Plain	Heat-killing temperature (°C)	Exposure time	References
Fruit			
Grapes (ripe)	63		Müller-Thurgau (see Huber, 1935)
Tomatoes	45		Huber, 1935
Apples	49–52		Huber, 1935
Pollen			
Red pine pollen	50	4 hr	Watanabe, 1953
Black pine pollen	70	1 hr	Watanabe, 1953

tures), grow and develop at temperatures of about 10° to 30°C. Any temperature above about 35°C may be a heat stress for them. (3) Thermophiles (heat lovers) may grow and develop at temperatures between 30° and 100°C. Only temperatures above 45°C (moderate thermophiles) or much higher (extreme thermophiles) are heat stresses for them. Thus a quantitative evaluation of high-temperature stress might be defined as the number of degrees above 15°C, since this is the approximate threshold of heat injury for the least heat-resistant group—the psychrophiles. Such a choice, however, is arbitrary; a species may be found with a threshold below 15°C. The alga *Koliella tatrae*, for instance, grows optimally at 4°C and any long-lasting temperatures above 10°C are lethal (Hindak and Komarek, 1968). No quantitative definition for heat stress is, therefore, possible. It can only be said qualitatively that the specific heat stress for any organism increases with the temperature above the lowest one that causes stress.

B. Limit of High-Temperature Survival

The high-temperature limit is generally lower for growing than for resting organisms (see below). Yet it may be astonishingly high for some growing organisms. There are records of growth of blue-green algae at temperatures as high as 93°–98°C, although 80°–85°C is usually accepted as the upper limit for their growth (Vouk, 1923; Robertson, 1927), and it is more commonly lower, e.g., 63°–64°C (Castenholz, 1969). According to Brock (1967), and Brock and Darland (1970) there are some extreme thermophilic bacteria for which no upper limit can be determined. This is

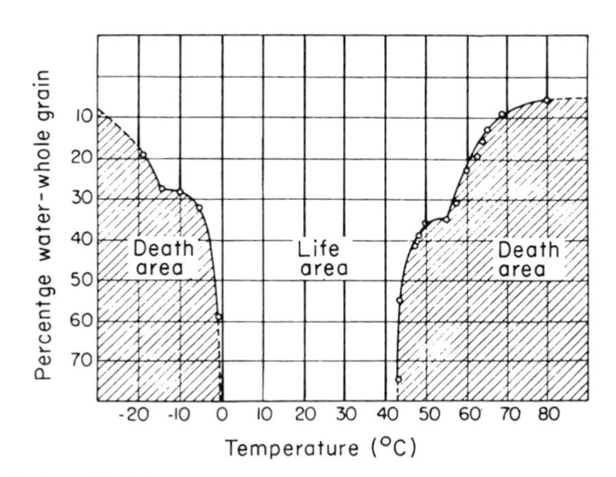

FIG. 11.1. Relation of killing temperature to water content of corn grains. *Left:* Hogue's yellow Dent exposed for 24 hr to temperatures below zero (Kiesselbach and Ratcliff). *Right:* Reid's yellow Dent exposed for 2 hr to high temperatures. (From Robbins and Petsch 1932.)

because they were able to grow and develop right up to the boiling point (92°–100°C), at which point the absence of liquid water brought growth to a stop. Nevertheless, they are still subject to heat stress and will be killed at some temperatures above 100°C. Thus, unlike low-temperature stress, the most extreme of which can be tolerated by living cells under certain conditions, there is a limiting high-temperature stress for all organisms. Exactly what this limit is for plants, in general, cannot be stated with certainty. By analogy with low temperatures, the heat-killing temperature for a plant may be defined as the temperature at which 50% of the plant is killed. As indicated above, this varies markedly from plant to plant (Table 11.2). Probably the highest recorded temperatures for a growing higher plant are 60°–65°C (Table 11.7; Biebl, 1962a). However, these temperatures were maintained for very short periods of time, usually in mid-afternoon, when growth probably had temporarily ceased.

In agreement with these older results, the heat-killing temperatures for some thirty-nine species of plants from August to September on the coast of Spain ranged from 44° to 55°C (Lange and Lange, 1963). The low limit for the higher, multicellular algae has been corroborated by Giraud (1958). At optimal illumination (5000-10,000 lux) the optimal temperature for growth of *Rhodosorus marinus* (a red alga) is 20°C. The cells die in a few hours above 29°C and in 2–3 days at 25°C.

Resting tissues in the dehydrated state have long been known to tolerate much more severe treatment than when active and fully hydrated (Just, 1877). Dry seeds are able to survive as high as 120°C, in contrast to highly hydrated tissues that are killed by temperatures below 50°–60°C (Table 11.2). When dried and heated for 16 min in a vacuum, they can survive 122°–138°C (35°C higher than without a vacuum; Ben-Zeev and Zamenhof, 1962). Of course, not all seeds survive such high temperatures. Some may be killed by 50°–60°C (Crosier, 1956), while others may survive boiling in water for several hours, provided that they do not swell during the boiling (Just, 1877). Dry barley and oat grains can be made to survive high temperatures for even longer times if dried further. In the ordinary, air-dry state they survived 100°C for only 1 hr without injury. However, if they were carefully dried for 9 days at 50°C, 2 days at 60°C, 2 days at 80°C, and finally transferred to 100°C for 3 days, more than 58% were still able to germinate (Just, 1877). Even hard-coated seeds that survive autoclaving at 120°C for one-half hour, are killed by boiling for 10 min if their seed coats are first filed (Schneider-Orelli, 1910). Presumably the filing permits the living cells to take up enough water during this 10-min period to lower their heat tolerance. The relationship between the heat-killing point of seeds and their moisture content and the analogous relation to low-temperature tolerance are clearly shown in Fig. 11.1. The seeds (maize) with less than 10% moisture had killing temperatures above 80°C, but when their moisture content rose to 75%, their killing temperature dropped to about 40°C. It is not surprising then that some apparently air-dry seeds are killed at 60°C, due to a high internal water content (Crosier, 1956).

Even in the case of nonresting tissues, the same relationship holds. Dallinger (see Lowenstein, 1903) showed that infusoria can survive 70°C if their water content is first reduced. Sea algae that succumb to seawater if kept at 35°C for 12 hr, survive 42°C for the same length of time if they are first dried on microscope slides (Biebl, 1939). *Nostoc muscorum* and *Chlorella* sp. in the lyophilized state survive 100°C for 10 min (Holm-Hansen, 1967).

C. The Time Factor

In contrast to the relatively minor role of exposure time in the case of freezing, the time subjected to high temperatures is of fundamental importance (Table 11.3). Not only does the heat-killing temperature vary inversely with the exposure time, but the relationship to time is actually exponential, so that Arrhenius plots give the expected straight-line relation between the log of the heat-killing rate and the reciprocal of the absolute temperature (Fig. 11.2). This has been thoroughly confirmed by

TABLE 11.3 RELATIONSHIP BETWEEN HEAT-KILLING TIME AND TEMPERATURE[a]

Temp. (°C)	Heat-killing time (min)				
	Tradescantia discolor	*Beta vulgaris*	*Brassica oleracea*	*Draparnaldia glomerata*	*Pisum sativum*
35				480	300–400
40	1300 (app)	>1500–2500	1100	80	32
45	725	420	577	7	2.2
50	243	90	45	1.2	0.27
55	44	4.3	3.8	0.32	0.095
60	7	0.7	0.8		
65	1.8				

[a] From Collander, 1924.

Alexandrov (1964). At intermediate temperatures, however, there is a break in the curve (Fig. 11.2), which is interpreted below.

According to Lepeschkin (1912):

$$T = a - b \log Z$$

where T = heat-killing temperature; Z = heating time; a and b are constants.

The agreement between the values calculated from this equation and

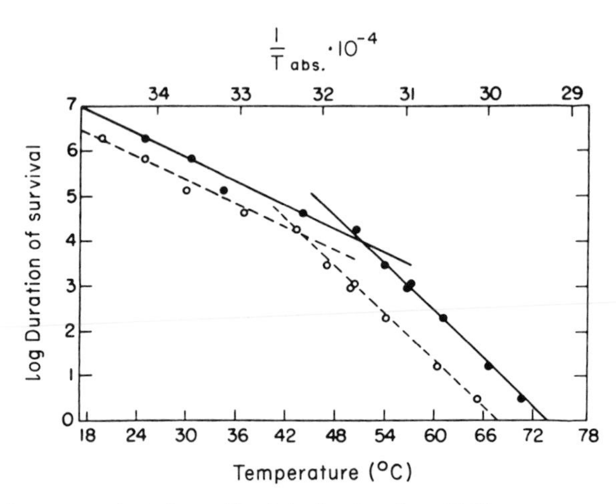

FIG. 11.2. *Upper curve*(–●–): modified Arrhenius plot of killing temperature in relation to log of time. *Lower curve* (–○–): analogous Q_{10} relation. Break in curve indicates two distinct processes in *Helodea canadensis*. (From Belehradek and Melichar, 1930.)

TABLE 11.4 HEATING TIME AND COAGULATION TEMPERATURE *Tradescantia discolor*.[a]

Heating time (min)	Coagulation temperature (°C)	
	Determined	Calculated[b]
4	72.1	
10	69.6	67
25	63.2	62
60	57.0	57.1
80	55.7	55.5
100	54.1	54.2
150	52.0	

[a] From Lepeschkin, 1912.
[b] $a = 79.8$ and $b = 12.8$. in above equation.

the measured values is very good (Table 11.4). Other empirical equations have also been used successfully (Porodko, 1926b; Belehradek, 1935). S-Shaped curves relating heating time to injury may sometimes be obtained (Porodko, 1926b). According to Belehradek (1935), these are probability curves and simply mean that the organisms of a single type possess dissimilar heat-killing temperatures, in accord with the laws of variability.

D. Occurrence of Heat Injury in Nature

All the above results were obtained with artificially induced high temperatures. The question is whether the plant is ever exposed to high enough temperatures to be injured under natural conditions. It is not enough to know the air temperatures exposed to, for it has long been known that the plant's temperatures may rise above that of its environment. Dutrochet (1839) showed that this occurs if the plant is kept in saturated air to prevent the cooling effect of transpiration. The higher the external temperature, the greater was the elevation of the plant's temperature. In the case of fleshy organs (Table 11.5) with high metabolic activity, the elevation may be as high as 11°C (Vrolik and de Vriese, 1839), or even 14°R in the spadix of *Arum* (Goeppert—see Dutrochet, 1840). Due to their small specific surface such fleshy organs are unable to transfer all the excess heat to their environment, and their temperature rises. Thin leaves, on the other hand, may actually be cooled below the air temperature due to transpiration (see Chapter 12). More commonly, however, the

TABLE 11.5 Elevation of Temperature above that of the Air, in the Spadix of *Colocasia odora*[a]

Air temperature (°C)	Spadix temperature (°C)
20.0	27.8
16.7	26.1
15.6	26.5
20.7	28.9

[a] Vrolik and de Vriese, 1839.

temperatures of leaves exposed to sunlight may be well above that of the surrounding air (Table 11.6), due to absorption of 44 to 88% of the total radiation received (Raschke, 1960). According to Dörr (1941), the leaf temperature increases with its color in the following order: yellow, green, orange, and red. The strong absorption by read leaves caused both a rapid uptake and rapid loss of heat. Wilted leaves were always a few degrees warmer than turgid leaves. Herbaceous and expecially woody stems reached temperatures as much as 12.2°C above those of the leaves. The temperature of the roots showed the closest agreement with the surrounding temperature. Different parts of the same leaf (or other plant part) may have different temperatures (Fig. 11.3), and those parts exposed to the most intense

TABLE 11.6 Plant Temperatures in Relationship to Temperatures of Surrounding Air

Plant	Plant part	Temp.	Air temp.	Reference
Sempervivum spp.	Succulent leaves	48–51°C	31°C	Askenasy, 1875
Tomato	Ripe fruit	100–106°F	80°–83°F	Hopp, 1947
Various	Leaves	44.25°C	36.5°C	Harder, 1930
Various	Leaves	37.6°C	24.7	Fritzsche, 1933
Conifer	Needles, twigs	9–11.8°C above air		Michaelis, 1935
	Stems (herbaceous and woody)	12.2°C above leaves		Dörr, 1941
Various	Thin leaves	6–10°C above air	20°–30°C	Ansari and Loomis, 1959
Herbaceous plants	Thick leaves	20°C above air	20°–30°C	Ansari and Loomis, 1959

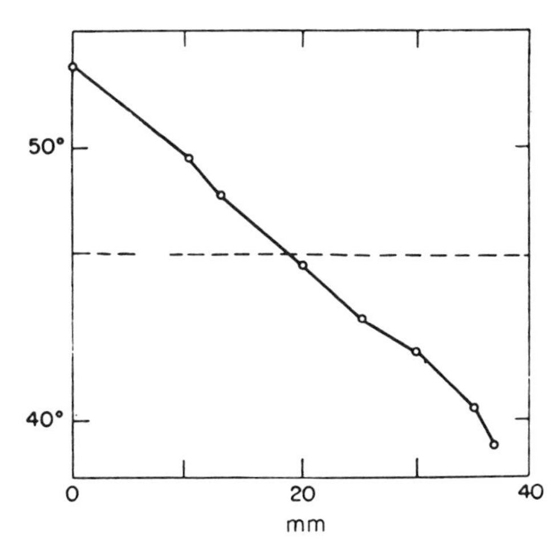

FIG. 11.3. Temperature (°C) gradient with depth (abscissa) in a joint of *Opuntia*. (From Huber, 1932.)

radiation reach the highest temperatures (Konis, 1950; Waggoner and Shaw, 1953). For this reason, position is also important. McGee (1916) found that *Opuntia* joints in the meridional position heat up more during a day in the sun than those in the equatorial position, but the heating above air temperature is great only in still air (Harder, 1930).

That dangerously high temperatures may occur under natural conditions seems obvious from all these observations. How high they actually do rise is shown in Table 11.7. In many cases they reach and may even exceed the 45°–55°C range that is usually accepted as the normal tempera-

TABLE 11.7 HIGH PLANT TEMPERATURES RECORDED UNDER NATURAL CONDITIONS

Plant	Highest temperature (°C)	Reference
Pine tree (south side cambium)	55	Hartig (Sorauer, 1924)
Opuntia	65	Huber, 1932
Sempervivum hirtum	50.2	Dörr, 1941
Globularia	48.7	Dörr, 1941
Tortula (dried out pulvinus)	54.8	Rouschal, 1938b
Arum italicum (fruit)	50.3	Rouschal, 1938b
Viburnum (leaves)	43.8	Rouschal, 1938b
Iris (rhizomes)	42.5	Rouschal, 1938b
Rhamnus alaternum	52.5	Konis, 1949
Fleshy fruit (various)	35–46	Huber, 1935

ture limit for most plants (Huber, 1935). Heat injury has, in fact, been described by many workers, especially in the case of bulky organs (see Sorauer, 1924; Huber, 1935). A well known type is the sun or bark burn that occurs on the south and southwest sides of thin-barked trees. This may be followed by drying and stripping of the bark, leading to defoliation and wood injury. The worst affected are the older, stronger stems; pole timber and branches are seldom injured in this way. The injury is thought to be due to overheating of the cambium. Hartig actually recorded a cambium temperature of 55°C on the southwest side of a spruce tree in the open at an air temperature of 37°C (Sorauer, 1924).

Burns have frequently been described in fleshy fruit—more commonly in grapes, cherries, and tomatoes, and less commonly in pears and gooseberries (Huber, 1935). As in the case of bark injury, they are usually confined to the most strongly heated southwest side (Huber, 1935). The burns may later dry up and become separated from the uninjured portion by a cork layer. Less often (e.g., in grapes) the whole fruit is killed. That true heat injury is involved follows from Müller-Thurgau's production of the same kind of injury by raising the fruit temperature artificially above 40°C (Huber, 1935). Szirmai (1938) was able to produce the "drought fleck disease" of paprika (which damaged 4–12% of the crop in Hungary) artificially by exposures to temperatures of 50° to 52°C when the surface was wet. In the case of dry fruit, 55°C was required to produce the injury. Direct exposure to sunlight produced the symptoms at air temperatures of 49°C.

Heat injury has been less often reported in leaves and other thin organs than in the above bulky plant parts (Sorauer, 1924; Huber, 1935). In practice, it has received the most attention when greenhouse-grown or shade plants are placed outdoors directly in the sun. Huber states, however, that it occurs most commonly on inland plains, but it is difficult to be sure that these are true cases of direct heat injury, since there is usually also a water deficiency. Thus, in spite of the many records of high temperatures in plants under natural conditions, Rouschal (1938b) concluded that the maximum temperature the leaves can reach in the Mediterranean region during summer cannot produce any heat injury. By enclosing shoots in blackened tubes, he exposed them to temperatures higher than any he was able to observe under natural conditions; 55°C was reached for a short time, and 47°C was maintained for 10 to 15 hr. Even after several days of such treatment, no injury occurred. Konis (1949) used the same method and came to the same conclusion in the case of maquis plants under natural conditions in Israel. The lethal temperature was several degrees higher than the highest temperature recorded in the field. On the other hand,

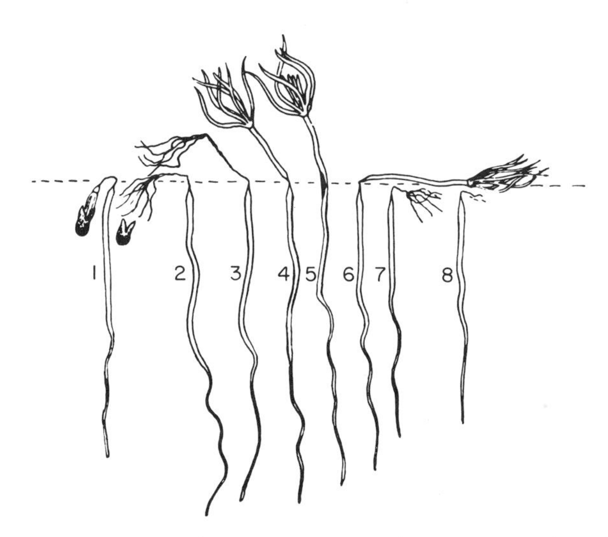

Fig. 11.4. Killing of pine seedlings at soil level, due to overheating of soil surface. (From Münch, 1913.)

when sorghum plants were exposed at head emergence to simulated heat waves for 5 days (107°–108°F by day, 90°F by night) most of the enclosed flowers were killed (Pasternak and Wilson, 1969). High temperatures were apparently responsible since humidity (41 versus 70% by day) had little effect.

The greatest danger of heat injury occurs when the soil is exposed to insolation, reaching temperatures as high as 55° to 75°C (Lundegårdh, 1949). One of the most serious seedling "diseases," according to Münch (1913, 1914) is the killing of a narrow strip of bark around the stem of young woody plants at soil level (Fig. 11.4) when soil temperatures exceed 46°C. Since the seedlings usually die, he calls this "strangulation sickness." In laboratory tests, pear seedlings were found to succumb within 3 hr at 45°C or within 30 to 60 min at 50°C. Baker (1929) also concluded that fatal temperatures are reached in nature only at the base of the stem of 1- to 3-month-old conifers. He points out that surface soil temperatures of 130° to 160°F have been detected in temperate climates and that injury may occur at as low as 120°F. Henrici (1955) showed that temperatures of prostrate plants may exceed 55°C, and for a short time 60°C during summer in South Africa, although trees and other plants seldom reach 36°C. Young coffee plants suffer from collar injury, which Franco (1961) was able to induce by a temperature of 45°C. At 50°C most of the plants died, at 51°–55°C all died. Soil surface temperatures in the field reached 45°–51°C and sometimes above. Therefore, high-temperature injury must occur in nature, but this depends on the bulk of the structures. Although Rouschal (1938b) measured maximum soil temperatures as high as 64°C, no injury occurred to the bulky iris rhizomes that he investigated, for their temperature did not rise above 42.5°C. According to Julander's (1945) observations, however, the much thinner stolons of range grasses are in definite danger of injury. He observed a soil temperature of 51.5°C when the air temperature was 36°C. Since he was able to produce definite injury to the stolons at 48°C, and since air temperatures as high as 43°C are not uncommon under severe drought conditions, the possibility of heat injury under natural conditions seems obvious.

Lange (1953) showed that lichen temperatures may be well above that of the surrounding atmosphere (Fig. 11.5) and as high as 69.6°C under natural conditions. He suggested that they may rise to 75°–80°C, since these values have been observed for surface temperatures of soils that are their natural habitats. Since one-half hour periods at 70°–100°C were sufficient to kill a large number of different kinds of lichens in the dry state, and since the high temperatures he observed were maintained for much longer than one-half hour (Table 11.8), he concluded that they must be injured sometimes by heat under natural conditions.

Most of the above investigations were not conducted in the hottest climates. Researches by Lange (1958) in the Sahara desert have clearly demonstrated the importance of heat injury in this region. Even the most successful plants commonly attained leaf temperatures within 2°–6°C of their heat-killing points and some natural heat injury occurred similar to that produced under artificial conditions. Some of the plants (e.g.,

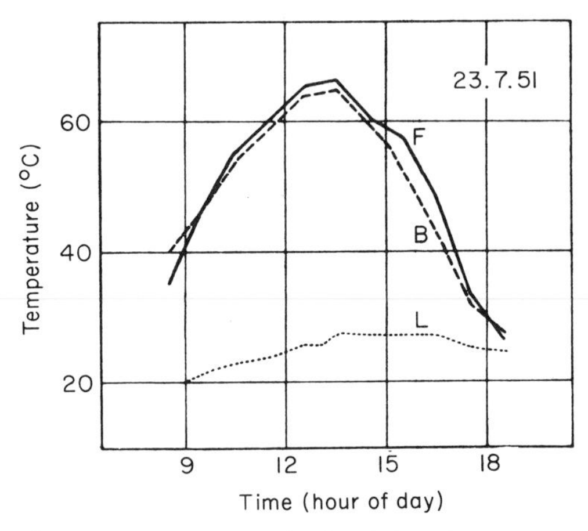

Fig. 11.5. Diurnal temperature course in the lichen *Cladonia furcata* var. *palomoea* (F), the soil (B), and the air (L). (From Lange, 1953.)

TABLE 11.8 TIME (MIN) DURING WHICH THE TEMPERATURE OF THE LICHEN REMAINED ABOVE THE LISTED TEMPERATURES (°C)[a]

Species	62.5°	65.0°	67.5°	Max. temp.
Cladonia pyxidata	270	225	150	69.6°
Cladonia subrangiformis	195	125	—	67.0°
Lecidia decipiens	50	—	—	64.3°

[a] From Lange, 1953.

species of *Citrullus*), in fact, survived only if sufficient water was available for high transpiration rates (see Chapter 12). Attempts to grow cultivated plants introduced in an oasis during summer failed, even with plentiful watering, due to heat injury. Even in temperate climates, Lange (1961) found that *Erica tetralix* owed its survival of summer heat to a marked summer rise in heat tolerance, without which its leaf temperatures would have been above the killing point (see Chapter 12). According to Khan and Laude (1969), heat stress during maturation of barley seed may aid in explaining differences in % germination at harvest of seed produced in successive years, or in different locations in the same year. The germination of the freshly harvested seed was depressed following a heat stress 7–10 days after awn emergence, but was enhanced by the same stress applied 3 weeks after emergence.

In view of all these observations, it cannot be denied that heat injury does occur in nature, though perhaps only on relatively rare occasions. By analogy with freezing injury, it may be expected only during relatively rare "test summers." It is most likely to occur when crop plants are grown in regions to which they are not adapted.

E. Nature of the Injury

A complicating factor in low-temperature injury is the change of the cell's water from the liquid to the solid state. At high temperatures, the analogous sudden change to vapor occurs only at temperatures that are not found under natural conditions. The slower vaporization at normal temperatures is, however, a possible cause of injury, but greater complexity is to be expected at high than at low temperatures, since all the reactions in the plant are already taking place rapidly, and a further rise in temperature might easily disturb the balance. Because of these complicating factors, the heat stress may produce direct or indirect stress injury as well as secondary stress injury.

1. Secondary Heat-Induced Drought Injury

Gäumann and Jaag (1936) measured the cuticular transpiration at temperatures of 20° to 50°C (Fig. 11.6). Their results show how pronounced the increase is at the higher temperatures. There are two reasons for the sharp rise in transpiration with the rise in environmental temperature: (a) the direct effect of temperature on the diffusion constant of water, and (b) the steepening of the vapor pressure gradient between the leaf and the external atmosphere. Curtis (1936a) points out that if the leaf temperature is 5°C above the atmospheric temperature, this is equivalent to a steepening of the gradient by a 30% lowering of the atmospheric r.h. (relative humidity). In other words, if the external atmospheric r.h. is 70%, the 5°C rise in leaf temperature would double the gradient and,

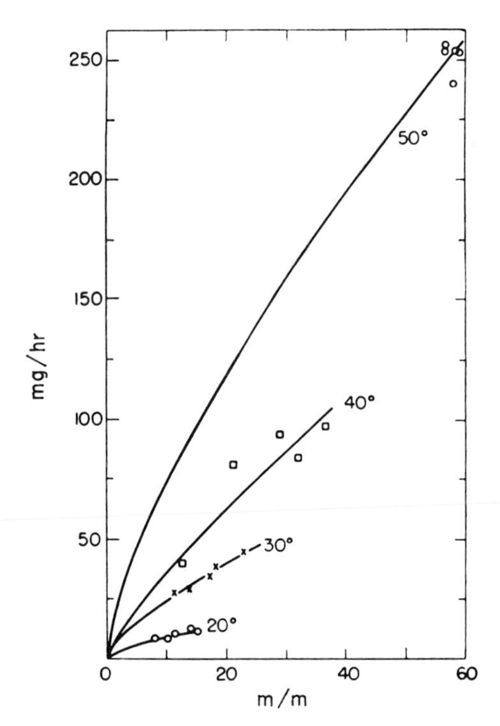

FIG. 11.6. Cuticular transpiration of *Quercus robur* at different air temperatures. (From Gäumann and Jaag, 1936.) m/m = physiological saturation deficit (mm Hg).

TABLE 11.9 Effect of Different Times of Exposure of *Penicillium* Cultures (1-Day Old) to 35°C on Subsequent Growth[a]

Exposure time (days)	Time for growth to recommence (days)
2	2
4	3
8	4
12	5
19	6
25	6
30	8
1 month	Dead

[a] From Hilbrig, 1900.

therefore, the evaporation rate, aside from the increase due to the increased molecular velocity at the higher temperature. A 10°C rise would have a proportionately even greater effect than expected from the doubling of the temperature gradient. The danger of drought injury under such conditions is obviously great, even without a deficiency in soil moisture.

The above described secondary drought injury will be considered together with the primary drought stress. It must be eliminated in any attempt to study the primary heat injury, which in its turn may be either direct or indirect. The existence of two kinds of primary heat injury is demonstrated by the break in the Arrhenius plot, giving two straight-line relations between the log of the heat-killing rate and the reciprocal of the absolute temperature (Fig. 11.2). This break has been confirmed in a number of plants by Lorenz (1939) and by Alexandrov (1964). The logical explanation is that the straight line in the lower temperature range is due to indirect heat injury, and in the upper range to direct heat injury.

2. Primary Indirect Heat Injury

The indirect injury is due to an elastic (reversible) metabolic strain that is converted to a plastic (irreversible) strain and, therefore, to injury. It has long been known that the growth of plants is stopped at temperatures that are not immediately fatal. Hilbrig (1900) showed that the injury at such temperatures is gradual. The longer the plants are exposed to the high temperatures, the longer it takes them to recommence growth (Table 11.9).

Even fungus spores finally die after 52 days if kept at temperatures too high for growth. Temperatures that are not quite high enough to stop growth completely, eventually may also be injurious (Hilberg, 1900). Results with higher plants (bean, pea, and cucumber) were similar, though the temperature zone in which growth was stopped without immediate injury was smaller. Growth stoppage (at 45°C) could never be maintained for more than 1 hr and 45 min without killing the seedlings. Dangeard (1951c) stated that cessation of cell elongation may be reversed after several days at room temperature.

The gradual injury produced at such high temperatures can be shown by respiration measurements (Table 11.10). In *Crepis biennis*, respiration decreased with time even at 30°C (Kuijper, 1910). The marked decrease in reserves as a result of the high respiration rate affected both the starch and the proteins. The production of CO_2 by yeast shows no depression up to 45°C, but at 46° and higher it decreases with exposure time (van Amstel and Iterson; Belehradek, 1935). The greater heat injury when submerged in water has actually been ascribed to an oxygen deficiency (Just, 1877). Other metabolic processes may be even more heat sensitive than respiration. A 1-hr exposure to 47°C in *Saccharomyces* and to 4°C in *Torula*, reduced oxygen consumption by 40 to 60% and nitrogen assimilation to zero (Van Halteren, 1950). After $2\frac{1}{2}$ hours at room temperature, nitrogen assimilation recommenced, though at a lower rate than the control. The longer the exposure time, the slower the recovery rate and the rate of nitrogen assimilation when it does recommence (Table 11.11). Uptake of phosphate was also stopped for 2 to 3 hr after the high temperature exposure. Four basic types of indirect heat injury have been proposed.

TABLE 11.10 CO_2 EVOLUTION BY PEA SEEDLINGS AT HIGH TEMPERATURES[a]

Temperature (°C)	CO_2 evolution					
	1st hr	2nd	3rd	4th	5th	6th
30	51.7	50.9	52.2	53.6	53.5	53.5
35	68.7	62.8	60.1	61.7	60.9	60.9
40	73.3	55.2	49.0	45.3	43.0	41.2
45	73.5	48.4	41.9	35.9	31.9	28.6
50	74.0	38.8	17.8	12.0	8.0	5.9
55	35.7	12.8	9.7	5.4		

[a] From Kuijper, 1910. Similiar results were obtained with wheat.

TABLE 11.11 DURATION OF HIGH TEMPERATURE, RECOVERY TIME AND RATE (ON RECOVERY) OF NITROGEN ASSIMILATION IN YEAST[a]

Duration of high temperature (hr)	Time for recovery of N assimilation (hr)	Rate of N assimilation after recovery ($\gamma/cm^3/2$ hr)
0	0	6.3
$\frac{1}{4}$	$\frac{1}{2}$	5.3
$\frac{1}{2}$	1	4.2
1	$2\frac{1}{2}$	3.8
$1\frac{1}{2}$	$2\frac{1}{2}$	3.3
2	5	2.5

[a] From van Halteren, 1950.

a. STARVATION

The simplest kind of metabolic injury is quantitative in nature. It is due to the higher temperature optimum for respiration than for photosynthesis (50° and 30°C, respectively in potato leaves; Lundegårdh, 1949). Assimilation reached zero at 45° to 50°C when measured over short periods of time and at 37° to 43°C when measured over longer periods. In the case of 25 species of plants, the temperature maximum for assimilation (36°–48°C) was from 3 to 12 degrees below the heat-killing temperature (44°–55°C; Pisek *et al.*, 1968). The temperature at which respiration and photosynthesis are equally rapid is called the *temperature compensation point.* Obviously, if the plant's temperature rises above the compensation point, the plant's reserves will begin to be depleted. A sufficiently long time at such temperatures would ultimately lead to starvation and death. Since the temperature compensation point drops with light intensity, this kind of injury can occur at relatively low temperatures in shade plants and may account for the low heat-killing points of some algae. This, however, would be the slowest kind of heat injury if the temperatures are moderate, and would normally occur only after one or more days during which the daylight temperature is maintained uninterruptedly at the supracompensation point. In the case of the higher compensation points of terrestrial sun plants, the process would be more rapid. As the temperature rises above the compensation point, respiration rate continues to increase and photosynthesis to decrease and, therefore, the starvation rate would increase exponentially. It is interesting that a thermophilic blue-green alga has no apparent compensation point, at least up to 65°C (Fig. 11.7) and is therefore in no danger of this kind of injury.

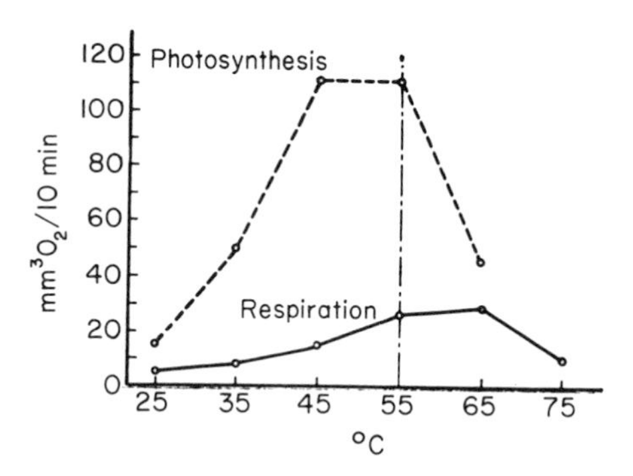

FIG. 11.7. Change in rate of photosynthesis and respiration with temperature. Alga acclimatized at 55°C. (From Marré and Servettaz, 1956.)

b. TOXICITY

A second and more rapid kind of metabolic injury could result if a specific disturbance in the normal process occurred at the high temperature, leading to formation of a toxic substance. This may explain the typical heat injury symptoms induced in French prunes by oxygen concentrations of 1 and 2.5% at 68°F, and the protection against such injury at 86°–100°F by oxygen concentrations of 60–100% (Maxie, 1957). Possibly the heat injury is, in this case, due to toxic products of anaerobic respiration. Petinov and Molotkovsky (1957) suggest that heat injury is due to the toxic effect of NH_3 produced at high temperatures and that this effect is counteracted by the respiratory production of organic acids. This may help to explain the heat tolerance of succulents, since they possess a highly acid metabolism. On the other hand, their acidity is at a maximum when danger of heat is minimal (at night) and at a minimum when the danger is maximal (in the afternoon). In seedlings of *Pennisetum typhoides*, after exposure to 48°C for 12–24 hr, ammonia N was found in detectable quantities (Lahiri and Singh, 1969), injuring the plants.

c. BIOCHEMICAL LESIONS

A third possibility is the production of biochemical lesions (Kurtz, 1958). If the accumulation of an intermediate substance necessary for growth is inhibited at high temperatures, growth inhibition and eventually injury may occur. Evidence of this kind of injury was offered in the case of *Neurospora crassa*. A mutant that was able to grow in standard medium only up to 25°–28°C was induced to grow at as high temperatures as the wild type (35°–40°C) by supplying it with riboflavin. Adenine similarly induced growth at higher temperatures in the case of other plants. This agrees with Galston and Hand's (1949) evidence of adenine destruction at high temperature. Langridge and Griffing (1959) have supported this concept in the case of some races of *Arabidopsis thaliana*. Eight races showed marked decreases in growth at 31.5°C, five of them showing morphological symptoms of high temperature damage. Three of these showed increased growth at this temperature when supplied with vitamins, yeast extract, or nucleic acid. In the case of two of these races, biotin was specifically effective and completely prevented heat lesions. The third race showed a partial alleviation on the addition of cytidine. Not only biotin, but even 1% sucrose may increase the growth of *Arabidopsis thaliana* at supraoptimal temperatures, although the former had a much smaller affect on plants grown at optimal temperatures, and the latter actually decreased their growth (Shiralipour and Anthony, 1970). Sherman (1959) also found a greater nutrient requirement for yeast growth at elevated temperatures. In fact, the occurrence of such biochemical lesions has been suggested as the cause of the slow growth of even the thermotolerant yeasts (Loginova and Verkhovtseva, 1963), since these tend to be more heterotrophic than the mesophiles. Loginova *et al.* (1962), in fact, were able to accelerate the growth of a thermotolerant yeast by addition of Tween 80, ergosterol, or oleic acid. Similarly, Starr and Parks (1962) found that sterol synthesis by yeast was inhibited increasingly above 30°C, becoming critical at 40°C. Death at 40°C was averted by adding oleic acid. In order for cell growth to occur, however, both oleic acid and ergosterol had to be added.

Neales (1968) showed that the vitamin requirement of the roots of one strain of *Arabidopsis thaliana* was greater at 31.5° than at 27°C. On the other hand, in the case of *Neurospora crassa*, riboflavin accumulated to a higher degree at a high temperature (37° and 40°C) than at 25°C (Ojha and Turian, 1968), even though injury occurred at 40°C (mitochondria shrunken, endoplasmic reticulum broken). Therefore, biochemical lesions are not the answer in all cases.

Just how these biochemical lesions arise is not known. One possibility has been suggested by Al' tergot (1963). He found that a temporary oxygen deficit decreased heat injury, while excess oxygen increased it. This is due to the thermostability of oxidases which, therefore, are able to destroy the readily oxidized substances. If these substances (e.g., ascorbic acid, glutathione), are essential, this destruction would produce a biochemical lesion.

There is, of course, no direct evidence, that added substances are producing their effect in the manner postulated, and other explanations are equally valid. In one case, in fact, this mechanism was excluded since the protection was obtained even if the substance was supplied after the heat treatment. Air-dry lettuce seed, were injured by a 1-hr exposure to 75°C as shown by a reduced subsequent germination. A concentration of 10^{-5} kinetin in the germination medium after the heat treatment prevented this injury (Porto and Siegel, 1960). This has been corroborated by Ben-Zeev and Zamenhof (1962).

d. Protein Breakdown

A fourth kind of metabolic injury (which may be considered as a special kind of biochemical lesion) is a net breakdown of protoplasmic proteins. Lepeschkin (1935) was probably the first to produce indirect evidence of this process and of the importance of synthesis or repair of heat injury. He found that interrupting the exposure to high temperature at the mid-point by 2 min at 20°C had no effect on the total time needed to produce killing, of *Spirogyra*. When the interruption was for $2\frac{2}{3}$ hr at 20°C, however, a longer total time at the high temperatures was needed for heat killing. From the brief heat break, he concludes that the protein denaturation is physically and chemically irreversible, but from the long break he concludes that it can be repaired by physiological activity. The same conclusion was arrived at by Allen (1950), who showed that, in the absence of nutrients, thermophilic bacteria die at 55°C just as rapidly as the mesophilic bacteria. The enzyme systems of the thermophiles were rapidly inactivated at this temperature. She concluded, therefore, that the thermophiles can synthesize enzymes and other cell constituents far faster than they are destroyed by heat, and that they have higher coefficients of enzyme synthesis than in the mesophiles. According to this concept, heat killing occurs when the speed of resynthesis of an indispensable component (e.g., an enzyme or an intermediate substance) is unable to compensate for its degradation. In support of this concept, some strains of *Neurospora crassa* are able to produce the hydrolytic enzyme cellulase in the presence of cellulose at 35°C (Hirsch, 1954). At 25°C they fail to produce the enzyme. It is perhaps possible that injury is due to a similar self-digesting proteolytic enzyme produced at high temperatures, perhaps from lysosomes.

On the other hand, the net loss may be due to a decreased synthesis rather than an increased breakdown. Several cases of uncoupling at high temperatures have been reported. In corn mitochondria, coupling between oxidation and phosphorylation began to decrease at 30°–35°C and at 40°C uncoupling was complete (Kurkova and Andreeva, 1966). The decreased phosphorylation would certainly lead to decreased synthesis of proteins as well as other substances. In *Chlorella* sp. K, when grown at 37°C and 43°C, accumulation of biological mass was uncoupled from cell multiplication, leading to hypertrophy (Semenenko *et al.*, 1969). Carbohydrates were synthesized at a high rate, and after 7 hr comprised up to 45% of the dry weight, while protein decreased to 18%. During the first hour the protein increased. Thermal uncoupling in chloroplasts may also occur at temperatures lower than those normally leading to heat inactivation (Emmett and Walker, 1969). Similarly, in the case of the psychrophile *Sclerotinia borealis* grown at 0°C, the maximum temperature for growth is 15°C, but the optimum for respiration is 25°C. It has, therefore, been postulated that growth becomes uncoupled from respiration above 15°C (Ward, 1968). In support of this explanation, the uncoupling agents, 2,4-DNP and dicoumarol stimulated oxygen uptake relatively more at 5° than at 25°C. In the case of peas, however, no change in the efficiency of respiration could be detected at 41°–43°C compared to 18°–20°C (Nikulina, 1969). In the case of wheat roots, DNP induced the same injury as a 45° heat shock (Skogqvist, and Fries 1970).

Uncoupling is not the only possible cause of decreased protein synthesis at high temperatures. When *Physarum polycephalum* was subjected to heat shocks at 40° for periods of 10 or 30 min (Schiebel *et al.*, 1969), the incorporation of amino acids into protein was decreased by approximately 40 or 70%, respectively. There was also a decrease in polyribosomes of more than 50%, which was postulated to be the cause of the decreased protein synthesis. In the psychrophile *Pseudomonas* sp., RNA synthesis was not detected at supramaximal temperatures (Harder and Veldkamp, 1968).

The occurrence of metabolic heat injury due to proteolysis, has now been established in higher plants by Engelbrecht and Mothes (1960, 1964), but it occurs in a different manner from that proposed by Allen for microorganisms. A 1 to 2-min exposure of a leaf of *Nicotiana rustica* to 49°–50°C produced a reversible, sublethal "heat-weakening". The leaf remained turgid after the heat treatment, but the normal yellowing (aging) occurred more rapidly than in the control. Amino acids were also translocated to the unheated half of the leaf only if the other half had been exposed to the heat treatment. When half the leaf was sprayed with kinetin, this prevented the above described (heat-induced) yellowing and the kinetin-treated leaf half accumulated amino acids, maintaining or increasing the protein content. This sublethal heat weakening could even be reversed by a kinetin treatment after the heating (confirming Porto and Siegel, 1960; see above), as well as by other treatments that checked the efflux of metabolites. A similar proteolysis occurred in 3-week-old plants of *Pennisetum typhoides* when exposed to 48° (±1°C) for durations up to 24 hr (Lahiri and Singh, 1969). Soil water stress was negligible, so this was solely due to the heat stress. Wheat roots exposed to a heat shock of 45°C were uninjured if this was followed by 25°C, but were injured if it was followed by 35°C (Skogqvist, and Fries 1970). Kinetin overcame or prevented the injury, and so did chloramphenicol. In the case of a thermal blue-green alga (*Aphanocapsa thermalis*), the optimal temperature for growth and photosynthesis was 40°C (Moyse and Guyon, 1963). At higher temperatures, which had a strongly inhibiting effect, both amino acid synthesis and their condensation to proteins were slowed down to a greater extent than was carbohydrate synthesis.

Alexandrov (1964) has produced evidence of a repair mechanism even during the heating. He observed cessation of cytoplasmic streaming after 90 min at 42.0°C, but after 200 min at the same temperature, some streaming recommenced, and streaming was practically normal at 6 hr. After a longer period, however, the cells eventually died.

This fourth kind of metabolic heat injury may therefore be due to a net loss of protoplasmic proteins, but it may occur after the heat stress has been removed. It may be suggested that a slightly more severe heat stress could perhaps produce the same kind of net protein loss during the heating. Engelbrecht and Mothes (1964) showed that this does not occur. The slightest increase in heat stress (1–2 min exposure to 50°–52°C) resulted in an "irreversible" heat injury. The normal degradation of proteins was actually inhibited, the heat damage occurred during the stress, and kinetin treatment failed to prevent the injury. Early evidence (Illert, 1924) had indicated that this kind of injury is not metabolic in nature for it was unaffected by CO_2 or oxygen supply. Yet Engelbrecht and Mothes (1964) were able to prevent it by preheating of the leaves at 46°C, i.e., by a hardening treatment. Heat tolerance (see Chapter 12), therefore, is induced by these hardening treatments against this metabolically "irreversible" rapid heat injury which is, therefore, a direct injury.

3. Primary Direct Heat Injury

All the above described four kinds of primary indirect heat injury are slow processes. If the injury is due to starvation, toxins, or biochemical lesions, it may require an exposure to the high temperature for hours or even days. If due to a breakdown of protoplasmic proteins, it may result from a brief exposure to high temperature, but it then appears only hours or days after the heat stress. Direct heat injury is induced by brief exposures (seconds to 30 min) to the heat stress and appears either during the heating or immediately after, though it may progress for 24 hr or more after cooling (see Chapter 12). In further contrast to the fourth kind of indirect injury, which leads to a translocation of raw materials for protein synthesis to unheated parts, direct heat injury may actually result in damage to unheated parts. Yarwood (1961b) demonstrated that heat injury could be translocated from one primary bean leaf killed by heating for 10 sec at 65°C to the opposite unheated one. If the heated leaf was removed quickly, the unheated leaf was uninjured. In spite of these differences, the fourth kind of indirect heat injury has at least two features in common with direct heat injury (see above). For these and other reasons, the two may for some purposes be classified together (see Fig. 11.10).

Three hypotheses have been proposed to explain the mechanism of direct heat injury.

a. Protein Denaturation

This was the earliest explanation (Belehradek, 1935) and is the commonly accepted one to this day. Brock (1967) has recently rejected this explanation because heat injury follows first-order kinetics. However, first-order kinetics are characteristic of monomolecular reactions, and protein denaturation is simply the unfolding of a protein molecule. Brock's objection is, therefore, actually a point in favor of denaturation as the cause of heat injury. Unfortunately, the kinetics of the heat denaturation process are difficult to measure, because of the difficulty in separating it from the subsequent aggregation (see below). The early evidence in favor of the denaturation concept was the commonly observed coagulation of protoplasm in cells heated under the microscope. Sachs (1864) describes a heat solidification of protoplasm that may be reversible on cooling. As the small epidermal strips from young leaves or flower buds of *Cucurbita pepo* were warmed, protoplasmic streaming in the hair cells speeded up, until it became very violent. At higher temperatures, strands were pulled vigorously into one larger protoplasmic mass. Finally, the protoplasm all lay at rest against the cell wall. Five to ten minutes after cooling, protuberances gradually began to form, and the network of strands was slowly regenerated. This heat solidification of the protoplasm occurred when the strips were plunged into water at 46° to 47°C for 2 min. Even after exposure to 47°–48°C, streaming recommenced within 2 hr of cooling. In air, higher temperatures had to be used, e.g., 25 min at 50°–51°C. The solidification was then reversed only after a 4-hr cooling. Lepeschkin (1912) states that all layers of the protoplasm including the plasma membrane coagulate simultaneously at the heat-killing temperature. Therefore, he determined protoplasmic coagulation by the time semipermeability of the plasma membrane was lost, i.e., when the pigments of the cell sap were observed to diffuse out under the microscope. He later (1935) states that protoplasmic coagulation of plasmolyzed cells was recognized first by a rapid decrease in protoplast volume, and that the injury worked inward from the outer protoplasm layers (1937). It required a higher temperature for the color to begin to leave the cell. Recovery was not possible after the chloroplasts had begun to coagulate, but the coagulation of the superficial protoplasm layer was reversible. In agreement with Lepeschkin's observations, soybean and *Elodea* exposed to sublethal temperatures showed a loss of chlorophyll and swollen chloroplasts (Daniell *et al.*, 1969). At the thermal death point, disorganization of the tonoplast, plasmalemma, and chloroplast membranes occurred. It was, therefore, concluded that the primary cause of the injury is disintegration of the cell membranes.

Other observations, however, indicated that injury could occur in the absence of observable coagulation. In opposition to Lepeschkin's earlier (1912) description, Bogen (1948) observed loss of color from *Rhoeo discolor* cells before any protoplasmic change could be detected. Lepeschkin's later (1935) observations seem to agree with Bogen. He describes four stages of heat coagulation in *Spirogyra* cells: (1) An imperceptible change in dispersion is detected by an increased permeability to water. The starch grains show just detectible swelling. (2) Starch swells significantly due to a greater increase in permeability, and coagulation of the protoplasmic surface begins. There is often a movement of the chloroplast ribbon toward the middle of the cell. (3) Complete heat swelling of the starch follows the complete coagulation of the chloroplast. (4) The proteins coagulate completely. Therefore, both investigators seem to agree that the first sign of injury is an increase in permeability.

Döring (1932) was able to detect heat swelling of chloroplast starch in living cells and suggests that this may actually injure the protoplasm. He concluded that the tonoplast is more heat tolerant than the rest of the protoplasm and that the changes in heat tolerance under different conditions may not be the same in these two protoplasmic components. Scheibmair (1937) observed the first signs of heat injury in the chloroplasts of mosses. They enlarged and became pale and irregular in contour. At the instant of death, the elaioplasts also changed observably and the whole protoplast contracted in apparent plasmolysis, which was easily distinguished from true plasmolysis by the angular form. Perhaps due to the chloroplast injury, the leaves of some plants (e.g., *Oxalis*) change from green to yellow on heat killing, though others (e.g., Polygonaceae) fail to show this color change (Illert, 1924).

Dangeard (1951a,b,c) attempted to find out how far disorganization by heat can proceed without causing death. He examined sections of fixed radicles after exposing them to heat-killing temperatures for various lengths of time. The chondriosomes (mitochondria) were rapidly destroyed, often by less than 1 min at 55° to 60°C. At lower temperatures, an hour or more was needed. They were rarely destroyed at 42°C or lower. In some cases their destruction was accompanied by the survival of a small percentage of the cells, judging survival by their appearance when fixed. Belehradek (1935) reviews many other changes observed by different workers in cells undergoing heat injury, e.g., the formation of granules, vacuolization, and protoplasmic contraction. Liberation of lipids has been recorded, even from the cell walls. The nucleus has been found the most heat-sensitive part of the cell, though more heat tolerant in the resting than the dividing state.

Since many proteins are denatured at the heat-killing temperatures for many plants, this has long been accepted as the explanation of the above observed cell coagulation, and of heat killing in general. Direct evidence in its favor is obtained from the reaction times. Lepeschkin (1912) points out that the same logarithmic relation to temperature has been found for the heat-killing time as for the protein coagulation time in solutions, although the former is more rapid.

The Q_{10} for heat killing of protoplasm is very high, though it varies markedly from plant to plant as well as with the conditions (Table 11.12). According to Belehradek (1935), it is usually low when the tissues are dry or semidry, or in general when heat tolerance is high, and also at temperatures slightly above the optimum. This lack of constancy has been stated to preclude a single cause of death (Illert, 1924). The determinations do not always involve simply heat killing. Sometimes, they really measure the acceleration at high temperatures of death that would normally occur at a slow rate even at room temperature, e.g., in the case of excised leaves (Belehradek and Melichar, 1930).

The only two processes known to have as high Q_{10} values as heat killing are the heat coagulation of proteins and the conversion of starch into paste (Lepeschkin, 1935). The actual values for heat killing usually fall within the range for heat coagulation of proteins. The Arrhenius formula gives values for μ of from 65,000 to 132,000, for the cortical parenchyma of several trees, which also fall within the range for heat coagulation of proteins (Lorenz, 1939). Christophersen and Precht (1952) even suggest that the death rate of cells follows the curve for a monomolecular reaction and therefore may be due to the breakdown of a single molecule. They admit that in the case of bacteria, deviations from the curve for monomolecular reactions are frequent.

Against the explanation of heat killing by protein denaturation is the long recognized fact that many organisms are killed at temperatures too

TABLE 11.12 TEMPERATURE COEFFICIENTS (Q_{10}) FOR HEAT KILLING OF PLANTS

Plant	Q_{10}	Temperature range (°C)	Reference
Barley grains (soaked 1 hr)	10	55–70	Goodspeed, 1911
Ceramium tenuisimum	37.6	28–38	Ayres, 1916
Tradescantia discolor	26	40–65	Collander, 1924
Beta vulgaris	71	40–60	Collander, 1924
Brassica oleracea	80	40–60	Collander, 1924
Elodea densa	31	35–55	Collander, 1924
Drapernaldia glomerata	43	35–55	Collander, 1924
Pisum sativum seedlings	118	35–55	Collander, 1924
Elodea canadensis	7.6	22–44	Belehradek and Melichar, 1930
	50	44–65	
American elm	19 (lower slope of curve)	55–68	Lorenz, 1939
	243 (upper slope)		
Catalpa	360	54–65	Lorenz, 1939
Northern white pine	63	54–66	Lorenz, 1939
White spruce	73 (lower slope)	53–69	Lorenz, 1939
	3.6 (upper slope)		Lorenz, 1939
Red pine	66	55–66	Lorenz, 1939
Monterey pine seedlings	28–100	49–55	Baker, 1929
11 Species	123–2150	43–52(?)	Alexandrov, 1964

low for the denaturation of known proteins. From such evidence, early investigators (Sachs, 1864; Just, 1877) concluded that protein coagulation cannot be involved. Collander (1924) points out that the temperature for denaturation of protoplasmic proteins is unknown. Furthermore, not all the cell proteins need be denatured in order to cause cell injury. There is always the possibility that a single sensitive protein may be responsible for the injury. Recent investigations lend support to this possibility. Alexandrov (1964) found that cytoplasmic streaming is halted at a lower temperature than other cell processes. He, therefore, concluded that the proteins responsible for cytoplasmic streaming are more sensitive to heat denaturation than are other cytoplasmic proteins. Other results from his laboratory showed specific differences between proteins. The thermostability of urease in leaf homogenates was lower in the mesophilic *Leucojum vernum* than in the thermophilic species, *L. aestivum* (Feldman and Kamentseva 1967). This was correlated with the thermotolerance of protoplasmic movement and respiration. The inactivation temperature for urease required much higher temperatures than the highest heat-killing temperature for higher plants. It required 40 min at 78°C to produce 50% inactivation in the less tolerant species, and the same time at 81°C in the more tolerant species. On the other hand, acid phosphatase (a much more heat-labile enzyme) was 50% inactivated at 48°–55°C (Feldman *et al.*, 1966).

Lepeschkin (1935) points to the effects of various factors on heat injury as proof of the protein denaturation theory. Small amounts of acid (e.g., nitric) lowered the heat-killing temperature of *Spirogyra*, and also the heat-coagulation temperature of proteins. Brock and Darland (1970) have obtained a similar effect of pH on the maximum temperature for the most extreme thermophilic bacteria found growing at boiling temperatures (92°–100°C). At pH 2–3, the upper limit was lowered to 75°–80°C. Conversely, Lepeschkin raised both the heat-killing temperature and the heat coagulation temperature by use of very dilute alkalis. Narcotics, such as alcohol, ether, chloroform, and benzol, lowered both temperatures. Even the protective effect of plasmolyzing solutions against heat injury (see below) is in accord with the fact that denaturation of proteins occurs at higher temperatures in concentrated than in dilute solutions.

He also points to the effect of salts in lowering the temperature for both processes. The effect of salts on heat killing (Table 11.13) is complicated and not agreed on by all workers. Kaho (1921, 1924, 1926) plasmolyzed epidermal strips in single salt solutions and observed them with a horizontal microscope while their temperature was being raised at a constant rate in a water bath. The plasmolyzed protoplasts were seen to expand at a speed depending on the salt used. Shortly before death, the swelling was especially striking. The temperature resulting in rupture of the cells with expulsion of their contents was taken as the heat-coagulation temperature. In some salt solutions, no such rupture occurred, and a less sharp coagulation temperature was obtained. The time from initiation of warming to the complete decolorization of the cells averaged 15 min.

The anions lowered the heat-coagulation temperature more than did the cations, and followed the lyotropic series. For potassium salts, the order was: $CNS^- > Br^- > I^- > NO_3^- > Cl^- >$ tartrate$^{2-} > CH_3COO^- >$ citrate$^{3-} > SO_4^{2-}$. The range of temperatures was from 67.5° (for KCNS) to 76.5°C (for K_2SO_4). The cations also had some effect in the order: K^+, $NH_4^+ > Na^+$, Li^+, $Ca^{2+} > Mg^{2+}$, Ba^{2+}, Sr^{2+}.

As a general rule, the most rapidly penetrating salts, judging by the rate of protoplast expansion, lowered the heat-coagulation temperature the most. This conclusion is also in agreement with the order of permeability found by other workers. Heat coagulation of egg albumin was also

TABLE 11.13 PROTECTION BY DEHYDRATING SOLUTIONS AGAINST HEAT KILLING OR SHOCK (INACTIVATION OF AN ENZYME OR METABOLIC PROCESS)

Plant	Solution	Killing or shocking temperature (°C)	Time exposed (min)	Reference
(a) Heat Killing				
Agave americana	Water	57	10	de Vries, 1871
	10% NaCl	58.2	10	
Hyacinthus orientalis	Water	47.5	10	de Vries, 1871
	10% NaCl	55.4	10	de Vries, 1871
Saxifraga sarmentosa	Water	44.0	10	de Vries, 1871
	10% NaCl	58.2	10	de Vries, 1871
Fucus eggs	Seawater	45	1	Döring, 1932
	Seawater + 1*M* sucrose	50	1	
Plagiochila asplenioides	Water	48	$4\frac{1}{2}$	Scheibmair, 1937
	$2\frac{1}{2}$ *M* sucrose	>52	9	
Hookeria luscens	Water	52	3	Scheibmair, 1937
	1 *M* sucrose	>52	8	
Rhoeo discolor	Water	60	20	Bogen, 1948
	$\frac{1}{8}$ *M* sucrose (hypotonic)	60	17	
	$\frac{1}{2}$ *M* sucrose (hypertonic)	60	36	
(b) Heat shock				
Torula utilis	Water	44	1 hr	Van Halteren, 1950
	1–3 *M* NaCl, glucose, sucrose	>44	1 hr	
Torulopsis kefyr	Water	54		Christophersen and Precht, 1952
	$\frac{1}{3} - \frac{2}{3}$ *M* maltose	60		

found to follow the same series, provided the solutions were acid. Magnesium sulfate gave exceptional results, owing (according to Kaho) to its strong hydrolysis and acidity, which enhance coagulation. That the salts are toxic to the cell even in the absence of an injurious high temperature was shown by determining the death rate of the cells in the salt solutions at temperatures from 0° to 36°C. Kaho cautions that this heat coagulation of the protoplasm cannot be taken as synonymous with heat injury, since it is only the last link in a chain of phenomena that may cause injury at high temperatures.

Kaho was actually able to raise the heat-coagulation temperature with Ca salts, if the correct anion was used. In agreement with these results, Scheibmair (1937) found that a one-half hour treatment in a hypotonic solution previous to heating in water resulted in a higher heat-killing temperature when $CaCl_2$ was used and a lower one when KCl was used.

Bogen (1948) plunged sections directly into the heated salt solution and determined the heat-killing time by loss of vacuole pigment. Monovalent cations of equimolar concentration lowered heat resistance according to their position in the lyotropic series, i.e., $Li^+ > Na^+ > Rb^+ > Cs^+$. Monovalent anions lowered the heat-killing temperature according to their order of adsorption, i.e., $SCN^- > NO_3^- > Br^- > Cl^-$, $C_2O_4^{2-}$. Multivalent cations did not follow the lyotropic series and he explains their behavior as due to the interference of the oppositely acting adsorption effect, i.e., a discharging of the colloid. Thus, when both monovalent and divalent cations are grouped together, the following is the order of injury: control $< Mg^{2+} < Cs^+ < K^+ < Na^+ < Li^+ < Ca^{2+} < Ba^{2+} < Al^{3+}$. When he combined the anion and cation that were both least injurious in their respective series (i.e., $MgSO_4$) he actually obtained an increased heat tolerance. This was the only salt tested that produced an increase. The change in heat tolerance was proportional to the salt concentration in an exponential manner, i.e., when the concentrations were plotted in a geometric series, a straight-line relation was obtained with killing time. Mixtures failed to produce any antagonistic effect. Thus, KCl plus $CaCl_2$ produced an additive effect, while LiCl plus $MgSO_4$ produced purely a $MgSO_4$ effect.

Lepeschkin (1912, 1935) found that mechanical agents lower the heat-killing temperature. Cutting the tissues produces a sensitization that may last up to 15 hr in the case of beets. Bending of *Spirogyra* filaments has a similar effect. The farther from the cut or bend a cell is, the less its heat-killing temperature is affected. In contrast to Lepeschkin, Scheibmair (1937) found that centrifuging actually raised the heat-killing temperature of the moss *Plagiochila*. In a few cases, however, it was lowered.

Lepeschkin also found that light increased the speed of heat killing. Low concentrations of narcotics raised the heat-killing temperature though not affecting protein denaturation. Scheibmair (1937) was unable to confirm this last result and was able to obtain only a lowering of the heat-killing temperature.

It is obvious that these early investigators failed to differentiate between denaturation and coagulation of proteins. Modern investigations of pure proteins have, however, provided the basis for a reevaluation of the protein denaturation concept of heat injury. Protein denaturation is due to an unfolding of the molecule, with consequent loss of the activity possessed in the native state. According to Haurowitz (1959), such unfolding occurs when the temperature is high enough to break the relatively weak H bonds (bond strength usually 2–3 kcal) which help to hold the folds in place in the native proteins. More recent evidence, however, has shown that the folds in the tertiary structure of the protein molecule are mainly held by hydrophobic bonds. As mentioned earlier, it is the weakening of these hydrophobic bonds that results in the reversible denaturation which occurs at low temperature. Recent evidence (Brandts, 1967) indicates that a similar reversible denaturation occurs at high temperatures. As the temperature rises, the conformational entropy favoring the denatured state increases more rapidly than the increase in strength of the hydrophobic bonds (Fig. 11.8) and a temperature is finally reached at which unfolding begins. The first effect of the high temperature on the proteins is, therefore, denaturation. This remains reversible, unless fol-

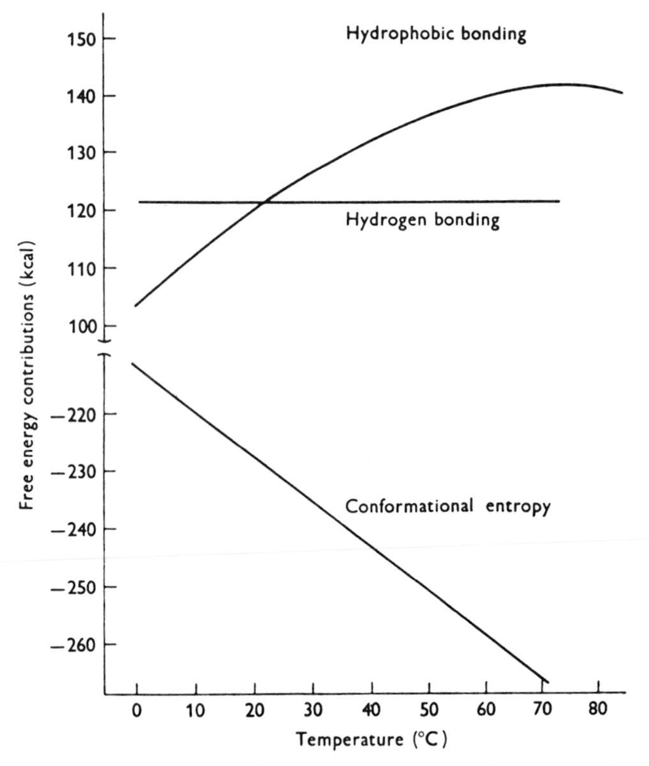

FIG 11.8. The relationship of hydrophobic bonding, hydrogen bonding, and conformational entropy to temperature. (From Brandts, 1967.)

lowered by aggregation:

$$N \underset{\text{normal } T}{\overset{\text{high } T}{\rightleftharpoons}} D \xrightarrow{\text{high } T} A$$

where N = native; D = denatured; A = aggregated proteins. Many proteins are converted so rapidly to the irreversible aggregated state that the reversible denaturation is difficult to detect. This is, no doubt, why it has been overlooked until recently. It also explains the observation of coagulation (an aggregation of proteins to form a gel structure) in heat-injured cells. Aggregation is irreversible only in the thermodynamic sense. Sachs' observation of reversal of heat coagulation may, perhaps, have been due to an enzymatic solubilization of the irreversible aggregate. On the other hand, it was reversible only for a short time, resembling the heat denaturation of proteins. It is, therefore, possible that the initial reversible coagulation was actually due to reversible denaturation, and that only the later irreversible coagulation was due to aggregation.

In the case of the larger protein molecules, aggregation may be preceded by disaggregation on heating; for the breaking of bonds may occur not only within a protein subunit, but also between the subunits of a large multimer. For instance, a highly purified enzyme complex from bakers' yeast possessed two different enzymatic activities and a regulating site for both activities. After heating at 50°C for 5 min, the complex was disaggregated into subunits $\frac{1}{4}$ the size of the original complex, and these possessed only one of the enzymatic activities (Lue and Kaplan, 1970).

The hydrophobic bond strength increases with temperature up to and beyond the heat-killing temperature of most cells; at heat-killing temperatures their strength is greater than that of the hydrogen bonds (Fig. 11.8). Therefore, unlike low-temperature denaturation, heat denaturation may be expected to involve the breaking of H bonds even before the hydrophobic bonds. This conclusion has been supported by analysis of heat denaturation with infrared spectrometry (Boyarchuk and Vol'kenshtein, 1967). The number of CO and NH groups increased in all the six proteins tested. This indicates the disruption of secondary and tertiary protein structures by the abolition of many hydrogen bonds.

These recent determinations of heat denaturation of proteins at high temperatures have demonstrated that the less stable enzymes denature reversibly at temperatures considerably lower than previously suspected (Fig. 11.9). In the case of chymotrypsinogen A, for instance, the measured rates of H exchange at pH 5 and 37°C were higher and the activation energies lower than expected from thermal cooperative unfolding (Rosenberg and Enberg, 1969). The heat stability of enzymes, in fact, covers a wide range of temperatures (see Tables 12.9 and 12.10). It is, therefore, not

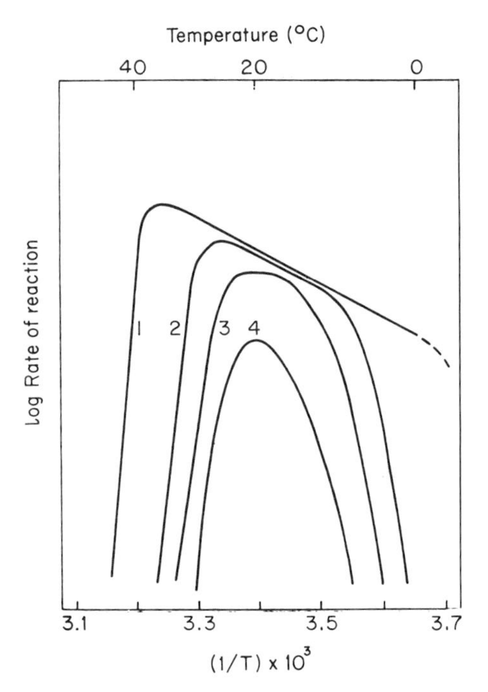

FIG. 11.9. Arrhenius plots for four enzymatically controlled reactions. (From Brandts, 1967.)

surprising that denaturation of some proteins can occur at low enough temperatures to fall within the heat-killing range.

Unfortunately, it is difficult to interpret results obtained with pure proteins or cell extracts in terms of the normal, living cell. The stability of proteins may be greater when in the cell due to the presence of protective substances. Evidence of this is the 10°C higher inactivation point for crude amylase than for the crystalline product (Roy, 1956). Nakayama (1963), in fact, has isolated and partially purified a small-moleculed substance from Japanese radish leaf which protects R-amylase from sweet potato against heat inactivation. The half-life of trehalase and invertase at 60° and 65°C was much greater when intact ascospores were heated than in the case of extracts of *Neurospora tetrasperma* (Yu *et al.*, 1967). Dialysis of the ascospores removed the protection. Similarly, complex formation with DNA (and also with CM-cellulose) prevented heat coagulation of serum albumin, ovalbumin, and chymotrypsinogen A (Hofstee and Bobb, 1968). Cystathionase and cysteine sulfinic acid decarboxylase were protected against heat denaturation by the cofactor pyridoxal phosphate

(Chatagner *et al.*, 1968). Even the artificial attachment of a dye molecule (methyl orange) to a protein (serum albumin) increased its thermostability by 2.5°C (Vitvitskii, 1969). Many other examples are to be found in the literature.

It is even possible to distinguish large differences between the thermostability of proteins in the pure state, which are not detectible under natural conditions. The protein of tobacco mosaic virus (TMV) is denatured at 40.5° in the vulgare form and at 27°C in the mutant (Jockush, 1968). However, this difference between the two proteins was difficult to detect in the whole virus particle, due to stabilization by coaggregation with RNA. Similarly, quaternary aggregation (at pH 5.0) shifts the denaturation temperature upward by 20°–25°C.

On the other hand, substrate decreased the heat stability of succinate cytochrome *c* reductase but increased the heat stability of succinate oxidase (Luzikov *et al.*, 1967). The thermal inactivation of yeast cytochrome b_2 is increased in the presence of flavin mononucleotide (FMN) (Capeillare-Blandin, 1969). Similarly, though the presence of other proteins may have a protective effect, the opposite is also possible. The ovomucin–lysozyme complex is heat denatured at a lower temperature than that observed for either protein separately (Garibaldi *et al.*, 1968).

It must also be realized that activity and conformation are not always changed simultaneously by heat, so that injury to the cell might be expected at a lower or a higher temperature than the denaturation temperature. Lysozyme, for instance, is stable up to 75°C, yet the activity reached a maximum at 50°C, decreasing markedly above this temperature (Hayashi *et al.*, 1968). The loss of activity at 60°–70°C was believed due to the formation of an enzyme–substrate complex difficult to hydrolyze.

In the case of psychrophiles, the relationship of heat injury to protein denaturation appears to be clear-cut. Pyruvate carboxylase was 50% inactivated after it was heated for 10 min at 35°C (Grant *et al.*, 1968). This agreed with the cessation of glucose fermentation by resting cells (and cell-free extracts) after 30 min at 35°C. Similarly, the amino acid incorporating system of the psychrophilic yeast *Candida gelida* was completely inhibited after incubation at 35°C for 30 min (Nash *et al.*, 1969). This was due in part to unusually temperature-sensitive aminoacyl-sRNA synthetases. Of the thirteen tested, seven retained less than 50% of their activity after 30 min at 35°C. Leucyl-sRNA synthetase was inactivated after only 7 min at 35°C. *Candida gelida* also possessed thermolabile soluble enzymes involved in the formation of ribosomal-bound polypeptide chains. Similar results were obtained with another psychrophile, *Micrococcus cryophilus* (Malcolm, 1968, 1969). It cannot grow above 25°C due to an inhibition of protein synthesis. Heating for 10 min at 30°C inactivates three species of tRNA synthetase of this psychrophile, but has no effect on those of a mesophile or thermophile. It is primarily the glutamyl-tRNA synthetase and the prolyl-tRNA synthetase that are inactivated above 25°C. The tRNA is still functional. In the case of the marine psychrophile *Pseudomonas* sp., RNA synthesis could not be detected at supramaximal temperatures (Harder ant Veldkamp, 1968). In contrast to *C. gelida*, the isoleucyl- and leucyl-tRNA synthetases from *Bacillus stearothermophilus* (a thermophile) and *Escherichia coli* (a mesophile) showed no striking difference in thermostability. They were able to amino-acylate the tRNA *in vitro* at temperatures above 70°C (Charlier *et al.*, 1969). It is, therefore, apparently only in the psychrophiles that this mechanism of heat injury functions. In these organisms, the mechanisms of indirect and direct injury are apparently identical; the maximum temperature for growth is due to cessation of protein synthesis (indirect heat injury) which in its turn is due to the direct heat inactivation of enzymes (synthetases).

From all these results, it must be concluded that protein denaturation or inactivation does, indeed, occur in the living cell at a low enough temperature to account for heat killing, at least in some cases. It must be remembered, however, that protein denaturation at high temperatures is a reversible process and does not become irreversible until followed by aggregation. It, therefore, does not seem likely that the rapid heat killing can occur unless the denatured proteins are also aggregated. What kind of bond formation may lead to this aggregation? Some evidence now points to a possible explanation. Just as in the case of freezing injury, heat injury results in an increased SS content of the proteins at the expense of SH

TABLE 11.14 EFFECT OF HEAT KILLING (15 MIN AT 58°C) ON SH CONTENT OF SUPERNATE FROM KHARKOV WHEAT AFTER ABSORBING 60 GM H_2O PER 100 GM GRAIN[a]

	SH	SH + 2SS	%SH (of SH + 2SS)
A. Unvernalized (3 days at 20°C)			
Control	0.25	0.85	30
Heat killed (1)	0.16	0.80	20
(2)	0.12	0.90	13
B. Vernalized (40 days at +3°C)			
Control	0.52	1.36	38
Heat killed	0.20	0.96	21

[a] From Levitt, 1962.

groups (Table 11.14). These results have been confirmed by Kotlyar *et al.* (1969). When germinated wheat was heated to 40°–80°C, the level of SH groups in the water-soluble proteins decreased, and the SS: SH ratio increased. Similarly, the proteins of soybean hypocotyl show maximum thermostability after a 4-hour incubation of the tissue with $10^{-5}M$ 2,4-D and this was correlated with a decrease in SH content (Morré, 1970) and, therefore, a decrease in ability to form intermolecular SS bonds on heating. The cysteine-cystine content of the flagellar proteins (as well as of the ribosomes; Friedman, 1968) of bacteria is very low, and does not differ significantly in thermophiles and mesophiles (Mallett and Koffler, 1957). Furthermore, the addition of 0.2 *M* sodium thioglycollate at pH 8 in 5 *M* urea had no effect on the thermostability. Therefore, SS bridges do not seem to form in these proteins. They may not, however, be the ones responsible for heat injury to bacteria, and SH groups have been found in their membrane proteins (see Chapter 10). Nevertheless, other bonds may conceivably be involved in heat-induced protein aggregation, for instance, hydrophobic bonds may be formed between hydrophobic groups unmasked by the heat-induced rupture of H bonds.

The inverse relationship between killing temperature and moisture content is also explainable by protein denaturation; there is a quantitative relation between moisture content and protein denaturation. This is because the unfolding can take place only in the presence of adequate water, which permits the necessary freedom of movement of the protein molecules. Thus, the rate of inactivation of a protein (e.g., sweet potato β-amylase at 63°C; Nakayama and Kono, 1957) increases with decreasing protein (i.e., enzyme) concentration. Consequently, dehydrated protoplasm can survive high temperatures without injury. In the case of some seeds, in fact, there is no injury until the temperature is high enough to break the valence bonds in the proteins and other protoplasmic substances, as shown by the charring after prolonged periods at these temperatures. The above two types of heat injury can therefore be called denaturation injury and decomposition injury, respectively.

Between the two extremes, a combination of the two types of injury may conceivably occur; because some purely chemical reactions (e.g., caramelization of sucrose) can also take place. That there are two types of injury is indicated by the break in the Arrhenius plot (Fig. 11.2) of the reciprocal of the killing temperature versus log of time (the reciprocal of the reaction rate). When this break occurs at relatively low temperatures (e.g., at 42°C; Fig. 11.2) it is explainable by indirect heat injury in the lower range and by direct heat injury in the upper range. When it occurs at a relatively high temperature (e.g., 60°C for white spruce; Lorenz, 1939) it is explainable by denaturation injury in the lower range and by decomposition injury in the upper range.

As mentioned above, heat denaturation of proteins has a temperature coefficient as high as that of heat killing. This is the strongest evidence in favor of the protein denaturation theory of heat injury in the intermediate heat-killing zone. In agreement with this, heat (decomposition) injury in the highest heat-killing zone does not have such high temperature coefficients (Belehradek, 1935). This conclusion has been corroborated for white spruce (Table 11.12), the Q_{10} dropping from 73 for temperatures below 60°C to 3.6 for temperatures above it (Lorenz, 1939). In terms of activation energy, the values were 94 and 28 kcal/mole, respectively. Siegel's (1969a) more recent results are in good agreement with these—93 and 19 kcal/mole. He found that beet root tissue differs in stability toward oxygen at moderately high (45–60°C) and very high (60°–100°C) temperatures.

b. Lipid Liquefaction

The observed liberation of lipids at high temperatures led Heilbrunn to suggest that heat killing may be due to liquefaction of protoplasmic lipids (see Belehradek, 1935). Many attempts have been made to support this theory, but without success (Belehradek, 1935; Campbell and Pace, 1968). Lepeschkin (1935) concludes that since the lipids occur in such thin layers in the protoplasm, they would liquefy instantly at a definite temperature and therefore could not account for the high Q_{10} values for heat killing. He believes that lipids are freed as the result, rather than the cause, of death. Support for this view is provided by modern concepts of membrane lipid attachment to proteins by either electrostatic or hydrophobic bonds. The liberation of membrane lipids at high temperatures would then require breaking of these bonds, and not simply a liquefaction of the lipids. The attachment to the proteins might also conceivably alter the liquefaction temperature. This does not happen in the case of *Mycoplasma laidlawi* (Reinert and Steim, 1970). The phase transition occurred at the same temperature whether determined (calorimetrically) in viable organisms, in isolated membranes, or in isolated membrane lipids. Furthermore, it occurred at a much lower temperature than the denaturation of the membrane protein—a transition temperature of 20°–45°C, compared to denaturation beginning slightly above 50°C. Since the lipid phase transition began at a temperature well below the temperature at which the organism grew, and the normal growing temperature (37°C) was well into the phase transition zone, the change in phase of the lipid can have no relation to heat injury in this organism. It was, in fact, possible by enriching the membranes in oleate to lower the phase transition temperature of the membrane lipids to −20°C. It must be concluded from these results that the intermolecular forces with the membrane proteins are strong enough to maintain the membrane lipids in their normal orientation even when they are completely liquid. In the case of a protozoan, when the temperature rose above the optimum for growth, there was a significant reduction in protein synthesis, but lipid biosynthesis was slightly stimulated (Byfield and Scherbaum, 1967). This would certainly oppose the lipid concept even as an explanation of indirect heat injury, since lipid loss would be more readily replaced than protein loss. It also indicates that the energy source was still adequate.

In the case of poikilothermic animals, Ushakov and Glushankova (1961) found no correlation between iodine number of protoplasmic lipids and cell thermotolerance. Ushakov (1964) was also unable to detect a significant relationship between the melting point of the lipids and the thermostability of the cells. On the other hand, he was able to establish a close relationship between cell thermotolerance and the denaturation temperature of the proteins. In 88.9% (57 species) of the pairs of allied, but distinct species tested, parallel differences were found in the thermotolerance of the homologous cells and thermostability of their proteins.

In contrast to Ushakov's results, a relationship between lipid properties and growing temperature has been found in the case of a thermophilic, unicellular algal eucaryote (*Cyanidium caldarium*) that grows in acid hot springs (Kleinschmidt and McMahon, 1970a). It can grow at temperatures from below 20°–56°C. When grown at 55°C, total lipid content decreased to one-half. The ratio of unsaturated: saturated fatty acids also decreased three times at the higher temperature. At 20°C, fully 30% of the fatty acid was linolenic, though none was detected at 55°C. The cells grown at 55°C were more heat tolerant than those grown at 20°C by 10°–15°C (Kleinschmidt and McMahon, 1970b). This was attributed to the higher saturation of the membrane fatty acids. The decrease in unsaturation at the higher temperature was explained by the decreased solubility of oxygen, which was apparently required for desaturation. A similar direct relationship between lipid saturation and heat tolerance was found in nine thermophilic and nine mesophilic species of seven genera of fungi (Mumma *et al.*, 1970). Total lipids varied between 8 and 54.1%, most falling between 8 and 18.3%. The predominant fatty acids were palmitic, oleic, and linolenic. The mesophiles contained 0–18.5% linolenic acid; the thermophiles inappreciable amounts ($< 0.5\%$). The fatty acids of the thermophiles were, therefore, more saturated than those of the mesophiles.

Even in the case of higher plants, Kuiper (1970) has shown that lipid saturation increases with the growing temperature (see Chapter 10). Heat injury may, therefore, involve both lipid and protein changes. This would seem to point to membrane damage as the cause of direct heat injury. A loss of semipermeability at high temperatures could then be due to either (1) excessive fluidity of the lipids, leading to disruption of the lipid layer, or (2) denaturation and aggregation of the membrane proteins, leading to "holes" in the membrane (see Chapter 10).

c. Nucleic Acids

Like proteins, nucleic acids can also be denatured by heat, and the reaction is again first order (Peacocke and Walker, 1962). It is, therefore, not surprising that attempts have been made to implicate them in heat injury. No relationship exists between the heat stability of DNA and thermophily, since both the base composition and the melting temperature are identical in thermophiles and mesophiles (Campbell and Pace, 1968). Similarly, the thermal denaturation of the sRNA's from a mesophile (*E. coli*) and a thermophile (*Bacillus stearothermophilus*) are virtually identical. Yet the ribosomes of the thermophiles are much more heat stable than those of the mesophiles. With few exceptions, the guanine and cytosine of the rRNA tended to increase, the adenine and uracil to decrease with increasing growth temperature. As in the case of DNA and sRNA, however, the rRNA is not significantly different in mesophiles and thermophiles, either as to its thermal denaturation or its gross base composition. Similar results have been obtained with a psychrophile (*Micrococcus cryophilus*). It is unable to grow above 25°C due to an inhibition of protein synthesis (Malcolm, 1968). This was not due either to an inability to synthesize mRNA or to a degradation of existing RNA. On the other hand, three enzymes (tRNA synthetases) were found to be temperature-sensitive. Similarly, none of the ten tested sRNA species from the psychrophile *Candida gelida* was temperature-sensitive (Nash *et al.*, 1969).

In the case of higher plants, there has been little work on the possible relationship between nucleic acids and heat injury. According to Baker (Jung, personal communication), the influence of high temperature on growth of grasses may occur through its influence on RNA. This conclusion was based on the decrease in RNA, which began at temperatures slightly lower than those necessary to decrease top growth.

In summary, there appears to be three zones of heat injury (Table 11.15), each controlled by a different mechanism. In the lowest heat zone, the injury is commonly metabolic (and, therefore, indirect), and the Q_{10} is low, at least in the higher plants. In the case of psychrophilic microorganisms, however, protein denaturation occurs even at these low tempera-

TABLE 11.15 THE DIFFERENT ZONES AND TYPES OF HEAT INJURY

High temperature zone (°C)	Kind of injury	Q_{10}
Lowest (e.g., 15°–45°)	Indirect or metabolic	2–3?
Intermediate (e.g., 45°–60°)	Direct due to protein denaturation	10–100 or higher
Highest (e.g., 60°–)	Decomposition injury	2–3?

tures and appears to be the cause of the heat injury. In the intermediate heat zone, there can be no doubt that protein denaturation is the cause of death, since these temperatures (45°–60°C) denature many if not most proteins, and the Q_{10} for heat killing is high, in agreement with the values for heat coagulation of proteins. In the highest heat zone, the zone for heat killing of dry tissues, the injury is due to chemical reactions, for the temperature coefficient is low and insufficient water is available for protein denaturation.

It now appears likely that both the direct and the fourth kind of indirect heat injury due to a net protein loss are caused by protein denaturation. Different proteins must, however, be involved. The net protein loss must be due to a denaturation and consequent inactivation of the protein synthesizing enzymes. This has been demonstrated in the case of psychrophilic microorganisms. The direct heat injury is more likely due to denaturation of the membrane proteins, causing immediate efflux of cell contents, and cell death. There are several lines of evidence indicating that this kind of injury is initiated in the membranes:

1. Direct observation by Lepeschkin.
2. The salt effects on the heat-killing temperature seem to parallel their effects on cell permeability.
3. Cells in sections near the cut edge are both more permeable and more sensitive to heat.
4. Narcotics have been found to raise or lower both cell permeability and heat tolerance.
5. Heat tolerance is proportional to lipid saturation in some organisms, indicating a relation between membrane properties and heat tolerance.

The fact that the indirect heat injury due to protein loss occurs at a lower temperature than the direct injury (see Mothes, above), is also explainable by protein denaturation. Since hydrophobic bond strength increases with temperature, the membrane proteins with their higher hydrophobicity (see Chapter 10) would have a higher denaturation temperature than the less hydrophobic protein synthetases.

On the basis of all the above evidence, the different kinds of heat injury are all initiated by four direct strains, induced by the primary heat stress (Diag. 11.1): (1) kinetic changes, or changes in metabolic reaction rates, (2) protein denaturation, (3) increased lipid mobility, and (4) chemical decomposition. The first three are all elastic (thermodynamically reversible) strains, and therefore noninjurious by themselves. It is not known, however, whether denaturation of the membrane proteins is indissolubly linked with loss in cell permeability, or whether this occurs only when denaturation is followed by aggregation (see Chapter 10). Even in the latter case, aggregation follows so quickly after denaturation that the two processes occur, to all intents and purposes, simultaneously. On the basis of speed alone, therefore, and regardless of the exact mechanism, heat-induced membrane injury must be called direct injury. The same argument holds for lipid mobility. Chemical decomposition is a plastic (thermodynamically irreversible) strain and therefore is a direct injury by definition.

The heat-induced kinetic changes, however, must involve indirect injury, since a simple, relatively equal increase in the rates of all the meta-

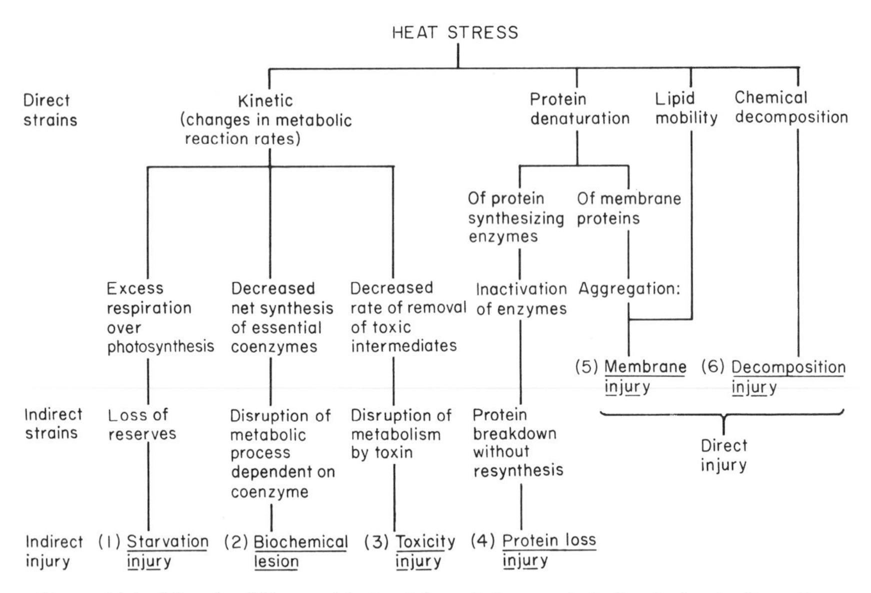

DIAG. 11.1. The six different kinds of heat injury and their relation to the primary, heat-induced strains.

bolic processes would produce no injury. The changes in rate must be different for different metabolic reactions in order to produce a metabolic imbalance. Furthermore, this imbalance must be continued long enough (hours or days), before a deficiency or excess is serious enough to injure the cells. The kinetic strains can, therefore, produce only the slow, indirect kinds of injury.

The fourth kind of indirect injury is actually due to a combination of strains: a protein denaturation followed by a metabolic imbalance which is not due to a direct effect of the heat stress on reaction rates (although this is also a possible factor), but to the inactivation of the protein synthetases as a result of denaturation. Thus, the mere dependence on protein denaturation does not, in itself, mean that the injury is direct.

F. Protective Substances

Some positive evidence (see above) has been obtained of the utility of protective substances such as kinetin in the case of the proteolytic kind of metabolic heat injury. The protection by the kinetin was due to an increased protein synthesis relative to hydrolysis according to Engelbrecht and Mothes (1960). Porto and Siegel (1960) ascribe it to antioxidant activity of the kinetin. In the case of decomposition heat injury no attempt has yet been made to use protective substances, due probably to the technical difficulties of applying substances to dry and sometimes impermeable tissues. Attempts have been made, however, to prevent denaturation injury by the use of protective substances.

Although the results of different investigators do not always agree, it has generally been found that nearly all the salt solutions tried, instead of protecting against heat injury actually lowered the heat-killing temperature (see above). In some cases, however, a rise in the heat-killing temperature has been reported (Table 11.13). An 18–24 hour soaking of millet and wheat seed in 0.25 *M* $CaCl_2$ increased heat tolerance as judged by yield (Genkel and Tsvetkova, 1955). Pretreatment with dilute and concentrated seawater had opposite effects on *Chaetomorpha cannabina* (an intertidal alga). The dilute solution lowered heat tolerance while the concentrated solution increased it (Biebl, 1969). Pretreatment with 0.2–0.4 *M* glucose solution in seawater had no effect. Heat tolerance of bacteria has been increased by salts (Ljunger, 1962). Ljunger (1962) has been able to induce gas production from bacteria at a temperature (37°C) that is above the maximum for normal production, by adding Mg, Ca, Ba, Sr, Na, or K chlorides to the medium. He explains this as a stabilization of the enzyme protein against the heat denaturation. Petinov and Molotkovsky (1961, 1962) have increased the yield of sugar beets, potatoes, sunflowers, and watermelons in hot climates, and therefore supposedly increased their heat resistance by foliar spraying with 0.05–0.08% $ZnSO_4$. The treated plants had twice the respiration rate of the controls. Infiltration with 0.07–0.1 *M* citric, malic, and other acids similarly protected them, due supposedly to combination with and detoxification of NH_3 formed by decomposition of protein.

A decrease in heat coagulability of proteins of growing pea stems has been reported due to application of the auxin 2,4-D (Galston and Kaur, 1959), but the addition of auxin to the homogenate had no affect (Galston *et al.*, 1963). The soluble pectin content of the homogenate was doubled by the auxin treatment, and this may be the cause of the protection. In support of this explanation, the addition of citrus pectin to the control homogenate stabilized the protein against heat coagulability. According to Morré (1970), the amount of homogenate protein (from soybean hypocotyl) coagulated by boiling was increased by low concentrations of 2,4-D, and decreased by concentrations supraoptimal for growth. Maximum heat stabilization occurred after 4 hr of tissue incubation with 10^{-5} *M* 2,4-D. The homogenates from tissue treated with high 2,4-D concentrations tended to have lower SH contents than those from the control untreated hypocotyls. Homogenates from tissues treated with low concentrations of 2,4-D tended to have higher SH contents. Morré, therefore, suggests that the 2,4-D protection against coagulation was due to a conversion of protein SH groups to intramolecular SS bonds which stabilized it against coagulation.

Feldman (1962) raised the heat-killing temperature 0.6–1.7°C in five different species by use of hypotonic or isotonic sucrose, glucose, or lactose solutions. Protection rose to a maximum in concentrations of 0.2 to 0.3 M. Galactose lowered the heat resistance. The increase was detected after 4 hr, reaching a maximum after 24 hr. On transfer to water, the resistance decreased to that of the control in 30 min.

Chloroethanol increased the heat resistance of wheat coleoptiles at 45°C (Miyamoto, 1963). Maleic hydrazide protected growing but not nongrowing leaves (Alexandrov, 1964). It also stopped growth. In opposition to these results, growth-promoting (IAA) and growth-inhibiting (maleic hydrazide) substances were able, respectively, to stimulate and completely inhibit the growth of wheat coleoptiles without affecting the heat resistance of the coleoptile cells (Gorbanj, 1968), unlike the correlation between natural growth inhibition and heat resistance. Morré (1970) also obtained no effect of 2,4-D when used at a concentration that stimulated growth.

Dipicolinic acid is present in the spores of two heat-resistant species of actinomycetes (Cross, 1968). It has also been found to enhance the heat stability of glucose dehydrogenase from the spores of *Bacillus subtilis* (Hachisuka *et al.*, 1967). In the case of wheat roots, casein hydrolysate prevented heat injury (Skogqvist and Fries, 1970). The protection was due to the histidine which was replaceable by a number of nonphysiological imidazoles and other ring substances.

CHAPTER 12

HEAT RESISTANCE

A. Heat Avoidance

1. Occurrence

Since the term thermotolerance has come into more or less general use in recent years, as synonymous with heat tolerance, the parallel terms, thermoresistance and thermoavoidance, may also be used for heat resistance and heat avoidance, respectively. As has already been mentioned, higher plants are poikilotherms. At first sight, this might seem to eliminate avoidance in the case of high temperature just as in the case of low-temperature injury. The problem is more complicated in the former. The intense absorption of radiant energy by leaves and other plant parts usually results in a much higher leaf temperature than the temperature of the transparent and practically nonabsorbing air (see Chapter 11). Even floating algae may possess a temperature in the middle of a mat up to 6.4°C higher than that of the surrounding water (Schanderl, 1955). Therefore, avoidance in the case of the heat-resistant plant does not require a plant temperature lower than that of the air, but simply a temperature lower than that of a control, less avoiding plant under the same environmental conditions. Plants A and B, for instance, may both have the same heat-killing temperature (and, therefore, tolerance), which is 7°C above that of the surrounding air. If the temperature of plant A's leaves rises to 10°C above that of the air, and the temperature of plant B's leaves rises only to 5°C above that of the air, A's leaves will be killed, but B's will not. Since there is no difference in tolerance, the greater heat resistance of plant B is obviously due to avoidance.

That this kind of avoidance does, indeed, occur, has been shown by

Lange and Lange (1963). Hard-leaved woody plants on the south coast of Spain attained leaf temperatures as much as 18.4°C above the air temperatures during August to September. The soft-leaved plants had leaf temperatures 10°–15°C below that of the hard-leaved plants, though still slightly above the air temperature. Since the hard-leaved plants attained leaf temperatures as high as 47.7°C and the killing temperature of the soft-leaved plants was as low as 44°C, the latter survived because of heat avoidance.

2. *Possible Mechanisms*

Heat avoidance may conceivably be due to several causes.

a. Insulation

Insulation cannot protect the plant whose temperature (due to absorption of radiant energy) is higher than that of its immediate environment. The insulation could only aggravate matters by preventing conductive loss of heat from the plant to the cooler environment. It can protect, however, if the plant part is in direct contact with a warmer environment above its killing temperature. This explains why mature tree seedlings with a good protective layer of bark are more heat resistant than immature seedlings with thin bark. The latter may be killed at soil level due to the high temperature of the soil surface. Of course, the soil temperature does not rise as high under the canopy formed by a mature seedling, so that protection of the stem by insulation may be superfluous.

b. Respiration

2. Respiration might conceivably have a harmful effect by contributing to the rise in temperature; a lowered respiration rate might conceivably induce avoidance. In the case of leaves, the quantity of heat released in this way is so much smaller than the radiant heat absorbed as to be insignificant (less than 10^{-5} and 0.7 kcal/cm^2 leaf/min, respectively). In the case of fleshy, insulated organs, however, such as the aroid inflorescence, respiration may be an important contributor to the temperature (Table 11.5) though this normally occurs at a time of year when there is no danger of heat injury. The major heat avoidance must, therefore, be due to the remaining two causes.

c. Absorption of Radiant Energy

A decreased absorption of radiant energy may be brought about in three ways.

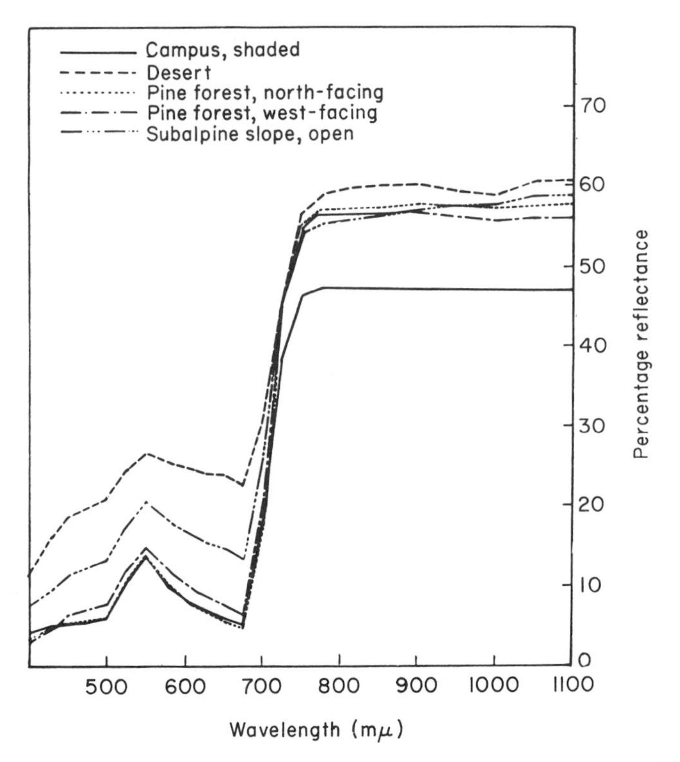

FIG. 12.1. Percentage reflectance from leaves of various species. (From Billings and Morris, 1951.)

(1) Reflectance of radiant energy by leaves varies with the wavelength and the type of leaf (Fig. 12.1). Measurements (Billings and Morris, 1951) indicate a greater reflectance by species of hot habitats. In general, the reflectance curves of green leaves showed values of about 5% at 440 mμ, rising to a peak of about 15% at 550 mμ, with a gradual slope down to 5 or 6% at 675 mμ. Above this, the curves rose steeply to a plateau of about 50% in the infrared region of 775–1100 mμ. On the average, desert species reflected the greatest amount of visible radiation, followed by subalpine, west-facing pine forest, north-facing pine forest, and shaded campus species, respectively (Fig. 12.1). In the infrared, the differences between groups were not so marked, but the greatest reflectance here also was shown by the desert species, with an average value of about 60%. The greatest infrared reflectance (almost 70%), was measured from the glabrous leaves of the desert peach, *Prunus andersonii*. The hairs or scales on desert and sub-alpine plants were correlated with higher reflectance in the visible but not

necessarily in the infrared region. The specific relationship depends on the particular range of infrared. By removing hairs from the upper surface of leaves of *Gynura aurantiaca*, Gausman and Cardenas (1969) showed that pubescence on young leaves increased the total and diffuse reflectance in the 750–1000 nm region but decreased both in the 1000–2500 nm region. Yet, according to Taleinsnik and Usol'tseva (1967), the hairy coverings are responsible for the heat resistance of Manchu cherry. On the basis of the above results, this is possible only in the visible and the short infrared.

On a clear day, considerably more than 50% (often as much as 65%) of the solar radiation incident at the earth's surface is in the infrared region (Gates and Tantraporn, 1952). The major effect on the plant's temperature must therefore be produced by reflectance in the infrared rather than in the visible region, the more so since the plant reflects so little of the latter. On the basis of Billings and Morris' results, shade leaves absorb about one-third more in the infrared than do leaves of desert plants. Their temperatures might therefore be expected to rise proportionately higher.

This assumes that the reflectance found by Billings and Morris for the short infrared (up to 1100 mμ) holds as well for the long infrared, which includes most of the incident infrared in the radiation from the sun (Gates, 1965). Gates and Tantraporn (1952) have measured the reflectance by leaves in the long infrared region (1–20 μ). They find that it is generally small—less than 10% for an angle of incidence of 65°, and less than 5% for 20°. Since the transmissivity of the leaves is zero in the infrared region beyond 1.0 μ, this means that the absorption varies from above 90 to nearly 100% of the incident infrared. Such small differences can have little effect on leaf temperatures. Furthermore, *Opuntia* leaves showed nearly the lowest reflectivity of the 27 species tested, and the hairy leaves of *Verbascum thapsus* failed to reflect any. On the other hand, *Citrus limonia* reflected the most. Shade leaves, in fact, reflected more than sun leaves of the same plant.

Therefore, in spite of the promising difference found in the short infrared, the results with the long infrared (including most of the infrared radiation), lead to the conclusion that plants have not succeeded in developing heat avoidance by increasing their reflectance of the incident radiation. Unfortunately, however, the major portion of the infrared used by Gates and Tantraporn is of negligible importance in normal solar radiation since the solar energy above 2000 nm (or 2 μ) accounts for only 5–10% of the total (see Gates, 1965). It would, therefore, be desirable to confine measurements of leaf infrared reflectance to the range of 750–2000 nm. It is conceivable, in view of Billings and Morris's positive results for the lower portion of this range (which accounts for a little less than half of the solar energy in the infrared), that significant differences may exist between some heat resistant and nonresistant plants. This conclusion seems to be supported by Lange's third group of leaf temperature behavior (see below).

(2) Transmissivity has long been known to vary markedly in leaves. It is obvious that a pale green leaf transmits much more visible radiant energy than a dark green leaf. Even the same leaf may show an increased transmissivity due to a change in orientation. Many plants are, in fact, known to turn their leaf edges to the sun, thus decreasing the absorption of radiant energy and the consequent rise in temperature. This adaptation might conceivably lead to avoidance of heat injury. The plastids have also been observed to turn edge-on toward the surface of the leaf when intensely illuminated. This characteristic has no apparent relationship to heat resistance (Biebl, 1955). This empirical result is to be expected from the above mentioned fact that most of the absorbed radiant energy is due to the infrared radiations, which are mainly absorbed by the water.

(3) Absorption by protective layers is possible since the external water layer protects the internal cells by filtering out most of the heat-producing infrared radiation. This would, however, require an external layer of cells with higher heat tolerance, in order to protect internal (chlorenchyma) cells of lower tolerance. In the case of succulents, this is a distinct possibility, not only because of the high water content of the cells, but also because measurements have revealed a marked negative temperature gradient from the surface of the organ to the center (Huber, 1932; Fig. 11.3). It is perhaps possible that a thick enough cuticle may protect in the same way.

d. Transpirational Cooling

Due to the difficulties inherent in the measurement of leaf temperatures, earlier records of the cooling effect of transpiration were frequently exaggerated (Curtis, 1936a). For instance, the elimination of transpiration by vaselining of leaves resulted in a warming of only 1°–3°C above the temperatures of the freely transpiring leaves at air temperatures of 30°C or less (Ansari and Loomis, 1959). Carefully performed measurements have even led to the conclusion that the maximum cooling is 2°–5°C, and calculations confirmed this conclusion (Curtis, 1936b). Even this amount, of course, could be a deciding factor in heat survival. More recently, however, fully reliable methods of measuring leaf temperatures have been developed, using the older thermocouples as well as thermistors (Lange, 1965b), and an infrared radiometer which does not touch the leaf (Gates, 1963). Gates found some sunlit leaves as much as 20°C above the air temperature (a leaf temperature of 48°C in *Quercus macrocarpa* at an air temperature of 28°C), while shade leaves averaged 1.5°C below the air temperature. He concluded that transpiration must play a relatively strong role in reducing leaf temperatures, and that convection is a relatively inefficient process. The earlier estimates did not agree with this conclusion; according to the above calculations (Curtis and Clark, 1950), the heat absorbed by the evaporation of water from a rapidly transpiring leaf at 40°–50°C could account for only about 15% of the radiant energy absorbed in full sunlight. More recent calculations (Wolpert, 1962) indicate that transpiration may be expected to remove 23% of the incoming heat during the midday hours. These and other recalculations (Raschke, 1956, 1960) have indicated that under extreme enough conditions of temperature and vapor pressure, transpiration can produce greater lowerings of leaf temperature than the above limit of 2°–5°C.

It is dangerous, however, to generalize from such calculations. Below 35°C, for instance, the average temperatures of *Xanthium* leaves were above the air temperature by an amount dependent on the wind velocity, an increase in wind velocity producing a decrease in transpiration (Drake *et al.*, 1970). At air temperatures above 35°C, on the other hand, the leaf temperatures were below the air temperatures and increasing wind markedly increased transpiration. A further complication not usually taken into account is the root temperature. Coffee plants transpired maximally at a root temperature of 33°C, with a drop in rate both above and below this temperature (Franco, 1958).

In view of all this more recent information, it is no longer surprising that Lange (1958, 1959) was able to observe pronounced lowerings of leaf temperatures at air temperatures of 40°C and higher. That his measure-

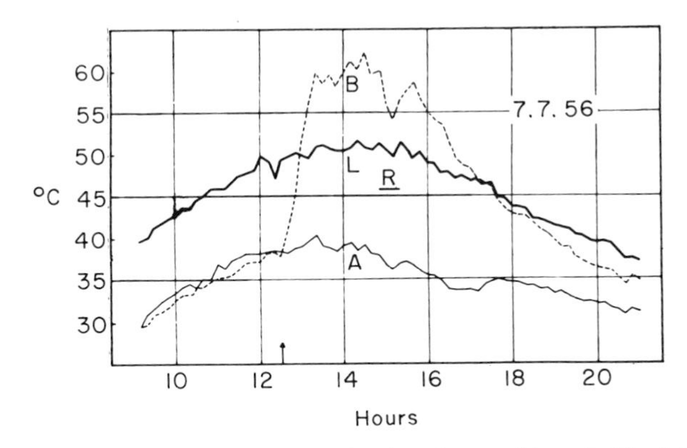

FIG. 12.2. Daytime temperature curves for two leaves (A and B) of *Citrullus colocynthis*. Leaf B excised at 12:30 p.m. L, air temperature; R, heat-killing temperature of the leaf. (From Lange, 1959.)

ments of leaf temperature were not in error was elegantly proved by parallel measurements of the heat-killing temperature of the leaf. The air temperatures were as much as 10°C above the heat-killing temperature. Since the leaves were alive, their temperatures must have been lower than the air temperatures by more than this 10°C difference. Lange found three types of leaf temperature behavior in the extreme heat of the desert and savannah regions of Mauretania:

Group a. The temperature of the horizontal leaves in full sunlight was markedly (as much as 15°C) below that of the surrounding air (Fig. 12.2), e.g., a leaf temperature of 42.5°C when the air temperature was 55°C. Such species that maintain leaf temperatures below the air temperature when in full sunlight, were called *undertemperature plants.*

Group b. The leaf temperature in full sun was above that of the surrounding air (Fig. 12.3), by as much as 13°C. Such species he calls *overtemperature plants.* If the temperatures were sufficiently extreme (under artificial conditions), even these overtemperature plants were able to make use of transpirational cooling. Thus, at temperatures of 30°–50°C, the undertemperature plant (*Cucumis prophetarum*) transpired at increasing rates with rise in temperature, nearly paralleling evaporation rates from an evaporimeter (Fig. 12.4). In the same temperature range, the overtemperature plant (*Phoenix dactylifera*) actually decreased its transpiration slightly with rise in temperature, due to closure of its stomata. Above 50°C, however, the stomata opened again and the overtemperature plant showed an even steeper rise in transpiration rate than the undertemperature plant, and a greater cooling effect per unit of water transpired (Figs. 12.4 and 12.5).

Group c. Transpiration had no effect on the leaf temperature. Yet even when strongly irradiated, the leaf temperature rose very little above the air temperature, perhaps due to greater reflectance of radiation (see above).

That the first group owed their low temperatures to transpiration was proved by cutting off the shoot, thus severing it from its water supply. This resulted in a sudden rise in leaf temperature above that of the air

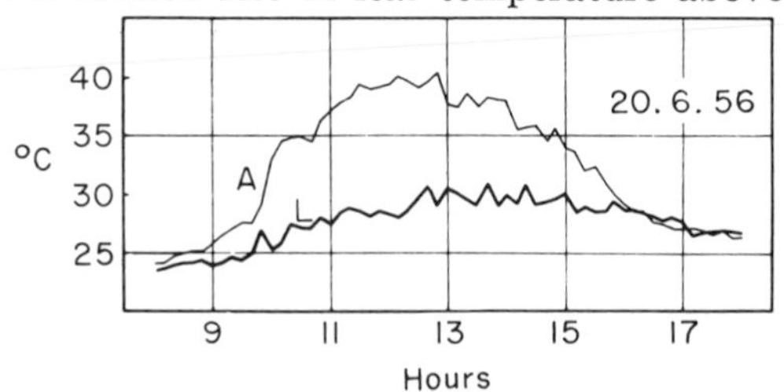

Fig. 12.3. Daytime temperature curve of a leaf (A) of *Zygophyllum fontanesii.* L, air temperature. (From Lange, 1959.)

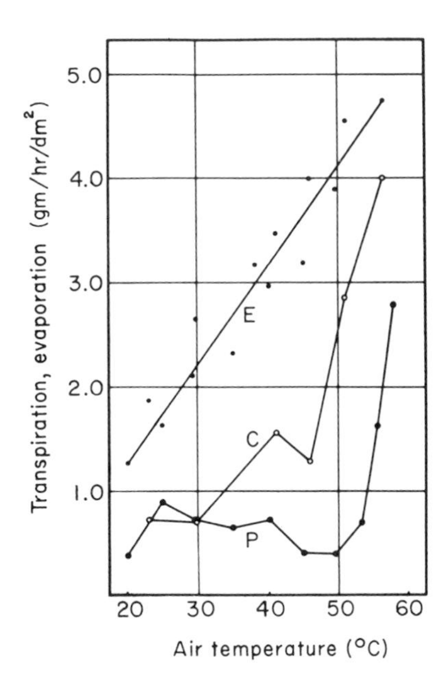

FIG. 12.4. Transpiration rates (gm/hr/dm²) of *Cucumis* (C) and *Phoenix* (P) compared to evaporation (E) in relation to air temperatures at 30% r.h. (From Lange, 1962a.)

(Fig. 12.2). On the other hand, the third group proved that other factors besides transpiration could control leaf temperature to some extent. These factors were not identified, but must have involved a decreased absorption of radiant energy (see above).

On the basis of measurements on 133 different species in Brazil, Coutinho (1969) observed the "De Saussure effect" (dark fixation of CO_2) in overtemperature plants, some of which developed leaf temperatures 10°–15°C above that of the surrounding air (e.g., a leaf temperature of 45°C at an air temperature of 30°C). It, therefore, seems likely that the most extreme overtemperature plants are to be found among those with an active dark CO_2 fixation system; they tend to keep their stomata closed during the day, thereby decreasing their transpiration rate.

The importance of transpirational cooling is well illustrated by the recent use of antitranspirants. These have resulted in a rise in leaf temperature of tobacco by as much as 9°F above the controls (Williamson, 1963). Similarly, when heat injury was determined in a temperature-controlled chamber, a doubling of the injury was obtained by raising the relative humidity from 50 to 75% or from 75 to 100%, when heated to 43°C for 8 hr (Kinbacher, 1969).

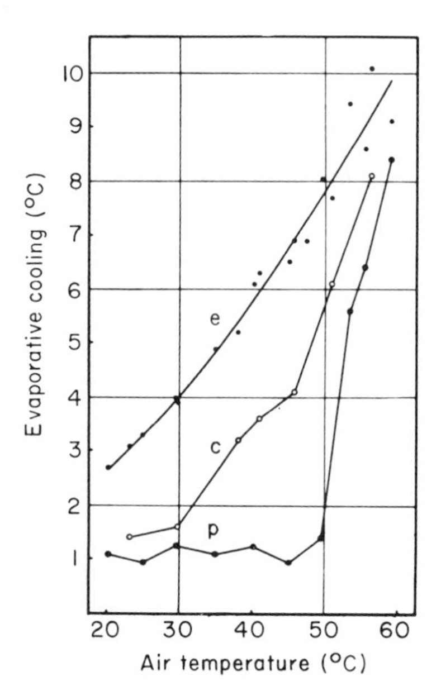

FIG. 12.5. Transpirational cooling (°C) of leaves of *Cucumis* (C) and *Phoenix* (P), and evaporation cooling of evaporimeter disc (e) in relation to air temperature at 30% r.h. (Evaporimeter cooling values reduced to half for purposes of comparison.) (From Lange, 1962a.)

3. *Measurement*

Although heat avoidance has never been measured, it should, at least be theoretically possible to do so, with the aid of modern control chambers. The temperature of the plant part would have to be measured under standard, steady-state conditions (e.g., an air temperature of 45°C and a soil temperature of 35°C, a light source of standard intensity and wave length distribution, a standard rate of air movement, and adequate soil moisture). The lower this temperature the greater the avoidance. A numerical value for avoidance might perhaps be obtained from a ratio of air temperature to leaf temperature at the steady state. It is obvious, however, that such measurements would give rise to many difficulties (e.g., due to the position of the leaf and the presence of others near it). Here again, however, a standardization of conditions could be developed. The heat avoidance should, of course, be measured at a temperature just below the heat-killing temperature.

B. Heat Tolerance

The terms thermotolerance and thermostability have sometimes been used in the literature interchangeably for heat tolerance of organisms. In conformity with the general stress terminology defined earlier, the term thermotolerance will be used here interchangeably with heat tolerance—the ability of an organism to survive a heat stress within its tissues. The term thermostability will be used only for the ability of *substances*, e.g., proteins or nucleic acids, to remain stable, native, or active (e.g., as a catalytic enzyme), after exposure to a heat stress. Lange and Lange (1963) found that avoidance was the main basis of the heat resistance in the woody plants characterized by mesomorphic leaf structure and a high transpiration rate. Their thermotolerance was low. On the other hand, the hard-leaved woody plants had a low transpiration rate and thermoavoidance, but a high thermotolerance. Unlike thermoavoidance, thermotolerance can be measured simply and with a high degree of precision in both of these types of plants. The main methods have remained essentially unchanged for over a century.

1. Methods of Measurement

As in the case of low-temperature tolerance, heat tolerance is measured by determining the temperature at which 50% of the plant or plant part is killed—in this case called the heat-killing temperature. The two methods originally used by Sachs (1864) are still the standard procedures for measuring the heat-killing temperature of plants, though more elaborate equipment may now be employed. With the first, potted plants are placed in a heat chamber (whose temperature can be controlled) and maintained at a constant temperature for a standard length of time. They are then transferred to the greenhouse for a period of a week or two, following which the degree of injury is observed. Heyne and Laude (1940), for instance, exposed corn seedlings to 130°F at a relative humidity of 25 to 20% for 5 hr. With the second method, potted plants are overturned (Sachs, 1864), or small pieces or sections of the plant are plunged into water at a known temperature for a standard length of time, and then observed for growth or microscopically examined for injury. Sometimes the plant is not allowed to come in contact with the water. Julander (1945), for instance, cut $1\frac{1}{2}$ inch pieces from stolons of range grasses, transferred them in lots of eight into stoppered glass tubes, which were then immersed in a constant-temperature bath at $48° \pm 0.1°C$ for periods of 0, $\frac{1}{2}$, 1, 2, 4, 8, and 16 hr. The stolon pieces were then planted and recovery was estimated after 4 weeks. Lange (1958) used this method in the field without removing the plant parts, by simply plunging them into the heated water in a thermos flask.

The first method has the advantage of using the whole plant, but there are definite objections to it. The actual temperatures attained by the tissues are not known, and may be far below those of the surrounding air due to the rapid transpiration at the high temperatures and low relative humidities usually employed. As evidence for this conclusion, Sapper (1935) found that the plants were able to survive air temperatures as much as 5°C higher in dry than in moist air; though wilted plants withstood higher air temperatures than turgid plants when heated in saturated air, in dry air the relationship was reversed. It is possible, therefore, to subject prairie grasses to hot winds at 135° to 145°F without injury, as long as soil moisture is available (Mueller and Weaver, 1942). Even if the plants are heated in still air, some time is needed for temperature equilibrium to be reached, and during this time the tissues are in danger of drought injury, especially at the low relative humidities (e.g., 25 to 30% in Heyne and Laude's or 30–35% in Laude and Chaugule's 1953 experiments) that usually prevail. Kinbacher (1963) demonstrated this in the case of eight varieties of winter oats. When subjected to 112°F (43°C) for 8 hr, only a heat stress resulted if the r.h. was maintained at 100%, but both a heat and a drought stress occurred if the r.h. was 50 or 75%. It is not surprising, therefore, that the "heat hardiness" of the plants tested by Heyne and Laude paralleled their field drought resistance. Consequently, in order to prevent drought injury during tests by the first method and to obtain plant temperatures identical with the measured air temperatures, Sachs' (1864) original precaution of maintaining 100% relative humidity should be adopted. Any injury produced at the high temperature will then be purely heat injury, as is always true of the second method. The two kinds of tolerance may, of course, be correlated, in which case a measurement of drought tolerance may yield a relative measurement of heat tolerance (see Chapter 21). This is not true in the case of some organisms, e.g., the green alga *Spongiochloris typica* (McLean, 1967). Even if the air is not saturated, the heat-killing point can be determined, provided that the actual temperature of the plant is used and not that of the surrounding air (Lange, 1967). Results obtained in this way do not differ by more than a degree or two from those obtained by the immersion method. Of course, as mentioned above, such tests should not be continued for long enough periods to permit drought injury.

It must be realized, however, that even when the above precautions are taken, the two methods may not always yield exactly the same results.

Sachs (1864) found that with the first method (in air), 51°C was the killing temperature, though the plants withstood 49°–51°C for 10 min or more without injury. The same plants, however, were killed if plunged in water at 49° to 51°C for 10 min. According to de Vries (1870), killing occurs at about 2°C lower in water than in air. Yet, in spite of these results, agreement is often surprisingly good. Sapper (1935), for instance, gives a heat-killing temperature of 40.5°C for *Oxalis acetosella* exposed for one-half hour by the first method; Illert (1924) gives a value of 40°C for 20 min using the second (immersion) method on the same species.

The use of tissue pieces enables determinations on individual tissues or even cells. It has the disadvantage of judging survival by cellular methods instead of by the ability of the plant to continue normal metabolism and growth. In the hands of an untrained observer, cellular methods are dangerous to judge by, particularly if only one criterion is used. Schneider (1925), for instance, was able to obtain plasmolysis-like contraction of the protoplasts in heat-killed cells, e.g., moss cells dipped in water at 80°C for half a minute, "plasmolyzed" in hypertonic sucrose. This observation was confirmed for guard cells by Weber (1926a), but he points out that the "plasmolyzed" cells look abnormal and are not difficult to distinguish from truly plasmolyzed living cells. Schneider, himself, was unable to deplasmolyze and replasmolyze these heat-killed cells, and found that the plasmolyzability was lost in about 10 min. As pointed out by Weber, mistaking heat-killed cells for living cells can be avoided by combining vital staining with plasmolysis. Therefore, the observations of men like De Vries (1871), who used several criteria for distinguishing injury, appear fully reliable. Döring (1932) objects, however, that even such methods may fail to distinguish between cells with living protoplasm and those in which only the tonoplast remains functional. Scheibmair (1937) concluded that even this danger can be avoided by combining vital staining, plasmolysis, observation for abnormalities, and finally deplasmolysis and replasmolysis. Even with perfectly reliable cellular methods of distinguishing living from dead cells, the second (immersion) method may be expected to yield slightly higher heat-killing temperatures than the first, since plants that are doomed to die after heating (by the first method) may retain full turgor and appear fully healthy for 24 hr (Sachs, 1864). To avoid overlooking such postheating changes, Scheibmair recommends leaving the sections for some time before testing. The actual differences obtained have been in the opposite direction (see above). This result must be at least partly due to the instantaneous rise of the cell temperature to that of the immersion water, the slower rise to the air temperature because of the far greater specific heat of water, and the larger plant bulk.

A modification of the first method, used in the open, is to enclose a shoot in a blackened box, permitting its temperature to rise due to the insolation (Rouschal, 1938b; Konis, 1949), but neither the temperature nor the time exposed to it can be controlled in this way. Besides the above mentioned methods of evaluating injury (plasmolysis, vital staining, observation of plants for recovery) several other criteria of survival have been used. Lange (1953) determined the temperature that reduced the subsequent equilibrium respiration of lichens to 50% of normal. The actual growth of the lichens after the heat stress gave similar results.

Alexandrov (1964) developed a technique for observing cytoplasmic streaming (of the spherosomes) in the epidermal cells of intact leaves, as a criterion of heat survival. He covers the leaf surface with silicone oil, thus eliminating surface reflections when observed under the microscope. The leaves are heated at a series of temperatures for 5 min, and the temperature that is just sufficient to stop cytoplasmic streaming is taken as the heat-killing temperature. This method has been extended to the leaf parenchyma, in which the phototaxis of the chloroplasts was used as a criterion of heat injury (Lomagin *et al.*, 1966). The method is precise to 0.1°C. One objection, pointed out by Alexandrov, is that the cell can recover from this effect of heat if cooled rapidly enough, as shown by Sachs (1864). This objection may be valid for the absolute killing temperature, however, Alexandrov obtained the same relative results by this method and by several others (e.g., plasmolysis, vital staining, respiration, photosynthesis).

Kappen and Lange (1968) object to Alexandrov's method because it is confined to the epidermal cells, which may not respond to stresses in the same way as the leaf as a whole, and because injury is estimated immediately after heating instead of one or more days later. They point out that Till (1956) observed a winter rise in freezing tolerance of whole plants or leaves of *Hepatica nobilis*, but Alexandrov's (1964) group failed to detect any rise by their method, perhaps due to the reversible effect of the temperature on cytoplasmic streaming in the epidermal cells. Similarly, Alexandrov's method failed to reveal the changes in heat tolerance detected by Lange's modification of Sachs' first method, and which were produced by climatic or developmental changes. Again, when they observed cytoplasmic streaming immediately after the heat treatment (Alexandrov's method), they were unable to detect the increase in heat tolerance following moderate drying of *Commelina africana*, which was established by their method. When, however, the cells were observed over a period of 200 hr, a faster recovery of protoplasmic streaming could be detected in the dried leaves than in those heated in the saturated state. In agreement with Lange, Biebl (1969) concluded that valid measurements of heat injury cannot be made earlier than 24 hr after heating *Chaetomorpha cannabina*, an intertidal alga. Alexandrov's *et al.* (1970) evaluation of these objections will be considered below.

Oleinikova (1965) used the conductivity method to estimate heat injury in wheat and millet varieties, in the same way as it has been used for estimating freezing injury. Feldman and Lutova (1963) estimated the heat injury in algae by vital staining and by the effect on photosynthetic rate.

As in the case of freezing tolerance, the relative heat tolerance of different species or varieties should be compared only when they are in the hardened

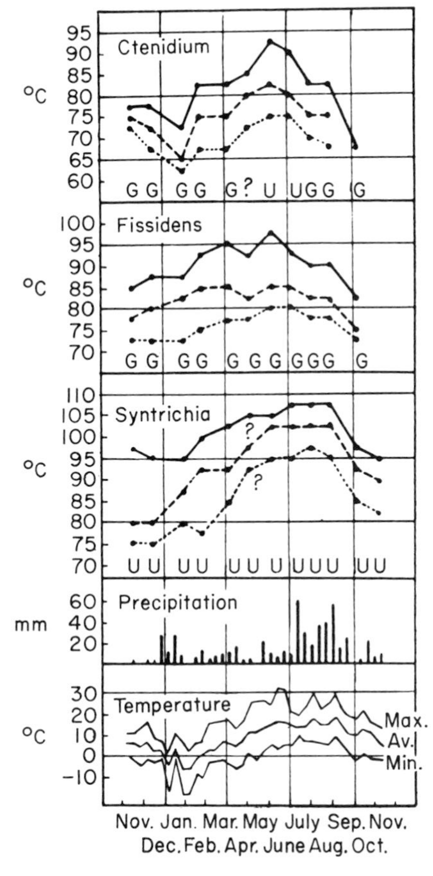

FIG. 12.6. Seasonal change in heat tolerance of mosses, measured by exposure in the dry state for one-half hour to the high temperatures. Complete killing (top curve), severe injury (middle curve), and slight injury (bottom curve). G, injured by 50 hr, exposure over P_2O_5, U, uninjured. (From Lange, 1955.)

condition, though few investigators take this precaution. Julander (1945) found little difference between range grass species when they were "unhardened," i.e., well watered and clipped and grazed. In the hardened condition, however, there were definite differences. Bermuda and Buffalo grass were the most tolerant; they were not killed by even 16 hr at 48°C. Bluestem was intermediate. Slender wheat, smooth brome, and Kentucky bluegrass were the least tolerant.

2. Effects of Environmental Factors

a. Temperature

As in the case of freezing tolerance, a relationship between the environmental temperature and heat tolerance is indicated by seasonal cycles in tolerance, although this was not clearly demonstrated until relatively recently. The heat tolerance of mosses in the dry state shows an annual rise to a maximum during summer and a drop to a minimum during winter (Fig. 12.6). Wintergreen and evergreen plants in their normally moist state also show a seasonal cycle (Lange, 1961), but there may be two maxima in heat tolerance, one during the heat of summer, and the other during the cold of winter (Fig. 12.7). The cyclic change occurred not only in the new leaves but also in the previous year's leaves and, therefore,

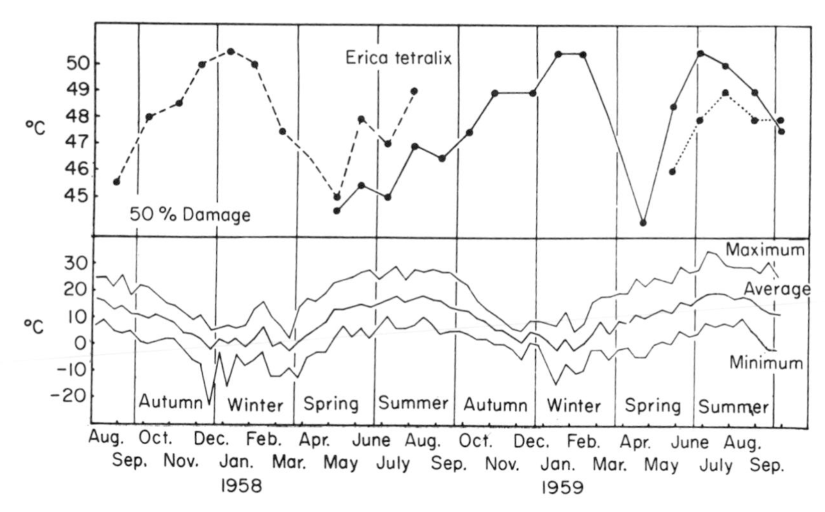

Fig. 12.7. Seasonal changes in heat tolerance of *Erica tetralix* leaves exposed to high temperatures for 30 min (50% killing). Lower curves: the corresponding air temperatures. (From Lange, 1965b.)

could not be explained simply by a change in growth and development. Furthermore, a higher heat tolerance was developed during the hot and dry summer of 1959 than during the cool summer of 1958. Although the differences are not as great as in the annual cycle of freezing tolerance, *Erica tetralix* showed a rise in heat tolerance of as much as 6.5°C during summer.

In other plants (e.g., grasses), the same winter maximum in heat tolerance was found, paralleling its freezing tolerance, but there was no summer maximum (Alexandrov *et al.*, 1959; Alexandrov, 1964). The third combination has also been reported. Feldman and Lutova (1963) observed a rise in heat tolerance and a drop in freezing tolerance during summer in the case of *Fucus* sp. In winter, the reverse occurred—freezing tolerance rose and heat tolerance fell. It is difficult to evaluate their measurements of freezing tolerance since these were obtained by plunging test tubes of the thalli into ice–salt mixtures at −30° to −20°C, and determining the length of time survived. This method may have measured freezing avoidance, or intracellular freezing avoidance, rather than freezing (i.e., freeze-dehydration) tolerance. Nevertheless, the results do agree with the seasonal changes found by others (Chapter 7).

The grasses that failed to show a summer maximum have a high degree of heat tolerance under any conditions, and this tolerance remains constant at growing temperatures of 22°–36°C (Alexandrov, 1964). This may explain the lack of a summer maximum; when they are exposed artificially to 38°–44°C, their heat tolerance increases markedly. Presumably, the summer temperatures to which they are exposed in nature do not attain such high values and, therefore, no hardening occurs.

The relationship between temperature and hardening is obviously not as straightforward as in the case of freezing tolerance. This is further shown by comparing the heat tolerance of higher plants in different habitats. Thus, plants of dry, hot environments are frequently more heat tolerant than those of moist, cool environments (Table 12.1). Yet, in contrast to these results, plants of the Sahara desert range in heat tolerance from 44°–59°C (Lange, 1959), many of them thus showing no higher tolerance than those of the native mid-European plants used by Sapper. Furthermore, though most of the species did show a general correlation between heat tolerance and habitat, there were several pronounced exceptions (Table 12.2). All the exceptions that survived the heat of the desert in spite of low heat tolerance were undertemperature plants and, therefore, owed their survival to heat avoidance. Consequently, if heat-avoiding plants are eliminated, the generalization still holds that the heat tolerance of most (but not all) higher plants varies directly with the temperature of their natural habitats. The relationship, of course, does not prove a direct hardening effect of high temperature on plants since they may have been selected for a stable heat tolerance.

In the case of mosses and lichens there is also usually a correlation between habitat and heat tolerance (Lange, 1953, 1955), although exceptions do occur. An unexpectedly high tolerance was found in two lichens in spite of their habitats (one in the shade and the other a hydrophyte). Similar relationships exist among algae, though the heat killing range is low (3 hr at 22°–38.5°C; Montfort *et al.*, 1957). The correlation with temperature changes under natural conditions may not necessarily prove a cause and effect relationship. *Chaetomorpha cannabina*, an intertidal alga, when cultivated in dishes for 5 consecutive days showed conspicuous fluctuations in heat tolerance in a "tidal rhythm" coinciding with the 6-hr intervals of ebb and flood (Biebl, 1969), and in the absence of any corresponding temperature cycle.

The rise in heat tolerance found by Lange during the heat of summer, suggests the possibility of artificial hardening by exposure to moderately warm temperatures. Early results, however, were the opposite of this expectation. Exposures to high, sublethal temperatures for extended periods actually reduced heat tolerance (Sapper, 1935). This was accompanied by a drain on the carbohydrate reserves at the high temperature. Conversely, low-temperature hardening that leads to an accumulation of reserves

TABLE 12.1 Heat-Killing Temperatures of Plants from Different Habitats[a]

Habitat or ecological type	Species	Heat-killing temp. (°C), ½-hr exposure
Submerged aquatics	*Helodea callitrichoides*	38.5–41.5
	Vallisneria species	
Shade	*Oxalis acetosella*	40.5–42.5
	Impatiens parviflora, etc.	
Partially shadded	*Geum urbanum*	
	Chelidonium majus	45–46
	Asplenium species	
Very dry and sunny	*Alyssum montanum*	
	Teucrium montanum	
	Dianthus species	48–50 or over
	Iris chamaeiris	
	Verbascum thapsus, etc.	
Succulents	*Sedum* species	48.5–50
	Other succulents	50–54

[a] From Sapper, 1935.

TABLE 12.2 RELATION BETWEEN HABITAT AND HEAT TOLERANCE[a,b]

Species	Habitat	50% killing temp. (°C)
Trichomanes erosum	rain forest	44
Piptadenia africana	rain forest	45
Acacia tortilis	dune	45–46
Acacia senegal	savannah	47
Heisteria parvifolia	rain forest	47
Tamarix senegalensis	dune	47
Commelina africana	rain forest	47
Geophila obvallata	rain forest	47.5
Cassia aschreck	desert	49
Ziziphus nummularia	desert	49
Adenium honghel	savannah	50
Palisota hirsuta	rain forest	52
Culcasia angolensis	rain forest	52
Capparis decidua	desert	53
Boscia senegalensis	desert	55
Aristida pungens	desert	55
Phoenix dactylifera	desert	58

[a] Adapted from Lange, 1959.

[b] All the undertemperature desert plants and the overtemperature nondesert plants have been omitted. Expected order of increasing tolerance: rain forest < dune < savannah < desert.

actually increases heat tolerance (Sapper, 1935). A cold treatment of pea plants has produced a similar effect (Highkin, 1959). These results were later corroborated by the maximum heat tolerance found during winter (Fig. 12.7).

Sapper's original failure to harden plants to heat has received some confirmation, but many have succeeded where he failed. Coffman (1957) found that exposure to high temperature increased the heat tolerance of winter oats, but not of spring oats. Alexandrov (1964) has not succeeded in hardening grass species by exposure for long periods to high but not injurious temperatures (22°–36.5°C). At 38°C and higher, however, a reversible hardening occurred (Alexandrov and Yazkulyev, 1961). He has, also, succeeded in hardening *Tradescantia fluminensis* by exposures to temperatures of 28°–36°C for 16–18 hr. A similar long period at 37.5°C killed the plant. Algae showed a direct relation between heat tolerance and growing temperature throughout the range tested (e.g., 10°–30°C for *Porphyra*). On the other hand, the green alga *Chlamydomonas* showed no capacity for heat hardening (Luknitskaya, 1967).

Lange (1961) succeeded in hardening *Commelina africana* and *Phoenix dactylifera* by cultivating them at high but not injurious temperatures. He (1962c) obtained a 4°C increase in heat tolerance (from 47° to 51°C) in the leaves of *Commelina africana* by growing the plants at 28°C for 5 weeks, compared to control plants grown for the same length of time at 20°C. Even tissue cultures (of Jerusalem artichoke) increase in heat tolerance as a result of an elevated growing temperature (Guern and Gautheret, 1969). It is possible to explain these discrepant results on the basis of the compensation point. Perhaps it is only those plants whose compensation point is near or below that of the hardening temperature used, that show a decrease in heat tolerance due to the consequent loss of reserves. Similarly, the absence of hardening at high but not injurious growing temperatures may be characteristic of plants (e.g., Alexandrov's grasses) that already possess so high a heat tolerance in the unhardened state that only a more severe heat shock (see below) can harden them.

During the past two decades, the Russian team under Alexandrov's direction has conclusively and repeatedly shown that brief (even a single second) but severe heat shocks can induce a pronounced heat hardening of higher plants (Table 12.3). They have also succeeded in hardening lower plants by exposure to supraoptimal temperatures for brief periods, e.g., in the case of *Physarum polycephalum* (Lomagin and Antropova, 1968). Similar results have been obtained independently by Yarwood (1961a, 1963). In general, the higher the heat tolerance of the unhardened plant,

TABLE 12.3 Influence of 1-Second Heat Hardening at 59°C on the Heat Resistance of Cells of *Campanula persicifolia*[a]

Number	Maximal $t°$ at which protoplasmic motion is retained after 5-min heating		Difference[b]
	Pretreated cells	Control cells	
1	46.6	45.4	+1.2
2	46.6	45.8	+0.8
3	45.8	44.6	+1.2
4	45.4	43.8	+1.6
5	46.2	44.2	+2.0
6	45.4	44.2	+1..2
7	46.6	45.0	+1.6
Average	46.1	44.7	+1.4

[a] Lomagin, 1961. From Alexandrov, 1964.
[b] For significance of the average difference, $P < 0.001$.

the higher the temperature required for hardening by heat shock (Alexandrov, 1964). Yarwood (1967) has succeeded in hardening bean, cowpea, corn, cucumber, fig, soybean, sunflower, and tobacco, as well as some rusts and viruses by exposure to air temperatures of 32°–40°C or water temperatures of 45°–55°C. The optimum exposure time was 20 sec at 50°C, with times at other temperatures corresponding to a coefficient of 50.

Even cells in tissue culture are capable of hardening by heat shock (Schroeder, 1963). Cultures grown from pericarp of avocado fruit when treated for 10 min at 50°C and then incubated at 25°C for 3 days, survived 10 min at 55°C and grew during a subsequent 4 weeks at 35°C. Nonshocked controls failed to grow after a similar exposure to 55°C for 10 min. This persistence of the hardening has also been shown by Alexandrov (1964), although dehardening was perceptible after 24 hr, and was complete within 6 days.

Yarwood (1962) has also shown that a brief heat shock may actually sensitize a plant so that it is more injured by a later heat shock. According to Wagenbreth (1965), hardening of beech requires a heat shock of 55°C, and high temperatures lower than this (e.g., 50°C) reduce its heat tolerance by increasing protein hydrolysis, accelerating yellowing, and increasing respiration. This may conceivably explain Yarwood's sensitization to heat.

Conversely, a moderate heat shock may increase heat tolerance only if this is measured some time after the heat stress (Lutova and Zavadskaya, 1966). They explain this by an increased ability to repair the damage, since no difference in heat injury could be detected immediately after the heat stress. Intermediate hardening periods of 3 hr at temperatures of 28°–50°C also increased heat tolerance, as measured by cytoplasmic streaming, chloroplast phototaxis, selective permeability, and respiration in leaves of *Tradescantia fluminensis* (Barabal'chuk, 1969). Maximum hardening occurred at 38°C.

b. Moisture

Sapper showed that dry cultivation increases heat tolerance by as much as 2°C (Table 12.4). Wilting alone increased hardiness as much as dry cultivation. This is in agreement with other results on the effect of dehydration (Chapter 11). Similarly, watered bluestem grass was killed by 4 hr at 48°C, and droughted bluestem only by 16 hr at the same temperature (Julander, 1945). The effect of moisture is also seen in the frequently observed burn injury to fleshy fruits after long rains when these are followed by hot, sunny weather (Sorauer, 1924; Huber, 1935). Similarly, fungal spores in younger developmental stages with higher water content are more sensitive to heat injury than those in later stages with lower water

TABLE 12.4 EFFECT OF DRY CULTIVATION ON HEAT HARDINESS[a]

Species	Heat-killing temperature (°C)		Osmotic value at incipient plasmolysis (*M* sucrose)	
	Sparingly watered	Kept saturated	Sparingly watered	Kept saturated
Melilotus officinalis	46.5	45	0.55–0.60	0.45–0.50
Taraxacum officinale	46	44.5	0.65	0.45
Avena sativa	44.5	<43.5	0.65	0.40–0.45
Hordeum distichum	46–46.5	<44	0.80	0.45–0.50
Hieracium pilosella	>50.5	48.5	0.65–0.75	0.35–0.55
Ceterach officinarum	100–120	47		

[a] From Sapper, 1935.

content (Zobl, 1950). The former become more heat tolerant on stepwise, careful drying, in agreement with results using seeds. Some species may fail to differ from each other in heat tolerance unless first exposed to drought hardening (Julander, 1945). The most extreme example of a relation between moisture and heat tolerance is, of course, the case of seeds (see above). The same phenomenon occurs in the vegetative state among lower plants. In the turgid state, mosses and lichens are killed by temperatures around 40°C; in the dry state (occurring regularly at midday), the killing temperature rises to 70°–100°C (Lange, 1955). Yet in opposition to this relation, succulent leaves have shown the highest heat tolerance of higher plants (Table 12.1). Lange (1958), however, found that the succulents of the Sahara desert are not especially heat tolerant. Their range of heat-killing temperatures (48°–53°C) falls in the middle of the range for all these desert plants. In fact, the plants with the highest tolerance were not the fleshy-leaved succulents but the hard-leaved species. In the case of *Kalanchoe blossfeldiana,* the strongly succulent leaves of the flowering plants have a higher heat tolerance than the weakly succulent leaves of vegetative plants. The heat-killing points are 50° and 47°C, respectively (Lange and Schwemmle, 1960). The two properties may, however, be separated, for a 10-day exposure to a short photoperiod was not enough to induce the succulence, yet it produced a significant rise in heat tolerance. Similarly, the differences in heat tolerance between leaves of different levels only partially paralleled succulence. These two investigators were therefore led to the conclusion that succulence is not in itself an indication of heat tolerance, but just as in the case of freezing tolerance, some plants possess much greater tolerance than is to be expected from their high water contents.

Even brief (6 hr) periods of dehydration leading to water saturation deficits of 22–27% were able to raise the heat-killing temperatures by about 3°C in the case of leaves of *Commelina africana*, *Hedera helix*, and *Phoenix reclinata* (Hammonda and Lange, 1962). Kappen and Lange (1968) showed that loss of water induced an increase in heat tolerance not only in lower plants and in xerophytic higher plants, but also in mesophytic higher plants. Zavadskaya and De'nko (1968) investigated twelve species of plants and found that the heat tolerance was higher when they were subjected to a water deficiency than when growing in humid habitats. As the leaves became saturated with water, heat tolerance gradually decreased.

c. LIGHT

The effect of light on heat tolerance is not so easily determined. Five days in the dark increased the heat tolerance of *Oxalis acetosella* (Illert, 1924). In agreement with these results, Sapper (1935) was unable to reduce the heat tolerance of plants by keeping them in the dark for as long as 3 days. Longer times than this, however, did reduce tolerance. Nevertheless, etiolated young seedlings were more tolerant than green, assimilating ones of the same age. In *Hordeum distichum* the heat-killing temperatures were 47° and 45°C, respectively. Sun plants were always hardier than shade plants of the same species, though both were kept thoroughly watered. However, watering alone is unable to maintain the same hydration in the leaves under such atmospheric conditions. According to Heyne and Laude (1940), even a 1-hr exposure to light increased the heat tolerance of plants previously kept in the dark for 12 to 18 hr; as mentioned above, they were probably determining drought rather than heat tolerance. Similarly, older bromegrass seedlings deprived of light showed greatly reduced resistance (Laude and Chagule, 1953). Nevertheless, bromegrass seedlings showed their highest heat tolerance immediately after emergence, even in seedlings deprived of light. Older seedlings deprived of light showed greatly reduced tolerance. When grown in a 12-hr day, *Kalanchoe blossfeldiana* shows a maximum heat tolerance at night in the middle of the dark period and a minimum at midday (Schwemmle and Lange, 1959a). These differences are not simply due to the direct effect of light, since the plant continues to show this "endogenous rhythm" for at least 2 days after transfer to continuous dark under constant temperature and moisture conditions. Furthermore, only the middle, actively growing leaves show this rhythm.

Oleinikova (1965) hardened wheat and millet plants by exposure to 33°–35°C. Although hardening occurred both in the light and in the dark, light accelerated the process, especially at the beginning of hardening. Some of the contradictions may be due to this increased hardening in the light as opposed to increased injury if exposed to light after the heat stress. In the case of leaves of *Tradescantia fluminensis*, exposure to light after short-term (5–10 min) heating induced injury at 10°C lower heating temperatures than if kept in darkness (Lomagin and Antropova, 1966). This injury in the light was accompanied by a considerable destruction of chlorophyll. In variegated leaves of *Chlorophytum elatum*, the light injured only the green parts of the leaf blades. The minimal light intensity causing injury was 1000 lux. The injury was sharply decreased if the air was replaced by nitrogen. The explanation suggested was a photooxidation sensitized by chlorophyll and occurring at the expense of energy not used in photosynthesis. It is, therefore, conceivable that light enhances heat hardening, but that it increases photooxidative heat injury.

d. Mineral Nutrition

The effects of salts on heat injury have already been discussed (Chapter 11). The effects of mineral nutrition have received little attention. Nutrient deficiency raised the heat-killing temperature of *Oxalis acetosella* (Illert, 1924). In agreement with these results, Sapper (1935) found that excess nitrogen or potassium reduced heat tolerance while deficiency very effectively raised it (by about 2°C). High nitrogen also reduced the heat tolerance of turf grasses (Carroll, 1943).

3. Relationship to Plant Characteristics

Age may markedly affect heat tolerance, although there are many apparent contradictions in the literature. In general, Sachs (1864) found that the blades of young, fully grown leaves were killed first. Younger leaves that were not fully grown and bud parts were more tolerant. The most tolerant of all were the old, healthy leaves. De Vries (1870) obtained the same results. According to Illert (1924), older leaves of *Oxalis acetosella* are less heat tolerant than younger ones; the significance of this observation is doubtful, since the leaves were already moribund. Seeds that were allowed to swell in water for 24 hr showed a decreased heat tolerance with age (Porodko, 1926a), even though no decrease in percentage germination of the unheated seeds could be detected. In fruit, heat tolerance increases with ripeness, e.g., from 43°C in unripe grapes to 62°C in ripe grapes (Müller-Thurgau, see Sorauer, 1924). Tree seedlings show an increased tolerance with unfolding of the cotyledons (Huber, 1935). Young leaves of *Helodea* are less tolerant than older ones, but the oldest are the least tolerant (Esterak, 1935). On the other hand, Scheibmair (1937) showed that young moss leaves are more tolerant than older ones, and younger cells more tolerant than older basal ones. According to Heyne and Laude (1940), 10- to 14-day-old corn seedlings are more tolerant than older ones. They also noted an exhaustion of the food material in the endosperm by about the fourteenth day. Bromegrass seedlings, on the contrary, decreased in tolerance about 14 days after planting, and increased with age after 28 days (Laude and Chaugule, 1953). As mentioned above, however, Heyne and Laude may actually have been measuring drought tolerance rather than heat tolerance. Sapper (1935) was unable to detect any difference in heat tolerance between seedlings and older plants. Baker (1929) concluded that the apparent increase in tolerance of conifer seedlings with age was not protoplasmic (i.e., true tolerance), but simply due to the development of mechanical protection (i.e., avoidance). Similarly, the age of tree seedlings and the associated maturity of the cortex and other tissues had little effect on heat tolerance, according to Smith and Silen (1963). According to Bogen (1948), however, younger leaves of *Rhoeo discolor* survive heating for a longer time than older leaves. The same was true of younger basal cells versus older tip cells.

Many other investigators showed similar tolerance gradients in a single organ that may not be related to age differences. The base of *Iris* and *Anthericum* leaves had higher killing temperatures than the tip (De Vries, 1870). In apples, the center may be killed at 50° to 52°C, and the outer layers only by temperatures above 52°C (Huber, 1935). According to Belehradek and Melichar (1930), the basal cells of *Helodea canadensis* leaves succumb first, the apical cells showing the greatest heat tolerance. This order was reversed at lower temperatures, and they point out that the gradient may actually be due to the injurious effect of cutting, which is perhaps propagated from the base to the tip. The accuracy of this suspicion was later demonstrated by Esterak (1935), who obtained a gradient in the reverse direction on the same material. The result apparently depended on whether the leaves were removed with forceps before testing, or by excising with scissors as done by Belehradek and Melichar. A real reversal was found by de Visser Smits (1926). Heat tolerance of beet roots decreased from the base to the tip, except at inception of the second growth period, when the gradient was reversed.

In the case of evergreen and wintergreen plants, Lange (1961) found the younger leaves to be much less heat tolerant than the older ones. Yet young leaves were more heat tolerant than mature leaves in the many species growing on the south coast of Spain (Lange and Lange, 1963). In the case of *Ilex aquifolium*, leaf no. 5 (a younger leaf) had a heat-killing point of 46°, and leaf no. 10, a heat-killing point of 48°C (Lange, 1965c).

In *Kalanchoe blossfeldiana,* however, leaf no. 4 showed maximum sensitivity, and tolerance increased progressively above and below this leaf.

As in the case of freezing tolerance, there seems to be an inverse relationship between growth rate and heat tolerance and this may complicate the above relations to age. The vegetative cone, youngest leaves, and fully grown older leaves of *Kalanchoe blossfeldiana* are more heat tolerant than the middle, not yet fully grown leaves (Schwemmle and Lange, 1959b). In summer, when growth was most intense, the growing cells of the leaves of *Zebrina* and *Echeveria* showed less heat tolerance than the mature cells (Gorbanj, 1962). A decrease in tolerance occurs in some plants on flowering (Henckel and Margolin, 1948; Henckel, 1964), though as mentioned above, *Kalanchoe blossfeldiana* shows the opposite relationship. Similarly, vernalization of peas markedly increased their heat tolerance, although this was independent of any floral initiation (Highkin, 1959).

There is some evidence of a relation to cell sap concentration. The effects of nutrition, low temperature, and moisture found by Sapper (1935) also paralleled the osmotic value (Table 12.4). When different species were compared, or even different parts of the same plant, this relation did not hold. Both with respect to their high water contents and low osmotic potentials, the very thermotolerant succulents are an exception. Julander (1945) found, however, that hardened grasses had about twice as high carbohydrate contents as the unhardened, though there was very little starch. Sucrose accumulated but reducing sugars did not. The substances that accumulated the most were the colloidal carbohydrates, especially levulosans. Since Julander heat-hardened his plants by exposure to drought, these results are not comparable to any obtained by heat hardening. Sapper (1935), on the other hand, found etiolated lower carbohydrate plants more tolerant than the normal green ones. Albino sunflower leaves, however, had a lower heat tolerance than that of the green plant (Avilova, 1962).

The relation of cell sap concentration to thermotolerance is also indicated by the opposite changes in heat tolerance that occur in the guard cells and subsidiary cells (Weber, 1926b). When the stomata are open in the light, the guard cells are more thermotolerant than the subsidiary cells; when they are closed in the dark, the subsidiary cells are more tolerant. In each case, the more thermotolerant cells are free of starch, but presumably contain sugar; the less thermotolerant contain starch, but probably little or no sugar.

The photosynthetic rate of cucumbers in a hothouse rose to a maximum at 40°C, dropping at 47°C to the same value as at 27°C in the open (Fedoseeva, 1966). Relatively more oligosaccharides were formed at the higher temperatures. It was, therefore, proposed that the oligosaccharides play a protective role, raising the thermotolerance of the chloroplasts.

Parija and Mallik (1941) incubated seeds at 40° to 60°C for 8 to 120 hr and recorded the percentage germination. Oily seeds survived the high temperature better than starchy seeds; the higher the oil content, the better was the survival. Linseed was an exception, since it showed greater heat tolerance than cottonseed, though the latter had a higher oil content. They ascribe this to the mucilaginous seed coat. According to Zobl (1950), protein content is a factor in the heat tolerance of spores. This is based on nitrogen analyses which showed that the protein content (6.25 × N content) was three times as great in the heat tolerant bacterial spores as in the less tolerant fungal spores.

Protoplasmal properties have received little attention with regard to heat tolerance. Scheibmair (1937) observed more rapid rounding up in the more thermotolerant, upper, marginal cells of mosses on plasmolysis than in the less thermotolerant basal cells. On the other hand, the tip cells, which were just as thermotolerant as the upper marginal ones, never rounded up at all. This latter result, however, must be due to something other than true viscosity—perhaps a firmer adhesion to the cell wall. Henckel and Margolin (1948) conclude that two important reasons for the heat tolerance of succulents are high cytoplasmic viscosity and high bound water. According to their measurements, both of these factors are developed to a higher degree in the succulents than in the other xerophytes, which are less heat tolerant. They state that the bound water in succulents is much higher than in mesophytes—as high as 70% in certain cacti. During flowering, viscosity decreased suddenly but bound water increased and to some extent compensated for the decreased thermotolerance resulting from viscosity change. These conclusions are obviously due to a misunderstanding of the meaning of bound water, since nearly all the water of succulents is free (see Chapter 14).

Swelling of mitochondria isolated from plants differing in heat tolerance differed at optimal (25°C) as well as at elevated (30°–50°C) temperatures. It was greater in mitochondria from pea plants and from nondrought hardened maize plants than from bean plants and drought-hardened maize (Andreeva, 1969). In the case of the heat-resistant bean and maize plants, shrinkage occurred on addition of ATP even at extreme temperatures (45°–50°C), while in the nonresistant pea plants the shrinkage was much less, indicating irreversible heat injury. This kind of mitochondrial injury is usually ascribed to membrane damage.

4. *Mechanism of Heat Tolerance*

Tolerance of the heat stress requires either strain avoidance or strain tolerance, but the strain may be either a direct or an indirect effect of the heat stress (see Diag. 11.1). The mechanism of heat tolerance, therefore, depends on which of these strains the plant must avoid or tolerate.

a. Avoidance or Tolerance of Indirect Strains

If indirect heat injury is due to starvation or any other metabolic abnormality, those plants that are capable of continuing a normal type of metabolism at high temperatures would be thermotolerant. Harder *et al.* (1932) have shown that many plants adapted to high temperatures have a much higher compensation point than unadapted plants. In contrast to potato and similar plants with low compensation points, these adapted plants are able to assimilate even at temperatures of 45°–53°C. Thermophilic blue-green algae show the same phenomenon (Bünning and Herdtle, 1946). Their respiration rises very slowly with temperature, and even at high temperatures, assimilation is in excess. According to Marré and Servettaz (1956) and to Prat and Kubin (1956), there is no compensation point in thermophilic blue-green algae, or at least it is above 55°C (Fig. 11.7). Though the temperature optima for photosynthesis are the same in tomato and cucumber, the latter is more thermotolerant, due to the less steep drop of its photosynthesis curve to zero at the maximum temperature (Lundegårdh, 1949). In four desert shrubs, it is the dark respiration that adapts, for it drops to about half the value in summer, as compared to the winter value when both are measured at 25°C (Strain, 1969). Desert plants, in general, have a low sensitivity to high temperature in comparison with cultivated plants (Fig. 12.8). For example, the apparent CO_2 absorption of apricot sinks to the compensation point at 35°C, though all the desert

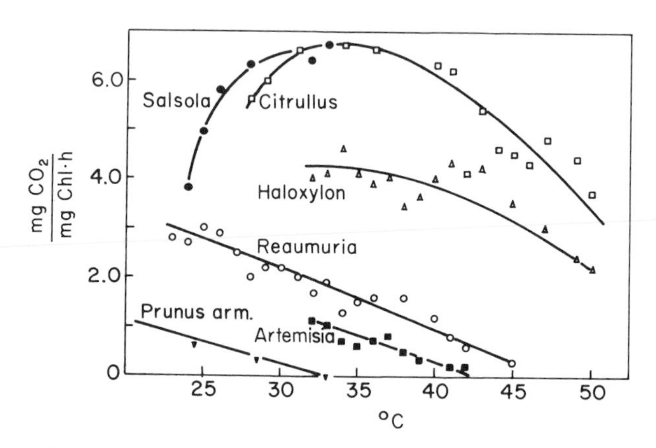

Fig. 12.8. Effect of high temperatures on net assimilation of cultivated plants (e.g, apricot—*Prunus arm*) and of desert plants (e.g., *Citrullus*). (From Lange *et al.*, 1969.)

plants still exhibit a pronounced assimilation at this temperature, and do not attain the compensation point until the leaf temperature rises above 40°C, and in some cases 50°C (Lange *et al.*, 1969). In the case of *Atriplex polycarpa*, even intraspecific differences in CO_2 exchange were correlated with the climates of origin (Chatterton *et al.*, 1970b). Furthermore, a daily temperature regime of 43°D/32°N for a week reduced the rate of dark CO_2 evolution; an equal exposure to 16°D/5°N increased it.

If the metabolic injury is due to a decreased rate of formation of essential intermediates at the high temperature leading to a biochemical lesion, tolerance would require a higher rate of production (see Chapter 11). Similarly, in the case of heat injury due to toxins, thermotolerance would require a decreased production of the toxin, or its detoxification. Neutralization by the synthesis of organic acids, has been suggested as the thermotolerance mechanism against NH_3 toxicity (Chapter 11). For the fourth kind of metabolic heat injury, thermotolerance due to a repair mechanism has already been indicated (Chapter 11). This may perhaps be the basis of some differences in tolerance among four grass species. The two species that are more heat sensitive contained less RNA and DNA per unit dry weight than the two more resistant species (Baker and Jung, 1970b). When subjected to a heat stress (35°C for up to 15 hr) the RNA content was reduced more in the two sensitive species. This could, of course, affect protein synthesis both during and after the heat stress.

TABLE 12.5 Temperature Differences in Certain Plants between the Temperature that stops Protoplasmic Streaming and the Temperature that Prevents Resumption of Protoplasmic Streaming[a]

Species	*t*° stopping motion after 5-min heating I	Maximum *t*° after which restoration is possible II	II–I	References
Tradescantia fluminensis	44.5	52.0	7.5	Alexandrov, 1955
Zebrina pendula	46.7	53.0	6.3	Gorban, unpub.
Campanula persicifolia	44.0	52.0	8.0	Alexandrov, 1955
Ficus radicans	45.8	54.6	8.8	Derteva, unpub.
Chlorophytum elatum	47.0	56.0	9.0	Liutova, unpub.
Catabrosa aquatica	44.7	48.5	3.8	Den'ko, unpub.
Oxilatoria tenuis	49.2	54.0	4.8	Mel'nikova, 1960
Phormidium autumnale	52.2	56.4	4.2	Mel'nikova, 1960

[a] From Alexandrov, 1964.

Further evidence that repair after exposure to the heat stress may be a factor has been produced by Alexandrov (1964). Epidermal cells of *Catabrosa aquatica* showed greater heat tolerance than those of *Campanula persicifolia* as judged by the temperature that stops protoplasmic streaming in 5 min (44.7° and 43.5°–44.0°C, respectively). The protoplasmic streaming can be restored if the heating is not excessive, and in contrast to the above result, *Campanula persicifolia* is better able to repair this injury, for it recovers its streaming even after heating to 51.5°C for 5 min, as compared with 49.0°C in the case of *Catabrosa aquatica*. The temperature difference between these two points varies from 3.8° to 9.0°C, depending on the species (Table 12.5). This repair of injury to cytoplasmic streaming may continue up to 18 days after the heat stress.

Metabolic injury of another kind has been suggested by Ljunger (1970). He found that thermophilic bacteria die at temperatures ordinarily employed for their cultivation, if the medium lacks Ca^{2+}. Heat tolerance also required K^{+}, phosphate ions, and glucose or another energy source. He, therefore, proposes that their heat tolerance is dependent on an active transport of Ca^{2+} from the environment into the cells.

b. Avoidance or Tolerance of Direct Strains

The four above kinds of heat tolerance require four different metabolic mechanisms. Where the direct effects of the heat stress are involved, the mechanism must again be different. Furthermore, a heat-tolerant plant does not necessarily develop the different kinds of tolerance to the same degree. Pisek *et al.* (1968), for instance, measured the tolerance of the direct effects of the heat stress by the rapid (30 min) heat-killing temperature (using Lange's modification of Sachs' method), and tolerance of the indirect effects by the maximum temperature for net assimilation. Among twenty-two species of higher plants, the difference between the two ranged from 3°–12°C, and the difference had no relation to the heat-killing temperature, which ranged from 43°–55°C.

Many attempts have been made to explain the heat tolerance that is due to avoidance or tolerance of the direct strain. In all cases, the direct strain is assumed to be protein denaturation. Therefore, in all cases, tolerance is assumed to be due to (1) thermostability of the proteins leading to avoidance of the denaturation strain, or (2) increased speed of resynthesis of the proteins, leading to tolerance of the denaturation strain. Several mechanisms of one or other of these two kinds of thermotolerance have been suggested.

i. Protective Substances. Molisch (1926) explained the thermotolerance of thermophilic organisms living in hot springs, by (a) presence in their protoplasm of substances that inhibit coagulation, just as in blood, or (b) increased alkalinity of the protoplasm (see also Illert, 1924).

ii. Decrease in Free Water. Bünning and Herdtle (1946) propose an increase in protoplasmic (presumably protein) stability by a decrease in free water, which would also reduce the rate of metabolic processes by slowing down diffusion. Christophersen and Precht (1952) expand this explanation and repropose the identical theory put forward much earlier by Gortner (1938) to explain freezing and drought tolerance. They suggest that metabolism and tolerance are inversely related to each other, the latter being directly related to bound water, or more correctly nonsolvent space. As the solvent space (which is nearly the same as free water) decreases, the transport of substances is decreased and with it metabolism. On the other hand, free water weakens the intermicellar bridges. The thermal oscillations induced by high temperatures can therefore readily break the H and S–S protein bridges. This frees the SH groups, leading to new intra- and intermicellar bonds and loss of the original specific structure. In this way, heat denaturation of the proteins occurs. Bound water, on the other hand, strengthens the protein bridges and opposes high-temperature denaturation. The same objections may be raised to this concept of thermotolerance, as in the case of freezing tolerance (Chapter 9). In further opposition to the above concept, the amount of water bound firmly enough to affect diffusion is such a small proportion of the total that the effect on total diffusion must be negligible. As shown by experiments with Thiogel (see Chapter 10) SS bonds are not broken even by boiling.

iii. Protein Bond Strength. Bogen (1948) attempts to explain his results on the basis of proteins, particularly the folding of proteins and the structural bonds between them. He assumes that heat tolerance is determined by the stability of the molecular form and that any molecular deformation or tearing of molecular bonds can lead to a structural disturbance and finally to death of the cell. Thus he concludes that heat tolerance is not due simply to the degree of hydration, but to the original order of charges and the hydration centers of individual molecules. Any disturbance of this order may lead to death, since the molecule is thereby deformed and the system of molecular bonds is exposed to tensions. In this process, the total water content may be increased, decreased, or unaltered. This, he thinks, may explain "the many contradictory results on the influence of hydration on heat tolerance" (but, as shown above, the results agree very well). Where heat tolerance and hydration do run parallel, it is because they are similarly affected by many other factors.

The effects of the monovalent cations, according to Bogen, can be explained only by adsorption which occurs without discharging the colloidal particle. As a result, the imbibition of the proteins is increased, the molecular bonds are loosened, and heat tolerance is lowered. The same is true for the monovalent anions, since they are effective in the order of their adsorbability. In both cases, however, Bogen states that it is not the hydration of the colloid as a whole that is involved but the hydration of individual molecules, and even then not equally over the whole molecular surface. Nor is it likely to involve a simple enlargement of the hydration shell about previously existing hydration centers. More likely, new dipole moments are induced under the influence of the ion and new hydration centers are created. Thus, both the charge distribution and the arrangement of the hydration centers are altered. The increased heat tolerance produced by $MgSO_4$ does not involve a change in hydration but is associated with strengthening of the structure. It may bind two protein molecules together and hold them in their original orientation, thereby stabilizing the system of molecular bonds. The effect of pretreatment with water, he explains, as due to the washing out of ions.

The basis for Bogen's theory is the effect of ions on the coagulation temperature of living cells. On the basis of an earlier and very extensive investigation of this relationship, Kaho cautioned against drawing any conclusions about heat tolerance from this last stage of heat killing. Furthermore, the order of effectiveness of the ions was basically the same as in Bogen's later results. Kaho pointed out that this order agreed with the order of penetration of the ions into the living cell. Nevertheless, Bogen chooses to conclude that the ion effects are purely due to adsorption on the protoplasmic proteins, which could be true only if they all penetrated at the same rate. Although Bogen's hypothesis is stimulating, the results on which it is based may be explained more simply. It has been shown (see above) that increased total hydration of protoplasm causes decreased heat tolerance, owing perhaps to a more ready unfolding (i.e., denaturation) of the proteins in the presence of ample water. The order of effectiveness of the ions on heat tolerance is the same as the order of their effectiveness on protein swelling (Scarth and Lloyd, 1930). A combination of this effect with the order of penetration of the ions leads to a clear understanding of the results. The protective effect of $MgSO_4$, for instance, follows from its lack of penetration and consequent osmotic dehydration of the protoplasm. Another factor which cannot be ignored is the toxic effect of the unbalanced ions (Kaho, 1926), which must become far more pronounced at the higher temperatures.

iv. ALEXANDROV'S HYPOTHESIS. The most detailed explanation of the mechanism of heat tolerance has emerged over the years (1940–1970) as a result of the many, original experiments of Alexandrov and his group of co-workers in Leningrad (Feldman, Lomagin, Lutova, and many others). This hypothesis has been enlarged over the years, to accommodate all the new evidence as it accumulated. It can be divided into several components, arranged in chronological order as they were developed.

(a) *General basis of thermotolerance.* Nassanow and Alexandrov (1940—see Alexandrov, 1960; Alexandrov *et al.*, 1961), proposed that thermotolerance is due to protein thermostability. Since the properties of the proteins are determined genetically, Alexandrov concludes that thermotolerance is a phylogenetic characteristic of the species, which depends on their temperature of origin. This conclusion was based on experimental evidence with higher plants, but it also agrees with results obtained by his colleague Ushakov with animals. As a result of a vast number of comparisons between closely related species of many animals (mostly poikilotherms but including some mammals), Ushakov (1966) concluded that interspecific differences in heat tolerance are due to differences in thermostability of their proteins. Purification of the proteins showed that the differences were not due to contamination. This absence of stabilizing factors has since been corroborated by measurements on mixtures of proteins from thermophilic and mesophilic organisms (Amelunxen and Lins, 1968). According to Ushakov's results, the thermotolerance of animals is fixed and they, therefore, cannot be hardened. In the case of higher plants, however, Alexandrov's group demonstrated repeatedly that hardening could be induced by heat shock (see above). Lower organisms (algae and protozoa), on the other hand, possessed a labile thermotolerance which varied directly with the growing temperature. No explanation is offered for this fundamental difference between lower and higher organisms.

(b) *Multiple nature of tolerance.* Alexandrov (1961) concludes that the cell has three mechanisms for increasing its heat tolerance: (i) increased rate of replacement of the altered proteins, (ii) an increased concentration of antidenaturing substances, especially sugar, and (iii) structural changes in the protein molecules, increasing their thermostability. On the basis of stresses and strains, the first of his three mechanisms of heat tolerance (repair) would involve tolerance of the denaturation strain, the second and third would confer avoidance of the strain. He later (1970) amplifies this concept by explaining the resulting injury as a balance between two destructive effects (the degree and kind of primary injury, and the destructive aftereffect; see below), and two constructive effects (adaptation and repair). The two constructive effects can again be recognized as strain avoidance and strain tolerance, respectively.

(c) *Two kinds of tolerance.* The sharp difference between the results obtained when thermotolerance is measured by his method and by Lange's (see above), led Alexandrov to propose two kinds of thermotolerance. (i) "Primary" tolerance is measured by his method, involving a brief (5 min) intense heat treatment followed by observation immediately after cooling. (ii) "General" tolerance is measured by Lange's method, involving a more moderate and prolonged (30 min) heating, followed by observation some time after heating. As demonstrated by Lange's group, the only important difference between the two methods was the last point—the time of observation. Alexandrov applies his above concept, and points out that Lange's "general" tolerance depends on a balance between the four destructive and constructive effects. Specifically, it would include both the destructive aftereffect and the repair, neither of which could be involved in Alexandrov's method. It appears, then, that the "primary" thermotolerance of Alexandrov, is a measure of the avoidance of the denaturation strain (due to the specific protein structure or the protective substances), whereas the "general" thermotolerance measured by Lange's method, includes both avoidance and tolerance of the denaturation strain. It further follows that avoidance of the denaturation strain is species specific and also can be changed by heat shock; tolerance of the denaturation strain can be changed by the growing temperature, by decreasing the water content, etc. (see above).

(d) *Proposed mechanism.* Alexandrov explains all the above results on the basis of a conformational flexibility of the proteins (Alexandrov *et al.*, 1970). In the less thermophilic species, the flexibility is supposed to be greater. Those adapted to higher temperatures have been selected for less flexible proteins, which would, therefore, have more stable structural bonds. This greater bond stability would be due to (i) genetically determined structure, and (ii) the properties of their medium and their ability to form complexes with other molecules. In support of this concept, Alexandrov's group demonstrated a parallelism in cell resistance to heat and to many other injurious agents, e.g., many toxic substances and hydrostatic pressure. This was taken as evidence of a general stability of the proteins.

Alexandrov (1964) postulated two stages in the denaturation. The first stage activates some groups in the protein molecules and can increase metabolism. The second stage is a stronger denaturation which results in injury. The former is reversible, the latter irreversible. This concept seems to be disproved by his own colleague Feldman (1966), who showed that heat hardening (Alexandrov's first stage) results in an increased thermostability but a *decreased* activity of the enzymes, in opposition to Alexandrov's prediction of activation. Feldman's results do, however, agree with his concept of heat hardening. Since he was able to harden higher plants only by a rather intense heat shock, Alexandrov concludes that heat hardening is a specific reaction to an injurious action of heat. As evidence, he points to a suppression of photosynthesis at the high temperature which induces hardening. This is not true, however, in the case of algae; he showed that the heat tolerance of algae increases with the temperature at which they are grown, throughout the temperature range for growth. The hardening mechanism is explained by an increase in the general stability of all the proteins, since the hardening process increased the thermostability of both the more labile and the more stable metabolic processes (ranging from 44° to 65°C). He, therefore, favors the formation of a factor that stabilizes all the different proteins, rather than a stabilization of the individual protein molecules. He suggests that there are stabilizing substances in an inactive or bound form, and that hardening activates or releases them, allowing them to interact with the proteins as stabilizing ligands.

Alexandrov's concept is an attempt to explain the mechanism of his "primary" tolerance, and therefore of the avoidance of denaturation strain. He considers the hardening of algae, which is directly related to the temperature of cultivation, and which is inversely related to freezing tolerance (unlike the direct relation in many higher plants) as a different process from the hardening by heat shock, and calls it "temperature adjustment."

v. Thermostability of Proteins. Alexandrov's proposed protein flexibility concept is essentially a variation of the above concepts (e.g., Bogen's), as it must be, since it is based on the same assumption—that thermotolerance is due to protein thermostability. In order to evaluate these theories, it is, therefore, necessary to obtain more direct information as to the thermostability of proteins.

As pointed out above, the earlier ideas were actually based on observations of coagulation rather than denaturation, and of whole cells rather than proteins. More direct evidence of the protein denaturation theory has since then accumulated by the numerous demonstrations of a relationship between the thermostability of proteins and the heat tolerance of the tissues or cells in which they occur (Table 12.6). That these results are due to the proteins themselves and not to the presence or absence of stabilizing factors was indicated by heat hardening of cucumber and wheat plants (Feldman *et al.*, 1966). The thermostability of their acid phosphatase increased by 0.5° to 3.0°C. Since this increase in thermostability persisted after dialysis, it does not depend on the accumulation of a dialyzable protective substance. In contrast to the increase in *thermostability* of the enzyme, its *activity* declines with adaptation to high temperature (Christophersen, 1963; Feldman, 1968). The activity of acid phosphatase was decreased by half due to the hardening treatment, although

TABLE 12.6 RELATIONSHIP BETWEEN THERMOTOLERANCE OF ORGANISMS AND THERMOSTABILITY OF THEIR PROTEINS[a]

Enzyme	Organism	Thermotolerance versus thermostability	Reference
TPNH-cytochrome c reductase	*Aphanocapsa thermalis* (t) vs. *Anabaena cylindrica* (m)	Enzyme from thermophile more resistant to heat inactivation	Marré and Servettaz, 1956
Eleven	*Bacillus stearothermophilus* (t) vs. *B. cereus* (m)	With two exceptions much greater thermostability in thermophile	Amelunxen and Lins, (1968)
Hexokinase	*Candida pseudotropicalis*	Thermostability increases with adaptation temperature	Christophersen, 1963
Urease, acid phosphatase, ATPase	Leaves of cucumber, wheat, Caragana	Heat hardening of leaves increased heat tolerance of cells and thermostability of enzymes	Feldman, 1968
Urease	*Leucojum aestivum* (t) vs. *L. vernum* (m)	40 min at 78° (m) vs. 81° C (t) for 50% inactivation	Feldman and Kamentseva, 1967
Malate dehydrogenase	*Typha latifolia*	Thermostability higher when obtained from plants native to a hot climate	McNaughton, 1966
Malic dehydrogenase	*Phaseolus acutifolius*	Thermostability higher when extracted from hardened plants	Kinbacher *et al.*, 1967
Cellulolytic enzymes	*Aspergillus fumigatus*	Activity maximal at 60° C in thermotolerant at 55° in mesophilic organism	Loginova and Tashpulatov, 1967
Pyruvate carboxylase	*Bacillus licheniformis* (t) vs. *B. coagulans* (m)	Enzyme lost activity at 45° (10 min) in (m) not in (t) even after 45 min	Sundaram *et al.*, 1969
Acid phosphatase	Cucumber	Hardening (29 hr at 39°C) increased thermostability 0.5° to 3.0°C	Feldman *et al.*, 1966

[a] t, thermophile; m, mesophile.

its thermostability increased 0.5° to 3.0°C (Feldman *et al.*, 1966). Urease, however, is an exception, since its activity does not decline with increase in thermostability. The concentration of protein also decreases as a result of heat hardening (Feldman, 1966).

Some enzymes have not shown a relation between their thermostability and the heat tolerance of the plant from which they are extracted. In the case of *Typha latifolia* (McNaughton, 1966), in contrast to malate dehydrogenase, glutamate-oxaloacetate transaminase was quite resistant regardless of origin, and aldolase was rapidly inactivated regardless of origin. This may be interpreted to mean that only certain specific enzymes are important in the heat tolerance of plants, or that some enzymes are not thermostable unless the plant is first hardened (see below). On the other hand, the thermostability of an enzyme in a living cell may be quite different from that in extracts of the cell, due to pH differences, presence of substrate, etc. Since, in the above experiments, the leaves were blended with 30 volumes of water and no buffer was used, only the most stable enzymes would be unaltered. Aldolase is an unstable enzyme (see above) and any differences between plants would therefore quickly disappear. According to Kurkova (1967), ATPase activity rose more with heating (40°C) in the unhardened than in hardened corn shoots. This difference explained the observed phosphorylation which continued in the hardened but not in the unhardened shoots at 40°C. This result agrees with Feldman's observation (see above) that although heat hardening increases the thermostability of enzymes, it decreases their activity.

Many other investigators have established conclusively that the thermostability of an enzyme (measured by the temperature at which 50% of their activity is lost) depends on the organism from which it is obtained. In the case of catalase, it varied from 48.1° to 67.1°C among 26 kinds of mammals tested (Feinstein *et al.*, 1967). As indicated above, these differences are not due to the presence of protective substances, but are characteristics of the enzyme molecules. The fundamental question is how such differences in thermostability of the molecule can arise. Some of the recent investigations provide the possible answer to this question.

The thermostability of collagens was correlated with their amino acid contents, e.g., the proline, hydroxyproline, and threonine contents (Rigby, 1967). Jockush (1968), in fact, has suggested that any pair of homologous proteins will differ in thermostability if there is at least one difference in amino acid sequence. He investigated the coat protein of TMV, whose tertiary structure is not complicated by covalent (SS) cross-links. At low ionic strength (0.02) and a slightly alkaline reaction (pH 7.5) the protein is mildly heat denatured and stays in solution. This mild denaturation consists of a partial unfolding of protein subunits leading to exposure of hydrophobic regions, which in turn causes limited aggregation, and at higher ionic strength or lower pH results in precipitation. The temperatures of half denaturation were 40.5° and 27°C for the vulgare form and the mutant, respectively. The following lowerings of the denaturation temperature resulted from amino acid replacement:

tyrosine by cysteine	40°C → 35°C
asparagine by lysine	30
asparagine by glycine	32
asparagine by alanine	31
proline by threonine	29
proline by leucine	27
threonine by isoleucine	32
proline by serine	32
isoleucine by threonine	37
proline by leucine	37

It is, of course, known that the thermostability of proteins depends not only on its amino acid composition, but also on its three-dimensional structure (Kasarda and Black, 1968). The enzyme D-amino acid oxidase, for instance, occurs in two forms—a low temperature and a high temperature conformation (Koster and Veeger, 1968). They have about the same activation energies but differ in activity due to different transition probabilities. It is a fact, however, that the primary structure (i.e., the amino acid arrangement) determines the secondary and tertiary structure. A full understanding of the effect of the replacement of one amino acid by another, as in the above experiments of Jockush (1968) may therefore perhaps depend on the location of the change in the molecule.

In some cases, the thermophile may possess a completely different kind of enzyme that performs a similar function but possesses a higher thermostability. Temperate Gramineae, for instance, show maximal net photosynthesis at 20°C and tropical Gramineae at 30°–35°C (Treharne and Cooper, 1969). RUDP carboxylase is primarily involved in the photosynthesis of the temperate Gramineae and shows maximum activity at around 20°C. PEP carboxylase is primarily involved in the photosynthesis of the tropical species and shows maximal activity at 30°–35°C.

There are other factors that may conceivably result in a difference in thermostability of proteins. The enzymes trypsin and papain when covalently coupled to porous glass show a marked increase in thermostability (Weetall, 1969). Obviously, the relationship of any protoplasmic protein to other components of the protoplasm may similarly affect its thermostability.

It must be realized that protein thermostability is required for prevention of indirect proteolytic heat injury as well as for prevention of direct heat injury; in order to resynthesize its protoplasmic proteins more rapidly at higher temperatures, the enzymes involved in protein synthesis must be fully active at these temperatures. Since these enzymes are proteins, the heat-tolerant plant can prevent heat injury only if its protein synthesizing enzymes are thermostable (see above). The thermotolerant plant must, therefore, possess the following two kinds of protein thermostability. (1) Its protein synthesizing enzymes must be thermostable, for prevention of indirect injury due to proteolysis. (2) The protoplasmic proteins whose denaturation and aggregation would cause instant death must be thermostable, for prevention of direct injury. These must be the membrane proteins, for they (and probably not the enzymes) can be expected to cause instant death if denatured and aggregated (see Chapter 10). This conclusion is supported by direct observation of cells during heat killing (see Chapter 11).

Lange (1961) has postulated that these two kinds of thermotolerance occur at two different times of the year. The greater rate of protein synthesis would account for the summer increase in thermotolerance that is not accompanied by an increase in freezing tolerance. He believes that the winter rise in freezing tolerance involves a general stabilization of the proteins, accounting for the accompanying rise in thermotolerance. Since such a general stabilization would also be expected to lead to increased rate of protein synthesis, Lange's hypothesis appears less likely than the above explanation—a winter increase in the thermostability only of the membrane proteins. This conclusion is supported by the lack of any change in freezing stability of the soluble proteins (see Chapter 10) which opposes Lange's concept.

C. Molecular Aspects of Thermotolerance

In spite of their differences, all the above concepts agree that protein denaturation is the cause of heat injury and that protein thermostability is the basis of plant thermotolerance. The negative evidence in the case of nucleic acids and lipids supports this conclusion. Some attempts have also been made (see above) to explain the thermostability of the proteins of thermotolerant plants. These attempts were made before adequate information was available on the molecular structure of proteins, and on the effects of temperature on this structure. The following is a more detailed molecular explanation on the basis of Brandts' concepts of native (N) proteins, their denaturation (D) and aggregation (A) (see Chapter 11):

$$N \underset{\text{normal } T}{\overset{\text{high } T}{\rightleftharpoons}} D \xrightarrow{\text{high } T} A$$

The thermostability of the proteins, on this basis, can only be due to the prevention of either the reversible denaturation (N $\rightleftharpoons$ D) or of the irreversible aggregation (D $\rightarrow$ A).

1. Prevention of Reversible Denaturation (N $\rightleftharpoons$ D)

This could be due to a strengthening of the protein bonds. Thus Koffler's *et al.* (1957) analyses of the amino acid content of the proteins led him to conclude that the thermostability of the proteins from thermophiles is due to their higher hydrophobicities. Similar results were obtained by Ohta *et al.*, (1966). This conclusion was supported by the disintegration of the flagella of the mesophile in 4–6 *M* urea and those of the thermophile only in 9 *M* urea (Mallet and Koffler, 1957). It is now known that these high concentrations of urea break the hydrophobic bonds. The higher concentration needed to disintegrate the thermophile's protein indicates a greater strength or number of hydrophobic bonds. They also demonstrated a greater resistance to sodium dodecylsulfate, an anionic detergent that breaks hydrophobic bonds. Similar results were obtained by Marré *et al.* (1958a). Acetamide and urea inhibited the activity of cytochrome *c* reductase to a lesser extent when the enzyme was extracted from a thermophilic blue-green alga (*Aphanocapsa thermalis*, than when extracted from a thermosensitive alga *Anabaena cylindrica*). All the above results are explained by Brandts' concept. Since the strength of hydrophobic bonds increases with the rise in temperature up to about 75°C (Brandts, 1967), proteins with a higher proportion of hydrophobic bonds would remain in the folded, native state at temperatures high enough to denature proteins with a lower proportion of hydrophobic bonds. In agreement with this concept, the β form of polylysine is the more stable form at 50°C and appears to owe a large part of its stability to hydrophobic interactions between the lysyl residues. The α-helical polylysine is stabilized largely by interamide hydrogen bonds and is the more stable form at 4°C (Davidson and Fasman, 1967).

Bigelow (1967) has calculated the hydrophobicities of the amino acid side chains and has used these values to obtain a quantitative comparison between those proteins of thermophilic and mesophilic organisms for which amino acid analyses are available. The phycocyanins of five mesophilic species yielded average hydrophobicities of 980–1110 cal/residue, compared to 1190 cal/residue for one thermophilic species. The α-amylases of four mesophilic species had hydrophobicities of 1020–1130 cal/residue, compared to 1210 for the one thermophile. Friedman (1968), on the other hand, has compared the amino acid analyses for the ribosome protein of a mesophilic and a thermophilic bacterium and has concluded that there is no obvious difference between the two. This conclusion was based solely on visual comparison. Calculation of the hydrophobicities, with the use of Bigelow's values, supports his conclusion, yielding average values of 1062 cal/residue and 1095 cal/residue, respectively (Table 12.7). However, some of the hydrophobicity values for the side chains are so low (0.45–1.70 kcal/residue) that they may not be able to form bonds strong enough to hold the protein folds in place. Furthermore, these differences in strength are exaggerated at high temperatures (Brandts, 1967). The strength of the weaker hydrophobic bonds rises only slightly with temperature to a maximum at 50°C, only 200 cal above the value ot 0°C. The strength of the stronger hydrophobic bonds rises much more steeply with temperature to a maximum at 80°C, which is 750 cal above the value at 0°C. If only side chains with values of 2 kcal/residue and above are compared, all five amino acid residues are found to be present in larger quantity in the thermo-

TABLE 12.7 Relative Hydrophobicities of Proteins from a Thermophile (*Bacillus stearothermophilus*) and a Mesophile (*Escherichia coli*)[a,b]

Amino acids with hydrophobic side chains	Hydrophobicity (kcal/res)	*B. stearothermophilus*			*E. coli*		
		Mole (%)	kcal	Av. cal/res	Mole (%)	kcal	Av. cal/res
Trp	3.00	—	—	—	—	—	—
Ile	2.95	6.64	19.55	—	5.51	16.25	—
Tyr	2.85	2.16	6.15	—	1.78	5.07	—
Phe	2.65	3.55	9.40	—	3.03	8.03	—
Pro	2.60	4.44	11.52	—	3.67	9.55	—
Leu	2.40	8.34	20.00	—	7.40	17.75	—
Val	1.70	9.07	15.40	*666*	9.63	16.35	*566*
Lys	1.50	6.30	9.45	—	9.01	13.50	—
Met	1.30	2.33	3.02	—	2.40	3.12	—
Cys/2	1.00	0.82	0.82	—	0.53	0.53	—
Ala	0.74	10.51	7.90	—	10.98	8.23	—
Arg	0.75	5.01	3.76	—	7.30	5.46	—
Thr	0.45	5.53	2.49	—	5.22	2.35	—
Totals		64.70	109.46	*1095*	66.46	106.19	*1062*

[a] Data from Bigelow, 1967 and Friedman, 1968. Numbers in italics indicate averages.
[b] From Levitt, 1969b.

philic than in the mesophilic protein (Table 12.7). On the basis of these five amino acids, the average hydrophobicity of the protein of the thermophile is 666 cal/residue compared to 566 cal/residue in the mesophile. In terms of numbers of strong hydrophobic bonds, there would be a maximum of 1 bond/8 residues in the thermophile, and only 1/9.3 residues in the mesophile. It is even conceivable that some of the weaker hydrophobic side chains may be effective in the proteins of the thermophile because of proximity to the one strong SS bond, which apparently occurs in the protein of the thermophile but not in the protein of the mesophile. A similar analysis of Matsubara's (1967) data for the heat-resistant protein thermolysin again reveals no higher value for total hydrophobicity (1010 cal/residue), but a very high value for the six groups having high hydrophobicities (706 cal/residue).

This mechanism could conceivably also arise due to hardening. If the proteins are broken down and resynthesized during this exposure to moderately high temperature, the newly synthesized proteins, though having the same amino acid sequence, would fold in a somewhat different manner due to the increased affinity of the hydrophobic groups for each other at the higher temperature and the weakening of the H bonds. They would therefore have a higher proportion of hydrophobic to hydrogen bonds and would be more stable at the high temperatures. This would account for the slow natural hardening during summer (see above). Similarly, even if the proteins are not broken down and resynthesized, if some of the hydrogen bonds are broken at the high temperature, this may permit the formation of new and stronger hydrophobic bonds, between groups previously separated sterically, and which now have a stronger affinity for each other because of the high temperature. This more rapid hardening would occur due to heat shock.

As mentioned earlier, the change in conformation may or may not alter the activity of the enzyme, depending on whether or not the active site is altered. Thus, heat-hardening of leaves, which leads to increased thermotolerance of their cells, increases the thermostability of their enzymes by 1.5°–7°C in the case of ATPase, acid phosphatase, and urease (Feldman, 1968). In the ase of the first two, this was accompanied by a decrease in activity, but the activity of urease was not affected by its increase in thermostability.

In agreement with this concept, a protein conversion during heat hardening has been observed by Jacobson (1968) in the case of *Drosophila*. An increase in thermostability accompanied the conversion of the electrophoretically slowest moving isoenzyme of alcohol dehydrogenase to the fastest moving form. This result is predicted by the above concept, since the increased strength of the hydrophobic bonds and the rupture of some of the hydrophilic bonds would lead to a greater proportion of internal hydrophobic bonds, and a greater proportion of external, free hydrophilic (polar) groups, leading to a greater surface charge and, therefore, greater electrophoretic mobility.

It does not necessarily follow, however, that all the thermostability of the proteins is due to a high proportion of hydrophobic side chains. The strongest of all intramolecular bonds responsible for the conformation of the proteins is the covalent SS bond. The importance of this bond in the thermostability of proteins can be seen by comparing thermostable with thermolabile enzymes. The thermostable are characterized by absence of SH groups and presence of SS bonds (Table 12.8). Similarly, when different lysozymes are compared, hen's egg lysozyme has four SS bonds and is the most thermostable, human lysozyme has three SS bonds and is less thermostable, and goose egg lysozyme has two SS bonds and is the least stable (Jollès, 1967). Similar results have been obtained with bovine lens proteins (Mehta and Maisel, 1966).

Although analyses of pure plant proteins have not yet been made,

TABLE 12.8 Properties of Heat-Stable Enzymes

Protein	Inactivation temperature (°C)	Time (min)	Inactivation (%)	Molecular weight	SH (groups/molecule)	SS (groups/molecule)
1. Pepsin	65 (in acid)	15	50	35,000	0	3
2. α-Amylase (*Bacillus subtilis*	65	30		48,700	0	0
3. Arginase	70	177	50	140,000	0	0
4. α-Amylase (thermophile)	90	60	10	15,000	0	2
5. Inorganic pyrophosphatase	90–100	—	—	63,000	Trace	
6. Cytochrome *c*	>100	—	—	13,000	0	0
7. Muramidase (lysozyme)	>100	—	—	15,000	0	4
8. Ribonuclease	>100 (in acid)	—	—	12,700	0	4
9. Trypsin	>100	—	—	24,000	0	6
10. Myokinase (adenylate kinase)	>100	—	—	21,000	2	0

[a] From Levitt, 1966a.

indirect evidence points to a possible role of SS bonds in the thermotolerance of plants. Fraction I protein was found to be more thermostable when isolated from the leaves of heat-hardened bean plants than from unhardened plants (Sullivan and Kinbacher, 1967). Blocking the SH groups with PCMB (parachloromercuribenzoate) did not change the thermostability of the protein from hardened leaves. On the other hand, cleavage of the SS bonds with ME and sodium sulfite decreased the thermostability of the protein from hardened plants to that of the protein from unhardened plants. No significant difference was found, however, between the number of SS bonds in the hardened and unhardened plants. It was therefore suggested that the hardening process led to a repositioning of the SS bonds in such a way as to increase protein stability. This agrees with the above concept of a change in conformation leading to a larger number of hydrophobic bonds.

2. Prevention of Irreversible Aggregation (D $\rightarrow$ A)

Aggregation could be prevented only by eliminating or protecting the chemical groups capable of interacting with each other to form intermolecular bonds. It is still a question as to which chemical groups can form such bonds. As in the case of intramolecular bonds, the strongest *intermolecular* bond that can be formed is the covalent SS bond. In agreement

TABLE 12.9 PROPERTIES OF HEAT-LABILE ENZYMES[a]

Enzyme	Inactivation temperature (°C)	Time at temperature (min)	Molecular weight	SH (groups/ molecule)	SS (groups molecule)
1. β-Galactosidase	55	1	750,000	12	2
2. L-Glutamate dehydrogenase	55	—	1,000,000	90–120	
3. α-Glycerophosphate dehydrogenase	55	1 (53% activity lost	78,000	15–16	
4. Glyceraldehyde-3-phosphate dehydrogenase	Room temperature (stabilized by ethylenediaminetetraacetate at 39)	—	120,000	11±2	0
5. Succinic dehydrogenase	Unstable even at 25	—	200,000	SH enzyme	
6. Xanthine oxidase	56	—	290,000	SH enzyme	
7. Glucose oxidase	Above 40	—	154,000	SH enzyme	

[a] From Levitt, 1966a.

with this, SH-containing enzymes are thermolabile (Table 12.9), presumably because the SH groups of adjacent molecules can combine to form intermolecular SS bonds. The thermostable enzymes, on the other hand, are mostly free of SH groups (Table 12.8). The most striking difference is between the same enzyme (α-amylase) from a thermotolerant and a thermosensitive species of bacterium. The former has two SS bonds per molecule, while the latter has none (Table 12.8). Similarly, the proteins of some thermophilic bacteria are free of any SH or SS groups (Koffler *et al.*, 1957; Ohta *et al.*, 1966; Matsubara, 1967).

Considerable evidence points to the conversion of SH groups of proteins to intermolecular SS bonds during freezing (see Chapter 10). Since, as mentioned above, an increase in freezing tolerance is accompanied by an increase in heat tolerance, the same chemical groups must be involved in protein aggregation at both temperatures. In favor of this conclusion, a decrease in protein SH and an apparent increase in SS groups has been, found during heat killing (Levitt, 1962). In agreement with these results, heat injury was decreased when oxygen was deficient and was increased when oxygen was increased (Al'tergot, 1963). These results were explained by the thermostability of the oxidases and the consequent oxidation of ascorbic acid, GSH, and tannins at high temperature. Porto and Siegel (1960) also implicate oxidation in heat injury. They explain the kinetin-induced restoration of heat-injured germination on the basis of anti-oxidant activity. Similarly, the results of Levy and Ryan (1967) indicate that heat inactivation of the relaxing site of actomyosin occurs because certain labile SH groups are oxidized to the SS form. Dithiothreitol led to both prevention and reversal of this effect. Mishiro and Ochi (1966) found that human serum albumin becomes turbid at 60°–95°C. This turbidity was completely prevented in 0.05 M solution in the presence of 10^{-3} M sodium dipicolinate, which decreased the number of SH groups.

However, even SH-free proteins of thermophilic bacteria are inactivated if the temperature is high enough. Even gelatin, which is denatured and free of SH groups, forms covalent cross-links between adjacent molecules when the water content falls below 0.2 gm/100 gm protein (Yannas and Tobolsky, 1967). It therefore seems reasonable to conclude that as the temperature rises, and more and more bound water is converted to free water, previously protected chemical groups become available for the intermolecular covalent bond formation. Consequently, at moderately high temperatures, aggregation may be primarily due to intermolecular SS bonding by oxidation of free SH groups (which do not bind water); at still higher temperatures, other groups may become available for intermolecular bonding. What these other groups are, is not known.

In the absence of SH groups (e.g., in the case of proteins from thermophiles), Awad and Deranleau (1968) suggest that intermolecular hydrophobic bonding is possible. Nearly all the hydrophobic groups are internal to the native molecule and therefore unavailable for intermolecular bond formation as long as the molecules remain native. When the proteins are denatured reversibly at the high temperature, this is due to the greater thermodynamic tendency to unfold than to remain bonded by the hydrophobic groups in the folded state (see Fig. 11.8). Although these hydrophobic bonds are too weak to hold the protein in the folded state, they may be strong enough to aggregate the unfolded protein molecules. On the other hand, the initial unfolding also involves the breaking of hydrophilic bonds (e.g., H bonds) which are weaker than the hydrophobic bonds at the high temperature. This may permit the formation of new hydrophobic bonds previously sterically impossible. It is, therefore, conceivable that as soon as the hydrophilic bonds are broken, the molecule quickly refolds to a conformation that is more stable at the high temperature. This would clearly explain the difference between rapid thermohardening, and thermoinjury. A brief exposure to a supraoptimal temperature would permit partial unfolding and immediate refolding to a more hydrophobically bonded, and therefore, more thermostable form. A longer exposure or a higher temperature would permit aggregation and, therefore, injury.

An apparently opposite relation has been suggested by Berns and Scott (1966). They conclude that the protein from a thermophilic member of the Cyanophyta (*Synechococcus lividus*) has a greater number of charged and polar groups than that of mesophiles. According to their interpretation, aggregation reversibly inactivates the enzymes at 25°C, and only by raising the temperature to 50°C would the smaller, active form of the enzyme occur. This explanation is difficult to accept, since the association of monomer protein subunits to form di-, tri- and multimers usually involves hydrophobic bonding which increases in strength from 5°–50°C (see Fig. 11.8). Association would, therefore, be more likely at the higher than at the lower temperature. Furthermore, it is usually the larger, associated protein molecule that is active, and the monomer that is inactive. Aggregation of a protein has also been reported for another thermophilic bluegreen alga (*Oscillatoria*) from a hot spring (Castenholz, 1967). Since these organisms survive as much as 30° higher temperatures than the most thermotolerant higher plant (see above), the mechanism is perhaps different.

On the basis of the above concepts, thermostability of proteins depends on their possession of one or more of the following characteristics: (1) a high proportion of strong hydrophobic bonds, (2) intramolecular SS bonds, and (3) absence of SH groups. Thermotolerance of plants is apparently due to protein thermostability and therefore presumably dependent on these same characteristics. Since soluble enzymes are less hydrophobic than membrane proteins (see Chapter 10), this concept also explains why the indirect metabolic injury due to protein loss occurs at a lower temperature than the direct (membrane) injury. The different kinds of thermotolerance are tabulated in Diag. 12.1. Although the evidence points to the proteins as the main factors in heat tolerance, this does not, of course, preclude a role for saturated lipids with high melting points.

D. Relationship between Thermotolerance and Low-Temperature Tolerance

1. Thermotolerance and Chilling Tolerance

The different kinds of indirect stress injury induced by the chilling (Chapter 3) and the heat (Chapter 11) stresses are apparently identical. It does not necessarily follow, of course, that the respective tolerances of these two stresses are correlated. If, for instance, the metabolic disturbance leading to the injury is due to a reversible denaturation and, therefore, an inactivation of certain enzymes, the proteins most readily denatured at a high temperature would be those with low hydrophobicities, and this would lead to a greater stability at low temperature. On this basis, thermotolerance and chilling tolerance of the indirect stress effects would be mutually exclusive. This prediction agrees with the known facts; chilling-sensitive plants are commonly those adapted to growth at high temperatures and therefore possessing thermotolerance.

Tolerance of the direct effects of the chilling stress is apparently related to the membrane lipids (see Chapter 3). The evidence indicates an increase in unsaturation of lipids in chilling tolerant and a decrease in thermotolerant plants. This again leads to the prediction that the two tolerances are mutually exclusive.

2. Thermotolerance and Freezing Tolerance

A relation between thermotolerance and freezing tolerance has already been indicated (see above); when a plant hardens in the autumn, it develops freezing tolerance and at the same time its thermotolerance rises to a maximum, paralleling the freezing tolerance (Alexandrov, 1964). In both cases, it is the tolerance of the direct effects—of the secondary, freeze-induced water stress, and of the heat stress, respectively. This correlation

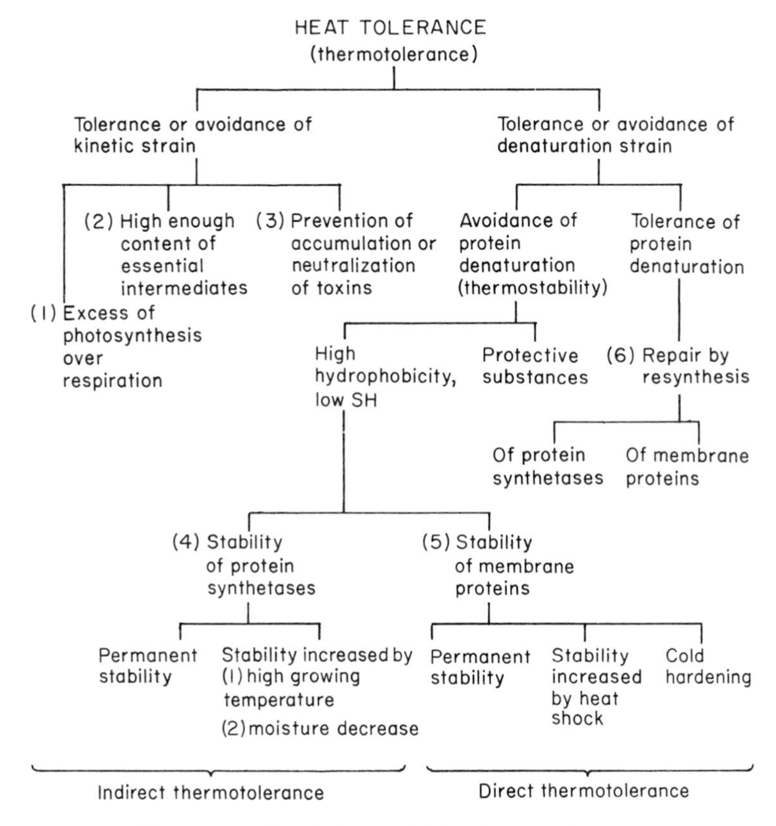

DIAG. 12.1. The different kinds of heat tolerance.

is paradoxical (see Precht, 1967), since heat tolerance during winter can play no role in nature. The freezing tolerance reaches a maximum when the plant is exposed to the lowest temperatures, but the thermotolerance reaches its maximum at the time when the plant is never exposed to a high-temperature stress. This parallel development of the two kinds of tolerance, therefore, cannot be an accidental correlation, due to parallel selections by the habitat. It must mean that those factors developed by the plant in the fall that lead to freezing tolerance, also confer thermotolerance on the plant, even though this is a useless character. Since freezing tolerance is due to a combination of avoidance and tolerance of the freeze-induced dehydration strain (see Chapter 7), the accompanying thermotolerance must be due to one or both of these characteristics. Avoidance of freeze-induced dehydration is due to accumulation of sugar in the cell vacuole (see Chapter 10). There is some evidence that sugar content may be a factor in thermotolerance. This may be due either to increased avoidance of protein denaturation or to a supply of metabolic energy for the repair mechanism and therefore an increased tolerance of protein denaturation. Tolerance of freeze-induced dehydration, according to the molecular concept of freezing tolerance (see Chapter 10), is due to a prevention of the irreversible aggregation of its proteins (D $\rightarrow$ A). This prevention depends on the absence or protection of chemical groups capable of interacting with those of adjacent protein molecules. It, therefore, confers both freezing tolerance and thermotolerance on the plant.

During the summer, on the other hand, the thermotolerance of a plant rises without any concomitant rise in freezing tolerance (see above). There are, therefore, two kinds of thermotolerance. This can be readily explained on the basis of the above molecular concept of protein thermostability. The thermotolerance developed at high temperatures (during summer) is due to a prevention of the reversible denaturation of its proteins (N $\rightleftharpoons$ D) when exposed to a high-temperature stress. Since the factor involved—an increase in hydrophobic bonding—leads to greater denaturation at low temperature, this kind of thermotolerance and freezing tolerance are mutually exclusive. Evidence of this mutual exclusion has been obtained in the case of bacteria. Conversion of mesophilic to psychrophilic bacteria lowered the minimum temperature for growth from 11° to 0°C. This was accompanied by a decrease in the maximum temperature for growth (Olsen and Metcalf, 1968).

Why should the normal (summer) thermotolerance be due to prevention of denaturation (N $\rightleftharpoons$ D), and freezing tolerance to prevention of aggregation (D $\rightarrow$ A)? It must be remembered that in order to survive the heat stress as it occurs in nature, a plant must prevent both direct and indirect (proteolytic) heat injury. The ability to prevent aggregation would not prevent the indirect (proteolytic) heat injury; denatured enzymes are usually inactive whether or not they are aggregated. Therefore, if the protein synthesizing enzymes were denatured, protein synthesis would cease even in the absence of aggregation. On the other hand, freezing injury could not be prevented simply by an increased resistance toward denaturation (N $\rightleftharpoons$ D), since the plasma membrane of the extracellularly frozen cell is subjected to a mechanical stress due to contraction of the freeze-dehydrated cell. The consequent stretch of the cell surface would itself induce a partial unfolding (as indicated by experiments with keratin; see Chapter 10). Therefore, the only safe mechanism of freezing tolerance available to the plant is prevention of aggregation (D $\rightarrow$ A) of the membrane proteins. The thermotolerant plant does not have to prevent irreversible aggregation (D $\rightarrow$ A) even in the case of direct heat injury involving its membrane proteins, since its cells remain fully hydrated when subjected to the heat stress, and no tension-induced unfolding of its membrane proteins occurs. Therefore, the thermotolerant plant's prevention of denaturation (N $\rightleftharpoons$ D) by the heat stress suffices, for a native protein does not undergo irreversible aggregation.

It must be pointed out that the denatured form is not as stable at high temperatures as at low temperatures. In fact, some proteins are so quickly converted from D $\rightarrow$ A at high temperatures, that the reversibility of the denatured state (N $\rightleftharpoons$ D) was not recognized until recently. Therefore, when the proteins in the thermosensitive plant are denatured on exposure to a heat stress, aggregation will quickly follow. When this occurs in the membrane proteins, semipermeability is lost and the cell dies as a result of direct heat injury.

On the basis of this concept, the similarities and differences between freezing tolerance and thermotolerance now become clear.

1. The development of freezing tolerance by hardening at low temperature also results in thermotolerance of the direct type because the hardened plant can prevent irreversible aggregation (D $\rightarrow$ A). In the case of the frozen plant, the membrane proteins are under tension and therefore incipiently unfolded; in the case of the heated plant the proteins unfold spontaneously and reversibly at the high temperature (N $\rightleftharpoons$ D). Both are, therefore, in danger of irreversible aggregation (D $\rightarrow$ A).

2. Low-temperature hardening is slow; high-temperature (heat shock) hardening is rapid. On the basis of the temperatures involved (5° and 45°C, respectively), the maximum ratio of the two rates, assuming a Q_{10} of 3, is $1/(3)^4 = 1/81$. In actual fact, full hardening requires weeks at low temperature and hours at high temperature. The minimum time for low-temperature hardening is about 24 hr, and for high-temperature hardening 1 sec, the ratio being 1/86,400. Therefore, heat hardening is three orders of magnitude larger than expected solely from the direct effect of temperature on reaction rates. This would be possible only if the two processes involved different kinds of reactions. On the basis of the above concept, heat hardening would be almost instantaneous since it involves only a conformational change in the protein. Low-temperature hardening would be much slower since it depends on chemical reactions.

3. Many enzymes of heat-tolerant plants have been shown to possess greater thermostability than the enzymes of less tolerant plants, but no enzymes of freezing-tolerant plants have been shown to possess greater cryostability than the corresponding enzymes of less tolerant plants. This is to be expected, since the reversible denaturation (N $\rightleftharpoons$ D) is also an

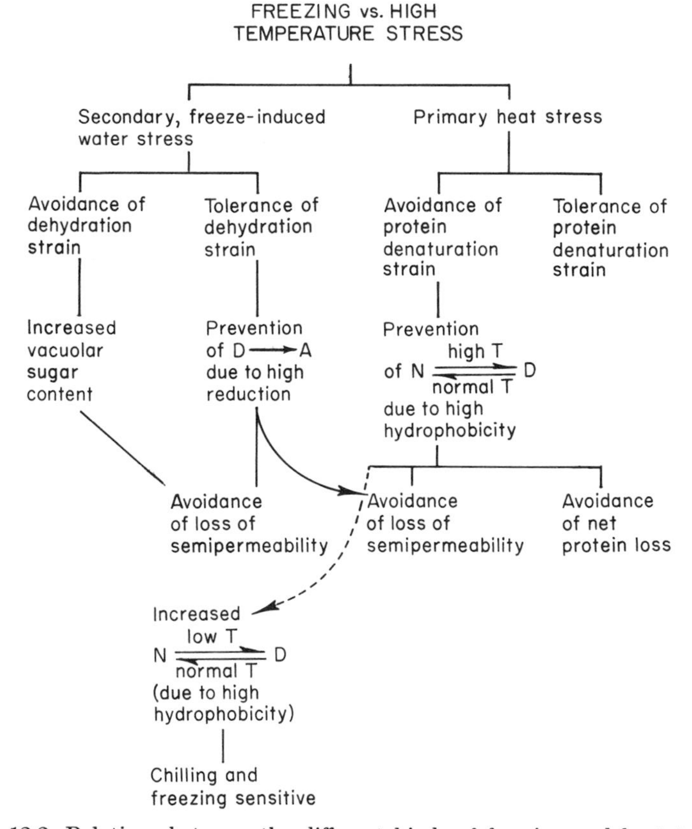

DIAG. 12.2. Relations between the different kinds of freezing and heat tolerance.

inactivation and therefore both are prevented in the thermotolerant plants. In the freezing tolerant plant, this reversible denaturation does not have to be prevented, since the subsequent irreversible aggregation of the membrane proteins is prevented.

4. Repair is a far more important factor in thermotolerance due to the more rapid rate of metabolism at the postheating than at the postthawing temperatures.

The proposed relations between freezing tolerance and thermotolerance are presented in Diag. 12.2.

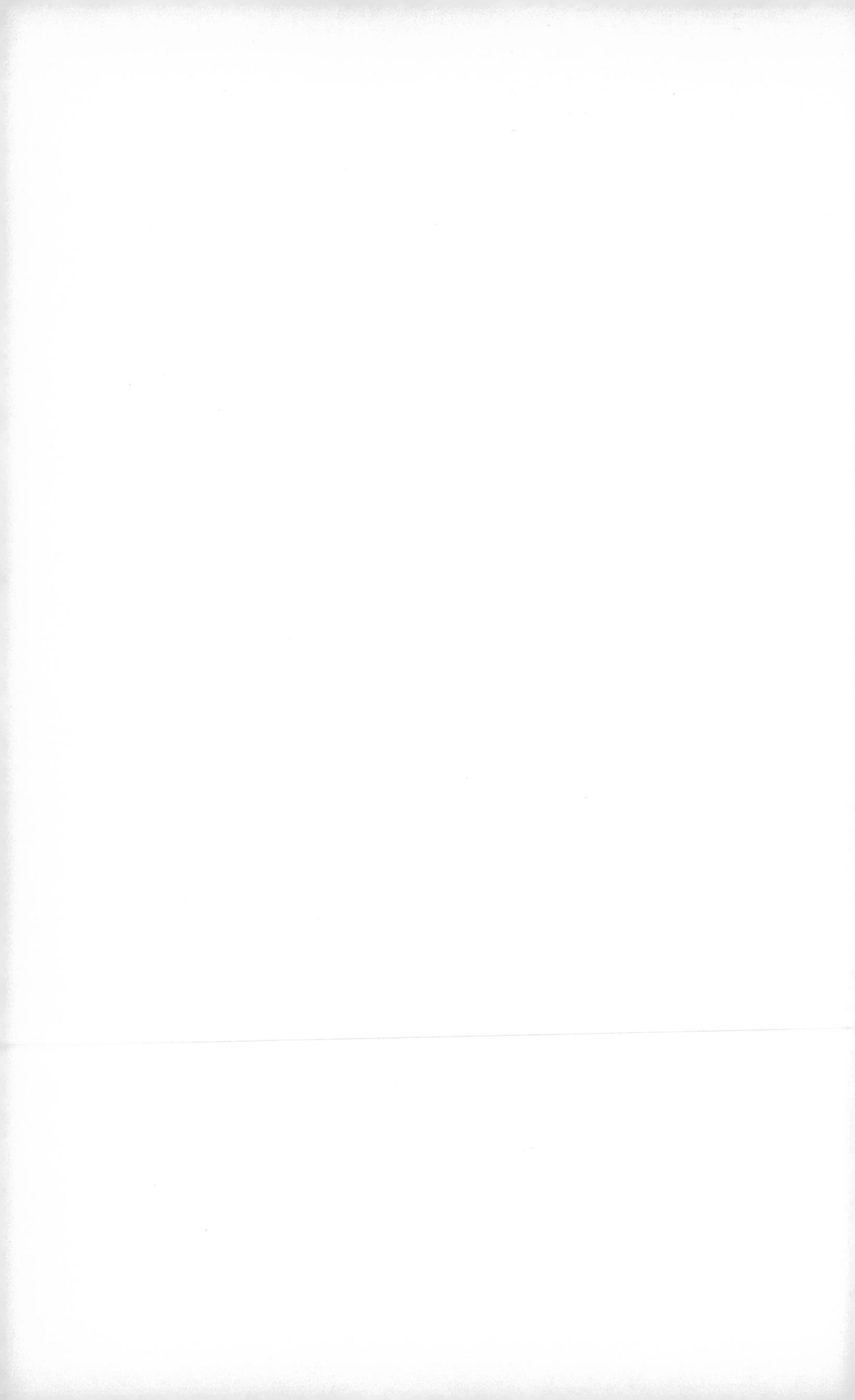

WATER STRESS

CHAPTER 13

WATER DEFICIT (OR DROUGHT) STRESS

A. Drought Stress

As in the case of temperature, a water stress may conceivably arise either due to a deficit or an excess of water. Since stresses due to the deficit are so much more common, the more correct but more clumsy term "water deficit stress" is usually shortened in the literature to water stress. The water deficit to which a plant is exposed in nature is called a drought stress. When the plant is subjected to an artificially induced evaporative loss of water, this is commonly called a desiccation stress. A stress that is capable of inducing a loss of water in the liquid state is called an osmotic stress.

The water deficit or drought stress may be measured by the vapor pressure deficit (v. p. d.):

$$S_d = p_0 - p$$

where S_d = drought stress; p_0 = vapor pressure of pure water at the temperature of the environment; p = vapor pressure of the environment.

With the general acceptance of water potential terminology, it is commonly expressed as:

$$S_d = -\psi_e$$

where ψ_e = the water potential of the environment (bars or atm).

Since ψ is always negative, the water stress becomes positive. There are, of course, objections to this definition and difficulties in its application. No one value can be assigned to a land plant because the water stress in the root environment (frequently called the soil moisture tension or stress) will always be different from that in the shoot environment. This is also true, of course, in the case of low- and high-temperature stresses, but it is much more important in the case of water stresses because of the rapid transfer of water from soil to roots to shoot. The water stress to which a plant is exposed is, therefore, not known unless both the stress in the shoot environment $(S_d)_s$ and that in the root environment $(S_d)_r$ are known. The two may not necessarily be equally important. Even in the vicinity of the chloroplasts, which are very close to the shoot environment, the water potential is chiefly controlled by the status of the soil moisture (Idso, 1968). Agricultural drought has, in fact, been defined on the sole basis of the root environment (Rickard and Fitzgerald, 1969). It is said to occur when the soil moisture in the root zone is at or below the wilting point (see below). On this basis, it was concluded that very severe periods of drought could be expected in New Zealand about every 5 years. Although this simplification may suffice under certain practical conditions, other conditions require a consideration of both shoot and root environments. According to Mack and Ferguson (1968), wheat yields were not as closely related to seasonal precipitation or potential evaporation (of the shoot environment) as to a function of absorption (from the root environment) minus transpiration (into the shoot environment). Another complication is the lack of uniformity within each of the two environments, particularly in the root environment. Thus, Evenari (1962, personal communication) found that desert succulents in the Negev grow although their roots occur in soil below the permanent wilting point (the water content below which essentially no water can be absorbed; see Chapter 14). Yet the plant remains turgid due to a few roots which find their way to soil protected from evaporation by overlying stones. These small pockets of soil have a moisture content well above the average for the soil, and, therefore, above the permanent wilting point. Similar observations have been made in Death Valley (Stark and Love, 1969). Water condensed on rocks from distillation in the soils was a major source of supply to the roots of *Peucephyllum schotii*, *Atriplex hymenelytra*, and *Larrea divaricata*. A range of soil temperatures of 10°C with a 0–3 cm soil maximum of 36°–40° is essential to provide the energy for distillation and condensation of significant quantities of water. Under nondesert conditions, however, the reverse relationship is more common. Water stresses may occur in the perirhizal zones even when the water stress in the bulk of the soil is essentially zero (Tinklin and Weatherley, 1968). Collis-George and Williams (1968) conclude that the influence of the soil system on seed germination can be wholly attributed to the isotropic effective stress in the solid framework of the soil and not to the free energy of the soil water, but this conclusion may perhaps be due to a complication such as the above nonuniformity of soil water potential.

The drought stress may produce a harmless, reversible, elastic dehydra-

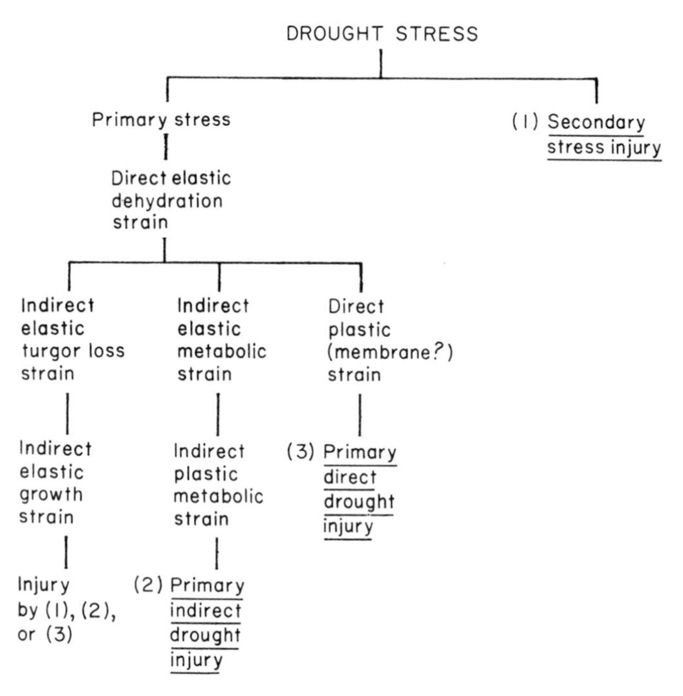

DIAG. 13.1. The four kinds of drought strains leading to the three kinds of injury.

tion strain, but it may also induce at least four different types of injurious strains: (1) elastic growth strain, (2) secondary stress-induced strain, (3) indirect plastic strain, and (4) direct plastic strain (Diag. 13.1).

B. The Direct, Elastic Dehydration Strain

1. Equilibrium versus Steady-State Strain

The direct effects of a temperature stress are on the molecular kinetic energy, body volume, and sometimes physical state. The direct effect of a water deficit stress is simply a transfer of water or dehydration. Since the transfer of mass is a slower process than the transfer of heat, equilibration with a water stress is always slow compared to the rapid equilibration with a temperature stress. Due to low diffusion gradients, permeability, and specific surface, and to large diffusion distances, large bodies may, therefore, require days, weeks, or months to come to equilibrium with the water potential of their environment. Even when the specific surface is large, the permeability high, and the total quantity small, the last traces of free water may be lost very slowly, e.g., requiring as long as a month in the case of freeze-dried chloroplasts (Levitt, 1958), due to the very low gradients.

Besides this quantitative difference between the rates of attainment of temperature and water potential equilibria, there is also a qualitative difference. Plants are poikilotherms and, therefore, tend to follow closely the temperature changes of their environment. In contrast, higher plants at least are homoiohydric (Walter and Stadelmann, 1968) and tend to maintain a steady-state water potential well above that of their environment.

This steady state can be maintained, of course, only so long as a water supply is available to the plant. If the plant as a whole is exposed to the stress, it will gradually approach and finally attain equilibrium with the external water potential. Even if equilibrium is finally reached with an environment of 0% relative humidity at room temperature, the tissues will still have measurable quantities of "bound" water that can be driven off only at higher temperatures. Proteins, for instance, dried to equilibrium at 100°C may lose more and more water as the temperature is raised further, and it may require 130°C to remove some of it (Gortner, 1938). This bound water differs from ordinary free water in several respects which are not understood too well. Several suggestions have been made of an icelike structure of the water in protoplasm (Gortner, 1938; Ling, 1968; Cope, 1969). The recent concept of polywater (Lippincott *et al.*, 1969) suggests another possible state of the water in living cells. Others (Rousseau and Porto, 1970; Kurtin *et al.*, 1970), now believe that polywater does not exist, and according to Abetsedarskaya *et al.* (1968) the structure of water in the hyaloplasm of both plant and animal cells is very similar or even identical to that of water *in vitro*, on the basis of NMR analysis.

2. Limits of Survival of Dehydration Strain

As long as the plant can prevent the attainment of equilibrium with its environment, there is no limit to the water stress it can survive. There may, however, be a limit to the dehydration strain. A general characteristic of protoplasm in the resting state is its ability to become air dry without loss of life, e.g., in seeds, spores, and other reproductive bodies. Even when their rest is broken and growth recommences, plants may still retain this ability for some time. Thus, seedlings, sporelings, and mosses may survive drying over concentrated sulfuric acid (Table 13.1). Not all plants show the same degree of tolerance. Among seedlings, the Gramineae are most tolerant, those from oily seeds less so, and legumes least of all; the

TABLE 13.1 SURVIVAL OF EXTREME DROUGHT BY VARIOUS PLANTS

Plant	Treatment survived	Reference
Germinated wheat, rye, and barley; radicle half the length of the grain	Dried *in vacuo* over H_2SO_4 for 6 months	de Saussure, 1827
Funaria, Grimmia pulvinata	18–22 weeks over H_2SO_4	Schröder, 1886
Bulbils of *Cystopteris bulbifera*	5 months of air drying plus 2 weeks over H_2SO_4	Heinricher, 1896
Fungal cultures (14 of 21 species)	Dried in air for 2½ years	Wehmer, 1904
Germinated spores of leafy mosses and some thallus cells	Dried over H_2SO_4	Rabe, 1905
Myrothamnus flabellifolia	Lost 93% of its water	Thoday, 1921; Genkel and Pronium 1969
Covillea glutinosa	Air dry	Maximov, 1929a
Pelvetia (brown alga)	Air dry	Isaac, 1933
Notochlaena marantae	Powder dry for 2 months at 50% relative humidity	Iljin, 1931
Cololeugenia calcarea *Madotheca platyphylla*	Dried over conc. H_2SO_4 for more than 1 month	Höfler, 1942a
Chamaegigas intrepidus	Herbarium specimens dried 4 years	Hickel, 1967

germinated spores of liverworts and ferns possess as little ability to survive water loss as when they have developed into mature plants (Rabe, 1905). It has been suggested (Schröder, 1886) that absolute dryness occurs only on death and that about 5% of the plant's water is needed to keep it alive (2–3% according to Ewart, 1898). Webb (1966) has recently adopted this bound water concept as an explanation of drought injury in bacteria. In at least two specific cases (*Sticta pulmonaria*—Schröder, 1886; *Notochlaena marantae*—Iljin, 1931) death did occur after this "minimum" water content of 5% was exceeded. Lower values were obtained by Oppenheimer and Halevy (1962). A small fern, *Ceterach offiicinarum*, had 3% water left (on a dry weight basis) even after drying over $CaCl_2$, and they concluded that this quantity is essential for life. After 5 days over 90% H_2SO_4, Rouschal (1938a) showed that fronds of this fern were still living, though their water content dropped to 1.77% of saturation. Over concentrated H_2SO_4 it dropped to 1.6% and they were damaged. The former lived after resaturation, the latter died. The lower limit has since been extended beyond this point. The lichen *Ramalina maciformis* can survive even after its water content drops to 1% (Lange, 1969). Furthermore, the plants can be kept in this extremely desiccated state for long periods without injury. Lichens kept in a desiccator over $CaCl_2$ for 12 months rapidly recovered their ability to fix $^{14}CO_2$ after wetting (Nifontova, 1967), as did Lange's lichen. Similarly, a diatom (*Stauroneis anceps*) stored at vapor pressure deficits of 11.9 mm Hg or higher, survived longer than 16 months, although when stored at 5.9mm Hg or lower vapor pressure deficit, the survival time was decreased (Hostetter and Hoshaw, 1970). In a few cases even the greater part of the bound water has been driven off without killing the cells. Thus the water contents of seeds of Kentucky bluegrass and Johnson grass was reduced to 0.1% without affecting germination (Harrington and Crocker, 1918). A further drying in a vacuum oven for 6 hr at 100°C failed to decrease germination. It should be pointed out, however, that the vigor of germination was reduced. In agreement with the minimum water concept, tobacco seed retained their viability after 15 years of storage over H_2SO_4 in a relative humidity of 10%, better than when stored at 0% relative humidity (over P_2O_5). The seed retained 2.9% H_2O (Franco and Bacchi, 1960). Perhaps careful remoistening by Pruzsinszky's method (see below) would have overcome the decrease in viability of the drier seeds. As Harrington and Crocker emphasize, such extreme drought resistance is not a characteristic of all seeds, for there are some that cannot survive even air drying (e.g., *Acer saccharinum*, willows, and many water plants).

As mentioned earlier, it is just as difficult to remove these last traces of

TABLE 13.2 The Water Loss Resulting in Death of Half the Leaves in Different Plants[a]

Species	Percentage of water content lost	Percentage of fresh weight lost
Kalmia latifolia	95	47
Ilex aquifolium	90	53
Helianthus annuus	87	70
Sambucus nigra	85	68
Fagus sylvatica	80	40
Myrica cerifera	77	41
Tropaeolum majus	75	60
Acer pseudoplatanus	71	45
Platanus orientalis	63	40
Polygonum cuspidatum	57	40
Impatiens parviflora	44	40

[a] From Schröder, 1909.

TABLE 13.3 THE WATER LOSS FOUND UNDER NATURAL CONDITIONS ON JULY 1 AND 2 (COLUMN 1), AND THE CRITICAL WATER LOSS (COLUMN 2), IN PERCENTAGE OF THE WATER CONTENT WHEN SATURATED[a]

Species	Natural sat. deficit	Critical sat. deficit
Hieracium pilosella	73.6	87.2
Potentilla arenaria	52.5	81.8
Anthyllis vulneraria	77.0	80.4
Globularia cordifolia	56.2	75.1
Thymus praecox	44.3	70.8
Teucrium montanum	57.2	70.2
Bupleurum falcatum	28.1	66.2
Anemone pulsatilla	33.3	66.0
Linum tenuifolium	57.2	65.0
Helianthemum canum	53.2	60.5
Aster linosyris	45.3	48.2

[a] From Höfler *et al.*, 1941.

water from nonliving as from living material. It is, therefore, still a question whether or not traces of water are essential for life. Even enzymes are believed to be inactivated by a detachment of bound water from the protein molecules, as judged by a partial (21–49% loss in activity when stored under a high vacuum (10^{-8}–10^{-10} mm Hg) for 72 hr (Imshenetskii *et al.*, 1968).

The injury due to the removal of the last traces of water is at least sometimes more apparent than real. Pollen grains, for instance, though uninjured by exposure to relative humidities as low as 2–15%, showed a drop to nearly zero germination after drying over concentrated H_2SO_4 (0.08% r.h.) if transferred directly to the germination solution (Pruzsinszky, 1960). If, however, they were transferred first to a relative humidity near saturation (97–100%) and later to the germination solution, a high percentage germination was obtained even after several weeks over concentrated H_2SO_4.

On the basis of all the above evidence, it must be concluded that resistant plant cells can lose essentially all of their water without loss of life. Whether or not the last traces can be lost without injury has not yet been established.

The above limits of drought survival are characteristic only of reproductive structures (seeds, pollen, bulbils, etc.), some lower plants, and a few so-called resurrection plants (Oppenheimer, 1960). Most higher plants are killed by a loss of 40 to 90% of their normal water content (Tables

13.2 and 13.3), or when they come to equilibrium with relative humidities of 92 to 97% (Iljin, 1931). Iljin states that survival of equilibrium relative humidities as low as 88% is not uncommon, although only the exceptional plant can tolerate values as low as 85%.

C. Drought Injury

1. Elastic Growth Strain

In the case of the freezing stress (Chapter 6), the primary, direct strain is the solidification and expansion of the protoplasmic water, which separates some protoplasmic constituents and aggregates others. It is, therefore, a plastic, irreversible, and always injurious strain. In the case of the drought stress, on the contrary, the primary direct strain is a cell dehydration, which is elastic and completely reversible up to a point, beyond which it is plastic, irreversible, and therefore injurious. The elastic growth strain, however, although not injurious per se, may be indirectly injurious, by limiting the ability of the plant to send out new roots in search of moister soil. It, therefore, must be considered in relation to the other causes of drought injury.

Although the higher plant is homoiohydric, it may be regularly subjected to moderate, elastic dehydration strains. This is shown by the diurnal expansion and contraction of leaves and fruits (Chaney and Kozlowski, 1969). The effect of such elastic dehydration on growth varies with the species. In any one case, however, when the dehydration is sufficient to eliminate turgor, growth ceases (Lawlor, 1969). Instead of measuring the dehydration strain, Walter (1955) measures the external water stress necessary to stop the growth of lower plants. Since they are poikilohydric, the external water stress is identical to their internal water stress. Those that grow only within a narrow range of water stresses he calls *stenohydric* plants. In extreme cases they cannot grow below a relative humidity of 99%. Those that can grow within a wide range of water stresses he calls *euryhydric* plants. Molds, for instance, can grow at relative humidities as low as 88%. By analogy with mechanical stresses, the stenohydric plants have a low modulus of elasticity (since a small water stress produces a large growth strain); the euryhydric plants have a high modulus of elasticity.

In the case of the homoiohydric, higher plants, the only external stress that can be used in applying Walter's classification is soil water stress, for it is only the fine roots of a higher plant that are likely to be close to equilibrium with the water potential of their environment. The internal water stress must be measured for all other parts of the plant. Corn is apparently a stenohydric plant, for although its seedling root growth continues down to a soil moisture stress of 12 atm, it is most sensitive to change between 1 and 3 atm (Gingrich and Russell, 1957). In the case of the whole plant, growth was slowed down by a soil moisture stress of less than one bar, and a leaf water potential above −3.5 bar (Shinn and Lemon, 1968; Hsiao *et al.*, 1970). In some cases, there may be an optimum below saturation. When pea seedlings were grown in sugar solutions, with the shoot atmosphere at the same relative humidity as that of the solution, elongation of the main root increased from 0 to −7 atm, then decreased steadily to zero at about −30 atm (Walter, 1963a). Shoot growth, however, decreased steadily from 100% at zero water potential to 0% at −25 atm. In bean, cotton, and maize, all leaf growth ceased when the root water potential was lower than −10 bar (Lawlor, 1969). Growth of bean actually stopped at −6 bar. Rye was more resistant and showed some growth even at −10 bar. Some partial recovery of growth was possible during the second week in solution, due to recovery of turgor.

Ungerminated seeds may be classified according to the water stress of the external medium, since they are small enough to be completely immersed in a single medium. In the case of ponderosa pine seeds, germination was greatly depressed by −7 bar (Larson and Schubert, 1969). The water stress that prevents germination is not necessarily a constant value for a specific kind of seed, unless other conditions are kept constant. Kinetin-treated lettuce seed germinate only within a range of 0 to −1.1 bar at 35°C, but show 30% germination in a medium of −8.0 bar at 15° or 25°C (Kaufmann and Ross, 1970). Wheat, on the other hand, germinated when exposed to a water stress of −8.0 bar and showed little effect of temperature. Bauman (1957) made use of this relationship by controlling irrigation so as to maintain the optimum water stress for maximum growth rate.

The effect of the water stress is not always only quantitative. Sometimes it is also qualitative—producing a growth retardation in the above ground part of the plant, exposed to the major water stress, and a corresponding (or even greater) increase in the subterranean portion, which is not exposed to a severe enough stress to inhibit growth (Simonis, 1947).

2. Secondary Stress Injury

As in the case of the temperature stresses, the water stress is capable of leading to injury due to the induction of a secondary stress. The one that has attracted the most attention is the phosphorus stress (i.e., deficiency), which may produce either an elastic or a plastic strain. A reduced growth (the elastic strain) of tomato plants was observed at a water stress of –10.4 atm. According to Greenway *et al.* (1969) this was not due directly to the water stress, but to the marked decrease in phosphorus. This was corroborated by measuring uptake with ^{32}P (Thorup, 1969). Other evidence, however, indicated that the phosphorus decrease was caused by an increased leakage from the cells and that ion uptake *per se* was not affected by the water stress (Greenway *et al.*, 1968). In the case of seeds of crested wheatgrass, ^{32}P was not incorporated into organic substances until they were moistened to a water potential of -130 atm (Wilson and Harris, 1968). In the case of *Litchi chinensis* (Nakata and Suehisa, 1969) leaf N and K were unaffected by a water stress but the P level in the leaf stem was lowered. The absorption of ^{32}P by root cells of maize and soybean was severely inhibited by water potentials below -12 to -15 bar (Dove, 1969). Uptake of ^{32}P by passive processes was increased slightly by exposure of the roots to air, but this did not compensate for the decrease in active absorption.

These growth effects are not due to a cessation of phosphate ester synthesis, since ATP and other phosphate esters are synthesized at water potentials far too low for seeds to germinate (Wilson, 1970). In the case of an osmotically induced water stress, the primary effect on maize roots is apparently a decrease in P translocation, and the inhibition of uptake is secondary (Resnick, 1970).

3. *Primary Indirect Drought Injury*

a. Starvation

A net loss of reserves must occur if respiration exceeds photosynthesis. According to Smith (1915), the effect of drought on respiration rate depends on the extent of the water loss. Up to a 30% loss, respiration increases gradually to a maximum rate which is maintained up to a loss of 50–60% of the plant's water. Any further loss is accompanied by a progressive decrease in respiration rate. Somewhat similar results have since been found by many others (Iljin, 1923b; Simonis, 1947; Domien, 1949; Montfort and Hahn, 1950), the effect varying with the plant (Table 13.4). Not all plants show such respiration increases on dehydration. No change in respiration rate could be detected in droughted *Helodea canadensis* (Walter 1929). Brix (1962) observed a gradual, steady decrease in the respiration rate of tomato plants with decreasing water potential from -12 to -36 atm, but no change from 0 to -12 atm (Fig. 13.1). In loblolly

TABLE 13.4 MAXIMUM RESPIRATION INCREASE ON DEHYDRATION[a]

Species	Percentage increase in respiration
Hedera helix	34, 67 (different experiments)
Triticum vulgare	6
Arum maculatum	44
Rumex acetosa	10
Phaseolus vulgaris	0

[a] From Domien, 1949.

pine, however, there was a sharp drop at −12 atm followed by a steep rise to a maximum at −36 atm, which was 150% of the original value. This was followed by a steep, steady drop to −48 atm, the lowest value investigated (Fig. 13.2). When rape plants were grown at a series of soil moistures, respiration rate (measured on pieces of leaf tissue) increased with decreasing moisture up to a maximum 2.3 times the rate at maximum soil moisture (Takaoki, 1962). In the root, it increased to 3.2 times. Even in the same plant a change may occur at one stage of development but not at another (Musaeva, 1957). Of course, a sufficiently severe dehydration leads to a pronounced decrease in respiration rate, but this decrease is usually found only after a degree of dehydration severe enough to cause direct drought injury.

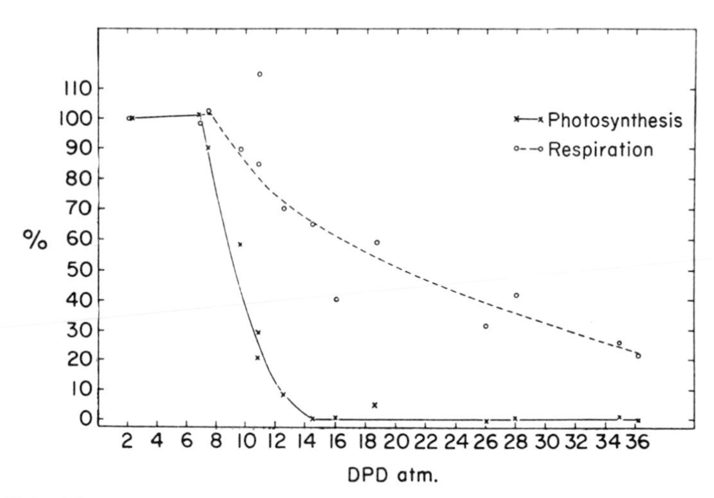

FIG. 13.1. The effect of water stress (DPD or $-\psi$) on the rates of photosynthesis and respiration in tomato plants (as percentage of rates with soil moisture at field capacity. (From Brix, 1962.)

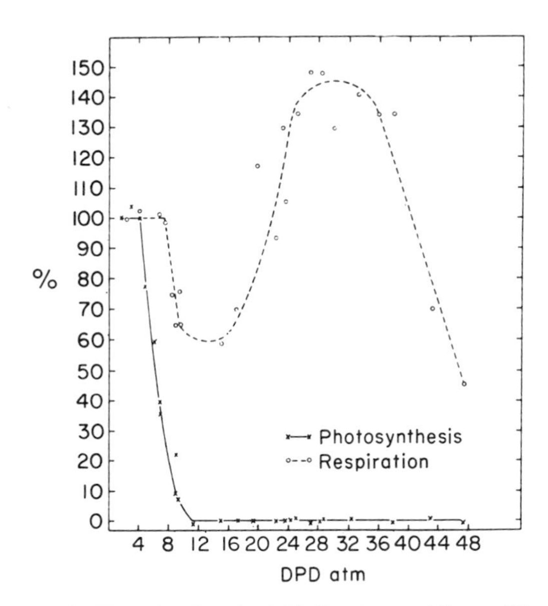

FIG. 13.2. Same as in Fig. 13.1, but for loblolly pine seedlings. (From Brix, 1962.)

It is unlikely that all the biochemical steps in the respiratory process are equally affected by a water stress. According to Zholkevich and Rogacheva (1968) the ratio P/O tends to increase under conditions of moderate water deficiency, but uncoupling (and, therefore, a decrease in the ratio) occurs at more severe water deficiencies. In the case of wheat, however, under drought conditions, the P/O ratio remained constant (Gordon and Bichurina, 1968). Similarly, although substrate oxidation was limited in soybean mitochondria when subjected to osmotically lowered water potentials (using sucrose or KCl), the ADP/O ratios were unaffected, except at high (470 m*M*) KCl concentrations (Flowers and Hanson, 1969). From results with respiratory inhibitors, Zholkevich and Rogacheva (1968) concluded that in some plants glycolysis is accelerated by moderate stress, whereas in others it is the hexose monophosphate shunt. In wheat plants subjected to drought (Lukicheva, 1968), the resulting increase in respiration rate was accompanied by a rise in catalase activity and ascorbic acid content, but a drop in peroxidase activity. When mitochondria of corn (*Zea mays*) root cells were exposed to a severe water stress, oxygen consumption was decreased, but the activity of cytochrome oxidase increased (Nir *et al.*, 1970). The structure of the cristae seemed to be changed, due apparently to loss of lipids.

The photosynthetic rate in higher plants decreases much more precipitously with increased water stress than does the respiration rate (Figs. 13.1 and 13.2). This is because the first effect of water reduction in leaves is a partial or complete stomatal closure (Iljin, 1923a). This markedly slows down the movement of carbon dioxide into the assimilating leaves, reducing the photosynthetic rate two to ten times, according to the amount of water removal and the sensitivity of the plant. In agreement with these earlier observations, stomatal closure was the primary factor limiting the exchange in cotton plants, and the mesophyll resistance did not vary with leaf water content down to 75% of saturation, but increased progressively as the water content dropped from 75 to 56% of saturation (Troughton, 1969). Similarly, stomatal behavior accounted for the difference in photosynthetic sensitivity to low leaf water potential, down to −16 bar in corn and soybean (Boyer, 1970b). Yet leaf enlargement was inhibited earlier and more severely than photosynthesis or respiration (Boyer, 1970a). Others, (see Walter, 1929) have shown a marked reduction in assimilation rate even in the absence of stomatal control. In very sensitive plants, such as *Helodea canadensis,* the rate is reduced at relatively slight degrees of dehydration, and in sucrose solutions above 0.5 M, CO_2 is actually evolved in full sunlight (Walter, 1929). Even free chloroplasts from pea and sunflower plants show an inhibition of oxygen evolution when isolated from leaves with water potentials below −12 or −8 bar, respectively (Boyer and Bowen, 1970).

Photosynthesis of intact cotton plants increased at first as soil water decreased and then decreased (Pallas *et al.*, 1967). Respiration decreased moderately, but increased again before watering. Similarly, slight water deficits increased CO_2 output by about 20% in whole potted wheat seedlings (Kaul, 1966b). More severe stress decreased it by about 50%. Due to the stomatal closure on exposure to a water stress, the decrease in turgor of coffee leaves was small, yet the reduction in photosynthesis was much more pronounced (Bierhuizen *et al.*, 1969). A sharp decline in photosynthesis of Douglas fir occurred when a soil moisture stress of 1 atm was reached (Zavitkovski and Ferrell, 1968). In the case of Monterey pine, a soil moisture stress of 0.70 bar affected transpiration and net photosynthesis to the same degree, due to changes in stomatal aperture (Babalola *et al.*, 1968). The sensitivity of photosynthetic rate to water stress is so great, that the net photosynthesis (and growth) of plants may be very substantially increased by keeping the soil water content high enough to maintain transpiration at its maximum potential rate.

Morris and Tranquillini (1969) transferred seedlings of *Pinus contorta* to nutrient solutions having osmotic potentials of −3, −6, and −9 atm due to added PEG 1000. After 48 hr the photosynthetic rate was measured.

It was found to decrease greatly at -3 atm and to cease at -9. This was due to a deterioration in the plant water balance. Results with PEG as the osmotic agent must, however, be accepted with caution. When the growth rate of "osmophilic" and "nonosmophilic" yeasts were compared, it was found that they were less tolerant of low water activities induced by PEG than by sugars (Anand and Brown, 1968). As a result, no differences between the two groups could be detected with PEG, though the differences were conspicuous in sugar solutions. However, the authors suggest that it may be a case of sugar tolerance rather than a tolerance of water stress. Some of the differences between the results of different investigators may be due to interactions with temperature. In the case of citrus leaves, a sharp optimum temperature for photosynthesis occurred between 15° and 20°C in air dried over $CaCl_2$, in contrast to an unchanged rate up to 30°C with slight reduction at 35°C in air with a relative humidity above 80% (Kriedemann, 1968). In the case of poikilohydric plants, there is no increase either in photosynthetic or respiration rate, but only a uniform decrease on dehydration (Genkel and Pronina, 1968).

These immediate effects of water loss on assimilation and respiration rates must not be confused with the gradual hardening effects of long-continued growth under conditions of drought (see Chapter 15). Simonis (1952) showed that this results in a marked increase in rate of photosynthesis, though respiration rate sometimes increased and sometimes decreased. Others have failed to obtain such an increase. Thus, in the case of Douglas fir (Zavitkovski and Ferrell, 1968), respiration and transpiration rates were higher for seedlings from a wet site than for those from a dry site, at both high and low soil moisture tensions. Photosynthesis, however, declined sharply after a stress of 1 atm, with no major difference between the seedlings from the two sources.

The starvation effect, however, may be indirect due to an effect on translocation rather than a direct effect on the photosynthetic rate. According to Zholkevich *et al.* (1958), an accumulation of photosynthetic products occurred in the leaves in spite of a slightly reduced rate of photosynthesis. This resulted from cessation of translocation out of the leaves to the roots (in sugar beets) or to the ear (in wheat). Similarly, when the roots of cotton plants were subjected to an osmotic stress, soluble sugars accumulated in the leaves and were proportionately decreased in the roots (Vieira-da-Silva, 1968b). In the case of *Lolium temulentum*, the transfer of labeled assimilation products from the photosynthetic tissue into the conducting tissue was delayed in water-stressed leaves (Wardlaw, 1969). This was apparently due to indirect effects of the water stress on translocation—presumably the decreased growth removed the sinks. The velocity of assimilation product translocation was only slightly affected, indicating that the translocation pathway is either resistant to water loss or capable of functioning efficiently when dehydrated.

On the basis of all these results, a water stress may be expected to inhibit photosynthesis and translocation of the photosynthetic product, and within certain ranges to enhance respiration. When these three changes do occur, the net result must be a decrease in reserve carbohydrates, i.e., a starvation effect. The question is whether this starvation occurs to a sufficient degree to account for drought injury. Iljin and Demidenko (see Tumanov, 1930) report a significant decrease in the dry matter of wheat and corn exposed to drought. Tumanov was unable to detect any loss during a 10-day wilting period, though he suggests that a more gradual and longer lasting wilting might have produced a different result. According to Schafer (see Tumanov, 1930) and Mothes (1928), however, the increased respiration occurs only at a water loss of not more than 20%, at which point the assimilation rate is quite adequate to compensate for the higher respiration rates. In spite of the above-described sensitivity of photosynthesis, recent results uphold this contention. In apple leaves, a marked reduction in photosynthesis and an increase in respiration may occur even before wilting has set in (Schneider and Childers, 1941). When the plant was definitely wilted, photosynthesis was reduced 87%. Maize leaves showed less than 10% the normal rate of photosynthesis only when the leaves were badly wilted (Verduin and Loomis, 1944). Since these were decreases in net photosynthesis (i.e., photosynthesis − respiration) the remaining rate was still positive. Consequently, no loss in dry weight could occur in the light. At the degrees of wilting necessary to produce injury, the rates of both respiration and photosynthesis are inhibited (Tumanov, 1930).

In further opposition to a starvation effect due to water stress, Carlier (1961) disagrees with those who found an increase in respiration rate during drying (e.g., Schneider and Childers). He found a steady decrease in respiration rate of apple leaves with decrease in soil moisture, all the way to the permanent wilting point. The decrease was as much as 50% and was more marked in the younger leaves. Unlike the above investigators who measured respiration rates on detached leaves which were dehydrated rapidly over short periods of time, Carlier used attached leaves that were dehydrated slowly under natural conditions over long periods of time. His results are, therefore, applicable to the normal plant; the results with detached leaves are not.

It is difficult, however, to generalize for all conditions of water stress. In some dry climates, photosynthesis may stop in the afternoon in all species (Eckardt, 1952). This results in a loss in dry weight (Shmueli, 1953; Slatyer, 1957b) and if this state is maintained for a long enough time, the plant's reserves will eventually be used up and death will result. A lichen, in fact, has been reported uninjured by the direct effect of drought, but injured only by this negative balance (Ried, 1960b). Those lichens that have a compact thallus may, however, show a rise in photosynthesis during the first stages of dehydration, reaching a maximum at 65% of maximum water content in extreme cases (Ried, 1960a). This is due to the hindrance to gas movement in the maximally swollen state. Mothes (1956) concluded that respiration rate is usually higher the higher the water content, even when there is no real water deficiency. When they are stressed further, the leaves may survive a larger (still reversible) water loss better than a smaller one, due presumably to the reduced respiration rate. As a result, in many herbaceous plants the leaves remain green when strongly wilted, but more weakly wilted leaves quickly turn yellow.

In general, on the basis of all the above results, we must conclude that drought injury usually occurs while the plant still contains large quantities of reserves, and starvation is not likely to be a cause of injury *per se*. A negative assimilation balance may, in fact, exist without causing injury. Thus *Myrothamnus*, which is among the most drought tolerant of all higher plants, and may be air-dried without injury, does not begin to release oxygen until a relatively long time after resaturation and illumination. Respiration, on the other hand, is induced immediately on resaturation (Hoffmann, 1968) and this must result in a considerable negative assimilation balance. However, starvation is more likely to be due to the prevailing high temperature rather than the water stress, and would probably not occur in a plant wilted at moderate temperatures.

b. Protein Breakdown

Metabolic drought injury of another kind is suggested by Mothes (1928). He allowed sunflower and tobacco plants to lose water until the lower (but not the upper) leaves wilted. If these lower leaves were allowed to regain their turgor, the recovery was only apparent, for they died sooner than the lower leaves on control, unwilted plants. Soluble substances, as well as water, moved from these lower leaves to the upper ones during the onset of wilting. Protein was converted to asparagine or glutamine which were translocated to the younger leaves and resynthesized there to proteins. The injury, according to Mothes, is therefore due not only to the water removal but also to the protein loss. He suggests that wilting may speed up the aging of leaves by decreasing the protein synthesizing power of the chloroplasts. He also suggests that proteolysis on wilting may be at least partly due to the increased enzyme concentration. As evidence, he points to the less rapid protein breakdown in leaves floated on weaker than in those on stronger sugar solutions. He proposes the following series of changes. Wilting causes proteolysis and the resulting amino acids activate diastase; this produces an increase in soluble carbohydrates which leads to increased respiration. This series is directly opposed by his own later admission that the proteolysis occurs much more slowly than the starch hydrolysis.

It has already been shown in Chapter 4, that one characteristic of metabolic chilling injury is its greater severity at moderate than at extreme low temperatures. For similar reasons, one would expect a metabolic drought injury to be more severe at moderate than at extreme desiccation. In agreement with this, Mothes states that intermediate aged leaves survive strong wilting better than moderate wilting, due presumably to the less intense respiration and proteolysis. According to Oparin and Kursanov (Mothes, 1956) syntheses in general decrease with decreasing water content and hydrolyses increase. Oparin distinguishes between adsorbed enzymes active in synthesis and soluble enzymes active in hydrolysis. Water loss, according to him, frees the adsorbed enzyme. On the basis of modern knowledge of enzymology, this is, of course, an oversimplification. Since Mothes' pioneering work, a great deal of further evidence has accumulated of protein breakdown (e.g., Dove, 1968) and of accelerated leaf senescence (e.g., in maize; Wilson, 1968).Feeding with ^{14}C-labeled serine showed that lack of water hindered the incorporation into the protein of grape leaves but increased its transformation into other free amino acids (Morchiladze, 1969). A single succulent leaf, which remains turgid at its base and even produces roots there may wilt at the tip and lose soluble nitrogen (i.e., amino acid) to the base (Sullivan 1962, unpublished). In perennial rye grass (Kemble and MacPherson, 1954), wilting resulted in degradation of protein to α-amino acid, volatile base, amide, or peptide. All the amino acids occurred in amounts smaller than expected from their occurrence in the protein, with the exception of proline which was greatly in excess of expectation. They suggest a possible connection between this synthesis of proline and the known formation of amide during starvation.

Many other investigators have reported an accumulation of proline as a result of water stress (Stewart *et al.*, 1966). An analysis of the amino acids in barley plants under conditions of soil water deficit revealed that proline was the only amino acid constantly present in large amounts in all of the organs (Savitskaya, 1967). This was also true of droughted wheat plants (Tyankova, 1967; Vlasiuk *et al.*, 1968). The proline content of wilting leaves of *Solanum laciniatum* was twelve times as high as that of the control, and this was due not to protein breakdown but to *de novo* synthesis (Palfi, 1968a). Kinetin, 2,4-D, and antimetabolites failed to inhibit this abnormal increase (Palfi, 1968b). The proline content of the leaves rose to several times normal after a water deficiency for only 2–3 days (Palfi and Juhasz, 1969). In wheat, according to Protsenko *et al.* (1968), an increase of proline and asparagine in leaves is one of the symptoms of adaptation of winter wheat plants to drought and other unfavorable conditions; it accumulates more in the drought-resistant varieties. Further evidence that the proline is not simply formed by hydrolysis of protein is the incorporation of label from glutamic acid and *N*-acetylglutamic acid into uncombined proline. Considerably more was incorporated by leaf disks under moisture stress than by normal leaf disks (Morris *et al.*, 1969).

Even in the highly drought-resistant creosote bush, total amino acid content more than doubled under moisture stress, predominantly due to proline, phenylalanine, and glutamic acid (Saunier *et al.*, 1968). Significant increases, however, also occurred in alanine, arginine, histidine, isoleucine, and valine. Changes in other amino acids were not significant. A possible function of proline was considered to be for NH_3 storage.

It is quite apparent from all these results that proline accumulates in water-stressed leaves, and that at least some of the accumulation is *de novo* synthesis rather than a direct result of protein breakdown. It may still, however, be an indirect result of protein breakdown, in the same way as the accumulation of amides, i.e., by conversion into proline of other amino acids which are primary products of protein hydrolysis. Proline may, therefore, simply be a more innocuous amino acid than some of the others (and than NH_3 which might otherwise accumulate), and can, therefore, accumulate without causing injury. It may also, perhaps, be a more readily translocated form. On the other hand, it may conceivably play a direct role in drought resistance (see Chapter 15).

c. Enzyme Inactivation

Todd and Yoo (1964) followed the changes in activity of specific enzymes, which may or may not have been due to breakdown or synthesis. Wheat leaves were desiccated over $CaCl_2$, and compared with controls kept over H_2O. The desiccated leaves lost 50% of their original water in 16 hr. Saccharase activity decreased rapidly to a low value but more rapidly in the turgid sample. Phosphatase activity disappeared to a lesser extent but more rapidly in the dried leaves. Peptidase activity decreased slightly in the dried but not in the turgid leaves. Peroxidase showed a marked increase in the turgid leaves but decreased with drying. A particulate dehydrogenase decreased slightly in the turgid and, at first, in the dried leaves, but this was followed by a sharp decline when the leaves had lost 60% of their water. Protein content decreased rapidly in both sets of detached leaves. It dropped to half when 87% of the water was lost. Lukicheva's (1968) droughted wheat plants also showed a drop in peroxidase activity, though there was a rise in catalase activity and ascorbic acid content. A rise in cytochrome oxidase activity occurred in corn mitochondria exposed to a severe water stress (Nir *et al.*, 1970). In three out of five cotton species, acid phosphatase activity increased under conditions of osmotic stress (Vieira-da-Silva, 1969), in opposition to the above decrease in wheat.

d. RNA Decrease

Kessler (1961) has produced evidence of the possible cause of protein hydrolysis in water-stressed leaves. Water stress led to an initial rise in RNA content followed by a drop over a 20-day period. As the RNA content decreased, RNase activity increased (after an initial decrease). He explains this by a liberation of the enzyme from the bound state. Adenine treatment prevented both of these changes and the treated plants were apparently much less injured by the drought. DNA synthesis was also apparently impaired under conditions of water stress. The water stress, according to him, impairs the nucleic acid system intimately correlated with protein synthesis.

Similar results were obtained by Gates and Bonner (1959). Tomato plants exposed to a brief period of water shortage showed a suppressed increase in RNA, even though the leaves retained their ability to incorporate ^{32}P-labeled phosphate into RNA. They, therefore, concluded that the rate of RNA destruction is increased. This conclusion was supported by the increased rate of RNA loss in excised tomato leaves subjected to osmotically induced moisture stress.

Others have since investigated the RNA effect in more detail. Genkel *et al.* (1967b) exposed maize and bean plants to drought. Although protein incorporation of ^{15}N decreased, RNA content did not. Polysomes, however, consisting of four to ten or more ribosomes in the controls, disappeared from the droughted plants. They concluded that RNase destroyed mRNA which binds the ribosomes into polysomes. In agreement with this conclusion, Chen *et al.* (1968) found that dehydration of wheat embryos inactivated mRNA and arrested protein biosynthesis. Similarly, Sturani *et al.* (1968) concluded that in germinating castor beans subjected to water stress, some factor affecting mRNA or its interaction with ribosomes is responsible for the cessation of protein synthesis. When barley plants were exposed to drought, the RNA content of the developing pollen was greatly decreased, reversibly if droughted prior to the appearance of the pollen mother cells, and irreversibly if after their appearance but prior to fertilization (Simonova, 1968).

There is more than one possible interpretation of all these results. Although the decrease in RNA is associated with and believed to be caused by increased RNase activity (see above), this increased activity was associated with and perhaps caused by injury, senescence, or both in the case of *Avena* leaf tissue (Wyen *et al.*, 1969). Similarly, the above described role of mRNA is opposed by more recent results (Hsiao, 1970). In agreement with the above results, the polyribosomes in the coleoptilar node regions of *Zea mays* did shift to the monomeric form when the water potential of the tissue began to decrease measurably (30 min after stress initiation). After about 4 hr of stress, resulting in a 5-bar decrease in tissue potential, most of the ribosomes were in the monomeric form. The shift, however, was not an artifact resulting from RNase-induced loss of mRNA (as suggested above), but occurred only if polypeptides could be terminated and released. Another possible cause of the observed protein hydrolysis has been proposed by Itai and Vaadia (1968). Exposure of the roots to a nutrient medium with increased osmotic stress resulted in decreased translocation of cytokinins from the roots. This decrease was reversible, and on return of the osmotic stress to normal, cytokinin activity in the root exudate increased, initially to more than the control level. They, therefore, suggest that the stress-induced decline in protein synthesis in the leaves is due to a deficiency of cytokinins.

It has been suggested that the protein breakdown injures the plants, not simply due to a protein deficit, but to the accumulation of a toxic product of protein breakdown, probably NH_3, as in the case of heat injury (see Chapter 12). In agreement with this concept, ammonia N could only be detected in wilted plants of *Pennisetum typhoides*, and it showed a sharp decrease on rewatering, with an associated increase in the level of amide N (Lahiri and Singh, 1968).

4. *Primary Direct Drought Injury*

Since a drought-induced decrease in reserves or in protein content takes a considerable time at normal growing temperatures, indirect drought injury must be relatively slow. It, therefore, cannot explain the very rapid kinds of drought injury. Similarly, metabolic injury is greater at moderate than at severe wilting (Mothes, see above). It, therefore, cannot explain the injury that occurs at tissue dehydrations that are severe enough to slow down the metabolic changes. It must be realized, however, that metabolic distrubances which are not severe enough to injure by themselves, may nevertheless amplify the other effects of the dehydration strain and, therefore, the injury. A drought-induced decrease in net photosynthesis, for instance, may not be sufficient to produce starvation injury; yet it may decrease translocation to the roots, reducing root growth, and with it the ability to extend into moister soil.

Drought injury may, therefore, be labeled direct (1) if it occurs too rapidly to be metabolic; (2) if it occurs at dehydration strains that are severe enough to prevent metabolic injury; or (3) if it occurs in spite of the retention of ample reserves and protoplasmic proteins in the injured cells. On the basis of these criteria, at least many (if not most) of the cases of drought injury must be direct rather than indirect in nature (Mothes, 1928).

a. Effect of Rate of Drying and Remoistening

As mentioned above, the rate of water loss in the case of higher plants subjected to drought, cannot ordinarily be as rapid as the rate of heat loss during cooling. This is largely due to a surface barrier that is more or less impermeable to water. In order to investigate the effect of high speeds of dehydration, it is therefore necessary to use sections of higher plant tissues (unprotected by the cutinized or suberized surface layer) or lower plants that do not possess an impermeable surface. At 50% relative humidity, for instance, the hanging moss, *Bazzenia stolonifera*, dries to 6% of is saturation weight within 55 min (Biebl, 1964a).

Schröder (1886) points out that slow water removal allows the plant to go over into a resistant, resting state, but if the plant is already in this state, he believes that rate of drying is not important. Thus, a shoot of *Grimmia pulvinata* was quickly made powder-dry by moving air previously dried over $CaCl_2$ and H_2SO_4 at 35° to 40°C. This drying was continued for 18 hr; yet, on remoistening, the shoot was completely alive. Similarly, he concluded that there is usually no difference in survival whether the water is added slowly or rapidly. If the plants were alive in the dried state, they remained alive whether moistened rapidly or slowly; if they were injured, slow remoistening failed to help. Under natural conditions, he points out, plants are rapidly remoistened by rain.

On the other hand, the rapid addition of water to dried seedlings has frequently been reported to be more harmful than gradual moistening, though different workers have obtained different results. Rabe (1905) explains the discrepancy by the harmful effect of fungus growth during the slow moistening. In the case of mosses (Irmscher, 1912), more rapid drying by placing immediately in a desiccator was more injurious than drying first in air and then in a desiccator; the same amount of injury, however, resulted, whether the addition of water was slow or rapid. Höfler (1942a) confirmed the protective effect on liverworts of slow drying and in some cases of slow remoistening. Allowing them to take up water slowly from saturated air enabled them to survive a relative humidity of 72 to 77%, although they were killed by relative humidities of 86 to 88% if remoistened directly in water, in agreement with Pruzsinszky (1960; see above). This was true of only two species; others showed little difference. Both slow and rapid drying give rise to the same differential response between species.

In the case of sections, Iljin (1933a) found that the dried cells were killed if plunged directly into water, but they survived if allowed to take up water slowly from a saturated atmosphere. Rapid drying (within a period of minutes or even seconds) immediately killed six out of ten species tested (Table 13.5). The remaining species died after longer periods of time in the dried state (1 hr to a few days). Gradual drying resulted in better survival in some cases. Best results were obtained when they were dried at intermediate rates (during 14 hr). The injury increased with the time in the dry state (Table 13.5) even after equilibrium had been at-

TABLE 13.5 Percentage of Living Cells after Different Methods of Drying[a]

Species	No. days kept dry	Dried in air		Dried in sugar solutions	
		Rapid	Gradual	Rapid	Gradual
Syringa vulgaris	3	0	45	85	100
	7	0	25	80	100
Centaurea scabiosa	3	0	0	20	100
	7	0	0	0	100
Ligustrum vulgare	3	0	0	100	100
	6	0	0	25	25
Buxus sempervirens	3	0	0	0	100
	7	0	0	0	5
Berteroa incana	2	50	100	100	100
	6	0	20	5	20
Bupleurum falcatum	3	65	100	100	100
	7	5	85	5	20

[a] From Iljin, 1935b.

tained. These results are far more significant than earlier reports (Rabe, 1905) which gave no evidence that equilibrium had been reached. Irmscher (1912) showed that mosses suffer greater injury from repeated drying and wetting than from a single continuous desiccation.

All these results, though far less numerous, agree admirably with those for freezing injury—even to the extent of the apparent discrepancies. Thus, not only are the same time factors operative in drought as in freezing injury, but the same qualifications must be added, e.g., some plants are so tolerant that they are uninjured by any kind of drought treatment and therefore the time factor does not apply to them. The moment of injury due to both freezing and drought may occur, therefore, during either drying or remoistening. Injury during remoistening was observed directly by Iljin in those cases in which pseudoplasmolysis occurred followed by deplasmolysis leading to cell rupture (Fig. 13.3). The possibility that, as in the case of postthawing, injury may occur after turgor is regained, has not, as yet, been adequately studied.

b. Observations of Drought-Dehydrated Cells

Drought plasmolysis is no more to be expected during the dehydration process than is plasmolysis during freezing (Chapter 7), since there is no solution to enter between the wall and protoplast. This conclusion is corroborated by the observations of dried cells described by many investigators. According to Schröder (1886), the cell loses its transparency on drying and air enters. Dried *Grimmia* leaves, when mounted in oil, showed many folds in the cell wall. In places where the wall was no longer able to follow the collapsing protoplast, air penetrated the cell. Immediately after remoistening, an air bubble could be seen in each cell. It rapidly rounded up, became smaller, and in a short time disappeared. The protoplasm refilled the cell and full turgor was quickly regained. In other plants the process took longer. Steinbrinck (1900, 1906) confirmed the above obser-

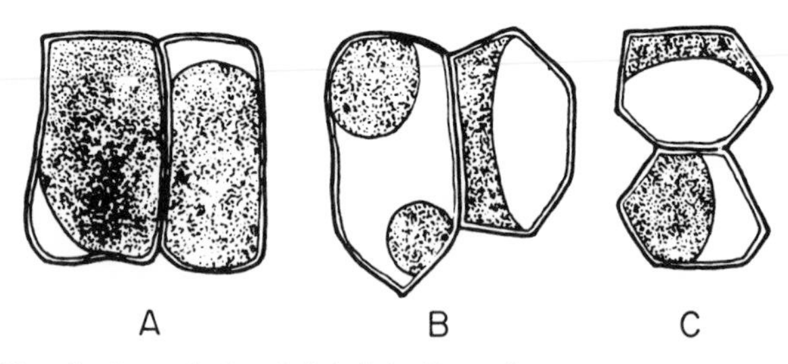

Fig. 13.3. Pseudoplasmolysis of dried (collapsed) cells on transfer to water. (A) *Cirsium canum.* (B) *Rhoeo discolor.* (C) *Tradescantia fluminensis.* (From Iljin, 1933a.)

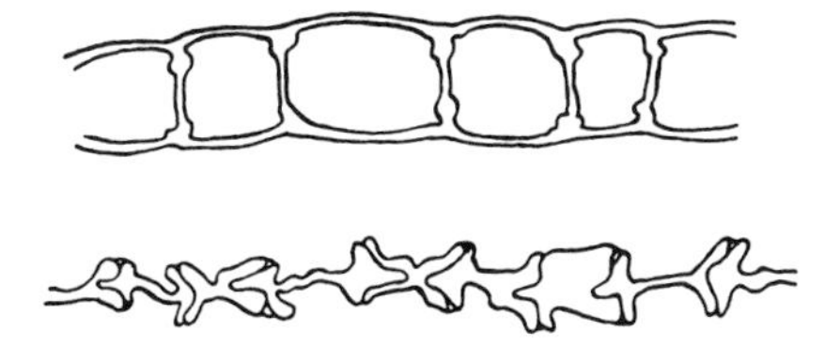

FIG. 13.4. *Top:* normal (turgid) cell shape. *Bottom:* collapsed cells in desiccated leaf of *Mnium punctatum.* (From Steinbrinck, 1903.)

vation that instead of the protoplasm separating from the wall during desiccation, the cell wall is pulled in with the cell contents. As a result, folds and concavities arise, and finally the opposite walls meet (Fig. 13.4). The addition of water returns the cells to their original volume. As shown by Schröder and others (Thoday, 1922) this may occur without injury. Folds do not occur in the case of thick-walled cells, such as in some mosses (Steinbrinck, Laue 1938). It is in such cells that the protoplasm apparently contracts from the wall in places, forming the air pockets earlier observed in *Grimmia* by Schröder (1886) and later by Holle (1915). Iljin (1927) also noticed a partial separation of protoplasm from cell wall, but nothing that he could call plasmolysis. Holle (1915) showed that plasmolysis may occur at the locus of a cut due to the concentration of sap exuded from the cut cells, which can then plasmolyze the uninjured cells. This may be the basis for the conclusion that drought plasmolysis occurs (Livingston, 1911; Caldwell, 1913). If the sections are first rinsed with water, plasmolysis is avoided (Holle, 1915). Some investigators may have used the term plasmolysis incorrectly for the above described cell contraction.

The actual degree of contraction normally suffered by the leaves of plants exposed to drought is clearly evident from Schratz's (1931) measurements on a small-leaved xerophyte (Table 13.6). On progressive drying,

TABLE 13.6 COMPARATIVE LEAF MEASUREMENTS ON NORMAL (DROUGHTED) AND CONTROL (WATERED) PLANTS OF *Covillea tridentata*[a]

Quantity measured	Control leaves	Normal leaves	Ratio, normal: control
Water content (% fresh wt)	58	32	0.55
Water content (% of dry wt)	139	47	0.34
Leaf thickness (mm)	0.206	0.120	0.58
Leaf surface (cm_2)	0.540	0.240	0.44
Leaf volume (cm^3)	0.0111	0.0029	0.26

[a] From Schratz, 1931.

the leaves turn yellow-brown, shrivel up, and become brittle. They are then in their rest period. Schratz compared these with the leaves of control bushes which had been continuously supplied with sufficient water so that the leaves had their maximum possible water content. The reduction in size of the individual cells on drying varies with the tissue and the kind of plant (Table 13.7). Moss cells and cells of storage tissue (e.g., in seeds) undergo only a 1.5-fold reduction in diameter. At the other extreme are the pith cells of *Pelargonium*. On drying, they undergo a ninefold reduction in diameter. Mouravieff (1969) attempted to observe the changes occurring during desiccation of epidermal cells of leaves. He stained them with tetracycline, then observed the fluorescence by phase contrast microscopy. The most pronounced changes were observed at the borders in contact with the cell wall. The protoplasm here became granular due to the accumulation of mitochondria, spherosomes, and part of the plastids. A plasmaschism was observed between a thin layer adhering to the wall and the rest of the protoplasm. This seems to indicate that the rupture observed by Iljin during rehydration may also occur during dehydration.

The main changes in the submicroscopic structure of maize roots on dehydration (a loss of 60–70% of their weight) were observed in the mitochondria, plastids, the plasma membrane, and the chromatin in the nucleus (Nir *et al.*, 1969). Numerous lipid droplets were found in the cytoplasm of the dehydrated cells, possibly due to displacement from the membranes. Cutting off the root tips caused disruption of polysomes into monosomes. On rehydration in a nutrient medium, restoration of the normal submicroscopic structure occurred, except for the polysomes. This was not always accompanied by restoration of function, e.g., growth. Rehydration of tissue previously dehydrated to the lethal point did not result in restoration of normal structure.

TABLE 13.7 Reductions in Cell Dimensions on Drying[a]

	Ratio of original dimension to dimension after drying		
Species	Diameter	Surface	Volume
Pelargonium stem	8.9–9.3		7.3–9.0
Red beet root	5.4–7.5	4.8–5.3	5.0–6.6
Pea seeds	1.4	1.5	
Bean seeds	1.5	1.7	
Moss leaves (5 species)	1.3–1.5	1.7–2.3	1.8–3.1

[a] From Iljin, 1930a.

c. Enzyme Inactivation

Changes in activity of proteins have been described, under conditions of slow dehydration at normal temperatures (see above). These may, therefore, have arisen due to differential rates of hydrolysis of the proteins, and may involve indirect effects of drought. Santarius (1969) (also Santarius and Ernest, 1967; Santarius and Heber, 1967) has observed changes in enzyme activities of chloroplasts dried rapidly (2 hr over $CaCl_2$ *in vacuo*) at +2°C. At this low temperature, protein hydrolysis obviously must have been negligible and any change in enzyme activity must have been a direct effect of the drought stress.

When the free chloroplasts of spinach or beet were 90% dehydrated by osmotic methods (concentrated sugar or lutrol solutions), the Hill reaction was not depressed, but cyclic photophosphorylation was decreased even at a less severe dehydration (Santarius and Ernst, 1967). The decrease was reversible on rehydration. On the other hand, when dehydrated more thoroughly, removing 98–99% of their water, both reactions were inactivated (Santarius and Heber, 1967). In this case, protein denaturation apparently occurred and the inactivation was irreversible. The two phenomena were, therefore, not comparable; in fact contrary to the above osmotic inhibition by a high concentration of sugars, low concentrations of sugars protected against the irreversible (desiccation-induced) inactivation. Furthermore, the critical dehydration for the irreversible inactivation by desiccation was near 10–15% of the total water content, and, therefore, in the same range of dehydration as the reversible osmotic inactivation.

It must be realized that desiccation does not necessarily inactivate enzymes. Even in the crystalline state, the resistant enzymes chymotrypsin, RNase, and papain may be completely active (Sluyterman and De Graaf, 1969). Similarly, ^{14}C was released from urea by urease at relative humidities as low as 60% (Skujins and McLaren, 1967). The minimum amount of water required for the reaction was 1.3 moles/mole of side chain polar groups of the urease protein. Finally, the activity of mitochondrial cytochrome oxidase actually increased as a result of water loss from roots of *Zea mays* (Nir *et al.*, 1970).

5. Protective Substances

a. Nonpenetrating Solutes

As in the case of freezing injury, drought injury can be prevented by protective solutions. Rabe (1905) showed that although the sporelings of certain fungi do not normally withstand drying, they survive month-long dehydration in concentrated sucrose and dextrose solutions (except when they contained considerable amounts of inorganic salt). The most thorough investigation of this phenomenon was made by Iljin (1927, 1930a, 1933a, 1935b). He found that the degree of drying tolerated by the tissues is proportional to the concentration of the protective solution used (Table 13.8). Similar results were obtained with several different plants, and salt solutions were just as effective, though (in agreement with Rabe) injurious after a long time. Otherwise, the length of time the cells were kept in the solutions had no effect on the results. The solutions plasmolyzed the cells, and, according to Iljin, the protective effect was proportional to the degree of plasmolysis. When epidermal sections were gradually transferred from weak sucrose solutions to stronger ones and then blotted, they could be exposed to lower humidities and even finally dried in a desiccator over concentrated H_2SO_4 without injury. They survived for weeks or even 4 months in this condition—the plasmolyzed protoplasts had a regular shape, and the sap retained its color. Iljin concluded that plant protoplasm can withstand complete drying in all except those cells having a high vacuolar content.

There is no question of the validity of Rabe's (1905) results since he actually grew the germinated spores after their drought treatment. Iljin, however, accepted plasmolysis (usually in 2.7 *M* sucrose) and dye retention as proof positive of survival. By this method, he even produced evidence (1933a, 1935a) that the protection could be obtained by adding the solutions to the already dried sections. It was shown earlier (Chapter 9) that his similar extreme protection against freezing injury could not be repeated by others and was interpreted by them as due to tonoplast survival. He states (1933a) that the outer protoplasm is more easily destroyed by desiccation than is the tonoplast, and even when the cells deplasmolyze

TABLE 13.8 Survival of Mesophyll Cells of Iris in Protective Solutions of Different Concentrations[a]

Solution (1*M* glucose)	Minimum relative humidity survived
0.0	>99
0.1	97
0.2	93
0.5	90

[a] Sections left in solutions for 24 hr, then kept for 2 days in chambers of different relative humidities. From Ijlin, 1927.

completely, he observed that death may occur after deplasmolysis is complete. He clearly demonstrated that the plasmolyzed cells are often abnormal, since sometimes pseudoplasmolysis persisted for 1 to 2 hr before deplasmolysis could occur. Obviously the protoplast was prevented from expanding by a rigid surface layer. As further evidence against this method, even those cells of *Helodea canadensis* that appear normal some time after deplasmolysis may still be injured, according to Walter (1929), and may die after a few days.

It can be seen from these facts that Iljin estimated survival on the basis of incompletely killed cells, and when he speaks of 100% survival he may really mean that 100% of the cells were incompletely killed. This may not detract from the validity of many of his conclusions, assuming that the degree of injury to the protoplasm parallels the degree of injury to the plant. In favor of this assumption are the following results. After very severe drying, the cells died immediately on transfer to water; after less severe drying they showed pseudodeplasmolysis and died shortly after swelling was complete; after intermediate severities of drying they died during pseudodeplasmolysis. This parallels the response of the plant as a whole to different degrees of drying, but more severe drying is required to kill the cell completely than to kill the plant completely. His results, therefore, apparently give an exaggerated picture of the drought tolerance of the plant and even of the protoplasm. His statement that the protoplasm of growing plants can withstand complete drying cannot be accepted in the sense that the protoplasm remains fully alive and functional and can resume normal growth. Thus, Kaltwasser (1938) was unable to obtain any protection against drought injury by Iljin's method of applying protective solutions to the dried tissues. He judged injury not by plasmolysis, but by the effect on assimilation rate which was followed for 8 days after the reabsorption of water. Iljin (1935a) showed that his method does not work with some species, so Kaltwasser's negative results may not detract from Iljin's positive results.

Oppenheimer and Jacoby (1963) were unable to corroborate Iljin's results. They were unable to detect any increase in drought survival of plasmolyzed tissue. They, therefore, concluded that he must have observed only tonoplast survival. More recently, Samygin and Matveeva (1968) verified the protective action of solutions during drying, but they were unable to confirm Iljin's observation of survival over concentrated sulfuric acid.

b. Penetrating Solutes

Kessler (1961) treated pea seeds and seedlings with purine and pyrimidine bases, each treatment comprising twenty-five plants. The plants were grown in the open and then subjected for 8 hr to 40°C at 17% r.h. All wilted strongly and only those treated with adenine and kinetin recovered. Adenine induced drought resistance only when applied at very early developmental stages and this may be due to its influence on DNA synthesis. In a few other cases, substances that promote RNA synthesis (caffein, uracil, xanthine, and, even more so, uridine triphosphate) were found to be effective, particularly in fully grown plants.

The growth inhibitors CCC (chlorocholine chloride) and Phosphon increased both the fresh weight and dry weight of bean plants compared to the untreated plants when both sets were subjected to drought, but not when they were adequately watered (Halevy and Kessler, 1963). Later tests (Plaut *et al.*, 1964), however, failed to confirm these results. In fact, the transpiration of the treated plants was in many cases higher than that of the untreated ones. However, it was suggested that the more extensive root system and the lower top: root ratio may contribute to their survival under conditions of water stress. Positive results were obtained with other plants. Both CCC and Carvadan increased the drought survival of gladiolus plants (Halevy, 1964). Two growth-retarding chemicals produced a pronounced increase in the dry weight and grain production of wheat plants after two drought cycles (Plaut and Halevy, 1966). The growth retardants increased the ratio of root:top in all plants, and CCC reduced the transpiration rate. Other retardants, however, either increased transpiration or had no effect. The increase was due to an increased ability of the treated plants to regenerate new shoots on rewatering. Another explanation for the increased drought resistance of plants treated with growth retardants was the apparent delay in leaf senescence (Halevy, 1967). The suggested cause was a delay in the breakdown of nucleic acids and proteins under stress. The resistance of rutabaga seedlings to desiccation was increased by proline, due supposedly to its retardation of cell growth (Hubac, 1967). The growth inhibitor, CCC, counteracted the effect of drought at ear emergence on wheat yield (Humphries *et al.*, 1967). In this case, the explanation given was the increased size of the root system. A similar increase in root growth was proposed to explain an increase in drought resistance of *Dolichos biflorus* and *Eleusine coracana* due to thiamine (Sastry *et al.*, 1968). On the other hand, phenylmercuric acetate (PMA) applied to spring wheat at heading or flowering stages reduced plant growth and water use, but not at other stages of growth (Brengle, 1968).

Santarius and Heber (1967) protected the enzyme systems of the Hill reaction and photophosphorylation from inactivation by desiccating them in the presence of sugars, soluble proteins, or polypeptides. More sugar was needed to protect the phosphorylating system than the Hill reaction system. Lutrol (polyethylene glycol) and glycerol were not protective.

Glycerol, in fact, prevented protection by sugar. Cysteine was also ineffective.

According to Chinoy *et al.* (1965) pretreatment of barley seed with ascorbic acid (25 mg/liter for 5–6 hr, then dried) was beneficial to growth and yield in a number of varieties when exposed to drought.

c. Antitranspirants

Recent attempts have been made to protect plants against drought injury by substances calculated to convert them into water savers (see Chapter 14). These substances are intended to decrease transpiration rate and are therefore called transpiration suppressants or antitranspirants. Some of these are intended to act in a purely physical manner as a barrier to diffusion of water from the leaf when they are deposited on the surface (e.g., in the case of long-leaf pine plantings; Allen, 1955). Hexadecanol (Roberts, 1961; Peters, 1963), vinyl acetate–acrylate esters (Gale, 1961), cetyl alcohol, oxyethylene docosanol, and S-600 (a plastic spray) are all examples of this group. According to Slatyer and Bierhuizen (1964) they all significantly reduced transpiration, but primarily due to stomatal closure. Cetyl alcohol was effective when applied in the rooting medium, but when sprayed onto the plants killed them (Kriedemann and Neales, 1963). A second group of antitranspirants act indirectly on the diffusion process by altering the physiology or biochemistry of the cell. According to Zelitch (1964), the monomethyl ester of alkenylsuccinic acid is effective in reducing transpiration, probably by altering the permeability of the guard cell membranes. Maximum stomatal closure in tobacco was obtained with dodecenylsuccinic acid or with the monomethyl ester of decenylsuccinic acid at low concentrations (5×10^{-5} M). The effect lasted for 5–7 days without any toxicity, and transpiration was inhibited more than photosynthesis. Shimshi (1963) obtained similar results with phenylmercuric acetate spray on tobacco and sunflowers. There was a 31% reduction, at first, in transpiration by tobacco, dropping to a 10% reduction after 11 days when new leaves had unfolded. Again, there was no adverse effect. These substances failed, however, to save the drought-sensitive *Thuja plicata* (Oppenheimer, 1967). According to Kuiper (1964), the alkenylsuccinic acids increase cell permeability by incorporation into the lipid layer of the cytoplasmic membrane. At 30°C, they increased cell permeability to eight times. As indicated previously Chapter 8), this increase in permeability is simply due to death of the cells.

According to Slatyer and Bierhuizen (1964) all these antitranspirants reduced photosynthesis, partly by increased resistance to CO_2 diffusion, and partly by acting as metabolic inhibitors. The only one that caused a proportionately greater reduction of transpiration than of photosynthesis was phenylmercuric acetate, which produced stomatal closure. Apparently, the closure was only partial even in the highest concentration. Transpiration reduction could be maintained for 25 days, but the control declined in transpiration more than the treated plants, so that finally its rate was lower than that of the treated plants.

Due to the several kinds of drought injury, and the failure of investigators to determine which of the kinds of injury they were dealing with, it is not possible to generalize from the above results. Many of the apparent contradictions as to the protective effect of individual substances against drought injury may simply be due to their ability to protect against one kind of injury but not against another. This relationship will become clearer after a consideration of the different kinds of drought resistance which have been developed by plants.

CHAPTER 14

DROUGHT AVOIDANCE

A. Classification of Adaptations to Water Stress

More than one basis has been used to classify the adaptations of plants to water stress.

1. Ecologists classify them according to the environmental water supply required for the normal completion of their life cycle. Those adapted to partial or complete submergence in free water are called *hydrophytes*. Land plants adapted to a moderate water supply are *mesophytes*. Those adapted to arid zones are *xerophytes*. There are, of course, all gradations between these groups and it is, therefore, not always easy to place a plant in one or the other group. It is even possible for a plant to fit into more than one group. *Chamaegigas intrepidus* grows in shallow water pans in southwest Africa, but during the dry season it exists in the air-dried condition (Hickel, 1967). Herbarium specimens which had been dried for 4 years came to life when immersed in water, but this plant is an obvious exception. Unlike the normally homoiohydric higher plants, it is a poikilohydric flowering plant, possessing few or no structural adaptations to desiccation.

2. Microorganisms have been classified into the same three groups (hydrophytes, mesophytes, and xerophytes; Walter, 1951). Since members of each group may grow submerged in water, the classification is based on water potential rather than on water supply. Furthermore, since even some of the hydrophytes may survive the maximum water stress (i.e., air drying) without injury, this classification must be based not on resistance to the irreversible, plastic or injurious strains, but on resistance to the elastic strain which leads to cessation of growth, or on Walter's euryhydry. According to Walter (1951), hydrophytic microorganisms can grow at relative humidities of 100–95%, mesophytes down to 95–90%,

and xerophytes at as low as 90–85%. In modern terminology, these would correspond to water potentials of approximately 0 to −75, −75 to −150, and −150 to −225 bar, respectively. The hydrophytes are, therefore, the least euryhydric, and the xerophytes the most euryhydric of the microorganisms. Because of the poikilohydric nature of microorganisms, this euryhydry must be based on their drought tolerance, for they are always essentially at equilibrium with the water potential of their environment. Furthermore, since the xerophytic microorganisms can grow at lower water potentials than the hydrophytic or mesophytic organisms, this kind of drought tolerance must be due to avoidance of the dehydration strain (see Chapter 15); because growth depends on the maintenance of cell turgor. Xerophytism among microorganisms is, therefore, simply due to increased solute concentration, which permits the attainment of low water potentials while avoiding dehydration.

3. Higher plants cannot be classified on the basis of the above elastic drought tolerance for several reasons. (a) Hydrophytes among higher plants are not able to grow beyond about −5 to −10 bar, mesophytes probably not beyond −20 bar, and xerophytes not far beyond −40 bar (Walter, 1951). Therefore, the differences between the classes may be too small for quantitative measurement. (b) Furthermore, one class of higher plant xerophytes falls within the range of the hydrophytes in this respect (see below). (c) The only consistent difference between the three groups among higher plants is the environmental water stress that members of each group can survive.

Regardless of the classification or the organism, drought resistance can be defined in the same way as any other stress resistance (Chapter 2). It is the water stress necessary to produce a specific plastic strain, for instance, death of 50% of the plant. Among the three groups, it can be expected to reach its peak in the xerophytes, since they are the group best adapted to water stress. It may, however, still be an important characteristic in mesophytes, and even in some hydrophytes. The xerophytes, themselves, are a highly varied group and have been divided by Maximov (1929b) into (1) ephemerals—plants that complete their life cycles in arid habitats during the brief periods free of water stress; (2) succulents or water-conserving plants, and (3) "true xerophytes"—the plants that survive severe water stresses without conserving water. On the basis of stress and strain terminology, xerophytes can be classified more logically (Diag 14.1). It can be seen that there are actually five different basic kinds of xerophytes. Four of these are drought resisting and the fifth (ephemerals) possessing no greater drought resistance than mesophytes. The xerophytic microorganisms differ from plants by possessing high dehydration avoid-

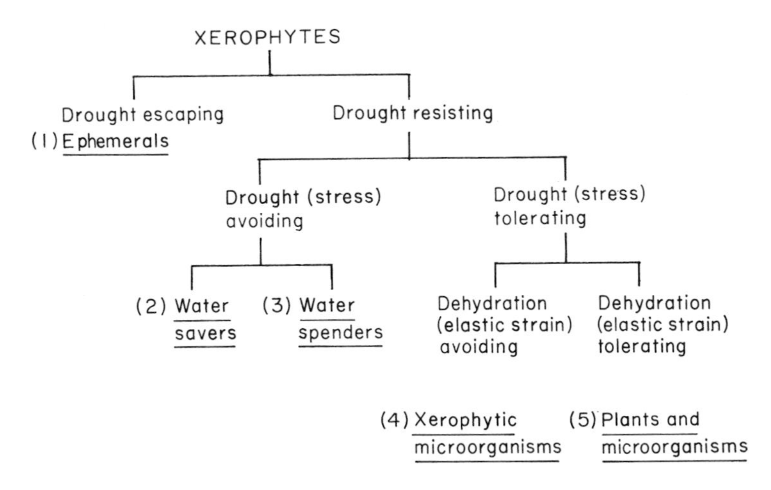

DIAG. 14.1. The nature of xerophytism.

ance. Dehydration tolerance, on the other hand, occurs both in xerophytic plants and in all groups of microorganisms (even some hydrophytes). As mentioned above, dehydration tolerance has even been found in a hydrophytic higher plant.

As will be seen below, some plants may belong to more than one subgroup. Since, as mentioned above, mesophytes as well as even some hydrophytes possess drought resistance to a moderate degree, the same classification may be applied to them.

B. Drought Avoidance

The freezing resistance of plants is solely or almost solely due to tolerance (Chapter 7). Heat resistance is usually due to tolerance, but in some cases, avoidance plays a decisive role (Chapter 12). In the case of drought resistance, avoidance comes into its own, and due to the many combinations between the different kinds of avoidance and tolerance, a wide variety of adaptations to drought has arisen.

By definition, a drought avoider must maintain a high water potential when exposed to an external water stress. Ephemerals are, therefore, not drought avoiders, since they survive the severe water stress only in the same way as all land plants of low drought resistance—in the drought-tolerant dormant, seed stage. They have, therefore, been called drought escapers. From the point of view of drought resistance, they are no different from mesophytes. Maximov's classification of xerophytes, however, points to two distinct kinds of drought avoiders: (1) the water savers, that avoid drought by water conservation, and (2) the water spenders, that avoid drought by absorbing water sufficiently rapidly to keep up with their extremely rapid water loss. Both kinds of adaptation maintain the plants in the turgid, high water potential (i.e., stress avoiding) state when exposed to water stress. Perhaps the most striking contrast between these two types is provided by Ferri's (1955) measurements in two different dry regions of Brazil. In one of these regions (the "cerrado"), water is stored in the soil equivalent to 3 years' rainfall. Most plants of this region, though exposed to very marked atmospheric drought, show no restriction of water expenditure either at the beginning or the end of the dry season. The much more arid region (the "caatinga") has no water reserves in the soil throughout the year. The rivers dry up during the 7-month dry season. Even when rain does occur, the runoff is heavy. The average temperature is very high, the relative humidity low, and strong winds persist all day and evening. Plants growing in this extreme drought markedly restrict their water loss—even during the rainy season.

Although these drought adaptations reach their highest development in the xerophytes, they are also found to a lesser degree among mesophytes, since drought resistance is a quantitative character. Mesophytic crop plants that succeed in relatively dry climates must have developed one of the adaptations of the xerophytes to a sufficient degree for survival. Wheat varieties, for instance, produce adequate yields in India only if they can complete their life cycles before the arrival of the annual drought (Chinoy, 1960). They are, therefore, to a slight degree ephemerals.

1. *Water Conservation by Water Savers*

In contrast to tolerance, drought avoidance is largely morphological–anatomical in nature (Shields, 1950; Parker, 1956), although the resistance develops due to the effects of drought on physiological processes. It was, in fact, the striking morphological characteristics of xerophytes that led Schimper to explain drought resistance on the basis of water conservation. The later discovery that some of these apparent conservers of water may actually lose water more rapidly than plants of moist habitats threw doubt on the validity of this explanation (Maximov, 1929b). It is now known, however, that both kinds of plants exist. The water savers may lose as little as $\frac{1}{4300}$ their weight per day, whereas the water spenders lose as much as five times their weight per hour (Kilian and Lemée, 1956). Thus water spenders may lose water as much as 500,000 times as rapidly as water savers. The water savers are able to restrict transpiration long before wilting occurs (Grieve, 1953). As mentioned above, they may succeed under conditions of more extreme drought than the water spenders. Yet in some cases, of course, both kinds may live side by side (Oppenheimer, 1953).

The succulents and some sclerophylls are examples of extreme water savers. Among more moderate xerophytes, or even some mesophytes, the superior drought resistance of a species or variety is also often due partly or even solely to water conservation. This adaptation commonly accounts for the differences in habitat of conifers (Parker, 1951; Satoo, 1956; Tazaki, 1960c; Oppenheimer and Shomer-Ilan, 1963), as well as the differences in yield of varieties of oats (Stocker, 1956), sugar cane (Naidu and Bhagyalakshmi, 1967), and peanuts (Gautreau, 1970) in dry climates.

Several adaptations permit the water savers to conserve their water supply.

(a) The stomata open during daylight only for a short time in the early morning and they are able to close very rapidly (Table 14.1). An extreme example is *Spondias tuberosa* whose stomata may close completely within 5 min and are open for only a few hours during the less severe morning (Ferri, 1955). Even the more drought-resistant variety of a mesophytic species may show a greater degree of stomatal closure than the less resistant variety, e.g., among oats (Stocker, 1956) and peanuts (Gautreau, 1970). The major control of water loss must, of course, be exerted by the stomata. In support of this conclusion, the resistance to water transport between the soil and the open stomata in a herbaceous perennial was negligible (Lake *et al*, 1969). When, however, the soil dried, changes in water vapor conductance of the stomata were linearly related to the changes in soil water potential (from -1 to -20 bar). The control mechanism appears to depend on an inhibitor of stomatal opening which accumulates in leaves after a period of water stress, or alternatively, a deficiency of a substance which promotes opening (Allaway and Mansfield, 1970). As a result, stomata of *Rumex sanguinea* showed an aftereffect of wilting, failing to open as widely as usual after recovery of turgor. The two substances appear to be abscisic acid and cytokinin, respectively. The cytokinin decreases in quantity in stressed plants (Itai and Vaadia, 1965; Vaadia and Itai, 1969), and is found to stimulate transpiration (Livne and Vaadia, 1965). The abscisic acid reduces transpiration (Little and Eidt, 1968) and produces stomatal closure (Mittelheuser and van Steveninck, 1969). Furthermore, it increases markedly (to as much as 40 times) 4 hr after wilting (Wright and Hiron, 1969) because of synthesis from a pre-

TABLE 14.1 Rate of Stomatal Movement and Cuticular Transpiration[a]

Species	Percentage reduction in initial transp. rate	Time (min)	Cuticular transpiration (% of total)
a. Cerrado			
Byrsonima coccolobifolia	50	30	15
Didymopanax vinosum	50	20	
Stryphnodendron barbatimam	50	20	
Kielmeyera coriacea	45	50	
Erythroxylum tortuosum	40	50	
Erythroxylum suberosum	20	50	
Anona coriacea	60	30	2
Andira humilis	40	20	33
Palicourea rigida			35
b. Caatinga			
Spondias tuberosa	55	2	2–5
Caesalpinia pyramidalis	50	10	15
Jatropha phylacantha	50	2	10–20
Maytenus rigida	50	10	15
Ziziphus joazeiro	50	5	15
Bumelia sartorum	50	3	10
Aspidosperma pyrifolium			15
Tabebuia caraiba			15
Capparis yco			10

[a] From Ferri, 1955.

cursor (Milborrow and Noddle, 1970) and continues to rise in leaves of osmotically stressed plants for at least 48 hr (Mizrahi *et al.*, 1970). Similarly, a wilty mutant of tomato, whose stomata resist closure, was prevented from wilting and showed a decreased resistance to stomatal closure as a result of treatment with abscisic acid (Imber and Tal, 1970).

(b) Once the stomata are closed, the cuticle on the leaf surface reduces the transpiration rate of the water savers to a much smaller fraction of the stomatal transpiration (e.g., $\frac{1}{5}$ to $\frac{1}{50}$; Table 14.1) than in the case of mesophytes ($\frac{1}{2}$ to $\frac{1}{5}$). This difference in cuticular transpiration accounts for the superior drought resistance of *Pinus halepensis* over *P. pirea* (Oppenheimer and Shomer-Ilan, 1963), of pine over mulberry seedlings (Tazaki, 1960c), and of *Quercus ilex* over *Q. pubescens* (Larcher, 1960).

It has long been known that some drought-resistant plants have very low rates of cuticular transpiration. The classic example is the cactus that holds onto a large fraction of its water for long periods, even if deprived of all contact with water. In the caatinga, the transpiration was actually less during the dry month of January than in the moist month of April, the former being mainly cuticular while the latter was largely stomatal (Ferri, 1955). In this case, in striking contrast to the plants of the cerrada, transpiration rate and evaporation curves form almost mirror images of each other (Fig. 14.1), the transpiration rate decreasing at midday when simple evaporation rises. Some Australian xerophytes behave in exactly the same way (Grieve, 1956) and the relative transpiration rate (transpiration rate/evaporation rate) is at a minimum during the driest part of the day.

Cuticular transpiration is not a constant quantity for a specific leaf, for incipient drying reduces it progressively, until the cuticle becomes practically impervious to water. The transpiration rate may then become so

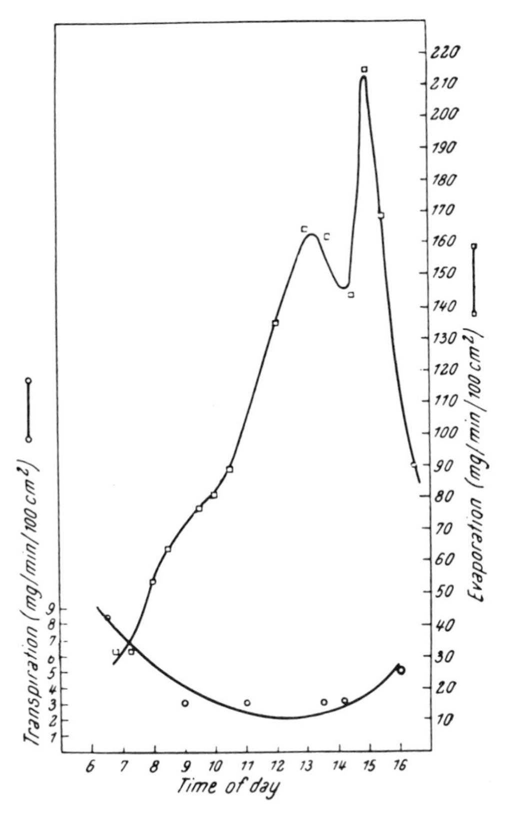

FIG. 14.1. Diurnal course of evaporation (measured by Piche evaporimeter) and of transpiration from *Jatropha phyllacantha*. (From Ferri, 1955.)

low as to be indetectable (Oppenheimer, 1947). This change must be one of the major factors responsible for the above described mirror-image relationship between transpiration and evaporation rate. Another factor may also be involved, since in some plants, the transpiration rate may fall even when the stomata are still open (Grieve, 1956).

A second cause of change in the cuticular transpiration from a specific leaf is an increase in the surface lipids. Even in relatively unadapted plants, exposure to moderate drought may result in a "pseudohardening" (i.e., an increase in avoidance but not in tolerance), due to a deposit of lipids on the leaves resulting in reduced cuticular transpiration (Table 14.2). Soybeans, in this way, may develop an ability to recover from wilting without an increased supply of water. It was even possible to raise the transpiration rate of the pseudohardened leaf to that of the control leaf by removing the soluble surface lipids (Fig. 14.2). In agreement with these results, analyses of cuticle removed from leaves (Skoss, 1955) showed that they contained a greater proportion of wax if obtained from plants undergoing water stress, and this wax was found to control the permeability of the cuticle to water. Similarly, rubbing away the bloom from leaves of *Brassica aleracea* increased the cuticular transpiration rate (Denna, 1970). The various leaf modifications responsible for low cuticular transpiration have been amply discussed and described (see Shields, 1950; Parker, 1956).

It was at one time suggested (Gortner, 1938) that the high pentosan content of many succulents was responsible for a high bound water and, therefore, slowed down the water loss, but it is easy to show that when the morphological protection (i.e., the cutinized epidermis) is removed, water loss is comparable to that from a free water surface. Thus an uprooted cactus was kept in diffuse light without any addition of water for 6 years (see Oppenheimer, 1960). It lost no more than 0.05% of its weight

TABLE 14.2 RELATIONSHIP OF CUTICULAR TRANSPIRATION RATE[a] TO SURFACE LIPIDS OF LEAVES OF SOYBEAN[b]

	Transpirational loss (gm/area/1½ hr)	Lipids removable by lipid solvent/cm² leaf
Nondrought hardened (low drought resistance)	0.119	0.0167
Drought hardened (higher drought resistance)	0.058	0.0289

[a] Water loss from excised leaves one-half hour after excision.
[b] From Clark and Levitt, 1956.

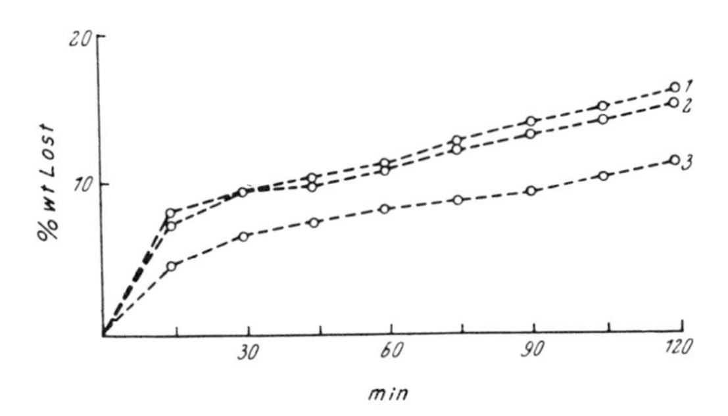

FIG. 14.2. Water loss (% of fresh weight) by excised leaves of drought-hardened soybean plants. Some lipids removed from leaves 1 and 2; leaf 3 untreated. (From Clark and Levitt, 1956.)

per day, at first, and 0.015% later. One that was peeled lost 87% of its weight in 48 hr. Similarly, a high cell sap concentration has been suggested as a method to reduce water loss by lowering the vapor pressure, but the reduction in vapor pressure is very slight at the solute concentrations found in plants, so this effect must be negligible.

(c) In spite of the generally lower cuticular transpiration of succulents, their water loss per unit area may sometimes be quite high (Kilian and Lemée, 1956). In these cases, the principal adaptation responsible for reduced water loss is the reduction in their specific surface. In the case of nonsucculents, rolling, folding, or shedding of leaves have long been recognized as aids to drought resistance (Maximov, 1929; Shields, 1950). Clements (1937a,b) could find no adaptive response to drought in the case of tomatoes and sunflowers other than the shedding of the lower leaves. Ferri (1953) gives a striking example of leaf folding in dry weather (Fig. 14.3). In the caatinga, when the drought reaches its maximum severity the leaves drop and the vegetation becomes whitish. This is responsible for the name of the region which means "white forest." Only a few of the most resistant species are able to retain their foliage during this period. In the case of *Corynephorus canescens*, a species of grass growing in sandy soils in Europe, the stomata-bearing epidermis is inside the rolled leaf, while the heavily cutinized epidermis is on the outside (Sebbah, 1967). Similarly, the dropping of leaves during severe droughts markedly decreases the evaporative surface. Thus, *Calamagrostis canescens* and *Koeleria cristata* had high rates of transpiration during early growth, but this abruptly decreased later due to death of the lower leaves (Sebbah, 1967). This mechanism was, in fact, more effective than the above described leaf rolling, for the water deficit in the remaining tissue was smaller in the leaf-dropping plants. Some reduction in relative surface may be achieved simply by a

Fig. 14.3. Seedlings of *Caesalpina pyramidalis*. Leaves open in high humidity (left), folded under dry conditions (*right*). (From Ferri, 1953.)

compact form of the foliage and crown. Leaves of apple trees with a dense pyramidal crown had a 20.2% higher water-retaining capacity than trees with a loose, dispersed crown (Sivtsev and Kabuzenko, 1967).

(d) In the case of many succulents, the shallow spreading root system functions by quickly absorbing the small amounts of water supplied by light rains. "Rain roots" may develop within a few hours after a shower, and as soon as the soil dries up, they disappear (Oppenheimer, 1959). Thus the root: top ratio may actually be smaller in some of these extreme water savers than in mesophytes, in this way also reducing the root surface from which water can be lost. It must be realized, of course, that though roots do not have stomata and are not exposed to the extreme drought suffered by the leaves, they must nevertheless also develop xeromorphic characters if the plant is to survive drought. As in the case of leaves, the plant must either cover its fine roots with a layer relatively impermeable to water, or it must separate and abscise them by a cicatrization layer in order to cut down the loss of water to the dry soil. The amazing ability of some succulents to produce roots can be demonstrated by suspending a leaf in air. Even when the relative humidity is low, they can send out roots into the dry air from the base, at the expense of water from the tip of the leaf, which shrivels up although the base remains turgid (Sullivan, unpublished).

(e) In the case of sclerophylls it has been suggested that the stiffness of the leaves prevents them from being blown around in the wind and therefore reduces their water loss (see Grieve, 1953).

(f) Genkel *et al.* (1967a) have suggested that the formation of metabolic water from intense respiration may be a mechanism for maintaining the plant's water content. In support of this concept, they state that drought-hardened plants respire more rapidly than nonhardened plants. This mechanism is at least a theoretical possibility in the case of the heterotrophic organisms that obtain their respirable material from the external environment. It is not possible in the case of autotrophic plants, since these synthesize their respirable substances from their own water. Respiration can, therefore, do no more than return to the plant the water used up photosynthetically. Some plants do, however, have water reserves that they store in "water cells" (Gessner, 1956b). These water cells give up their water to assimilating cells during the gradual drying of the leaves.

(g) Those adaptations that favor the conservation of water by stomatal closure, are, of course, unfavorable to photosynthesis since the stomata may be closed during much of the daylight hours, preventing CO_2 from entering the leaves. The most extreme water savers therefore have the most markedly reduced CO_2 absorption (Eckardt, 1952). The succulents have succeeded in partially counteracting this deficiency by the development of an active dark CO_2 assimilation, accompanied by the opening of the stomata at night when danger of water loss is at a minimum. As a result their respiratory quotients at night are actually negative. The assimilated CO_2 is stored in the form of organic acids. During daylight, these are then converted to phosphoglycerate and directly metabolized to carbohydrates (Bruinsma, 1958). Even some nonsucculents adapted to arid regions (e.g., *Salvadora persica* and *Prosopis juliflora*) may possess this acid metabolism (Gaur, 1968).

In the case of tropical grasses as well as some dicotyledons there is a special C_4-dicarboxylic acid pathway of photosynthetic CO_2 fixation. This is believed to be an adaptation for efficient rapid carbon fixation in environments where water stress frequently limits photosynthesis (Laetsch, 1968). Some plants actually store up CO_2 in the form of cystoliths, and it has been suggested that these may conceivably serve as a supply for photosynthesis during long periods of closed stomata (Pireyre, 1961).

It is obvious that these adaptations result only in making the best of an unfavorable situation. The rate of photosynthesis in the water savers will still be very slow as compared to that of well-watered mesophytes that have stomata open during most of the daylight hours and that usually possess a much larger specific surface for absorbing the photosynthesis-producing radiations. Even under conditions of adequate water supply, the water savers may close their stomata during part of the day and, therefore, restrict photosynthesis (Grieve, 1953). As a result of this slow rate of photosynthesis in the extreme water savers, there will be relatively little accumulation of carbohydrates and, therefore, a very slow growth rate. This may explain their inability to send roots deeper into the soil where there may be more moisture. It also explains the low osmotic potentials in succulents—sometimes the lowest found in plants (5–10 atm). There are exceptions, however, and some succulents (e.g., pineapple) are highly efficient and produce large yields. *Agave americana* showed no CO_2 uptake during the first 8 hr of daylight, rising to 4 mg/dm^2/hr during the remaining 6 hr of daylight. At night, however, it absorbed steadily at a rate of 8–12 mg/dm^2/hr (Neales *et al.*, 1968). Thus, its dark CO_2 assimilation at night, compared favorably with the daylight CO_2 assimilation of two mesophytes (sunflower and tobacco), which absorbed at rates of from 7–15 mg/dm^2/hr.

2. Accelerated Water Absorption by Water Spenders

It may appear contradictory to speak of water spenders under conditions of drought. Walter (1963b) explains the apparent paradox. He points out that the individual plant in a desert has at its disposal as much water as the individual mesophyte in its normal habitat. The reason is that the desert plants are so much farther apart that each can use the rain that falls on a far larger soil surface. Furthermore, desert plants often occur in depressions that receive the runoff from a considerable soil area. In support of his conclusion, Walter has, in fact, found that the yield per unit area is directly proportional to the rainfall for arid areas with rainfall up to 500 mm.

In contrast to the water savers, the water spenders are able to maintain higher water contents than unadapted plants in spite of higher transpiration rates (Fig. 14.4), by extracting larger quantities of water from the soil per unit time and leaf surface. As a result, they can keep their stomata open through most of the day in spite of their exceedingly high rates of transpiration (as high as nearly 5 gm/gm/hr; Oppenheimer, 1951). This permits a larger accumulation of photosynthetic product per day than in the case of water savers, and, therefore, also a more rapid rate of growth. Evidence of this relation between photosynthesis and transpiration is the parallelism between yield and transpiration rate (Robelin, 1967). Relative transpiration rate in fact provided a useful growth index. The water spenders are, therefore, often more successful than the water savers as long as some water reserves remain in the deeper soil. There are many adaptations that aid in this rapid supply of water to the leaves (Maximov, 1929).

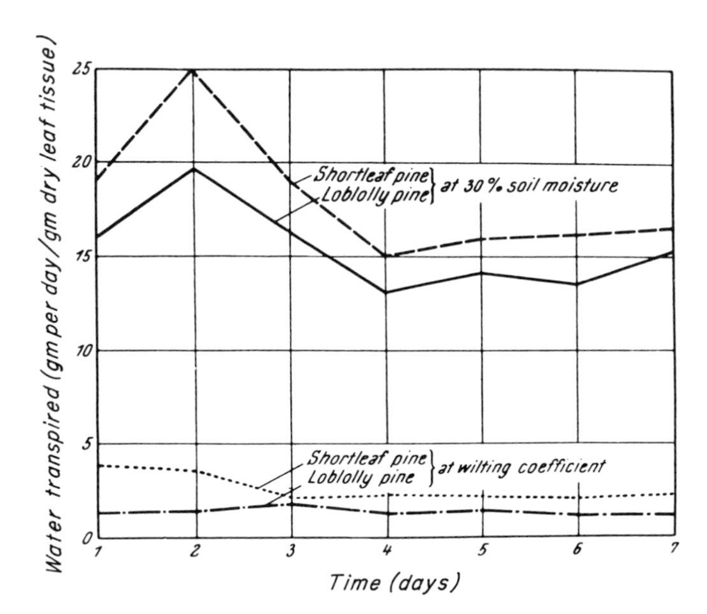

FIG. 14.4. Higher transpiration rate in shortleaf than in loblolly pine at 30% soil moisture and at the wilting coefficient. Water content of shortleaf pine remained higher than that of loblolly pine. (From Schopmeyer, 1939.)

a. A Larger Proportion of Conducting to Nonconducting Tissues

The veins are closer together than in mesophytes and the length of the network per unit leaf area is greater (Table 14.3). Consequently, more water can be delivered per unit leaf surface, leading to a higher transpiration rate even if the rate of flow in the veins is less (Table 14.4). Among the several tree species that Huber (1931) studied, the most drought sensitive (*Fagus*, *Picea*, *Tilia*) showed the most difficulty in supplying water to their upper parts. This was indicated by the marked reduction in transpiration rates of their leaves with increased height (Fig. 14.5). In the most drought-resistant species (*Pinus*), there was a marked increase in transpiration rate with height. Huber, therefore agreed with Maximov in assigning a minor role in drought resistance to reduced transpiration. The frictional resistance to water movement was further shown by cutting off shoots and putting them in water. This doubled the transpiration rate of shade shoots, though that of the sun shoots rose insignificantly. Further evidence of the large frictional resistance to water movement in a mesophyte was produced by Tennant (1954). Even at a high relative humidity of 88%, more water was lost by shoots of the moss *Thamnium alopecurum* than could be taken up from a potometer. Thoday (1931) concluded that the major significance

TABLE 14.3 COMPARISON OF THE ANATOMICAL STRUCTURE AND INTENSITY OF TRANSPIRATION OF STEPPE AND FOREST SPECIES OF *Asperula* AND *Galium*[a,b]

Species	Habitat	Length of network of Veins per unit area	Number of stomata per unit area	Intensity of Transpiration		
				I	II	III
Asperula glauca (=*A. galioides*)	Steppe	100	100	100	100	100
A. odorata	Forest	30	14	31	46	56
Galium verum	Steppe	100	100	100	100	100
G. Cruciata	Forest	38	21	33	46	53

[a] According to Keller and Leisle.
[b] From Maximov, 1929.

of reduction in leaf size is the reduced frictional resistance to flow of water. This follows from the shorter distances of the mesophyll cells from the main channels of supply. Oppenheimer (1949) found the rate of transpiration of oaks related to the width of vessels, in agreement with the earlier results of Huber and Rouschal (see Oppenheimer, 1951).

b. A High Root: Top Ratio

Kausch and Ehrig (1959) showed by root and leaf pruning experiments that the surface area of the root system is the regulator of water expenditure. Their artificial increase in root: top ratio (by leaf pruning) increased the transpiration rate per unit leaf area. This ratio must therefore be one of the most important factors contributing to the high transpiration rates of water spenders. Obviously, the plant with the larger root system will explore a larger volume of soil and therefore will have removed more total

TABLE 14.4 RELATIVE RATES OF WATER FLOW IN SUN AND SHADE BRANCHES OF OAK[a]

	Transpiration rate (mg/dm²/hr)	Cross-sectional area of conducting cells per unit of leaf surface (mm²/dm²)	Volume of water (ml) crossing 1 cm² of xylem per unit time
Sun branches	75.7	0.42	18.0
Shade branches	45.9	0.20	22.5

[a] From Huber, 1924; see Maximov, 1929.

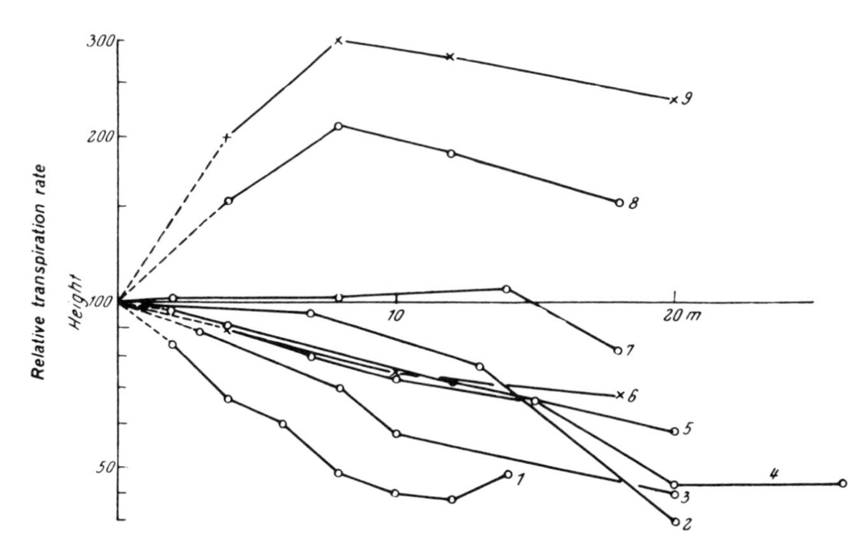

FIG. 14.5. Transpiration rates of some trees at different heights (meters). 1. *Sequoia gigantea*. 2. *Fagus sylvatica*. 3. *Tilia grandifolia*. 4. *Picea excelsa*. 5. *Acer pseudoplatanus*. 6. *Quercus sessiliflora*. 7. *Fraxinus excelsior*. 8. *Larix leptolepis*. 9. *Pinus austriaca*. (From Huber, 1931.)

water when the wilting coefficient is reached (Kramer and Coile, 1940; see Parker, 1956). That drought increases the ratio of root to top has long been known (Maximov, 1929). Even differences in varietal or species resistance may be accounted for by the size of the root system (Misra, 1956; Devera *et al.*, 1969; Parsons, 1969). Simonis (1952) has found that if this increase in root system is taken into account, the total dry weight of plants grown under dry conditions may actually exceed that of similar ones grown with full moisture, though the above ground part is considerably reduced in size. Satoo (1956) has shown that this factor may lead to a reversal in the relative drought resistance of conifer species. *Cryptomeria japonica* was more resistant than *Chamaecyparis obtusa* when the roots were permitted to penetrate freely into deeper layers of soil. This was associated with its higher ratio of root: top. When, however, the depth of root penetration was limited by planting in shallow containers, the order of resistance was reversed. On the other hand, *Pinus densiflora* was the most resistant of the three under both conditions, and it had the highest root: top ratio. The root: top ratio may reach values as high as 30–50 (Oppenheimer, 1959). Besides providing a high ratio of water-absorbing to water-evaporating surface, such highly developed root systems may be capable of reaching the deeper water reserves of the soil. Plants that are capable of reaching the water table are called *phraeatophytes* (Meinzer, 1927). An extreme case is *Prosopis farcata,* the roots of which extend 15 m to the water table in the Dead Sea area (Oppenheimer, 1951). The phraeatophytes are so successful in utilizing the water reserves that they may actually lower the, water table and are, therefore, a menace in certain dry regions (Robinson 1957).

Associated with this high root: top ratio is an ability to continue sending out new roots into unexplored soil during a drought that stops root growth in unadapted plants. Slatyer (1956), for instance, found that the roots of privet continued to increase in dry weight even when the top was already decreasing in dry weight due to the drought. In the less drought resistant tomato, both roots and top decreased in dry weight simultaneously. Similarly, Dastane (1957) found that sugarcane varieties all reduced the soil moisture to the same value at the death point, but the greater drought resistance was associated with a greater capacity of the roots to grow from a region of high to a region of low moisture tension. Similarly, four species of *Cucurbita* are adapted to the hot, arid regions of the Sonoran desert primarily through modifications of their root systems (Bemis and Whitaker, 1969).

c. A Higher Water-Absorbing Potential

Even without any difference in ratio of root: top, the root system of one plant with a lower osmotic potential may conceivably continue to remove water from a soil that is too dry for absorption of water by the roots of another plant with a higher osmotic potential. This possibility has led to measurements of the wilting coefficient or permanent wilting percentage of the soil, i.e., the water content of the soil at the point of "permanent wilting" of the plant. By definition, a plant can recover from "temporary wilting" if the transpiration is stopped. Temporary wilting is therefore due to a greater rate of water loss than water absorption, when both are taking place simultaneously. A plant can recover from "permanent wilting" only if it is supplied with more water. It is, therefore, permanent only if no water is added. Permanent wilting occurs, theoretically, only when water absorption stops. The wilting coefficient or permanent wilting percentage of a soil should therefore be the point at which a plant stops absorbing water, and the determination of this value should reveal whether or not different plants have different water-absorbing potentials. Early determinations by Briggs and Shantz (see Maximov, 1929) yielded the surprising result that the wilting coefficient of any one soil is a constant (within the limits of error of the method used) and is independent of the kind of plant (Table 14.5). This result has been confirmed by more recent investigations

TABLE 14.5 Relative Wilting Coefficients for Different Plants[a,b]

Plants	Numbers of obser-vations	Average value	Probable error of average value	Probable error of single ratio
Corn	75	1.03	±0.003	±0.042
Andropogon	66	0.98	±0.008	±0.062
Chaetochloa	48	0.97	±0.006	±0.035
Wheat	653	0.994	±0.002	±0.049
Oats	46	0.995	±0.007	±0.047
Barley	60	0.97	±0.006	±0.047
Rye	19	0.94	±0.011	±0.049
Rice	21	0.94	±0.012	±0.054
Various Gramineae	77	0.97	±0.005	±0.040
Various Leguminosae	138	1.01	±0.005	±0.059
Various Cucurbitaceae	17	0.99	±0.016	±0.068
Tomato	20	1.06	±0.009	±0.040
Colocasia	19	1.13	±0.005	±0.060
Hydrophytes	8	1.10	±0.037	±0.105
Mesophytes	35	1.02	±0.010	±0.058
Xerophytes	16	1.06	±0.008	±0.032

[a] According to Briggs and Shantz.
[b] See Maximov, 1929.

(Veihmeyer, 1956). It would seem to eliminate the water-absorbing potential of the roots as a factor in drought avoidance.

Unfortunately, these investigators never thoroughly tested the "water spenders," although a group of so-called xerophytes was averaged by Briggs and Shantz (Table 14.5). Any inclusion of water savers with the water spenders would, of course, defeat the whole purpose of such an investigation, since the two would cancel each other out. Furthermore, they were so impressed by the large differences between the permanent wilting points of different soils, that they overlooked the possible value of differences between plants too small to be detected by the methods they used. Thus, the difference between osmotic potentials of different plants will probably not be more than about 30 atm and if water absorption stops when the root cells wilt, this would correspond to only 1–2% soil moisture (Slatyer, 1957a). Yet even a difference of 0.5% in permanent wilting percentage, which Veihmeyer admits is within the limits of error and therefore "insignificant," would certainly be significant as far as the plant is concerned. Calculations reveal (Levitt, 1958), that this small difference could supply a plant with enough water to keep it alive for 6 full days. In some cases, at least, this could easily mean the difference between survival and death.

Of course, it does not follow that water absorption stops completely at zero turgor of the root cells. Water moves according to diffusion gradients, and therefore will continue to be absorbed by roots as long as the gradient favors such movement. Conduction from the roots to the tops may conceivably be interfered with until a certain degree of recovery of root cell water content. In some cases, absorption may seem to stop at soil moisture tensions as small as 6 atm (Oppenheimer and Elze, 1941); in others (e.g., corn roots) it may continue at a slow rate at 12 atm judging from the continued though greatly retarded growth (Gingrich and Russell, 1956). In any case, there is now ample evidence that different plants do lower the soil water to different permanent wilting percentages (Slatyer, 1957a), and that this depends on the osmotic potential of the plant (Table 14.6). It is also possible for plants to continue removing water from the soil until they lower the soil water content well below the permanent wilting percentage (Slatyer, 1957b). Here again, the limiting value varies with the plant (Table 14.6), and Slatyer states that arid zone plants may dry soils to still lower values, equivalent to a stress of several hundred atmospheres.

Thus, the whole concept of the permanent wilting point as a magic point at which absorption stops is theoretically incorrect. Yet it may often appear to be true in practice. The absorption of water is slowed down long before the permanent wilting percentage, and is, in fact inversely related to soil moisture tension practically from zero tension (Gingrich and Russell, 1957). The growth of the plant may be reduced significantly, by as little as 0.3 to 1.5 atm soil moisture tension (Sands and Rutter, 1959).

TABLE 14.6 Relationship of Permanent Wilting Percentage of a Soil to the Osmotic Potential of the Plant[a,b]

Plant	Permanent wilting percentage (% water in soil)	Water or osmotic potential (atm)	Total soil moisture stress (atm)	Final soil water content (%)
Standard	12.2	−15		
Tomato	11.8	−18	20	9.8
Cotton	10.2	−38	38	7.0
Privet	9.7	−47	48	6.9

[a] The soil was covered and lost water only to the still living plants.
[b] From Slatyer, 1957a.

The range, however, varies with the plant. A pepper plant (*Capsicum frutescens*) transpired at a constant rate down to a soil water potential of −6 to −8 bar (Rawlins *et al.*, 1968). Below this value, transpiration decreased linearly with drop in soil water potential to −37 bar. At this time the plant's water potential was −50 bar, and the transpiration rate was near zero. After irrigation, the plant regained full turgor. Other plants have also been shown to differ with respect to water-absorbing potential. At or below the permanent wilting point, oats, wheat, and barley showed little penetration of the soil, but two range grasses (side-oats grama and lovegrass) penetrated the soil rather extensively (Salim *et al.*, 1965). Roots of the highly drought resistant Mediterranean species have a higher potential for removing water from the soil than those of the less resistant Monterey pine (Oppenheimer, 1967). One of the Mediterranean group, the Aleppo pine, had water potentials as low as −20 to −39 atm while still able to absorb water. The resistant species were therefore able to use much more soil water before becoming damaged. Even in the case of corn, a more resistant variety was able to remove more water from the soil than a less resistant variety (Barnes and Woolley, 1969).

It is obvious from the above that the soil water level as well as the plant's should be expressed not as percentage values, which are meaningless when comparing different systems, but as water potentials. It can then be seen that even a mesophyte is able to lower the soil water level well below the previously accepted values (e.g., of −15 atm) for the permanent wilting point (Table 14.6).

On the basis of the newer information, what then is the significance of the permanent wilting point to the plant? The absorption of water slows down suddenly when a plant wilts, due to stomatal closure and the consequent sharp drop in the rate of movement of the transpirational stream. Transfer of the plant into a saturated atmosphere (standard procedure in determining the permanent wilting percentage) will, of course, bring transpiration to a stop and therefore will at least slow down absorption. Under normal conditions of drought, a diffusion gradient is maintained between the plant and its environment and both transpiration and absorption will continue slowly as long as this gradient persists from soil to plant and from plant to air. The permanent wilting point is the point at which there is not enough water in the soil to permit recovery of turgor by the leaves even in the absence of transpiration. Since the roots will normally have a lower cell sap concentration than that of the leaves, it follows that they will also be flaccid and therefore unable to grow. Consequently, what the permanent wilting point really reveals is that the plant is unable to send out new roots into the deeper, moister soil, and therefore absorption for practical purposes stops (Slatyer, 1960). Even this conclusion is not valid in all plants, since some succulents can produce roots even though their leaf tips are strongly wilted (see above). It is, however, obvious that determinations of permanent wilting percentages must be made with extreme care and must be interpreted in relationship to other information. This is especially true since poor water absorbers may actually be better water conservers due to better suberization of the root surface.

d. Dew Absorption

Several investigators have recently suspected that the leaves of plants might absorb enough dew at night to increase their drought resistance appreciably. That water can be absorbed by leaves has long been known (Gessner, 1956a). Thus, in the majority of 25 species of horticultural plants tested, 2–4 hours' absorption through the leaf surface enabled wilted leaves to regain their full turgor after a loss of 10–15% of their fresh weight (Brierley, 1934). This absorption occurred either when immersed in water or when the surface was wetted and suspended in a saturated atmosphere.

Some experiments in the laboratory at first indicated that such absorbed water might actually be pumped down the plant against a gradient and excreted by the roots into the surrounding medium (Stone, 1957; see Slatyer, 1960). These early results were soon shown to be simply due to condensation of water from the roots onto the sides of their container because of a temperature differential. When the temperature was kept constant, no such movement against a gradient occurred. However, movement along a gradient certainly can occur both downward and upward (Slatyer, 1956), and leaves certainly can absorb some water from the dew drops deposited on them (Arvidsson, 1951). The average maximum values of condensed dew for the best dew collectors among the many species investigated both in Sweden and Egypt by Arvidsson (1958) were 0.10 mm or about $\frac{1}{3}$ of the saturation leaf weight. Among the species investigated were two main groups. The Chenopodiaceae, being halophytes, condensed a greater amount of dew (in mm) than the other species. However, because of their thick leaves, the amount absorbed in percentage of leaf weight was not so great as in a second group of thin-leaved plants (e.g., potato, barley). In Sweden (at Olands), condensation of dew on the leaves must have been possible for half of the nights from mid-June to mid-October. In Egypt it must have been possible during a third of the nights of a whole year. Just how important such absorption is in drought resistance, is another question. Indirect evidence indicates that it may be sufficient to account for the existence at dawn in strongly wilted leaves of higher water potential values than in the soil (e.g. −80 atm versus −110 atm; Slatyer, 1957b); although when the saturation deficits were not as strong (40 atm), no such difference occurred. The amounts are centainly very small and seem to be unrelated to the drought resistance of the plant. In carefully conducted experiments, the maximum uptake was hardly sufficient to account for water loss equal to 20–30 min transpiration during daylight (Waisel, 1958a) and it never led to full recovery of turgor. An extreme example is *Tamarix aphylla* (Waisel, 1960). Under field conditions the twigs become wet almost nightly, but little water is absorbed—not enough to equal the saturation deficit of the plant. Nevertheless, water is precipitated by the trees from the atmosphere (from fog or dew) in quite large quantities, and this might be a significant water source, especially in regions with frequent humid nights and under conditions of sublethal water deficits. Tamarix, however, is a salt excreter, and due to its hygroscopic nature the salt can absorb water vapor even at relative humidities of 80%. This mechanism is, therefore, not available to plants in general. Yet, when transpiration drops to extreme low values, it may conceivably account for a very significant extension of life. Artificial dew prolonged the life of permanently wilted conifers from 20–72 days (Stone, 1957). The fog-catching by plants in the high parts of the Canary Islands is said to be large (Morey and Gonzalez, 1966), and may conceivably prolong their life in the same way. *Pinus radiata* has a relatively high transpiration rate and an ability to absorb water directly into the foliage from precipitation (Leyton and Armitage, 1968). It, therefore, appears to be well adapted to the moist atmospheric conditions of its habitat in the "fog belt" region of California. Slatyer (1960), however, suggests that in most such cases the main benefit may be from the mere presence of surface water on the leaves, which prevents transpiration. On the other hand, if the dew absorption permits photosynthesis to proceed for one-half hour longer each morning, before the stomata close, this could also have significant survival value. In some arid climates, the annual dew increments are equal to about 1.5 inches of rain, and this may be sufficient to induce some growth of the meristems (Talli and Durgham, 1968).

Although atmospheric moisture is apparently of relatively minor importance to higher plants, it is of foremost importance in the case of lichens growing in the Negev desert (Lange *et al.*, 1968; Lange, 1969). When their water content drops to less than 5% of their dry weight, there is no measurable gas exchange. Water uptake from the air, however, is very rapid leading to a rapid reactivation of both respiration and photosynthesis. The hydration compensation point occurs at a water content of about 20% when the thallus temperature is 10°C at an illumination of 10,000 lux. This compensation point is reached even when in equilibrium with a relative humidity of 80% (a water potential of −287 atm). The net photosynthesis increases with increasing hydration up to 60% of the thallus dry weight. Dew-fall may, therefore, activate photosynthesis to a high or even optimal degree, and it may continue for 2 hr after sunrise.

e. Conversion to Water Savers

Because of their increased water absorption, water spenders are characterized by very high transpiration rates—higher than those of plants adapted to moist habitats, but this is true only as long as the absorption rate is sufficient to keep up with the transpiration rate. When the water supply becomes so limited that this is not possible, the water spenders then become water savers (Maximov, 1929). This is due to their development of some of the characteristics of the latter, e.g., a well-developed cuticle, though not to the same degree as in the water savers. Once the stomata have closed, the previously rapidly transpiring shoots now lose water less rapidly than those of mesophytes in the same condition (Table 14.7). Huber (1931) showed this clearly by comparing the water loss from the more drought resistant sun shoots and the less resistant shade shoots (Fig. 14.6). In spite of their earlier more rapid transpiration rate, the sun shoots became air dry a full 2 days later than the shade shoots. This emphasizes the fact that it is not always possible to draw a sharp line between the water spenders and the water savers. Thus *Phalaris tuberosa* (a Mediterranean grass) survives under conditions of drought that kill the related annual species *P. minor*. This is partly due to the dormancy of the culms, which presumably conserve water, but also due to the deep root system (McWilliam and Kramer, 1968), which absorbs more water. It is not always possible to generalize even as to the morphological characteristics of savers and spenders. The sclerophylls (hard-leaved plants) of the Mediter-

TABLE 14.7 Relative Transpiration Rates of a Xerophytic and a Mesophytic Shoot[a,b]

Species	Day 1	Day 2	Day 3	Day 6	Day 22	Percentage of total water at end
Stachys sp. (xerophyte)	5	$2\frac{1}{2}$	$\frac{1}{4}$		0	8.9
Vinca sp. (mesophyte)	1	1	1	0		2.2

[a] Measured at 1–22 days after removed from the plant.
[b] From Migahid, 1938.

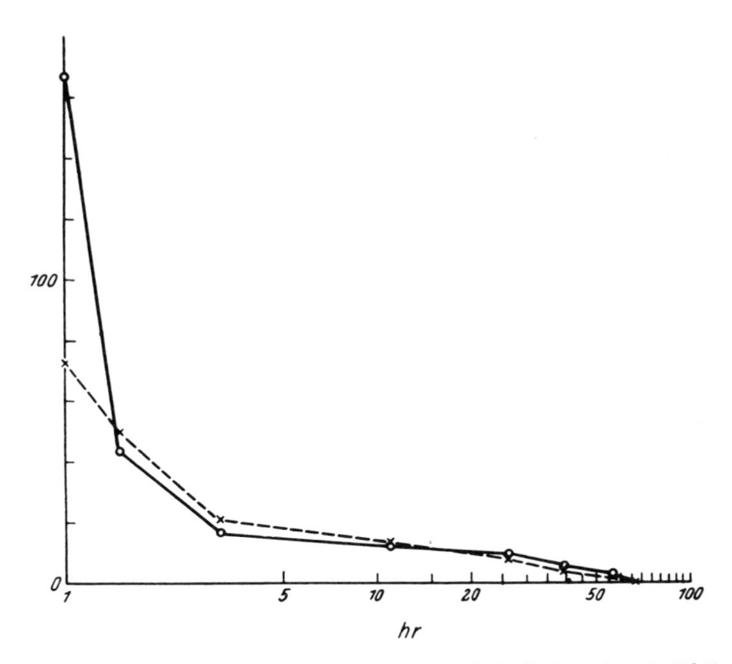

FIG. 14.6. Transpiration rate (mg/gfw/hr) of excised shade branches (solid line) and sun branches (broken line) of *Quercus pedunculata*. Time in hours on logarithmic scale. (From Huber, 1931.)

ranean-type climates are obviously water savers even before extreme water deficiency develops, although Maximov considered them all to be water spenders (Grieve, 1953). The transpiration of such Maquis type plants is usually not above 5 mg/gm/min, and the rate of movement of the transpiration stream is very slow—about $\frac{1}{10}$ of that of mesomorphs (e.g., a maximum of 4.5 as compared with 45 m/hr) and there is no reliable difference between their stomatal frequency and that of mesophytes. Consequently, the drought resistance of such sclerophylls to some extent resembles that of the succulents rather than that of his "true xerophytes" (the water spenders).

3. *Xeromorphy and Xerophily*

The above mentioned morphological characteristics of plants showing drought avoidance are part of what is known as xeromorphic structure. Some of them lead to water saving, others to water spending. Consequently, only some may be found in water savers, whereas most are likely to be found among the water spenders as a group, since they may change from spenders to savers (see above).

Xeromorphic structure is a quantitative character and is developed to different degrees in different plants. It may be hereditarily fixed, or it may vary with the environment. Thus, the same plant when grown under conditions of moderate drought will develop it to a greater degree than when grown with ample moisture (Pyykko, 1966). This is supposedly why it also occurs in a relative sense in sun leaves as compared with shade leaves, and in the upper leaves of plants, in general, as compared to the lower leaves. Xeromorphic structure has therefore been explained as the result of leaf growth under conditions of reduced turgor (Maximov, 1929). The lower leaves of the plant are more mesomorphic than the upper leaves because they are nearer to the water supply (i.e., the roots) and develop earlier in the history of the plant when the water supply is better. For this reason, it is suggested that the upper leaves must complete their growth in the presence of a somewhat reduced turgor pressure compared to the earlier growth of the lower leaves. Since cell enlargement depends on the stretching force of turgor pressure, a smaller force will result in a smaller stretch. Thus the upper leaves, when mature, will have (1) smaller cells and intercellular spaces than the lower leaves and (2) the stomata and also the conducting tissues will be closer together. Although the turgor pressure is reduced, the cells will still retain enough turgor to maintain the stomata open and to continue photosynthesis at its maximum rate per unit area. The smaller cell enlargement will leave more of this photosynthetic product unused. In other words, the extra cellulose that would have been used for intussusception (i.e., cell enlargement) in the larger cells of the lower leaf is now usable in the upper leaf only by apposition, leading to (3) thicker cell walls. A similar excess of photosynthetic product may conceivably lead to (4) formation of more of the thick-walled mechanical cells, and (5) thicker cuticle and wax. Under more extreme conditions, the growth of leaves may be so greatly slowed down that the excess photosynthetic product is mainly translocated to the roots, resulting in more of the growth there, and therefore a greater ratio of root:top. Other characteristics of xeromorphic structure (e.g., a greater development of palisade cells) may be due to light intensity rather than water stress (Oppenheimer, 1960).

Physiological differences accompany these morphological differences. The upper leaves of a plant have (1) a lower water content than the lower leaves. This is to be expected from the smaller cell size which is accompanied by a smaller proportion of the aqueous vacuole and a larger proportion of the less hydrated and thicker cell wall. (2) Sugars accumulate and, therefore, (3) the cell sap concentration is higher. During wilting, however, (4) the upper leaves draw water from the lower ones. (5) They also draw amino acids from the lower leaves (Mothes, 1928). (6) The transpiration rate per unit area is greater in the upper leaves as long as they are turgid. (7) When exposed to drought, however, this relationship may be changed, partly because of the lower cuticular transpiration of the upper leaves, but partly also because of the senescent nature of the lower leaves which, therefore, lose the capacity to close their stomata. Tazaki (1960a) calls these leaves "dull" because of the insensitivity of the closing mechanism. The stomata may therefore, remain open 24 hr a day in the lower leaves of mulberry, and also in those of poplar (Tazaki and Ushijima, 1963). The transition to the "dull" state due to ageing was more gradual in the poplar, in contrast to the sudden change in the mulberry plant. All these morphological and physiological characteristics of plants growing under conditions of moderate drought are together known as xerophily.

There are, of course, extreme hereditary differences between the abilities of different plants to develop xerophilic characters. There are plants (e.g., hydrophytes) that cannot be made to develop them under any conditions of water supply, and there are those that develop them without being exposed to a water deficiency (e.g., some sclerophylls; Grieve, 1953). Thus a dry habitat may be relatively poor in xeromorphic plants, and conversely xeromorphic plants may close their stomata for long periods of the day even during the rainy season (Ferri, 1960). Similarly, the water savers may continue to conserve their water even when irrigated, e.g., *Agave sisalana* (El Rahman *et al.*, 1968). Under artificial conditions, mesquite (*Prosopis glandulosa*) plants also failed to respond to differences in the environment (Wendt *et al.*, 1968). When kept at four moisture levels with water potentials of -15.0, -6.3, -1.5, and -0.1 bar, respectively, there was no significant difference between the transpiration rates of the plants at these widely different soil water potentials, but this lack of a difference may be more apparent than real, due to a mutual cancellation of two factors. In 32 species of trees (Gindel, 1969a), the soil water deficit due to lack of irrigation led to a significant increase in density of stomata. In the case of cotton, maize, and wheat plants, it led to both an increase in stomata and a decrease in leaf area (Gindel, 1969b). In spite of the maximal number of stomata per unit area, full turgor was maintained in the driest periods. Such morphological changes may conceivably prevent a difference in transpiration rate at different soil water potentials.

Due to the existence of such extremes, it is perhaps not surprising that some investigators have failed to find a relationship between xerophilic characters and the dryness of the habitat. Greb (1957) concluded that xeromorphy is a secondary phenomenon due to any deficiency that inhibits leaf development. He suggests that the term xeromorphy should be used only when the structural changes result from water deficiency, and that peinomorphy should be used whenever the structural change is due to some other deficiency, e.g., of nutrients. It should first be realized, however, that the development of xeromorphic characters in the absence of drought may, nevertheless, confer some drought resistance, for it has long been known that rapidly growing plants (e.g., due to excess N) may be less drought resistant than slowly growing plants. Conversely, it must be realized that drought resistance is possible without the development of xeromorphic characters, if the resistance is due to tolerance (see below).

In short, xerophily (and in particular xeromorphy) is not necessarily always related to the aridity of the environment. Nevertheless, the evidence has amply shown that drought-resistant plants frequently do owe part or all of their resistance to xerophily.

CHAPTER 15

DROUGHT TOLERANCE

The above described drought avoidance (Chapter 14) permits the plant to avoid all four types of drought injury (Chapter 13); the drought-avoiding plant maintains a high internal water potential in spite of the low environmental water potential to which it is exposed. Therefore, (1) it also avoids any secondary, drought-induced stress. By maintaining a high cell water potential, (2) it also maintains cell turgor and growth. Finally, in the absence of a low cell water potential, there can be no dehydration-induced (3) indirect metabolic or (4) direct injury. There is, however, one exception. The adaptation that leads to drought avoidance may, in turn, introduce another potentially harmful strain. Stomatal closure, for instance, effectively confers drought avoidance, but it simultaneously prevents photosynthesis, and this may lead to a drought-induced starvation injury. The water savers are able to prevent starvation by their specialized dark carbon assimilation described above. This permits photosynthesis to proceed at an adequate rate during daylight in spite of stomatal closure. To some extent, stomatal control may occur even among mesophytes. Thus, the more hardy of two oat varieties opens its stomata more rapidly in the early morning when drought is at its minimum and photosynthesis can, therefore, occur with the least loss of water (Stocker, 1956). The water spenders, on the other hand, can avoid starvation only by keeping their stomata open during daylight. Their high transpiration rate is, therefore, a result of this adaptation.

In opposition to the general protection conferred by drought avoidance against all kinds of drought injury, drought tolerance must be highly specific in nearly all cases. There are, in fact, a minimum of six possible types of drought tolerance (Diag. 15.1). Of these, the secondary, drought-induced stress will not be considered, since this is a problem of nutrient stress, which is a field in itself and outside the scope of this monograph.

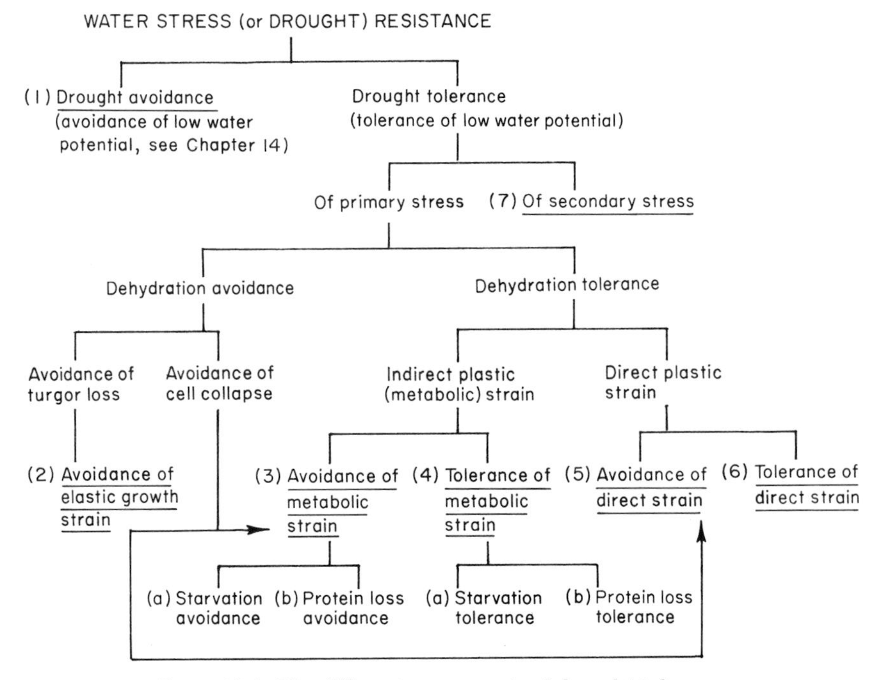

DIAG. 15.1. The different components of drought tolerance.

The remaining five types of drought tolerance fall naturally into three main groups.

A. Dehydration Avoidance (Drought Tolerance due to Avoidance of Elastic Growth Strain; Walter's Euryhydry)

There is only one possible mechanism of tolerating the drought stress without a concomitant cessation of growth—by dehydration avoidance. The drought-tolerant cells that continue to grow when exposed to a drought stress, must possess sufficient dehydration avoidance to retain their turgor (the force responsible for cell growth). This is possible only if they accumulate sufficient solute to produce a lower osmotic potential than that of their environment. In this way, cell turgor can be maintained in spite of the drought stress, and cell division and enlargement can continue.

The following are the quantitative relationships:

$$\psi_c = \pi_c + P_t$$

where ψ_c = cell water potential; π_c = cell osmotic potential; P_t = cell turgor pressure.
A cell possessing maximum drought tolerance is in equilibrium with its environment, so that:

$$\psi_c = \psi_e$$

where ψ_e = environmental water potential. Therefore:

$$\psi_e = \pi_c + P_t$$

and

$$P_t = \psi_e - \pi_c$$

Since both π_c and ψ_e are negative quantities (i.e., at a lower energy level than pure water), P_t can be positive only when π_c is lower (a larger negative quantity) than ψ_e.

This is the mechanism used by the extreme euryhydric molds. It is successful only at relatively moderate droughts: -200 to -300 bar, at the extreme. In the case of higher plants, this tolerance mechanism is possible only at an extreme of about -40 bar, for they are stenohydric organisms (see Chapter 13).

Just as drought avoidance is a general kind of drought resistance, so too is dehydration avoidance a general kind of drought tolerance; in the absence of an elastic dehydration strain, which is the precursor for all the plastic drought-induced strains (see Fig. 13.1), there can be no plastic drought injury, direct or indirect. The only kind of injury against which it may not protect is the secondary stress injury, for as long as there is a primary drought stress there is always the possibility of an accompanying, drought-induced secondary stress. Even a less pronounced dehydration avoidance, which permits loss of turgor, may still confer avoidance of both the indirect (metabolic) plastic strain and the direct plastic strain (Diag. 15.1), although there are also other methods of achieving these two types of drought tolerance (see below).

B. Primary Indirect Drought Tolerance (Drought Tolerance due to Avoidance or Tolerance of Drought-Induced Metabolic Strain)

As mentioned above, the simplest method of avoiding metabolic strain is by dehydration avoidance (see Diag. 15.1); the greater the elastic dehydration strain, the greater the danger of a resulting plastic metabolic strain, at least up to a specific degree of dehydration, beyond which all metabolism is too slow to injure. Even in the absence of dehydration avoidance, however, the plant may be able to develop mechanisms of avoiding an accompanying plastic metabolic strain. There are three possibilities. (1) If dehydration affects the kinetics of all reactions equally, there will be no metabolic unbalance and, therefore, no injury. This is very unlikely, not only because of the large number of reactions involved, but also because different reactions occur in different organelles, and it is not likely that all organelles will be equally dehydrated. If, on the contrary, the dehydration does not affect all reactions equally, there are two remaining mechanisms for avoiding metabolic injury. (2) The metabolic unbalance does not produce a deficiency of an essential metabolite. (3) The metabolic unbalance does not lead to an accumulation of a toxic substance.

1. Starvation

a. Avoidance of Starvation Strain

A great deal of information is available on the temperature and the CO_2 compensation points. These are the temperature and CO_2 concentrations, respectively, that are just high enough to induce equal rates of photosynthetic CO_2 absorption and respiratory CO_2 evolution, yielding a net photosynthesis of zero. The corresponding hydration compensation point has received very little attention. Yet it does exist, and it may be more important than the other two compensation points when the plant is exposed to a drought stress.

The greater the dehydration a plant can tolerate and still keep its stomata open, the lower will be its dehydration compensation point, and the greater will be its avoidance of the starvation strain. This kind of drought tolerance will, therefore, be due to a specific dehydration avoidance of the guard cells, permitting them to remain turgid although all other leaf cells are wilted. In general, xerophytes manage to continue photosynthesis under conditions of drought that bring it to a stop in mesophytes. In many, if not most cases, this is due to drought avoidance (see Chapter 14). There is a good deal of evidence, however, that a moderate degree of drought tolerance may be due to the specific dehydration avoidance of the guard cells. According to Huber (1931), the most drought resistant of the species he investigated (*Teucrium montanum*) was the last to close its stomata in the morning (at 7–8 o'clock) during periods of severe drought, and therefore continued to photosynthesize for a longer time than the less drought-resistant species. Shade plants are known to possess very low drought resistance (see below). They close their stomata with a very small water loss (3–10% of their saturation water content), whereas the more drought-resistant plants of sunny, dry regions undergo far greater water loss (5–33%) before closing theirs (Pisek and Winkler, 1953). Eckardt (1953) found that a "mesophytic xerophyte" continued to photosynthesize at a high rate though transpiration rate was relatively low. Apparently, the stomata remained open but the transpiration rate dropped due to the resistance to water flow in the thin roots responsible for water absorption from deep soil during drought. Conifers can apparently withstand a little more dehydration before closing their stomata than can broad-leaved trees—about 8 and 3% of the saturation water content, respectively, for the beginning of closure (Pisek and Winkler, 1953), and 30 and 10%, respectively, for closure in spruce and aspen (Jarvis and Jarvis, 1963a). In some cases, they may permit photosynthesis to continue by keeping their stomata open even after general leaf turgor is lost (Larcher, 1960).

Because of stomatal control, it has generally been found that assimilation or gain in dry weight parallels the plant's transpiration rate, e.g., in the tomato (Wassink and Kuiper, 1959). Exceptions have, however, been reported. Thus Jarvis and Jarvis (1963b) obtained maximum assimilation at soil water potentials of about -1.5 bar, dropping to 50–80% of this value at -3.0 to -3.5 bar in aspen, birch, pine, and spruce. The ratio of the assimilation (as measured by increase in dry weight) to transpiration remained about constant for spruce, aspen, and birch at -1.0 to -2.5 bar, but in pine it rose to $3\frac{1}{2}\times$ at -2.5 bar.

A possible mechanism of avoiding stomatal closure has been suggested by Henrici (1946). Although alfalfa wilted at 3% water loss, Karoo bush did not wilt even at 30% water loss, though the leaves became brittle. One possible explanation of such results is a much greater elastic extensibility of the cell walls in the drought-resistant plants. In soybeans, however, no increase in elastic extensibility of the walls was detected as a result of "pseudohardening" by noninjurious wilting (Clark, 1956). Therefore, although Henrici's concept is intriguing, it needs experimental confirmation. In the absence of corroborative evidence, the known properties of the cell walls make this postulated mechanism extremely unlikely. A far more likely mechanism is an increase in guard cell solutes, leading to dehydration avoidance of the guard cells.

It is obvious from the above examples that higher plants possess only a very modest avoidance of starvation strain, i.e., they can tolerate only a very small dehydration strain and still keep their stomata open. It appears likely that an ability to maintain their stomata open at a greater degree of dehydration would do more harm than good. Not only would this lead to a more rapid approach to direct injury (due to a higher transpiration rate), but it would also probably fail to increase photosynthesis markedly; due to the greater dehydration, translocation would no doubt quickly halt photosynthesis (see Chapter 13).

Lichens, on the other hand, have no stomata and no long distances for transport. They have developed avoidance of starvation strain to a maximum degree. The lichen *Ramalina maciformis*, growing in the Negev desert, assimilates CO_2 maximally at 80% r.h., the rate dropping on both sides of this value to zero on the low side at 15–20% r.h. Respiration also decreased steadily, from a maximum at 100% to zero at 15% r.h. The dry thalli have their CO_2 exchange reactivated by exposure to unsaturated air (Bertsch, 1966b). In the case of some, the hydration compensation point is reached between 80–85% r.h., or at a hydration of 20% (Lange *et al.*, 1969). Similarly, an aerophilic green pleurococcoid alga (*Apatococcus lobatus*) has an unwettable thallus but achieves vapor pressure equilibrium with the air (Bertsch, 1966a). Fully 50% of its maximum photosynthetic capacity is attained at a water potential of −142 bar (90% r.h.) and 10% at −370 bar (76% r.h.). The limit of CO_2 uptake is reached at about −520 bar (68% r.h.). On drying, respiration decreases and an increase in apparent CO_2 uptake is observed. Even some of the ferns show extreme adaptation. In *Polypodium polypodioides*, photosynthesis was proportional to water content all the way from 100 to 20% of its saturation water content (Stuart, 1968).

Even in the case of higher plants, when stomatal control and long-distance translocation are eliminated by using leaf pieces, the results may be very different from the normal plant. Thus even the loss of up to 80% of the water content of beet leaves does not completely suppress photosynthesis, and the rate is still comparable to the respiratory rate. With still greater losses, photosynthesis drops below the respiratory rate, but the latter is also reduced to a very low value (Santarius, 1967). Photosynthesis apparently came to a complete halt somewhere between 88 and 99% water removal.

Some evidence exists of a hardening effect of dehydration on the assimilation rate. According to Simonis (1952), long continued growth under conditions of drought leads to a marked increase in rate of photosynthesis. Periodic droughting has also been reported to increase photosynthesis (Stocker, 1961). Others, however, have failed to detect any such change (Petrie and Arthur, 1943; Eckardt, 1953). Since net assimilation was measured in these cases, if a change did occur, it may have been due either to the photosynthetic or the respiratory process.

A possible case is the lower respiratory rate in varieties of sugarcane that are more resistant in the field (Raheja, 1951). Some plants appear to achieve this low respiration rate by entering a rest period at the height of the drought (Grieve, 1953).

b. Tolerance of Starvation Strain

Even when below its hydration compensation point, a plant in its rest period will survive longer than an actively metabolizing plant. It will, therefore, possess a greater tolerance of dehydration strain. This may also be one reason for the success of succulents under conditions of drought. Due to their extremely low dry matter contents and their fundamental metabolism per cell their carbohydrate needs must be very low. They can, therefore, tolerate a low net rate of photosynthesis that would be fatal to plants with a higher fundamental metabolism.

2. Protein Loss

a. Avoidance of Protein Loss

Since avoidance of protein loss is a net result of two processes, drought tolerance of this kind could be due to a change in either of them—a decreased rate of protein hydrolysis or an increased rate of synthesis. That this kind of tolerance does exist, has been suggested by Mothes (1928). He ascribes the greater drought resistance of young leaves to their higher protein content, well above the minimum necessary for life. Kessler (1961) suggests two possible methods of drought resistance which may prevent the net loss of RNA, and the consequent inability of the plant to synthesize proteins. (1) RNA synthesis may be intensified. (2) The increase in RNase activity, which he observed during drought, and which led to a decrease in RNA, may not occur due to stabilization of the inactive (complexed?) form of RNase. He was actually able to reverse the hydrolysis of RNA during drought by spraying the plant with solutions of purines. Results of Chen *et al.* (1968) favor the first method. They found that a 48-hr dehydration of germinating wheat embryo in both the drought-tolerant (germinated 24 hr) and the drought-sensitive (germinated 72 hr) stages inactivated mRNA and halted protein synthesis. In the sensitive embryos RNA actually disappeared while in the resistant ones it was preserved but rendered inactive as in the ungerminated embryo. Rehydration of the drought-tolerant embryos resulted in the formation (by transcription) of complementary RNA which resembled normal mRNA. Rehydration of the drought-sensitive embryos resulted in the formation of false and inactive RNA.

Confirmation of these concepts has been obtained in winter wheat plants (Stutte and Todd, 1968). Water stress resulted in a reduction of the total RNA content, with the nonresistant variety showing a greater decrease. The (G + C)/(A + U) ratio increased with water stress in both varieties. Under more severe drought, however, the nonresistant variety yielded higher ratios than the more resistant variety. Temperature also affected the ratios, making it more difficult to assess the effect of water stress alone.

All of these results attempt to explain avoidance of protein loss by an increased rate of synthesis. A decreased rate of protein breakdown has not been indicated.

b. Tolerance of Protein Loss

If the loss is not sufficient to kill the plant, it is conceivable that it could be rapidly repaired on rehydration. This possibility does not seem to have been investigated. There is, however, another possible source of injury due to the protein loss. Long before the protein deficit is severe enough to injure the plant, products of protein hydrolysis may accumulate to a

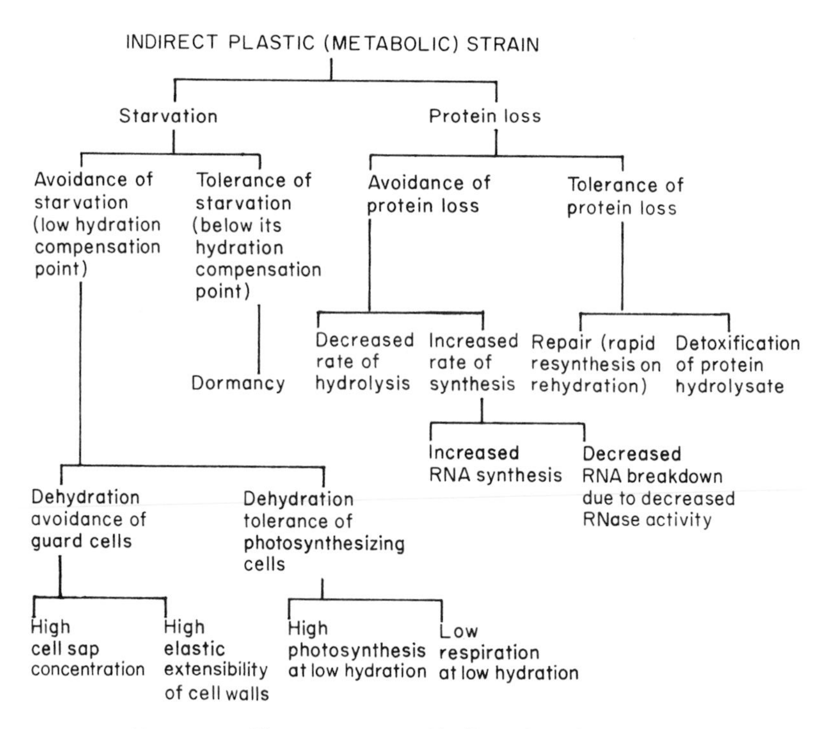

Diag. 15.2. The components of indirect drought tolerance.

sufficient degree to be toxic. Tolerance of protein loss could then be induced by some method of decreasing the toxicity. This may conceivably explain the accumulation of proline (see Chapter 13) that appears to be associated with drought tolerance (Palfi and Juhasz, 1969, 1970). In *Carex pachystylis*, a very drought-tolerant plant, both proline and to a smaller degree γ-aminobutyric acid accumulated during a 21-day drought, at the expense of glutamic and aspartic acid in the aerial parts, and of arginine in the subterranean parts (Hubac *et al.*, 1969). However, as pointed out by Hubac *et al.* (1969), there are many other reports of an accumulation of proline under conditions that would not be expected to lead to an increase in drought resistance. The relationship between the above described possible kinds of indirect drought tolerance is shown in Diag. 15.2.

C. Primary Direct Drought Tolerance (Avoidance or Tolerance of Direct, Drought-Induced Plastic Strain)

As in the case of the indirect, metabolic strain, the simplest method of avoiding a direct plastic strain is by dehydration avoidance (Diag. 15.1)—either the more extreme avoidance of turgor loss, or the less extreme avoidance of cell collapse after loss of turgor. Again, however, other mechanisms may be capable of accomplishing the same objective in the absence of adequate dehydration avoidance, but first it must be established that drought tolerance of this kind does, indeed, exist.

1. Existence of Primary, Direct Drought Tolerance in Nature

In the case of the higher plants, it is not always possible to decide whether a drought effect reported in the literature is direct or indirect, or even whether the resistance involved is tolerance or avoidance. The decision is much simpler in the case of the lower plants that Walter classifies as poikilohydric, since they possess maximum drought tolerance and minimum avoidance. Plants of this type cannot, of course, grow and metabolize when air dry, but they are capable of becoming air dry without injury (Chapter 13), and they can grow and metabolize actively again when the drought is over. Lichens and mosses, for instance, owe their drought resistance to tolerance, since they have neither a significant barrier to water loss, nor any means of supplying water to the cells or tissues exposed to drought. Yet a marked difference in tolerance occurs among them (see below). Even among the pteridophytes (the fern family), which are usually homoiohydric, there are several species known as resurrection plants, which commonly become air dry in the vegetative state without injury (Oppenheimer, 1959). The high degree of tolerance survived by some ferns is illustrated by the epiphytic fern, *Polypodium polypodioides*, which was not damaged by a loss of 97% of its normal water content (Stuart, 1968).

FIG. 15.1. Comparison between the highest average natural water deficit (broken lines) and the critical water deficit (solid lines) for the leaves of different species. Expressed in percentage of maximum water content. (From Arvidsson, 1951.)

As shown above, in the case of vegetative higher plants (which are mainly homoiohydric; Walter, 1950), the most extreme droughts are usually survived (particularly over long periods of time) due to avoidance. As a result, their water deficits under natural conditions are small and fall within a relatively narrow range that is well below the critical saturation deficit (Fig. 15.1). Even in the case of these, however, drought tolerance may be in the same order as drought resistance, as Satoo (1956) found to be true of three conifer species, but this relationship held only when the root system was restricted. In other words, it is possible for a drought-resistant species to overcome a lack of drought tolerance by the development of one or more drought-avoidance factors. When a number of species are compared, the natural water deficits of the drought-tolerant cover a wider range and reach higher values than in the case of the drought-avoiding plants under the same conditions (Fig. 15.1). It is obvious that all gradations and combinations occur in different species. The actual deficit tolerated usually increases from early to late summer (Fig. 15.2).

Some higher plants can also become air dry without injury (see Table 13.1). One of these species with extreme drought tolerance in higher plants is the creosote bush (Duisberg, 1952). In spite of its mesophytic morphology, it is the dominant plant of the hottest and driest plains and deserts of Mexico and southwest United States. Under conditions of extreme drought the more mature leaves die; but the immature leaves and buds dry out and turn brown. When more favorable moisture conditions exist, the immature leaves and buds continue growth. These drought-tolerant immature leaves have lower contents of nordihydroguaiaretic acid and resin and a higher protein content than the mature, nonhardy leaves, but it is not known whether these differences are related to the difference in drought resistance.

Even grasses may show little restriction of transpiration during drought, and this may result in marked saturation deficits in the leaves—often 20% of the fresh weight and sometimes 50% without suffering any damage (Henrici, see Oppenheimer, 1959). An extreme case is grama grass (*Bouteloua gracilis*) which may lose 98.3% of its free water without injury (Whitman, see Oppenheimer, 1959).

2. Drought Hardening

The term "hardening" has been used for an exposure to a sublethal stress that results in resistance to an otherwise lethal stress. Most commonly an increase in tolerance is implied. Nevertheless it must be recognized that an increase in avoidance may also occur. Growth under con-

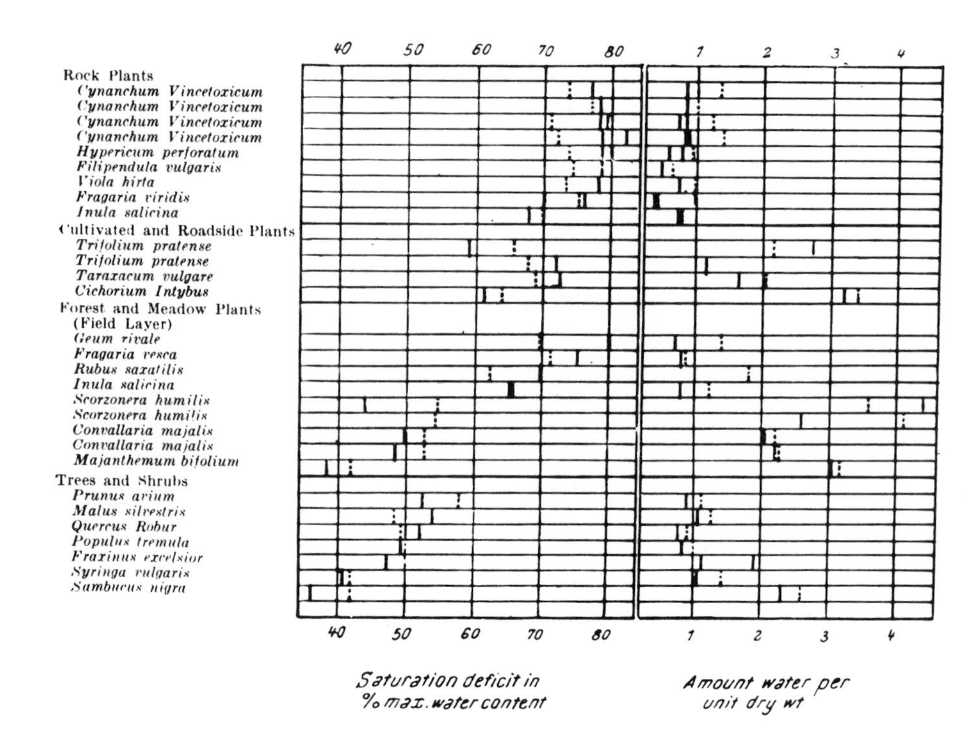

FIG. 15.2. Early (broken line) and late (solid line) summer average critical saturation deficits (left) and water per unit dry matter at the critical saturation deficit (right). (From Arvidsson, 1951.)

ditions of moderate drought has already been shown to lead to xeromorphy in many plants (Chapter 14). Two simple mechanisms leading to a decrease in cuticular transpiration were described: (1) drying of the leaf surface, and (2) a deposit of lipids on the leaf surface, both of which decreased the permeability of the cuticle to water. Such decreases in transpiration rates and, therefore, increases in avoidance, have been reported for trees (Dobroserdova, 1968), as well as for cereals (Salim *et al.*, 1969), as a result of exposure to drought. There is also some evidence of an increased avoidance due to improved water absorption as a result of growth under conditions of moderate drought.

In at least some cases (e.g., the deposit of lipids on the leaf surface of soybeans; Clark and Levitt, 1956) the increase in avoidance is unaccompanied by any increase in tolerance, and the protoplasm itself is not any "hardier" to drought. Any increase in avoidance without a concomitant increase in tolerance will therefore be called a "pseudohardening." The term "hardening" will be applied only if there is an increase in tolerance.

Hardening may conceivably involve an increase in the indirect as well as the direct drought tolerance. Simonis (1952), for instance, reports an increase in photosynthetic rate, which might imply an increased ability to prevent starvation effects. Unfortunately, little information is available on this kind of tolerance. It is, therefore, necessary to confine our attention to direct drought tolerance. Many plants are incapable of hardening to drought. An extreme example are hydrophytes. They are homoiohydric and, in general, die when exposed to the moderate drought needed for hardening. At the other extreme, many lower plants have a "built-in" drought tolerance even in the absence of any hardening treatment. Some lichens and mossess, for instance, become air dry within a few hours without suffering any injury (Lange, 1953).

Most higher plants (including those of agricultural importance) fall between these two extremes. They are neither homoiohydric nor poikilohydric in the extreme senses of the terms, but their water contents show

TABLE 15.1 Drought Tolerance of Unhardened Plants and of Plants Hardened by Droughting for 2–6 Weeks[a]

Species	Relative humidity causing 50% killing	
	Unhardened	Hardened
Millet	92	92
Cabbage (*Brassica oleracea*)	96	92
Sempervivum glaucum	94	94
Wheat (Thorne)	98	98
Wheat (Seneca)	97	94
Barley (B475) long-day	96	90
Barley (B475) short-day	96	94

[a] From Levitt *et al.*, 1960.

daily fluctuations within relatively narrow limits and greater fluctuations during periods of drought. They can therefore be dehydrated to the moderate degree required for hardening. The standard method of hardening is to withhold water for some days, allowing the plant to undergo temporary and even permanent wilting (Tumanov, 1927). If the wilting is not too severe, the plants recover their turgor on watering. Some plants harden (i.e., become more drought tolerant) as a result of such treatment, others do not (Table 15.1). Under natural conditions, the former will, of course vary in hardiness with their environment, the latter will not. As a result of the predroughting, plants survive for longer periods in a drought chamber (Oppenheimer, 1967). Even a water stress too small to induce wilting may also lead to some hardening (Table 15.2).

Exposure to low temperature, so as to induce frost hardening, may also confer drought hardiness, and sometimes to a greater degree than drought hardening itself (Table 15.3). This has also been found true in nature, and Mediterranean plants may actually become more drought tolerant in winter than in summer (Fig. 15.3). Epiphytic mosses may show the same phenomenon (Hosokawa and Kubota, 1957). According to Oppenheimer (1959), this is not surprising since the northern conifers may be subjected to more severe water stress during the frost of winter than during summer, when some water is still supplied by the roots.

Henkel and his co-workers (see Henckel, 1964) have reported increases in drought resistance due to pretreating seeds before sowing. They allow grain to take up water to the extent of 30% of their dry weight (40–50% in the case of some seeds). They are then left for 24 hr at 10°–25°C, then air dried. Best results are obtained if this is done 2–3 times. The exact treatment is purely empirical and varies with the species. Another method is to allow the imbibition of water from a 0.25% $CaCl_2$ solution for 20 hr.

TABLE 15.2 EFFECT OF MODERATE WATER STRESS ON DROUGHT TOLERANCE[a]

Species	Water stress (bar)	Drought tolerance (bar)
Birch	0.1	34
	1.0	50
	2.0	60
Aspen	0.1	48
	0.5	56
	1.0	58
	4.0	75

[a] Adapted from Jarvis and Jarvis, 1963b.

TABLE 15.3 Drought Tolerance of Unhardened Barley Plants and of Plants Hardened by Exposure to low Temperature[a]

Variety	Relative humidity causing 50% killing	
	Unhardened	Hardened
B475	98	90 (Grown outdoors)
	94	90 (Greenhouse-grown)
Tennessee winter (nonhardy variety)	97	90 (Field-grown)
Kearney (hardy variety)	97	85 (Field-grown)

[a] From Levitt *et al.*, 1960.

Borates have also been used. In all cases, the effect on drought resistance is judged by yield. Increases of 10–25% are usually reported, though it is sometimes greater and there is sometimes no response. Parija and Pillay (1945) obtained increased survival of severe wilting in the case of rice plants subjected to such presowing treatment. Chinoy (1960) reported small increases in yield of wheat after such treatments, but ascribed these to earlier maturing of the crop, and the consequent escape from the later more extreme drought. In none of the above cases was drought tolerance measured. Four-year-long tests of Henckel's method on two wheat varieties,

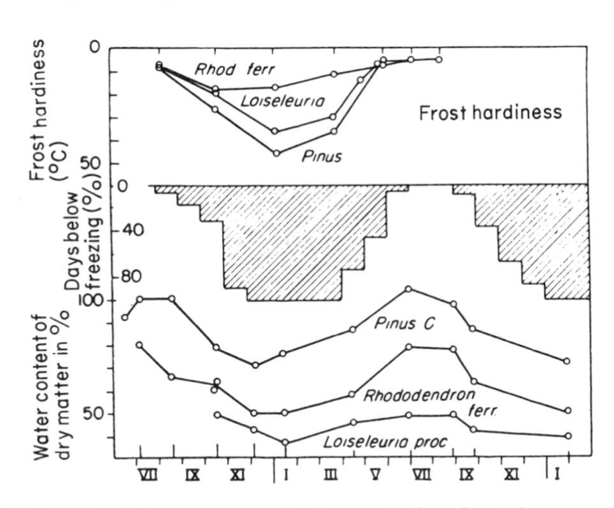

Fig. 15.3. Parallelism between seasonal changes in freezing tolerance or frost hardiness (upper curves) and in critical water removal (lower curves) in three evergreens. (From Pisek and Larcher, 1954.)

resulted in higher yields by one variety, with optimal water supply and even more so with periodic water deficits. The other variety showed a decrease in yield in some cases (Gej, 1962). Waisel (1962b) was unable to obtain any significant yield increases, or any increase in drought tolerance as a result of presowing treatments. Other measured factors also failed to show any significant change. Waisel ascribes Genkel's results to a selecting effect of the treatment on a genetically impure seed sample. Jacoby and Oppenheimer (1962) also failed to affect the drought tolerance of sorghum plants by grain pretreatment. There was, however, an increase in resistance to water loss. They soaked sorghum seed in distilled water for 18 hr, and then dried them at room temperature. This was repeated twice before sowing. After germination to a height of 8 cm, the plants were droughted by exposure to 30% of field capacity.

Husain *et al.* (1968) used Genkel's method on barley seeds as did Genkel, and concluded that the evidence failed to support Genkel's claim, although they obtained a 15% increase in grain size. Salim and Todd (1968) were unable to generalize as to the effects of the presowing treatment, since the response depended on the treatment and on the variety used. The Russian investigators, however, continue to support the concept. Bozhenko (1968) used the presowing treatment on sunflower seeds. Aluminum nitrate and cobalt nitrate raised the RNA and DNA contents of the growing parts under water deficiency conditions, and lowered RNase activity. They offer this as a possible cause of the increased drought resistance. According to Keller and Black (1968), however, the presowing treatment induces emergence of the seed (of crested wheatgrass) 40 hr ahead of untreated seeds. This seems to agree with Chinoy's explanation. Examination of the treated seed has revealed how this earliness is achieved. Carrot seed were hardened for 24 hr after the addition of 70% of their weight in water and then dried (Austin *et al.*, 1969). Three cycles of this treatment produced embryos 51% longer than in the controls, mainly due to cell division during hardening. The hardened seed imbibed water more quickly and the seedlings emerged in the field 3–4 days earlier than untreated seed. Although relative growth rates during early growth were similar in the treated and control plants, at harvest, 14–21 weeks after sowing, the mean yield of roots was 64.0 for the hardened compared to 59.2 tons/hectare for the unhardened seed.

Since in the two cases when tolerance was measured, no increase could be detected, any effects of the presowing treatment on drought resistance can only be due to avoidance, and effects on yield not due to avoidance must be due to drought escape by earlier maturation. The ability of seeds to germinate in solutions of higher concentrations of solutes has, in fact,

TABLE 15.4 PERCENTAGE OF SURVIVAL OF BAKERS' YEAST AFTER THREE DAYS' EXPOSURE TO GIVEN RELATIVE HUMIDITIES[a]

Rel. hum. (%)	Control culture		Control culture 1.1 *M* sorbitol added before exposure to rel. hum.		Grown in culture plus sorbitol	
	1	2	1	2	1	2
60	41	32	49	36	94	91
40	0	4	5	4	48	63
20	0	0	1	0	24	35
0	0	0	0	0	4	15

[a] From Füchtbauer, 1957a.

been tried as a measure of drought tolerance, but the results are conflicting (Manohar *et al.*, 1968).

In contrast to the delayed avoidance effect of solutes on germinating seeds, a direct effect on the drought tolerance of some microorganisms has been obtained. Among the several species of bacteria and fungi tested by Burcik (1950), addition of salt solution to the substrate raised the threshold relative humidity for four species, lowered it for two species, and failed to change it for five species. Among the alga and two fungi tested by Füchtbauer (1957a,b) only bakers' yeast and the alga (*Chlorella pyrenoidosa*) showed a significant increase in tolerance with the addition of sorbitol to the medium. The increased tolerance was not due to any purely physical effect of the sorbitol, since the organisms had to be grown in its presence before developing resistance (Table 15.4). The sorbitol did not

TABLE 15.5 PERCENTAGE OF SURVIVAL OF BAKERS' YEAST[a,b]

Rel. hum. %	Control cultures			Sorbitol cultures		
	1	2	3	1	2	3
90	100	100	100	100	100	100
80	84	80	85	96	100	97
70	87	79	72	99	100	95
60	54	39	31	80	63	73

[a] Sorbitol cultures washed three times with distilled water, then suspended in distilled water before exposure to given relative humidities.

[b] From Füchtbauer, 1957a

have to be present in the surrounding solution during the drying process. Thus, after growth in the presence of sorbitol, the cultures could be washed with NaCl or other solutions or even with distilled water and left there for 3 hr before drying without loss of the increased drought tolerance (Table 15.5).

3. Relationship of Tolerance to Development

De Saussure (1827) clearly showed that the drought tolerance of seedlings decreases with the progress of development. Similar results were obtained by Rabe (1905) and Milthorpe (1950). The relationship of developmental stage to drought tolerance was shown by Sun (1958) in the case of the excised pea seedlings (cotyledons removed). The smallest ones, 7–12 mm long, recovered completely after water losses amounting to 77% of their fresh weight (final moisture content 27%). Intermediate size seedlings (13–20 mm long) recovered only after water losses of 74.4% (final moisture content 48.8%). Large seedlings (21–30 mm long) were still more sensitive.

Another drop occurs in later stages, from a minimum osmotic potential of −63 and −76 atm for oats and wheat in the early stages of growth to −32 and −40 atm at heading time (Binz, 1939). Brounov (see Maximov, 1929) applied the term "critical period" to the stage of development at which the plant is particularly susceptible to injury. This period occurred during the rapid internode growth that preceded heading in cereals, as was later corroborated by Campbell (1968). Moliboga (1927—see Maximov, 1929) showed that the critical degree of wilting for this period did little harm in earlier or later stages. In wheat, both the flowering phase and the stage of grain filling and maturation are more sensitive to drought than the vegetative period of growth (El Nadi, 1969). Similar sensitivities during the flowering stage have been found in maize (Wilson, 1968), barley (Singh and Singh, 1966), and pepper plants (De Lis *et al.*, 1969). Finally, a sharp rise in drought tolerance occurs during seed maturation, equaling the sharp drop during germination (Maximov, 1929).

During the development of seeds of *Phaseolus lunatus*, as the cells mature, the polysomes disappear and free ribosomes appear in the cytoplasm (Klein and Pollack, 1968). It was suggested that these changes may be necessary to prepare the cells to withstand the desiccation during seed maturation.

Within a single plant, the younger tissues are more tolerant than the older ones, e.g., in the plumules of seedlings (Rabe, 1905), the dormant buds of mosses (Irmscher, 1912), and the younger parts of liverworts (Höfler, 1942a,b). Pringsheim (1906) found a movement of water from older to younger parts on wilting. This protected the younger parts longer and permitted them to develop further. In many cases, the water in the stems moved directly to the growing points and the fully developed leaves were sacrificed, sometimes even before drying up (e.g., *Sedum*, *Erica*, *Bryophyllum*, *Euphorbia*). Christ (1911) observed the same phenomenon in beech, oak, and ash trees which retained small rosettes on the ends of twigs. Tumanov (1930) and Clements (1937a,b) describe it in crop plants. Thus, the better survival of drought by younger tissues cannot be accepted as evidence of greater drought tolerance so long as they are still attached to the older tissues. Direct determinations have shown that well-ripened, 1 to 2-year-old leaves of evergreens are more drought tolerant than both underdeveloped and overage leaves (Pisek and Larcher, 1954). The effect of age is undoubtedly indirect. Thus growth inhibition by accommodation to dry soil led to enzyme activities in the leaves of cowpea (*Vigna sinensis*) characteristic of young leaves (Takaoki, 1968).

There are, of course, some age differences due to avoidance. Exposure of young cembran pine plants to 15% r.h. at 25°C and with an air flow of 10 m/sec showed that the needles dried out most rapidly at shooting time and were completely killed within 2 days (Tranquillini, 1965). Even the older needles survived only 12 hr longer. The needles became increasingly resistant throughout the summer, and by fall (October) they were not killed until 7 days' treatment under the above conditions. These changes may be related to photoperiod. Thus, Vaartaja (1960) has found that short-day treatment of 3-month-old spruce (*Picea glauca*) seedlings induces dormancy and markedly increases survival of artificial drought, but he admits that it may be due to tolerance, to avoidance, or both.

Artificial treatments may also affect both development and drought resistance simultaneously. For instance, those doses of γ-rays (from ^{60}Co), which delayed the growth of peppermint plants (*Mentha piperita*), increased the drought resistance (Savin and Stepanenko, 1968). It is unknown, however, whether tolerance was involved. It appears more likely to be due to avoidance, since transpiration rate and water deficit were decreased. Attempts to improve drought hardiness by preventing growth with naphthalene acetic acid have not been successful (Maki *et al.*, 1946). Gibberellic acid, on the other hand, reduced the resistance of pea and fodder beans to low moisture levels (Fedorova, 1967), as would be expected since it is a growth promoter. Similarly, allyl alcohol has been found to increase the growth of Monterey pine seedlings and to decrease their drought resistance (Wilde *et al.*, 1957). The effect of mineral nutrients on drought resistance is undoubtedly complex. There may be an effect on avoidance, as shown by an increase in transpiration rate following nitrogen fertilization, which Biebl (1962a) ascribed to an improved root system, or there may be no effect (Bierhuizen *et al.*, 1959). These differences may perhaps be explained by Sands and Rutter's (1958b) observation that nitrogen deficiency increases leaf water deficit (and, therefore, decreases avoidance) at low soil moisture tensions, but has the reverse effect at high moisture tensions. Drought tolerance, however, may be lowered by excess nitrogen as in the case of freezing tolerance, due perhaps to the increased growth and the consequent exhaustion of soluble carbohydrates. According to Singh and Singh (1966), boron increased drought resistance of barley at tiller initiation and shooting stages. A rise in K:Ca ratio occurred in the press sap of *Vicia faba* as a result of dry culture (Simonis and Werk, 1958). In agreement with these results, drought hardening on dry soils resulted in higher K, a little more P, and less Ca than in plants grown on wet soil (Takaoki, 1966).

4. *Factors Related to Tolerance*

Those factors that have received the most attention in connection with freezing tolerance have also been investigated for possible roles in drought tolerance, although the investigations have been fewer in number. Unfortunately, many of the results are difficult to interpret, since drought tolerance was not measured. Sometimes field performance is recorded, but as already emphasized, such differences may be completely unaccompanied by differences in drought tolerance.

Morphological factors have been intensively investigated in relation to xerophily (Maximov, 1929), but have received relatively little attention in connection with drought resistance *per se*. Maximov (1929), however, has drawn attention to the extensive work of Kolkunov (1905–1915) which showed a pronounced inverse relationship between drought resistance and cell size when different varieties of crop plants were compared (wheat, beets, corn). Before attempting to interpret these results, Maximov cautions that the early maturation of small-celled plants enables them to escape the drought of late summer, and therefore the relationship may be coincidental. Kolkunov explains the small size as a means of conserving and improving the water supply of the plant. Maximov discusses the errors on which this explanation is based, and points out that a general ability to endure injurious conditions usually accompanies reduction in cell size. Many others have shown that plants have smaller cells when grown with a reduced water supply (Maximov, 1929). Iljin (1930a) was the first to determine the drought tolerance of the plants whose cells were measured. He concludes (1931) that, in general, drought-tolerant plants, such as xerophilic mosses and resting organs of higher plants, have cell volumes of 100–1000 μ^3 compared with an average of 10,000 (and as high as 1–2 million) μ^3 in mature cells of higher plants. The ratio of volume/surface is an expression of this factor. The smaller the ratio the greater the tolerance (Table 15.6) In xerophilic mosses it is 1 to 2 or even lower. In sensitive cells (killed by relative humidities of 99 to 97%) it is near 20. Cells of intermediate tolerance have ratios of 5 to 10. He points out that small elongated cells with a long vacuole and a visible protoplasm layer are also hardy. When his data (Table 15.6) are examined, however, an inverse relationship between cell size and osmotic value is evident. Since a relationship between osmotic value and drought tolerance has been repeatedly found (see below) and may occur independently of any difference in cell size, the inverse relationship between cell size and drought tolerance may be coincidental. In agreement with Iljin, however, Whiteside (1941) has shown that when wheat plants are grown with low moisture, their cells are smaller and they are more drought tolerant. This is also true of *Aspergillus niger* (Todd and Levitt, 1951). Kisselew (1935) found that small leaved forms of *Scorzonera* survive a greater loss of water than do the broad-leaved forms.

Iljin also considers cell structure important. Thus, elimination of the vacuole by contraction or thickening of the protoplasm (e.g., in *Fuligo*) or by filling with nondrying substances (e.g., in seeds and spores) accom-

TABLE 15.6 RELATIONSHIP BETWEEN AVERAGE CELL DIMENSIONS AND DROUGHT HARDINESS[a]

Species	Organ	Drought-killing relative humidity (%)	Osmotic value (moles sucrose)	Volume (μ^3)	Volume/ Surface
Begonia maculata	Stem	99	<0.2	1,690,000	21.4
Clerodendron fragrans	Stem	97	0.3	720,000	16.2
Pelargonium zonatum	Stem	93	0.4–0.5	885,000	17.2
Nerium oleander	Leaf	90	<0.5	3,600	2.3
Dianthus sp.	Leaf	87.5	0.7–0.8	21,660	4.4
Hedera helix	Leaf	85	0.85	18,100	4.2
Buxus sempervirens	Leaf	85	1.0	6,350	3.1
Mnium hornum	Thallus	0	0.7	3,350	2.7

[a] From Iljin, 1930a.

panies the development of extreme drought tolerance. Many other drought-tolerant cells, e.g., those of leafy mosses and of the meristems of higher plants, have very small vacuoles.

There are many reports of an increased osmotic value on exposure to drought (see Levitt, 1956), although the degree of response varies with both the species and organ (Table 15.7), and all the tissues do not even behave in the same way (Beck, 1929). There is a marked and rapid response to drought in the palisade cells, the conducting, and the spongy parenchyma; a less regular one in the guard cells. The epidermal cells respond very irregularly, and the responses may be completely masked by the effects of other factors such as temperature; low temperature increases and high temperature (38°C) decreases the osmotic value under conditions of drought (Beck, 1929). Besides the effect of drought on the osmotic value, there is also an inherent difference between plants that are adapted and those that are not adapted to drought even when growing under the same conditions (Tables 15.6 and 15.8).

Walter and Schall (1957) were unable to corroborate earlier evidence of a sharp rise in osmotic value during the first hours of wilting of excised leaves, followed by a sudden drop. Instead, they observed a steady rise in osmotic value that was solely due to the increase in water deficit. The above results by Iljin and by Beck and others, however, were obtained plasmolytically. They, therefore, cannot be due to a passive loss of water, but must indicate an increase in solute content. Some of the differences may, however, be exaggerated, since KNO_3 was used as the plasmolyte. It is now known that cells are somewhat permeable to KNO_3, and too high values are therefore obtained (Tadros, 1936). Since hardy cells are more permeable than unhardy cells (see Chapter 8), the differences might be

TABLE 15.7 RELATIONSHIP OF CELL SAP CONCENTRATION TO SOIL MOISTURE IN DIFFERENT PLANTS AND PLANT PARTS[a]

Soil moisture (%)	Osmotic value (moles sucrose)				
	Roots			Leaves	
	Maize	Wheat	Barley	Wheat	Barley
32	0.240	0.192	0.240	0.200	0.440
27	0.264	0.264	0.272	0.390	0.490
22	0.264	0.304	0.304	0.630	0.588
19	0.272	0.450	0.400	0.784	0.644

[a] From Iljin, 1929a.

TABLE 15.8 Osmotic Values in Leaves of Mesophytes and Xerophytes[a,b]

Mesophytes	
Erodium ciconium	0.3
Papaver strigosum	0.3
Hirschfeldia adpressa	0.3
Senecio vernalis	0.4
Succulents	
Sedum maximum	0.15
Sedum oppositifolium	0.15
Xerophytes	
Artemisia maritima	0.5
Gypsophila acutifolia	0.5
Kochia prostrata	0.6
Centaurea ovina	0.7
Parietaria judaica	0.8
Zygophyllum fabago	0.8
Dianthus fimbriatus	0.9

[a] Expressed as molarity of KNO_3 used as plasmolyte.
[b] From Maximov, 1929.

spurious. That differences nevertheless do exist is proved by those investigators who used nonpenetrating solutes (such as sucrose) for the plasmolyte (Beck, 1929; Iljin, 1929a, 1930a; Schmidt, 1939; Whiteside, 1941; Bartel, 1947). Bartel even found the osmotic increase during the droughting of four varieties to parallel their accepted differences in drought resistance, in agreement with Schmidt *et al.* (1940). In sugar beets, however, the osmotic value was lower in the more resistant variety (Schmidt *et al.*, 1940). Höfler *et al.* (1941) also found that drought-resistant species, in general, have high osmotic values, although there are exceptions.

Many of the above results must be accepted with caution, however, since most of the investigators falied to measure drought tolerance as distinct from field drought resistance. That osmotic value definitely is a factor, however, is evident from the astonishingly close correlation with Iljin's desiccation resistance (Tables 15.6 and 15.9). On the other hand, Höfler (1942b) later showed that younger parts of liverworts are more drought tolerant than older parts, though they frequently have lower osmotic values. It seems, therefore, that in drought tolerance, just as in freezing tolerance, cell sap concentration is a factor although not always the deciding one.

A reason for the correlation was already suggested by Rabe's (1905) observations that the drought tolerance of seedlings and sporelings decreases with the progress of germination and exhaustion of reserves. He

TABLE 15.9 Relationship of Osmotic Value of Plant Cells to Drought Tolerance[a,b]

Species	Osmotic value (moles)	Osmotic potential (atm)	Drought killing	
			Rel. hum. (%)	Osm. pot. (atm)
I. *Coleus hybridus*	<0.2	<−4.7	99	−14
II. *Malachium aquaticum*	0.26	−6.2	97	−40
Rumex acetosa	0.28	−6.7	97	−40
Bidens tripartitus	0.35	−8.2	96	−52
III. *Scrofularia nodosa*	0.40	−10.0	94	−79.5
Plantago media	0.45	−11.4	92	−110
Dorycnium germanicum	0.50	−12.8	92	−110
Iris pseudacorus	0.55	−14.2	92	−110
IV. *Ranunculus repens*	0.60	−15.8	92	−110
Clematis vitalba	0.60	−15.8	94	−79.5
Centaurea rhenana	0.60	−15.8	90	−142
V. *Aster trifolium*	0.65	−17.2	90	−142
Tetragonolobus siliquosus	0.75	−20.7	90	−142
Linaria genistifolia	0.80	−22.4	90–88	−142–173
Plantago maritima	1.00	−30.0	90	−142
Red beet	0.8	−22.4	87.5	−181
Hedera helix	0.85	−24.3	85	−220
Buxus sempervirens	1.0	−30.0	85	−220

[a] R.h. causing 50% killing.
[b] From Iljin, 1930a.

also found that the storage regions were the most tolerant parts of the seedlings. Wilting was soon shown to result in the disappearance of starch (Lundegårdh, 1914; Neger, 1915; Molisch, 1921; Horn, 1923; Ahrns, 1924; Iljin, 1927; Henrici, 1945), accompanied by a sugar increase (Table 15.10). Even after a 2-hr wilting there was a significant increase in sugars, and the greater the water loss, the greater the sugar increase (Ahrns, 1924). By 24 hr the starch had completely disappeared and sugars reached their maximum. A reabsorption of water by the wilted leaves resulted in a sugar decrease, though no starch formation could be detected. In mesophyll cells, high temperature may produce starch hydrolysis, but water loss is more effective (Kisselew, 1928). Lundegårdh (1914) made the interesting observation that osmotic dehydration has the same effect (Table 15.11). Leaves floated on 40% sugar solutions showed a starch → sugar conversion which was reversed when they were transferred to 10% sugar. By this method, Iljin (1930b) showed that the degree of dehydration needed to produce the starch→sugar conversion differs with the plant (Table 15.12). Many others reported an increase in sugar content on exposure to drought (Rosa, 1921; Iljin, 1927, 1929b; Vassiliev and Vassiliev, 1936; Clements, 1937a; Miller, 1939; Julander, 1945). Iljin (1929b) also showed that when plants are grouped ecologically, their sugar content increases with the dryness of the habitat. Although these differences were unaccompanied by measurements of drought tolerance, the conditions were the same as had previously been shown to increase drought tolerance. The sugar increase is easily understood, since the reduced moisture stops or at least decreases growth, without having as much effect on photosynthesis (see above). Consequently, carbohydrates may increase (Willard, 1922; Clements 1937a, b; Grandfield, 1943; Eaton and Ergle, 1948), but in most cases, there is a clear increase in sugars without any increase in other carbohydrates, or else the sugar increase exceeds the increase in other carbohydrates (Table 15.13). In the case of loblolly pine subjected to drought, there was a marked increase in reducing sugars, nonreducing sugars, and total carbohydrates, and an approximately equivalent decrease in starch (Hodges and Lorio, 1969). The sugar increases may not occur in the whole plant. In *Pennisetum*

TABLE 15.10 STARCH → SUGAR CONVERSION IN DETACHED LEAVES ALLOWED TO WILT FOR 24 HR IN THE DARK COMPARED WITH SIMILAR LEAVES KEPT MOIST FOR THE SAME TIME[a]

Species	Treatment	Water content	Starch (parts per thousand)	Total sugar (parts per thousand)
Tropaeolum majus	Control	85.92	62.87	101.1
	Moist	87.2	40.74	117.4
	Dry	79.8	3.28	149.3
Vitis vinifera	Control	71.03	47.76	70.33
	Moist	72.06	17.80	99.38
	Dry	69.75	7.54	110.7
Phaseolus vulgaris	Control	79.51	128.1	38.49
	Moist	81.7	30.24	130.6
	Dry	70.4	12.82	115.9
Helianthus annuus	Control	84.8	41.27	52.49
	Moist	86.0	16.20	54.85
	Dry	77.3		85.45
Pisum sativum	Control	82.6	40.65	124.9
	Moist	84.2	30.65	127.4
	Dry	77.7	13.09	170.4

[a] From Ahrns, 1924.

TABLE 15.11 EFFECT OF OSMOTIC DEHYDRATION ON STARCH CONTENT OF LEAVES OF *Homalia trichomanoides*[a]

Conc. of glucose in external solution (%)	Starch content after 24 hr	Plasmolysis
15	Much	None
20	Some	Insignificant
25	Traces	Slight
30	None	Considerable
35	None	Strong
40	None	Strong

[a] From Lundegårdh, 1914.

typhoides leaf sugar content increased at low soil moisture levels, but the sugar content of the stems decreased (Naidu *et al.*, 1967). Even a difference in drought resistance between two rice varieties was found to be correlated with sugar concentration (Murty and Srinivasulu, 1968).

The speed of wilting is important. In rapidly wilted disks of pelargonium leaves, the content of nonreducing sugars increases at the expense of the starch (Deutsch and Carlier, 1965). The increase does not occur when the wilting is slow, or the initial starch content is low. On rehydration, the nonreducing sugars decrease back to their original value within 12 hr.

Some have failed to find any effect of drought on carbohydrates (Magness *et al.*, 1932; Schneider and Childers, 1941), and wilting has even been reported to produce a decrease in sugars (Barinowa, 1937). This decrease, however, has been shown to occur only when wilting is severe enough to produce injury (Vassiliev and Vassiliev, 1936; Henrici, 1946) or when respiration is increased sufficiently to counterbalance the reduced rate of photosynthesis. Tadros (1936) found that though the freezing-point lowerings of Egyptian desert plants are large, this is mainly due to electrolytes,

TABLE 15.12 CONCENTRATION OF SUCROSE JUST SUFFICIENT TO PREVENT STARCH FORMATION IN LEAF PIECES FLOATED ON SOLUTIONS (CONTAINING 0.06 *M* DEXTROSE) FOR 36 HR IN DIFFUSE LIGHT[a]

Aquatic plants	0.46–0.96 *M*
Meadow plants	0.76–0.96 *M*
Forest plants	0.66–1.06 *M*
Sand plants	1.06–1.46 *M*
Rock plants	1.06–1.26 *M*

[a] From Iljin, 1930b.

TABLE 15.13 EFFECT OF 18 DAYS' DROUGHT ON TOTAL CARBOHYDRATE AND SUGAR CONTENT OF WHEAT PLANTS[a]

Variety	Condition	Water content (relative)	Total carbohydrates (mg)	Total sugars (mg)
Kitchener	Check	100	118.8	29.9
	Droughted	69	98.5	34.7
Gemtchoujina	Check	100	97.8	22.6
	Droughted	63	98.7	24.9
Caesium—0111	Check	100	97.5	17.8
	Droughted	62	108.0	29.0
Sarroubra	Check	100	90.3	18.3
	Droughted	65	107.9	51.4
Erithrospermum—341	Check	100	85.1	14.8
	Droughted	67	105.2	32.8

[a] From Vassiliev and Vassiliev, 1936.

but these plants grow under special conditions. The annual rainfall is less than 2 inches, and as a result most of the plant body is subterranean. Consequently, the relatively small amount of photosynthetic products formed per plant must be quickly used up by the large root system, and no accumulation of sugars can occur. Furthermore, in many cases the plants were growing on somewhat saline soils. In the case of loblolly pine, moisture stress resulted in a sugar increase, but it was primarily due to the decrease in growth rather than to starch hydrolysis (Hodges and Lorio, 1969).

There are, obviously, exceptions to the commonly found correlation between water stress and sugar accumulation. Whether or not the correlation is found, drought tolerance is not normally measured. A role of sugars in drought tolerance has not, therefore, been proved. Nevertheless, attempts have been made to explain this role. Fitting gives two explanations of the role of high cell sap concentration in drought resistance: (1) an increased water-absorbing power, and (2) a reduced water loss. The erroneous bases of these explanations are fully discussed by Maximov (1929), although he does not completely reject them. Such roles would, of course, have no bearing on drought tolerance, with which osmotic value seems definitely correlated. Maximov (1929) suggests two other explanations: (1) the accumulation of substances might protect the protoplasm from coagulation and desiccation, and (2) the high concentration might prevent visible wilting for a long time, in spite of an increasing water deficit. Thus, xerophytes wilt only after loss of 30 to 40% of their water content, while delicate shade plants (e.g., some of the Balsaminaceae) after loss of 1 to 2%. The latter difference must depend on differences in physical properties of the cell wall.

Increased bound water has been correlated with drought tolerance (Rosa, 1921; Tumanov, 1927; Newton and Martin, 1930; Calvert, 1935; Barinowa, 1937; Migahid, 1938; Grandfield, 1943); but Whitman (1941) failed to obtain any consistent differences. The same objections may be raised as against the correlations between bound water and freezing tolerance (Chapter 8). A good example is given by Migahid (1938). He compared the bound water in two mesophytes (*Vinca rosea* and *Withania somnifera*) with that in four xerophytes (*Zygophyllum coccinium*, *Peganum harmala*, *Pulicaria crispa*, and *Stachys aegyptiaca*). His method was to measure the water lost from the air-dry shoots (intact or ground) on oven drying. According to his data, the water remaining may be four times as much in the xerophytes as in the mesophytes. Unfortunately, aside from his neglect to measure drought tolerance, his results are due to the same error in calculation as was made in investigations of freezing tolerance (Chapter 8); the bound water is expressed as a percentage of the original water. When, however, the bound water per gram dry matter is calculated from his figures, it is found to be 0.15–0.17 gm for the mesophytes and 0.12–0.27 gm for the four xerophytes. His large differences are obviously due to the much lower original water contents of the xerophytes—as low as 51.9% compared with 87–87.5% in the mesophytes.

Rao *et al.* (1949) show a relationship between drought resistance and the amount of water taken up by the dried leaf tissue from an atmosphere of 0.5 r. h. No attempt, however, was made to measure drought tolerance and the relationship with rate of growth was just as good. As in the case of freezing tolerance, the only evidence that appears to apply to the protoplasm rather than the nonprotoplasmic contents is the previously described increase in the water-holding power of the dry matter of *Aspergillus* mycelium with the concentration of the medium in which the organism is grown (Chapter 8). Northen (1943) arrived at the same conclusion from indirect evidence. When species of *Mnium* and *Bryum* were dried for periods up to 50 min over anhydrous $CaSO_4$, thus producing "incipient drought," there was a decrease in structural viscosity which he ascribed to protein dissociation. He concluded that this assumed protein dissociation induced an increased protoplasmic swelling pressure. Füchtbauer (1957b) has failed to detect any increase in bound water of yeast cells when their drought tolerance increased due to addition of sorbitol to the medium, but his method of measuring bound water is open to objection (Levitt, 1958). The bound water of the cytoplasmic proteins from wheat leaves has recently been calculated from measurements of the dielectric constant (Sedykh and Khakhlova, 1967). The values obtained were found to increase upon subjection of the plants to drought.

Protoplasmic factors have not received much attention in recent years. Levitt and Scarth (1936) found an increase in permeability to urea of more than 100% in cells of *Spartium* plants unwatered for 14 days. Whiteside (1941) observed an increase to three times the original permeability in two wheat varieties droughted for 12 days. During this time he showed that the drought tolerance of the cells increased from a threshold humidity corresponding to a water potential of −45 to −60 atm to one of about −100 atm.

The Larmstadt group (Schmidt, 1939; Schmidt *et al.*, 1940) obtained permeability changes that are in agreement with the above. They never measured tolerance, however, and those measurements that they did make established differences in avoidance. Their results are, therefore, not comparable to the above (see Levitt, 1958). They also concluded that the water stress leads to an increase in protoplasmic viscosity. Increases in both protoplasmic viscosity and osmotic concentration were also obtained by Simonis (1952) in the case of droughted plants. Five varieties of summer wheat showed the same relationship between tolerance and protoplasmic viscosity save that the most tender variety had the highest viscosity. However, these varieties were not drought-hardened previous to testing, and their tolerance was not determined. They were simply graded according to the field experience of the grower, and may have differed only in avoidance. Similar results have been reported by Beysel (1957), and by Chuikov and Skazkin (1968).

Some investigators have attempted to relate enzyme properties to drought tolerance. The activity of glutamate decarboxylase from wheat embryo increased for 54 hr during germination, and the stability decreased (Nations, 1967). The responses to desiccation by active protein preparations following defatting with *n*-butanol were measured. These responses suggested that the stability of the enzyme may reside in a capacity for dissociation into smaller protein components under conditions of water stress, and that this capacity may depend on the presence of lipids. In agreement with this conclusion, wheat plants subjected to water stress showed a decrease in the larger mol. wt. fraction of the proteins and an increase in the smaller mol. wt. fraction (Stutte and Todd, 1967). However, the more drought-resistant variety maintained a higher percentage of the larger mol. wt. proteins, indicating perhaps that the increase in the smaller mol. wt. fraction was a result of injury rather than a tolerance factor. Phosphorylase and phosphomonoesterase were more active in cowpea

(*Vigna sinensis*) plants raised in dry soils than in plants raised in wet soils (Takaoki, 1968). The activities of α-amylase, β-amylase, and catalase were highest in plants raised at optimum soil moisture, and decreased on both sides of this level. Peroxidase activity showed no definite relationship to soil moisture. Lowering the osmotic potential of the nutrient medium increased the phosphatase activity in *Gossypium thurberi* (Vieira-da-Silva, 1968b). Severe wilting of wheat leaves (by withholding water for 4–8 days) increased the number and quantities of iron-containing proteins, as determined by acrylamide gel electrophoresis (Stutte and Todd, 1969), but decreased the number of lactic dehydrogenase bands. Certain peroxidase bands disappeared and new ones appeared. Total peroxidase activity per unit of protein increased, while total soluble protein decreased drastically with increasing water stress. Since the wilted leaves were able to regain turgor, these changes are presumably related to tolerance rather than injury.

The above results obviously fail to agree as to any general relationship between drought tolerance and enzyme activity. It does not, of course, follow that those factors that have been correlated with drought tolerance are necessarily causally related. In most cases, tolerance was not measured directly. Therefore, some of the factors may conceivably be related to avoidance rather than tolerance. The leaves of droughted soybean plants, for instance, increased in osmotic value to a higher value than in the control, undroughted plants (Kaku, 1963). Since the values were determined plasmolytically, this must have been due to an increase in solutes. The greater the number of times the plants were wilted (up to 4–8 times), the greater was the increase in osmotic value. Yet soybeans fail to increase in drought tolerance when droughted (Clark and Levitt, 1956).

D. Mechanisms of Drought Tolerance

No serious attempt has been made to explain the mechanism of drought tolerance in the case of indirect injury. How, for instance, does a drought-tolerant plant exposed to a water stress, manage to lower its hydration compensation point or to increase the dehydration avoidance of its guard cells so as to prevent starvation injury? Some attempts have been made (see above) to explain the mechanism by means of which a net loss of protein on exposure to water stress may be prevented in drought-tolerant plants, but these attempts only go so far as to relate the process to the nucleic acids, which obviously must be involved if it is a matter of protein synthesis. No explanation is attempted for the mechanism of protein loss or for prevention of this loss in the tolerant plant. It is conceivable, however, than when a mechanism is discovered, it may reveal a closer relationship between the direct and the indirect drought effects than at first suspected. This possibility will be reexamined after considering the hypothetical mechanisms of direct drought injury and tolerance.

1. Iljin's Mechanical Stress Theory

As in the case of freeze-induced dehydration injury, Iljin has explained drought injury on the basis of mechanical stress. Early implications of mechanical injury can be found in Schröder's (1886) statement that resistance must depend on specific properties of the protoplasm. He considered the presence of oil and other reserve substances useful because they prevented too strong cell collapse. Iljin (1927, 1930a, 1931, 1935a) adopts a completely mechanical explanation of drought injury. He states that it is not the water loss itself that kills, but the mechanical injuries that accompany drying and remoistening. When a plant part dries, the cells collapse. If the cell wall is sufficiently rigid it opposes the collapse and thereby subjects the protoplasm to a strong tension that may lead to destruction. If the wall is thin and soft, it is pulled in together with the protoplast and forms folds and wrinkles, Before regaining their normal size and shape, cells that survive drying are subjected to new mechanical stresses on remoistening, which may lead to death. Iljin's mechanical theory of drought injury is supported by the following facts:

(1) Cell contraction does occur on desiccation (Fig. 15.4), and the wall is pulled in with the collapsing protoplast. The cell diameter may be reduced to $\frac{1}{5}$ or $\frac{1}{10}$ of normal (Iljin, 1931).

(2) Tensions have been measured and were shown to reach high values. Chu (1936) measured the "suction force" of conifer leaves. The threshold

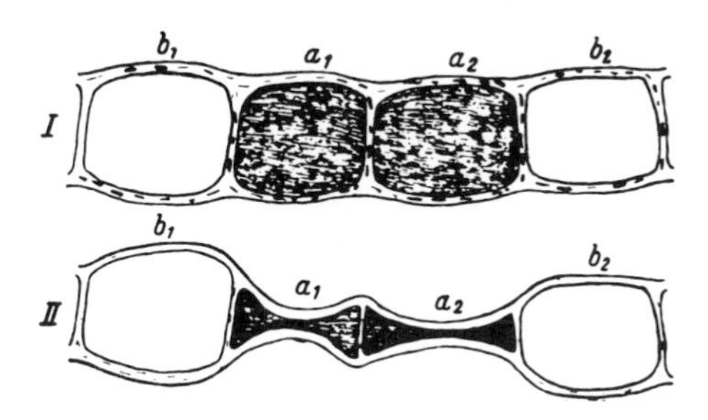

FIG. 15.4. Cell collapse in desiccated cells of *Mnium undulatum* (II) compared with turgid state (I). (From Iljin, 1930a.)

for death occurred at values of 50 to 75 atm in summer and 300 atm in winter. Since "suction force of the cell contents" (which is numerically equal to their osmotic potentials) never exceeded 50 atm, negative pressures (or tensions) must have been responsible for the high values. The largest of Blum's (1937) values (17.9 to 73.4 atm) in dry regions must also have been due to negative wall pressures. Holle (1915) had previously recorded a value of 20 atm for leaves of *Catherinea undulata;* in the epidermis of *Rochea falcata,* the tensions during pronounced wilting split and lifted the cuticle.

(3) Pieces of torn protoplasm may actually be seen attached to the cell wall and a single protoplast may be torn into two protoplasts during the pseudoplasmolysis following remoistening (see Fig. 13.3).

(4) Plasmolyzing solutions release the walls and prevent tensions. They also prevent drought injury in cells that are normally killed instantaneously by the same drought (see Chapter 13).

(5) Factors that are associated with drought tolerance can be readily explained by this theory. According to Iljin (1930a), the amount of tension depends on:

a. The quantity of water lost. The higher the cell sap concentration, the less the degree of cell collapse, but according to him (1931), this factor operates only at relative humidities above 85%, since the vacuole reaches its minimum volume at this point.

b. The extent of interface between protoplasm and wall, which is controlled by cell size and structure. Iljin (1931) concludes that the small cell size or large specific surface of hardy plants is one of the most important methods of avoiding the tear during desiccation and remoistening, since the greater the specific surface of any one pressure, the less the tension per cell. For the same reason, the smaller the vacuole, the less the tension on its surface and the greater the drought tolerance. Small cell size is the prevailing factor in cases of extreme hardiness (e.g., mosses), and cell sap concentration plays no significant part since all the free water is given up when they become air dry.

c. Accumulation of carbohydrates and other substances. If these are soluble, the explanation is already given in (a). Insoluble carbohydrates or other substances would also occupy space and reduce cell shrinkage. Iljin (1931) describes an exceptional example of this. The vegetative cells of the fern *Notochlaena marantae* can lose 94% of their water content without injury. Though powder dry, they can regain their normal state within 24 hr of remoistening. He ascribes their extreme hardiness to the fact that the cells retain nearly the same volume when dry as when fully swollen (Fig. 15.5). The protoplasm is therefore not subject to severe stresses.

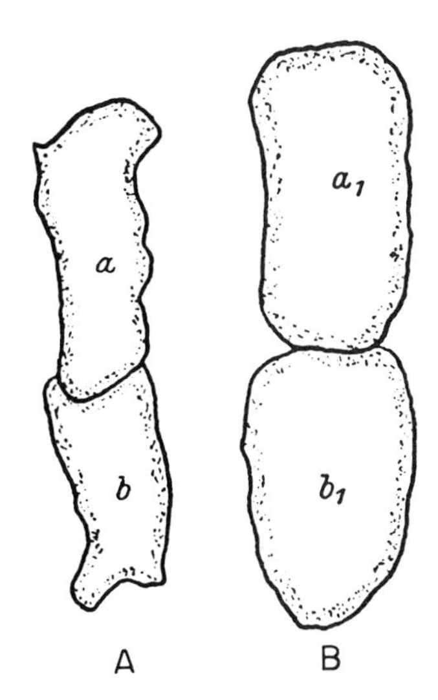

FIG. 15.5. Relatively small change in cell volume of *Notochlaena marantae* when severely desiccated (A) as compared with the turgid state (B). (From Iljin, 1931.)

Another epiphytic fern, *Polypodium polypodioides* has recently been discovered with similar properties. Its cells have vacuoles that solidify when dehydrated (Stuart, 1968). This could conceivably protect the cells against injury by Iljin's mechanical stress, and would, therefore, explain its high degree of drought tolerance (loss of 97% of its water content without injury). Even the extreme drought tolerance of seeds can be explained in this way. Their cells do not undergo marked cell contraction because solid reserve material gradually replaces the water, the two processes occurring simultaneously. The net result is a maintenance of cell volume essentially unchanged.

Iljin's experimental results have not been corroborated by Oppenheimer and Jacoby (see Chapter 13). Nevertheless, Iljin's theory seems capable of explaining all the known facts about drought injury and hardiness. Consequently, most investigators have accepted it as a working hypothesis. Iljin later (1935) concluded that some protoplasm has a greater ability to withstand desiccation than others, even when the stresses are the same. Earlier (1930a), he pointed out that plasmolyzed moss protoplasts show free, ameboid movement inside the cell wall. He suggests that this may explain their ability to tolerate extreme desiccation. Higher plant protoplasts are joined on all sides by plasmodesmata, which hold them fast to the cell wall and lead to a tear on plasmolysis.

As in the case of freezing tolerance, drought tolerance on the basis of

Iljin's concept must be twofold in nature, consisting of (1) avoidance of cell contraction, due largely to a high vacuolar sugar concentration, and (2) tolerance of protoplasmic dehydration. Höfler (1942a) attempts to eliminate the first of these, and to determine whether there can be a "protoplasmic drought tolerance" as distinct from "cell drought tolerance." In order to avoid the mechanical injuries described by Iljin, he chose the leafy Jungermanniales. The cell walls of these liverworts are elastically extensible; instead of being collapsed in folds when the tissues wilt, they closely follow the contracting protoplasts (Laue, 1938). Höfler, therefore, believed that in these cells there is no danger of the protoplasm tearing loose. The wide differences that he found between different species (Table 15.14) led him to conclude that differences in protoplasmic drought tolerance do exist.

There is little reason to doubt that the absence of plasmodesmata and of folds in the dried cells of liverworts reduces the stresses below those in desiccated cells of higher plants, but it is highly doubtful that stresses can be completely absent. Höfler (1942a) admitted that at least some of these

TABLE 15.14 Drought Tolerance of Different Species of Liverworts Determined by Survival in Chambers of Different Relative Humidities[a]

Species	Minimum desiccation causing complete killing: H_2SO_4 (%)	Minimum desiccation causing complete killing: Relative humidity (%)
Lophozia incisa	10	95.0
Aplozia riparia	15	91.3
Calypogeia neesiana	20	87.0
Chiloscyphus rivularis	25	81.9
Chiloscyphus palescens	30	75.0
	35	66.2
Aplozia caespiticia	40	56.5
Metzgeria pubescens, Lophozia quinque-dentata, Scapania nemerosa	50	36.4
Plasiochila asplenioides-maior, Ptilidium pulcherrimum	60	15.4
	80	1.0
Frullania dilatata	100	0.0
Modotheca platyphylla		
Cololeugenia calcarea	Not killed even at 0% relative humidity	

[a] From Höfler, 1942a, b.

liverworts must be injured by mechanical stresses, since speed of drying and remoistening had such a pronounced effect on their survival. In a later paper (1950) he is careful to point out that since 1943 he has been remoistening the cells gradually by allowing them to take up water from moist air, in order to avoid danger of mechanical injury. In those cases where the speed of drying and remoistening had little effect, he concluded that the constancy of the critical relative humidities for each species indicated that protoplasm tolerance was involved. Other experiments (1945) showed, however, that these values are not so constant, but vary with the hardening conditions. Höfler's "hardening" consisted of artificial predrying at relative humidities of 87%. This is equivalent to attempting to harden against freezing by exposure to about −15°C. Since nearly all the water is removed at this relative humidity, it seems obvious that no hardening process can take place. It is more plausible to suggest reduced injury due to less sudden and intense mechanical stresses, as he explains in the case of his other experiments on speed of drying and remoistening. It would, in fact, seem impossible to remove completely this kind of injury unless naked protoplasm such as the plasmodia of myxomycetes, is used. There are, however, many soft-walled cells (e.g., many animal cells) that are just as readily killed by desiccation as are the hard-walled cells. It does not seem possible to explain drought injury in such cells by Iljin's theory, nor does Iljin's theory explain how the mechanical stress injures firm-walled cells at normal rates of desiccation and rehydration, which cannot lead to the rupture observed in the laboratory under conditions of extremely rapid rehydration.

As in the case of freezing injury, Iljin's theory of drought injury must be either expanded or replaced, in order to obtain a modern, molecular concept of drought injury and tolerance.

2. Protein Aggregation or SH Hypothesis

Since drought injury is due to cell dehydration, it is subject to explanation by the same molecular hypothesis as injury due to freeze-induced dehydration. Whether or not it is initiated by Iljin's mechanical stress, it may be explained by protein aggregation due to intermolecular bonding of the dehydrated and therefore more closely packed protein molecules. It is more difficult to evaluate this concept in the case of drought injury because of the much smaller body of available evidence. Kaloyereas (1958) obtained a correlation between SH content and drought resistance. Subbotina (1959) observed a decreased reducing power in wilting leaves, as measured by E_h. An observed increase in the ratio of oxidized: reduced ascorbic acid (Stocker, 1961) may also have some bearing on this mechanism.

Unlike freezing injury, however, drought injury is not preceded by a reversible denaturation of the proteins, since the cell is not necessarily exposed to either a low or a high temperature. It is, therefore, more probable that the mechanical stress itself subjects the protein to a strain, leading to an incipient unfolding or denaturation. At the same time, the dehydration brings the molecules close enough together to permit the formation of intermolecular bonds between the sensitive groups which have become exposed ("unmasked") by the incipient denaturation. Again, as in the case of freezing injury, it is the plasma membrane proteins which are most likely to undergo these changes, primarily because of their attachment to the cell wall and, therefore, their direct subjection to the mechanical stress. Direct evidence has, in fact, indicated that intermolecular SS bonding does arise at the dehydration that causes injury (Table 15.15), and the bonds are to be found in the insoluble "structural" proteins which presumably include the membrane proteins (Gaff, 1966). The partial unfolding due to the mechanical stress, would also unmask some of the hydropho-

TABLE 15.15 Disulfide Content of Desiccated and Undesiccated (Control) Half Leaves of Cabbage[a,b]

Weight of desiccated leaf[c] (as % of initial fresh wt.)	Water potential[d] (atm)	Disulfide content (as % of total titratable sulfur): "Structural protein" Desiccated	"Structural protein" Control	Soluble protein fraction Desiccated	Soluble protein fraction Control	Soluble nonprotein fraction Desiccated	Soluble nonprotein fraction Control
88	−13	26.7	24.1	42.1	36.5	83.1	86.0
88	—	Wilting commences					
75	−20	22.6	23.5	49.2	45.3	77.6	78.4
50	−41	27.5	30.9	50.5	37.1	91.5	79.9
42	−76	20.9	17.8	52.9	41.6	85.7	84.0
—	−86	50% of cells cease accumulation of neutral red					
28	−94	49.3	23.3	51.4	36.7	87.0	83.5
28	—	Approx. 65% of tissue fails to recover turgor in water					

[a] Each value is the mean of the data of three experiments.
[b] From Gaff, 1966.
[c] ±2%
[d] ±5 atm.

bic groups formerly within the native molecule. These would now be available for intermolecular bonding between the dehydrated and, therefore, closely adjacent molecules. Consequently, protein aggregation can also be induced by hydrophobic bonding, as occurs in the freeze-dehydrated, denatured BSA molecules on warming to room temperature (Goodin and Levitt, 1970). In one preliminary investigation it has been possible to protect cabbage cells against direct drought injury by low concentrations (10^{-2} *M*) of mercaptoethanol (Paricha and Levitt, 1967). Independent observations by other investigators support the protein aggregation concept. A constituent of the stroma of *Phaseolus* plastids aggregates when leaves are subjected to dehydration by high-speed centrifugation, by plasmolysis, or by wilting (Gunning *et al.*, 1968). It was suggested that this may be due to aggregation of Fraction I protein. A possible similar result is the change in conformation of cytoplasmic proteins induced in leaves of spring wheat by drought (Gusev *et al.*, 1969). In two out of three isolated protein fractions, these changes are described as an increased asymmetry, looseness, and flexibility of the molecules.

Results obtained by Santarius and Heber (1967) apparently oppose the above concept. They observed a doubling in SH content of free chloroplasts as a result of removal of 98–99% of their water over $CaCl_2$. They interpreted this as an unmasking due to protein denaturation. At the same time both the Hill reaction and phosphorylating systems were inactivated. Sugars prevented both the denaturation and the inactivation. However, the SH increase was prevented by sugar concentrations too small to protect the two enzyme systems from inactivation. Although SS groups were not measured, on the basis of the increase in SH groups, they concluded that none were formed. Their results do, however, support the above hypothesis to the extent that unfolding apparently occurred as a result of the drought stress. Since it requires only one SS group to bond two protein molecules, this may be enough to produce the proposed irreversible increase in cell permeability (Chapter 10). Therefore, if the unfolding unmasked several SH groups and only one formed an SS bond, a large increase in protein SH would still result, as found by Santarius and Heber. They did not investigate the possibility of hydrophobic bonding. Unfortunately, however, there is no evidence that changes during the desiccation of free chloroplasts (which are already injured due to the preparative procedure), are in any way related to changes during the desiccation of living cells.

Other results point to a possible relationship between SH and SS groups of proteins and drought tolerance. According to Khokhlova *et al.* (1969), the soluble amino acid-activating enzymes of wheat leaves showed higher activities when obtained from plants subjected to partial dehydration. There was a positive correlation between the enzyme activity and their SH and SS content. It was suggested that this biochemical response increased the drought tolerance of the cells.

As in the case of freezing tolerance, attempts have been made to explain drought tolerance, whether achieved by hardening or by use of protective substances, on the basis of bound water. Santarius and Heber (1967), and Santarius (1969) found that although the above phosphorylation and electron transport systems of spinach chloroplast fragments were inactivated by desiccation, various soluble enzymes were not affected. They were able to prevent enzyme inactivation during desiccation of free chloroplasts by use of sugar solutions. As they point out, glucose and disaccharides are unable to form H bridges with proteins, whereas the smaller molecules, such as pentoses, can and do form sugar–protein complexes. They concluded that the mere retention of water by the sugar molecules must be sufficient to protect the chloroplast structure from complete dehydration in their experiments.

Santarius (1969) found that the inactivation of photophosphorylation and electron transport due to desiccation was identical to that due to high concentration of electrolyte. He, therefore, suggested that the desiccation produces inactivation by disorientation of the structural water due to the high concentration of the electrolytes at the membrane surface. It is difficult to understand, however, why this effect should be confined to the phosphorylating and electron transporting enzymes, and should not affect the soluble enzymes. Proteins, in general, should be all affected by such a process, and since the inactivation is known to be associated with destruction of the membrane, there would be no barrier between the electrolytes and these soluble enzymes.

In order to understand their results, the method used to desiccate the chloroplast fragments must be considered. Desiccation and rehydration were both extraordinarily drastic. The fragments were dried at +2°C *in vacuo* over $CaCl_2$ for 3 hr, resulting in removal of 98–99% of the water. They were then rehydrated (instantaneously!) in distilled water. The successive high-speed dehydration and instantaneous rehydration would produce osmotic rupture of the membrane in a manner not possible during the normal dehydration and rehydration of plants droughted in nature. It has been shown that rupture of these thylakoid membranes does occur and that the above enzyme systems are inactivated in such ruptured thylakoids (Uribe and Jagendorf, 1968). Nonpenetrating solutes (e.g., sucrose) would prevent the rupture and inactivation osmotically, while penetrating solutes (e.g., salts) could lead to rupture. The effectiveness of a salt in producing rupture, and therefore inactivation, would depend on how rapidly it penetrates the thylakoids as its concentration increases during the 3-hr desiccation, and whether or not it could diffuse out of the thylakoids rapidly enough to prevent osmotic rupture during rehydration.

It is, unfortunately, unknown at this time, whether or not there is any relationship between the effects of dehydration on free chloroplast fragments and on the plant as a whole. The fact that sugars, proteins, or polypeptides in such small quantities (40/μmoles) protect against the injurious effects of dehydration in the case of the free chloroplast fragments would seem to eliminate the possibility of such injurious effects in the living cell, which always contains quantities of these substances. Similarly, if SS formation really does not occur in the free chloroplast fragments, this may be due to the absence of enzyme systems normally present in the living cell, which control the oxidation of SH $\rightarrow$ SS.

The present state of uncertainty as to the mechanism of drought injury makes it doubly difficult to postulate a mechanism of tolerance and an explanation of how this is achieved during the hardening process. It is, of course, logical to adopt the mechanism suggested for the freezing stress, since both involve dehydration. It is interesting to note that, just as inhibitors have been found to play a role in the normal hardening process during the fall (see Chapter 8), there is now evidence of the same factor in the case of drought hardening. When detached leaves of cotton, pea, or bean plants were allowed to wilt in a stream of air until they lost 9% of their fresh weight, and were maintained in this condition for 4 hr at 22°C in the dark, abscisic acid was produced (Wright, 1969; Wright and Hiron, 1969). It was, therefore, suggested that this inhibitor might play a role in the physiological changes induced by the water stress.

It is even possible to explain tolerance of the indirect effects of the water stress on the basis of the protein aggregation concept. If tolerance involves an increase in reduction potential, this would (1) require a higher rate of photosynthesis to produce this high reduction potential, and therefore would carry with it a prevention of starvation injury, and (2) possibly lead to a reduction of the SS groups in RNase, thus inactivating the enzyme and preventing the protein hydrolysis that leads to indirect drought injury.

E. Relative Importance of Avoidance and Tolerance

Since either avoidance or tolerance may be responsible for a plant's drought resistance, it is essential to know whether they are equally important, or whether one is far more important than the other. Obviously, the answer must depend on the specific plant. In the case of lower plants (such as algae, lichens, mosses), since they are poikilohydric, and therefore, have not developed efficient mechanisms for conserving water or for absorbing it as rapidly as it is lost to the environment, water savers and water spenders are essentially nonexistent. Drought resistance in lower plants is, therefore, to all intents and purposes solely due to tolerance. Even those that are never exposed to arid climates may possess a far greater tolerance than the most tolerant higher plants. For instance, the hanging mosses of the humid "mossy forest" and those of the constantly moist soil of the virgin forest survive drying to relative humidities of 10–40% (Biebl, 1964b).

In the case of higher plants, the situation is more complicated. They may be (1) intolerant avoiders, (2) tolerant avoiders, or (3) tolerant non-avoiders. Group (1) consists of two main types: (a) The water savers such as the succulents owe their resistance mainly or solely to avoidance. There are also less extreme examples with low transpiration rates and low saturation deficits such as *Ruscus*, *Iris*, and *Smilax* (Rouschal, 1939). Many cases of superior drought resistance belong in this group [see Table 15.16(a)]. (b) Water spenders with low saturation deficits also belong to this group. They maintain a high water supply due to a well-developed root system, e.g., *Pistacia lentiscus*, *P. terebinthus*, and *Rubus* (Rouschal, 1939). The most extreme examples of this type are the phraeatophytes. There are also cases of superior drought resistance that belong in this group [see Table 15.16(b)]. As indicated previously (Chapter 14), there are plants that combine some of the characteristics of both types of group (1).

Group (2) consists of two subtypes: (a) The drought-tolerant water savers such as the sclerophyllous maquis plants with low transpiration rates under almost all conditions, yet with the ability to survive water saturation deficits of 70% and more (Oppenheimer, 1951; Grieve, 1953). (b) The drought-tolerant water spenders that do not restrict transpiration except under extreme water deficiency.

The importance of a combination of factors in the drought resistance of the plant is illustrated by the moderately drought-resistant Monterey pine (*Pinus radiata*). According to Oppenheimer (1968), the needles close their stomata quickly when exposed to drought, but this adaptation is insufficient to prevent early death because (1) cutinization is much less effective than in the xeromorphic pines (*P. halepensis* and *P. pinea*), and (2) drought tolerance is low, death occurring when only ¼ to ⅓ of the saturation water is lost. In agreement with this evidence of the importance of tolerance in this group, thirteen species of the Mediterranean Midi showed differences

TABLE 15.16 Examples of Species or Varietal Differences in Resistance due to Avoidance Rather than to Tolerance

Species in order of decreasing resistance	Avoidance factor correlated with resistance	Reference
(a) Avoidance due to water conservation[a]		
1. Ponderosa pine, Douglas fir, Western arbovitae	Decreased rate of water loss	Parker, 1951
2. Grain sorghum, cotton, peanut	Lower transpiration rate	Slatyer, 1955
3. Two oat varieties	More rapid stomatal closure	Stocker, 1956
4. *Pinus densiflora* versus *Chamaecyparis obtusa* and *Cryptomeria japonica*	Cuticular transpiration inversely related to resistance	Satoo, 1956
5. *Quercus ilex* versus *Q. pubescens*	Ten times as great a drop in transpiration on stomatal closure	Larcher, 1960
6. Sugarcane varieties	More rapid stomatal closure	Naidu and Bhagyalakshmi, 1967
7. *Pinus halepensis* versus *P. pirea*	Lower cuticular transpiration	Oppenheimer and Shomer-Ilan, 1963
8. Pine versus mulberry seedlings	Lower cuticular transpiration	Tazaki, 1960c
9. Six clones of blue panic grass	Fewer stomata	Dobrenz *et al.*, 1969
10. Six wheat cultivars	Higher percentage water retention on desiccation at 81% r.h.	Salim *et al.*, 1969
(b) Avoidance due to greater water absorption		
1. Short-leaf versus loblolly pine	Higher transpiration rate	Schopmeyer, 1939
2. Grain sorghum, cotton, peanut	Greater water uptake	Slatyer, 1955
3. *Pinus densiflora* versus *Chamaecyparis obtusa* and *Cryptomeria japonica*	Greater root development	Satoo, 1956
4. Varieties of corn	Larger root system	Misra, 1956
5. Pine species	More water removed from soil	Oppenheimer, 1967
6. Two-eared versus single-eared corn	Extracted 1–2% more water from soil	Barnes and Woolley, 1969
7. Spring wheats	Greater root development	Devera *et al.*, 1969
8. *Eucalyptus socialis* versus *E. incrassata*	Higher root-shoot ratio	Parsons, 1969

[a] Measurements of tolerance when made revealed no difference in Nos. 1, 6, and 10; differences in tolerance found in No. 4 correlated with drought resistance.

in drought tolerance during the dry season, as measured by Iljin's method (Mouravieff, 1967). In the case of plants belonging to group (2), the hardening process may lead to a combination of "pseudo-hardening" (an increase in avoidance) and true hardening (an increase in tolerance). Barley and wheat, for instance, become more drought tolerant on exposure to a moderate moisture stress (Table 15.1). They also show a 50% reduction in transpiration (Wassink and Kuiper, 1959). In agreement with these results, Salim *et al.* (1969) found that survival was greater in barley and wheat than in oats after the tissue had reached a given water level, indicating better tolerance in the former two. On the other hand, among five wheat and one rye cultivars, resaturation values were strongly correlated with relative water content after drying, indicating that avoidance is more important than tolerance among these plants. Similarly, two wheat and one barley cultivars exhibited pseudohardening by increasing their water retention on exposure to drought. Resistant Bajra varieties also appeared to possess both avoidance and tolerance (Bhargava, 1967).

The xerophytic *Quercus ilex* is apparently more drought resistant than *Q. pubescens* because of both avoidance and tolerance of both the direct and indirect effects of water stress (Larcher, 1960). It is able to close its stomata at a smaller water saturation deficit (17 versus 20%) and it has a higher sublethal saturation deficit (71 versus 50%). It also shows a higher ratio of assimilation to transpiration between 10–16% water saturation deficit.

Finally, since water spenders may become water savers when the water stress becomes more severe, it is possible for a plant with superior drought resistance to depend on all three i.e., it is then a tolerant water saver and spender. As an example, *Pinus densiflora* is more drought resistant than *Chamaecyparis obtusa* and *Cryptomeria japonica* because of (1) greater drought tolerance of both tops and roots, (2) greater root development, and (3) lowest transpiration rate when available soil is exhausted (Satoo, 1956).

Among the xerophytes, those that belong to group (3) possess no greater avoidance than ordinary mesophytes. The relative number of species in group (3) (see Biebl, 1962a) is, however, undoubtedly far less than the number in the first two groups (see Oppenheimer, 1959). Some members of the fern family belong to this group. Platycerium (a heliophilic-epiphyte) is definitely poikilohydric, and its prothallus survives the loss of 99% of its water (Boyer, 1964). The fronds are slightly less resistant, surviving the loss of 80–95% of their water content. Many of the ferns that can become air dry without loss of life have been called resurrection plants. Few flowering plants can be placed in this group. Even such an extremely tolerant species as the creosote bush is not without avoidance, since it reduces its evaporative surface by dropping its mature leaves, and has a very large root system. Oppenheimer (1959) lists several resurrection plants among spermatophytes: *Carex physoides, Ramonda nathaliae, Myrothamnus flabellifolia,* and *Chamaegigas intrepidus.* However, at least in the case of the first of these, in the author's experience, it is only the tiny buds that remain alive and they are, threrfore, essentially acting as dormant, vegetative, reproductive structures analogous to seeds. Some epiphytes (e.g., *Tillandsia* sp.) are poikilohydric according to Biebl (1964b), and are able to survive natural deficits of as much as 68–70%. However, even these plants retain relatively large reserves of water even after several rainless days. Nevertheless, they do demonstrate that tolerance may be an important factor in the drought resistance of higher plants. It may, in fact, be the deciding factor. *Acacia aneura,* for instance, recovered from a soil water potential of −130 bar without any injury, while −45 and −90 bar were fatal to tomato and privet plants, respectively (Slatyer, 1960). However, it is dangerous to judge tolerance from field experience. Although *Acacia aneura* was previously considered to represent an extreme of drought tolerance among Australian arid zone plants, it proved to be less tolerant of desiccation in the laboratory than bigelow (*Acacia hapophylla;* Connor and Turnstall, 1968).

Among the xerophytic higher plants, therefore, group (3) is the most poorly represented. Among economically important plants, tolerance also seldom seems to be the deciding factor in their drought resistance. According to Satoo (1956) the most important characteristics in the drought resistance of the three conifers he investigated are the cuticular transpiration and the development of the root system. Tolerance paralleled their drought resistance only when their root system was restricted. When allowed to develop their root system normally in the field, the order of resistance was not the same. Waisel (1962a) obtained similar results, but even the drought resistance of the shoot (without roots) is commonly due to avoidance. Thus, among nineteen species tested, pine possessed the highest drought resistance of the shoot and mulberry the lowest, yet the difference between these two was solely due to the difference in cuticular transpiration (Tazaki, 1960b). This agrees with Parker's (1951) earlier results. Larcher's (1960) measurements also point to avoidance factors as the main cause of the greater drought resistance of one species of *Quercus,* than another. In this case, however, a difference in tolerance may also exist, since the critical water saturation deficit was also greater in the more resistant species.

Parker (1968) was unable to recognize any difference in drought tolerance of the root bark tissue in such diverse trees as sugar maple (*Acer saccharum*), white ash (*Fraxinus americana*), and red oak (*Quercus rubra*). The fact that practically no higher plants are able to survive air drying proves that drought tolerance can be a factor only if accompanied by drought avoidance. It is apparent that just as animals have evolved in the direction of less low-temperature tolerance and more low-temperature avoidance, thus becoming homoiotherms, so plants have evolved in the direction of less drought tolerance and more drought avoidance, thus approaching homoiohydry. The evolution of the latter character in plants, however, is not as complete as found in animals. Therefore, higher plants must possess appreciable drought tolerance. It must be concluded that drought avoidance is more important in higher plants while drought tolerance is more important in lower plants.

This must mean that avoidance is of greater survival value than tolerance, in the case of the water stress. Why should this be so? It is obvious that a plant with well-developed avoidance can not only survive the drought, but will also be able to continue metabolism, growth, and development in the presence of an extended water stress. Drought avoidance is, therefore, a more complete adaptation to arid climates than is tolerance, since it carries with it an avoidance of all the strains that may be produced by a water stress—elastic as well as plastic, direct as well as indirect, and secondary as well as primary. Avoidance alone, if well enough developed, may therefore permit completion of the life cycle of the plant while subjected to the water stress.

The plant with only tolerance, on the other hand, can do no more than survive in the presence of the water stress, for the tolerant plant will have no positive turgor pressure when subjected to the water stress, and therefore cannot grow. It, therefore, is unable to prevent elastic (reversible) strain. Tolerance merely permits the plant to survive until the water stress, (and therefore, the elastic strain) is removed, when it can grow once more and complete its life cycle. Furthermore, tolerance of the indirect effects may require a different mechanism from tolerance of the direct effects, and tolerance of the secondary stress is even more likely to be different from tolerance of the primary stress. Therefore, several kinds of tolerance may have to be developed even for mere survival.

Of course, since the plant is not perfectly homoiohydric, it cannot survive the water stress without any drought tolerance. Similarly, no higher plant can exist without any avoidance, because of the presence of separate water-absorbing and water-evaporating organs. Therefore, it is not surprising that plants depending mainly on avoidance may grow side by side with others that depend mainly on tolerance, with astonishingly similar success. Thus, Oppenheimer (1953) found four species were able to reduce their transpiration to essentially zero during the hot hours of the day, whereas another species (*Phillyrea media*) maintained its stomata open and therefore its transpiration high. As a result, the former maintained osmotic potentials as high as −14 atm, while the latter's dropped to about −50 atm and it had very little sap, and it can in fact become practically air dry without injury.

On the basis of the above analysis, this success of tolerance in competition with avoidance is unexpected. It favors the conclusion that the plant does not have to develop three kinds of tolerance, but that the tolerance of the direct effects of the water stress carries with it also a tolerance of the indirect effects. This is precisely what the protein aggregation or SH hypothesis predicts (see above).

In spite of the necessity of both avoidance and tolerance in higher plants, most plants tend to develop mainly the one or the other rather than an equal development of both kinds of resistance. According to Bornkamm (1958), none of the species with low transpiration rate and low daily fluctuation in water content possessed high drought tolerance. There are several reasons for this. (1) If a species successfully avoids a decrease in its water content when exposed to a water stress, the possession of tolerance will have no selection value. (2) If tolerance depends on hardening by a sublethal decrease in water content, the plant that possesses avoidance will never be able to harden. Both of these reasons assume that avoidance and tolerance are compatible. Actually, however, at least some of the factors that give rise to these two types of resistance are mutually exclusive. A simple example is provided by high water content and high osmotic potential. The former is an avoidance factor, for plants with higher water content (e.g., succulents) are capable of losing more water before reaching the same water potential as plants with lower water contents. Consequently, if both groups possess the same drought tolerance, the one with the higher water content will be more drought resistant. On the other hand, the development of a high osmotic potential leads to greater tolerance and a lower water content. Therefore, the plant that has evolved with a higher tolerance will have a lower avoidance and vice versa.

In some cases, even the two kinds of drought tolerance (avoidance of elastic growth strain and dehydration tolerance) may not be correlated. The two are inversely related in *Filipendula vulgaris* and *Saxifraga hypnoides* (Jarvis, 1963) and in two conifers (pine and spruce) as opposed to two broad-leaved (birch and aspen) trees (Jarvis and Jarvis, (1963b). This inverse relation is not universal, for it did not occur in the case of *Prunus padus* or *Thelycrania* (*Cornus*) *sanguinea* (Jarvis, 1963). The reason for the independence of survival and growth at moderate stress is that the two quantities depend on different cell properties. Prevention of injury depends mainly on protoplasmic properties. Ability to grow when subjected to a drought stress is limited by vacuole and cell wall properties; the maximum water stress at which growth can continue, increases with the turgor pressure and the plasticity of the cell wall in the cells that are at the stage of elongation. These values have never been measured in relation to water stress, and therefore nothing is known about the relationship between water stress and cell growth. One possible explanation of the above results, however, is stomatal closure. Aspen and birch closed their stomata much sooner (at water potentials of −5 and −10 bar, respectively) than spruce and pine (−37 and −15 bar, respectively), and therefore may be expected to maintain a higher turgor pressure. The ability to continue assimilation, of course, would be decreased by this adaptation.

It must be realized, of course, that since different plants owe their drought resistance to different factors, one may be the most drought resistant under one set of conditions, another may be the most resistant under a different set of conditions (Satoo, 1956).

CHAPTER 16

THE MEASUREMENT OF DROUGHT RESISTANCE

A. Yield and Other Indirect Measurements

A major barrier to progress in our understanding of drought resistance has been the inadequacy of the methods of measurement. Indirect methods (e.g., the measurement of xerophilic characters or individual factors in drought resistance) have failed to produce results in agreement with field experience (Asana, 1957). As in the case of freezing resistance, it is possible to measure drought resistance qualitatively or semiquantitatively by field survival, but the methods used in the two cases are different. Freezing resistance may be evaluated in the field long (2–5 months) after injury has occurred to the dormant plants, but before any appreciable subsequent growth. Drought injury, on the contrary, occurs during the growing season, and the plants are allowed to grow and produce a crop before, during, or after the drought. Drought resistance is then assumed to parallel the yield. When evaluated in this way, resistance is a combination of many things, since the yield depends not only on the ability of the plant to survive the drought, but also on its ability to grow and complete its development before, during, or after the drought. In India, for instance, the only wheat varieties that produce good yields are those that complete their development before the drought (Chinoy, 1960). In this case, yield is not a measure of drought resistance, but merely of drought escape. Sometimes an apparent difference in drought resistance may result from the lack of uniformity of field methods. Thus, a variety of tobacco considered to be drought resistant from field experience, failed to show any superiority from more direct tests. The field behavior was explained by the fact that only half as many plants were grown per acre as in the case of the "less resistant" variety and, therefore, the soil moisture was not reduced as rapidly to the danger level (Bliss *et al.*, 1957). From all these considerations, field determinations of yield cannot be relied upon to give a true measurement of drought resistance as distinct from the other properties of the plant.

Other indirect measurements have also proved incapable of determining drought resistance. Even those characteristics that must be in some way related to drought resistance may give contradictory results, e.g. transpiration rate is directly correlated with drought resistance in some cases, and inversely correlated in other cases (see Maximov, 1929). It is not surprising then that other empirical methods have failed. The ability of seeds to germinate in media of high osmotic concentration (low osmotic potential) has been used to measure their drought resistance (see McGinnis bibliography). Recent attempts to use this method have failed (McGinnis, 1960). Kaloyereas (1958) obtained correlations between drought resistance and heat stability of chlorophyll, "bound water," and SH content. In all these cases, the drought resistance of the plants is judged from field experience and therefore any correlation is of doubtful significance to true drought resistance. Consequently, most investigators now attempt to determine drought resistance directly on the basis of survival of drought.

B. Survival Time

The earliest method of measuring drought resistance on the basis of survival was simply to withhold water from plants in the open and to determine how long they survived, or the percentage survival after an arbitrary time (e.g., 2 weeks) in the unwatered condition (Tumanov, 1927). A variation of this method is to count survival time from the time when the plant reaches the permanent wilting point. This would undoubtedly increase the importance of tolerance in the overall drought resistance measured, but avoidance would not be completely eliminated since both loss and even uptake of water will continue during the period measured. The first results with summer wheats gave good agreement with field experience (Table 16.1), but later results with other plants showed wide differences (Waisel, 1959).

The highly variable field conditions soon led investigators to attempt an evaluation of drought resistance by survival under the controlled conditions of a drought chamber. This has given less satisfactory results than the freezing chamber method of measuring freezing resistance, for the following reasons.

(1) It is far easier to obtain an accurate control of temperature than

TABLE 16.1 DROUGHT RESISTANCE OF SUMMER WHEATS ACCORDING TO FIELD SURVIVAL AND SURVIVAL OF PERMANENT WILTING[a]

Variety	Line	Field survival (Pisarew)		Survival of permanent wilting (Tumanov)
		1919	1920	
Ferrugineum rossicum	Tulun 81/4	72.6	67	90
Ferrugineum rossicum	Tulun 120/32	77.4	75	94
Ferrugineum rossicum	Tulun 916/4	50.0	67	50
Pseudohostianum	Prélude	8.1	43	49
Lutescens	Marquis	3.1	43	77
Anglicum	Pusa 4	5.5	20	23
Ferrugineum rossicum	Tulun 324	2.9	—	42

[a] From Tumanov, 1927.

of relative humidity, even when no plants are present. In the presence of plants, control of relative humidity becomes much more difficult due to the continuous evaporation of water from the plants.

(2) In order for relative humidity to indicate a specific vapor pressure, temperature must also be controlled. This is again more difficult in a drought chamber than in a freezing chamber because the plant is both absorbing radiant energy and cooling itself by transpiration. The leaf and even the air temperature may therefore fluctuate much more than in a freezing chamber.

(3) In freezing chambers, the time at a given temperature is only 1–2 hr since this is long enough for the leaf temperature to reach equilibrium with that of the surrounding air. In the case of the drought chamber, relative humidity equilibrium would take many days and, in any case, must not be allowed to occur since it would almost invariably lead to death, and would defeat the purpose of the chamber by eliminating avoidance. The time chosen, must therefore be a purely arbitrary one.

(4) The rate of air movement must also be controlled (again at an arbitrary speed) in order to control the rate of water loss from the plant. Since all these conditions are purely arbitrary, no two investigators are likely to choose the same conditions. Therefore, their results are not directly comparable with each other.

It was easily shown, of course, that artificial droughts in such chambers can kill plants. Differences occurred between species (or even between varieties) in abilities to survive such artificial droughts, but the order of survival frequently failed to agree with field experience (see Levitt, 1956).

More recent attempts with this method have proved no more successful. Drought resistance of potted tree seedlings as determined in a drought chamber (25°C, 15% r.h., 10 m sec $^{-1}$, 5000 lux daylight) failed to agree with field survival (Tranquillini and Unterholzer, 1968). Oppenheimer (1967), in fact, obtained best survival by the least drought-resistant species, due apparently to their poorer development and therefore slower exhaustion of the soil moisture.

In these early experiments with drought chambers, no attempt was made to cope with a fifth deficiency in this technique. The effect of a drought on plants is due, at least as much to its effect on the root system as on the top of the plant. Yet only the relative humidity of the atmosphere in contact with the top was actually controlled. This is easily understood, since controlling the relative humidity of the root environment is much more difficult, and no truly satisfactory method has as yet been developed. In the case of small seedlings investigated for short periods of time, it may be sufficient to control the water content of the medium (Gingrich and Russell, 1956), since the amount removed by the seedling during the time of the experiment may be negligible and therefore the vapor pressure will remain essentially constant. This is not true in the case of a plant with a well developed root system. The problem may be simplified by using solutions instead of soil as the root medium, since it is easy to measure the freezing point of solutions and to keep their concentrations essentially constant. However, besides being nontoxic the solute must be nonpenetrating, in order not to alter the drought resistance of the plant. Unfortunately, all solutes seem to be taken up, and even though the rate is often slow, the time of exposure to a drought must be considerable, and the penetration of the solute may easily be excessive during this time. Thus, even when the high molecular weight (30,000–90,000) polymer PVP (polyvinylpyrrolidone) is used, it accumulates to a sufficient degree to injure some of the leaves within 1 or 2 days (Krull, 1961). Carbowax (lutrol, polyethylene glycol or PEG) is taken up very slowly and may cause little or no injury even after several days or weeks. Kaul (1966a) measured the drought resistance of grain seedlings by determining their relative growth rates on exposure to water stress. They were grown in water culture and subjected to stress by the addition of polyethylene glycol. The lowest water potential obtainable by this method was −7.7 atm. It is obvious, therefore, that this method cannot be used to investigate the extreme water stresses experienced by plants in nature. Furthermore, as emphasized by Kaul (1966a), due to the slow diffusion rate of such large molecules, the water potential in the rhizosphere may soon become very different from that a short distance away. To eliminate this problem, Kaul used an automatic periodic draining system, which kept the solution well mixed and uniform.

A field method of evaluating the maximum time of drought survival ("Ausdauer"), yielding values from 2.1 to 166 hr, has been developed by Pisek and Winkler (1953). They calculate it from two other quantities as follows:

$$\text{Ausdauer} = \frac{\text{available water content at complete stomatal closure}}{\text{cuticular transpiration}}$$

The "available water content" was calculated in its turn by subtracting the water content at the drought-killing point from the water content in the state of complete stomatal closure. The "Ausdauer" can be a true measure of survival time only if the cuticular transpiration rate remains constant from the time of stomatal closure until death. Actually, the vapor pressure of the plant must drop steadily as the concentration of its cell sap increases, and as its turgor pressure becomes more and more negative. Both of these factors as well as the drying of the cuticle lower the transpiration rate. Cuticular transpiration has been found to decrease on drying to as little as $\frac{1}{100}$ its original value (Oppenheimer, 1960). If, however, these complicating factors do not differ much among the plants being investigated, even though true survival time cannot be predicted in this way, the values may still be in the approximately correct relative order.

Levitt *et al.* (1960) measured survival time by exposing shoots to a moving stream of air at 15% r.h. (by passing it through a saturated solution of LiCl) under standard conditions of light and temperature in a Plexiglas chamber. At predetermined intervals, shoots were removed and placed in water to determine their survival of the treatment. Among the seven species tested, six ranged in survival from 5 to 53 hr, the seventh survived 32 days (Table 16.2). Since only shoots were used, this failed to include the contribution of the roots to drought resistance. Tazaki (1960a) has used a similar method and defines his "dehydration resistance" as the time, after detachment from the plant, to kill a plant part due to water loss under standard conditions. In agreement with Pisek and Winkler, Tazaki points out that survival time depends on two factors: (1) the amount of water that must be lost before the plant is injured, and (2) the rate of water loss, which, in the case of detached shoots, is primarily dependent on the rate of cuticular transpiration, but also on the specific surface. Survival time is, therefore, really a measure of total drought resistance, due to both tolerance and avoidance. However, since only the shoot was used, this method cannot measure the overall drought resistance of the plant. In order to understand the problems involved in any attempts to measure

TABLE 16.2 DROUGHT RESISTANCE OF SHOOTS OF DIFFERENT PLANTS[a,b]

Species	Time for 50% killing
Oxalis sp.	5–6 hr
Lycopersicon esculentum (tomato)	11 hr
Mentha (*citrata?*)	23 hr
Hordeum vulgare (barley)	24 hr
Helianthus annuus (sunflower)	31–34 hr
Brassica oleracea (cabbage)	48–53 hr
Setcreasea striata	32 days

[a] Measured at 15% r.h. and 30°C, under continuous light (4–40 W fluorescent lamps at 2 ft.)
[b] From Levitt *et al.*, 1960.

overall drought resistance, the methods used to measure the two components must be examined.

C. Avoidance

Efficiency of Water Utilization

One of the earliest measurements was developed at a time when water conservation was considered to be the basic, if not the sole cause of resistance. Measurements were, therefore, made of the amount of water lost per unit of dry matter produced. The following are the terms used and their meanings (Maximov, 1929):

Water requirement (w. r.)
(or transpiration coefficient) = water lost/dry matter produced (1)
Efficiency of transpiration = dry matter produced/kg water lost (2)

The former quantity is a coefficient, i.e., a number without units. It was believed, at first, that the water requirement of all plants was about 300. It was soon found, however, that cereals could be roughly divided into more mesophytic ones with a high water requirement, and more xerophytic ones with a low requirement. On the other hand (Table 16.3), some still more xerophytic grass species had the highest water requirement of all—even higher than that of rice, which may be considered a partial hydrophyte. More recent measurements have yielded even more extreme values, ranging from <200 to >2000 (Slatyer, 1964) for most plants.

Exceptional values as low as 25–50 have been recorded for succulents such as pineapple. These results clearly establish the two main avoidance types among xerophytes—the water savers with a low water requirement (or high efficiency of transpiration) and the water spenders with a high water requirement (or low efficiency of transpiration). The majority of crop plants are mesophytes and fall between these two groups. Nevertheless, even among crop plants, just as among native plants of dry habitats, two kinds of drought avoidance exist. The water savers have a higher efficiency, the water spenders a lower efficiency of water usage, though the differences are less extreme than in the case of native xerophytes.

Another drought-resistance factor is also involved in measurements of water requirement: resistance to the indirect effects of the water stress, specifically starvation. A plant that adapts simply by closing its stomata, in this way decreasing transpiration, would decrease photosynthesis simultaneously. The result might be no change in the water requirement, for the decrease in water loss would be compensated for by the corresponding decrease in photosynthetic products accumulated. It was, perhaps for this reason that, as mentioned above, all plants were at first thought to have the same water requirement. The lower water requirement of the one group of xerophytes, therefore, must depend as much on their ability to increase the rate of photosynthesis under conditions of water stress (relative to the rate in a mesophyte) as on their ability to decrease the water loss. Similarly, the second group of xerophytes undoubtedly transpire more rapidly even than indicated by their high water requirement,

TABLE 16.3 RELATIONSHIP OF WATER REQUIREMENT (W.R.) TO XEROPHYTISM[a]

Plant	Xerophytism	Water requirement
Wheat, Oats, Barley, Rye	Mesophytic	350–470
Millet, Sorghum, Corn	Moderately xerophytic	168–196
Agropyron sp., *Bromus* sp.	Most xerophytic	977–1035
Rice	Hydrophyte (in seedling stage)	682

[a] From Maximov, 1929.

because the latter value is lowered by the simultaneous high rate of photosynthesis.

It must be realized, of course, that the water requirement is not a constant for a species or variety, but will vary with the environment. It is normally higher the greater the water stress of the atmosphere, provided that sufficient soil moisture is available (Table 16.4). In cotton leaves it increased linearly with the vapor pressure differences between leaf and air (Bierhuizen and Slatyer, 1965). The w.r. values ranged from 100–500 as the vapor pressure difference increased from 5 to 25 mm Hg. This is a natural consequence of the increase in evaporation rate with the vapor pressure gradient. In other words, when the water requirement is measured under field conditions, it is the environment as well as the plant that is being measured. Only if measured in a controlled chamber under constant conditions, can the value be taken as a measurement of the plant's properties. The water requirement can also be altered by changes in rate of photosynthesis. Increasing CO_2 concentration therefore lowered the water requirement of cotton to below 100 when the concentration of CO_2 no longer limited photosynthesis. It is also affected by other environmental factors (Maximov, 1929), and, therefore, may fluctuate markedly from year to year, e.g., from 300 to 650 for a single oat variety, and from 140 to 440 for a variety of corn during 1911–1917.

Neales *et al.* (1968) measure the transpiration ratio (T) as

$$T = (\text{net efflux of water vapor/time})/(\text{net influx of } CO_2\text{/time}) \qquad (3)$$

This can be simply converted to water requirement by multiplying by 1.5, since the dry matter produced in photosynthesis is about 0.67 times the weight of the CO_2 absorbed. When this conversion is made, they obtained water requirements of about 250 for the mesophytic sunflower and tobacco and only about 75 for two succulents (pineapple and *Agave americana*).

Of what value are these measurements of water requirement in relation to avoidance? They may be used to identify the kind of avoidance, since high values during drought are characteristic of water spenders and low values of water savers. As pointed out above, they may also measure one kind of drought tolerance due to the plant's avoidance of the indirect metabolic strain induced by the water stress, specifically starvation.

Many measurements have been made of the osmotic potential of the plant in relation to drought (Walter, 1931; Oppenheimer, 1953). In order to have any significance in connection with avoidance, the freezing point of the juice (or some related measurement) must be used. Plasmolytic determinations are useless since they permit water uptake by the flaccid cells. Theoretically, the osmotic potential may give too low values in

TABLE 16.4 THE EFFECT OF DIFFERENCES OF AIR MOISTURE ON THE WATER REQUIREMENT[a,b]

Plant	Water requirement	
	Humid air	Dry air
Kubanka wheat	826	1052
Hannchen barley	758	1037
Spring rye	875	1100
Burt oats	760	1043
Honduras rice	585	743
Grimm alfalfa	906	1378
Kursk millet	267	386
Red Amber sorghum	223	297
Dent corn	210	263

[a] According to Briggs and Shantz.
[b] From Maximov, 1929.

relation to the actual state of the water in the flaccid cells since it ignores the effect of wall tension on water potential. For this reason, Weatherley and Slatyer (1957) measured the leaf's diffusion pressure deficit (DPD), nowadays expressed as water potential (ψ), by allowing leaf disks to come to equilibrium above a graded series of solutions over a period of 32 hr. The DPD (or negative water potential) was that of the solution that failed to lose water to or gain it from the leaf disks. In practice, the difference between the osmotic and the water potential may sometimes be negligible (Slatyer, 1957b). Since these quantities measure the energy state of the plant's water, they should be closely related to drought avoidance, but they cannot be used by themselves as a measure of plant avoidance. They must be related to similar measurements of the plant's environment.

A quantitative evaluation of drought avoidance by definition, must depend on the relationship between the energy level of the plant's water and that of its environment (Levitt, 1963; Jarvis and Jarvis, 1963b). The farther apart these are, the greater will be the plant's drought avoidance. This does not mean, however, that it can be measured by the difference between the water potential of the plant and that of the environment ($\psi_p - \psi_e$); a true measure of avoidance must relate the steady state decrease in the plant's water potential to the water removing potential of the environment. A quantitative definition of drought avoidance is, therefore, the water removing potential of the environment required to lower the water potential of the unstressed plant by one unit. This can be calculated from the ratio of a specific water removing potential ($\psi_0 - \psi_e$) to the steady state decrease that it produces in the plant below the saturated state ($\psi_0 - \psi_p$):

$$A_d = \frac{\psi_0 - \psi_e}{\psi_0 - \psi_p} \quad \text{or} \quad \frac{\psi_e}{\psi_p} \tag{4}$$

where A_d = drought avoidance; ψ_0 = water potential at saturation; ψ_e = water potential of the environment; ψ_p = water potential of the plant in the steady state when exposed to ψ_e.

Avoidance is, therefore, a dimensionless coefficient, and a value of 1 means that the plant is unable to avoid reaching equilibrium with its environment. The values will, therefore, range from 1 to infinity.

Since avoidance means a lack of equilibrium, its true measure is possible only under steady-state conditions, i.e., when the rate of water absorption exactly equals the rate of water loss. In order to attain this state, the plant must be exposed to a specific, constant environmental drought long enough for its internal water potential to reach a constant value. In practice this may not be possible, partly because (at least in a plant with low avoidance) the curve will level off gradually and the end point may be difficult to identify, and partly because it may not be possible to maintain the water potential of the root medium constant for a long enough time. There are ways of circumventing such difficulties, especially where extreme exactness is not needed.

The plant's avoidance will not be constant at different environmental droughts. If, for instance, the water stress increases step by step from zero water potential (i.e., 100% r.h.), the plant may retain full or nearly full turgor at first, and avoidance will be at or near infinity. Furthermore, one plant may conceivably have a greater avoidance than another plant at a slight environmental drought, but a lower avoidance at a more severe one (if, for instance, they close their stomata at different degrees of drought). What avoidance should, then be measured? As far as survival of the plant is concerned, it is obvious that the avoidance at an environmental water potential just above the drought-killing point (i.e., just sufficient to prevent injury) is the critical one. This, of course, would be at different external droughts for different plants, and can be determined only by measuring drought tolerance. The avoidance coefficient, however, may perhaps not be appreciably altered between the permanent wilting point and the drought-killing point, since the stomata are closed and the cuticle surface has become dry, so that the barrier limiting water movement will be constant. Equation (4) then becomes:

$$A_{d50} = \frac{\psi_{e50}}{\psi_{p50}} \tag{5}$$

where A_{d50} = the critical drought avoidance of the plant; ψ_{e50} = the environmental water potential producing 50% killing; ψ_{p50} = the water potential of the plant at the 50% killing point.

D. Tolerance

1. Dehydration (Strain) Avoidance

Stocker (1929) defines the "water saturation deficit" (w.s.d.) of the plant at any moment (at which its fresh weight is referred to as its natural weight) by

$$\text{w.s.d.} = \frac{\text{saturation wt.} - \text{natural wt.}}{\text{saturation wt.} - \text{dry wt.}} \times 100 \quad (6)$$

Weatherley and Slatyer (1957) define "relative turgor" (r.t.) as:

$$\text{r.t.} = \frac{\text{natural wt.} - \text{dry wt.}}{\text{saturation wt.} - \text{dry wt.}} \times 100 \quad (7)$$

These two quantities actually measure the same property since one is 100 minus the other, i.e., they measure the percentage of the plant's maximum water content lost or retained respectively at a particular moment (Fig. 16.1). The techniques of obtaining the saturation water content are a little different in the two cases. For w.s.d., shoots are usually allowed to become saturated by standing them in water. This method does not work in the case of some exceptional plants that actually wilt under these conditions (Oppenheimer, 1960). For r.t., leaf disks are floated on water. This method may give spurious results due to cell enlargement and other complications, but improved methods seem to have overcome these difficulties (Catsky, 1960). A new method for measuring water deficit is to determine the amount of water necessary to infiltrate a leaf. The increase over the control-saturated leaf is a measure of the water deficit (Czerski, 1968).

The term "relative turgor" is an unfortunate one since most of the values are obtained when the plant is below zero turgor pressure. Even in the turgid plant, water content is not a measure of turgor. The correct term now adopted at Walter's (1963) suggestion is "relative water content." A true measure of relative turgor has, in fact, been used by Shardakov (1957), which he calls the "index of saturation."

$$H = \frac{T}{P} \times 100 \quad (8)$$

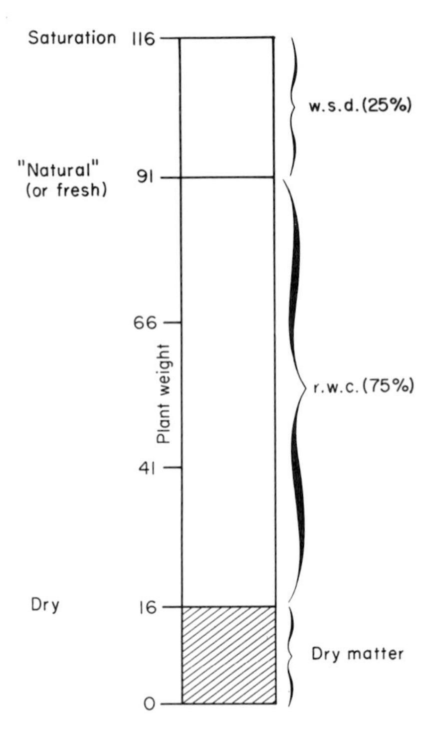

FIG. 16.1. Relationship between water saturation deficit (w.s.d.) and relative water content (r.w.c.). Ordinate represents the weight of the plant, and of its dry matter and water components. If the sample contains exactly 100 gm H_2O, the above water components in grams would represent the w.s.d. and r.w.c., respectively.

where H = the index of saturation or the relative turgor; T = actual turgor pressure; P = osmotic pressure or maximum potential turgor pressure.

Shardakov uses his index of saturation as an indicator for time to irrigate.

On the basis of stress and strain concepts, it is now clear that w.s.d. is a measure of the dehydration strain induced by the drought stress, and r.w.c. (relative water content) is a measure of dehydration avoidance. Neither of these quantities can measure drought (stress) avoidance, since this quantity, by definition, measures the plant's ability to maintain a thermodynamic disequilibrium with its environment. This is illustrated in Table 16.5. Obviously, though both plants have the same w.s.d. and r.w.c., and, therefore, the same dehydration avoidance, the succulent plant maintains its water at a higher energy level and therefore has the greater drought avoidance. Of course, if the w.s.d. of one plant remains essentially constant during a drought and that of a second plant rises markedly, this usually reveals a greater drought avoidance in the former. The differences must be quite large before such conclusions can be accepted categorically, for there is no single proportionality factor between w.s.d. or r.w.c. and water potential for all plants (see Table 16.7). Therefore, the only reliable method of measuring drought avoidance is to determine the energy level of the plant's water when subjected to a water stress.

2. Dehydration (Strain) Tolerance

If a shoot is allowed to dry until the drought-killing (or "critical") point is reached, the "lethal" (Oppenheimer, 1932) or "critical" (Höfler *et al.*, 1941) w.s.d. may be determined. Similarly, the critical relative water content (c.r.w.c.) may also be determined. Just as w.s.d. and r.w.c. are measurements of dehydration strain and dehydration avoidance, respectively, so also c.w.s.d. is a measurement of dehydration tolerance. When both are measured, they may yield qualitative information as to the kind of stress resistance made use of by the plant. A plant whose water saturation deficit on exposure to drought is far above its c.s.d. must owe its

TABLE 16.5 Assumed Steady State Values for w.s.d., r.w.c., and Osmotic Potential in Relation to Drought Avoidance in a Succulent and a Halophyte[a]

State of plant	Atmosphere	Succulent	Halophyte
a. Fully turgid			
r.h. (%)	100		
w.s.d. (%)		0	0
r.w.c. (%)		100	100
Osmotic potential (atm)		−10	−50
Water potential (atm)	0	0	0
b. Droughted			
r.h. (%)	60		
w.s.d. (%)		40	40
r.w.c. (%)		60	60
Osmotic potential (atm)		−17	−83
Water potential (atm)	−600	−17	−83

[a] Exposed to an r.h. of 60%. Succulent retains a water potential much higher than that of its environment and therefore possesses greater avoidance.

drought resistance primarily to avoidance. If, however, its c.s.d. is very low but its w.s.d. approaches close to it, the plant must owe its resistance primarily to tolerance (see Table 13.3). Nevertheless, just as w.s.d. and r.w.c. cannot measure drought avoidance, so also c.w.s.d. and c.r.w.c. cannot measure drought tolerance. For instance, in the case of the same imaginary succulent and halophyte referred to above, the succulent may have the larger c.w.s.d. (and therefore a larger dehydration tolerance), yet the halophyte may have the lower osmotic potential at its c.w.s.d. and, therefore, the greater drought tolerance (Table 16.6).

3. Drought (Stress) Tolerance

Dehydration avoidance and tolerance are measures of the drought-induced strain. The former measures the elastic resistance, the latter the yield point of the plant. Drought tolerance, on the other hand, is a measure of the drought stress at the limit of dehydration tolerance (or yield point) of the plant. The drought tolerance of the plant must, therefore, be measured by the decrease in free energy of its water (relative to the saturated state) which is just sufficient to produce 50% killing:

$$T_d = \psi_0 - \psi_{p50} \quad \text{or} \quad -\psi_{p50} \tag{9}$$

where T_d = drought tolerance; ψ_{p50} = the water potential of the plant at the 50% killing point.

The earliest measurements of drought tolerance were expressed in units of relative humidity. Iljin (1927) used a graded series of solutions and allowed sections of plant tissue to come to equilibrium with the atmosphere above each (24–48 hr). The relative humidity just insufficient to injure the cells (determined by plasmolysis, vital straining, etc.) was taken to be inversely related to the "desiccation resistance" of the plant. Since r.h. is directly related to the energy level of water, the r.h. of the plant at the drought-killing point should give a direct measure of drought toler-

TABLE 16.6 Assumed c.s.d., c.r.w.c., and Osmotic Potentials Calculated for the Same Succulent and Halophyte as in Table 16.5

	Succulent	Halophyte
c.s.d. (%)	75	50
c.r.w.c. (%)	25	50
Osmotic potential (atm)	−40	−100

TABLE 16.7 RELATIONSHIP BETWEEN RELATIVE WATER CONTENT (R.W.C.) AND WATER POTENTIAL (ψ) IN SEVERAL PLANTS[a]

Species	r.w.c. at ψ leaf = −15 bar	ψ at 50% r.w.c. (bar)
Tomato	73	−22
Lupin	68	−24
Birch	86	−35
Privet	87	−38
Aspen	86	−44
Eucalyptus	84	−51
Pine	84	−53
Acacia	92	−60
Spruce	90	−65

[a] From Jarvis and Jarvis, 1963b.

ance. Large differences between plants have been found (see Table 15.9). The "maximum osmotic pressure" (Walter, 1931) attained by the plant at the drought-killing point may also be used as a measure of tolerance. However, as mentioned above, this may not be equal to the water potential, and therefore may approach, but not equal, drought tolerance.

Drought tolerance can be determined from measurements of c.s.d. or c.r.w.c., if the relationship between these quantities and the water potential of the tissue is known (Jarvis and Jarvis, 1963b). Of course, this relationship differs for each species or tissue and must be determined experimentally. The values for a number of plants are given in Table 16.7.

Attempts have been made to improve the methods of evaluating the injury quantitatively. The efflux of salts (e.g., chlorides) or of metabolites from the tissue after exposure to a specific drought was inversely related to tolerance (Gessner and Hammer, 1968; Shmat'ko and Rubanyuk, 1969).

E. Total Drought Resistance

Survival time can, at best, give only a relative measure of drought resistance and no comparison is possible between the results of different workers using different, arbitrary conditions of drought. More important, even the relative survival times of two plants may be reversed, depending on the particular conditions of droughting (as will be illustrated below). It is, therefore essential to develop an absolute system of measuring resistance. This can be done by measuring avoidance and tolerance separately and calculating the total drought resistance of the plant from these two measurements (Jarvis and Jarvis, 1963b; Levitt, 1963).

By definition, the drought resistance of a plant is the water stress that is just sufficient to produce 50% killing. This is expressed as an equation:

$$R_d = -\psi_{e50} \tag{10}$$

where R_d = drought resistance; $-\psi_{e50}$ = water stress just sufficient to produce 50% killing.

This definition does not include the time factor because killing time depends on the rate of net loss of water, whereas the above measurements must be made under steady state conditions, i.e., when the water content of the plant remains constant due to equal rates of loss and absorption. Whether or not drought resistance, as defined above, is capable of predicting survival time will be discussed below. It should be possible to calculate overall drought resistance without measuring it directly, if the two components (drought tolerance and critical drought avoidance) have been measured. From Eqs. (9) and (5):

$$T_d \times A_{d50} = -\psi_{p50} \times \frac{\psi_{e50}}{\psi_{p50}} = -\psi_{e50} \tag{11}$$

Since from eq. (6):

$$R_d = -\psi_{e50}$$

$$R_d = T_d \times \mathrm{A}_{d50} \tag{12}$$

This equation should be very useful since R_d is very difficult to measure in the field and is impossible to measure correctly in the growth chamber for field conditions. The two components may, however, be measured separately in the field, leading to the possibility of measuring total drought resistance under field conditions (Tables 16.8 and 16.9). What information does this calculated value of drought resistance yield? It indicates that the higher the drought resistance of a plant, the lower the water potential (i.e., the more severe the drought) that it must be exposed to before being injured. More important, these calculated values of drought resistance are based on two measurements and, therefore, determine whether the superior resistance is due to avoidance, tolerance, or both.

The measurements themselves will involve some difficulty. Drought tolerance is relatively easily measured by use of Iljin's "desiccation resistance" method (1927). Sections of tissues are allowed to come to equilibrium with atmospheres of known water potential (Iljin expressed it as

TABLE 16.8 DIRECT MEASUREMENT OF DROUGHT AVOIDANCE IN RELATION TO SOIL WATER STRESS[a]

Soil moisture stress (atm)	DPD just before sunrise (atm) $-\psi_p$			Drought avoidance ψ_e/ψ_p		
$-\psi_e$	Privet	Cotton	Tomato	Privet	Cotton	Tomato
90	59			1.5		
110	70			1.6		
80		62			1.3	
107		77			1.4	
45			41			1.1
40			37			1.1

[a] Adopted from Slatyer, 1957b.

relative humidity) and cell survival is then determined (e.g., by plasmolysis, vital staining, streaming movement). The water potential (or relative humidity) resulting in death of 50% of the cells is taken as the drought-killing point and drought tolerance is $-\psi_{p50}$.

Drought avoidance is more difficult to measure: (1) because it involves the steady state in contrast to the equilibrium state of tolerance, and (2) because both root and top environments have to be controlled and at different degrees of drought. The problem is to determine the water potential of an atmosphere in which the plant shoot would maintain a constant (measurable) water potential not far above the drought-killing point. The latter value would require an instantaneous determination without changing appreciably the absorption or loss of water by the plant, the leaf temperature or state of opening of the stomata, the cuticular hydration,

TABLE 16.9 TOTAL DROUGHT RESISTANCE, TOLERANCE, AND AVOIDANCE IN TOMATO AND PRIVET[a]

Plant	Drought resistance (measured $-\psi_{e50}$ in atm)	Drought tolerance $-\psi_{p50}$ (in atm)	Drought avoidance $-\psi_{e50}/-\psi_{p50}$	Drought resistance [calculated from $-\psi_{p50} \times (-\psi_{e50}/-\psi_{p50})$]
Tomato	45	41	1.1	45
Privet	110	70	1.5	105

[a] Adapted from Slatyer, 1957b.

etc. Though difficult, such a measurement may nevertheless be possible. An absolute determination of avoidance in a controlled chamber would be necessary, and it would have to permit controls not at present available in such chambers, e.g., of the root medium. This may, perhaps, not be as difficult as at first suspected, since the steady state may possibly be reached in a relatively short time, and the root medium may be controlled for such short periods. In the field, the water potential of the leaf would have to be measured while it is still attached to the plant. A method of accomplishing this has been developed (Tinklin, 1968). Such measurements must be made, however, without altering the environmental conditions. Even field determinations now in use may come close to an evaluation of absolute avoidance, since Slatyer (1957b) showed that the leaf at sunrise may be close to equilibrium with the soil. It is, therefore, conceivable that the plant may be close to the steady state at certain times of the day. In any case, such field determinations should give a relative measure of avoidance, if not an absolute one. Thus, Slatyer (1957b) measured the water stress (in this case the soil water stress) at sunrise, and, on this basis, privet showed an avoidance about 15% higher than cotton (Table 16.8), which is in agreement with its more xerophytic nature. Unfortunately, no measurement of drought tolerance was made in the case of cotton, otherwise the total relative drought resistance of the two plants under these conditions could be calculated. Privet, however, was shown to be more than twice as drought resistant as tomato, due partly to tolerance and partly to avoidance (Table 16.9).

Jarvis and Jarvis (1953b) eliminated the problem of measuring water potential in the field by determining r.w.c. and using a predetermined conversion table (see Table 16.7) for calculating the water potential. Unlike Slatyer, who determined the soil water stress, they measured the atmospheric water stress (Table 16.10). Consequently, their avoidance measurements are about 3–9 times those obtained from Slatyer's data. Among the four species investigated, the order of tolerance was not the same as the order of avoidance. Thus, although the avoidance of birch was about 10% higher than that of spruce, the latter was nearly twice as drought resistant, on the basis of tolerance $\times$ avoidance.

If the shoot is removed from the plant, and its avoidance determined, it consists of at least two components—avoidance of stomatal water loss and avoidance of cuticular water loss. It is interesting that among four species tested by Jarvis and Jarvis (1953b), the order of the former was almost the reverse of the overall avoidance, for the water potentials at which stomatal closure occurred were: spruce -37, pine -15, birch -10, and aspen -5 bar, respectively. In other words, aspen closed its stomata

TABLE 16.10 RELATIONSHIP BETWEEN DROUGHT TOLERANCE, SHOOT AVOIDANCE, AND TOTAL DROUGHT RESISTANCE OF TREES[a]

Species	Drought tolerance		Shoot avoidance at 85% r.h. (−220 bars)			Drought resistance
	r.w.c. (%)	$-\psi_1$ (bar)	r.w.c. (%)	ψ_2 (bar)	ψ_{air}/ψ_2	$\psi_1 \times \psi_{air}/\psi_2$
Spruce	40	95	76	−32	6.9	655
Pine	38	70	73	−23	9.6	672
Birch	40	50	63	−29	7.6	380
Aspen	56	40	27	−73	3.0	120

[a] From Jarvis and Jarvis, 1963b.

at the smallest water stress, and therefore had the greatest avoidance of stomatal water loss, but its cuticular transpiration was so high that it had the smallest overall avoidance. This is shown by the rates of cuticular water loss: spruce 0.02, pine 0.04, birch 0.20, and aspen 0.49% of r.w.c./min.

Drought resistance as measured in the field cannot usually be expected to involve either equilibrium or steady-state conditions, but simply the conditions of the moment. It cannot, therefore, be a constant for a given plant under different conditions. In fact, the relative order for a series of plants may conceivably change with the conditions. For instance, a deep rooting plant may be expected to show a greater drought resistance than a shallow rooted plant if ground water is available, but perhaps lower resistance if it is not present. This, of course, shows the flexibility of the concept and its validity in revealing the changing drought-resisting properties of the plant.

Table 16.11 illustrates how this might work out in practice. In spite of the absence of avoidance (being a poikilohydric plant) and due to its high tolerance, plant B shows a higher drought resistance than plant A, which possesses both avoidance and a slight tolerance. With the prevailing soil moisture stress and air temperature, plant A is killed by an atmospheric water potential of −130 atm (or a relative humidity of 90%), and plant B only when it drops to −600 atm (or 60%). Plant C, on the other hand, possesses as high a drought resistance as plant B, though it is killed when its own water potential drops to −150 atm and its tolerance is therefore only one-quarter of the latter's. This is, of course, due to its high avoidance. If the soil moisture stress is high, plant C's high drought avoidance would

TABLE 16.11 ILLUSTRATION OF DROUGHT RESISTANCE QUANTITIES IN THREE IMAGINARY PLANTS

Plant	Critical avoidance (of atmospheric water stress) (A_{d50} or ψ_e/ψ_p)	Tolerance (T_{d50} or $-\psi_{p50}$) (bars)	Total drought resistance (R_d or $-\psi_{e50}$) (bars)
A (mesophyte)	2	65	130
B (rootless or shallow-rooted, poikilohydric plant)	1	600	600
C (deep-rooted xerophyte)	4	150	600

have to be due to water conservation. If the soil moisture stress is relatively low (at least at considerable depth), plant C's avoidance could be due to a high rate of water absorption and transport to the leaves.

The above results also point to the necessity of measuring both tolerance and avoidance, rather than solely drought resistance. Even though plants B and C have the same total drought resistance under the above conditions, it is obvious that plant B will never be killed by the direct effect of drought at relative humidities above 60%, whereas plant C will eventually die at any relative humidity below 95%, if it comes to equilibrium with it. On the other hand, as long as adequate ground water is available to their roots, both plants C and A will have higher avoidance than at the critical point and may conceivably live indefinitely at relative humidities of 60% or even less, though plant B will eventually die at this relative humidity due to lack of avoidance. These comparisons further emphasize the inability of measuring drought resistance adequately by survival time, since one plant may survive longer than another in one drought, and the reverse may be true in a different drought. They further indicate that, in order to characterize a plant's behavior completely, it is necessary to determine not only its critical avoidance, but also its avoidance above this point, e.g., with its stomata open.

What of the limiting case when no water is being absorbed but some is steadily being lost? Nothing like a steady state would exist, and it would seem, at first sight, that avoidance measurements would be useless. In practice, however, determinations may have some value even under these conditions. Succulents, for instance, may remain alive for long periods of time without absorbing any water, due to their extremely low transpiration rate. Suppose one is kept at an environmental relative humidity of 40% (a water potential of −900 atm) for 1 year and loses half its water during this time. If its original potential in the turgid state is -10 atm, its final value will be -20 atm. Its drought avoidance during the year would, therefore, be 67.5 ± 22.5. Though this variation is numerically large, it may be relatively unimportant since most plants have much lower values (see Tables 16.8 and 16.10). Consequently, even in the absence of steady-state conditions (e.g., under natural conditions) determinations of avoidance may be very useful. Accurate measurements of avoidance, however, must be made under controlled conditions.

It must be realized that this measurement of drought resistance includes only drought avoidance and direct drought tolerance. Strictly speaking, it does not measure the other components of drought tolerance—euryhydry or indirect drought tolerance. Of course, if a plant has a sufficient degree of drought avoidance, there will be no need for these other components of drought tolerance. What of a plant that does not possess sufficient drought avoidance? In such a case, it is possible that the measurement of direct drought tolerance carries with it a measurement of these other components of drought tolerance. Euryhydry and avoidance of cell collapse both increase with cell sap concentration, and so does direct drought tolerance (see above). The only remaining components of indirect drought tolerance are avoidance and tolerance of starvation or of protein loss. If these are all due to the prevention (or repair) of the denaturation and aggregation of specific proteins, and if this is also the molecular mechanism of direct drought tolerance (see Chapter 15), then indirect and direct drought tolerance may be correlated. It is, therefore, conceivable that the above measurement of "total" drought resistance does, indeed, include a measurement of all the components of drought resistance. However, it will not be possible to decide this until the molecular mechanisms of drought tolerance are fully understood.

F. Identification of the Individual Factors Responsible for Drought Resistance

Although the calculated values of drought resistance are based on direct measurements of avoidance and tolerance, they do not reveal the factors on which each depend. Many attempts have been made to measure these individual factors, sometimes simply by measurements of morphological characters (e.g., Kilen and Andrew, 1969), of which transpiration has long been one of these. Larcher (1960) has attempted to standardize such measurements by determining transpiration under controlled, standard conditions (22°C, 10,000 lux, and 0.24 cm^2/hr by a Piche evaporimeter). He also determined earliness and speed of closure of stomata as drought resistance factors. Such factors were found to be better developed in the species of *Quercus* that appeared to be better adapted to drought from field experience. However, as emphasized by Larcher, such field observations may be related to adaptations to the indirect effects of water stress, e.g., the ability to maintain stomata open and therefore to continue assimilation under conditions of slightly unfavorable water balance. Tazaki (1960a) has set up an equation relating the drought resistance of the shoot to three individual shoot factors. By applying this equation, he can determine whether the greater drought resistance of the shoot of one species than that of another is due to cuticular resistance, high water content, or drought tolerance.

The role of the root has also been measured by determinations of the root water potential at different soil moisture stresses (Table 16.8). However, this is only one aspect of the root factor and no method of measuring the overall contribution of the root to the plant's drought resistance has been developed. Waisel and Pollak (1969) have developed a method of estimating water stresses in the active root zone of plants growing in natural habitats.

RADIATION STRESSES

CHAPTER 17

RADIATION STRESS—VISIBLE AND ULTRAVIOLET RADIATION

A. Visible Radiation (Light)

1. Light Deficit or Shade

Although it is conceivable that a light deficit may lead to a secondary (e.g., temperature or water) stress under certain conditions, this possibility is remote and does not seem to have been investigated. Similarly, the daily exposure of plants to the dark without injury, also eliminates the possibility of primary direct (i.e., rapid) light deficit injury. As in the case of temperature and water, however, a level of illumination below the light compensation point can lead to a slow, indirect injury, due to starvation. Shade plants are adapted to low illumination and are therefore more "shade resistant" (i.e., resistant to light deficit) than sun plants.

a. Avoidance of Light Deficit

Avoidance of light deficit can refer only to the illumination at the photochemically active centers in the cell, relative to that in a less avoiding plant. Since higher plants are not motile, and have no light-concentrating lens, they cannot receive a higher light intensity than that incident at their aerial surface. This kind of avoidance, can therefore, be due solely to (1) a decreased reflectance and absorbance of light by the part of the plant external to the photochemically active centers and (2) an increased absorption of light by the photochemically active centers. The light-deficit avoiding shade plants are, therefore, characterized by a much less reflecting leaf surface than in the nonresistant sun plants, due to a poor development of cuticle, and the absence of wax and hairs. Furthermore, chloroplasts may occur in the epidermal cells, unlike sun plants in which the epidermal layer is a light-absorbing barrier between the chloroplast-containing mesophyll cells and the light source. As a result of these adaptations, a higher illumination is incident on the chloroplasts of the shade plant than on those of the sun plant exposed to the same low light intensity. The chloroplasts can also absorb a larger fraction of the weak illumination due to a higher concentration of chlorophyll or other photochemically active pigment. An extreme example of this kind of avoidance of light deficit is shown by the red algae. Some of these organisms live and photosynthesize actively as deep as 100 m below the surface of the water. At this depth nearly all the red and blue radiations have been absorbed by the water. The green radiation, however, is absorbed to a much lesser degree and is fully 50% of its intensity at the water surface. By means of their red pigment, the red algae are able to absorb strongly the green radiations and to transmit the absorbed energy to the chlorophyll pigment for use in photosynthesis. This is shown by their relative rates of photosynthesis in light of different colors (Table 17.1).

b. Tolerance of Light Deficit

A net CO_2 assimilation of zero requires a minimum illumination at the reaction centers, in order to compensate for the respiratory CO_2 evolution. When the illumination incident on the leaf surface is no greater than this minimum (the light compensation point), even a theoretically perfect avoidance of light deficit cannot result in a positive net photosynthesis. The resistance of shade plants to light deficit is, therefore, also due to tolerance. They are characterized by lower compensation points (Table 17.2) and can, therefore, accumulate photosynthetic products at lower light levels than can sun plants (e.g., at as little as 0.5% of full sunlight). On the other hand, they show a light saturation at lower light levels than in the case of sun plants (Table 17.2). It has long been a question whether the low

TABLE 17.1 Relative O_2 Production by a Red Alga and a Green Alga in Light of Different Colors[a]

	Relative O_2 production by algae in		
Species	Blue light (435.8 nm)	Green light (546 nm)	Red light (620–660 nm)
Schizymenia (red alga)	48–53	288–340	100
Ulva (green alga)	94	46	100

[a] From Haxo and Blinks, 1950.

TABLE 17.2 LIGHT COMPENSATION POINTS AND LIGHT SATURATION LEVELS OF SUN VERSUS SHADE PLANTS[a]

	Light compensation pt. (ft. c)	Light saturation (ft. c)
Eight (8) sun species	100–150	2000–2500
Five (5) shade species	50	400–1000

[a] From Böhning and Burnside, 1956.

compensation points of shade plants are due to a higher rate of actual photosynthesis at the low illumination, or to a lower rate of respiration, or to both. A higher actual rate of photosynthesis could conceivably be due to avoidance, as a result of a higher absorption of the incident radiation by the chloroplasts. It could also conceivably be due to a greater efficiency of utilization of the absorbed radiation, and therefore a greater tolerance of light deficit, for instance, by possession of more efficient (or a higher concentration of) photosynthetic enzymes. At low light intensities, however, enzyme activity would not be limiting. In fact, shade plants have lower activities of the enzyme carboxydismutase (Björkman, 1968a), and this explains their lower light saturation values than for sun plants (Table 17.2). It, therefore, does not seem likely that shade tolerance can be due to a greater efficiency of utilization of the absorbed radiation.

Tolerance of light deficit could, however, be due to a low respiration rate, which would yield a net photosynthetic product even though the actual rate of photosynthesis was no greater in the shade plant than in the sun plant. Recent evidence indicates that respiration is, indeed, the important factor, and also demonstrates that a plant can become adapted (i.e., can "harden") to low light levels. In the case of white clover, lowering the illumination from 70 to 11 W/m^2 (i.e., from 6 to 1% of full sunlight) resulted in a gradual drop of the respiration rate to less than half its original value (McCree and Troughton, 1966). Raising the illumination to the original level brought the respiration rate back to its original value. The adaptations occurred within 24 hr of the light change and permitted the plants to grow at light levels which were below the compensation point for the unadapted plants, but not below that of the adapted plants. The compensation point in the adapted plants was about 10 W/m^2, and in the unadapted plants about 18 W/m^2.

It must be realized, however, that the light compensation point is markedly dependent on temperature. In the case of the lichen (*Ramalina maciformis*), the light compensation point was as low as 200–300 lux between $-5°$ and $+2°C$, and as high as 8000 lux at 27°C (Lange, 1969). Shade tolerance is, therefore, inversely related to temperature.

2. Excess Light

a. Secondary Stress Injury

High light intensity, particularly if accompanied by high infrared, as in sunlight, may give rise to two types of secondary stress injury. (1) Sunscald is a secondary, radiation-induced heat stress injury (2) Drought injury occurs mainly under conditions of high insolation and may be classified as a secondary, radiation-induced stress, since it is due to the evaporative effect of energy absorbed from the sun. These two types of stress injury have already been discussed (Chapters 6, 11, and 13). Shade plants are particularly sensitive to both (see Chapters 12, 15, and 16).

b. Primary Direct Light Injury

The direct effect of radiation of any kind is a rise in energy. If it is a rise in molecular kinetic energy, the result is the above mentioned secondary heat stress. If it is a rise in chemical reactivity, the result is a photochemical strain. If this photochemical strain is injurious, it must be a primary direct injury. This type of injury is commonly induced by a light-absorbing photosensitizer. The photosensitizer may be introduced artificially, by staining vitally with erythrosin or fluorescein. Exposure to light kills these stained cells. This is the so-called *photodynamic effect*, which has been used to kill bacteria. In the case of *Saccharomyces*, the dye attaches itself to important components of the cell, probably DNA or RNA, absorbing light energy and transferring it to these attached components (Lochmann and Stein, 1968). When rose bengal is the photosensitizer, the membrane enzymes suffer photooxidative damage (Duncan and Bowler, 1969), particularly those associated with active transport. The normal plant possesses its own pigments. It may, therefore, be asked whether or not any of these pigments are capable of producing a photodynamic effect. This does not seem likely in the case of higher plants, for the living protoplasm is protected from direct irradiation by the cell wall, cuticle, wax, suberin, etc. Under normal conditions, therefore, these protective layers lead to a light avoidance and the plant is resistant to excess light.

If, however, the plant is grown at low light intensitities, the protective layers are poorly developed. The leaf may have no wax and little or no cutin. Such plants may be injured by exposure to high light intensities.

Even plants grown in the greenhouse during the dark days of winter may be injured if transferred to growth chambers. The much higher intensities in the short-wave regions are apparently responsible for the injury (van der Veen and Meijer, 1959). The younger leaves, however, soon become "hardened" to the higher light intensities, due perhaps to the development of the protective surface materials.

Shade plants are particularly sensitive to strong light. When shade ecotypes of *Solidago virgaurea* are grown in strong light and then transferred to weak light, they use this weak light less efficiently than if grown from the beginning in weak light. According to Björkman (1968b) this is due primarily to damage to the photosystems or to a site close to them. The evidence seemed to indicate that photosystem II was affected more than photosystem I. Shade grown ecotypes of *Solanum dulcamara* show photo-inhibition on transfer to strong light (Gauhl, 1970), presumably due to similar damage.

Such injury to shade plants due to excess light is usually relatively slow and the plant may recover and become fully adapted. In some cases, however, injury is rapid and fatal. The most extreme sensitivity to light is shown by the shade plants belonging to the lower groups of plants—the algae and mosses. Red algae from the deep sea may be killed by as little as 1–2 hr of direct sunlight, whereas algae of the tidal zone (*Ulva lobata, Cladophora trichotoma, Porphyra perforata*) are not injured even by a 5-hr exposure to direct light (Biebl, 1952c, 1956b, 1957). This is an example of what Biebl calls "ecological resistance," i.e., resistance which parallels the stress of the natural habitat. The light injury can be detected by a rounding up and swelling of the rhodoplasts of the red algae (Biebl, 1956b, 1957). The cells that are killed turn red-violet in color and their cell walls swell. Some (e.g., *Microcladia*) may bleach completely in the sun (Biebl, 1952c, 1956b). In the branched red algae, there is a gradient of light resistance from the tip of the thallus to the main stem (Biebl, 1956b), the young tips dying first.

Among mosses, *Mnium serratum* (which grows in a weakly illuminated environment with as little as 26 lux) was the most light-sensitive species. It was partially killed by 1½ hr of sunlight. Five hours of direct sunlight killed all the mosses investigated, either partially or fully (Biebl, 1954). The injury was due to the visible radiations, and the UV of sunlight (300–400 nm) produced no injury. Even when the light was filtered to yield only the wavelengths above 425 nm, a 7-hr exposure killed most of the mosses. In the case of red algae (*Petrochondria Woodii* and *Callophyllis marginifructa*), wavelengths longer than 500 nm may kill parts of the plants (Biebl, 1956b, 1957).

c. Primary Indirect Light Injury

In the case of other stresses (e.g., low temperature and water deficit), a growth inhibition in the absence of injury is clearly an elastic strain. It is produced instantly and is instantly reversible by removal of the stress. Growth inhibition by light, on the other hand, does not appear instantly on exposure to the light and is not instantly reversed on removal of the stress. It is, therefore, a plastic strain, and, at least in some cases, due to an injury. The photoinhibition of stem elongation by full sunlight in the case of bean (*Phaseolus vulgaris*) is apparently due to a decrease in the amount of available gibberellin (Lockhart, 1961a,b). In the case of sunflower seedlings, it is due to an increased transport of a growth inhibitor from the leaf to the stem and the consequent inhibition of IAA transport (Shibaoka, 1961). The two mechanisms may, of course, be interrelated. In general, then, light appears to inhibit growth by disturbing the balance of the growth regulators. Whether or not the disturbed balance is dependent on transport, the growth inhibition is an indirect strain following an unknown direct photochemical strain.

Other indirect metabolic effects of light may, however, be more ob-

TABLE 17.3 Enzymes Showing Decreased Activity in Illuminated Plants

Plant	Enzyme	Light used	Reference
Lemna gibba	NAD^+-dependent glyceraldehyde-3-phosphate dehydrogenase	White	Muller and Ziegler, 1969
Wheat seedlings	α-Galactosidase α-Galactokinase β-Galactosidase Ascorbic acid oxidase Peroxidase	White	Wolf, 1968
Oat (stimulated first leaf)	Peroxidase	White	Parish, 1969
Oat seedlings	Nitrate reductase	White	Chen and Ries, 1969
Euglena gracilis	Malic enzyme decreased to 1/10		Ammon and Friederich, 1967
Chlorella protothecoides	dTMP kinase	Blue most effective, red least	Sokawa and Hase, 1968
Watermelon seedlings	Isocitrate lyase	White	Hock, 1969
Spinach leaves	Catalase	406 nm	Björn, 1969
Nicotiana tabacum (aurea mutant)	Glycollate oxidase	Blue	Schmid, 1969

viously injurious. Blue light and near UV inhibited the respiration of the colorless alga *Prototheca zopfii*, and this was accompanied by destruction of cytochrome *a* (Epel and Butler, 1969). These radiations also destroyed cytochrome oxidase of yeast and cytochrome a_3 of beef heart mitochondria. Several of the plant's substances can be photooxidized, e.g., chlorophyll, IAA, and ascorbic acid. Light also has been reported to lower the activities of many enzymes (Table 17.3). In some cases (e.g., catalase; Björn, 1969), it seems to be due to a direct photoinactivation. In the case of glycollate oxidase, the light-induced excited state of the coenzyme (flavin mononucleotide) inactivates the enzyme (Schmid, 1969). There are many more cases of increased than of decreased activity due to light, but this is to be expected, due to the increased protein synthesis that accompanies photosynthesis.

There is one kind of light injury that has been observed only under artificial conditions of growth. Tomato plants when grown in a growth chamber have an absolute requirement for a daily periodicity (Ketellapper, 1969). If exposed to continuous light, they become chlorotic. Yet this chlorosis may be prevented by a daily period of drastically lowered light intensity. Complete darkness is not necessary. Therefore, the chlorosis appears to be due to a net destruction of chlorophyll by the continuous light.

B. Ultraviolet Radiation

Since UV radiations play no essential role in the physiology of the plant, there can be no stress due to a deficit. A stress due to excess radiation is more easily achieved than in the case of visible radiations due to the higher quantum value of the UV radiation and therefore the larger energy values for inducing radiochemical reactions. The physiological effects on higher plants have been reviewed by Dubrov (1968).

1. UV Injury

a. Absorption by the Plants

As in the case of visible radiations, only the absorbed UV radiations can have any effect on the plant. Measurements on algae have shown that 80–95% of the incident UV (from 400 to 200 nm) is absorbed (Biebl, 1952b). Higher plants are also highly absorptive. The leaf epidermis of

Vicia faba shows maximum transparency at 250 and 310 nm,and minimum transparency at 220, 280, and 350 nm (Lautenschlager-Fleury, 1955).

Many substances present in the plant participate in this absorption. Water absorbs UV much more intensely than visible radiations. Nevertheless, it would require about 100 cm H_2O to absorb half of the UV present in sunlight. Consequently, other substances must be responsible for the major absorption by plants. One such substance is anthocyanin, which absorbs strongly in the UV. Since this pigment accumulates in the cell sap, and is usually located in the epidermal and subepidermal cells, it has long been suspected to protect the mesophyll cells by filtering out the excess UV. Other related but colorless substances may, of course, be just as effective as the anthocyanins.

Substances that occur in the protoplasm of plant cells may also absorb UV radiations. Unlike absorption by substances in the vacuole, such absorption may be expected to injure the protoplasm by the above mentioned photodynamic effect. According to Gilles (1940), destructive effects are produced by radiations at 290 nm, but not at higher wavelengths. Injury, of course, has long been known to be produced by radiation of 253.7 nm, e.g. in soybeans (Tanada and Hendricks, 1953). These two wavelengths are near the peaks of maximum absorption for proteins and nucleic acids, respectively (see McLaren, 1969). The photoinactivation of proteins by UV irradiation depends on the temperature in the 5°–35°C region, due apparently to changes in molecular conformation (Konev *et al.*, 1970).

b. Injury in Nature

In spite of these specific destructive effects of UV irradiation to proteins and nucleic acids, there is little evidence of UV-induced injury to plants in nature. On the other hand, there is good evidence of resistance to UV injury in nature (see below). This apparent paradox is, however, readily explained. The shorter UV wavelengths (below 300 nm) that are most damaging to proteins and nucleic acids are present in sunlight in sufficient intensity to induce DNA damage in microorganisms (Saito and Werbin, 1969). The intensity is usually too low, however, to damage these substances in the cells of higher plants, due to protection by the epidermal layer with its cutin, cell wall, and vacuole water and solutes. Other injuries may, however, occur due to the higher intensities of the longer UV wave lengths (above 300 nm). Chlorophyll, for instance, may conceivably be destroyed by the UV of sunlight and this may perhaps account for the lighter green color of sun than of shade plants. UV irradiation of chloroplasts at high altitudes reduces the intensity of the Hill reaction, NADP reduction, and both cyclic and noncyclic photophosphorylation (Nikitina *et al.*, 1969).

Cyclic phosphorylation is the most sensitive. Again, under natural conditions the chloroplasts are not likely to receive a high enough intensity of UV irradiation to produce these inhibitions.

c. PRIMARY INDIRECT UV INJURY

A plastic (irreversible) growth inhibition has long been known to result from UV irradiation. This has, in fact, been suggested as one reason for the small size of alpine plants, which grow at high elevations and, therefore, are exposed to more intense UV irradiation. The UV-induced growth inhibition can be ascribed, at least partly, to auxin effects. Epinastic (growth movement) responses to auxin can, for instance, be prevented by UV irradiation (253.7 nm). According to de Zeeuw and Leopold (1957), this is due to the prevention of the plant's ability to respond to auxin, since the effect is obtainable by exposure of the plant to the UV irradiation before applying the auxin. Other investigators, however, have concluded that there is either a direct destruction of auxin or of an auxin precursor. Exposure of timothy roots to UV irradiation (253.7 nm) resulted in the production of unusually long cells, due apparently to inhibition of cell division (Brumfield, 1953). In contrast to auxin, scopolin and chlorogenic acid increased in UV-irradiated tobacco and sunflower plants (Koeppe *et al.*, 1969).

d. PRIMARY DIRECT UV INJURY

In general, a 1–4 min irradiation with UV (230–310 nm) produces killing within 24 hr (Biebl, 1952c), e.g., in the case of onion epidermis (1 min for inner, and 3–4 min for outer epidermis). Fatal injury to the nucleus precedes injury to the cytoplasm (Biebl and Url, 1958). Therefore, even when these nuclei are dead, the cells may still be able to plasmolyze. The tonoplast is especially resistant. The mitochondria are the most sensitive of the cytoplasmic organelles. The plastids are a little less sensitive, and the spherosomes are the least sensitive. Two types of inactivation occurred when the green alga *Platymonas* (Volvocales) was irradiated in the far UV (223–300 nm). One type was an immediate immobilization (Halldal, 1961). The action spectrum of this kind of inactivation followed the absorption curve of proteins. The second type of immobilization was observed after 1 week, and gave an action spectrum which followed the absorption spectrum of DNA. Even the longest UV (or shortest visible) radiation may produce some damage. In the case of bacteria, irradiation with 360 nm destroys endogenous quinones without disruption of organelle structure or dissociation of enzymes and coenzymes necessary for oxidative phosphorylation (Brodie, 1969).

UV irradiation (<310 nm) induced an increase in permeability to water in the cells of the inner epidermis of onion when measurements were made 24 hr after irradiation (Biebl and Url, 1958), but the permeability to glycerol decreased.

2. UV Resistance

a. UV Avoidance

Since the UV intensity increases with the elevation, the stunting of alpine plants has long been ascribed to UV radiation. Similarly, it has long been assumed that alpine plants survive the higher UV intensities of their environment by a filtering out of the harmful UV radiations as they pass through the epidermal layer of the leaf. This kind of resistance would be due to avoidance. In support of this concept, the epidermal layers are sometimes highly colored by anthocyanin pigments, which absorb intensely in the UV. Actual measurements have shown that the upper epidermis is less transparent than the lower epidermis (Lautenschlager-Fleury, 1955). The lowest transparency was found in alpine plants, the highest in shade plants. These results support the suggestion that the epidermis of alpine plants protects the assimilating tissue against UV irradiation.

Caldwell (1968) tested this concept by comparing different elevations from 125–4350 m. The UV irradiation from 280–315 nm increased with elevation to a lesser degree than expected, because the component from the sky (as opposed to the direct irradiation from the sun) actually decreased with elevation. In agreement with the above results, by means of high intensity monochromatic irradiation in the laboratory, Caldwell established an inverse relationship between UV-B (280–315 nm) filtration capacity of the epidermis and sensitivity to UV damage. On the other hand, UV filtration capacity of alpine plants was no greater than that of plants from lower elevations. In the case of *Oxyria digyna*, however, populations from the arctic, where the UV radiations are particularly attenuated, possessed a higher transmission than populations from alpine areas. The differences were small, but this may have been due to the fact that the tested plants had been grown under laboratory conditions.

Further evidence that avoidance may be an important component of resistance was the existence of a kind of hardening (or "pseudohardening;" see Chapter 14). Plants grown under UV-absorbing filters on snow-free tundra often possessed a decreased UV filtration capacity. If they were then exposed suddenly to unfiltered solar radiations, there was no apparent damage, but the UV filtration capacity usually increased very rapidly, with a concomitant reddening of the plants.

b. UV Tolerance

The plant may also possess UV tolerance due to a repair mechanism called photoreactivation. This was originally discovered in microorganisms. Inactivation by far-UV was reversed by light of short wavelength, e.g., by 360–490 nm in the case of yeast (Townsend, 1961), and by 320–460 nm in the case of other organisms, with a maximum effectiveness at 405 nm (Saito and Werbin, 1969).

Photoreactivation has been demonstrated in soybeans, by reversing the harmful effects of UV irradiation (253.7 nm) with visible radiations below 450 nm (Tanada and Hendricks, (1953). A similar photoreactivation of a UV-irradiated green alga (*Platymonas*) occurred with an action spectrum resembling the absorption spectrum of flavoprotein (Halldal, 1961). The same alga, however, could also be completely reactivated in light wavelengths which normally do not affect photoreactivation (500–800 nm). This occurred when, unlike the above photoreactivations, the UV wavelengths used were absorbed only insignificantly by DNA (223 and 238 nm). In this case, reactivation to 60% of normal occurred even in the dark. This reactivation was explained by a destruction of proteins by the far UV, which produced immediate immobilization, followed by a resynthesis of the protein when DNA synthesis was not damaged (Halldal, 1961).

The photoreactivating capacity is higher in young daughter cells than in mature cells of chlorella, due perhaps to variations in activity of a photoreactivating enzyme (Taube, 1970). The primary photoproducts persist in yeast cells up to 6 hr after UV irradiation (Kiefer, 1970). The photoreactivation may, therefore, include an avoidance of post irradiation injury, as well as a repair of injury during irradiation. This conclusion is supported by the observation of a "simultaneous reactivation"of UV damage in *Xanthium* leaves (Cline *et al.*, 1969). It was much smaller, however, than photoreactivation due to a post-UV irradiation. Photoreactivation was demonstrated for several plants irradiated with damaging doses of UV-B (280–315 nm). The importance of this photoreactivation process in nature was demonstrated by subjecting the plants to a simulated alpine day of precisely defined UV-B irradiation in the absence of longer wavelengths (Caldwell, 1968). Tissue destruction was evident in a few individuals. In nature, where photoreactivation can occur, such severe tissue destruction is not normally observed.

In opposition to the photoreactivation by longer wavelengths after irradiation with short UV, the injurious effect of the long-wave UV of daylight increases or even first arises when combined with longer light waves (Biebl, 1956b). Thus the long UV (300–400 nm) which is harmful in mixed light has no effect by itself even at doses six times that in sunlight

(Biebl, 1957). Similarly, the UV of sunlight (mainly 300–400 nm), does not begin to injure even the most light-sensitive red algae until they have been exposed to it for 15 hr a day, for 4 days (Biebl, 1956b, 1957). Furthermore, this UV does not add to the light injury, for if a sublethal dose of sunshine (40 min) was followed by 4 hr of filtered UV light (300–400 nm) no injury occurred, though 1 hr of full sunlight killed the algae.

Unlike resistance to light, which is of the "ecological" kind, resistance to short-wave UV (230–310 nm) differs among members of the same ecological group of algae and is therefore "constitutional" (Biebl, 1952a). This was also true of the mosses (Biebl, 1954). For instance, the two mosses *Mnium serratum* and *Trichocolea tomentella* are very sensitive to sunlight but relatively resistant to short-wave UV which is not present in sunlight.

Biebl (1961) reported a protection of the inner onion epidermis against UV radiation by thiourea. The protection increased with concentration from 0.01 to 0.5 M (the highest used). Later investigations (Biebl et al., 1961) proved, however, that this was a purely physical effect due to absorption of the UV radiations by the thiourea solutions. Solutions of urea, glucose, KCl, $CaCl_2$, or glycerol, gave no protection. Several other substances were as effective as thiourea—*S*-methylthiourea, thiocarbohydrazide, aminoethylisothiourea, *N*-acetylthiourea, and thioacetamide. In all cases the protection was due to absorption of the UV by the solutions.

c. Mechanism of UV Resistance

The above investigations establish the fact that UV injury can be induced by wavelengths present in sunlight, and that the plant can at least "pseudoharden" to UV stress, but that in nature, UV injury is not likely to occur, because of the presence of both UV avoidance and tolerance. Nevertheless, the mechanism of UV resistance is at least of theoretical interest. The existence of UV avoidance is established, though no attempts have been made to discover how the exposure to UV irradiation sets in motion the chemical reactions that lead to this kind of pseudohardening. Since both primary direct and primary indirect injury occur, two kinds of UV tolerance may be expected. However, only the one kind, repair by photoreactivation, has been established. According to Saito and Werbin (1969), this kind of tolerance is probably due to two photoreactivating enzymes—one repairing UV damage to DNA, and the other to RNA.

The existence of another kind of tolerance is suggested by a correlation between resistance both to visible and to UV radiations and factors previously found to be correlated with tolerance of other stresses. (1) The light-resistant sun leaves of the copper beech have a higher proline content than the shade leaves, in spite of the higher concentration of the majority of the amino acids in the light-sensitive shade leaves (Haas, 1969). Besides proline, however, valine, histidine, and arginine were also higher in the sun leaves. (2) Inactivation of enzymes involves the breaking of bonds, some in the primary process of photon absorption and some following energy transfer, e.g., from excited aromatic residues to the disulfide amino acid cystine (McLaren, 1969). Grossweiner and Usui (1970) explain the UV inactivation of enzymes by the generation of a hydrated electron from amino acids—specifically by reduction of cystine and oxidation of cysteine in the presence of indole (e.g., from tryptophan). According to Shibaoka (1961) growth inhibition due to excess light seems to involve inactivation of SH groups. (2) According to Mantai (1970b), the injurious effect of UV irradiation on chloroplasts appears to be due to a general breakdown of membrane structure, rather than an inactivation of specific sites in the electron transport chain. Light-induced photodynamic membrane injury has also been reported (Duncan and Bowler, 1969). All of these results are highly suggestive, in view of the apparent involvement of sulfhydryl groups and membranes in other kinds of stress resistance. More evidence is needed, however, before attempting to explain either light or UV resistance on the basis of these factors.

CHAPTER 18

IONIZING RADIATIONS

The ionizing radiations that have been most commonly investigated as environmental stresses are X-rays and γ (gamma)-rays (photons), α (alpha)-rays (positive helium nuclei), and, in a few cases, β-rays (negative electrons). Other ionizing radiations have so far received little attention. Ionizing radiations can transfer their energy to any matter with which they collide. The collision may result in ionization of the matter by ejection of an electron or of a proton, or by capture of the particle by the nucleus of the atom. Free radicals may also be formed, which are extremely reactive due to the unpaired electrons, and may, therefore, also lead to ionization. The radiation-induced strain is obviously the addition or removal of electrons from cell molecules. This so activates the molecule that it quickly participates in a chemical reaction, converting the theoretically elastic (reversible) electron displacement into an irreversible and, therefore, plastic strain. If the molecules that are altered chemically are essential components of protoplasm, injury can be expected.

A. Radiation Injury

Injury due to ionizing radiations is, to all intents and purposes, absent in nature, due to the presence of essentially negligible quantities of ionizing radiations in the plant's normal environment. The increasing usage of such radiations has, however, aroused a tremendous amount of interest in their effects under artificial conditions. Furthermore, the quantities in nature have increased due to contamination from artificial sources. It is, therefore, conceivable that injury due to ionizing radiations in nature may be a problem of the near future.

Injury to plants from artificially applied ionizing radiation has been amply reported. The α-rays from polonium were first shown to retard or inhibit three processes in the germination of the spores of *Pteris longifolia*: (1) cracking of the cell wall, (2) development of chlorophyll, and (3) cell division (Zirkle, 1932). The number of α-particles necessary to produce a 50% effect for these three processes was 33,000, 13,000, and 4000/protoplast, respectively. Johnson (1948) exposed the tiny plantlets from the leaves of *Kalanchoe tubiflora* to X-rays in two doses of 2300 R each. The final size attained by the treated plants was smaller than in the controls. Injury was detected by an increased efflux of electrolytes from X-irradiated (3–100 kR) potato tuber tissue (Higinbotham and Mika, 1954). Injury is not always accompanied by growth inhibition. Lettuce plants exposed to 6–7 kR of γ-radiation formed tumors (Bankes and Sparrow, 1969). Either an increase or a decrease in growth may be caused by the same radiation dose, depending on the daylength (Feldmann, 1969). Numerous other observations of injury by ionizing radiations can be found in the scientific literature of the past four decades.

1. Limits of Survival of Ionizing Radiations

Most plants can tolerate much higher doses of radiation than those fatal to animals (Hluchovsky and Srb, 1963). The average lethal dose for man is around 500 R, whereas some plants tolerate 20,000 R (Kuzin *et al.*, 1958), 47,000 R (inner epidermis of onion scale; Biebl and Pape, 1951), or even 72,000 R (Bacq *et al.*, 1957—see Vasiliev, 1962). Among the mosses and liverworts tested by Biebl and Url (1963b), the α-irradiation which produced partial or complete killing within 24 hr ranged from 1460 mCi/sec (*Plagiochila asplenioides*) to 21,900 (*Hookeria lucens*). The lethal dose for a given plant at the same time of the year was relatively constant even when the ^{210}Po strength was varied. The blue-green algae are particularly tolerant. Three categories of resistance were found among twenty-three species of these cyanophycean algae (Kraus, 1969): (1) low resistance, with an LD_{90} of less than 400 krad, (2) moderate resistance, equal to or better than the resistant bacterium *Micrococcus radiodurans*, with an LD_{90} of between 400 and 1200 krad, and (3) high resistance, with an LD_{90} greater than 1200 krad.

There are some reports of plants that are as sensitive as animals to ionizing radiations. Plants of *Soja hispida* showed abnormally small, stiff, and crumpled leaves if given as little as 350 R/day of γ-irradiation from ^{60}Co (Biebl, 1956a). Plants of *Brassica oleracea* were the most resistant among those tested, becoming dwarfed and markedly succulent as a result of 2000 R/day. Yet even after this treatment they were able to flower.

2. *Secondary, Radiation-Induced Heat Injury*

Some of the stresses already discussed can injure by inducing a secondary environmental stress. Both high temperature and high visible and infrared radiation stresses, for instance, can induce a water deficit which in turn can then produce the injury. It does not seem possible for ionizing radiations to lead to a similar secondary stress injury, for it would require a dose of 400,000 R or rads to raise the temperature of the plant by a mere 1°C, due to an increase in energy of 1 cal/gm tissue. Since this dose is much higher than those usually sufficient to produce injury, it is obvious that the usual doses of ionizing radiations cannot injure due to a secondary high-temperature stress. Extreme doses producing immediate injury (see below), on the other hand, undoubtedly do produce secondary stress (i.e., heat) injury. It is, therefore, logical to conclude that the slower injury, due to the commonly used doses of ionizing radiations, is primary in nature, caused by injurious radiochemical reactions in the plant. As in the case of other stresses, this primary radiation effect may be either direct or indirect (Vasiliev, 1962).

3. *Primary Indirect Radiation Injury*

a. Plastic (Irreversible) Growth Inhibition

It has long been known that the nucleus is much more sensitive to radiation injury than is the cytoplasm, and that chromosome breaks are among the earliest effects of the radiation (Read, 1959). In the case of wheat seedlings, mitotic breaks occur at relatively low doses (< 6.2 kR; von Wangenheim, 1969), although the cells may continue a delayed development even after much higher doses (as high as 660 kR), perhaps because the extranuclear organelles are much less sensitive than the functional activity of the nucleus. These nuclear changes explain the slow injury to plants exposed to small doses of irradiation. In the case of seeds, for instance, the dose may be large enough to prevent mitoses and, therefore, cell division, but too small to prevent the meristematic cells from expanding and differentiating. The irradiated seed may, therefore, produce a small plant (Vasiliev, 1962). No further growth will, however occur after about 3 days, when all the cells have completed cell expansion, since due to the injury no further cell division can occur. Such plants may remain turgid and green for several days after the irradiation. In fact, the only obvious change may be the cessation of growth, but this may be followed by a state of rest, after which the plant dies (Vasiliev, 1962). The growth that does occur is not merely due to water uptake, but is sustained by metabolism, and the seedlings increase in dry matter, protein, and RNA, and they photosynthesize (Haber *et al.*, 1961). In the case of onion root meristem exposed to X-irradiation, DNA synthesis was actually stimulated even though cell proliferation was inhibited (Das and Alfert, 1961). The coleoptiles of γ-irradiated (500 kR) wheat seeds grow without cell division or DNA synthesis, and may have their growth promoted by kinetin and gibberellic acid when 2 mm long, and by IAA when 8 mm long (Rose and Adamson, 1969). Massive γ-irradiation of dry wheat grains (0.5 Mrad) prevents cell division but permits growth without any gross acceleration of senescence. Foard and Haber (1970), therefore, conclude that the radiation is not lethal in the physiological sense although lethal in the genetic and proliferative senses.

Growth may also be stimulated by ionizing radiations. The cells in the quiescent zone of roots of *Allium sativum* respond to X-rays by entering into mitosis due to stimulation of DNA synthesis immediately after irradiation (Clowes, 1970). When injury does occur due to γ-irradiation of leaf buds (of apple and peach), the most actively dividing cells are not necessarily the most sensitive to radiation (Lapins and Hough, 1970).

Growth inhibition due to low doses of radiation obviously differs in two ways from growth inhibition due to other stresses. (1) Only cell division is inhibited. Cell enlargement is apparently unaffected. (2) The inhibition of cell division is irreversible, and is, therefore, a plastic injurious strain.

b. Radiation Capture by Water Molecules

Since all substances absorb ionizing radiations approximately equally per unit mass, and since water accounts for by far the largest part of the normally hydrated cell, nearly all the ionizing radiations that penetrate the cell must be captured by the water molecules. It has long been assumed that the most common type of indirect injury is due to interaction between the particles produced from the radiation-capturing water molecules and molecular oxygen. This would lead to the formation of the hydroperoxyl radical HO_2, and finally to hydrogen peroxide (H_2O_2):

$$2\,H_2O + 2\,h\nu \rightarrow H\cdot + OH\cdot + H^+ + OH^-$$
$$H\cdot + O_2 \rightarrow HO_2\cdot$$
$$2\,HO_2\cdot \rightarrow H_2O_2 + O_2$$

The H_2O_2 can, of course, induce injurious oxidations. The irradiated water molecules also may release other particles (Phillips, 1968), which may conceivably result in injurious chemical reactions, due perhaps to involvement of the nucleoproteins, lipoproteins, or other essential protoplasmic substances. Since this type of injury is due to secondary reactions, and since the substances undergoing the chemical change receive the radiation energy indirectly, injury produced in this way is indirect.

In the case of microorganisms, attempts have been made to measure the indirect injury by exposing them to high concentrations of radical scavengers during the irradiation. These scavengers would not alter the direct effect (the capture of the particles of radiation by essential protoplasmic molecules), but would react quickly with the radicals produced by the breakdown of those water molecules that capture the ionizing radiation. They would, therefore, prevent reactions between these radicals and the essential molecules of the living cell and, consequently eliminate indirect injury. Unfortunately, the high concentrations needed for efficient action as radical scavengers are too drastic for higher plants or animals. Furthermore, their effect is not as unambiguous as desired, for although they interfere primarily with the action of water radicals, they may also, to some extent, influence the direct effect (Sanner and Pihl, 1969).

Other lines of evidence, however, clearly point to the occurrence of this kind of primary indirect radiation injury. The greater radiosensitivity of fermenting yeasts than respiring yeasts may be explained by their deficiency in catalase and, therefore, their inability to destroy the H_2O_2 that accumulates as a result of radiation capture by water molecules (O'Brien, 1961). In the case of higher plants, the oxidation–reduction potential of the plant is increased by irradiation (Vasiliev, 1962), as would be expected from the above described formation of H_2O_2. The plant is able to reverse this rise after the irradiation. Furthermore, the injury is more severe under aerobic than anaerobic conditions. Several other investigations have pointed to a relationship between oxygen and radiation injury. The relationship between water content and radiation sensitivity has been explained in this way. In dry barley seeds, the response to oxygen can be maintained for extended periods, but in moist seeds the effect can be demonstrated during but not after irradiation (Nilan *et al.*, 1961). Within a range of only 0.3% water content (from 10.7 to 11.0%), the response of irradiated barley grain to oxygen changed by a factor of more than 2.5 (Conger *et al.*, 1969). Similarly, exposure of barley grain to oxygen immediately after irradiation in its absence, intensifies the effects of irradiation (Peisker, 1967; Gudkov and Grodzinskii, 1968). If, however, the irradiated seeds are stored in a vacuum at low temperatures (-78 to -196°C) for several hours, the irradiation effect disappears. Further evidence of an oxygen effect is the increase in radiation damage to wheat seeds with an increase in oxygen pressure (Nor-Arevyan, 1968). The decreased injury to maize and barley seeds when soaked in D_2O (compared to H_2O) prior to irradiation, also depended on the presence of oxygen (Gaur *et al.*, 1969). There was no effect of the D_2O when seeds were irradiated anaerobically. DNP stimulated oxygen uptake and increased radiation damage to barley seeds (Gaur *et al.*, 1970). Higher concentrations of DNP decreased the damage.

The increase in radiation injury due to the presence of oxygen has become so well established that it is measured by the oxygen enhancement ratio (OER). Oxygen (20%) enhanced the sensitivity of barley roots to X-rays by a factor of 2.8 (Ebert and Barber, 1961) and of barley seeds by 5–8 (Nilan *et al.*, 1961). In the case of π-mesons (30 rad/hr) the OER was 1.35 and 1.5 respectively for the decrease in growth of roots of *Vicia faba* at room temperature and 4°C (Raju *et al.*, 1970). Oxygen-dependent damage was measured in barley seeds by exposure *in vacuo* to 25 krad of ^{60}Co γ-rays, followed by hydration in oxygen-free versus oxygenated water (Conger *et al.*, 1969). Development of the oxygen-dependent damage occurred faster than the rate at which oxygen-sensitive sites (measured by EPR) were lost during the postirradiation hydration process. Radical components appeared to remain even after 16 hr of hydration. Similar free radicals (of the narrow type as determined by EPR) were induced in the embryo and endosperm of barley kernels by γ-radiation with ^{60}Co at 77°K (Ehrenberg *et al.*, 1969).

The interaction between the irradiated water molecules and the oxygen can explain another well known result of the irradiation. In spite of the approximately equal absorption of ionizing radiations by all substances, some substances in the plant are much more readily inactivated by the radiations than others. In at least some cases, this can be explained by oxidation of the substances by the H_2O_2 produced in the primary radiation-induced reactions between water molecules and oxygen. Thus, two oxidase systems of potato tubers are relatively insensitive to γ-radiations. In contrast, the auxin concentration of plants is decreased within minutes by X-ray dosages of less than 100 R. This may occur (1) by preferential destruction of auxin, or (2) by inactivation of the molecules or systems responsible for auxin turnover. Gordon and Weber (1955) were unable to detect a direct effect on auxin. Gordon (1957), therefore, favors the second possibility, i.e., that radiation inhibits auxin synthesis. These indirect effects of irradiation on auxin content may explain some noninjurious effects on growth, but do not explain radiation-induced injury. Of course, growth effects due to cell division are not due to auxin decreases, e.g., the inhibition of root formation in a number of commelinacean plants by an irradiation of 24,000 R (Biebl, 1959).

c. Direct Capture by Metabolites

Of course, not all radiation injury is due to interaction with oxygen. In opposition to results with gramineous seeds, the radiosensitivity of *Vicia* seeds was not dependent on oxygen availability during soaking (Klingmuller, 1961). Other evidence also pointed to the absence of any relationship between damage and free radical content. Some kinds of injury, in fact, are related to the absence of oxygen. Hall and Cavanagh (1969) demonstrated that recovery of *Vicia* seedlings from sublethal radiation damage was substantially lowered if a state of hypoxia was maintained between the dose fractions. The state of oxygenation during X-ray exposures was relatively unimportant. In such cases, it is difficult to determine whether the injury is due to oxygen-independent effects of the products of breakdown of water molecules, or to direct capture by the metabolites.

In contrast to auxin, which is decreased by irradiation, the water content of leaves has been found to increase (Biebl, 1958b), as judged by the increase in succulence. Biebl relates this to the earlier observations by Cooke (1953—see Biebl, 1958b) that ascorbic acid decreases during the first 2–3 days of irradiation, but then increases at least up to 10 days' irradiation. Similarly, on the basis of investigations of some forty species of plants, Sparrow *et al.* (1968) concluded that enhanced pigment (presumably anthocyanin) formation is a common response of higher plants to ionizing (X-ray or γ-ray from ^{60}Co) radiations. Maximum increases approached twenty times that in the untreated plants. Large increases were generally indicative of lethal or near lethal exposures.

Irradiated leaves of bean, pea, and barley showed an appreciable increase in lipid peroxide after very low doses (100 R; Budnitskaya and Borisova, 1961). Within the first hours following irradiation, linoleic, oleic, and other unsaturated acids disappeared completely. This was an enzyme effect, since no such change occurred when the lipids themselves or when steam-inactivated leaves were irradiated. Radiosensitive yeast cultures showed an increase in lipids (sterols) at lower irradiation doses (10–25 kR) than in the case of the radioresistant cultures (Larikova and Gal'tsova, 1968). At higher doses, twice as much lipid was produced in the sensitive as in the resistant cultures.

In some cases, these radiation-induced changes in specific substances require higher doses of radiation than are sufficient to initiate other changes in the plant. Thus, in the case of X-irradiated sunflower seeds, growth was markedly reduced by smaller doses than required to inactivate the lipoxidase enzyme (Surrey, 1965), but these two effects involved different parts of the plant. Normal hypocotyl growth was inhibited at a minimal level of about 60 kR, while nearly seven times as much radiation was required to stop the initiation of lipoxidase metabolism in the cotyledons.

In view of these selective effects of ionizing radiations on specific plant substances, it is not surprising that effects on metabolic processes as well as on enzymes have been reported. Exposure of spinach chloroplasts to 80 krad during 20 min caused a measurable inhibition of oxygen evolution. Treatment with digitonin increased the sensitivity of the Hill reaction to irradiation (Perner *et al.*, 1961). Photophosphorylation was found to be more radiosensitive than the Hill reaction—50% inactivation at 130 and 530 krad, respectively (Füchtbauer and Simonis, 1961). Gamma-irradiation of tomato fruit induced an immediate stimulation of respiration (Abdel-Kader *et al.*, 1968). Similarly, irradiation of broad beans doubled the activity of cytochrome oxidase on the 7th day of germination (Babaeva and Seisebaev, 1968). The respiratory rate of darkened leaves of *Pelargonium zonale* was temporarily increased by X-irradiation (750–24000 rad), but it delayed the respiratory rise associated with senescence of starved leaves (Carlier, 1968). Mitochondria from irradiated tomato fruit had lower P/O ratios than from nonirradiated fruit (Padwal-Desai *et al.*, 1969). The effects of radiation on the proteins of barley seeds agreed with the effects of radiation on proteins *in vitro* (Shapiro and Bocharova, 1961). In both cases there were two kinds of change, one controlled by the presence of oxygen and the other not. In the case of potato tubers, some enzymes of carbohydrate metabolism were inactivated by γ-irradiation *in vitro* but not *in vivo* (Jaarma, 1968). In fact, according to Kuzin and Ivanitskaya (1967), the allosteric center of aspartokinase from yeast has a much greater radiosensitivity than its active center, and this may conceivably lead to an increased activity of the enzyme in the irradiated cell. Exposure of peanut seed to 500 kR of X-rays followed by 6 weeks storage before planting resulted in a severe reduction of amino acid incorporation by ribosomes (Huystee *et al.*, 1968). They concluded that the irradiation reduced the capacity of the cells to produce biologically functional mRNA.

There is some evidence of the formation of "radiotoxins." Water-soluble substances extracted from irradiated plants inhibited the growth and development of seedlings by 22–30%, and mitotic activity by 38% (Kuzin *et al.*, 1959). Heavily γ-irradiated endosperm depressed the growth in both the root and shoot of transplanted unirradiated embryos (Meletti and D'Amato, 1961). Similarly, DNA synthesis was suppressed by placing the roots of 3-day-old pea seedlings in extracts from potato tubers which had been irradiated with 15 kR of γ-rays from a ^{137}Cs source (Tokarskaya *et al.*, 1966). Photosynthetic $^{14}CO_2$ assimilation, however, was practically unchanged by the "radiotoxins." Further evidence of the formation of radiotoxins was obtained by X-irradiation (500–1500 R) of only one-half of the root system of *Vicia faba.* The shielded half developed symptoms of injury due to the presence of toxic metabolites which presumably produced injurious changes in the protoplasm (Muller, 1969).

Gamma-irradiation (10 kR) of the seeds of horsebean, pea, phaseolus bean, and cotton resulted in an increased content of amino acids in the subsequent 5 to 15-day-old sprouts relative to the control sprouts from nonirradiated seed (Fomenko and Medvedev, 1969). The "increase" was actually due to the rapid drop in amino acid content of the controls, indicating an interference in amino acid ultilization as a result of the previous irradiation. There was no effect of the radiation on barley, sunflower, and cabbage seeds. In the case of grapefruit peel, a linear relationship was found between the γ-irradiation dose and inhibition of protein synthesis (Riov and Goren, 1970), with almost complete inhibition at 400 krad.

The metabolic effect may sometimes be due to an effect on another process. Seedlings of *Vicia faba,* for instance, showed a decrease in $^{14}CO_2$ assimilation 8–24 days after X-irradiation (250 R). This was correlated with a reduced degree of stomatal opening (Roy and Clark, 1970). Similarly, γ-radiation of potato disks enhanced their ethylene content, which in turn was partly responsible for stimulated peroxidase development (Ogawa and Uritani, 1970).

Just as in the case of growth, enzyme activity may actually be stimulated by irradiation, though this may be reversed later. Rye seeds that were γ-irradiated with 50–500 krad showed a marked decrease in α- and β-amylase activity during subsequent germination (Bancher *et al.*, 1970a). Up to the third day, however, the activity was increased. Similarly, phosphorylase activity was higher after irradiation with 500 krad than after irradiation with 50 krad (Bancher *et al.*, 1970b). This was true only immediately after irradiation and on the first day of storage.

4. Primary Direct Radiation Injury

In the case of other stresses, a simple method of distinguishing between direct and indirect injury was by the greater speed of the former. This was because direct injury resulted from an instantly fatal plastic physical strain, e.g., a change in state leading to loss of semipermeability. Indirect injury was due to a slower chemical, metabolic strain, that required the buildup of an excess or a deficiency of a substance before becoming injurious. Biebl and his co-workers (Biebl and Url, 1963b; Biebl and Hofer, 1966; Biebl and Url, 1968) have attempted to apply this criterion to radiation injury. They have produced rapid injury (within 24 hr) in the epidermis of onion scales by high doses of radiation, and slow injury (after some days) by low doses. They call the rapid injury "direct," assuming that it damages the cytoplasm directly and the slow injury "indirect,"assuming that it injures the nucleus, leading to a gradually increasing injury to the cell's metabolism. Radiochemical reactions, however, are so extremely rapid that even an injury that appears only 24 hr later is likely to be indirect in nature.

Another possible criterion is on the basis of the radiation dose needed to produce the injury. In order to kill a plant immediately, very high doses must be used—several million R according to Vasiliev (1962), and 460 kR or higher in the case of small wheat seedlings (von Wangeheim, 1969), although larger seedlings may survive as high as 660 kR. A slower killing (after several days) may be produced by a few hundred to more than 10,000 R (Woodwell, 1965) depending on the plant. The same plants may remain alive but show slight growth inhibition when irradiated by 200–5000 R (Woodwell, 1965). In the case of large wheat seedlings, survival time may vary from 20 days to 3 months (von Wangenheim, 1969).

Is it possible to decide, however, on the basis of dose alone whether the injury is direct or indirect? Low doses might be expected to cause direct injury if the radiations were captured specifically by the essential components of protoplasm, e.g., proteins, nucleic acids, or lipids. This cannot happen because ionizing radiations are not absorbed specifically since all substances absorb them essentially equally per unit mass. Therefore, small doses are likely to be captured mainly by the water, because of its great excess in the living cell. Only the much larger doses are likely to get through to the essential components of protoplasm and to produce direct injury by changing them. This is analogous to other stresses—frost, heat, and drought stresses, for instance, all produce direct injury rapidly when the stress is severe and indirect injury slowly at less severe stresses. It, therefore, appears reasonable to classify rapid radiation injury due to large doses as direct injury, just as in the case of other stresses. This agrees with Biebl's classification, since the large doses were required to produce their direct cytoplasmic damage and the small doses produced only the indirect nuclear damage. Similar results have been obtained with yeast (Fritz-Niggli and Blattmann, 1969). It also agrees with the classification of nuclear injury as indirect due to the irreversible growth inhibition (see above).

The most likely type of direct injury, both on the basis of speed and the large dose required to induce it, is a loss of semipermeability. Even if caused by peroxides produced from the radiochemical splitting of water molecules, this would produce an immediate injury. A similar oxidation of enzymes, cofactors, or nucleic acids on the other hand, could cause injury only some time later as a result of metabolic disruption.

A loss of semipermeability in barley seeds, due to X-irradiation, is indicated by the large number of inorganic and organic substances leached from the seeds within 15 min of seed soaking (Kamra *et al.*, 1960). It was greater in nitrogen than in oxygen, and the inorganic substances were reabsorbed later, coinciding with apparent recovery.

Biebl (1956a) was unable to detect any difference between strongly and weakly γ-irradiated plants in permeability to urea or glycerol, as determined by deplasmolysis rate. In the case of onion epidermis, completion of plasmolysis in KNO_3 solutions was always delayed by X-irradiation (Hluchovsky and Srb, 1963; Srb and Hluchovsky 1963, 1967), whereas in sucrose solution it was shortened. They, therefore, concluded that the irradiation increased the permeability of cells to water and electrolytes, but decreased it to nonelectrolytes. The latter result, however, is undoubtedly solely due to an increase in permeability to water, for any decrease in permeability to sucrose would be indetectible, since the cells are essentially impermeable to it. The difference occurred immediately after irradiation and was retained for 17 days. On the 21st day, the cells returned to normal. The permeability to urea by onion was lowered by small doses of α-irradiation and increased by larger doses (Wattendorf, 1964, 1965). The increase occurred suddenly and immediately after the high irradiation (Wattendorf, 1965). A medium dose also gave rise to a sudden increase in permeability, but it gradually returned to nearly normal. Weaker doses also gave rise to a slight, temporary increase, but this was soon followed by a decrease below the original value.

Many other earlier publications have also reported effects of ionizing radiations on cell permeability. Stadelmann and Wattendorff (1966), however, have pointed out that there are difficulties in interpreting such results. They were unable to detect an effect of radiation on the urea permeability of onion epidermis, when the measurements were made some hours after irradiation. Later experiments showed an immediate response to irradiation as in Wattendorff's earlier experiments. Large doses of α-radiation (550–700 krad) produced an immediate 50% increase in urea permeability; 2–5 krad of β-radiation produced an immediate 50% increase in water permeability (Stadelmann, 1968). On the other hand, small doses of radiation (< 80 krad) immediately decreased urea permeability and β-radiation similarly decreased water permeability. Even the small doses of β-radiation (250–500 rad) which failed to produce an immediate effect, led to a decrease in permeability 24–72 hr later. After α-irradiation with 500 krad, the inner epidermal cells of the leaf sheaths of *Convallaria* showed an average threefold increase in permeability to water (Wattendorf, 1970). Similarly permeability to malonamide increased in most cells about 1 hr after irradiation. The radiation dose was not lethal even after several days. In the case of *Nitella flexilis* the increase in rate of penetration of ^{22}Na, following X-irradiation (80–120 kR at a rate of 100 kR/min), is apparently due to inactivation of the Na pump in the plasmalemma, and is, therefore, actually a decrease in efflux rate rather than an increase in rate of influx (Ozerskii, 1969). Increases in passive cation permeability have been obtained in the case of red blood cells (see below). Exposure of the ciliate *Tetrahymenia pyriformis*, to ^{60}Co (470 kR/hr) induced a decrease in permeability to dyes by as much as one-half (van de Vijver, 1969).

Permeability changes, of course, are evidence of membrane damage. More direct evidence is the occurrence of characteristic membrane complexes in wheat seedlings up to 6 hr after irradiation with 180 kR (but mainly with higher exposures), similar to those found during anoxia (von Wangenheim, 1969). The unusually high radiation dose and the rapid appearance of injury suggest that it is direct, as opposed to the indirect, oxygen-induced injury. Oxygen may actually be necessary for repair of such radiation injury (see below).

5. Radioprotectants

The radiochemical reactions are zero order, and therefore yield a straight line when plotted against time. In some cases, however, the reaction products may protect a radiosensitive solute, leading to an exponential inactivation curve. Similarly, the mere presence of another substance may decrease the yield per unit irradiation to as little as $\frac{1}{700}$ of that in pure solution. Inactivation of auxin may be lowered to this value by the presence of protein in as low an amount as one hundredth the concentration of the auxin (Gordon, 1957). Coenzymes may also protect enzymes against inactivation by ionizing radiations. Ascorbic acid and GSH, for instance, reduce the potentially harmful oxidized products formed by radiolysis of water molecules in the cell (Vasiliev, 1962). As a result, the total ascorbic acid is practically unaltered by radiation, although the reduced form is used up (Proctor and O'Meara, 1951—see Vasiliev 1962). Similarly, the reduced pyridine coenzymes protected trypsin against inactivation by γ-irradiation (^{60}Co) to a greater extent than did the oxidized coenzymes, both in dilute solution and in the dry state (Sanner, 1965). The protective effect of the oxidized coenzymes was almost independent of oxygen, whereas the protective effect of the reduced coenzymes was much greater in the absence, than in the presence of oxygen. The data are compatible with the view that $NADH_2$ and $NADPH_2$ may participate in the repair of target radicals in irradiated cells by a hydrogen transfer mechanism. This explanation, of course, implies that injury is indirect due to radiation capture by water molecules.

When ribonuclease or thiolated gelatin are X-irradiated in the dry state at 77°K *in vacuo*, secondary S radicals arise in the RNase, and the doublet-type radicals in the thiolated gelatin, primarily by an intermolecular mechanism (Copeland *et al.*, 1968). The presence of high concentrations of adenine during irradiation prevented both of these changes. Since the large adenine molecules are very unlikely to react with the radicals located in the interior of the protein molecules, the most plausible explanation is that they act as radical scavengers, preventing radical intermediates from interacting with protein molecules. The formation of the secondary sulfur radicals occurs on heat treatment (e.g., at 60°C) by intermolecular reactions mediated by diffusible reactants, primarily H atoms which can be scavenged by adenine. In the case of trypsin in solution, however, the results strongly indicate that the OH radical is mainly responsible for inactivation (Sanner and Pihl, 1967). Protection of *E. coli* B at 0°C against this same effect of X-irradiation, increased with the increasing ability of the protectants (e.g., ethanol, glycerol) to react with OH radicals (Sanner and Pihl, 1969). There was no similar relationship to their ability to react with the hydrated electron (e^{-}_{aq}).

A similar protection against the radiochemical disintegration products of water (free radicals, peroxides, ions) was adopted as the explanation for the effects of another group of radioprotectants. Thiourea, thioacetamide, and other thiourea derivatives in relatively high concentrations (e.g., 0.5 *M* thiourea) when applied one-half hour in advance, protected onion epidermis against injury by ionizing radiations (Biebl and Url, 1963a,c, 1968). A second group of substances, however (glucose, fructose, sucrose, glycerol, pentaerythrite, $CaCl_2$, and KCl), protected in a completely different manner, since they decreased the injury when applied after the irradiation.

Other miscellaneous substances have also been reported to act as radioprotectants. Methionine and lysine decreased the frequency of tumors in fern prothalli from X-irradiated spores, while glutamine and arginine had the opposite effect (Partanen, 1960). DMSO definitely protects seeds against initial X-ray damage, rather than stimulating recovery (Kaul, 1970). The antibiotic trichothecin, if applied prior to irradiation, acted as a radioprotectant for wheat against radiation doses below 2000 R (Il' na *et al.*, 1968). A similar presowing, soaking treatment of irradiated (40 kR γ-radiation) wheat seeds with IAA, tryptophan, and zinc induced recovery from the growth inhibition (Abrol *et al.*, 1969). Even animal hormones (hydrocortisone and prednisolone), have acted as radioprotectants for *Vicia faba* (Lozeron and Jadassohn, 1967). Plants from irradiated seed showed a favorable effect due to treatment with either 5-phenyl-2-imino-4-oxo-oxazoladine or procaine hydrochloride (Schwanitz and Schwanitz, 1968). Such results, however, must be interpreted with caution, since in some cases, control treatments with these substances (in the absence of irradiation) produced enhanced germination and better growth of the young plants.

The radioprotectants that have received the most attention are the sulfhydryl compounds (mercaptans or thiols). In 1949, Patt *et al.* discovered that cysteine could protect mice against the lethal action of ionizing radiation. In 1951, the protective action of cysteamine and its S-guanido derivative AET (2-β-aminoethyl-isothiourea) was discovered. Due to the many investigations of these substances, it is now established beyond question that injection of cysteine or cysteamine into animals affords some protection against X-ray damage (Dewey, 1965). On the other hand, several hundred compounds have been tested, yet only those which conform to the cysteine–cysteamine group are active (Eldjarn, 1965). These results with animal cells have been corroborated by the few investigations of plants. Cysteine decreased the radiation damage to onion roots to one-half or one-third (Forssberg and Nybom, 1953). Cysteamine protected the tissue of Jerusalem artichoke against the effects of X-irradiation (Jonard, 1968). Even postradiation soaking of seeds in solutions of cysteine (and ATP) leads to a reduction in radiation injury (Miller, 1968). Similarly, treating barley seeds with the radioprotectant 2,β-aminoethylisothiuronium bromide hydrobromide, before, during, and after α-irradiation (20–30 kR from ^{137}Cs) decreased the injury (Panoyan, 1969).

6. Mechanism of Injury

In the case of the temperature and water deficit stresses, it is not immediately clear why the respective strains (e.g., dehydration) should lead to injury, since the strain is elastic and, therefore, reversible. The general mechanism of injury due to the radiation stress is much more obvious. The radiation stress also produces an elastic strain—ionization or radical formation, but these are unstable products and might be expected to disappear immediately, either by reforming the original molecules, or by forming new stable substances. The latter change would be a (thermodynamically) irreversible and, therefore, plastic strain. If the substances that undergo the chemical change are essential protoplasmic components (e.g., nucleic acids, proteins, or lipids) injury will occur which may or may not be subsequently repaired. Many investigations (see above) have shown that the unstable intermediates are surprisingly long-lived, and therefore the elastic strain may persist for an unexpectedly long time.

The only real question in relation to the mechanism of injury is, therefore, which of the several possible types of chemical change is (or are) actually responsible for the radiation injury.

a. Mechanism of Indirect Injury

The discovery of the radioprotectant action of thiols has provided a new method of attack on this aspect of the mechanism of radiation injury. Four main hypotheses have been proposed to explain the protective effect of the thiols (Eldjarn, 1965): (1) Radical scavenging. More recently, the scavenging of Cu or Fe ions has been suggested (Foye and Mickels, 1965), thus preventing cellular oxidations normally initiated by radiation. (2) Repair of the ionized target molecules by H transfer. (3) Anoxia. Most of the early results may simply be due to oxidation of the CSH, leading to a reduced oxygen tension and a corresponding decrease in radiosensitivity (Dewey, 1965). In recent work, however, evidence has been produced which cannot be accounted for by each of these three hypotheses (Eldjarn, 1965). (4) Mixed disulfide formation. This hypothesis was first proposed by Eldjarn and Pihl (1956—see Eldjarn, 1965).

According to this concept, the small-moleculed protective thiols form mixed disulfides with the target (protein) molecules by combining with the SH and SS groups of these larger molecules. This binding is temporary, as is the protection, due to the normal presence of a GSSG reductase which will eventually split the mixed disulfides, releasing the thiols and, therefore, eliminating their protective effect. The freed thiols will then be excreted. Extensive formation of such mixed disulfides would temporarily interfere with the enzyme activities and, therefore, with cell metabolism. This would lead to a pronounced toxicity by these radioprotectants. This theory would also explain the moderate degree of sensitization to radiation injury by radioprotectants (Bridges, 1969) in those cases where naturally occurring protective agents (CSH, GSH) would be stripped off the proteins by the artificial radioprotectants. In agreement with the mixed disulfide hypothesis, it has been shown that the SH and SS groups of proteins attract a greater share of radiation damage than that corresponding to their mole fraction. It is against this part of the radiation damage that the mixed disulfide may offer partial protection.

Investigations of pure proteins have also supported the concept. Pihl and Sanner (1963a,b) investigated papain, the most radiosensitive enzyme studied. When its single essential SH group was converted to a mixed disulfide with cysteamine and the enzyme was then irradiated, the dose reduction factor (DRF) was extremely high (8.2). This demonstrated unequivocally that papain in the form of the mixed disulfide with cysteamine was effectively protected against the direct action of ionizing radiation. In the case of rabbit muscle phosphofructokinase, Chapman *et al.* (1969) found that destruction of SH groups plays an important role in the loss of catalytic activity induced by X-irradiation. Even in the case of fructose-1,6-diphosphatase, an enzyme stimulated by SS formation, the results indicated that SH groups play an important but not exclusive role in the X-ray inactivation (Little *et al.*, 1969). Irradiation of the enzyme in solution resulted in an initial stimulation followed by a subsequent exponential inactivation. The disulfide form of the enzyme was inactivated exponentially without an initial stimulation. In the native enzyme, the allosteric activity was much more radiosensitive than the catalytic activity. This was also true of phosphorylase *b* (Damjanovich *et al.* 1967). The X-ray inactivation of the SH groups of this enzyme was sufficient to account for the inactivation of the enzyme. Blocking two SH groups with pCMB decreased the radiosensitivity of the enzyme by a factor of 2. Both the X-ray inactivation and that due to SH reagents was largely an effect on allosteric sites.

By use of the ESR technique, Henriksen and Sanner (1967) were able to demonstrate the formation of sulfur radicals from the thiol penicillamine when irradiated at 77°K in a dry mixture with Sephadex. This was due to the secondary reactions resulting in a transfer of radiation energy from the Sephadex to the penicillamine. An appreciable transfer of unpaired spins was found to occur from the macromolecule to the thiol as a result of the X-irradiation (Sanner *et al.*, 1967).

All the above results supporting the mixed disulfide hypothesis have been produced by the Norwegian school of investigators, but support has come from other laboratories as well. Gorin (1965) points out that reactions do, indeed, occur between the applied mercaptan or disulfide radioprotectants and the body proteins. Myers and Church (1967) showed that the effect of dithiothreitol (DTT) on stromal enzymes mimics the effects of X-irradiation. Similarly, some SH reagents may sensitize the cells to radiation injury, leading to as much as a fourfold increase (Biarichi *et al.* 1965). Among these, at least the *N*-ethylmaleimide (NEM) appeared to produce its effect by reaction with SH groups. The action of iodoacetate (IA) and the other SH reagents, however, seemed to be independent of such alkylation.

From all these results, it must be concluded that sulfhydryl groups do, indeed, play a role in radiation damage. However, the most convincing evidence of a relationship between protein SH and radiation injury has been obtained in the case of red blood cells, which do not possess nuclei. What about the radiation injury produced by weaker doses in nucleated cells, which have been shown to involve nuclear damage? Inhibition of cell division is one of the most sensitive effects of radiation, and convincing evidence has shown that in microorganisms this inhibition is due to radiation-induced damage to DNA proper (see Sanner and Pihl, 1968). Several investigators have shown the effectiveness of SH substances (CSH, GSH, 2,3-dimercaptopropanol) as radioprotectants against nuclear (chromosome) damage (Vasiliev, 1962). Furthermore, CSH and cysteamine are protective even after the irradiation. Consequently, they are not acting simply as scavengers, but can induce repair. According to Bacq and Goutier (1968), the mixed SS theory (as well as the others mentioned above) is unable to explain protection by S-containing radioprotectants. These substances delay mitosis and DNA synthesis, permitting biochemical changes which increase the repair or which protect against radiation damage. In agreement with this explanation, treatment with β-mercaptoethanol prolonged the interval between X-irradiation and the first postirradiation cell division of diploid yeast cells, and resulted in increased survival (James and Werner, 1969). With haploids, no significant increase in survival was detected.

These results seem to eliminate the mixed disulfide theory as an explanation of nuclear radiation injury, but not as an explanation of cytoplasmic radiation injury. Even in the case of nuclear injury, however, a role for SH groups is not excluded, for according to Sanner and Pihl (1968), SH enzymes may be involved in the repair mechanism (see radioresistance below).

The actual chemical mechanism of injury via the SH groups still remains unexplained. In the case of small-moleculed thiols, like CSH and cysteamine, almost all the water radicals resulting from irradiation react with the SH groups (Sanner and Pihl, 1968). Under certain conditions, chain reactions may occur, leading to an even greater yield of reacting molecules. The SH groups of proteins do not participate in such chain reactions, and therefore, proteins are not as sensitive to radiation as are the small thiols. Furthermore, other residues may react with the water radicals, in this way protecting the SH groups of the proteins.

Nevertheless, the effects of irradiation on the small thiols may be instructive. When CSH is irradiated under anoxic conditions, the disulfide form, cystine, and alanine are the main products (Sanner and Pihl, 1968). In the presence of oxygen, sulfinic and sulfonic acids are also formed. Less

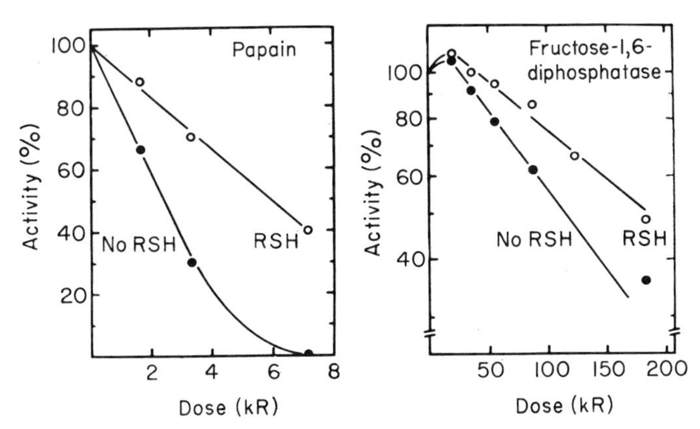

FIG. 18.1. Activation of catalytic activity by addition of cysteamine after irradiation. (From Sanner and Pihl, 1968.)

is known of the radiation products of the protein SH groups. In certain cases, the radiation-induced inactivation of SH groups is partly reversed by adding thiols after exposure (Fig. 18.1).

b. MECHANISM OF DIRECT INJURY

That permeability changes are the cause, and not the result of radiation induced injury, is indicated by the many observations of an increase in permeability and a loss of leachates followed by a reversal coinciding with recovery (see above). In the case of *Neurospora* conidia, X-irradiation led to a postirradiation leakage of ^{32}P, which was largely prevented by including calcium gluconate in the distilled water in which they were irradiated (Weijer, 1961). Less leakage occurred after irradiation with 7200R in the presence of calcium gluconate than in its absence after irradiation with 240 R. Since calcium has long been known to decrease permeability, this seems to indicate that the sole cause of injury is the radiation-induced increase in permeability, and that this injury can be prevented by the calcium induced protection against the injurious increase in permeability.

In the case of erythrocytes, there is strong evidence that this radiation-induced increase in permeability is due to changes in protein SH groups. That the change is truly in passive permeability and not in active absorption, is indicated by its occurrence when erythrocytes are irradiated (2000–20,000 R) at 4°C (Shapiro *et al.*, 1966). A similar increase in permeability was induced by the radioprotectant mercaptans, GSH and MEG, and by the SH-blocking agents PHMB, chlormerodrin, and NEM. Radiation and PHMB effects, in fact, were additive up to a point. The radioprotectants decreased the effect of both radiation and PHMB, and one of them (MEG) even reversed the effects when added after irradiation. Shapiro *et al.* (1966), therefore, suggest that the protein SH groups on the cell surface and inside the cell are the radiation targets, and that MEG may reverse the radiation effect by reducing protein disulfide formed by radiative oxidation of protein SH groups.

More direct evidence of this concept was obtained by Sutherland and Pihl (1968). Both in the whole erythrocytes and their isolated "ghosts" (membranes), a decrease in protein SH occurred at radiations as low as 40 kR (Fig. 18.2). Lipid peroxidation has also been shown to occur as a result of radiations (see above), and this was corroborated by Sutherland and Pihl (1968), but little or none occurred in the erythrocyte cells at less than 100 kR (Fig. 18.3), a dose far larger than required for the SH loss. The membrane SH groups consist of two classes with widely different sensitivities (Sutherland and Pihl, 1968), the more radiosensitive fraction constituting about 10% of the total number of SH groups. Factors expected to influence the structure of the membrane altered the percentage of radiosensitive SH groups. The loss of SH groups was greater when isolated ghosts were irradiated than when intact cells were exposed (Fig. 18.3). SS formation accounted for the major part of the SH loss on irradiation (Table 18.1). Sanner and Pihl (1968) point out that steric hindrance between the SH groups within a single molecule makes intramolecular reac-

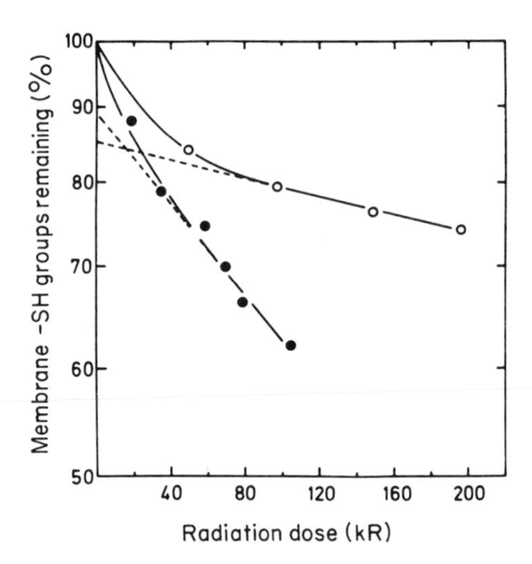

Fig. 18.2. Effect of X-rays on erythrocyte membrane SH groups. The disappearance of titratable membrane SH groups is plotted on a log scale as a function of radiation dose after exposure of cells (○——○) and ghosts (●——●). (From Sutherland and Pihl, 1968.)

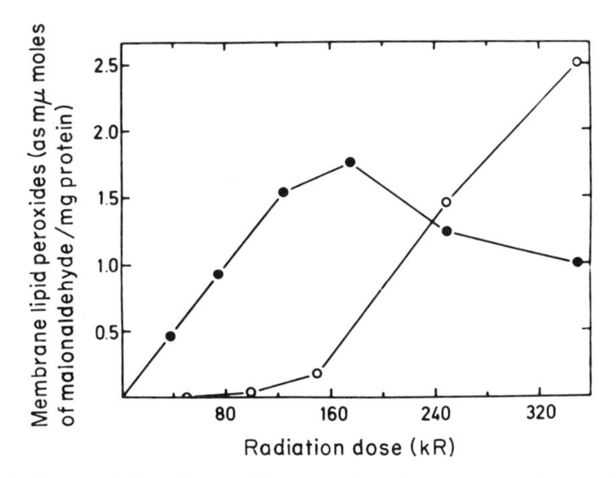

FIG. 18.3. Radiation-induced peroxidation of erythrocyte membrane lipids. Suspension of cells (○——○) and of ghosts (●——●) were irradiated with increasing doses of X-rays. (From Sutherland and Pihl, 1968.)

tions unlikely. However, no such hindrance would interfere with *intermolecular* SS formation.

In agreement with Shapiro *et al.* (1966), incubation of erythrocytes with SH-blocking agents resulted in an increase in passive permeability to cations, similar to that induced by x-irradiation (Sutherland *et al.*, 1967a). The radiation-induced loss of membrane SH groups was partly reversed

TABLE 18.1 RADIATION-INDUCED FORMATION OF MEMBRANE DISULFIDE GROUPS[a]

Preparation irradiated[b]	Dose (kR)	—SH lost (m/moles/mg protein)	—SH converted to —SS—[c]	
			m/moles/mg protein	% of —SH lost
Ghosts	35 (1)[d]	5.1	3.3	64.7
	70 (2)	7.1	4.2	59.2
	105 (2)	12.4	7.0	56.5
	196 (2)	20.5	11.0	53.7
Cells	105 (1)	10.3	7.4	71.8
	196 (6)	12.3	10.4	84.6

[a] From Sutherland and Pihl, 1968.

[b] The samples were kept in an ice bath during irradiation.

[c] Calculated on the basis of —SH titrations before and after reduction with sodium borohydride. The —SH titrations were carried out in the presence of 8 *M* urea and 0.2% lauryl sulfate.

[d] The figures in parentheses denote the number of experiments.

when the erythrocytes were allowed to metabolize after exposure under conditions known to give partial repair of radiation-induced permeability changes. A 60% recovery of SH groups occurred after 1 hr at 37°C in the presence of glucose, while none occurred at 4°C (Sutherland and Pihl, 1968). Direct measurements of membrane SS content revealed that the reappearance of SH groups after exposure was due to reduction of radiation-induced SS formation.

A similar association of repair with SH groups is indicated in the case of the fungus *Phycomyces blakesleeanus*. A decrease in growth occurred as a result of X-irradiation (100 R), which was reversible in 10 min (Forssberg *et al.* 1967). No cells suffered any injury. During this 10-min repair period, free SH increased, reaching a maximum 2–4 min after irradiation, coincident with a minimum in intracellular oxygen concentration.

From all these results, it must be concluded that (1) the radiation-induced increase in permeability is due to a loss of membrane SH. (2) The loss of membrane SH is due to oxidation to SS groups. (3) These SS groups form intermolecularly, inducing aggregation of membrane proteins (4) Repair of the radiation-induced damage caused by the increase in permeability is due to reduction of these intermolecular SS groups to SH.

What evidence is there that a similar mechanism may operate in plant cells? As in the case of erythrocytes, plant cells have been shown to increase in permeability as a result of irradiation (see above). Furthermore, this increase has been shown to disappear in some cases after the irradiation, as in the case of the erythrocytes. Finally, SH reagents have been shown to disappear in some cases after the irradiation, as in the case of the erythrocytes. Finally, SH reagents have been shown to alter the permeability of plant cells (see Chapter 10). Therefore, the same mechanism may be logically assumed to operate in plant cells as in erythrocytes.

B. Radiation Resistance

The method of exposure of the plant to ionizing radiations is quite simple, and resistance is measured by the dose sufficient to produce certain degrees of injury. The dose producing 50% killing is commonly used for animals, but most determinations of plant injury are based on 100% killing (Vasiliev, 1962).

1. Avoidance

Since all substances absorb ionizing radiations essentially equally per unit mass, avoidance must be proportional to the fraction of nonliving material in the irradiated tissues. This fact undoubtedly explains the far greater resistance in plants than in animals. Due to the relatively thick cell walls and large vacuoles of mature plant cells, the proportion of living protoplasm per cell is much lower than in mature animal cells. The proportion of any specific dose of radiation captured by the living protoplasm will, therefore, be much less in the plant than in the animal cell. This also explains the lower resistance of meristematic (largely protoplasmic) cells than mature cells, and the direct relationship between radioresistance of vegetative cells and their degree of differentiation (Vasiliev, 1962).

A special example of avoidance is the direct relationship between nuclear volume and radiation damage which has been established by Sparrow and his co-workers (Sparrow and Woodwell, 1962), and corroborated by many others. In diploid species, they found a clear relationship between the average nuclear volume of apical meristem cells and resistance to chronic γ-radiation (from ^{60}Co). The larger the nuclear volume, the greater the sensitivity of the nucleus, and ultimately of the whole plant (Sparrow *et al.*, 1961). Among the different taxa investigated, resistance differed by a factor of at least 1000 (Sparrow *et al.*, 1961). Since nuclear volume, amount of DNA per nucleus, and cell size are all related, it can be concluded that radiosensitivity is also directly related to these quantitities and to average chromosome size. This last relationship was confirmed in the case of eleven species of *Linum* (Bari and Godward, 1969). Both polyploidy and increasing chromosome number appeared to increase radioresistance (Sparrow *et al.*, 1961). On the other hand, Sparrow *et al.* (1970) observed an inverse relationship between interphase chromosome volume and radiosensitivity for a single 16-hr exposure.

According to the above results, a constant or nearly constant number of ionizations per nucleus is required to produce growth inhibition or death (Sparrow *et al.*, 1961). Measurements of nuclear volume, therefore, provide a convenient method of distinguishing between avoidance and tolerance. As an example, the apical cells of the gemmae of *Lunularia cruciata* have a nuclear volume larger than that of the other vegetative cells and are the most radiosensitive (Miller, 1968). The difference, however, disappears when energy absorption by the nuclei at the 100% lethal exposure is compared. Therefore, there is no difference in tolerance between the two types of cells and the difference in radioresistance is solely due to a difference in avoidance.

Avoidance of the above type allows a plant to avoid both the direct and the indirect radioinjuries. A specific avoidance of indirect injury in the absence of avoidance of direct injury is also possible. If the primary reaction induced by the radiation is the formation of radicals and ions from water molecules, then avoidance of this indirect injury can be simply achieved by loss of water. A direct relationship between radiosensitivity and the water content of seeds has, indeed, been reported by many investigators (e.g., Petri, 1921—see Vasiliev, 1962). The optimum water content at which radioresistance is at a maximum, was found to be at about 11.2% for wheat grains and 12.9% for barley grains (Biebl and Mostafa, 1965). Since much of this first absorption of water is adsorbed by the cell wall and starch grains, rather than by the living protoplasm, these nonliving cell components would, therefore, protect the protoplasm by absorbing both the radiation and their secondary products. Some reports, however, indicate an increase in resistance as the water content inincreases from 7 to 20% in the first stage of absorption (see Vasiliev, 1962). In agreement with these results, mungo beans (Quastler and Baer, 1949) and wheat grains (Biebl and Pape, 1951; Biebl, 1959a) exhibit their highest resistance when air dry (36,000 R for inhibition of germination). Resistance of the wheat grains decreases during the first 24 hr of imbibition until at the most sensitive stage they are inhibited by 500 R. Resistance rises again with splitting of the scutellum and emergence of the primary roots and coleoptile. Thus, the drop in radioresistance which parallels the absorption of water may be explained by a decrease in avoidance of indirect injury. The subsequent rise in resistance may perhaps be due to cell enlargement and the consequent increase in nonliving mass per cell (Fig. 18.4).

Changes in water content cannot explain all the changes in resistance. The relation breaks down if the partially germinated seed are redried to the air-dry state; they are then no longer as resistant as the original air-dry

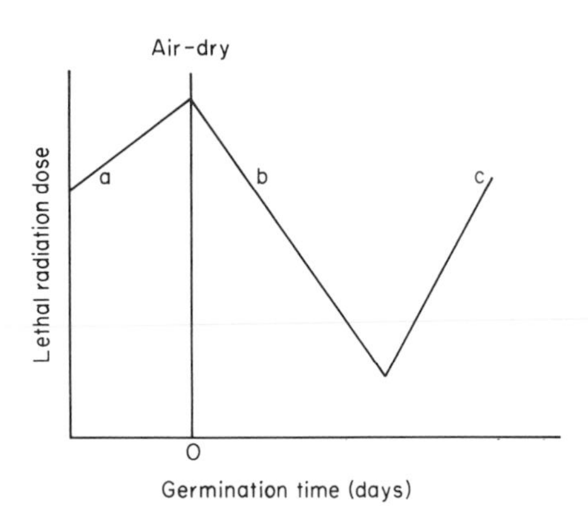

Fig. 18.4. Changes in radioresistance due to changes in radioavoidance during imbibition and germination of seeds. (a) Avoidance increases with water uptake (to 10–20%). Water mainly in walls and starch grains. (b) Avoidance decreases with water uptake. Water mainly going into the protoplasm. (c) Avoidance increases with cell differentiation. Proportion of protoplasm to cell decreases.

seed. Another possible type of avoidance may also be involved in this stage. Since the oxygen must diffuse into the tissues from the external atmosphere, avoidance of oxygen induced radiation injury would be inversely proportional to the permeability to oxygen of the external surface of the seed. Many seed coats are known to be impermeable to oxygen, and the initial swelling may conceivably decrease the permeability, which will increase suddenly when cracks develop. The three main types of radioavoidance are therefore due to: (1) a small nuclear mass or large proportion of nonprotoplasmal mass, (2) the major part of the water localized in nonliving parts of the cell, and (3) impermeability to oxygen.

2. Tolerance

In most cases of published differences in radiation resistance, insufficient information is available for a distinction between avoidance and tolerance. For instance, seeds of *Andropogon* from uraniferous soils when X-irradiated, showed a higher percentage germination and developed stalks and roots of greater average length than similarly irradiated seed from otherwise similar nonuraniferous soils (Mewissen *et al.*, 1959). In the absence of further information it is not possible to decide whether the difference in radioresistance is due to avoidance or tolerance. According to Sparrow *et al.* (1961), a constant or nearly constant number of ionizations per nucleus is required to produce a growth inhibition of meristems or death. This would seem to preclude any true tolerance. Nevertheless, evidence of tolerance can be found. The best evidence is the existence of a radiorepair mechanism.

Hardening of the plant is apparently possible, by exposing it to small doses of X-rays before subjecting it to the larger doses. Vasiliev (1962) believes that this is due to development of a repair mechanism. This explanation is supported by an increased resistance when exposed to room temperature after the irradiation than at 0°C which slows down metabolism and, therefore, any repair mechansim. This hardening would prevent the application of the Product Law where small doses are involved, for equal low intensity doses of prolonged irradiation and high intensity doses of short duration would not then produce the same amount of injury. This fact may require a reinterpretation of some results.

The "reversal phenomenon" (Palenzona, 1961) may also be due to hardening. It occurs in maize, barley, and wheat, with X-ray doses up to 240 kR. After a point of maximum sensitivity to a dose there is a diminution. Sparrow's group has discovered what appears to be a clearcut case of slow hardening. Eleven species of woody plants exposed to chronic

γ-irradiation (20 hr/day) for 8 years, showed a decline in the rate of death with increasing exposure time (Sparrow *et al.*, 1970).

Further evidence of tolerance is the existence of exceptions to Sparrow's rule of a direct relation between radiosensitivity and nuclear volume. In the case of two winter wheats which showed an extreme difference in radiosensitivity, there was no difference either in DNA content per nucleus or in nuclear volume (von Wagenheim and Walther, 1968). The difference in radioresistance between the two winter wheats was, therefore, presumably due to a difference in tolerance.

The existence of radiotolerance is also supported by investigations of rhythmic (both seasonal and diurnal) changes in radioresistance. Many plants are known to show such changes. A marked diurnal rhythm in radioresistance occurs in both higher and lower plants (Biebl and Hofer, 1966). Maximum resistance to α-irradiation occurred at noon in the case of onion epidermis, and at midnight in the case of a moss (*Mnium*). Minimum resistance occurred at 8 p.m. in the onion, and between noon and 4 p.m. in *Mnium*. The changes in nuclear size paralleled these diurnal changes in radioresistance, which were therefore *not* due to avoidance.

An annual (seasonal) rhythm in resistance also occurs in both the higher and lower plants. The bryophytes attain their maximum radiation resistance in June and their minimum in December. In the higher plants, the relationship is reversed—a seasonal increase in the radiosensitivity of winter wheat plants from February to May (Biebl, 1959b) and a summer decrease in the radioresistance of the outer epidermis of red onions (Biebl and Url, 1963b). The nuclei of onion epidermis also show an annual rhythm in size, attaining their maximum size in summer when radioresistance is lowest and their minimum size in winter when resistance is highest. In opposition to the diurnal change, the annual change in radioresistance of onion epidermis is, therefore, apparently due to a change in avoidance.

Biebl and Hofer were puzzled by the parallel diurnal changes in radioresistance and nuclear size, since these results "seem to contradict the observations of Sparrow and co-workers." They, therefore, investigated a wide range of radiation doses. In the case of doses causing death within 24 hr, nuclear size and radiation resistance ran parallel. In the case of doses producing injury after 1–3 weeks, cells with smaller nuclei were more resistant than those with larger nuclei. The resistance to large doses of radiation which cause rapid injury is apparently not due to avoidance and, therefore, must be due to tolerance. The resistance to small doses of radiation which can induce only a much slower injury, is apparently due to avoidance. Tolerance therefore seems to be characteristic of direct injury and avoidance of indirect injury.

Other rhythmic changes have been found in rape seed. Radiosensitivity increased in proportion to soaking time, but periodical (3-hr periods) fluctuations occurred (Shnaider and Vakher, 1969). These changes may be related to the quantity of nuclear RNA and the rate of protein synthesis.

Resistance varies with the plant organ. In the case of wheat, the coleoptile, leaves, primary roots, and lateral roots have relative radiosensitivities of 1:6:16:18, respectively (Vasiliev, 1962). Age or stage of development may also affect radioresistance. Younger seed were much more radiosensitive than older seed (Saric, 1961a). It is generally found that ionizing radiations produce more of an effect on dividing than on nondividing cells (see above). This relation has been called the "Law of Bergonie and Tribondeau." According to Haber and Rothstein (1969), however, the "law" is not universally valid, for the sensitivity of disks from tobacco leaves to γ-radiation was independent of cell division. Similarly, wheat grains attain their highest radiosensitivity after imbibition for 24 hr, just before nuclear division (Biebl and Pape, 1951). In the case of mosses, the older leaflets are frequently more resistant to radiation than the younger ones, though this did not seem to be due to age (Biebl and Hofer, 1966). Similarly, the resistance of onion epidermis to α-radiation decreased from the outer, more fully developed scales inward (Biebl and Url, 1963b). In the case of germinating seeds, radioresistance drops to a minimum after 1 to 2 days (Biebl and Pape, 1951; Biebl, 1958a). As mentioned above, this may be largely due to the increase in water content. All these differences would be more meaningful if separate measurements were made of avoidance and tolerance.

3. *Mechanisms of Radioresistance*

Since radioresistance may be due to either avoidance or tolerance, no one mechanism can explain all the different types of resistance. The three possible mechanisms of avoidance have already been mentioned. Tolerance may also conceivably be due to three mechanisms (Diag. 18.1). However, these mechanisms may not necessarily exist to any marked degree in the plant, since it has not been exposed to ionizing radiations in nature in sufficient doses to lead to a selection of adaptive mechanisms.

The evidence is strongest for the existence of the third radiotolerance mechanism—tolerance of the radiation strain due to a repair mechanism. Yeast cells recover from radiation damage induced by 30,000 R of γ-irradiation (^{60}Co) if kept in sterile water for 24 hr. Viability is then 70% as compared with 30% immediately after irradiation (Korogodin *et al.*, 1959). This is apparently a true recovery that occurs during mitotic rest. Such

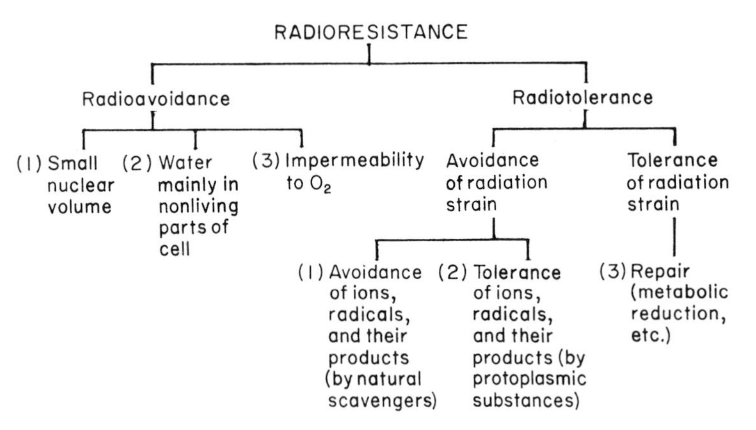

DIAG. 18.1. The six possible mechanisms of radioresistance.

"recovery," however, is not always due to repair. For instance, desiccated seed (2% H_2O) show apparent postirradiation recovery if stored under conditions of low moisture. This is explained by the existence of long-lived, radiation-induced radicals (Wolff and Sicard, 1961), which decay in the dry seed but continue to produce injury if the seeds are allowed to imbibe water and germinate. In the case of the slime mold *Dictyostelium discoideum*, the vegetative cells are extremely resistant to γ-radiation from ^{60}Co (Deering, 1968), having a 10% survival dose of 300 krad in air. Dose fractionation experiments indicated that this resistance is to a large degree due to repair of sublethal damage. Similar split-dose experiments with vegetative, stationary-phase cells of *Oedogonium cardiacum*, showed that part of the X-ray damage was repaired in a 2-hr interval (Howard, 1968). The ability of the cells to repair was independent of the presence of oxygen during the irradiation but did not occur unless more than 0.01 μmoles O_2/liter was present in the water surrounding the cells during the recovery interval. This supports the hypothesis that recovery is a metabolic process which requires an energy source in the cell. The metabolic repair (or "compensatory response") leads to *in vivo* recovery from radiation damage in the mitochondria of preclimacteric pear fruit but not in those at or near the climacteric peak (Romani *et al.*, 1968). In the case of Jerusalem artichoke, kinetin is a "radiorestorer" (Jonard, 1968). Pea seeds were subjected to 800 R of γ-irradiation for different lengths of time following the beginning of soaking (Shishenkova, 1967). Recovery processes played a substantial role in the differential radiosensitivity of the stages—the longer the time between the irradiation and fixation, the fewer the cells damaged. In the case of *Chlorella* cells, restoration following irradiation was inhibited at 0°C, but occurred at 25°C (Gilet and Terrier, 1969). In pear fruit cells, an increase in radiation dose that inhibited ribosomal synthesis also resulted in a transition from reparable to irreparable mitochondrial damage (Romani and Ku, 1970).

The repair mechanism can be explained on the basis of the mixed disulfide hypothesis of Eldjarn and Pihl (Eldjarn, 1965; Sanner and Pihl, 1968). Cells possess mechanisms for reduction of disulfides, restoring the biological activity of the proteins. Thus erythrocytes were capable of repairing damage caused by radiation-induced oxidation of membrane SH groups, when incubated after irradiation at 37°C in the presence of glucose for 1 hr. No repair occurred at 0°, or at 37°C in the absence of glucose (Sanner and Pihl, 1968). Furthermore, when cells are treated with SH-blocking agents prior to irradiation, the radioresistance is markedly lowered (Sanner and Pihl, 1968).

Not all radiotolerance is due to repair. Furthermore, the ability to repair the damage must depend on the radiotolerance of the enzymes themselves, i.e., on the second kind of radiotolerance (Diag. 18.1). Sanner and Pihl (1968) have shown that the relation between loss of enzyme activity and radiodestruction of enzyme SH groups may differ in different enzymes (Fig. 18.1). Furthermore, SH enzymes are not necessarily more sensitive to ionizing radiations than non-SH enzymes. Among the fifteen that they list, the destruction of SH groups can account for radiation-induced loss in activity (entirely or in part) in the case of only six. The reason is that different numbers of SH groups have to be destroyed in order to inactivate different enzymes, and that different SH groups differ greatly in radiation sensitivity.

Since such differences occur in radiosensitivities of different enzymes, it seems obvious that some of the differences in radiotolerance of different organisms must be due to differences in radiosensitivities of their proteins. This has not, however, been demonstrated in the case of any two organisms differing in radiotolerance. A role of lipids in tolerance has been suggested by Cervigni *et al.* (1969). Seeds of some nine species of plants were exposed to 20 kR and glow curves measured. The more resistant species had the higher thermoluminescence. The active component appeared to be the lipids.

The first kind of radiotolerance—avoidance of the ionizing and other products of the radiochemical reaction, due to the action of natural scavengers—seems the most logical kind of tolerance for the plant to develop. The lack of evidence for this kind of tolerance may be due to the lack of a search for it, or to the above mentioned absence of selection of plants for radiotolerance.

Interrelations between radioresistance and resistance to other stresses have received little attention. Gamma-radiation increased the "chemical" or salt resistance to $ZnSO_4$, $MnSO_4$, $VOSO_4$, and partly to H_3BO_3 but not to $Cr(SO_4)_3$ (Biebl, 1956a); it also increased the drought resistance of peppermint and sunflower (Savin and Stepanenko, 1967). On the other hand, cold, reduced light intensity, or dryness had no effect on radioresistance (Biebl and Hofer, 1966). The winter minimum in radioresistance of mosses was, therefore, presumably not due to the low-temperature stress. It may have been due to the short photoperiod, since this was found to lower their radioresistance markedly.

It should be emphasized, that even if the chemical mechanism of injury due to the radiation stress is the same as that due to the other stresses, resistance to the different stresses would not necessarily be correlated. According to the above hypothesis both radiation injury and freezing injury are due to aggregation of proteins (e.g., in the plasma membrane) by intermolecular SS formation. In the case of freezing injury the reaction producing the injury is due to dehydration, leading to a close enough approach of the molecules for the reaction to take place, and any factor capable of even slightly decreasing the dehydration would increase the freezing resistance. In the case of radioinjury, the reaction is due to activation of the irradiated molecules, leading to formation of highly reactive S radicals. Only a nearly complete dehydration could induce resistance, by eliminating the water radicals.

SALTS AND OTHER STRESSES

CHAPTER 19

SALT AND ION STRESSES

The problem of ion deficiency is a field in itself and will not be considered here. The terms salt or ion stress, as used here, therefore, refer only to an excess. The stress must be measured in energy units (as in the case of other stresses): chemical potential, activity, or more simply concentration. Although the effects of a salt are due to its ions, a distinction will be made between a salt stress and an ion stress. If the salt concentration is high enough to lower the water potential appreciably (0.5–1.0 bar), the stress will be called a salt stress. If the salt (or acid or base) concentration is not high enough to lower the water potential appreciably, the stress will be called an ion stress. In practice, salt stresses are commonly due to much higher concentrations than the above lower limit, and the ion stress is commonly due to a much lower concentration (10^{-3} M or lower). Salt stresses have received far more attention than ion stresses.

A. Salt Stress

1. Calcium Salt Stress

Although the Ca salts present in nature are of low solubility, the activity of the Ca^{2+} in the soil is much higher than that of ions responsible for ion stresses. Under artificial conditions, it has been investigated as the soluble salt ($CaCl_2$) at concentrations that definitely qualify it as a salt stress. Calcareous soils may injure, kill, or simply inhibit the growth of some plants. These are called *calcifuge* plants. Plants that are able to grow and develop on calcareous soils are called *calcicoles*. The distinction exists not only between species but also between ecotypes of the same species. A population of *Hypericum perforatum* found growing on calcareous soil grew better at high Ca levels than a population found growing on acid sites (Ramakrishnan, 1969). At low Ca levels, the reverse was true.

The strain produced in the plant is usually not due to the primary Ca stress, but to the secondary stresses induced by the Ca (see Kinzel, 1963)—pH, mineral deficiency, etc. (Diag. 19.1). The associated anion is particularly important. Calcareous soils have high quantities of bicarbonate as well as calcium. The root growth of the calcifuge (Ca-sensitive) species of grasses was more strongly inhibited by bicarbonate (up to 10 mEq/liter) than was the root growth of the calcicole (Ca-resistant) grasses (Lee and Woolhouse, 1969). The Ca may, therefore, injure the plants primarily because of the accompanying HCO_3^-. Even when large quantities of Ca are taken up by the plant, they can usually be precipitated in the cell sap, commonly as oxalates or carbonates, or neutralized in the soluble form in the cell sap as malates or citrates (Iljin—see Kinzel, 1963). Iljin has, therefore, classified plants on a phyiological basis (in contrast to the above ecological classification) as *calciophiles* (or calciotrophs) and *calciophobes*. The calciophile contains large quantities of soluble calcium in the cell sap, while the calciophobe contains no (or very little) soluble Ca in its sap but more or less large quantities of insoluble Ca. In general (though not in all cases), calciophiles are calcicoles and calciophobes are calcifuges (Diag. 19.1).

Jeffries and Willis (1964) postulate that calcicole plants are more selec-

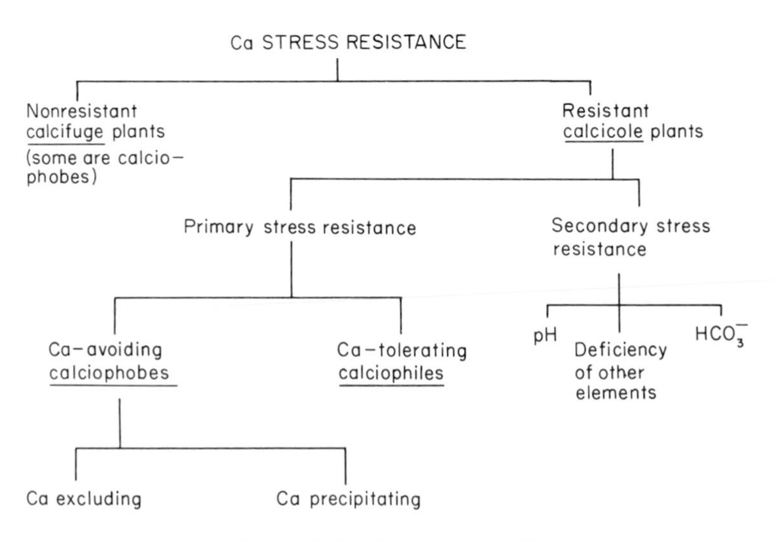

DIAG. 19.1. Kinds of resistance to Ca stress.

tive than calcifuge plants with respect to the ions absorbed. In the presence of excess Ca, the calcifuges would absorb considerable Ca at the expense of other ions, and would therefore suffer from a deficiency. This implies that the Ca resistance of the calcicoles is due to avoidance of a secondary deficiency stress. Their own data, however, fail to support this hypothesis, since the concentration of Ca in the calcicoles was at least as high and frequently higher than in the calcifuges. Iljin (see Kinzel, 1963), in fact, has found much higher Ca concentrations in calciophiles which succeed on calcareous soils than in calciophobes which do not. This was true whether they were growing in high Ca or low Ca soils. It must, therefore, be concluded that although plants differ in their reaction to Ca salts, there is no evidence that injury is due to a primary salt stress. The several possible kinds of resistance are shown in Diag. 19.1.

2. Sodium Salt Stress

Most of the salt stresses in nature are due to Na salts, particularly NaCl. The term *halophyte* literally means salt plant, but is used specifically for plants that can grow in the presence of high concentrations of Na salts. Plants that cannot grow in the presence of high concentrations of Na salts are called *glycophytes* ("sweet" plants). Because of the wide range of resistance found among halophytes, they have been subdivided into the extreme euhalophytes and the moderate oligohalophytes (Takada, 1954). Many halophytes are able to grow perfectly normally in low or nonsaline environments (Ungar *et al.*, 1969) and are, therefore, facultative halophytes. Other cannot, and are, therefore, obligate halophytes (Weissenböck, 1969).

a. Salt Injury

i. Limits of Salt Stress Survival. Some halophytes survive extremely high salt concentrations, compared to the low concentrations that injure nonhalophytes (Table 19.1). This difference is found, of course, both under natural and artificial conditions. Halophytes can grow on soils containing up to 20% salt, although most grow on soils with 2–6% salt (Strogonov, 1964), and they can grow in solution culture with very high salt concentrations. They may accumulate large amounts of salt, e.g., a 10.1% solution in the tissues of *Salicornia* (Keller 1940—see Strogonov, 1964). Leaves of *Nitraria schoberi* (the most tolerant among woody plants) may contain 14% of their dry matter in the form of NaCl, 57% as total salts (Strogonov, 1964). Growth of strand dune plants was decreased at

TABLE 19.1 LIMITS OF SURVIVAL OF SALT CONCENTRATIONS BY SOME HALOPHYTES AND NONHALOPHYTES[a]

Plant	Salt concentration survived	Reference
(a) Nonhalophytes (glycophytes)		
Snap beans	Injured by 1000 ppm NaCl	Barnes and Peel, 1958
Potatoes	Growth at more than 2000 ppm NaCl	Barnes and Peel, 1958
Barley	Chlorosis commences at 500 mEq Cl/liter	Greenway, 1965b
Soybeans	Stems, less than 15,000 ppm Leaves, less than 30,000 ppm	Abel and Mackenzie, 1964
Beta vulgaris and Spinacea oleracea	−4 atm NaCl caused only a slight reduction in yield	Nieman, 1962
Phaseolus vulgaris, *Pisum sativum*, and *Allium cepa*	−2 atm NaCl suppressed yields 45–61%	Nieman, 1962
(b) Halophilic microorganisms		
Penicillium notatum	Selected strain grows on saturated calcium acetate	Siegel *et al.*, 1967
Blue-green algae (growing in saline lagoons)	Begin to die at 1700 m*M* NaCl	Krishna, 1955
Phormidium tenue	Survives in 1700 m*M* NaCl	Krishna, 1955
Dunaliella tertiolecta	Grew at 3.75–12.0% NaCl	McLachlan, 1960
Dunaliella salina	In 5 *M* NaCl growth reduced to 1/6 of optimal value (in 1.25 to 2.5 *M* NaCl)	Marré *et al.*, 1958b
Debaryomyces hansenii (yeast)	Respiratory activity in 24% NaCl 10% of normal	Norkrans, 1968
Scrophulariopsis parvula (fungus)	Grew in media with saturated NaCl	Chen, 1964
(c) Halophytes (higher plant)		
Atriplex nummularia	Optimum growth at 100–200 mEq/liter of Na^+, K^+, or Mg^{2+} Cl; good growth at 300 mEq/liter	Greenway, 1968
Atriplex vesicaria	Seedlings established in 1 *M* NaCl	Black, 1960
Atriplex halimus	Growth retarded below −14 atm	Blumenthal-Goldschmidt and Poljakoff-Mayber, 1968
Succulent halophytes *Suaeda depressa*, *S. linearis*, *Salicornia europaea*, *Spergularia marina*	Germination impeded at 5% NaCl but uninjured	Ungar, 1962

TABLE 19.1 (Continued)

Plant	Salt concentration survived	Reference
Suaeda fruticosa	0.7 *M* NaCl or 0.2 *M* KCl	Shah, 1967
Halogeton glomeratus	1.4 molal highest survival	Repp *et al.*, 1959
Halophytes (several)	Germination in up to 1.0 *M* salt	Waisel, 1958b
Tamarix jordanis	Growth at −34 bar	Waisel and Pollack, 1970
Arthrocnemum glaucum	Growth at −45 bar	Waisel and Pollack, 1970

[a] Values in atm or bars refer to osmotic potentials of salt solutions.

salt concentrations higher than $\frac{1}{32}$ of sea water; growth of halophytes was maximal at $\frac{1}{8}$ to $\frac{1}{2}$ of seawater (Tsuda, 1961). Of course, the limit survived varies among the halophytes themselves (Tsuda, 1961; McMillan and Moseley, 1967). This is also true of the varieties of a single species (Alexander and Woodham, 1968). The difference was pronounced between ecotypes of one species from saline and nonsaline habitats, but not of another (Choudhuri, 1968).

Some families tend to show either high or low limits of salt survival. The limit is low in legumes (pea, beans), medium in cereal grasses (rye, oats, wheat, barley, sainfoin), and high in some forage and "technical" plants (Sudan grass, alfalfa, sunflower, sugar beet, forage beet; Maianu *et al.*, 1965). The limit may be indicated by a cessation of growth or by an actual killing of the tissues (Strogonov, 1964) in the form of a necrosis or a marginal burn (Ehlig and Bernstein, 1958), followed by a loss of turgor, falling of leaves, and finally death of the plant (Kovalskaia, 1958).

Bernstein (1964) classifies salinity effects as osmotic, nutritional, and toxic. In terms of stress terminology, the first two are secondary salt-induced stresses, while the third is a primary salt injury.

ii. Secondary Salt-Induced Stresses. When a plant is exposed to a salt stress, this means that the chemical potential, activity, or more simply the concentration of the salt is higher outside the plant than some arbitrary normal value. If the salt follows its diffusion gradient and penetrates the plant cells, two changes must accompany the internal salt stress or increase in concentration: (1) a change in ionic balance, and (2) a decrease in water potential. If the salt does not enter the cell, then only the second change will occur. These are elastic strains which would be immediately reversible on removal of the external salt stress. Can they give rise to other elastic strains, or only to plastic (irreversible and therefore injurious) strains?

In the case of some stresses (e.g., low temperature), decreased growth by itself may be a completely reversible, elastic strain. In the case of others (e.g., radiation) it is only partly reversible at best. Salt-induced growth inhibition seems to belong to the second group. Salt-stressed roots, for instance, deliver less hormone to the leaves, disrupting the leaf's hormone balance and increasing cell wall rigidity (O'Leary, 1970). Therefore, even if sufficient water is delivered to the leaf, cell expansion is decreased. The salt-induced increase in cell wall rigidity would, presumably, be irreversible. Nevertheless, as in the case of other stresses, if the meristematic tissues are uninjured, growth will recommence on removal of the stress, although not necessarily at the same rate. To what degree it is an elastic, reversible strain will depend on whether it is due to the primary or the secondary salt-induced stresses.

(a) *Osmotic stress.* If the salt lowers the external water potential below that of the cell, it exposes the cell to a secondary water deficit stress. To distinguish this from the drought stress, and because it leads to an osmotic dehydration strain, it is called an osmotic stress. It has also been called "physiological drought." Evidence of this water stress is the depression of growth and yield in rice by high concentrations of salt, regardless of whether it was NaCl, $CaCl_2$, or Na_2SO_4 (Ehrler, 1960), and the decrease in transpiration of bean plants in proportion to the salinity (Meiri and Poljakoff-Mayber, 1970). Because of this osmotic effect, halophytes show a quantitative relationship between the osmotic value of their sap and that of the root medium. This is usually (though not always—see below) due to the absorption of large quantities of salt from the external medium, achieving a higher total osmotic concentration inside the cell than outside. The salt concentration inside the cell will, nevertheless, be slightly lower than that outside, due to the presence of organic solutes in the cell (Marré and Servettaz, 1959). The osmotic value of the leaves of salt bush (*Atriplex*), for instance, varied directly with that of the solution in which the plants were growing (Ashby and Beadle, 1957; Black, 1960), showing a constant *absolute* value about 12 atm greater than that of the solutions throughout the range. As a result, their osmotic pressures ranged from thirty times that of the nutrient solution for control plants, to only two times for plants growing in solutions having 600 mEq/liter. This relationship was also true in nature. In soils with chloride contents varying from 19 to 9250 ppm, the leaves of the respective *Atriplex* plants had chloride contents ranging only from 9.6 to 13.6% of the dry weight (Beadle *et al.*, 1957). In the case of *Monochrysis lutheri*, the osmotic balance is mainly, if not solely, achieved by changes in cell content of cyclohexanetetrol (Craigie, 1969). The concentration was approximately 0.3 M in cells growing in seawater medium. Addition of 0.5 moles NaCl/liter of seawater medium caused an 80–90% increase in cyclitol concentration within 4 hr. Dilution of the medium reduced the cyclitol to the new steady-state level within 10 min.

Even in the case of extremely salt-sensitive plants, such as strawberry, the osmotic effect of the salt was the principal factor limiting its growth (Ehlig and Bernstein, 1958). Onion was unable to adjust osmotically to Cl^- salinity, while bean and cotton leaves did (Gale *et al.*, 1967). Even the bean plants, however, show a lag in adjustment, due to a rapid drop in water potential and leaf turgor, followed by a slower increase in cell sap concentration (Puscas *et al.*, 1966; Meiri and Poljakoff-Mayber, 1969). The increase in concentration was due to both inorganic and organic substances, but the former played the greater role, due primarily to accumulation of chloride (or of Na_2SO_4 according to Puscas *et al.*, 1966). As the cell sap concentration increased, turgor pressure was gradually restored. Even in the very salt-resistant yeast *Debaryomyces hansenii*, the total salt level is not sufficient to counteract the osmotic potential of the medium, and additional substances must be involved (Norkrans and Kylin, 1969).

Some microorganisms are obligate osmophiles, and can be grown only in media of high concentration. *Eremascus albus* is capable of growing in 2.4 *M* sucrose, but will not grow if the concentration is lowered to 5% (Paugh and Gray, 1969). Growth can be induced, however, by substituting an inert carbon source (1,2-propanediol) or a salt such as $CaCl_2$ for part of the sucrose.

(b) *Deficiency stress.* Even when the osmotic stress was eliminated, K caused a marked improvement in the growth of *Phaseolus vulgaris*, *Pisum sativum*, and *Citrus aurantium* (Giorgi *et al.*, 1967). Similarly, *Hordeum vulgare* was injured by NaCl only in the lowest of three nutrient concentrations (Greenway, 1963). Even leaf slices of the halophyte *Atriplex* when placed in a solution of NaCl, took up the Na largely in exchange for K lost from the vacuole (Osmond, 1968).

Solov' ev (1969a) observed a K deficiency in the lower parts of pumpkin (*Cucurbita pepo*) and sweet clover (*Melilotus alba*) plants subjected to NaCl salinization. Further experiments (1969b) led him to conclude that the main cause of NaCl-induced growth inhibition is the difficulty in uptake of mineral nutrient due to competition with the Na^+. In favor of this conclusion was the restoration of K supply and normal growth by placing an isolated root strand in a nutrient solution. When placed in pure water, growth was not restored. A split-root experiment with bean (*Phaseolus vulgaris*) and barley (*Hordeum vulgare*), on the other hand, revealed that the degree of osmotic adjustment and the growth rate, were functions of the proportion of the root system exposed to saline conditions, versus pure water (Kirkham *et al.*, 1969).

Potassium is not the only element that may be deficient as a result of a salt stress. Both chlorides and sulfates caused a decrease in content of total and inorganic P in tomatoes (Zhukovskaya, 1962). Resistance to this kind of secondary, salt-induced stress is shown by *Glycine falcata*, which has the ability to maintain a high P content in the presence of a salt stress (Wilson *et al.*, 1970).

iii. Primary Salt Injury. (a) *Distinction from osmotic injury.* In the above cases of secondary stress injury, plants are able to protect themselves against osmotic dehydration by accumulating the salt (and sometimes a soluble metabolic product) until the total osmotic concentration is in excess of that in the root medium. In the case of primary salt injury, on the other hand, the more the salt is accumulated (in the absence of tolerance—see below) the greater is the injury. Raspberries, for instance, accumulated Cl^- ions more rapidly than boysenberry and blackberry and were killed or severely injured (Ehlig, 1964). The Cl^- was accumulated by the plants from NaCl solutions more rapidly than the Na^+ and, therefore, Cl^- injury occurred earlier and was more severe than Na^+ injury. Tagawa and Ishizaka (1963) also concluded that the primary cause of injury to rice on transfer to 1.0% NaCl was the Cl^- accumulation in the shoots.

The most common method for distinguishing between secondary osmotic and primary salt injury is to compare the effects of isotonic solutions of the salt with those of organic substances. Carbowax (PEG with mol. wt. of 20,000), for instance, permitted larger yields from bean plants than from those grown in isosmotic solutions containing ions (Na, Ca, Mg chlorides; Lagerwerff and Eagle, 1961). Glycophytic, freshwater algae are injured less by sucrose than by NaCl, whereas halophytic algae are resistant to both (Kovarik, 1963). Similarly at equal osmotic concentrations, NaCl depresses the germination of alfalfa seeds much more than does mannitol (see Strogonov, 1964). Excised roots showed a similar effect—five times as much growth in mannitol as in NaCl of the same osmotic concentration. The toxic effect, is apparently due to the anion, rather than the cation. More recently, Strogonov and Lapina (1964) preferred dextran as a measure of osmotic injury because it is practically unabsorbed, it is nontoxic, and it maintains the osmotic value of the nutrient solution at a given level. However, it cannot be used by adding to the salt solution since the toxicity of the salt increases in dextran solutions (Lapina, 1968). At isosmotic concentrations (4.5 atm) they found that NaCl, and especially Na_2SO_4 significantly lowered growth and development compared to dextran. This, of course, indicates a specific inhibitory effect of the salts, over and above any osmotic effect. By use of this method they found some plants relatively insensitive to the osmotic effect, and some to the primary salt effect. Sodium chloride injury to maize, for instance, was primary, due to the penetration of excessive (injurious) amounts of the salt into the cells (Lapina, 1966). The same result was obtained with wheat seedlings, using sucrose as the purely osmotic agent (Bhardwaj, 1962). In this case, there was also an initial osmotic salt injury due to retarded water uptake.

Further evidence of a specific injury can be obtained by comparing the effects of different salts: NaCl, $CaCl_2$, and Na_2SO_4 depressed the growth of red kidney beans to the same degree at concentrations yielding the same osmotic potentials (−1.5 to −4.5 atm) relative to the control solution (−0.5 atm), while $MgCl_2$ and $MgSO_4$ produced a more pronounced retardation and definitely injured the plant (Gauch and Wadleigh, 1944). Similarly, very young sprouts of *Sophora japonica* required twice as much Na_2SO_4 or Na_2CO_3 (0.1 *N*) as NaCl (0.05 *N*) to inhibit them, in spite of the lower osmotic effectiveness of the NaCl (Alekperov and Khrzhanovskaya, 1961). Irrigation of Baccara roses with water containing chlorides was more detrimental than with water containing nitrates at the same level of salinity (Yaron *et al.*, 1969). Such differences can only be explained by a greater primary, injurious effect of the Mg^{2+} than the Na^+, and of the Cl^- than the SO_4^{2-}, CO_3^{2-}, or NO_3^-. Others have also concluded that Cl^- produces a specific injury, e.g., in the case of cotton (Strogonov *et al.*, 1963). This injury depended not on the total salt present but on the Cl^-: SO_4^{2-} ratio. In the case of rice, wheat, and barley grown in sand culture, NaCl was added up to amounts approaching the critical level for growth (Oga and Nishikawa, 1959). Even the largest additions of NaCl lowered the leaf water content only by 1.5%. The effect was, therefore, not due to osmotic dehydration. The distinction between a secondary osmotic and a primary salt effect is not always clearcut. Cl^- and SO_4^{2-} salinity, for instance, produced different effects on the water relations of cotton (Strogonov *et al.*, 1963).

(b) *Primary indirect salt injury.* As in the case of other stresses, the injury due to the primary salt stress may be either direct or indirect. Salt injury is usually preceded by a growth inhibition. According to Nieman (1965), the leaves of *Phaseolus vulgaris* were stunted by a salination of 72 mEq/liter because of fewer cells, judging by DNA content. However, Meiri and Poljakoff-Mayber (1967) arrived at the opposite conclusion from direct observation—the reduction in leaf area was due to a reduction in cell size. In any case, the growth inhibition is accompanied by one or more metabolic disturbances. This suggests, as mentioned above, that the growth inhibition is not simply a reversible elastic strain, but is one compo-

TABLE 19.2 METABOLIC CHANGES ASSOCIATED WITH SALT STRESS

Plant	Effect of salt	Reference
(a) Respiratory rate		
Cotton	Decreased by NaCl	Boyer, 1965
Algae (brown)	Inhibited by salinity	Munda, 1964
Bean	Salt increased number of mitochondria, perhaps an adaptive process	Siew and Klein, 1968
	Tended to increase in leaf but not in root	Nieman, 1962
Wheat and gram seedlings	Na_2SO_4 decreased it	Sarin and Rao, 1958
Wheat and gram seedlings	0.6% NaCl lowered it 0.2% Na_2CO_3 lowered it	Bhardwaj and Rao, 1960
Glycophytes	SO_4^{2-} increased it at medium concentration	Pokrovskaia, 1958
	Cl^- always decreased it at all concentrations	
Pea	Increase in NaCl content enhanced it by about $\frac{1}{3}$ in leaves, 10–15% in roots. Potential for oxidative phosphorylation appears to increase	Livne and Levin, 1967
Millet leaves	Cl^- and SO_4^{2-} stimulated it at first, later reduced it	Maksimova and Matukhin, 1965
Wheat and gram	0.6% Na_2SO_4 depressed it (max. depression 65% at 60 hr)	Sarin, 1961c
Pea root tips	NaCl increased percentage PPP Na_2SO_4 no effect	Porath and Poljakoff-Mayber, 1968
Dunaliella salina	Unchanged from 2 to 4 *M* NaCl, decreased from 4 to 5 *M* NaCl	Mironyuk and Einor, 1968
Barley	HMP pathway increased	Goris, 1969
(b) Photosynthetic rate		
Cotton	Decreased by NaCl	Boyer, 1965
Onion	Decreased by NaCl due to stomatal closure	Gale *et al.*, 1967
Bean	As above but later by affecting the light reaction	Gale *et al.*, 1967
Cotton	Decreased by NaCl mainly by affecting light reaction	Gale *et al.*, 1967
Algae	Stimulated by salinity	Munda, 1964

TABLE 19.2 (Continued)

Plant	Effect of salt	Reference
Algae	Depressed by salinity	Ogato and Matsui, 1965
Chlorella	No evident decrease within broad range (tolerant)	Trukhin, 1967
	not appreciably affected by NaCl	Nieman, 1962
Sudan grass	Increased by salt	Pokrovskaia, 1958
Dunaliella salina	Decreased with increase in NaCl from 2 to 5 *M*	Mironyuk and Einor, 1968
Marine plants (algae and phanerogams)	Lowering salinity (by dilution) decreases photosynthesis	Hammer, 1968
(c) Protein synthesis		
Phaseolus vulgaris	72 mEq NaCl/liter decreased the rate but prolonged it	Nieman, 1965
Pea seedlings	Salinization disturbs amino acid metabolism; formation of diamines and alkaloids, not normally present, causes necrosis	Prikhod'ko and Klyshev, 1963
Broad bean leaves	Retardation in synthesis of cysteine and methionine	Shevyakova, 1966
Pea	Inhibition of synthesis of basic proteins and increase in acidic proteins by NaCl or Na_2SO_4	Rakova *et al.*, 1969
Grape leaves	Salinity disturbed protein metabolism and inhibited growth	Saakyan and Petrosyan, 1964
Glycine javanica	Salinity up to 240 mEq/liter impaired synthesis	Gates *et al.*, 1966a
Pea roots	NaCl favored hydrolysis $NaHCO_3$ favored synthesis	Klyshev and Rakova, 1964
(d) Nucleotide synthesis		
Phaseolus vulgaris	72 mEq NaCl/liter reduced the rate of RNA synthesis but prolonged it	Nieman, 1965
Phaseolus vulgaris	No effect on DNA synthesis	Nieman, 1965
Soybean	Na_2SO_4 decreased synthesis in salt-sensitive variety, less in resistant variety	Rauser and Hanson, 1966
Grape leaves	Salinity disturbed NA metabolism and caused growth inhibition	Saakyan and Petrosyan, 1964

TABLE 19.2 (Continued)

Plant	Effect of salt	Reference
Barley	Cl increased content of AMP, ADP, and UDP in leaves, SO_4 increased GP, ADP, and ATP. Less increase in roots	Matukhin *et al.*, 1968
Pea leaves	0.2–0.4% NaCl did not interfere with free NA synthesis	Morozovski and Kabanov, 1968
Pea root tips	NAD + NADH decreased by NaCl or Na_2SO_4, NADPH + NADP increased only by NaCl	Hason-Porath and Poljakoff-Mayber, 1970
(e) Enzyme activity		
Wheat and gram (*Cicer arietinum*)	0.6% NaCl depressed catalase in gram, not in wheat. 0.2% Na_2CO_3 increased catalase activity in both	Bhardwaj, 1964b
Glycophytes	Catalase increased with brackishness. Polyphenol oxidase and peroxidase activity did not always parallel respiration rate	Pokrovskaia, 1958
Wheat, corn, pea	0.01–0.5 *N* NaCl, Na_2SO_4: catalase, peroxidase, and polyphenol oxidase all inhibited at high salt conc.	Vasile, 1963
Pea roots	NaCl resulted in 3rd MDH isoenzyme. NaCl, Na_2SO_4 increased mitochondrial MDH when NADP coupled, depressed when NAD coupled	Hason-Porath and Poljakoff-Mayber, 1969
Wheat	Salinization decreases amylase activity by inhibition (no effect on synthesis)	Sarin and Narayanan, 1968
Millet leaves	Cl^- and SO_4^{2-} lower activity of Cu enzymes, increase flavoprotein oxidases Cl^- suppressed polyphenol oxidase. SO_4^{2-} suppressed ascorbic oxidase	Maksimova and Matukhin, 1965
Pea root tips	NaCl or Na_2SO_4 decreased glucose consumption and C_6/C_1 ratios, increased G-6-phosphate dehydrogenase, decreased MDH	Porath and Poljakoff-Mayber, 1965

TABLE 19.2 (Continued)

Plant	Effect of salt	Reference
Pea root tips	NaCl and Na_2SO_4: most of glycolytic enzymes depressed, except G–P isomerase: not affected by SO_4^{2-}, increased 10X by Cl^-	Porath and Poljakoff-Mayber, 1968
Barley seedlings	Decreased activity of cytochrome oxidase, polyphenol oxidase, and ascorbate oxidase. Increased activity of flavoprotein oxidase	Goris, 1969
Pea roots	NaCl and Na_2SO_4 increased peroxidase activity	Rakova *et al.*, 1969 (also see Strogonov, 1964)
(f) Miscellaneous		
Pea and maize seedlings	Oxyacid content sharply dropped	Prikhod'ko, 1968
Spartina alterniflora (facultative halophyte)	NaCl decreased rate of dark $^{14}CO_2$ assimilation	Webb and Burley, 1965
Barley	Total lipid and phospholipid tended to increase slighty with increasing NaCl	Ferguson, 1966
Dunaliella salina	Chlorophyll decreased, carotene increased from 2 to 5 *M* NaCl	Mironyuk and Einor, 1968
Porphyridium sp.	Increased ratio of chlorophyll *b*: phycoerythrin in 2–3 times seawater	Leclerc, 1969
Bryophyllum pinnatum	0.04 *M* NaCl stimulates organic acid synthesis	Karmarkar and Joshi, 1969

nent of an overall plastic, irreversible strain. In order to understand the nature of this plastic strain, it is therefore necessary to find out just what these metabolic changes are. There have been many reports of a salt-induced decrease in several metabolic processes such as respiration, photosynthesis, protein synthesis, and nucleic acid synthesis (Table 19.2). Enzyme activity has also been reported to decrease in some cases and to increase in others (Table 19.2). An interference has been reported in carbohydrate metabolism (Bhardwaj, 1959), and in synthesis of chlorophyll and carotene (Kim, 1958). It is, therefore, usually assumed that salt injury is not due to a direct effect of the salts, but to the indirect effects of one or more of these metabolic disturbances. Strogonov (1964) has singled out the

N metabolism as the source of the injury. The salt-induced growth retardation leads to an accumulation of unused substances (Gauch and Eaton 1942). Synthesis of new proteins in cotton seedlings therefore decreases, while hydrolysis of the storage proteins is increased by the salt. As a result, amino acids accumulate. Strogonov points out that some of these may be toxic, in the following order of increasing toxicity: phenylalanine, aspartic acid, glutamic acid, proline, lysine, threonine, leucine, isoleucine, tyrosine, serine, and valine (Harris—see Strogonov, 1964). According to Strogonov, the actual toxic substance varies from species to species, depending on the metabolism of each species. In cotton leaves, NH_3 accumulated to two times the normal amount when weakly salinized and to ten times as much when strongly salinized, and may, therefore, be the cause of the salt injury. Putrescine accumulated in *Vicia faba* as a result of salination, but not in barley or sunflower. Strogonov actually produced the symptoms of salt injury in *Vicia faba* by the application of putrescine. Strogonov explains the accumulation of putrescine in some species and its absence from others on the basis of a single enzyme. He observed an increase in the cyclic amino acids, tyrosine and proline, in all plants as a result of salinity, both in the leaves and the roots. In the case of the leaves, nearly all the amino acids accumulated, but in the roots most were unchanged or decreased. Strogonov suggests that putrescine may be deaminated in the presence of diamine oxidase, liberating NH_3 and leading to the formation of proline as an end product. Injury due to accumulation of putrescine would then be the result of a salt-induced decrease in activity of diamine oxidase, which was shown to occur in cotton leaves. The proposed metabolic relationships are shown schematically by Strogonov (Fig. 19.1).

Strogonov's putrescine injury may indicate that it is a secondary salt-induced deficiency stress rather than a primary Na salt stress. Recent results (Smith, 1965) have shown that K deficiency leads to an accumulation of putrescine. If excess Na interferes with K absorption, it could lead to the accumulation for the same reason. Other effects of salts on amino acid metabolism have been described by Shevyakova and Komizerko (1969). In a culture of callus tissue from cabbage leaves, they found that salts slow down amination and amidation, as well as the transfer of S from sulfides to S-containing amino acids. They suggest that an increase in derivatives of S-amino acids may be one reason for the unfavorable effect of salts on the synthesis of amino acids in callus tissue.

Strogonov also observed a large increase in peroxidase activity in NaCl injured leaves. This may have played a role in the oxidation of the accumulated substances, leading to melanin formation from tyrosine in the necrotic areas. Catalase activity also increased, indicating a toxic accumu-

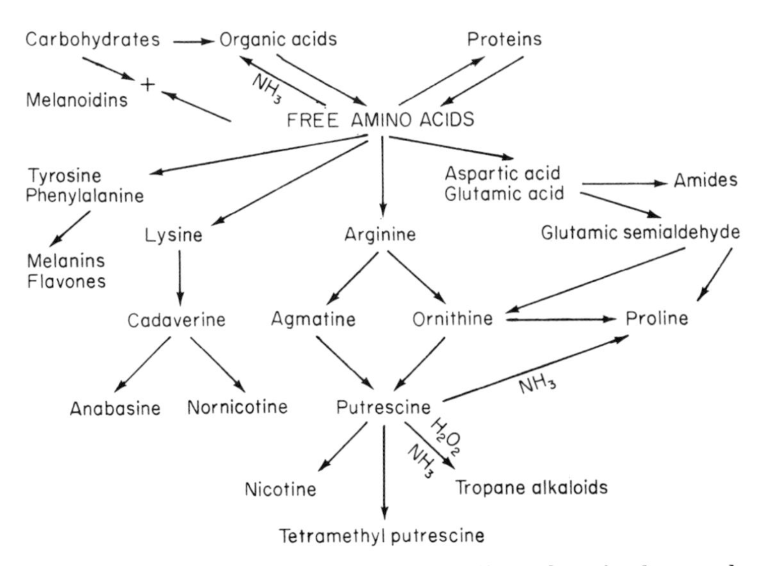

FIG. 19.1. Proposed accumulation of some metabolic products in plants under saline conditions. (From Strogonov, 1964. Reproduced by permission of the Israel Program for Scientific Translations, Jerusalem.)

lation of H_2O_2 according to Strogonov. He also found very toxic chlorine-containing substances in the salinized plants.

Since both a water-deficit stress and a nutrient deficiency stress are known to induce indirect metabolic injury (see Chapter 13 and above), it may be objected that the above indirect injury is just as likely to be due to secondary, salt-induced osmotic injury as to primary salt injury. It is, therefore, essential to eliminate secondary salt-induced stresses as the causes of injury (e.g., by control tests with nonsaline solutes) before concluding that any of the above indirect strains are due to the primary salt stress. Unfortunately, this precaution is not usually taken, and the metabolic disturbances can be ascribed to the primary salt stress only when they differ distinctly from those known to be induced by the secondary stresses.

(c) *Primary direct salt injury.* Although a primary indirect salt injury has apparently been established, the existence of a primary direct salt injury is still in question. In the case of other stresses, such direct effects are identified by their speed of action. A rapid salt injury may also occur—the salt shock. Even most halophytes, for instance, are sensitive to direct salt effects produced by a sudden salt shock (Repp, 1958). In habitats with marked fluctuations in salt concentration of the root medium, only the species with high shock resistance (e.g., *Salicornia* sp.) can survive.

Furthermore, the young plants that are still low in salt content are more shock sensitive than the older plants. Extreme resistance to salt shock is shown by the alga *Dunaliella salina* (Marré *et al.*, 1958b). The osmotic concentration of the medium was increased by 44 atm due to additions of NaCl, KCl, KNO_3, urea, or sucrose. In each case, cell movement ceased and the cells suffered severe shrinkage. This was followed by rapid recovery which was almost complete in less than an hour. The initial reaction was milder and the recovery was more rapid in the case of cells previously cultured in high salinity (3.75 M NaCl) than in those cultured in low salinity (1.25 M NaCl). In sucrose, the recovery was exceedingly slow.

Salt shock obviously requires such high concentrations of salt, that the secondary osmotic effect cannot be eliminated as the cause of the injury. A primary direct salt injury has, therefore, not been proved. Under strictly laboratory conditions, cells can be shown to suffer primary direct salt injury by immersing sections of tissue in high but hypotonic, single salt solutions. In the case of monovalent salts, this produces "vacuole contraction" or more correctly, protoplasm swelling which may lead to death in a few hours. The nature of this direct salt injury has not been clearly demonstrated. It appears to be due to a salt-induced increase in permeability, and therefore an effect on the plasma membrane. This implies an injurious change in either the lipids or the proteins of the membrane. The different possible kinds of salt injury are shown in Diag. 19.2.

iv. Effects of Environmental Factors. The degree of salt injury produced may be affected by a number of environmental factors. Kaho (1926) showed that the toxicity rises with the temperature (e.g., from 0 to 33°C), and this was corroborated by Sergeev (see Strogonov, 1964).

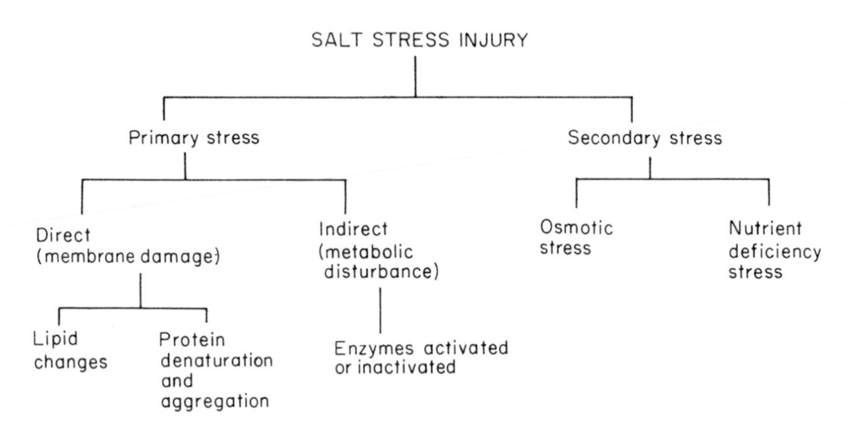

Diag. 19.2. The different possible kinds of salt stress injury (Na salts).

Salt resistance is also greater in the shade than in the light (Strogonov, 1964), although this may simply be due to a decreased transpiration leading to a decreased salt accumulation in the plant. It may also be due to the well-known photostimulation of ion absorption. Similarly, the presence of other ions may be expected to affect the absorption of Na salts and therefore the injury produced by them. The most important of these is Ca. It has long been known that the Ca^{2+} antagonizes the injurious effects of the Na^+, and that this effect is at least largely due to the decrease in cell permeability produced by Ca^{2+}, the increase produced by Na^+ (Scarth and Lloyd 1930). This important relationship has recently been reinvestigated in connection with salt injury. Wheat grains pretreated with water, germinated only to the extent of 8% on 1% NaCl, while grains pretreated with 1% $CaCl_2$ germinated 90% (Chaudhuri and Wiebe, 1968). The Ca pretreatment resulted in a 25% reduction in ^{22}Na uptake. Irrigation with saline water showed a similar effect (Deo and Kanwar 1969). Calcium-rich water decreased the uptake of Na but not of K. Bean plants (*Phaseolus vulgaris*) suffered no damage for 1 week in a NaCl concentration equal to $\frac{1}{10}$ seawater (50 mM) if the Ca concentration was 1 mM or higher (La Haye and Epstein, 1969). At lower concentrations, damage was severe, apparently due to a massive transport of Na into the leaves. In the presence of adequate Ca, the Na content fell to $\frac{1}{3}$ in the roots and to $\frac{1}{16}$ in the leaves. Bernstein (1970) points out that a distinction must be made between saline soils (containing enought Ca to meet the nutritional requirements of the plant), and sodic soils, in which the concentration of exchangeable Na is greater than 15%. The latter cause an accumulation of Na in the tops of bean plants. There are, however, differences between plants. Rice accumulated more Na and Ca than either barley or wheat (George, 1967), and is less selective between the two. Wheat accumulated more Na and less Ca in the roots. Silicon absorption has also been reported to increase salt resistance (Yoshida, 1965). Even the pretreatment of seed with other salts or their acids (e.g., 0.01 N boric acid or $AlSO_4$, has stimulated the subsequent germination in 0.1 and 0.2 N NaCl (Kustova, 1964).

v. Effects of Applied Substances. Growth regulators have also been reported to protect against salt injury. Soaking wheat seeds in IAA overcame the depression of root growth by a solution of Na_2SO_4 (Sarin, 1961b), and increased the yield by 31% when the growth was inhibited by adding Na_2SO_4 (0.15%) to the soil (Sarin, 1962). Even spraying the plants with 5 ppm IAA increased the yield 100% in such salinized soils (Sarin and Rao, 1961). The same IAA treatments had no effect on nonsalinized soils. Less promising results have been obtained with gibberellic acid. The growth of cotton plants was stimulated by spraying with 50 mg/liter, more pronouncedly on salinized soils (Agakishiev, 1964). The stimulation was greater when the salinization was due to chloride than when due to sulfate. Nieman and Bernstein (1959), however, concluded that GA is unable to overcome the salt-induced suppression of the growth of bean plants. The growth was increased at low levels of salinity (up to 1.5 atm) but not at high levels (3.0 and 4.5 atm). Kinetin is reported to promote the uptake and incorporation of amino acids into proteins in NaCl stressed tissue (Kahane and Poljakoff-Mayber, 1968), although inhibiting it in non-stressed and Na_2SO_4 stressed pea roots. The plant growth retardant B995 increased the salt resistance of kidney beans to toxic levels of NaCl and $(NH_4)_2SO_4$ (Ota, 1964).

Even a halophyte, *Salicornia*, is adversely affected by saline soils, since its chlorophyll content is greatly decreased. Weekly fertilization with 1% urea spray overcame this effect of the salt (Hoffmann and Sachert, 1967). Conversely, seed treatment with 0.06% urea reduced the resistance of wheat seedlings to high salt concentrations (Miyamoto, 1962). Yet a variety of other substances (0.1% chloroethanol, 0.2% LiBr, or 0.2% tannin) increased the salt resistance when the seeds were similarly pre-treated (Miyamoto, 1963). Inosine-5-phosphate greatly enhanced the growth of mutant *Penicillium* in both KCl and H_3BO_3 media (Siegel, 1969a). A reverse protection has been reported in the case of a salt-habituated form of *Penicillium notatum*. It was unable to grow normally in the absence of salt, but its growth was restored to near normal by yeast extract, RNA, or RNA hydrolysate (Siegel *et al.*, 1967). An even greater effect was obtained by adding adenylic, guanylic, or cytidylic acids singly or in combination. Uridylic acid was ineffective.

B. Salt Resistance

Plants differ widely in salt resistance (Table 19.1). The most resistant are the obligate halophytes. They are irreversibly inhibited at low salinities, e.g., two laminarian algae required salinities *above* 9 and 5%, respectively (Norton and South, 1969). Some of the facultative halophytes (e.g., *Salicornia rubra*) are found at the highest salinities, yet are capable of growing normally in low to nonsaline environments (Ungar *et al.*, 1969). *Atriplex* species have long been known for their remarkably high salt resistance (Gates and Muirhead, 1967). Some twenty-six salt-resistant species of fungi (mainly species of *Aspergillus* and *Penicillium*) were isolated from

TABLE 19.3 The Salt Tolerance of Agricultural Crops in Different Irrigated Areas of Central Asia[a,b]

Serial number of the crop	Central ameliorative station Zolotaya Orda (Golodnaya steppe)	Bukharian experimental field	Experimental field at Fedchenko (Fergana)	According to general farm experience (Khorezm)
1	Sunflower	Sunflower	Fodder beet	Beet root
2	Sugar beet	Millet	Cotton	Turnips
3	Cotton	Wheat	Sunflower	Millet
4	Wheat	Maize	Sugar beet	Sorghum
5	Alfalfa	*Phaseolus aureus*	Barley	Konzhut (*Sesamum indicum*)
6	Maize	Oats	Wheat	Cotton
7	Barley	Sugar beet	Oats	Melon
8	Millet	Barley	Sorghum	Potatoes
9	Oats	Cotton	Flax	Most of the vegetables
10	—	Sorghum	—	Wheat
11	—	Alfalfa	—	Alfalfa
12	—	—	—	Carrots

[a] The crops are arranged in order of decreasing salt tolerance.

[b] From Strogonov, 1964. Reproduced by permission of the Israel Program for Scientific Translations, Jerusalem.

soil by means of agar media containing 20 and 30% NaCl (Chen, 1964). Only two of the species were obligately osmophilic.

Glycophytes, in general, grow well only under nonsaline conditions. Yet even though most crop plants are glycophytes, there is a rather wide range of salt resistance among them, from a maximum in beet roots to a minimum in carrots (Strogonov, 1964). Among grains, barley was more resistant than oats which were more resistant than wheat (Ballantyne, 1962), as judged by the conductivity of the root medium. Among a large number of crop plants, tested, the most salt resistant were date palm, cotton, lucerne, sweet clover, asparagus, beets, leeks, and radish (Simonneau and Aubert, 1963). The least resistant were citrus, strawberry, and beans. Unfortunately, the order of resistance is not the same in all soils (Table 19.3). This is at least partly due to the fact that relative resistance of a given species is not the same for different salts (Sergeev—see Strogonov, 1964). Most plants (e.g., cotton) are less resistant to NaCl than to Na_2SO_4, but some (e.g., phaseolus, guayule, flax) show the reverse relationship. Sodium

TABLE 19.4 THE DEGREE OF TOXICITY OF VARIOUS SALTS FOR DIFFERENT PLANTS IN DECREASING ORDER[a]

White lupin	Alfalfa	Wheat	Maize	Sorghum	Oats	Cotton	Sugar beet
$MgSO_4$	$MgSO_4$	$MgSO_4$	Na_2CO_3	$MgCl_2$	$MgSO_4$	$MgSO_4$	$MgSO_4$
$MgCl_2$	$MgCl_2$	$MgCl_2$	NaCl	$MgSO_4$	$MgCl_2$	$MgCl_2$	$MgCl_2$
Na_2CO_3	Na_2CO_3	Na_2CO_3	$NaHCO_3$	Na_2CO_3	Na_2CO_3	Na_2CO_3	Na_2CO_3
$NaHCO_3$	Na_2SO_4	$NaHCO_3$	Na_2SO_4	$NaHCO_3$	$NaHCO_3$	Na_2SO_4	$NaHCO_3$
Na_2SO_4	NaCl	Na_2SO_4	$MgCl_2$	Na_2SO_4	Na_2SO_4	NaCl	Na_2SO_4
NaCl	$NaHCO_3$	NaCl	$MgSO_4$	NaCl	NaCl	$NaHCO_3$	NaCl

[a] From Strogonov, 1964. Reproduced by permission of the Israel Program for Scientific Translations, Jerusalem.

carbonate is almost always more toxic than the above two salts (Table 19.4). Algae showed higher resistance to Na_2CO_3 than mosses and flowering plants (Url, 1959). Among sixteen plant species, the resistance to NaCl was lower than to Na_2CO_3, which was lower than to Na_2SO_4 (Ballantyne, 1962). Corn seedlings, on the other hand, were more resistant to Cl^- than SO_4^{2-} at equal osmotic concentrations (Kaddah and Ghowail, 1964). These relations are, however, complicated, for NaCl depressed the growth of the corn plants more than $CaCl_2$ or $MgCl_2$ at low concentrations but the relation was reversed at higher concentrations. Even among the species of a single genus, wide differences in salt resistance may exist, e.g., among fourteen species of *Agropyron* (Dewey, 1960).

1. Avoidance

The plant can use any one of three methods to avoid the salt stress of its environment. (1) It can exclude the salt passively. (2) It can extrude it actively. (3) It can dilute the entering salt. All three methods have been reported. When varieties of barley (Greenway, 1962), were treated with 125 or 250 mEq NaCl/liter, the less resistant variety accumulated a higher content of Cl^- and Na^+ and a lower K^+ content than the two resistant varieties. The differences were particularly large in the inflorescences (Greenway, 1965b). Both the passive and the active uptake of Cl^- were higher in the less resistant variety. Even at high transpiration rates, the ascending sap attained only 1.5–4% of the concentration in the medium, showing that most of the water flowed through regions of low salt permeability (Greenway, 1965a). The Cl and Na contents of a more halophytic species, *Agropyron elongatum*, were considerably lower than those of a very resistant variety of *Hordeum vulgare*, and were low even when grown on highly saline sites (Greenway and Rogers, 1963). Similar results have been obtained with other plants. When the salinity of their medium was increased progressively, the four varieties of berseem clover showed little variation in the sum of total cations in their tops (Kaddah, 1962). The less resistant varieties of soybean accumulated larger amounts of Cl^- (Abel and Mackenzie, 1964). Among cultivars of *Glycine wightii*, one group in particular was more resistant to salinity stress than the others (Gates *et al.*, 1970), excluding Na, and to a lesser degree Cl, from the plant tops to a greater degree than the other cultivars. In the case of date (*Phoenix dactylifera*), seedling growth was depressed nearly linearly from 3000–24,000 ppm salt (mostly chlorides Furr and Ream, 1968), but the uptake of Na^+ and $Cl^-$was not proportional to the concentration of the external solution. The salt was actually excluded with increasing efficiency as the salinity increased. As a result, the accumulation of Na^+ and Cl^- in the tissues was not much greater at high than at low salinity levels and there were no visible symptoms of injury.

Indirect evidence also points to the existence of a salt exclusion mechanism in the halophilic alga *Dunaliella viridis* (Johnson *et al.*, 1968). Four of its enzymes were inhibited *in vitro* by NaCl concentrations far lower than that in its growth medium (3.75 *M*). The exclusion may function only for a specific ion. Blue-green algae growing in saline lagoons may resist up to 1700 m*M* chloride (Krishna, 1955). The resistance is due to exclusion of Cl^-. In its place, there is a large accumulation of SO_4^{2-} together with K^+, Ca^{2+}, and Mg^{2+} ions.

Salt exclusion has been found in the case of salts other than NaCl. *Helianthus bolandei* is endemic on serpentine (high Mg) soils and requires higher levels of Mg than *H. annuus* (Madhok and Walker, 1969). In artificial culture, the yield of *H. bolandei* increased steadily up to 10 m*M* Mg; in the case of *H. annuus* it increased only up to 0.25–2.0 m*M* Mg. The resistant species survived concentrations as high as 50 m*M*, with only a 30% decline in yield. The reason for this high degree of Mg resistance was because it required these unusually high levels of external Mg to bring the internal Mg of *H. bolandei* to the normal range. In other words, its resistance was due to Mg exclusion.

In most of the above cases, it is not possible to decide whether the Na salts have been excluded passively or actively. In the case of yeasts, the ratio of K:Na is much higher in the cells than in the medium, and higher in the salt-tolerant *Debaryomyces hansenii* than in the less tolerant *Saccharomyces cerevisiae* (Norkrans and Kylin, 1969). The difference between the two species was due to a better Na extrusion and K uptake in *D.*

hansenii. The authors, therefore, conclude that salt resistance is partly dependent on the ability to mobilize energy to extrude Na from the cells and to take up K, i.e., on avoidance. The K:Na ratio was greater than one in *Ipomoea pescaprae*, a succulent plant growing on saline shores (Dawalkar and Joshi, 1962). This also seems to imply an active extrusion process, since a plant able to exclude Na passively would have a low permeability for K as well as Na.

The active extrusion mechanism apparently confers avoidance not only of the primary Na stress but also of the secondary Na-induced K deficiency, since NaCl concentrations up to and including 500 m*M* (higher than in seawater) failed to interfere with the absorption of K from a 10 m*M* solution by the halophytic mangrove *Avicennia marina* (Rains and Epstein, 1967).

Some plants of intermediate salt resistance may succeed in excluding the salt only from the shoot. A large part of the Na^+ absorbed by their roots is retained in the roots and accumulates in the vacuoles of the root cells (Solov' ev, 1967). In these cases, the regulatory system must occur within the roots, preventing the translocation of the salt, rather than at the root surface where absorption is prevented. An example of such root-induced shoot avoidance is shown by *Eragrostis tenella* (Satyanarayana and Rao, 1963). The exclusion mechanism is apparently located in or near the translocatory system of the stem as well as in the root in the case of *Phaseolus vulgaris*; the Na was retained in the basal parts, whether roots of intact plants or stem bases of derooted plants were in contact with the labeled NaCl solution (Jacoby, 1964). Even the halophyte *Prosopis farcata* accumulates Na only in the root and hypocotyl, and excludes it from the upper part of the shoot (Eshel and Waisel, 1965). In the case of extreme halophytes, however, such as *Atriplex vericaria*, the roots mainly function in absorbing the salts and transporting them to the leaves (Black, 1958).

The avoidance mechanism may be achieved as well by salt excretion as by salt exclusion, and the two mechanisms may exist in two closely related plants. The mangrove *Rhizophora mucronata* excludes salt from its xylem more efficiently than the salt-secreting mangrove *Aegialitis annulata*. The respective concentrations were 17 and 85–122 mEq Cl^-/liter sap (Atkinson *et al.*, 1967). The concentration in the salt-secreting species may, therefore, reach as high as 0.5% NaCl (Scholander *et al.*, 1962). This highly concentrated xylem sap delivered 100 mEq Cl^-/day to each leaf of *A. annulata*, yet the leaf concentration remained unchanged due to an equivalent excretion (mainly of NaCl) from the salt glands. In the case of *Tamarix aphylla*, the salt secreted by the glands showed no apparent selectivity between Na^+ and K^+ ions (Berry and Thomson, 1967). Even Rb, in fact, was taken up from the nutrient solution and subsequently secreted by the glands (Thomson *et al.*, 1969). Electron micrographs of the glands showed accumulation of electron-dense material (presumably Rb) in the microvacuoles of the secretory cells. They concluded that the Rb was subsequently secreted by fusion of these microvacuoles with the plasmalemma.

Several *Atriplex* species possess epidermal bladders with high concentrations of ions. Chloride was secreted from the solution or the lamina to the bladders, against a concentration gradient (Osmond *et al.*, 1969). The vacuole of the bladder was highly electronegative with respect to the bathing solution, suggesting an active uptake into the bladders. Autoradiography with $K_2{}^{35}SO_4$ and $K^{36}Cl$ showed that the anion was concentrated in the stalk cell and the peripheral cytoplasm of the large vacuolated bladder cell. The stalk cell has the submicroscopic characteristics of a salt gland, and may secrete ions from the leaf symplasm to the bladder cell. The vesiculate hairs of *Atriplex halimus* remove salt from the remainder of the leaf, preventing a toxic accumulation of salt in the parenchyma and vascular tissues (Mozafar and Goodin, 1970). In this way, a nearly constant salt content is maintained in the leaf cells other than the hairs. Similar results were obtained with *A. polycarpa* (Chatterton *et al.*, 1970a). The excretion process is highly selective for Na^+ in *Aeluropus litoralis* (Pollack and Waisel, 1970), K^+, and Ca^{2+} being retained in the leaves to a greater degree than Na^+. There was an antagonism both between K^+ and Na^+ and between Na^+ and Ca^{2+}.

Rhizophora mucronata does not excrete salt, yet its growing leaves also retained a constant concentration (510–560 m Eq/liter) though receiving 17 mEq Cl/day. Due to the growth, water was also absorbed in a sufficient amount to prevent an increase in concentration. This "dilution" of the cell sap due to growth has also been found in some moderately salt-resistant nonhalophytes. Barley rapidly increases its NaCl concentration during early tillering, but shows little further change until grain formation due to the rapid growth (Greenway *et al.*, 1965). During senescence, when growth decelerates, there is a marked increase in Cl^- and Na^+ concentration, and at any one time the ion concentrations are higher at low than at high growth rates (Greenway and Thomas, 1965). Even the varietal salt resistance of twenty-two accessions of *Glycine javanica* seemed to be directly related to growth rate (Gates *et al.*, 1966b). This is also true of gram and wheat. The slow-growing varieties suffered more than the rapidly growing varieties from toxic concentrations of NaCl (0.8%) during early seedling growth (Sarin, 1961a).

Halophytes use another dilution mechanism. Under natural conditions, the salt content of plants increases gradually with increasing age, and may eventually reach a toxic concentration. Many halophytes avoid this increase in concentration by an increase in succulence. The cells (especially the parenchyma) enlarge due to an increase in water content, which prevents an excessive concentration of salts in the cell sap (Repp, 1958). This mechanism is well developed in *Atriplex* species (Greenway *et al.*, 1966). Succulence must not, however, be judged solely by leaf thickness. In the case of *Bryophyllum pinnatum*, chlorides produce thicker leaves, but these have more organic matter and less moisture than those of plants not exposed to chlorides (Karmarkar and Joshi, 1969). If, however, measured per unit leaf area, an increase in succulence will be accompanied by an increase in both water and dry matter (Weissenböck, 1969).

Avoidance of salt injury by salt exclusion is dependent on temperature. Thus, the optimum temperature for growth of chrysanthemum dropped with an increase in salinity (Lunt *et al.*, 1960). This was explained by the increase in accumulation of Na^+, Ca^{2+}, and Cl^- with increase in temperature. Sodium ions tended to be excluded from the upper leaves unless the temperature was high. Similarly, rice suffered more salt injury at 30.7°C and 63.5% r.h. than at 27.2°C and 73.4% r.h. (Ota and Yasue, 1959). This was explained by the greater intake of salt. The effect of temperature is presumably due mainly to the increased transpiration rate rather than the increased absorption rate, for such small changes in temperature (and relative humidity) can effect large changes only in the former process. In agreement with this conclusion, three (nontranspiring) fungi showed the opposite relation—low salinity favored growth at low temperature, while high salinity (up to 4.7%) favored growth at high temperature (Ritchie, 1957).

2. *Tolerance*

All the above results emphasize the importance of avoidance in salt resistance. Moderately resistant, nonhalophytic plants owe their resistance primarily to avoidance, e.g., barley (Greenway, 1965b). Nevertheless, Greenway points out that when halophytes and nonhalophytes are compared, ion accumulation (and therefore tolerance) appears to be a superior mechanism for growth in a saline habitat. The relatively salt-resistant beet, for instance, had a range of osmotic concentrations (8.8–20.6 atm) considerably broader than in the more salt sensitive climbing bean (12.6–16.1 atm) and maize (9.5–11.9 atm; Mercado, 1970). Another example is the halophyte *Atriplex halimus*, which unlike the salt-avoiding halophilic microorganisms discussed above, shows no inhibition of Na^+ absorption in the presence of high concentrations of K^+ (Mozafer *et al.*, 1970). On the contrary, absorption of K^+ was inhibited by excess Na^+, unlike the behavior of glycophytes. The growth of the mangrove *Avicennia marina* in nutrient culture was actually stimulated by NaCl, although all levels of KCl and $CaCl_2$ suppressed it (Connor, 1969). The optimum level of NaCl was 1.5%, one-half the concentration of seawater. Furthermore, at least one of the above avoidance mechanisms adopted by halophytes depends on the salt tolerance of some of its cells—the salt-excreting cells. Similarly, even though the above described mangroves possessed either a salt exclusion or a salt excretion mechanism, the leaves, nevertheless, did maintain a high salt concentration (510–560 mEq/liter sap) and, therefore, must also possess a high degree of tolerance. In some cases, in fact, it is not easy to draw a sharp line between tolerance and avoidance. For instance, an increase in bound Cl^- has been suggested as one cause of resistance (Matikhin and Boyko, 1957; Strogonov, 1964). Although such plants would be tolerating high *contents* of the chloride, high actual *concentrations* would be avoided. Possible examples are the halophytes *Suaeda fruticosa* and *Haloxylon recurvum* which grow luxuriantly in saline soil. Intracellular granular structures were found in various regions of the hypodermis, cortex, and pith, containing salts of Na, Ca, and Mg, impregnated with quartz and organic matter which renders them insoluble (Ahmad, 1968). Even in the case of nonhalophytes, avoidance cannot always provide the full explanation for salt resistance. Wheat plants, for instance, were more severely injured by salt under anaerobic than under aerobic conditions (Ogo, 1956). Nevertheless, Na accumulation in the leaves was less under anaerobic conditions. In other words, the salt tolerance of the plants was greater under aerobic than under anaerobic conditions. Tolerance is, therefore, unquestionably an important component of salt resistance, especially in the case of halophytes which markedly increase their cell sap concentration by storing salts. A significant correlation has, in fact, been found between the salt concentration tolerated by leaf cells and the salinity of the habitat in which the species occurs (Table 19.5).

Unfortunately, the term "salt tolerance" has been used in the literature for any plant possessing salt resistance, simply on the basis of the salinity in the external medium. This would, of course, include avoidance. It is, therefore, necessary to consider the methods of measurement used, before deciding whether or not true tolerance is involved.

a. Measurement

Several methods supposedly measure "tolerance."

(1) The specific conductance of a soil saturation extract corresponding

TABLE 19.5 Salt Tolerance of a Number of Species[a,b]

Species	NaCl molal conc. which killed 50% of cells
Halogeton glomeratus	1.6
Salsola kali	1.6
Agropyron elongatum seedlings	1.0
Beta vulgaris	0.7
Lotus tenuis var. Los Banos	0.5
Lotus corniculatus var. Granger	0.5
Trifolium repens var. Ladino	0.5
Ervum ervillea	0.4
Melilotus alba var. low coumarin	0.4
Medicago sativa var. Lahontan	0.4
Medicago sativa var. Province	0.4
Abutilon avicinnae	0.4
Vicia cracca	0.4
Medicago sativa var. Brigham Young	0.3
Lens culinaris	0.3
Vicia faba	0.2
Phacelia tanacetifolia	—

[a] Determined by immersing sections in a graded series of NaCl concentrations for 24 hr and then determining cell survival plasmolytically. (From Repp *et al.*, 1959.)
[b] First four species are halophytes.

to a standard percentage decrease in yield (Lagerwerff, 1969). This is, at best, a measure of total salt resistance, including avoidance as well as tolerance. It also assumes that salt injury is directly proportional to the electrical effect, and is therefore valid only for a single salt solution. Finally, it measures yield, which depends on many factors besides salt resistance.

(2) The osmotic adjustment of the plant. This has been measured in different ways. Bernstein (1961, 1963) measured the length of time required by the plant to adjust to a specific increase in osmotic concentration of the root medium. Beans, for instance, adjusted to an increase of 1 atm in 24 hr, while pepper plants incompletely adjusted to a 1.5 atm increase in 48 hr. Depending on the salt added to the root medium, the osmotic adjustment can be determined for a specific ion.

(3) Rate of uptake and translocation of Na^+. This was found to be the most important difference between corn and sugar beet (Marschner and Schafarczyk, 1967a). The Na^+ was rapidly translocated from the root to

the shoot and vice versa in sugar beet. In corn, translocation from the root to the shoot was very slow, and the reverse was not detectable.

(4) The ratio of Na:K in the plant. On the basis of this ratio, Sakazaki *et al.* (1954) arranged the tested plants in the following order of decreasing salt tolerance: atriplex > beet > tomato > metasequoia > corn. According to Takada (1954) plants can be classified into three groups on the basis of this ratio: euhalophytes, oligohalophytes, and nonhalophytes.

(5) Survival of sections of tissue in salt solutions. Repp *et al.* (1959) immersed the tissue sections in a graded series of NaCl solutions for 24 hr, and then determined the proportion of living cells by plasmolysis with hypertonic glucose (Table 19.5). This method would seem to yield the best measure of salt tolerance since it is based on survival rather than on growth or yield. The results, in fact, did agree well with salt resistance in the field, and with the degree of leaf succulence when grown on saline soils. The plasmolytic method also gave good agreement in the case of twenty-eight species of woody and herbaceous plants grown in salinized media (Monk and Wiebe, 1961). Ten were salt resistant, four were somewhat resistant, and fourteen were not resistant. On the other hand, the salt resistance of beet, bean, and maize plants could not be determined unequivocally in all cases by the plasmolytic method (Mercado, 1970).

Of the five listed methods, (1) measures total salt resistance (avoidance plus tolerance), while the other four measure tolerance. Methods (2) and (3) are designed to measure secondary osmotic tolerance since they measure the rate of achievement of osmotic balance. Nevertheless, they must also measure primary salt tolerance since the salt uptake and translocation is possible only by a plant with primary salt tolerance. Method (4) measures tolerance of the secondary, salt-induced deficiency stress, as well as primary tolerance. Secondary, osmotic tolerance is not, however, excluded. Method (5) probably measures primary, direct salt tolerance, due to its relative speed. There is no evidence, however, that 24 hr is a short enough time to completely eliminate the other two tolerances. In most cases the first method or one similar to it has been used, and the salt "tolerance" measured is really overall salt resistance.

b. Hardening Methods

In general, plants do not develop salt tolerance unless they are grown in saline media (Strogonov, 1964). This means that they must be hardened to the salt stress. The Soviet scientists have developed a method of salt hardening by presowing soaking of seeds in salt solutions (Genkel—see Strogonov, 1964). Maize seeds treated with 3% NaCl and 0.2% $MgSO_4$ before sowing, grew into plants with a lower leaf content of monosaccharides, a higher content in the roots and stems, as well as an increase in sucrose (Azizbekova and Babaeva, 1962). Bhardwaj (1961) obtained better growth and maturity of wheat plants after presowing soaking of the seed in salt solution, provided that water content was favorable throughout the life cycle. In the case of gram, however, the seed treatments were definitely harmful. When Na_2CO_3 was used to pretreat the seed, the results were not promising even for wheat (Bhardwaj, 1964a). Even when hardening by this method is said to occur, the increase in yield is only about 10% (Pang *et al.*, 1964). Furthermore, the effect of the salt stress is measured by the decreased growth or yield, which depends on many factors.

Increases in tolerance or avoidance of the plastic strain have been reported only as a result of a gradual exposure to increasing salt concentrations. Tomato and cotton plants are injured by a sudden increase in salt concentration, but if they are exposed to a gradual increase in salinization over a long period, the plants become salt tolerant (Kovalskaia, 1958). Irrigation with salt solutions increased the tolerance of pea, maize, and sunflower during the germination period (Imre, 1968). The effect was most marked when NaCl was used and least marked in the case of Na_2CO_3. Similar results were obtained with *Glycine javanica* (Gates *et al.*, 1966a). Even tissue cultures of carrot were able to adapt to a medium with high NaCl content (3.5 atm.), but not to an equal concentration of Na_2SO_4 (Babaeva *et al.*, 1968). Diatoms can also be hardened by a gradual, stepwise increase in NaCl concentrations, from a threshold tolerance of 0.45 to 0.55 *M* NaCl (Kovarik, 1963). This effect was presumably osmotic, since sucrose was equally effective. Low temperature has been reported to increase the salt tolerance of germinating cottonseed (Agakishiev, 1962).

3. Relationship of Resistance to Age and Development

Salt resistance is low in young tomato and cotton plants, becomes much higher by the bud stage, and decreases during flowering (Kovalskaia, 1958; Penskoy, 1956). Rice shows a similar increase in resistance with age (Pearson and Bernstein, 1959). Salinity at tillering was twice as inhibitory as at heading. In the case of barley plants, varietal differences increased during plant development (Greenway, 1965b). Another case of a change in relative tolerance with development has been found among *Tamarax* species. There was no direct relationship between the salt tolerance of these halophytes during germination (in salt solutions up to 1.0 *M*) and the salinity of their respective habitats (Waisel, 1958b). All the other species tested did show a correlation. In the case of soybeans, there is also no apparent relationship between the salt resistance of a variety during germination and during later growth (Abel and Mackenzie, 1964). Two sugarcane varieties differed in salt resistance, again only in the stage of germination and early growth (El Gibaly and Goumah, 1969). After 3 months no negative effect on yield, growth, and sugar content was found following watering with salinized water (6000 mmhos/cm). All the species tested by Choudhuri (1968) were less salt resistant at germination than in the seedling stage. Even the mechanisms for resistance were different in seedlings and mature plants (Hunt, 1965). In the case of fruit trees, salt resistance increases during initial growth but decreases as the plants grow older, dropping abruptly during the period of fruiting (Devyatov, 1962).

In the above cases, no distinction is attempted between avoidance and tolerance. In the halophyte *Suaeda vulgaris*, the Na content decreased with plant age from about 10–16% in the early stages (April–June) to about 1–3% of the dry matter at the end of the growing season (mid-September; Binet, 1963). This indicates an increase in salt avoidance with age. During the aging of bean leaves, both Na influx and Na efflux increased (Jacoby and Dagan, 1969).

4. Factors Associated with Salt Resistance

Many effects of salinization on the plant have been recorded. In most cases, unfortunately, it is difficult to determine whether the effect is due to injury or to a hardening process. Therefore, many of the factors listed, may not be associated with salt resistance.

Chloride salinization increased the nucleoside monophosphates, ADP, and uridyl diphosphates in barley plants (Matukhin *et al.*, 1968). Sulfate salinization increased the guanosyl phosphates and di- and triphosphates of adenylic acid, particularly in the leaves. Total lipid and phospholipid concentration in barley roots tended to increase slightly with increasing NaCl (Ferguson, 1966). This indicates that suppression of growth by NaCl is not due to suppressed lipid synthesis. There was, however, a change in distribution of the phospholipids. In general, polar phospholipids increased relative to the less polar classes, as NaCl concentration increased. In the halophytes *Suaeda macrocarpa* and *Aster trifolium* proline is one of the most important amino acids (Goas, 1966). The proline level is related to the NaCl dose. In fact, when transferred to a NaCl-free medium, *Aster trifolium* developed more slowly than in the presence of the salt and the normal high content of proline decreased rapidly (Goas, 1968). Ascorbic acid has also been reported to increase in the leaves of cabbage and radish, with increases in salt concentration (Kim, 1958).

The protoplsamic property that has been most commonly related to salt tolerance is cell permeability. In the case of higher plants, it has been suggested that the tonoplast must possess a low permeability to salts, in order to act as a barrier between the internal salt-containing cell sap and the protoplasm (Repp, 1958). Salt intrability (the permeability of the external protoplasmic plasma membrane—the plasmalemma), however, is higher than the permeability of the cell as a whole and therefore than the tonoplast. Salt-tolerant microorganisms (which possess no tonoplast) appear to be highly permeable to salts, in opposition to salt-avoiding plants (due to exclusion), which must possess a low salt permeability (see above). An osmiophilic yeast (*Saccharomyces rouxii*) showed a striking increase in permeability when cultivated in saline medium (Onishi, 1959). The increase did not occur in concentrated sugar and must, therefore, be a primary rather than a secondary effect. The cells with increased permeability were not able, however, to remain viable in saline medium unless allowed to metabolize actively with the aid of external substrate. As mentioned above, *Dunaliella salina* recovers its turgor within less than an hour after collapsing severely in a strong salt solution (Marré *et al.*, 1958b). This is possible only if the cell is highly permeable to the salt. Direct measurement of ^{24}Na transfer proved that it moved in and out of the cells freely (Marré and Servettaz, 1959). Another species of the halophilic microorganism, *Dunaliella parva*, is freely permeable to sucrose, inulin, starch, and even PVP (polyvinylpyrrolidone) with a mol. wt. of 30,000–40,000 (Ginzburg, 1969). It is impermeable, however, to dextran (mol. wt. 80,000). These results led to the conclusion that the cell membranes possess large pores.

The Soviet investigators have reported changes in the colloid-chemical properties of the protoplasm in connection with changes in salt resistance (Strogonov, 1964; Slonov, 1966). High protoplasmic viscosity is associated by them with salt resistance, due supposedly to an increase in hydrophily and bound water induced by Na^+ and Cl^- ions (Kovalskaia, 1958).

According to Kessler *et al.* (1964), the ribosomes of a salt-sensitive citrus species were less stable than those of a more resistant species, but were able to acquire an increase in stability.

A relationship between SH content and salt resistance has been reported by Shevyakova and Loshadkina (1965). A relatively weak concentration of Na_2SO_4 (0.6%) increased the SH content of horse bean leaves, apparently due to increased synthesis. Salinization with Cl^- also induced some accumulation of SH compounds. When salinization was severe enough to cause leaf necrosis, they proposed that this was due to a change in the ratio of SH:SS. Infiltration of the leaves with GSH prevented necrosis. In the case of corn leaves, the changes were not as pronounced or reproducible.

In kale, mustard, and broad beans, oxidation of methionine to sulfate was greatly enhanced by sulfate (Shevyakova and Popova, 1968). They concluded that Na_2SO_4 increases the oxidation of S-amino acids to sulfate.

5. *Mechanism of Salt Resistance*

a. Avoidance Mechanism

The mechanism of salt avoidance depends, of course, on whether the avoidance is due to exclusion, excretion, or dilution (Diag. 19.3). In the case of plants possessing the exclusion mechanism, the roots may show an impermeability to salts up to a point, followed by a "burst" of salts causing poisoning and sometimes death (Rikhter, 1927—see Strogonov, 1964). The salt resistance of such plants depends on maintenance of impermeability to the salt in the presence of high external concentrations. It has long been known (Scarth and Lloyd, 1930) that maintenance of the normal differential permeability of the cell depends on a balance (about 10:1) between monovalent (K^+, Na^+) and divalent (mainly Ca^{2+}) cations. When this balance is disturbed by too high a concentration of monovalent cations, the permeability increases, leading to injury. Therefore, a plant with salt avoidance due to exclusion must possess a low permeability to Na salts even in the presence of relatively high salt concentrations. This perhaps explains Iljin's (1932) discovery of halophytes that remain alive after drying up in a solution of pure NaCl, yet that die in a relatively weak Ca solution. Glycophytes showed the reverse behavior. This would seem

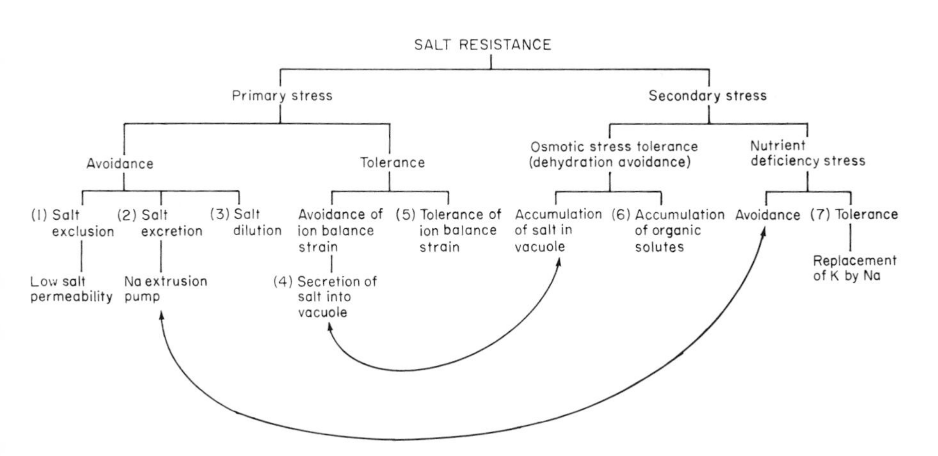

Diag. 19.3. The different possible kinds of salt resistance (Na salts).

to indicate a higher permeability of the halophytes to Ca^{2+} than to Na^+, whereas the reverse relationship is found in the glycophytes. The salt-excluding cell may, therefore, conceivably maintain the normal balance in the presence of high concentrations of monovalent cations, perhaps by a high, preferential adsorption of Ca^{2+} on the plasma membrane. In opposition to Iljin, however, Waisel (1958b) obtained equal germination of halophytes in solutions of NaCl and $CaCl_2$. It is, of course, essential to choose salt-excluding avoiders, in order to test this hypothesis.

Salt avoidance due to excretion requires a Na extrusion pump that continues to operate in the presence of high external concentrations of salt. On the basis of the published evidence, it is usually not possible to decide whether an exclusion mechansim (due to passive impermeability) or an active excretion mechanism is operating.

The third type of salt avoidance—due to dilution— is dependent on the succulent mechanism. Until this mechanism is understood, the dilution avoidance cannot be explained. Perhaps it depends on the maintenance of thin, plastically extensible cell walls, permitting continuous cell expansion by a water uptake sufficient to balance every salt increment in the cell.

b. Osmotic Tolerance Mechanism

Since the osmotic stress is one type of water deficit stress, tolerance can be of two types: (1) dehydration avoidance, and (2) dehydration tolerance. Evidence of the latter is the recording of high water tensions (180, 240, and 320 bar) in the leaves of a desert halophyte *Reaumuria hirtella* (Whiteman and Koller, 1964). At such high tensions, these leaves must have been strongly dehydrated. In order for growth to continue, however, the cells must remain turgid and this is possible only if they possess dehydration avoidance. When the plant is exposed to a hypertonic salt solution, there are two mechanisms by means of which it can retain its turgor: (1) by absorbing the salt, and (2) by increasing its concentration of organic solutes. Arnold (1955) concludes that salt resistance depends on the first of these—salt absorption. In the case of microorganisms, the high internal concentration appears to be due simply to a high cell permeability (see above), but it also requires the absence of a Na extrusion pump. Kylin and Gee (1970) showed that a halophytic mangrove (*Avicennia nitida*) possesses a Na extrusion pump, but it is inactivated by high concentrations of NaCl (0.2 to 0.4 *M*). This result is supported by Horovitz and Waisel (1970). They found that NaCl stimulated ATPase activity in a glycophyte (*Phaseolus vulgaris*) but inhibited it in a halophyte. The osmotic mechanism would require a highly specialized tonoplast system capable of rapidly excreting the excess protoplasmal salt into the vacuole by an active process. The tonoplast would, nevertheless, have to be highly impermeable to the salt even when present in high concentration, in order to prevent back diffusion into the protoplasm. In higher plants the situation is more complicated than in microorganisms, partly due to the large transfer distances from the root to the top of the plant. Thus, the most important difference between corn and sugar beet is in the rates of uptake and translocation of Na^+ These rates are much greater in the more salt-tolerant beet than in corn (Marschner and Schafarczyk, 1967a). The Na^+ is rapidly translocated from root to shoot and vice versa in sugar beet (Marschner and Schafarczyk, 1967b). In corn, it is very slowly translocated from root to shoot but the reverse is not detectible. They conclude that the translocation systems in the symplasm of the root cells and in the phloem have a high affinity for Na^+ in sugar beet and a low affinity in corn. The absorbed Na^+, therefore, accumulates in the corn roots.

In the case of halophytes the absorption and translocation of Na salts is even more pronounced than in the moderately tolerant sugar beet. The osmotic values for the leaf saps of a number of herbaceous halophytes (*Atriplex subcordata*, *Suaeda maritima*, *Salicornea herbacea*, and *Aster Tripolium*) ranged from 25 to 75 atm when growing in salt marshes (Yabe *et al.*, 1965). The osmotic adjustment is mainly due to Na^+ absorption; although most plants absorb about the same amount of K from a specific root medium, the above halophytes absorbed twice as much Na as K from the soil solution. This may, perhaps, be explained by the above described inhibiting effect of Na salts on the ATPase activity of halophyte cells, since this enzyme is part of the "ion pump" that excretes Na^+ from and absorbs K^+ into the cells of both glycophytes and halophytes. The inhibition of the pump would, therefore, also inhibit active K^+ absorption. Therefore, the halophyte presumably must be able to utilize the passively absorbed Na^+ in place of the deficient K^+.

According to Kessler *et al.* (1964), high osmotic pressures induce conformational changes in proteins, as well as changes in base composition of copolyribonucleotides synthesized from equimolar concentrations of ADP, GDP, CDP, and UDP. He, therefore, suggests that osmotic stress is coded into a changed composition of polynucleotides which may lead to adaptive reactions, for instance, by serving as primers for RNA-dependent RNA polymerase.

The osmotic mechanism, of course, requires an ability of the salt-resistant plant to absorb Na salts more rapidly than the nonresistant plant. Until the mechanism of salt absorption is fully understood, the osmotic mechanism of salt tolerance will also be only partially explained.

c. Primary Salt Tolerance Mechanism

There are two possible mechanisms of primary tolerance. (1) The plant may avoid the ion-balance strain by excreting the absorbed salt into the vacuole. (2) The plant may be able to tolerate the ion-balance strain. This would require protoplasmic substances and organelles capable of maintaining their normal properties in the presence of the salt-induced change in ionic balance. The first mechanism is identical with the osmotic tolerance mechanism. The second mechanism is, of course, the only one available to cells with very small or no vacuoles, e.g., microorganisms and the meristematic cells of higher plants. Primary indirect tolerance has been explained by a rapid (presumably adaptive) change in the plant's metabolism (Protsenko, 1956—see Strogonov, 1964). An example of this kind of tolerance is provided by yeasts isolated from a marine environment which have the same rates of respiration and fermentation in 0 as in 4% NaCl (Norkrans, 1968). They showed a higher NaCl tolerance for these two processes than for growth. This was not simply osmotic tolerance, since they were able to maintain 10% of their normal respiratory activity at lower water activities when produced by sugars than when produced by the electrolytes. According to Kessler *et al.* (1964) the incorporation of leucine into leaf proteins was enhanced by low concentrations of NaCl in the case of the salt-resistant Cleopatra Mandarin roots, while inhibited in the salt-sensitive variety (Sweet lime). However, the difference in resistance was not solely due to tolerance, since the more resistant variety also excreted NaCl more rapidly, and, therefore, possessed a greater degree of avoidance.

Until the mechanism of primary salt injury is understood, the specific substances that must be adapted to the ion-balance strain will not be known. A number of investigators favor the proteins. For instance, the extremely halophilic bacteria, which require very high concentrations of NaCl for growth, possess enzymes that are irreversibly inactivated in the absence of high concentrations of neutral salts. *Halobacterium salinarum* excretes a protease into the growth medium which it was possible to chromatograph only in the presence of 3.4 *M* NaCl (Norberg and Hofsten, 1970). Further evidence of specific properties of the proteins in another extremely halophilic bacterium (*Halobacterium cutirubrum*) is their unusually low isoelectric point (pH 3.9; Bayley, 1966). Another specific property of their proteins was the absence of half-cysteine residues, in opposition to the nonhalophilic *E. coli* which has 0.6 moles half-cysteine per 100 moles amino acid. Strogonov (1964) suggests that the plant adapts to high salinity by binding of the mineral element. The amount of bound Cl^- in leaves increased with salinity, due supposedly to an increase in content of soluble proteins (albumins) which he showed were able to bind Cl^-. The plants, however, were small and retarded in development, so that the adaptation was apparently not very successful. The presumed stability of proteins in the presence of high salt concentrations would prevent the breakdown of proteins, which according to Strogonov (1964) leads to accumulation of toxic products. On the basis of his suggested mechanism of injury by putrescine, salt tolerance would require a stability of the enzyme diamino oxidase in the presence of salts.

If a primary direct salt injury occurs, due to membrane damage, the tolerance mechanism would have to depend on either the properties of the membrane lipids or of the membrane proteins, or both. Unfortunately, no information is available on this possibility.

6. Relative Importance of Salt Avoidance and Salt Tolerance

In the case of other stresses, avoidance is a sufficient mechanism by itself, in the absence of tolerance. This is also true for the secondary salt-induced nutrient deficit; if the plant excludes Na^+, it can absorb adequate amounts of K^+. The same would be true for Cl^- versus essential anions. If, however, the salt stress is sufficient to eliminate cell turgor, then the plant must possess osmotic salt tolerance in order to grow and develop, and salt avoidance would be useless or even damaging. Even among the moderately salt-resistant plants tolerance may be a factor. Among varieties of avocado, for instance, there was a close correlation between Cl^- content and leaf scorch, except for one variety which showed the highest Cl^- content in the leaves but its damage was among the lowest (Kadman, 1963).

In the case of halophytes, tolerance is definitely the main factor, since the salt content of the cells approaches that of the external medium. The result is an unusually high cell sap concentration—38.7 to 46.5 atm in the case of leaf epidermis of mangrove trees (Biebl and Kinzel, 1965). Yet these same trees do separate essentially fresh water from their medium by a simple, nonmetabolic ultrafiltration process (Scholander, 1968). They obviously combine avoidance with tolerance.

In view of the different kinds of salt injury, the mutual exclusion of some kinds of resistance, and the double effects of others, the following summarizes the theoretical possibilities (Diag. 19.3).

(1) Salt avoidance due to salt exclusion and, therefore, to low salt permeability.

(2) Salt avoidance due to salt excretion by an active ion extrusion pump. This would confer resistance to both the primary stress and the secondary salt-induced deficiency stress (avoidance of nutrient deficiency).

(3) Salt avoidance due to dilution, perhaps depending on a high plastic extensibility of cell walls.

(4) Salt tolerance of the primary stress due to avoidance of the ion balance strain. This would be accomplished by secreting the excess absorbed ions from the protoplasm into the vacuole. This mechanism would also confer tolerance of the secondary salt-induced osmotic stress, by helping to maintain cell turgor.

(5) Salt tolerance of the primary stress due to tolerance of the ion balance strain. The protoplasmic organelles and their substances would have to possess special properties, permitting normal functioning, although subjected to an increase in ion concentration and a change in balance.

(6) Salt tolerance of the secondary, salt-induced osmotic strain, due to accumulation of organic solutes.

(7) Salt tolerance of the secondary, salt-induced nutrient deficiency strain due to a replacement of the deficient nutrient element by the excessively absorbed element (e.g., K by Na).

C. Ion Stress

1. Ion Injury

Spencer (1937) divides the elements into three groups on the basis of their toxicity to plants: (1) nontoxic, (2) toxic in medium concentrations, and (3) toxic in low concentrations (5 ppm or less). On this basis, the third group would include elements such as Hg, Se, I, Cd, Co, Ni, and Tl. The relative toxicities of a large number of ions was determined for myxomycete plasmodia by Seifriz (Table 19.6). The order, however, is not quite the same in the case of germinating tilletia spores (Burström, 1929), wheat seedlings (Serrano, 1956), or *Chlorella vulgaris* (Den Dooren de Jong, 1965). The differences may be due to differences in specific ion resistance between species, or to complications due to the presence of other ions (see below).

TABLE 19.6 Relative Toxicities of Ions for the Protoplasm of Slime Molds[a]

$Li^+ < Na^+ < K^+ < Rb^+ = Cs^+$

$Ca^{2+} < Mg^{2+} < Sr^{2+} < Ba^{2+}$

$La^{3+} < Pb^{2+} < Au^{3+} < Ag^+ < Th^{4+}$

[a] From Seifriz, 1949.

The toxicity of heavy metals has received considerable attention, partly due to its occurrence in nature. Algae have long been known for their extreme sensitivity to heavy metals, such as Cu, Ni, and Co. These elements completely inhibit the growth of *Chlorella vulgaris* in doses as low as 4×10^{-6} gm atom/liter (Den Dooren de Jong, 1965). Eight species of marine phytoplankton (dinoflagellates, diatoms, and blue-green algae) were completely inhibited by concentrations ranging from 0.03 to 0.265 μg Cu/ml (Mandelli, 1969) and were, therefore, 15–100 times as sensitive as chlorella. However, a significant decrease in growth of *C. pyrenoidosa* was obtained with 1 μg Cu/liter, or $\frac{1}{250}$ the above amount for complete inhibition of *C. vulgaris*. This inhibition was unaccompanied by any injury, and on transfer to ordinary medium they started to grow again (Steemann Nielsen and Kamp-Nielsen, 1970). In fact if the initial steps of division had taken place, the cells continued the division in the presence of Cu. Obviously, growth inhibition in the presence of Cu is an elastic, reversible strain. Higher plants are more resistant than are the algae, but they are nevertheless injured by relatively low concentrations of heavy metals. Copper produces a necrosis in higher plants (Repp, 1963), and mine workings are toxic to them due to the presence of several heavy metals (Gregory and Bradshaw, 1965).

A reasonable explanation for the Cu effect is provided by the Cu inhibition of yeast β-fructofuranosidase (invertase). Inactivation of one active enzyme center was caused by reaction with one Cu^{2+} ion (Myrback, 1965). The dependence of inhibition on pH was explained by competition with H^+ for two essential groups. One of these appears to be an SH group. Further evidence explainable in this way was obtained with *Chlorella vulgaris* (McBrien and Hassall, 1967). When allowed to absorb Cu anaerobically, the subsequent aerobic respiration, photosynthesis, and growth were all severely inhibited. This was not true when the Cu was absorbed aerobically. Under anaerobic conditions, the Cu was bound to sites not available under aerobic conditions. They suggest that these may be SH sites. In agreement with this interpretation, Gross *et al.* (1970) were able to prevent all Cu-induced changes in *Chlorella* by means of glutathione. They point out that Cu^{2+} catalyzes the oxidation of SH groups with the formation of SS bridges (Gurd and Wilson, 1956—see Gross *et al.*, 1970), and suggest that Cu leads to rupture of membrane barriers. The glutathione would neutralize the Cu before it could reach the membrane.

The cause of Cu injury is apparently different in pure solutions from the cause in balanced solutions. If a single salt is used ($CuSO_4$), Cu penetrates immediately into the protoplasm of *Chlorella* cells, reducing the rate of photosynthesis (Steeman-Nielsen *et al.*, 1969). In a balanced solution, the injurious effect of Cu is not due to a marked penetration of the ion into the protoplasm, but to a binding to the cytoplasmic membrane, which prevents the cells from dividing. Other cations compete with Cu for the active site on the membrane.

This may perhaps explain the inability of the plasmolytic method to measure the injurious effect under natural conditions (see below). Perhaps if the method is modified, so that the injurious ion is applied in a balanced solution, it would given a better measure of the plant's resistance.

Other ions,of course, may act as stresses for other reasons. In the case of Al-induced inhibition of cell division in barley roots (Sampson *et al.* 1965), DNA synthesis continued but it had an unusual base composition and was metabolically labile. Morgan *et al.* (1966) suggest that Mn toxicity in cotton is due to increased IAA oxidase activity leading to auxin deficiency. According to Williams (1967) the inhibition by metals may be due to chelation. Enzymes are inhibited by heavy metals other than Cu. Aldolase from leaves of *Zea mays* seedlings was completely inhibited by Zn^{2+} or Co^{2+} at 0.04 M, 60% by Fe^{2+}, and 20% by Mn (Clark, 1966). ATP sulfurylase is inhibited by SeO_4^{2-} or MoO_4^{2-} (Ellis, 1969).

Many other ions have been shown to injure higher plants. Fluorine injured peach, buckwheat, and tomato at concentrations of 10, 25, and 50 ppm, respectively (Leone *et al.*, 1948). All concentrations of Co above 0.006 ppm induced chlorosis in Biloxi soybean leaf. Necrosis occurred at 0.015 ppm and above (Pushpalata, 1967). Thallium produced chlorosis of the youngest leaves, and symptoms of the so-called frenching disease of tobacco (Spencer, 1937). The lowest toxic concentration was 0.007 ppm in water cultures. Arsenic was lethal to nutsedge tubers following repeated treatments, but this was evidently due to a depletion of food reserves as a result of sprouting, and not to an accumulation of a specific level of As in the tubers (Holt *et al.*, 1967).

2. Ion Resistance

The toxic effect of a specific ion depends on many other factors. It may be completely eliminated by the presence of another ion due to antagonism (Table 19.7). On the other hand, it may be enhanced by the H^+. In fact, Al and Mn toxicities may be the main factors in injury due to soil acidity (Vose and Randall, 1962). Resistance in this case is associated with low cation exchange capacity. When measuring the toxicity of an element to a plant, it is, therefore, essential to maintain the pH and other factors constant (Seifriz, 1949). Steemann-Nielsen and Kamp-Nielsen (1970)

TABLE 19.7 ANTAGONISM OF ION TOXICITY

Toxic ion	Toxicity antagonized by	Reference
Se	SO_4	Hurd-Karrer, 1939
As	P	Hurd-Karrer, 1939
Rb	K	Hurd-Karrer, 1939
Sr	Ca	Hurd-Karrer, 1939
I	Cl	Lewis and Powers, 1941
Cu	Al	Liebig *et al.*, 1942
Mn	Fe	Shive, 1941
Fe	Mn	Shive, 1941
Mg	Ca	Proctor, 1970
Mo	SO_4	Ramaiah and Shanmugasundaram, 1962
Cu	Mn	Nollendorf, 1969

point out that in media ordinarily used for culturing *Chlorella*, the effect of Cu is slight due to the extraordinarily high concentrations of Fe. The Cu is adsorbed to the negative charges on the micelles of $Fe(OH)_3$, created in the alkaline medium. They also point out that the concentration of algal cells in the medium is of fundamental importance when determining Cu toxicity. Furthermore, the H^+ competes with the Cu and, therefore, the Cu effect is slight at pH 5 as compared with pH 8. Because of these many complications, it is very difficult to develop a standard method for determining ion toxicity, and, therefore, for measuring ion resistance.

Nevertheless, attempts have been made to measure the quantity. The ion resistance of a plant is usually evaluated by determining the concentration of the ion just sufficient to stop growth. This, of course, is a measure of the elastic strain, since the growth inhibition is reversible on transfer to an ion-free medium (see above). Biebl and Rossi-Pillhofer (1954) have attempted to measure the resistance on the basis of the plastic (irreversible) strain, by use of a cytological method. Epidermal layers of leaves, petioles, or stems are immersed in a graded series of solutions of the ion to be tested (supplied usually as a salt). After 48 hr they are tested plasmolytically for viability. There are difficulties, however, in the use of this method. Url (1957) found that cells live in the lowest concentration range, die in the intermediate range, but remain alive in the relatively high range. This upper "life zone" may begin with solutions which are still hypotonic or it may occur only with hypertonic (plasmolyzing) solutions. The cells which remain alive in the high concentration range, die when transferred to the intermediate concentration range. This injury is not due to deplasmolysis, because it occurs even if the upper zone concentration is still hypotonic.

In spite of the difficulties in measurement, there is ample evidence that plants do differ in ion resistance. The alga *Dunaliella tertiolecta* survived Cu concentrations near saturation for seawater (0.6 μg Cu/ml) although others failed to grow at $\frac{1}{20}$ this concentration (Mandelli, 1969). Resistance of higher plants to heavy metals is established by the existence of 436 angiosperm species on Cu soils and 302 on Ni soils (Wild, 1970). Most of the latter occurred on soils containing 2000 ppm Ni or more. *Becium homblei* (Labiatae) grows on soils with high levels of Cu and may accumulate more than 100 ppm Cu in its leaves and roots (Reilly, 1969). Much of the accumulated metal is tightly bound within the plant, and is not in free solution. Selected populations of *Agrostis tenuis* grow on mine workings, which are toxic to most plants (Gregory and Bradshaw, 1965). When tested in culture they showed a remarkable resistance to metals found in high quantities in the soils of their habitat (Cu, Ni, Pb, Zn). Reclamation of toxic metalliferous wastes has, in fact, been suggested by use of Cu and Zn tolerant grass populations of *Festuca rubra*, *Agrostis stolonifera*, and *Lolium perenne* (Smith and Bradshaw, 1970). However, factors other than toxicity of a single heavy metal apparently limit the plant growth in "mine spoils" (Gadgil, 1969). The fungus *Polyporus palustris* has a particularly high tolerance of Cu (as high as 0.32 *M*, or 8% $CuSO_4 \cdot 5\ H_2O$; Osborne and Da Costa, 1970). The tolerance of the dikaryotic mycelium was substantially greater than that of its constituent homokaryons.

Cruciferae (radish, turnip, kohlrabi) possessed a marked ability to develop resistance, i.e., to harden (to $MnSO_4$, $ZnSO_4$, $Cr(SO_4)_3$, and H_3BO_3). Carrot and potato showed less hardening, while tomato showed practically none. The hardening was partly due to age, and partly to seasonal changes to which the plants were exposed. Copper-hardening occurs on Cu-rich soils (Repp, 1963) and some plants showed a high degree of resistance.

There are other ions besides the heavy metals, against which plants can develop resistance. Several species of higher plants have developed a high level of resistance to selenate, which occurs in high concentrations in some soils. It has been suggested that the synthesis of selenoamino compounds in *Astragalus bisulcatus* may be related to their ability to assimilate large quantities of Se without suffering injury (Nigam and McConnell, 1969). The unicellular green alga, *Anacystis nidulans* increased its resistance to Na selenate by about twelve times during ten transfers to increasing concentrations of Na selenate and chloramphenicol (Kumar, 1964). This resistance was partially lost after the resistant cells were given the same number of transfers in selenate-free medium.

Varieties of cotton differ in tolerance of an acid soil high in exhangeable Al (Foy *et al.*, 1967). Similarly, selection of winter barley for Al tolerance from a mixed population growing on acid, Al-toxic soil was demonstrated (Reid *et al.*, 1969).

Clones of *Agrostis canina* growing on serpentine soils were tolerant of higher concentrations of Mg than clones of *Agrostis stolonifera* from non-serpentine soils (Proctor, 1970). Ten ppm Mg had no effect on the former, but it decreased the roots to $\frac{1}{6}$ normal in the latter. This was true only in single salt solutions. As little as 0.5 ppm Ca was enough to counteract the toxicity of 10 ppm Mg.

Resistance to Al toxicity is also evident in the genus *Agrostis*, ranging from 0.2 m*M* for the least resistant, to 1.6 m*M* for the most resistant (Clarkson, 1966). Two wheat varieties also showed different degrees of Al resistance (Foy *et al*, 1955). This was definitely a case of avoidance, for the less resistant variety lowered the pH around its roots, while the more resistant one raised the pH. The difference between the two was as much as 0.7 pH.

Borate resistance was obtained in *Penicillium notatum* by isolating a strain after passing the organism through a nutrient solution saturated with calcium acetate (Roberts and Siegel, 1967).

Biebl and Rossi-Pillhofer (1954) concluded that ion resistance is not a general property of the cell for all ions, but varies with the ion. This conclusion was based on the pronounced increase in resistance to Mn with age of the plant, during a period when resistance to boric acid remained essentially constant, and resistance to $ZnSO_4$ increased slightly. These results varied somewhat with the plant. Gregory and Bradshaw (1965) also concluded that resistance to one heavy metal (among Cu, Ni, Pb, Zn) was not accompanied by resistance to another, except in the case of Zn and Ni. A similar linked resistance has, however, been reported for Li and Na (Bingham *et al.*, 1964). Of eleven species tested, those sensitive to low substrate concentrations of Na were also sensitive to low additions of Li (e.g., avocados, sour oranges, beans). Grasses, cotton, beets, and corn were relatively resistant to both. However, "low" toxic concentrations of Li and Na are much higher than in the case of the heavy metals, and may, therefore, involve secondary stresses (see salt stress). Furthermore, the halophilic alga *Dunaliella viridis* requires high external levels of NaCl for optimum growth, yet its growth is inhibited at low levels of Li, K, and Cs (Johnson *et al.*, 1968).

In most of the above cases, the ion resistance of plants has been called "tolerance," in the sense of tolerance of the external stress. In some cases, however, an attempt was made to determine whether the resistance was due to a true (internal) tolerance or to an avoidance of the stress. Ouellette and Dessureaux (1958) discovered that a clonal resistance of alfalfa to Mn and Al toxicity was due to a retention of the metals in the roots. These resistant clones suffered the same amount of injury as the less resistant clones when the metallic ions reached the same concentrations in the tops of the plant. The shoot of the resistant clone, therefore, owed its resistance to avoidance. Vose and Randall (1962) associated resistance to Al and Mn toxicities with a low cation exchange capacity, perhaps due to an increased uptake of monovalent cations at the expense of Al and Mn. In this case, too, the resistance would be due to avoidance. Other cases are more complicated.

Autoradiography of *Agrostis tenuis*, after supplying it with ^{65}Zn, failed to reveal any difference between the Zn-resistant and the nonresistant populations (Turner and Gregory, 1967). The Zn ions showed a similar distribution pattern in both cases. Consequently, the mechanism was not due to a differential uptake of the ion. Subsequent biochemical analyses of the root homogenate revealed that the cell wall is an outstanding accumulator of the metal ion and is probably involved in the mechanism of resistance.

Even in this case, resistance seems to be due to avoidance rather than tolerance; retention in the cell wall is avoidance as far as the protoplasm is concerned. In the case of selenate, however, resistance must involve tolerance since high concentrations of the ion are accumulated in the resistant plant.

It is obvious from the above brief review, that there is insufficient information on ion injury or the resistance of the plant to it, to propose a mechanism for either.

CHAPTER 20

MISCELLANEOUS STRESSES

Although the above temperature, water, radiation, and salt stresses have received the most attention, and produce the most profound effects on plants in nature, many other stresses have been investigated, and some are becoming more important due to the many new conditions being imposed on plants by man. These stresses may be divided into two main groups: (1) minor (mainly physical) natural stresses, and (2) man-made (chemical) stresses.

A. Minor Natural Stresses

1. Pressure Stress

To the physicist, the term "pressure stress" is repetitive and meaningless, since a mechanical stress has the dimensions of a pressure. When, however, stress is used in the biological sense (an environmental factor that is potentially injurious to the plant), the term is not repetitive and has a definite meaning. It has long been known that plants (and protoplasm, in general) can survive high pressures without injury, provided that the pressures are applied symmetrically. The best examples in nature, of course, are organisms living at great depths in the ocean, which are therefore exposed to symmetrical pressures of 1 atm/30 ft (or per 10 m). Red algae, for instance, may grow and photosynthesize at depths of as much as 100 m and, therefore, at pressures of about 10 atm above atmospheric. It was formerly thought that algae and sea animals cannot survive more than 600 atm, and, therefore, beyond an ocean depth of 6000 m. Yet life has been found at 7800 m, and more recently even at 10,400 m (see Gessner, 1952), and therefore at a pressure of over 1000 atm.

Even the protoplasm of land plants in the turgid state, is constantly subjected to a pressure of several atmospheres, in the case of halophytes to perhaps as much as 100–200 atm. In opposition to ocean pressure, which is an external, inwardly directed hydrostatic pressure, turgor pressure is an internal, outwardly directed hydrostatic pressure, leading to danger of cell rupture. Land plants, which are exposed to these high turgor pressures possess much thicker, more lignified walls than aquatic plants, which have much lower turgor pressures. Land plants, therefore, possess greater turgor pressure tolerance than aquatic plants.

Under artificial conditions, leaves and shoots can be subjected to symmetrical nitrogen gas pressures inside a bomb, of over 100 atm without injury (Scholander *et al.*, 1966). Nevertheless, gas pressures have been shown to injure plants. Shoots of the aquatic plant *Hippuris vulgaris* immersed in water and subjected to gas pressure in the air above it, showed no growth effects up to 0.5 atm above atmospheric pressure (Gessner, 1952). Pressures above 0.5 atm, however, completely stopped growth (Fig. 20.1), although assimilation was not affected. Similar results were obtained with three other species of aquatic plants. Observations of a doubling of the loss of solutes from the plants to the surrounding water led Gessner to suggest that the growth stoppage may have been due to an increase in permeability.

In view of the far greater pressures survived by plants at great depths in the ocean, this kind of result may conceivably mean that plants differ greatly in pressure resistance. On the other hand, the normal survival of turgor pressures far above 0.5 atm, and the survival of high artificial ni-

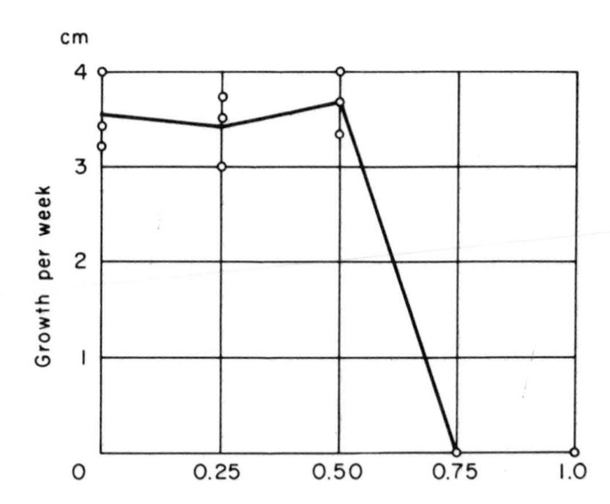

Fig. 20.1. Weekly growth in cm (ordinate) of *Hippuris vulgaris* shoots in relation to pressure (atm) above water. (From Gessner, 1952.)

trogen gas pressures without loss of semipermeability, indicate that the injury at low pressures is not due to the primary pressure stress. The existence of secondary stresses due to increased oxygen pressure has been indicated by Barker (1960). He obtained a decrease in respiratory rate at 3 atm pressure, but this decrease was practically eliminated when he provided CO_2 compensation. He concluded that the inhibition of respiration was due, at least mainly, to the high partial pressure of oxygen rather than the pressure itself.

These results and conclusions have been supported by others. Externally applied pressure markedly depressed the germination of corn grains (Johnson and Henry, 1967). As pressure on the seed increased, the optimum temperature for germination was lowered. This is presumably a reversible growth retardation due to the counteraction of turgor pressure. Increasing the partial pressure of oxygen from 711 to 1478 mm Hg decreased the growth of *Chlorella sorokiniana* linearly (Richardson *et al.*, 1969). Under 2 atm of oxygen pressure growth ceased after 10 to 12 hr. There was no permanent injury, however, and growth resumed when the oxygen partial pressures were returned to normal. Since the inhibition occurred under both autotrophic and heterotrophic conditions, it was not due solely to an inhibition of photosynthesis.

The damage due to moderate gas pressures is, therefore, largely if not solely caused by a secondary oxygen stress. This cannot be a factor in the case of hydrostatic pressures, since the gas pressure is unaltered. It, therefore, follows that plants adapted to growth at great ocean depths must possess resistance to the primary pressure stresses. Such resistance would have to be tolerance since hydrostatic pressures are transmitted uniformly throughout the aqueous medium, which continues through the cell wall and into the protoplasm and vacuole of the plant cells. Tolerance of high hydrostatic pressures is not, however, confined to organisms growing at great ocean depths. Ordinary land plants can survive hydrostatic pressures of well over 1000 atm, which is as high as the greatest pressures at which ocean organisms have been found. Nevertheless, pressure tolerance does change seasonally, paralleling the changes in freezing tolerance (see Fig. 21.3). Alexandrov (1964) has, in fact, shown that the pressure tolerance of plants can be increased, both by natural and by artificial hardening to other stresses (e.g., heat and freezing).

Although land plants possess as high a tolerance of the primary pressure stress as do organisms from great ocean depths, this is undoubtedly true only of the plastic irreversible strain. They undoubtedly cannot grow at the high pressures under which ocean organisms normally complete their growth and development. The difference between the two kinds of plants

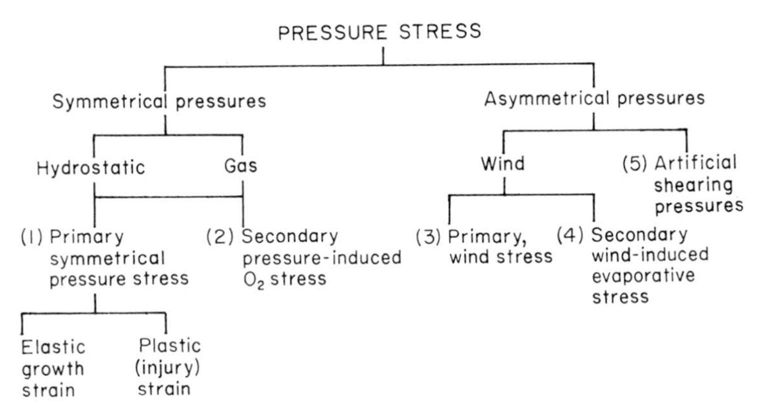

DIAG. 20.1. Five kinds of pressure-induced stresses.

is, therefore, presumably the greater pressure tolerance of the plants from great ocean depths, with respect to the elastic (reversible) strain.

Although plants may be subjected to large symmetrical pressures (in the absence of secondary gas stresses) without injury, small asymmetrical pressures may readily injure them. It is only necessary to press lightly on a glass cover slip with the finger, in order to kill cells between it and a microscope slide. This injury is easily understood, since the asymmetrical pressure may produce a direct physical, shearing effect on the protoplasm. Nevertheless, indirect injury is also possible. Thus, mechanical injury of leaf tissues by rubbing the surface with carborundum, or rapidly infiltrating with water resulted in an enhanced RNase activity (Bagi and Farkas, 1968).

In nature, of course, asymmetrical pressures due to wind may produce a sufficient mechanical stress to break plants. Similarly, plants may be torn from their roots by heaving due to the freezing of wet soils in winter. Besides the primary stress effects of wind, there are also secondary stress effects due to the increased evaporative effect. Somewhat surprisingly, however, although increasing wind velocity increases evaporation, transpiration of all the woody species tested lagged behind evaporation. The heavier the wind the larger was the difference (Tranquillini, 1969). Such trees apparently possess avoidance of this secondary evaporative stress induced by wind. Even photosynthesis was affected little, conifers maintaining 70–90% of the initial value when subjected to heavy winds above 7 m/sec. The slight injury that was detected was due to the secondary, wind-induced water stress leading to local desiccation of the leaf surface.

Five different kinds of stress produced by pressure are shown in Diag.

20.1. There is essentially no information as to the precise nature of primary pressure injury or of pressure tolerance.

2. *Anoxia or Oxygen-Deficit Stress*

FLOODING

Although a water deficit is a clear-cut stress, since it produces a potentially injurious dehydration strain, a water excess is not in itself a stress, since saturation of a cell with water is not injurious, but is a perfectly normal state. Flooding does, however, give rise to two secondary stresses and one tertiary stress (Diag. 20.2) which may, in their turn, give rise to an injurious strain. Even the turgor pressure stress that results from turgor and may conceivably be injurious only in rare cases, is itself a secondary stress. The secondary, oxygen-deficit stress, however, occurs under other conditions as well as in flooded plants, and must, therefore, be considered as a separate stress.

Although plants are capable of anaerobic respiration, it has long been known that most plants die in a nitrogen atmosphere within 24–48 hr. There are some exceptions, particularly among bulky organs that are protected from gas exchange with their environment, e.g., rhizomes, large roots such as turnips, and probably some large fruits. This oxygen deficit stress most commonly arises in nature as a secondary stress due to a primary excess water stress, leading to flooding injury. The secondary, flooding-induced oxygen stress may, in its turn, injure due to a tertiary ionic stress. At the low oxidation-reduction potential resulting from the oxygen deficit, manganous and ferrous ions may accumulate to toxic levels (Conway, 1940). On the other hand, the oxygen stress may also produce a primary indirect injury, due to a metabolic strain, leading to an accumulation of toxic intermediates of anaerobic respiration.

The injury may be of more than one kind. The oxygen deficit may lead to a decrease in water uptake and to wilting of tobacco subjected to poor aeration (Willey, 1970). According to Burrows and Carr (1969), the severe

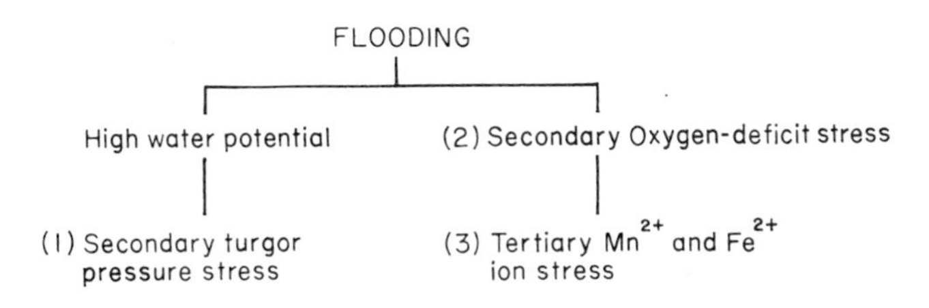

DIAG. 20.2. Secondary and tertiary stresses produced by flooding.

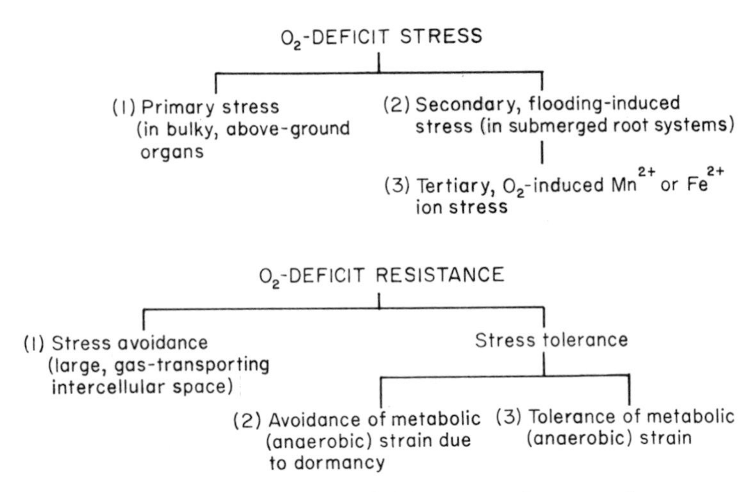

DIAG. 20.3. Kinds of oxygen-deficit stress and stress resistance.

chlorosis in the lowest leaves of flooded sunflower plants may be due to a reduction in import of cytokinins from the flooded roots. In the case of barley roots (De Wit, 1969) polysaccharide synthesis was apparently inhibited by low oxygen tension, but protein synthesis was not. Aquatic plants possess flooding resistance due to avoidance of the secondary oxygen-deficit stress, since they possess very large intercellular spaces (as much as 70% of the shoot volume, versus about 20% of the leaf volume of land plants). This permits the translocation of the oxygen down the plant from leaves at or above the water surface (Conway, 1940). Even the most sensitive plants can be grown in water cultures if the solutions in which their roots are submerged are continuously well aerated, i.e., under conditions of artificial oxygen-deficit avoidance. On the other hand, the above described rhizomes, large roots, and fruits must possess oxygen-deficit tolerance, for they survive in spite of a pronounced oxygen deficit in their tissues. Part of their oxygen-deficit tolerance may be due to their dormant state and, therefore, their very slow oxygen utilization. This tolerance would be due to strain avoidance. On the other hand, they must also possess strain tolerance, i.e., a tolerance of anaerobic respiration. The three kinds of injury and of resistance are shown in Diag. 20.3.

3. *Electric, Magnetic, and Sound Stresses*

Many attempts have been made in recent years to influence the growth rate of plants by exposure to electrical or magnetic stresses. Electrical fields have been reported to inhibit growth (Murr, 1965a, 1966; Sidaway, 1966), to affect growth orientation (Bentrup, 1967), and to increase respiration (Sidaway and Asprey, 1968). Murr (1964) explains cell damage to seedling orchard grass in an electrostatic field by (1) a biochemical enzyme stimulation acting as a trigger for plant destruction, and (2) enzyme oxidation by ozone breakdown. He later (1965b) concluded that leaf damage was associated with direct field or corona-type stresses induced in the leaf epidermis, and that there is no need for a physiological response to accompany the field stress as previously proposed, since the dynamic field intensities at the time of leaf damage are more than sufficient to cause epidermal deterioration by a field evaporation of organic molecules composing the outer cells. The injury, therefore, appears direct, although whether it is produced by the primary electrostatic field stress or by some secondary stress has not been established. The problem of resistance to electrostatic field stress does not seem to have been investigated.

The biological effects of magnetic fields have been investigated for some time (Barnothy, 1964). Many effects on plants have been reported: (1) increased seedling growth (Chao and Walker, 1967; Chuvaev, 1967; Leisle and Nikulin, 1967), (2) growth abnormalities (Chao and Walker, 1967), (3) an increase in the number of female flowers in cucumber (Abros'kin and Zadonskii, 1968), (4) a decrease in dry weight, (5) inhibition of respiration (Tarakanova *et al.* 1965), and after prolonged exposures (6) wilting and death (Novak and Valek, 1965). Enzyme–substrate reactions were unaffected by very high magnetic fields up to 170,000 G (Rabinovitch *et al.*, 1967). In the case of catalase, however, appreciable increases in activity occurred (Haberditzl, 1967). The inhibition has been explained by the influence of the magnetic field on the diffusion of charged particles (Liboff, 1965). Insufficient information is available for understanding the nature of the injury or whether plants differ in resistance.

Air ions have been reported to stimulate ATP metabolism of chloroplasts and to accelerate growth (Kotaka *et al.*, 1968; Kotaka and Krueger, 1968) when due to O_2^- or O_2^+, but to inhibit growth when due to CO_2^- or CO_2^+, and to accelerate chlorosis by excess use of Fe in synthesis of cytochrome *c*.

Many results have been published indicating an enhancing or inhibiting effect of sound on growth (Maslova *et al.*, 1965; Weinberger and Measures 1968, Measures and Weinberger, 1970). The reported effects of musical sounds on growth have been opposed by consistently negative results (Klein and Edsall, 1965). Sound waves have such low energy values, that it is difficult to conceive of a significant strain produced in the plant by them. Ultrasonic waves, on the other hand, have much higher energy levels, but are not as yet a factor in nature. They have been reported to increase, but also to decrease germination (Ghisleni, 1956; Limar, 1961a,b; Kochkar, 1961) and to induce callose synthesis at 850 kc of low intensity (Currier and Webster, 1964). The callose synthesis was reversible if there was no gross petiolar injury and cells remained alive, and there was no inhibition of subsequent growth. Yeast cells ceased synthesis when the ultrasonic oscillator was operated at a critical low power level (Burns, 1964). Synthesis was resumed immediately when the oscillator was turned off. The ultrasonic stress is, therefore, able to induce both an elastic (reversible) and a plastic (irreversible) strain.

B. Man-Made Stresses

1. Gas (Excess) Stresses

a. SO_2

Since the beginning of this century, sulfur dioxide from industrial plants has been recognized as an air pollutant, and as a source of injury to plants (Webster, 1967). Depending on the amount present, leaves may be partially injured or killed, the symptoms of injury being specific and different from those produced by other pollutants. The specificity of the damage is confirmed by the increased sulfur content of the leaves. There is apparently no decrease in growth or in the photosynthetic rate at concentrations that fail to injure the leaf. Alfalfa, for instance, grew and photosynthesized normally when exposed to 0.1 to 0.2 ppm continuously for 45 days (Thomas and Hill, 1949). When fumigated for 4 hr on 5 successive days with 0.43–0.46 ppm, there were only traces of leaf injury, and photosynthetic activity was slightly lowered (to 86–93% of normal). The SO_2 stress clearly produces a plastic (irreversible) strain but no elastic (reversible) strain. Furthermore, the plastic strain is apparently direct, since there is no evidence of an indirect metabolic disturbance. The nature of the injury should, therefore, be relatively easy to establish.

Some investigators have used much higher concentrations of SO_2 and it is, therefore, difficult to compare their results with all of the above. Lichens are known to be sensitive to atmospheric pollution, yet they were exposed to the much higher doses of 5 ppm SO_2 for 24 hr (Rao and Leblanc, 1966). This resulted in abnormal bleaching of the algal chlorophyll, permanent plasmolysis, and brown spots on the chloroplasts. The absorption spectrum of the chlorophyll indicated a degradation of phaeophytin *a*. Even higher concentrations were used by Taniyama and Arikado (1968). Rape and barley leaves were injured on exposure to 20 ppm SO_2, the injury increasing with the time exposed. In the vegetative (but not the flowering) stage, rape was more severely injured in the dark than in the light (20 klux). Barley was always more severely damaged in the light. The SO_2 was absorbed by the leaves at all times, and the sulfur content increased with the time exposed. Injury was correlated with sulfur content.

On entrance into the leaf, the SO_2 must dissolve to sulfurous acid, forming sulfite salts. The plant can presumably oxidize it to the sulfate. Since sulfates are normal constituents of plants and, in fact, are essential mineral nutrients, no injury can be expected if all the SO_2 is converted to this form. It is unoxidized sulfite that must cause the injury. According to Bersin (1950), the reaction with the sulfite is as follows:

$$RSSR + H_2SO_3 \rightarrow RSSO_3H + RSH$$

The disulfides are split forming thiosulfuric acid and thiols. L-Cystine would, in this way, be converted to cysteinthiosulfurous acid. Similarly, disulfide proteins would be split. This could, of course, inactivate disulfide enzymes, but it is difficult to see how such inactivation would kill the cell without first disturbing the metabolism and decreasing the growth. In view of the initial water-soaked appearance of the damaged marginal and interveinal areas (Webster, 1967), it seems more likely that injury is due to a loss of semipermeability, resulting from an alteration in the membrane proteins, as in the case of radiation injury (see Chapter 18). This would seem to indicate that the SO_2 stress induces primary direct injury.

There are marked differences in susceptibility of plants to SO_2 injury. Thomas and Hendricks (1956—see Webster, 1967) divide twenty-five species into three groups: sensitive, intermediate, and resistant. Differences in resistance have been found even between clones of a single species, e.g., *Pinus strobus* (Cotrufo and Berry, 1970). There are discrepancies between investigators as to the relative resistance of some of the species (Webster, 1967). This may be due partly to differences in the stages of growth, partly to environmental conditions. In most cases, injury increases as the young leaf unfolds and expands, reaching a maximum when fully expanded, and then decreasing. This relationship depends on the plant as well as the investigator (Webster, 1967). The degree of injury sometimes appears to be decreased by fertilization with N, P, and K (Cotrufo and Berry, 1970).

Any environmental condition that favors stomatal opening, increases the absorption of SO_2 and, therefore, the injury (Katz *et al.* 1939; Thomas and Hendricks, 1956 see Webster, 1967). On the other hand, 1 ppm of SO_2 may itself increase the degree of stomatal opening throughout the day and night (Majernick and Mansfield, 1970). It is, therefore, obvious that one kind of SO_2 resistance is due to avoidance by keeping the stomata closed. The differences between species are not, however, likely to be due to this factor, since all or nearly all crop plants grown under normal conditions open their stomata during daylight. Furthermore, the plant can survive injurious exposures to SO_2 if exposed to them intermittently (Webster, 1967). This demonstrates an ability to recover, presumably by metabolic activity. Even the ability to oxidize the sulfite to sulfate must vary from plant to plant, from one stage of growth to another, and from one environmental condition to another. The essential absence of injury in the dark, and the increased injury with light intensity, although largely due to avoidance (stomatal closure in the dark), may also depend on the high reducing activity due to photosynthesis in the light, and the high oxidizing activity due to respiration in the dark.

b. Photochemical or Oxidant Smog

This kind of gas stress is mainly due to two substances—ozone (O_3) and peroxyacetyl nitrate (PAN), although other injurious substances may also be present (Webster, 1967). In some respects, the actions of these two substances are directly opposite to the action of SO_2. They are oxidizing substances while SO_2 is a reducing substance. They injure the lower surface of the leaf while SO_2 injures the upper. Although PAN injury is increased by high light intensity and greatly decreased by darkness (as in the case of SO_2) its absorption is unaffected by the opening and closing of stomata. Finally, both O_3 and PAN may inhibit plant growth, as well as increase the respiratory rate and decrease the rate of photosynthesis. Thus, unlike SO_2, they can induce both elastic and plastic strains, and can produce both direct and indirect injury.

The elastic (reversible) growth strain is apparently due to interference with cell wall metabolism, according to Ordin and his co-workers (see Webster, 1967). In agreement with this conclusion, 2,4-D levels that are normally superoptimal for growth of *Avena* coleoptile sections, led to greater growth in PAN-treated sections than in the absence of 2,4-D (Ordin *et al.*, 1970). This seemed to be due to a decreased inhibition of wall metabolism. Both PAN and O_3 inactivate some enzymes, possibly by attacking the SH or histidine groups (Ordin and Skoe, 1964).

The plastic (irreversible) strain is first observed as minute blisters on the undersurface of the leaf, due to swelling of guard cells and epidermal cells. This is followed by cell collapse, producing water-soaked spots. On dehydration, these finally become silvered or bronzed (Webster, 1967).

Ozone injured bean leaves in two phases (Thomson *et al.*, 1966). In the first phase, the chloroplast stroma became granulated. In the second phase, there was a general disruption of cell membranes and organelles, and the cell contents aggregated in the center of the cells. These changes were identical with those previously observed in cells damaged by peroxyacetyl nitrate, and were probably related to the oxidative properties of both molecules.

Besides these direct types of injury, the above mentioned effects on rates of respiration and photosynthesis indicate an indirect injury. In agreement with these results, O_3 inhibited oxidative phosphorylation and O_2 uptake by mitochondria of tobacco (*Nicotiana tabacum*) leaves (Lee, 1967). The phosphorylative system was more sensitive than the respiratory system. After treatment for 1 hr with 1 ppm ozone, uncoupling of phosphorylation occurred without any detectable change in rate of respiration. There was no apparent change in optical density (at 520 mμ) of the mitochondria, therefore mitochondrial swelling was apparently not involved. Sucrose and glucose fed to detached leaves tended to raise the phosphorylative activity of the mitochondria.

Resistance is apparently not due to avoidance, since stomatal opening has no effect on the injury produced (Dugger *et al.*, 1962). It must, therefore, be assumed that resistance is due to tolerance. The resistance changes that parallel the stages of development are in agreement with this conclusion. The cells are somewhat resistant until they are fully developed. They then possess a short period of high sensitivity, and finally become more resistant again. The effects of O_3 are essentially confined to fully expanded leaf tissue, while PAN is most toxic to younger, expanding tissue (Dugger *et al.*, 1962; Webster, 1967).

A considerable body of evidence points to SH groups as a probable point of injury. As mentioned above, even the elastic strain has been explained by inactivation of SH enzymes. PAN reacts with all the SH groups of human hemoglobin at pH 4.5, but at pH 7.2 the reaction is slower and is limited to 2–3 SH/molecule (Mudd *et al.*, 1966). It does not react with the SH of native ovalbumin. Whether or not its reaction with SH groups explains its physiologcal effect, could not be decided. PAN inhibited cyclic photophosphorylation of isolated chloroplasts (Kuokal *et al.*, 1967). The evidence indicated that PAN can oxidize SH groups in enzymes necessary for photophosphorylation. Since, however, the inhibition of photophosphorylation by PAN could not be reversed by GSH, it was concluded that the oxidation of the SH group proceeded beyond the SS stage.

Ozone damage has also been related to SH. Ozonation lowers the SH content of bean, spinach, and tobacco plants (Tomlinson and Rich, 1968).

Treatment with IA yielded similar symptoms. The tobacco variety resistant to ozone was also more resistant to damage by SH reagents. The older the leaf, the more severely it was damaged by SH reagents and by ozone. The younger, more ozone-resistant leaves have higher concentrations of SH. Leaves of the resistant variety, however, contain lower concentrations. It has even been possible to reduce ozone injury to bean plants by spraying with CSH and GSH as well as other reducing agents (Das and Weaver, 1968). The severity of the damage was increased by PCMB and NEM. They, therefore, suggest that protein SH groups are the primary site of ozone injury.

c. NH_3

Ammonia has long been known for its toxicity to plants, and was suggested as the cause of injury to starving (darkened) leaves due to protein breakdown (Chibnall, 1939). The Soviet investigators have, therefore, explained the primary indirect injury produced by some of the above stresses (e.g., high temperature and water stresses) on the basis of a similar protein breakdown leading to NH_3 toxicity.

Increasing the supply of nitrogen [as $(NH_4)_2SO_4$ or KNO_3] to rice plants leads to unfavorable development (Varga and Szoldos, 1963). Root growth was especially inhibited due to the accumulation of NH_3. The ammonia inhibited the activity of IAA oxidase, presumably resulting in an anomalous increase in IAA and a consequent growth inhibition. Investigations of root IAA supported this suggestion. The injury is at least partly due to the secondary pH stress, and can be prevented in some cases by maintaining the pH near neutrality by the addition of $CaCO_3$ (Maynard and Barker, 1969). One kind of resistance to this secondary, NH_3-induced pH stress has long been known. Plants have, in fact, been divided into two classes on this basis (Prianishnikov, 1951). (1) The acid plants are NH_3 resistant, due to neutralization of the NH_3 by their large reserve of organic (N-free) acids stored in the vacuole. As much as 0.1 N NH_4 salt is tolerated by begonia without injury. The amide plants do not possess a reserve of organic (N-free) acids and, therefore, the NH_3 can only form amides with the small quantity of acidic amino acids (aspartic acid or glutamic acid) present in the protoplasm. The acid plants are, therefore, NH_3 resistant due to avoidance of the secondary pH stress, while the amide plants are not resistant due to the absence of this avoidance mechanism. The great majority of plants are of the amide kind and, therefore, are not NH_3 resistant.

In spite of this well-known toxicity of NH_3, *Allium* seed germinated up to 15% under an atmosphere containing NH_3 together with equal volumes of nitrogen or air (Siegel and Daly, 1966).

d. NONTOXIC GASES

Prolonged exposures of hydrated rye seed to acute anoxia progressively lowered viability (Latterell, 1966). For equal exposures of 9 days or longer, the mortaility was significantly higher in helium than in nitrogen.

2. Nongaseous Chemicals

The commonest nongaseous chemicals to which plants are artificially exposed are fertilizers, herbicides, fungicides, and insecticides. The literature on these substances is so voluminous that they must be treated as separate problems and cannot be considered here. Nevertheless, it must be recognized that the same principles apply to them as to the other stresses. They can induce both primary and secondary stresses. The strains produced may be elastic (reversible, e.g., growth inhibition) or plastic (irreversible). Resistance may be due to avoidance (e.g., very waxy leaves prevent penetration of some substances), or to tolerance (e.g., by metabolizing the substances in the cell). It is, therefore, possible for a plant to possess a general resistance to a number of these chemicals. It is even possible for a resistance to some chemical stresses to provide resistance to other stresses as well (see Chapter 21).

There are many other environmental stresses that have not been considered, particularly the soil stresses, such as acidity and alkalinity. This neglect is largely because of the lack of information relative to plant injury and resistance. Perhaps investigators of these stresses will be encouraged to attempt an attack on them from the point of view of the responses discussed in this monograph.

INTERRELATIONS

CHAPTER 21

COMPARATIVE STRESS RESPONSES

It has frequently been suggested that the plant may have a general resistance to several or even all environmental stresses, which has been variously called "physiological" (Maximov, 1929), "ecological" (Biebl, 1952a), or "environment" resistance (Levitt, 1956, 1958). It has already been shown, however, that even the avoidance and tolerance of a single stress require completely different mechanisms (Chapter 15), and that there may even be two kinds of tolerance of a single stress (Chapter 12). Therefore, there can be no single resistance against all stresses. It is, however, conceivable that one or a few types of resistance may protect a plant against several or even most stresses. The comparative stress responses of plants are, therefore, an important practical problem. They are also fundamentally significant in a search for the resistance mechanism; since if it is known for one type of stress resistance it may also explain a correlated stress resistance. Three comparisons may logically be made: (a) primary, secondary, and tertiary stresses, (b) the strains produced, and (c) the resistance mechanisms. It must be emphasized, however, that these comparisons can be made only on the basis of the, at present, inadequate evidence. Any conclusions arrived at, must therefore be accepted only as working hypotheses to be tested by further experiment.

A. Comparative Primary, Secondary, and Tertiary Stresses

One kind of general stress resistance may be predicted due to the occurrence of secondary and tertiary stresses. If a number of primary stresses produce the same secondary or tertiary stress, then the plant should possess a single resistance to this common stress, and therefore to one effect of the primary stresses producing them. The best example of this is the water stress, which may be produced in seven ways by six different primary stresses:

1. The secondary, chilling-induced water stress due to an excess of transpiration over water absorption by the chilled roots.
2. The secondary, freezing-induced water stress due to transpiration in the absence of translocation through the frozen xylem (leading to "freeze-desiccation").
3. The secondary, freezing-induced water stress due to intercellular ice formation (leading to "freeze-dehydration").
4. The secondary, heat-induced water stress.
5. The primary, drought-induced water stress.
6. The tertiary (visible and infrared) radiation-induced, secondary heat-induced water stress.
7. The secondary, salt-induced osmotic water stress.

The seven water stresses produced by all the above six primary stresses must expose the plant to the same kind of strain and, therefore, the same kind of injury. Another example of a common secondary stress is the nutrient deficit stress, which may be induced by both the water stress and the salt stress. The ability of the plant to resist the common secondary or tertiary stresses by a single avoidance or tolerance mechanism will be discussed below.

A resistance mechanism that succeeds against a secondary or tertiary stress, cannot be expected to succeed against the strains produced by the primary stress. On the other hand, any stresses that produce the same strain may be expected to give rise to the same resistance response. It is, therefore, important to compare the strains produced by different stresses.

B. Comparative Primary Strain Responses

The primary stresses may induce either elastic or plastic strains, and these may be either direct or indirect. In general, the direct elastic strains must differ for different stresses. The chilling stress, for instance, may produce a reversible retardation of the metabolic reactions and a reversible solidification of lipids (Table 21.1). The heat stress, on the contrary, may reversibly accelerate the metabolic reactions and liquefy any solid lipids present. Reversible protein denaturation, on the other hand, may be induced by these same two different stresses—chilling and heat. All three types of radiation stress, of course, induce a reversible photo- or radio-

TABLE 21.1 Comparative Strain Responses Induced by the Primary Environmental Stresses

Primary stress	Strains: Elastic (direct)	Direct plastic	Indirect plastic
Chilling	1. Lipid solidification 2. Protein denaturation 3. Metabolic retardation	Loss of semipermeability	1. Deficiency 2. Toxicity
Freezing (intracellular)	1–3 As above 4. Solidification of water (and expansion)	1. Loss of semipermeability 2. Disruption of protoplasm	None
Heat	1. Lipid liquefaction 2. Protein denaturation 3. Metabolic acceleration	1. Protein aggregation 2. Loss of semipermeability	1. Deficiency 2. Toxicity
Water	1. Dehydration 2. Concentration of solutes 3. Cell collapse	1. Protein aggregation 2. Loss of semipermeability	1. Deficiency 2. Toxicity
Radiation (visible and infrared)	Photosensitization	Photochemical reaction (photodynamic effect)	Deficiency
Radiation (ultraviolet)	Radiosensitization	Radiochemical reaction	Deficiency
Radiation (ionizing)	Radiosensitization	1. Radiochemical reaction 2. Protein aggregation 3. Loss of semipermeability	1. Deficiency 2. Toxicity
Salt	Ion unbalance	Loss of semipermeability	1. Deficiency 2. Toxicity
Ion	Ion exchange	1. Combination with protein SH 2. Loss of semipermeability	Enzyme inactivation (leading to deficiency or toxicity?)
Pressure	1. Reversal of turgor pressure 2. Cessation of cell enlargement		None

TABLE12 21.1 (Continued)

Primary stress	Strains: Elastic (direct)	Direct plastic	Indirect plastic
Oxygen deficit	Lowering of O–R potential	1. Formation of toxic intermediates 2. Loss of semipermeability	Toxicity
SO_2	1. Sulfite formation 2. Lowering of O–R potential	1. Splitting of protein SS	Enzyme inactivation (leading to deficiency or toxicity?)
Oxidizing pollutants (PAN, O_3)	1. Raising of O–R potential	1. Oxidation of protein SH to SS 2. Loss of membrane semipermeability	Enzyme inactivation (leading to deficiency or toxicity?)
NH_3	1. NH_4^+ formation 2. pH rise	1. Amide formation 2. Changes due to pH rise	

sensitization, but the type will be specific for each radiation. Similarly, more than one pollutant is capable of raising or lowering the oxidation-reduction potential, although the actual potential produced by each must be different.

Since the elastic strains are reversible, they are not injurious by themselves. Those stresses that induce the same kind of reversible elastic strain may, however, lead to an irreversible and potentially injurious plastic strain. It is even conceivable that different elastic or plastic strains may induce a similar kind of injury. A loss of semipermeability due to membrane damage, for instance, has been indicated as the direct injury in the case of fully nine stresses (Table 21.1). Even a biotic, phytotoxin-caused stress appears to injure via membrane damage (Johnson and Strobel, 1970). It, therefore, seems likely that the plasma membrane is the Achilles' heel of the cell, and that the direct injury produced by many, if not most, stresses is membrane damage.

The simplest kind of membrane damage (Table 21.2) is, of course, the membrane laceration by the expansion of intracellularly formed ice crystals, piercing the membranes in many places. A second simple method is suggested by investigations of the membranes of microorganisms, which

TABLE 21.2 FIVE TYPES OF STRAIN (AND THE CORRESPONDING NINE INDUCING STRESSES) CAPABLE OF LEADING TO INJURY BY MEMBRANE DAMAGE AND THE CONSEQUENT LOSS IN SEMIPERMEABILITY

Membrane-damaging strains	Stresses capable of inducing them
(a) Crystallization and expansion of protoplast water (membrane laceration)	(1) Freezing (primary, intracellular)
(b) Lipid solidification	(2) Chilling
(c) Lipid (per) oxidation	(3) Radiation
	(4) Salt
	(5) Ion
(d) Protein aggregation	(6) Freezing-induced water stress
	(7) Drought
	(8) Heat
	(9) Ionizing radiation
	(10) Oxidizing pollutants (PAN, O_3)
(e) Protein SH binding	(11) Ion

indicate that membrane lipids must be in the liquid state for normal membrane function (Overath *et al.*, 1970). The phase transition temperature is apparently unaffected by the protein component of the membrane (Melchior *et al.*, 1970). The evidence (Chapter 4) indicates that chilling injury may be due to this factor—lipid solidification at chilling temperatures leading to membrane damage and, therefore, to loss of semipermeability. Plants that are not chilling sensitive must, therefore, possess membrane lipids with melting points well below the chilling temperatures. Lipid solidification can also conceivably be induced by another method—oxidation of the unsaturated, low-melting point lipids to saturated, high-melting point lipids. Chaotropic agents, including a number of ions (e.g., NO_3^-) destabilize the native structure of submitochondrial particles and microsomes, rendering them susceptible to lipid oxidation by molecular oxygen (Hatefi and Hanstein, 1970). It is, therefore, conceivable that some salt or ion stresses may injure by means of a similar oxidation of membrane lipids, leading to loss of semipermeability. Even UV-irradiation has been suggested to injure chloroplasts by a structural disruption of the lamellar membrane (Mantai, 1970b).

A fourth possible mechanism of membrane injury is by an effect of the stress on the proteins. This effect may be initiated by protein denaturation at chilling or high temperatures, or by a dehydration strain leading to a mechanical stress on membrane proteins. In either case, aggregation of proteins by intermolecular bonding via SS or hydrophobic bonds may be favored. There is evidence of this kind of strain in the case of five stresses (Table 21.2). A sixth possible stress is the synthetic toxin, atrazine. It may be detoxified in corn leaves by conjugation with GSH (Shimabukuro *et al.*, 1970), perhaps in this way protecting the important SH groups in the membrane proteins. Finally (Table 21.2), these SH groups may combine with heavy metal ions (and perhaps others), again leading to "holes" in the membrane and loss in semipermeability (see Chapter 10).

There are, therefore, not only nine different stresses apparently capable of injuring plants by destroying their semi-permeability, but there are actually five different kinds of strains produced by them that can lead to this single common kind of injury. These are all cases of direct injury. Is there also a kind of indirect injury that may be common to more than one stress?

The indirect plastic strains are all metabolic in nature, and, therefore, the indirect injuries produced by the different stresses must resemble each other to some degree. They may be classified as (1) deficiencies of an intermediate or a terminal metabolite, or (2) toxicities (Table 21.3). Even the more specific kind of deficiency injury may be common to two or more stresses (Table 21.3). Some of these similarities, however, involve specifically different substances, for instance, in the case of the toxins.

TABLE 21.3 Examples of Identical Indirect Metabolic Injuries Induced by Different Primary Stresses

Indirect metabolic injury	Stresses producing them
(1) Deficiencies	
(a) Starvation	Chilling Heat Water Light deficit
(b) Protein breakdown	Chilling Heat Water Salt
(c) Biochemical lesion	Chilling Heat
(2) Toxicities	Chilling Heat Drought Salt

Several conclusions emerge from the above comparisons of injury responses:

1. Different primary stresses may be producing the same kind of injury in the plant if the strain is due to identical primary secondary or tertiary stresses. This is particularly common in the case of the water stress.
2. Many stresses induce the same kind of primary direct injury—a loss of semipermeability. This injury may be due to direct membrane-lipid strain or direct membrane-protein strain.
3. All stresses may produce a primary indirect injury due to a metabolic disturbance. These disturbances are of two main types—a deficiency of an essential metabolite or an excess of a toxic metabolite. End-product deficiencies (carbohydrate, protein) are common effects of several stresses.

There are thus three sources of similarity between the injuries induced by different stresses, and several sources of differences. Are the resistance mechanisms similar for the similar stress effects, and different for the different stress effects?

C. Comparative Resistance Responses

Some plants possess resistance to more than one stress. This does not, of course, prove that a single mechanism controls the correlated resistances. It may simply be due to parallel selection of two or more distinct adaptations. A plant native to a dry, hot climate, for instance, must have been selected both for heat and drought resistance, although the two mechanisms of resistance may be completely different. This kind of evidence must obviously be checked very carefully before accepting it as proof that the resistance to different stresses is due to a single mechanism.

Avoidance and tolerance mechanisms are generally very different, and may, in fact, be mutually exclusive (Chapter 15). It is, therefore, of fundamental importance when comparing the resistances to different stresses, to compare either the avoidances of the different stresses with each other, or the tolerances, but not the avoidance of one with the tolerance of another. Nor should the total resistances be compared with each other, so long as they are due to combinations of avoidance and tolerance.

1. Avoidance

Avoidance of any one stress must, of course, carry with it an avoidance of any secondary or tertiary stress that it is capable of inducing. Thus, visible and infrared radiation avoidance, also leads to avoidance of the secondary, radiation-induced heat and the tertiary, radiation-induced water stress.

The avoidance mechanisms developed by the plant against different stresses, have little if any relationship to each other (Table 21.4). Although the general mechanism is basically similar in the case of visible, IR, and UV radiations, the substances and structures that reflect, transmit, or absorb one wavelength of radiation will be ineffective against another wavelength. Consequently, no one structure or substance can confer avoidance of all radiations. Similarly, although freezing and drought both induce dehydration of the plant cell, the avoidance factors capable of decreasing the dehydration in the case of drought (e.g., stomatal and cuticular control) are completely incapable of affecting dehydration by freezing. Even within a single group of related stresses, e.g., ion stresses, there may be little relationship between avoidance mechanisms, for ion absorption is a highly specific process. Therefore, no one carrier, for instance, can function in the exclusion or excretion of different ions. A general low permeability may prevent the passive absorption of many or all ions, but since ions are to a large extent actively absorbed, this exclusion is not a very effective resistance mechanism. The avoidance mechanisms for intracellular and extracellular freezing strains are completely unrelated to other avoidance mechanisms (and to each other), and there is essentially no avoidance of

TABLE 21.4 Avoidance Mechanisms against Different Stresses

Primary stress	Avoidance mechanism
1. Chilling temperature	None
2. Freezing temperature	None
(a) Intracellular freezing strain	High cell permeability
(b) extracellular freezing strain	Small size, low water content, high cell sap concentration
3. Heat	High transpiration rate
4. Water	(a) Water conservation
	(b) Rapid water absorption (and high transpiration rate)
5. Visible and IR radiation	High reflection, transmission, and absorption (by protective substances)
6. UV radiation	High reflection, transmission, and absorption (by protective substances)
7. Ionizing radiation	High proportion of nonliving:living mass
8. Salt	Exclusion, excretion, dilution
9. Ion	Exclusion, precipitation
10. SO_2	Stomatal closure

chilling. The only remaining possibility of a relationship is between the water and heat stresses.

Heat avoidance is primarily due to a high transpiration rate (Chapter 12). Since one kind of drought avoidance—that due to high water-absorbing ability—is also characterized by a high transpiration rate, the mechanism is the same as for heat avoidance. On the other hand, the kind of drought avoidance that is due to water conservation cannot lead to heat avoidance. On the contrary, it prevents the development of heat avoidance by decreasing the evaporative cooling, and plants of this type can develop only heat tolerance. This was clearly shown by Lange (1959) in the case of those species native to the hot Sahara desert. He found that all "undertemperature" species (i.e., those showing heat avoidance) possess low-heat tolerance, all "overtemperature" species (without avoidance) possess high tolerance. It is not surprising, therefore, that succulents commonly have high heat tolerance. Yet, in spite of this correlation, Lange (1961) has shown that the increase in a plant's succulence does not necessarily lead to increased heat resistance. This is to be expected, for the correlation between succulence (the avoidance of water stress) and heat tolerance that is found in nature is simply due to parallel selection by the high temperatures of the dry climates inhabited by these plants. Since their drought resistance is due to water conservation, they cannot develop heat avoidance. Thus, only those succulents that possess high heat tolerance survive in these climates. In agreement with Lange, *Rhagodia baccata*, a plant of arid and semiarid regions of Western Australia exerts marked control over water loss and has high heat tolerance (Hellmuth, 1968).

What of the avoidance of secondary and tertiary stresses? If different primary stresses induce a common secondary or tertiary stress, can a plant develop a single avoidance mechanism against the several common secondary or tertiary stresses? The best example of a common secondary or tertiary stress is the water stress. Since it is usually transmitted to the plant via a dehydration, the avoidance mechanism must operate by blocking this dehydration strain. Unfortunately, the dehydration is not induced in an identical manner in all cases. The chilling-, heat-, drought-, and radiation-induced water stresses all arise by a net diffusional loss in the vapor phase from the shoot. On the other hand, the salt-induced osmotic stress arises by a net diffusional loss in the liquid phase from the root. Finally, the freeze-induced water stress arises by a net diffusional loss in both phases from the cell to the intercellular spaces. It is not surprising, therefore, that no one avoidance mechanism provides resistance against all seven kinds of water stress. Some may confer avoidance of two, three, four or even five types of water stress (Table 21.5), but there is no mecha-

TABLE 21.5 EFFECTIVENESS OF SOME AVOIDANCE MECHANISMS AGAINST THE SEVEN DIFFERENT WATER STRESSES

	Avoidance mechanism			
Water stress	Water impermeable surface	Rapid water absorption	Stomatal control	Freezing avoidance
(1) Primary drought-induced	x	x	x	
(2) Secondary chilling-induced	x	x		
(3) Secondary heat-induced	x	x	x	
(4) Tertiary radiation-induced	x	x	x	
(5) Secondary freezing desiccation	x			x
(6) Secondary freezing dehydration				x
(7) Secondary salt-induced (osmotic)				

nism for avoiding the osmotic stress, since this would involve impermeability of the cell to water.

There are two conclusions from the above comparisons of avoidance responses:

1. Each avoidance mechanism is specific for a single primary stress and is ineffective against others, except in the case of rapid absorption and loss of water, which confers avoidance of both the water and heat stresses.

2. A single mechanism may confer avoidance of several primary, secondary, and tertiary types of a single stress, but may be completely ineffective against other kinds of the same stress.

2. *Tolerance*

a. WATER STRESS TOLERANCE—PRIMARY, SECONDARY, AND TERTIARY STRESSES

In contrast to avoidance, water stress tolerance cannot be affected by the diffusion phase, since it is a measure of the actual water stress in the protoplasm which the plant can survive, regardless of how the water stress is achieved. Tolerance of one kind of water stress should, therefore, carry with it a tolerance of all the others. The available evidence supports this expectation.

1. The actual degrees of dehydration suffered under natural conditions due to freezing and drought are similar (Table 21.6).

TABLE 21.6 DEHYDRATION BY FREEZING AND BY NATURAL DRYING IN AIR[a]

Species	Percentage of water removed from the cells	
	By freezing at −15°C	By drying in air on a windy day (max. loss)
Fucus vesiculosus	82	91
Chondrus crispus	74	63
Ulva lactuca	69	77

[a] Kanwisher, 1957.

2. When the tolerance of freeze-induced water stress increases, there is a parallel increase in tolerance of both the drought and osmotic stresses (Fig. 21.1).

3. When species or varieties are compared, their relative tolerances of freeze-induced, osmotic, and drought stresses both in the unhardened and the low-temperature hardened conditions run parallel (Fig. 21.2).

4. Similar physiological changes occur during hardening to freezing and to drought (Table 21.7).

5. In several cases, a single species has shown the same injury when exposed to approximately equal water stresses, one induced by drought, and the other by freezing (Table 21.8).

It has even been suggested that cells, in general, have the same tolerance

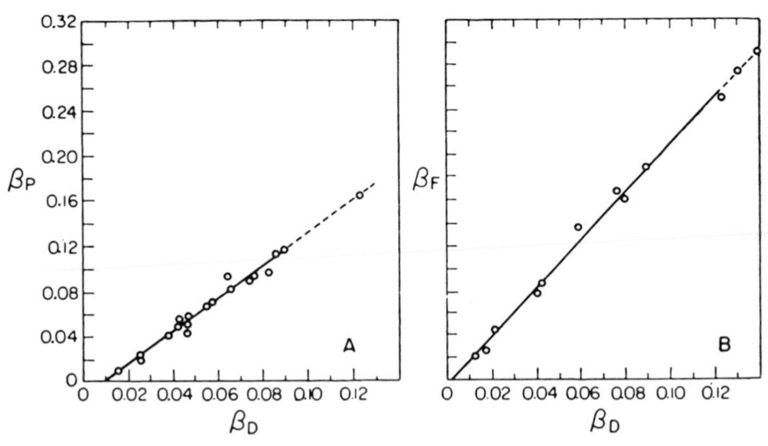

FIG. 21.1. Direct relationship between (A) tolerance of osmotic stress (β_P) and tolerance of drought stress (β_D), and (B) tolerance of freezing stress (β_F), and tolerance of drought stress (β_D). (From Siminovitch and Briggs, 1953a.)

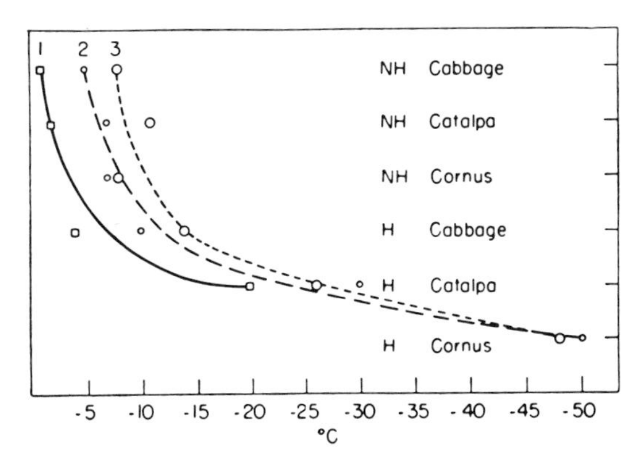

FIG. 21.2. Parallelism between the tolerances of three species of plants in the hardened and unhardened states. (1) Osmotic stress, (2) freezing stress, and (3) drought stress. Expressed as (1) the freezing point of the solution in contact with the cells, (2) freezing temperature, and (3) the freezing point of the solution in equilibrium with the air in contact with the cells, causing 50% killing of the cells exposed to the osmotic, freezing, and drought stresses respectively. (From Scarth, 1941.)

of dehydration (Meryman, 1967), but the above data (Table 21.8) show that this is not so. The osmotic water stress apparently does not agree too well with the other two stresses. This is not surprising, however, since the salt inducing the osmotic water stress may penetrate the cell in small or large quantities, depending on the plant, and, therefore, involves greater dehydration avoidance than in the case of freezing or drought. On the other hand, it may induce ion injury as well as osmotic injury. The first complication would lead to greater tolerance, and the second to less.

In spite of all the above qualitative and quantitative evidence of a close correlation between freezing and drought resistance, there are also several cases in the literature indicating a lack of correlation (Levitt, 1956). Pineapple, for instance, is drought resistant but has no freezing resistance. Its drought resistance, however, is due to avoidance as in the case of succulents, in general, and therefore cannot involve a tolerance mechanism, such as is required for freezing resistance. The nonsucculent millet is also drought resistant due to avoidance, but its drought tolerance is very low (see Chapter 16) and it is not freezing tolerant. The reverse situation is illustrated by the olive. In contrast to other evergreens (conifers, Ericaceae, ivy), the leaves of the olive survive the same drought stress in summer and winter, but a 4°–5°C lower freezing stress in winter than in summer (Larcher, 1963). This greater freezing resistance is solely due to avoidance, because of a lower freezing point. The olive leaves are killed by slight

TABLE 21.7 Some Physiological Factors Reported to be Correlated with Resistance to Different Stresses

	Physiological factors						
Stress	Decreased H_2O content	Perme- ability	Osmotic potential	Sugars	Proline	Ascorbic acid	SH
Chilling temp.		x					
Freezing temp.	x	x	x	x	x	x	x
Drought	x	x	x	x	x	x	x
Heat	x		x	x			x
Radiation:					x		x
Light							
UV							
Ionizing	x					x	x
Salt					x	x	x
Ion							x
SO_2							x
PAN, O_3							x

freezing both in summer and winter, and therefore show no difference in freezing tolerance.

On the other hand, Irmscher (1912) reports that some mosses with the lowest drought resistance have the highest freezing resistance. Since mosses possess essentially no avoidance, this difference must have been due to tolerance. In contrast to Irmscher, however, epiphytic mosses in winter were more resistant to both stresses than in summer (Hosokawa and Kubota, 1957). Another possible exception are tropical intertidal sea algae. They possess high drought and osmotic resistance and they are also chilling resistant, but they do not tolerate freezing (Biebl, 1962b,c). It is perhaps possible that due to the lack of an intercellular space system (such as occurs in higher plants) or to a low water permeability, these organisms cannot freeze extracellularly. In this case, no correlation would be expected between freezing resistance and water stress resistance.

b. Tolerance of Stresses Producing the Same Primary Direct Injury

The above evidence establishes the existence of a single tolerance of different primary stresses inducing a single (water) stress, which in its turn may be a primary, secondary, or tertiary stress. Can there also be a single tolerance of different primary stresses producing the same primary direct injury in the absence of any secondary or tertiary stresses?

The evidence points to a loss of semipermeability (and therefore membrane damage) as the primary direct injury produced by nine different stresses (Table 21.2). This suggests the possibility of a single tolerance mechanism for all nine stresses, due to an avoidance of the primary direct strain that induces the membrane damage. However, although the injury is the same, there are apparently five different direct strains capable of inducing it (Table 21.2). Therefore, at least five different kinds of tolerance are conceivable. It seems obvious, for instance, that chilling tolerance due to a low melting point of membrane lipids cannot confer tolerance of any

TABLE 21.8 QUANTITATIVE COMPARISONS BETWEEN THE CRITICAL WATER STRESSES OR THE CRITICAL DEHYDRATIONS INDUCED BY FREEZING, DROUGHT, AND OSMOTIC STRESSES, RESPECTIVELY[a]

Organism or tissue	Stress	Critical freezing temperature (°C)	Critical water stress (atm)	Critical dehydration (%)	Reference
Hedera helix	Freezing	−18.5	239		Kessler, 1935
(hardened)	Drought		225		Iljin, 1930a
Enteromorpha	Freezing	−15	195		Biebl, 1956a
clathrata	Drought		235		
	Osmotic		155[b]		
Muscle	Freezing	−2		78	Moran, 1929 (see
	Drought			78	Salt, 1955)
Red blood cells	Freezing	−3		65	Meryman, 1967
Mytilus edulis	Freezing	−10		66	Meryman, 1967
Venus mercenaria	Freezing	−6		65	Meryman, 1967
Kale	Freezing	−6			Samygin and
(unhardened)	Drought		−5.6°C[c]		Matveeva, 1962
	Osmotic (sucrose)		−11.2		
Kale	Freezing	−17			
(hardened)	Drought		−14.5		
	Osmotic (sucrose)		−18.6		

[a] Conversion values from Levitt, 1956, Table 73.

[b] Assuming a freezing point of −2.15°C for seawater, due solely to NaCl, which would then be 0.655 M. Five times seawater was maximum survived.

[c] Freezing point of solution in vapor pressure equilibrium with the plant.

of the other eight stresses, which do not injure due to a low temperature-dependent change of phase. This agrees with the known facts. Potato tubers are fully chilling tolerant, since they can be stored at 0°–5°C essentially indefinitely without injury. Yet they are killed by the slightest freezing and possess low drought tolerance. Similarly, tropical plants which are exposed to high temperatures and, therefore, must be heat tolerant, are the most chilling sensitive of higher plants. This is also true of tropical algae (Biebl, 1962b,c). Therefore, tolerance of the chilling stress does not confer tolerance of the freezing, drought, or heat stresses.

The second strain capable of leading to membrane damage, lipid (per) oxidation, is almost solely speculative. The two stresses that may conceivably injure due to this primary direct strain are radiation and salt stresses. Unfortunately, there is apparently no information as to the ability of the one tolerance to confer a tolerance of the other stress (see below).

There remains one tolerance mechanism that may conceivably be common to five or even six stresses—tolerance due to avoidance of membrane protein aggregation (and perhaps of SH binding by ions). Tolerances of two of these stresses (freezing and drought) have already been shown to be correlated, as long as they both injure via a water stress. Evidence of a correlation between heat tolerance and freezing tolerance has already been discussed (Chapter 12). The evidence for a correlation between all three tolerances (see Levitt, 1956, 1958) may be summarized as follows:

1. Environmental conditions (low temperature, drought, light, low N, etc.) that harden (i.e., increase tolerance) are effective against the freezing, drought, and heat stresses. As an example, light increases both the heat and freezing tolerance of guard cells (Weber, 1926a).
2. At least some of the factors that are frequently correlated with freezing and drought tolerance are also correlated with heat tolerance (e.g., low moisture and high sugar content and stage of development).
3. Heat tolerance rises and falls seasonally in parallel with freezing tolerance (Fig. 21.3), which in turn is paralleled by drought tolerance (Fig. 15.2),
4. The order of tolerance of different cells or tissues is the same for freezing and heat (Weber, 1926a,b), and for all three tolerances (Levitt and Nelson, 1942).
5. Species and varietal tolerances may be correlated. The three ecological groups of marine algae show the same order of tolerance of drought and heat (Biebl, 1952a). Twelve oat varieties show the same order for freezing and heat tolerance (Table 21.9). Species of *Fucus* growing in the upper (drier) littoral zones had higher tolerance of heat and freezing than those growing in the lower (moister) zones (Feldman and Lutova, 1963).

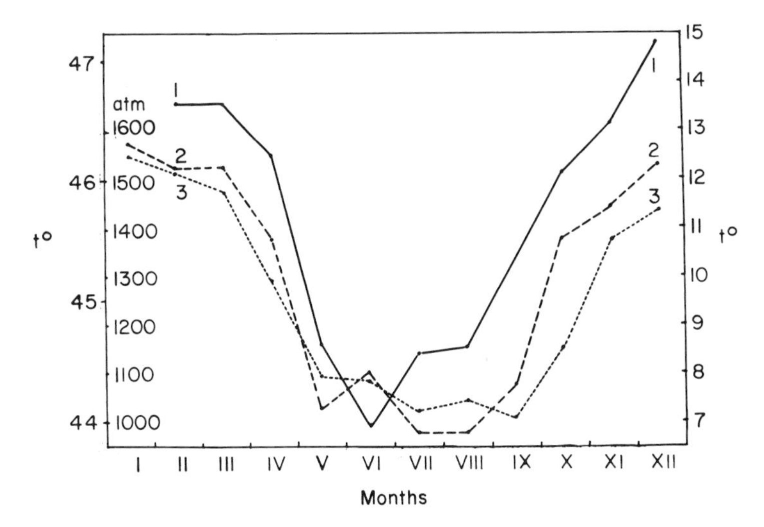

FIG. 21.3. Seasonal changes in tolerance. Epidermal cells of leaf sheath of *Dactylis glomerata*. Curve 1, freezing; curve 2, heat; curve 3, hydrostatic pressure. (From Alexandrov, 1964.)

As pointed out earlier (Chapter 12), there is also a heat tolerance that is *inversely* related to freezing tolerance, attaining a maximum in summer when freezing tolerance is minimal. As indicated previously (Chapter 12), this is perhaps due to avoidance of the reversible protein denaturation that may occur at high temperature, rather than of the protein aggregation. This summer heat tolerance is, therefore, presumably due to a property (high hydrophobicity) that cannot confer tolerance of other stresses. Avoidance of indirect heat injury may also be a part of the summer heat tolerance. It is not surprising, therefore, that negative comparative heat and freezing responses have also been obtained. This is particularly true in the case of some algae. Biebl (1958c), for instance, found that intertidal algae survive lower freezing temperatures than deep sea algae (−8°C versus −2° to −3°C), respectively. Yet the heat resistance of the two groups did not differ much—a critical temperature of about 30°C. The superior freezing tolerance of the intertidal algae was, however, correlated with a greater osmotic tolerance than that of the deep sea algae (up to three times seawater versus a maximum of 1.8 times seawater, respectively). Nevertheless, pretreatment with dilute and concentrated seawater (presumably decreasing and increasing osmotic tolerance) lowered and raised the heat tolerance of the intertidal algae, respectively (Biebl, 1969). One possible reason for the above observed lack of correlation between heat tolerance and the other two tolerances was the long exposure time (24 hr) to the heat.

TABLE 21.9 RELATIVE WINTER HARDINESS AND HEAT RESISTANCE OF TWELVE VARIETIES OF OATS[a,b]

Type and variety	*Avena* species	Winter survival (%)	Heat survival (%)
	True winter		
Fulwin	*A. byzantina*	137.6	123.4
Bicknell	*A. sativa*	130.1	124.7
Hairy Culberson	*A. sativa*	129.8	122.2
Pentagon	*A. byzantina*	128.2	103.1[c]
Tech (V.P.I. No. 1)	*A. sativa*	125.1	105.5[c]
Winter Turf	*A. sativa*	120.1	111.2
Lee	*A. sativa*	116.8	100.0[c]
Culred	*A. byzantina*	108.5	120.0
	Intermediate winter		
Appler	*A. byzantina*	100.3	99.8
Fulghum (check)	*A. byzantina*	100.0	100.0
	Spring		
Bond	*A. byzantina*	63.8	68.2
Victoria	*A. byzantina*	46.8	43.0

[a] Fulghum check, 100%.
[b] From Coffman, 1957.
[c] Additional data, not fully comparable with those shown, reveal that this variety is somewhat more heat resistant than indicated here.

Biebl was, therefore, probably measuring the tolerance of indirect rather than direct strain effects. Indeed, later tests (1969) showed that the intertidal algae survived 36°–37°C for 5 min to ½ hr, the usual exposure times for measurements of direct heat injury (as opposed to 30°C for 24 hr). Another apparently exceptional group of algae are the lake balls (*Aegagrophila Sauteri*). They are highly freezing tolerant, surviving −20°C for 24 hr without injury, but are easily killed by exposure to open air at room temperature (Terumoto, 1959). This injury was perhaps due to the drought stress, but may also have been indirect heat injury, since algae, in general, are very sensitive to long exposures to "high" temperatures as low as 15°–20°C (see Table 11.1).

Other lower plants, show as good a correlation between the three tolerances as do the higher plants. Lange (1953, 1955) observed a high tolerance of both heat and drought in the case of lichens and mosses of dry, hot regions, and a low tolerance of both in cool, moist regions. Even the exceptions agreed, for a shade lichen that possessed an unexpectedly high drought tolerance also had a high heat tolerance. The correlation was, therefore, not simply due to parallel selection. As mentioned above, a second kind of heat tolerance was not correlated with drought tolerance (Lange, 1961). Ferns also show a correlation between heat, freezing, and drought tolerance (Kappen, 1966). Heat and freezing tolerance both increased as the water content of the leaves decreased.

There have been very few attempts to correlate the tolerance of freezing, drought, and heat, with that of the other apparently protein aggregating stresses. Sergeev and Lebedev (1936—see Strogonov, 1964) concluded that the resistance of wheat and rye seedlings to salt solutions (NaCl, Na_2SO_4, Na_2CO_3) paralleled their freezing tolerance. Unfortunately, the salt resistance was apparently avoidance, due to salt exclusion. Biebl (1952) found no correlation between the tolerances of different ions (H_3BO_3, $ZnSO_4$, $VOSO_4$). A comparison between salt, radiation, and drought resistance also yielded negative results, in the case of four halophyte and three glycophyte species (Biebl and Weissenböck, 1968). The resistance was shown to be partly due to avoidance, since all three stresses induced succulence in the four halophytes, but the root system was decreased by salt and radiation, although increased by drought. Opposite effects on cell size and stomatal density were also observed, drought decreasing cell size and increasing stomatal density, while radiation had the opposite effect.

Gloeococcum bavaricus, an alga with much higher heat tolerance than other algae, showed an extraordinarily high resistance against Na, V, and Cr sulfates, but no greater resistance against Cu than the other algae (Url and Fetzmann, 1959). Resistances may, in fact, be mutually exclusive. A calcicole population of *Hypericum*, for instance, showed greater Ca resistance but less P, Al, and Mn resistance, judging by the higher yields at high Ca levels and the toxicity symptoms at high P, Al, and Mn levels (Ramakrishnan, 1969). Similarly, the algae *Porphyra* and *Ulva* were more resistant to salinity and drying than *Gelidium* and *Zostera*, but the latter two were less sensitive to injury by H^+ concentration (Ogato and Matsui, 1965).

A correlation between salt tolerance and high temperature (32°C) tolerance of germination has been found in the case of the halophyte *Salicornea europaea* as opposed to the less tolerant *Medicago sativa* and *Spergularia marinia* (Ungar, 1967). An interaction between stresses may also occur. As the temperature increased, the water stress became more critical in the germination of mesquite (Scifres and Brock, 1969). Similarly, as salinity levels increased (within a low to moderate range), the optimum growing temperature for chrysanthemum (a relatively tolerant plant) decreased (Lunt *et al.*, 1960). This may have been related to the increased accumulation of Na, Ca, and Cl with increase in temperature.

There is obviously insufficient evidence to decide whether or not a general correlation exists between the three correlated tolerances of freezing, drought, and heat, and the remaining stresses that appear to induce the same kind of primary direct injury.

c. Tolerance of Stresses Producing Different Kinds of Primary Direct Injury

There would seem to be no reason to expect a correlation between the tolerances of stresses that do not appear to induce the same kind of primary direct injury. Nevertheless, a single mechanism may conceivably protect a plant against more than one strain. As an example, two unrelated chemical substances, alcohol and urea, denature α-chymotrypsin, and hydrocinnamate protects the enzyme against denaturation by both (Friedberg *et al.*, 1969). If a stress disrupts the aerobic respiration it may lead to accumulation of alcohol from anaerobic respiration. Another stress that disrupted the protein or nucleic acid metabolism could lead to urea accumulation. Resistance to both stresses could then result from production of hydrocinnamate. Some positive results have, indeed, been reported. Alexandrov's group of investigators in the USSR have been particularly active in this field. Tolerance of hydrostatic pressure paralleled the seasonal changes in heat and freezing tolerance of epidermal cells of leaves (Fig. 21.3) reaching a maximum in winter. Resistance to ethyl alcohol showed similar changes (Alexandrov, 1964). Artificial heat-shock hardening also increased tolerance of both heat and hydrostatic pressure. In partial agreement with Alexandrov's group, the cells of *Bacillus cereus* were tolerant of octyl alcohol when they were heat tolerant, but the reverse was not necessarily true (Gupta *et al.*, 1970). Finally, a number of factors have been reported to be correlated with different kinds of stress tolerance (Table 21.7).

d. Tolerance of Stresses Producing the Same Indirect Injury

The indirect injury is apparently metabolic in all cases (Table 21.1), though the specific metabolic disturbance is commonly different for different stresses. There are several stresses, however, that may produce the same kind of metabolic disturbance, for instance, starvation (Table 21.3). Four different stresses may lead to this excess of respiratory breakdown over photosynthesis. The chief resistance mechanism is starvation avoidance by either a decrease in respiration or an increase in photosynthesis. The existence of two such mechanisms indicates that no single mechanism can be expected for all four stresses. In the case of high and low temperatures, in fact, the adaptations must be mutually exclusive, since any modification that leads to a rise in the optimum temperature for a process will usually also lead to a rise in the minimum and maximum temperatures. An enzyme, for instance, that is active at heat stress temperatures is likely to be inactive at chilling stress temperatures. It is still possible that a plant may adapt to the indirect metabolic strain produced by three of the four stresses (heat, water, and light deficit) due to a decrease in respiration. It does not seem likely, however, than an adaptation to water and light deficit at normal temperatures would succeed at high temperatures.

The same reasoning applies to protein breakdown. On the basis of the artificially induced tolerance by application of kinetin (Engelbrecht and Mothes, 1960), it may be proposed that a natural increase in cytokinin could confer tolerance of this indirect metabolic strain whether due to the heat, water, or salt stress. Another possible common mechanism on the basis of Kessler's (1961) results is an increase in RNA, perhaps due to inactivation of RNase.

There is less chance of a common mechanism that could prevent toxin injury. However, if a single toxin (e.g., NH_3) is produced by different stresses, a single tolerance mechanism could suffice (e.g., accumulation of organic acids). In the case of biochemical lesions a single mechanism is excluded, since the only two stresses apparently involved are chilling and heat, and again the adaptations would be mutually exclusive.

e. Repair Mechanisms

In all the above cases, stress tolerance is due to an avoidance of the injurious strain. Tolerance of the injurious strain is possible, within limits, if there is a repair mechanism. Its existence has been demonstrated unambiguously only in the case of the heat and radiation stresses. Even in these cases, the mechanism has been little investigated. It is conceivable that the plant may possess a repair mechanism against other stresses as well, and that a common mechanism may exist for several stresses. A possible indication of a common mechanism for repair of heat and radiation damage has been obtained in the case of the fungus *Penicillium expansum* (Baldy *et al.*, 1970). The spores were able to repair heat injury

(54°C for 1 hr) if held for 3 days in aerated water at 23°C before plating. Survival was twenty times that of spores plated immediately. Restoration of γ-radiation resistance accompanied the recovery of viability.

Since membrane repair of mechanical damage can occur (Prothero *et al.*, 1970), it is conceivable that this is also the repair mechanism for other kinds of stress injury causing membrane damage. In the case of radiation injury to red blood cells, evidence of this mechanism has been obtained (see Chapter 18). The photorepair mechanism for UV damage may also conceivably operate in this manner (see below).

Several conclusions emerge from the above comparisons of tolerance responses.

1. A single tolerance mechanism is apparently effective against all kinds of water stresses, whether primary, secondary, or tertiary.
2. A single tolerance mechanism may be effective against all primary stresses that produce the same kind of direct injury (e.g., membrane damage), provided that the direct strain produced by the different stresses is the same. An exceptional kind of tolerance may, nevertheless, occur due to a mechanism of preventing the direct strain which is effective only against one kind of stress (e.g., the heat stress).
3. There is some evidence of a common tolerance of some stresses producing different kinds of direct injury.
4. A single kind of tolerance may be effective against several kinds of stresses producing the same kind of indirect injury, in some cases, but not in others.

TABLE 21.10 Hardening Treatments and Types of Resistance They are Capable of Producing

Hardening treatment by sublethal doses of stress	Resistance conferred
1. Chilling temperatures	Chilling, freezing, drought, salt-induced osmotic, direct heat (perhaps pressure, alcohol) tolerance
2. Heat	Heat, pressure, alcohol tolerance
3. Drought	Drought, freezing, heat tolerance, drought avoidance (pseudohardening)
4. Radiation	
Light	Light avoidance and tolerance
UV	UV avoidance

5. A single repair mechanism may conceivably be effective against different stresses. This is presumably due to a common kind of injury.

D. A General Tolerance Hypothesis

Specific conclusions have been listed above for the comparative injuries produced by different stresses and the comparative resistances developed by the plant against them. Due to the existence in some cases of a single tolerance common to several stresses, it is not surprising that a hardening treatment consisting of a sublethal dose of a single stress, frequently confers tolerance of other stresses as well (Table 21.10). No one such mechanism can succeed against all the kinds of injury, produced by all stresses. Nevertheless, one stands out because of its apparent ability to protect the plant against the majority of stresses. Since these stresses are so diverse, there must be some basic injury mechanism common to all of them. It is postulated that this most common mechanism involves oxidation-reduction or exchange reactions with membrane proteins, via their SH and SS groups, and leading to a loss of semipermeability. The following is a summary of the evidence:

1. The oxygen stress and the three pollutant stresses are all due to reducing (oxygen deficiency and SO_2) and oxidizing (PAN and O_3) substances. The limited evidence points to their ability to act on the SS and SH groups of proteins, respectively.
2. Ionizing radiations are also known to produce oxidizing substances (peroxides) in the plant, and to oxidize membrane protein SH groups to intermolecular SS bridges in red blood cells. Even the lowest injurious doses which injure via nuclear damage may conceivably be explained by similar reactions. According to Dounce (1971), the nuclear DNA units are attached at both ends to a residual protein consisting of two SS-linked polypeptides. Reduction of these SS bridges depolymerizes the gel. Consequently, it may be postulated that the weaker radiation doses damage by reductive cleavage of the nuclear gel, preventing cell division and, therefore, growth; the stronger doses damage by oxidative aggregation of membrane proteins leading to loss of semipermeability.
3. Some ions (particularly heavy metal) may combine with the SH groups of the membrane proteins. In the case of divalent ions (Cu^{2+}, Hg^{2+}, etc.) protein aggregation may result due to combination of one ion with two SH groups.

4. The heat stress, though not specificially affecting oxidations, would accelerate all reactions. Since both the optimum and maximum temperatures for respiration are normally higher than those for photosynthesis, oxidative processes would be favored, including the oxidation of protein SH to intermolecular SS bridges. This process would be further aided at high temperatures by the reversible protein denaturation which would expose normally protected SH groups.

5. All seven of the primary, secondary, and tertiary water stresses lead to dehydration, which favors chemical reactions by bringing the solute or hydrated structural molecules closer together. Secondary aids to these reactions are (a) low and high temperature-induced reversible protein denaturation which exposes previously protected SH and other groups, and (b) mechanical stress induced by dehydrative cell collapse and exerted primarily on the surface membrane proteins. This mechanical stress has been shown to activate the SS bonds of proteins, leading to SH $\leftrightarrows$ SS or even SS $\leftrightarrows$ SS interchange reactions and, therefore, inducing intermolecular SS bonding and protein aggregation even in the absence of oxidation.

The plant can protect itself against this kind of stress injury by preventing or repairing such protein aggregations. It does this in more than one way.

1. It can develop a specific reducing system capable of maintaining the protein SH groups in the reduced state, thus preventing aggregation. This may occur via GSH, which in its turn would be maintained in the reduced state, probably by ascorbic acid or NADPH produced photosynthetically or by respiration. This mechanism must be clearly distinguished from the long known high SH content of actively growing tissues (see Pilet and Dubois, 1968), which frequently possess minimum stress tolerance. It has been shown that the basic characteristic of stress tolerant tissues is not their SH content per se, but their ability to prevent its oxidation to SS (see Schmuetz, Chapter 10). On exposure to a stress, the nontolerant actively growing tissues are apparently unable to prevent this oxidation, due presumably to inactivation of the reducing systems by the stress.

2. It can prevent the reversible protein denaturation and can therefore, maintain its sensitive SH and SS groups protected within the native molecules. This is accomplished by possession of highly hydrophobic proteins in the case of heat-tolerant plants, and proteins of low hydrophobicity in plants tolerant of chilling and freezing.

3. It possesses a repair mechanism capable of splitting the intermolecular

SS bonds reductively, and therefore, returning the protein aggregates to their native state. This again may occur via the above GSH system. The best evidence in favor of this repair system is in the case of radiation injury.

This mechanism of preventing or repairing protein aggregation via intermolecular SS bonding, may, therefore, protect the plant against direct injury by thirteen stresses (Tables 21.7, and 21.11), but it may also protect against at least some kinds of indirect injury, by preventing the inactivation of some SH enzymes, and perhaps even by preventing the activation of SS enzymes. The indirect injury due to protein breakdown, for instance, has been related to increased RNase activity. Since RNase is an SS enzyme, it may perhaps be retained in the inactive state by reductive cleavage of the SS groups.

It is, therefore, possible that a plant may be able to protect itself against a surprisingly wide spectrum of stresses by means of a single tolerance mechanism. One aspect of this mechanism, however, involves mutually exclusive properties; if a plant is protected against the heat stress by high

TABLE 21.11 Stresses against which the Plant may Conceivably Develop Resistance[a]

a. Oxygen stress
b. Pollutants (reducing and oxidizing)
  1. SO_2
  2. PAN
  3. O_3
c. Radiations
  4. Ionizing. Both membrane proteins and nuclear residual proteins
  5. UV Photorepair
d. Ions
  6. Heavy metals
e. 7. Heat
f. Water stresses
  8. Secondary, chilling-induced water stress
  9. Secondary, freezing-induced water stress
  10. Primary, drought-induced water stress
  11. Secondary, heat-induced water stress
  12. Tertiary, radiation (visible and infrared)-induced, secondary heat-induced water stress
  13. Secondary, salt-induced osmotic water stress

[a] Due to avoidance or tolerance of two kinds of strains: (1) splitting of essential SS groups, and (2) membrane protein aggregation, via intermolecular SS bridges.

hydrophobicity, it must be sensitized to the chilling and freezing stress, and vice versa. Furthermore, in some cases the mechanism may not be sufficient by itself. Chilling tolerance, for instance, apparently requires membrane lipids with low melting points; heat tolerance, on the other hand, requires high melting points. Similarly, the mechanism may be ineffective against elastic growth stresses, and it may not even be used by some plants with the highest resistance (e.g., due to avoidance).

BIBLIOGRAPHY

Abdel-Kader, A. S., Morris, L. L., and Maxie, E. C. (1968). Physiological studies of gamma irradiated tomato fruits. I. Effect on respiratory rate, ethylene production, and ripening. *Proc. Amer. Soc. Hort. Sci.* **92,** 553–567.

Abel, G., and Mackenzie, A. J. (1964). Salt tolerance of soybean varieties (*Glycine max* L. Merrill) during germination and later growth. *Crop Sci.* **4,** 157–161.

Abetsedarskaya, L. A., Miftakhutdinova, F. G., and Fedotov, V. D. (1968). The state of water in living tissues (results of NMR) nuclear magnetic resonance-spin echo studies. *Biofizika* **13,** 630–636.

Abrol, Y. P., Sirohi, G. S., and Sinha, S. K. (1969). Reversal of inhibitory effects of γ-rays on the seedling growth of wheat by the application of IAA, tryptophan, and zinc. *Indian J. Exp. Biol.* **7,** 114–116.

Abros'kin, V. V., and Zadonskii, P. G. (1968). The effect of orienting cucumber sprouts in the earth's magnetic field. *Zap. Voronezh. Sel'skokhoz Inst.* **34,** 86–91. Trans. from Ref. *Zh. Biol.* 1969, No. 4G28.

Adir, C. R. (1968). Testing rice seedlings for cold water tolerance. *Crop. Sci.* **8,** 264–265.

Agakishiev, D. (1962). Accelerating effect of low temperatures on germination of cotton seeds in salinized soils. *Fiziol. Rast.* **8,** 505–506.

Agakishiev, D. (1964). Effect of gibberellin on cotton under conditions of salinization. *Fiziol. Rast.* **11,** 201–205.

Ahmad, R. (1968). The mechanism of salt tolerance in *Suaeda fruticosa* and *Haloxylon recurvum. Plant Soil* **18,** 357–362.

Ahring, R. M., and Irving, R. M. (1969). A laboratory method for determining cold hardiness in bermudagrass. *Cynodon dactylon* (L). Pers. *Crop. Sci.* **9,** 615–618.

Ahrns, W. (1924). Weitere Untersuchungen über die Abhängigket des gegenseitigen Mengenverhältnisses der Kohlenhydrate im Laubblatt vom Wassergehalt. *Bot. Arch.* **5,** 234–259.

Akad. Nauk SSSR. Nauch. Sov. Prob. Fiziol. Biokhim. Rast. (1968). The physiology of the resting state in plants. 267 pp. Moscow. Ref. *Zh. Biol.* 1968. No. 12G125K.

Akazawa, T., and Sugiyama, T. (1969). Subunit structure of carboxydismutase. *Abstr. 11th Int. Bot. Congr. Seattle,* p. 1.

Akazawa, T., Saio, K., and Sugiyama, N. (1965). On the structural nature of fraction I. protein of rice leaves. *Biochem. Biophys. Res. Commun.* **20,** 114–119.

Åkerman, Å. (1927). "Studien über den Kältetod und die Kälteresistenz der Pflanzen," pp. 1–232. Berlingska Boktryck. Lund, Sweden.

Alekperov, S. A., and Khrzhanovskaya, T. E. (1961). The effect of salts on certain physiological characteristics of seeds and sprouts of *Sophora japonica. Izv. Akad. Nauk. Azerb. SSR Ser. Biol. Med. Nauk.* **12,** 3–15.

Alexander, D. McE., and Woodham, R. C. (1968). Relative tolerance of rooted cuttings of four Vinifera varieties to sodium chloride. *Aust. J. Exp. Agr. Anim. Husb.* **8,** 461–465.

Alexandrov, V. Ya. (1961). Die Bedeutung der Denaturations-theorie (Eiweisstheorie) der Verletzung und Erregung für die Untersuchung der Zelladaptation an die Wirkung von Verletzungsagenzien. III. *Humboldt-Symp. über Grundfragen der Biol. Berlin,* pp. 261–266. Gustav Fischer. Jena.

Alexandrov, V. Ya. (1964). Cytophysiological and cytoecological investigations of heat resistance of plant cells toward the action of high and low temperature. *Quart. Rev. Biol.* **39,** 35–77.

Alexandrov, V. Y., Lutova, M. Y., and Feldman, H. L. (1959). Seasonal changes in the resistance of plant cells to the action of various agents. *Cytologia* **1,** 672–691.

Alexandrov, V. J., Ouchekov, B. P., and Poljansky, G. I. (1961). La mort thermique des cellules par rapport au problem de l'adaptation des organismes à la température du milieu. *Pathol. Biol.* **9,** 849–854.

Alexandrov, V. Ya., Lomagin, A. G., and Feldman, N. L. (1970). The responsive increase in thermostability of plant cells. *Protoplasma* **69,** 417–458.

Alexandrov, V. Ya., and Shukhtina, G. G. (1964). The state of the protoplasm of plant cells in winter. A criticism of the theory of P. A. Henkel and E. Z. Oknina "the separation of protoplasm." *In* "Cytological Aspects of Adaptation of Plants to the Environmental Factors" (V. Ya. Alexandrov, ed.). Acad. Sci. USSR. Moscow. Pp. 136–154.

Alexandrov, V. Ya., and Yazkulyev, A. (1961). Heat hardening of plant cells under natural conditions. *Tsitologiya* **3,** 702–707.

Algera, L. (1936). Concerning the influence of temperature treatment on the carbohydrate metabolism, the respiration, and the morphological development of the tulip. I–III. *Kon. Ned. Akad. Wetensch. Amsterdam Proc.* **39,** 1–29.

Allaway, W. G., and Mansfield, T. A. (1970). Experiments and observations on the after effect of wilting on stomata of *Rumex sanguineus. Can. J. Bot.* **48,** 513–521.

Allen, M. B. (1950). The dynamic nature of thermophily. *J. Gen. Physiol.* **33,** 205–214.

Allen, R. M. (1955). Foliage treatments improve survival of long leaf pine seedlings. *J. Forest.* **53,** 724.

Al'tergot, V. F. (1963). The action of high temperatures on plants. *Izv. Akad. Nauk. SSSR Ser. Biol.* **28,** 57–73.

Al'tergot, V. F., and Bukhol'tsev, A. N. (1967). Induced resistance to cold in corn shoots. *Izv. sib. otd. Akad. Nauk. SSSR Ser. Biol. Med. Nauk.* **2,** 49–56.

Amelunxen, R., and Lins, M. (1968). Comparative thermostability of enzymes from *Bacillus stearothermophilus* and *Bacillus cereus. Arch. Biochem. Biophys.* **125,** 765–769.

Amin, J. V. (1969). Some aspects of respiration and respiration inhibitors in low temperature effects of the cotton plant. *Physiol. Plant.* **22,** 1184–1191.

Amirshahi, M. C., and Patterson, F. L. (1956). Development of a standard artificial freezing technique for evaluating cold resistance of oats. *Agron. J.* **48,** 181–184.

Ammon, R., and Friedrich, G. (1967). Enzyme activity in *Euglena gracilis.* I. *Acta Biol. Med. Ger.* **19,** 669–672.

Anand, J. C., and Brown, A. D. (1968). Growth rate patterns of the so-called osmophilic yeasts in solutions of polyethylene glycol. *J. Gen. Microbiol.* **52,** 205–212.

Anderson, J. W., and Rowan, K. S. (1967). Extraction of soluble leaf enzymes with thiols and other reducing agents. *Phytochemistry* **6,** 1047–1056.

Anderson, L. G., and Kneebone, W. R. (1969). Differential responses of *Cynodon dactylon (L.) Pers.* Selections to three herbicides. *Crop Sci.* **9,** 599–601.

Andersson, G. (1944). "Gas Change and Frost Hardening Studies in Winter Cereals," 163 pp. Håkan Ohlssons Boktryckeri, Lund, Sweden.

Andreeva, I. N. (1969). Effect of temperature on the swelling and shrinkage of isolated mitochondria. *Fiziol. Rast.* **16,** 221–227.

Andrews, J. E., and Roberts, D. W. A. (1961). Association between ascorbic acid concentration and cold hardening in young winter wheat seedlings. *Can. J. Bot.* **39,** 503–512.

Andrews, J. E., Horricks, J. S., and Roberts, D. W. A. (1960). Interrelationships between plant age, root-rot infection, and cold hardiness in winter wheat. *Can. J. Bot.* **38,** 601–611.

Andrews, S., and Levitt, J. (1967). The effect of cryoprotective agents on intermolecular SS formation during freezing of Thiogel. *Cryobiology* **4,** 85–89.

Angell, C. A., and Sare, E. J. (1970). Vitreous water: identification and characterization. *Science* **168,** 281.

Angelo, E., Iverson, V. E., Brierley, W. G., and Landon, R. H. (1939). Studies on some factors relating to hardiness in the strawberry. *Minn. Agr. Exp. Sta. Tech. Bull.* **135,** 1–36.

Ansari, A. Q., and Loomis, W. E. (1959). Leaf temperatures. *Amer. J. Bot.* **46,** 713–717.

Antoniani, C., and Lanzani, G. A. (1963). Influence of germination in the cold on the peroxidases of *Triticum vulgare* embryos. *Vitalstoffe Zivilisationskrankh.* **8,** 39–40.

Aoki, K. (1950). Analysis of the freezing process of living organisms. I. The relation between the shape of the freezing curve and the mode of freezing in plant tissues. *Low Temp. Sci.* **3,** 219–227.

Apelt, A. (1907). Neue Untersuchungen über den Kältetod der Kartoffel. *Beitr. Biol. Pflanz.* **9,** 215–262.

Arnold, A. (1955). "Die Bedeutung der Chlorionen für die Pflanze," 148 pp. Gustav Fischer. Jena.

Arnon, D. I. (1969). Regulation of ferredoxin-dependent photosynthetic activity of isolated chloroplasts. *Abstr. Int. Bot. Congr. Seattle,* p. 5.

Aronsson, A., and Eliasson, L. (1970). Frost hardiness in Scots pine. 1. Conditions for test on hardy plant tissues and for evaluation of injuries by conductivity measurements. *Stud. Forest. Suec.* **77,** 1–30.

Arreguin, B., and Bonner, J. (1949). Experiments on sucrose formations by potato tubers as influenced by temperature. *Plant Physiol.* **24,** 720–737.

Arvidsson, I. (1951). Austrocknungs- und Durreresistenzverhältnisse einiger Repräsentanten öländischer Pflanzenvereine nebst Bemerkungen über Wasserabsorption durch oberirdische Organe. Oikos Suppl. **1,** 1–181.

Arvidsson, I. (1958). Plants as dew collectors. *Compt. Rend. et Rapports—Assemblé Générale de Toronto 1957.* **2,** 481–484.

Asahi, T. (1964). Sulfur metabolism in higher plants. IV. Mechanism of sulfate reduction in chloroplasts. *Biochim. Biophys. Acta* **82,** 58–66.

Asahina, E. (1954). A process of injury induced by the formation of spring frost on potato sprout. *Low Temp. Sci. Ser.* **B11,** 13–21.

Asahina, E. (1956). The freezing process of plant cell. *Contrib. Inst. Low Temp. Sci. Hokkaido Univ.* **10,** 83–126.

Asahina, E. (1958). On a probable freezing process of molluscan cells enabling them to survive at a super-low temperature. *Low Temp. Sci. Ser.* **B16,** 65–75.

Asahina, E. (1961). Intracellular freezing and frost resistance in egg-cells of the sea urchin. *Nature (London)* **191,** 1263–1265.

Asahina, E. (1962a). Frost injury in living cells. *Nature (London)* **196,** 445–446.

Asahina, E. (1962b). A mechanism to prevent the seeding of intracellular ice from outside in freezing living cells. *Low Temp. Sci. Ser.* **B20,** 45–56.

Asahina, E. (ed.). (1967). Freezing injury in egg cells of the sea urchin. "Cellular Injury and Resistance in Freezing Organisms," pp. 211–230. Inst. Low. Temp. Sci. Hokkaido Univ., Sapporo, Japan.

Asahina, E., and Tanno, K. (1963). A remarkably rapid increase of frost resistance in fertilized egg cells of the sea urchin. *Exp. Cell Res.* **31,** 223–225.

Asahina, E., Aoki, K., and Shinozaki, J. (1954). The freezing process of frost-hardy caterpillars. *Bull. Entomol. Res.* **45,** 329–339.

Asahina, E., Shimada, K., and Hisada, Y. (1970). A stable state of frozen protoplasm with invisible intracellular ice crystals obtained by rapid cooling. *Exp. Cell. Res.* **59,** 349–358.

Asana, R. D. (1957). The problem of assessment of drought resistance in crop plants. *Indian J. Genet. Plant Breed.* **17,** 371–378.

Ashby, W. C., and Beadle, N. C. W. (1957). Studies in halophytes. III. Salinity factors in the growth of Australian saltbushes. *Ecology* **38,** 244–352.

Askenasy, E. (1875). Ueber die Temperatur, welche Pflanzen in Sonnenlicht annehmen. *Bot. Z.* **33,** 441–444.

Atkinson, M. R., Findlay, G. P., Hope, A. B., Pitman, M. G., Saddler, H. D. W., and West, K. R. (1967). Salt regulation in the mangroves *Rhizophora mucronata* Lam. and *Aegiolitis annulata* R. Br. *Aust. J. Biol. Sci.* **20,** 589–599.

Augusten, H. (1963). Der Einfluss der Temperatur auf einige Stoffwechselgrossen bei Bulbillen und Wurzelknollen von *Ficaria verna* Huds. *Beitr. Biol. Pflanz.* **38,** 421–424.

Austin, R. B., Longden, P. C., and Hutchinson, J. (1969). Some effects of "hardening" carrot seed. *Ann. Bot. (London)* **33,** (N.S.) 883–895.

Avilova, L. D. (1962). The thermostability of epidermal cells of albino, green, and etiolated sunflower plants. *Tsitologiya* **4,** 73–76.

Avundzhyan, E. S., Marutyan, S. A., Dogramadzhyan, A. D., and Petrosyan, A. (1967). A study of the bleeding sap from *Vitis vinifera. Fiziol. Rast.* **14,** 405–414.

Awad, E. S., and Deranleau, D. A. (1968). Thermal denaturation of myoglobin. I. Kinetic resolution of reaction mechanism. *Biochemistry* **7,** 1791–1795.

Ayres, A. A. (1916). The temperature coefficient of the duration of life of *Ceramium tenuissimum. Bot. Gaz. (Chicago)* **62,** 65–69.

Azizbekova, Z. S., and Babaeva, Zh. A. (1962). Carbohydrate metabolism in maize affected by pre-sowing hardening on soil salts of various qualities. *Izv. Akad. Nauk. Azerb. SSR Ser. Biol. Med. Nauk.* **6,** 3–8.

Babaeva, M. K., and Seisebaev, A. T. (1968). Activity of catalase and cytochrome oxidase in plants with different radiosensitivity. *Dokl. Akad. Nauk. SSSR* **178,** 1198–1201.

Babaeva, Zh., Butenko, R. G., and Strogonov, B. P. (1968). Effect of salinization of the nutrient medium on growth of isolated carrot tissues. *Fiziol. Rast.* **15,** 93–102.

Babalola, O., Boersma, L., and Youngberg, C. T. (1968). Photosynthesis and transpiration of Monterey pine seedlings as a function of soil water suction and soil temperature. *Plant Physiol.* **43,** 515–521.

Babenko, V. I. (1968). Some features of the metabolism of soluble carbohydrates and free acids in winter wheat induced by negative temperatures. *Fiziol. Rast.* **15,** 844–851.

Babenko, V. I., and Gevorkyan, A. M. (1967). Accumulation of oligosaccharides and their significance for the low temperature hardening of cultivated cereals. *Fiziol. Rast.* **14,** 727–736.

Bacq, Z. M., and Goutier, R. (1968). Mechanisms of action of sulfur-containing radioprotectors. *Brookhaven Symp. Quant. Biol.* **20,** 241–262.

Bader, H., Wilkes, A. B., and Jean, D. H. (1970). The effect of hydroxylamine mercaptans, divalent metals and chelators on (Na^+ + K^+)-ATPase. A possible control mechanism. *Biochim. Biophys. Acta* **198,** 583–593.

Bagi, G., and Farkas, G. L. (1968). A new aspect of the anti-stress effect of kinetin. *Experientia* **24,** 397–398.

Baker, B. S., and Jung, G. A. (1970a). Effect of environmental conditions on the growth of four perennial grasses. *Crop Sci.* **10,** 376–378.

Baker, B. S. and Jung, G. A. (1970b). Response of four perennial grasses to high temperature stress. *Proc. 11th Int. Grassland. Congr.* 499–502.

Baker, F. S. (1929). Effect of excessively high temperatures on coniferous reproduction. *J. Forest.* **27,** 949–975.

Baldy, R. W., Sommer, N. F., and Buckley, P. M. (1970). Recovery of viability and radiation resistance by heat-injured conidia of *Penicillium expansum* Lk. ex Thom. *J. Bacteriol.* **102,** 514–520.

Ballantyne, A. K. (1962). Tolerance of cereal crops to saline soils in Saskatchewan. *Can. J. Soil Sci.* **42,** 61–67.

Bamberg, S., Schwarz, W., and Tranquillini, W. (1967). Influence of day length on the photosynthetic capacity of stone pine (*Pinus cembra* L.). *Ecology* **48,** 264–269.

Bancher, E., Washuttl, J., and Stachelberger, H. (1970a). The influence of γ-irradiation on the activities of several starch degrading enzymes in resting and germinating rye seeds. I. The effect on α- and β-amylase. *Mikrochim. Acta* **2,** 413–418.

Bancher, E., Washuttl, J., and Stachelberger, H. (1970b). The influence of γ-irradiation on the activities of several starch degrading enzymes in resting and germinating rye seeds. II. The effect on phosphorylase. *Mikrochim. Acta* **2,** 419–422.

Banga, O. (1936). Physiologische Symptomen van Lage-Temperatuur-Bederf. *Lab. Tuinbouwplant.* (*Wageningen*) **24,** 3–143.

Bankes, D. A., and Sparrow, A. H. (1969). Effect of acute gamma irradiation on the incidence of tumor-like structures and adventitious roots in lettuce plants. *Radiat. Bot.* **9,** 21–26.

Barabal'chuk, K. A. (1969). Reaction of thermolabile and thermostable functions on the plant cell to heat-hardening action. *Tsitologiya* **11,** 1021–1032.

Bari, G., and Godward, M. B. E. (1969). Influence of chromosome size on the radiosensitivity of Linum species. *Can. J. Genet. Cytol.* **11,** 799–802.

Barinowa, R. A. (1937). The dynamics of the carbohydrate colloid complex as a factor of drought resistance in sugar beet. *Izv. Akad. Nauk. SSSR Ser. Biol.* **1,** 255–270.

Barker, J. (1960). High pressures of oxygen and respiration; an improved technique and a demonstration that high pressures as such may affect the rate of CO_2 output from plant tissues. *J. Exp. Bot.* **11,** 86–90.

Barnes, D. L., and Woolley, D. G. (1969). Effect of moisture stress at different stages of growing. I. Comparison of a single-eared and a two-eared corn hybrid. *Agron. J.* **61,** 788–790.

Barnes, W. C., and Peele, T. C. (1958). The effect of various levels of salt in irrigation water on vegetable crops. *Proc. Amer. Soc. Hort. Sci.* **72,** 339–342.

Barnothy, M. F. (1964). "Biological Effects of Magnetic Fields. Plenum Press, New York.
Bartel, A. T. (1947). Some physiological characteristics of four varieties of spring wheat presumably differing in drought resistance. *J. Agr. Res.* **74,** 97–112.
Bartetzko, H. (1909). Untersuchungen über das Erfrieren von Schimmelpilzen. *Jahrb. Wiss. Bot.* **47,** 57–98.
Bates, C. G. (1923). Physiological requirements of Rocky Mountain trees. *J. Agr. Res.* **24,** 97–164.
Bauer, H., Huter, M., and Larcher, W. (1969). Der Einfluss und die Nachwirkung von Hitze-und Kältestress auf den CO_2 Gaswechsel von Tanne und Ahorn. *Ber. Deut. Bot. Ges.* **82,** 65–70.
Baum, H., and Gilbert, G. A. (1953). A simple method for the preparation of cystalline potato phosphorylase and Q-enzyme. *Nature* (*London*) **171,** 983–984.
Bauman, L. (1957). Über die Beziehungen zwischen Hydratur und Ertrag. *Ber. Deut. Bot. Ges.* **70,** 67–78.
Bayley, S. T. (1966). Composition of ribosomes of an extremely halophilic bacterium. *J. Mol. Biol.* **15,** 420–427.
Beadle, N. C. W., Whalley, R. D. B., and Gibson, J. B. (1957). Studies in halophytes. II. Analytic data on the mineral constituents of three species of Atriplex and their accompanying soils in Australia. *Ecology* **38,** 340–344.
Beck, W. A. (1929). The effect of drought on the osmotic value of plant tissues. *Protoplasma* **8,** 70–126.
Becquerel, P. (1907). Recherches sur la vie latente des grains. *Ann. Sci. Nat. Ser.* **9,** 193–311.
Becquerel, P. (1932). L'anhydrobiose des tubercules des Renoncules dans l'azote liquide. *C. R. Acad. Sci. Paris* **194,** 1974–1976.
Becquerel, P. (1949). L'action du froid sur la cellule végétale. *Botaniste* **34,** 57–74.
Becquerel, P. (1954). La cryosynérèse cytonucléoplasmique jusqu' aux confins du zero absolu, son role pour la végétation polaire et la conservation de la vie. *8th Congr. Int. Bot. Rapports et Communications.* **11,** 269–270.
Belehradek, J. (1935). Temperature and living matter. *Protoplasma Monogr.* **8,** 277 pp.
Belehradek, J., and Melichar, J. (1930). L'action différente des températures élevées et des températures normales sur la survie de la cellule végétale (*Helodea canadensis*, Rich.). *Biol. Gen.* **6,** 109–124.
Beletskaya, E. K. (1967). The deposition of fat in the external layer of the endosperm of corn grains differing in their resistance to cold. *Rost. Ustoich. Rast.* **3,** 223–226.
Bemis, W. P., and Whitaker, T. W. (1969). The xerophytic cucurbita of Northwestern Mexico and Southwestern United States. *Madrono* **20,** 33–41.
Benko, B., and Pillar, J. (1965). Relation of sugars to the frost resistance of apple trees. *Ved. Pr. Vyzk. Ustavu Rastlinnej Vyroby Piestanoch.* **3,** 291–298.
Benson, A. A., Calvin, M., Haas, V. A., Aronoff, S., Hall, A. G., Bassham, J. A., and Weigl, J. W. (1949). C^{14} in Photosynthesis. *In* "Photosynthesis in Plants" (J. Franck and W. E. Loomis, eds.), pp. 381–402. Iowa State College Press, Ames, Iowa.
Bentrup, F. W. (1967). Plant cell morphogenesis in an electrical field. *Naturwissenschaften* **54,** 620–621.
Ben-Zeev, N., and Zamenhof, S. (1962). Effects of high temperatures on dry seeds. *Plant Physiol.* **37,** 696–699.
Bereznickaja, N. I., and Oveckin, S. K. (1936a). The enzymatic activity in winter wheat during winter. *Zb. Pr. Agrofiziol.* **1,** 186–197.

Bereznickaja, N. I., and Oveckin, S. K. (1936b). Activity of enzymes in relation to winter hardiness and phasic development of winter wheat. *Zb. Pr. Agrofiziol.* **2,** 67–82. (Herb. Abstr. **8,** 1860, 1938).

Berns, D. S., and Scott, E. (1966). Protein aggregation in a thermophilic protein: Phycocyanin from *Synechoccus lividus*. *Biochemistry* **5,** 1528–1533.

Bernstein, L. (1961). Osmotic adjustment of plants to saline media I. Steady state. *Amer. J. Bot.* **48,** 909–918.

Bernstein, L. (1963). Osmotic adjustment of plants to saline media. II. Dynamic phase. *Amer. J. Bot.* **50,** 360–370.

Bernstein, L. (1964). Effects of salinity on mineral composition and growth of plants. *Plant Anal. Fert. Probl. Collcq.* **4,** 25–45.

Bernstein, L. (1970). Calcium and salt tolerance of plants. *Science* **167,** 1387.

Berry, W. L., and Thomson, W. W. (1967). Composition of salt secreted by salt glands of *Tamarix aphylla*. *Can. J. Bot.* **45,** 1774–1775.

Bersin, T. (1950). Die Phytochemie des Schwefels. *Adv. Enzymol.* **10,** 223–323.

Bertsch, A. (1966a). CO_2 Gaswechsel und Wasserhausahlt der aerophilen Grünalge *Apatococcus lobatus*. *Planta* **70,** 46–72.

Bertsch, A. (1966b). Über den CO_2-Gaswechsel einiger Flechten nach Wasserdampfaufnahme. *Planta* **68,** 157–166.

Bewley, T. A., and Li, C. H. (1969). The reduction of protein disulfide bonds in the absence of denaturants. *Int. J. Protein Res.* **1,** 117–124.

Beysel, D. (1957). Osmotic value and protoplasmic viscosity in diploid sugarbeets. *Ber. Deut. Bot. Ges.* **70,** 109–120.

Bhardwaj, S. N. (1959). Physiological studies on salt-tolerance in crop plants. VI. Effect of NaCl and Na_2CO_3 on grain quality in wheat. *Proc. Nat. Acad. Sci. India Sect.* **B29,** 168–173.

Bhardwaj, S. N. (1961). Physiological studies on salt-tolerance in crop plants. XI. Inducing tolerance to NaCl by pretreatment of seeds. *Proc. Nat. Acad. Sci. India Sect.* **B31,** 160–171.

Bhardwaj, S. N. (1962). Physiological studies on salt-tolerance in crop plants. XVII. Influence of isosmotic concentrations of NaCl and sucrose on early seedling growth of wheat. *Agra. Univ. J. Res. Sci.* **11,** 69–74.

Bhardwaj, S. N. (1964a). Physiological studies on salt-tolerance in crop plants. XIX. Effect of sodium carbonate on growth and maturity of wheat and gram and inducing tolerance by pretreatment of seeds. *Proc. Nat. Acad. Sci. India Sect.* **B33,** 453–459.

Bhardwaj, S. N. (1964b). Physiological studies on salt-tolerance in crop plants. XXV. Effects of chloride and carbonate of sodium on catalase activity in the seedlings of wheat and gram (*Cicer arietinum*). *Sci. Cult.* **30,** 236–237.

Bhardwaj, S. N., and Rao, I. M. (1960). Physiological studies on salt-tolerance in crop plants. IX. Effect of sodium chloride and sodium carbonate on seedling respiration and growth of wheat and gram. *Indian J. Plant Physiol.* **3,** 56–71.

Bhargava, B. D. (1967). The diurnal variation in the internal water stress of Bajra varieties at the wilting stage. *J. Birla Inst. Technol. Sci.* **1,** 146–151.

Biarichi, M. R., Boccacci, M., Quintiliani, M., and Strom, E. (1965). On the mechanism of radiosensitization by iodoacetic acid and related substances. *Progr. Biochem. Pharmacol.* **1,** 384–391.

Biebl, R. (1939). Über die Temperaturresistenz von Meeresalgen verschiedener Klimazonen und verschieden tiefer Standorte. *Jahrb. Wiss. Bot.* **88,** 389–420.

Biebl, R. (1952a). Ecological and nonenvironmental constitutional resistance of the protoplasm of marine algae. *J. Marine Biol. Ass. U. K.* **31,** 307–315.

Biebl, R. (1952b). Ultraviolettabsorption der Meeresalgen. *Ber. Deut. Bot. Ges.* **65,** 37–41.

Biebl, R. (1952c). Resistenz der Meeresalgen gegen sichtbares Licht und gegen kurzwellige UV-Strahlen. *Protoplasma* **41,** 353–377.

Biebl, R. (1954). Lichtgenuss und Strahlenempfindlichkeit einiger Schattenmoose. *Oesterr. Bot. Z.* **101,** 502–538.

Biebl, R. (1955). Tagesgange der Lichttransmission verschiedener Blätter. *Flora* **142,** 280–294.

Biebl, R. (1956). Zellphysiologisch-Ökologische Untersuchugen an *Enteromorpha clathrata* (Rath) Greville. *Ber. Deut. Bot. Ges.* **69,** 75–86.

Biebl, R. (1956a). Morphologische, anatomische, und zellphysiologische Untersuchungen an Pflanzen vom "Gamma-Feld" des Brookhaven National Lab. (USA). *Oesterr. Bot. Z.* **103,** 400–435.

Biebl, R. (1956b). Lichtresistenz von Meeresalgen. *Protoplasma* **46,** 63–89.

Biebl, R. (1957). La résistance des algues marines à la lumière. *Coll. Int. Centre Nat. Rech. Sci.* **81,** 191–203.

Biebl, R. (1958a). Radiomorphosen an *Soja hispida. Flora (Jena)* **146,** 68–93.

Biebl, R. (1958b). Strahlensukkulenz. *Atompraxis* **11,** 411–416.

Biebl, R. (1958c). Temperatur-und osmotische Resistenz von Meeresalgen der bretonischen Küste. *Protoplasma* **50,** 217–242.

Biebl, R. (1959a). Strahlenempfindliche Keimungsphase und Dauerbestrahlung. *Oesterr. Bot. Z.* **106,** 104–123.

Biebl, R. (1959b). Strahlenempfindliche Entwicklungsstadien. *Ber. Deut. Bot. Ges.* **72,** 202–211.

Biebl, R. (1961). Protection against the effect of short-wave UV-rays on plant cells. *Progr. Photobiol.* **3,** 283–286.

Biebl, R. (1962a). Protoplasmatische Ökologie der Pflanzen. Wasser und Temperatur. *Protoplasmatologia* **1,** 1–344.

Biebl, R. (1962b). Kälte-und Warmeresistenz tropischer Meeresalgen. *Ber. Deut. Bot. Ges.* **7,** 271–272.

Biebl, R. (1962c). Temperaturreresistenz tropischer Meeresalgen. *Bot. Mar.* **4,** 241–254.

Biebl, R. (1964a). Austrocknungsresistenz tropischer Urwaldmoose auf Puerto Rico. *Protoplasma* **59,** 277–297.

Biebl, R. (1964b). Zum Wasserhaushalt von *Tillandsia recurvata* L. und *Tillandsia usreoides L.* auf Puerto Rico. *Protoplasma* **58,** 345–368.

Biebl, R. (1967a). Temperaturresistenz einiger Grünalgen warmer Bache auf Island. *Botaniste Ser.* **L,** 33–42.

Biebl, R. (1967b). Temperaturresistenz tropischer Urwaldmoose. *Flora (Jena)* **157,** 25–30.

Biebl, R. (1967c). Kurtztag-Einflüsse auf arktische Pflanzen während der arktischen Langtage. *Planta* **75,** 77–84.

Biebl, R. (1967d). Ueber Wärmehaushalt und Temperaturresistenz arktischer Pflanzen in Westergroenland. *Flora (Jena)* **157,** 327–354.

Biebl, R. (1969). Studien zur Hitzeresistenz der Gezeitenalge *Chaetomorpha cannabina* (Aresch.) Kjellm. *Protoplasma* **67,** 451–472.

Biebl, R., and Hofer, K. (1966). Tages und Jahresperiodik der Strahlenresistenz pflanzlicher Zellen. *Radiat. Bot.* **6,** 225–250.

Biebl, R., and Kinzel, H. (1965). Leaf structure and salt balance of *Laguncularia racemosa* and of other mangrove trees in Puerto Rico. *Oesterr. Bot. Z.* **112,** 56–93.

Biebl, R., and Mostafa, L. Y. (1965). Water content of wheat and barley seeds and their radiosensitivity. *Radiat. Bot.* **5,** 1–6.

Biebl, R., and Pape, R. (1951). Röntgenstrahlenwirkungen auf keimenden Weizen. *Oesterr. Bot. Z.* **98,** 361–382.

Biebl, R., and Rossi-Pillhofer, W. (1954). Die Änderung der chemischen Resistenz pflanzlischer Plasmen mit dem Entwicklungszutstund. *Protoplasma* **44,** 113–135.

Biebl, R., and Url, W. (1958). UV-Strahlenwirkungen auf Zellen von *Allium cepa,* besonders deren Chondriosomen und Plastiden. *Protoplasma* **49,** 329–352.

Biebl, R., and Url, W. (1963a). Untersuchungen über Strahlenschutz an Pflanzenzellen. *Protoplasma* **56,** 670–700.

Biebl, R., and Url, W. (1963b). Wirkungen von α-Strahlen auf die Pflanzenzelle. *Protoplasma* **57,** 84–125.

Biebl, R., and Url, W. (1963c). Chemical protection against the effects of alpha-rays and of thermal neutrons in plant cells by pre- and post-treatments. *Radiat. Bot.* **3,** 67–73.

Biebl, R., and Url, W. (1968). Chemical radiation protection against damages in the cytoplasm and the nuclei in plant (*Allium cepa*) cells by α- and γ-radiation. *Atompraxis* **14,** 302–305.

Biebl, R., and Weissenböck, G. (1968). Vergleichende Untersuchung der Salz-Trocken- und Strahlensukkulenz bei einigen Wild-und Kulturpflanzen. *Osterr. Bot. Z.* **115,** 229–254.

Biebl, R., Url, W., and Janecek, G. (1961). Untersuchungen über Strahlenschutz an Pflanzenzellen. II. Schutzwirkung verschiedener schwefelhaltiger Verbindungen gegen kurzwellige UV-Strahlen. *Protoplasma* **56,** 263–306.

Biel, E. R., Havens, A. V., and Sprague, M. A. (1955). Some extreme temperature fluctuations experienced by living plant tissue during winter in New Jersey. *Bull. Amer. Meteorol. Soc.* **36,** 159–162.

Bierhuizen, A. A., Rahman, Abd El, and Kuiper, P. J. C. (1959). The effect of nitrogen application and water supply on growth and water requirement of tomato under controlled conditions. *Mededeel. Landbouwhogesch. Wageningen* **59,** 1–8.

Bierhuizen, J. F., and Slatyer, R. O. (1965). Effect of atmospheric concentration of water vapor and CO_2 in determining transpiration-photosynthesis relationships of cotton leaves. *Agr. Meteorol.* **2,** 259–270.

Bierhuizen, J. F., Nunes, M. A., and Plocoman, C. (1969). Studies on productivity of coffee. II. Effect of soil moisture on photosynthesis and transpiration of *Coffea arabica. Acta Bot. Neer.* **18,** 367–374.

Bigelow, C. C. (1967). On the average hydrophobicity of proteins and the relation between it and protein structure. *J. Theoret. Biol.* **16,** 187–211.

Billings, W. D., and Mooney, H. A. (1968). The ecology of arctic and alpine plants. *Biol. Rev. Cambridge Phil. Soc.* **43,** 481–529.

Billings, W. D., and Morris, R. J. (1951). Reflection of visible and infrared radiation from leaves of different ecological groups. *Amer. J. Bot.* **38,** 327–331.

Binet, P. (1963). Sodium and potassium in *Suaeda vulgaris. Physiol. Plant.* **16,** 615–622.

Bingham, F. T., Page, A. L., and Bradford, G. R. (1964). Tolerance of plants to lithium. *Soil Sci.* **98,** 4–8.

Binz, E. (1939). Untersuchungen über die Dürreresistenz verschiedener Getreidesorten bei Austrocknung des Bodens. *Jahrb. Wiss. Bot.* **88,** 470–518.

Björkman, O. (1968a). Carboxydismutase activity in shade-adapted and sun-adapted species of higher plants. *Physiol. Plant.* **21,** 1–10.

Björkman, O. (1968b). Further studies on differentiation of photosynthetic properties in sun and shade ecotypes of *Solidago virgaurea. Physiol. Plant.* **21,** 84–99.

Björkman, O., Nobs, M. A., and Hiesey, W. M. (1970). Growth, photosynthetic, and biochemical responses of contrasting *Mimulus* clones to light intensity and temperature. *Carnegie Inst. Washington Yearb.* **68,** 614–620.

Björn, L. O. (1969). Photoinactivation of catalases from mammal liver, plant leaves and bacteria. Comparison of inactivation cross sections and quantum yields at 406 nm. *Photochem. Photobiol.* **10,** 125–129.

Black, R. F. (1958). Effects of sodium chloride on leaf succulence and area of *Atriplex hastata* L. *Aust. J. Bot.* **6,** 306–321.

Black, R. F. (1960). Effects of NaCl on the ion uptake and growth of *Atriplex vesicaria* Heward. *Aust. J. Biol. Sci.* **13,** 249–266.

Bliss, L. C., Kramer, P. J., and Wolff, F. A. (1957). Drought resistance in tobacco. *Tobacco* **145,** 20–23.

Blum, G. (1937). Osmotische Untersuchungen in Java. II. Untersuchungen in Trockengebieten Östjavas. *Ber. Schweiz. Bot. Ges.* **47,** 400–416.

Blumenthal-Goldschmidt, S., and Poljakoff-Mayber, A. (1968). Effect of substrate salinity on growth and on submicroscopic structure of leaf cells of *Atriplex halimus* L. *Aust. J. Bot.* **16,** 469–478.

Bobart, J. (1684). "Philosophical Transactions and Collections to the End of the year 1700," (Abridged and disposed under general heads) Vol. 2, 4th ed., pp. 155–160, 1731.

Böhning, R. H., and Burnside, C. A. (1956). The effect of light intensity on rate of apparent photosynthesis in leaves of sun and shade plants. *Amer. J. Bot.* **43,** 557–561.

Bogdanov, P. (1935). Photoperiodism in species of woody plants. Preliminary contribution. *Exp. Sta. Rec.* **73,** 22.

Bogen, H. J. (1948). Untersuchungen über Hitzetod und Hitzeresistenz pflanzlicher Protoplaste. *Planta* **36,** 298–340.

Bolduc, R. J., Cherry, J. H., and Blair, B. O., (1970). Increase in indoleacetic acid oxidase activity of winter wheat by cold treatment and gibberellic acid. *Plant Physiol.* **45,** 461–464.

Bornkamm, R. (1958). Habitat conditions and water economy of semiarid turfs of brome grass (Mesobromion) in the upper limits of its range. *Flora (Jena)* **146,** 23–67.

Boroughs, H., and Hunter, J. R. (1963). Effect of temperature on the growth of Cacao seeds. *Proc. Amer. Soc. Hort. Sci.* **82,** 222–224.

Borzakivś ka, I. V. (1965). Changes in the pigment system of the bark of some trees in winter. *Ukr. Bot. Zh.* **22,** 19–25.

Borzakivs'ka, I. V., and Motruk, V. V. (1969). Phosphorus metabolism in woody plant seedlings in connection with their winter hardiness resulting from the effect of variable temperatures on the seeds. *Ukr. Bot. Zh.* **26,** 67–72.

Bouwkamp, J. C., and Honma, S. (1969). The inheritance of frost resistance and flowering response in broccoli (*Brassica oleracea* var. *italica*). *Euphytica* **18,** 395–397.

Boyarchuk, Yu. M., and Vol'kenshtein, M. V. (1967). The effects of protein conformation on hydrogen bonds between peptide groups. *Biofizika* **12,** 341–343.

Boyer, Y. (1964). Contribution à l'étude de l'écophysiologie de deux fougères epiphytes: *Platycerium stemaria* (Beauv.) Desv. et *P. angolense* Welw. *Ann. Sci. Nat. Bot. Paris.* **5,** (*Ser.* 12), 87–228.

Boyer, J. S. (1965). Effects of osmotic water stress on metabolic rates of cotton plants with open stomata. *Plant Physiol.* **40,** 229–234.

Boyer, J. S. (1970a). Leaf enlargement and metabolic rates in corn, soybean, and sunflower at various leaf water potentials. *Plant Physiol.* **46,** 233–235.

Boyer, J. S. (1970b). Differing sensitivity of photosynthesis to low leaf water potentials in corn and soybean. *Plant Physiol.* **46,** 236–239.

Boyer, J. S., and Bowen, B. L. (1970). Inhibition of oxygen evolution in chloroplasts isolated from leaves with low water potentials. *Plant Physiol.* **45,** 612–615.

Bozhenko, V. P. (1968). Effect of aluminum and cobalt on the nucleic acid content and ribonuclease activity in the growing points of the sunflower plant under water deficiency conditions. *Fiziol. Rast.* **15,** 116–122.

Brandts, J. F. (1967). *In* "Thermobiology" (A. H. Rose, ed.), pp. 25–72. Academic Press, New York.

Brandts, J. F., Fu, J., and Nordin, J. H. (1970). The low temperature denaturation of chymotrypsinogen in aqueous solution and in frozen aqueous solution. *In:* "The Frozen Cell" (G. E. W. Wolstenholme and M. O'Connor, eds.), pp. 189–208. Ciba Foundation Symposium, J. & A. Churchill, London.

Brengle, K. G. (1968). Effect of phenylmercuric acetate on growth and water use of spring wheat. *Agron. J.* **60**; 246–247.

Bridges, B. A. (1969). Sensitization of organisms to radiation by sulfhydryl-binding agents. *Adv. Rad. Biol.* **3,** 123–176.

Brierley, W. G. (1934). Absorption of water by the foliage of some common fruit species. *Proc. Amer. Soc. Hort. Sci.* **32,** 277–283.

Brierley, W. G., and Landon, R. H. (1946). A study of cold resistance of the roots of the Latham red raspberry. *Proc. Amer. Soc. Hort. Sci.* **47,** 215–218.

Brierley, W. G., and Landon, R. H., (1954). Effects of dehardening and rehardening treatments upon cold resistance and injury of Latham raspberry canes. *Proc. Amer. Soc. Hort. Sci.* **63,** 173–178.

Brix, H. (1962). The effect of water stress on the rates of photosynthesis and respiration in tomato plants and loblolly pine seedlings. *Physiol. Plant.* **15,** 10–20.

Brix, H. (1969). Effect of temperature on dry matter production of Douglas Fir seedlings during bud dormancy. *Can J. Bot.* **47,** 1143–1146.

Brock, T. D. (1967). Life at high temperatures. *Science* **158,** 1012–1019.

Brock, T. D., and Darland, G. K. (1970). Limits of microbial existence: Temperature and pH. *Science* **169,** 1316–1318.

Brodie, A. F. (1969). Biological function of terpenoid quinones. *Biochem. J.* **113,** 25pp.

Brown, J. H., Bula, R. J., and Low, P. F. (1970). Physical properties of cytoplasmic protein-water extracts from roots of hardy and nonhardy *Medicago sativa* ecotypes. *Cryobiology* **6,** 309–314.

Brown, H. T., and Escombe, F. (1897). The influence of very low temperatures on the germinative power of seeds. *Proc. Roy. Soc.* **62,** 160–165.

Bruinsma, J. (1958). Studies on the Crassulacean acid metabolism. *Acta Bot. Neer.* **7,** 531–590.

Brumfield, R. T. (1953). The effect of ultraviolet irradiation on cell division and elongation in timothy roots. *Proc. Nat. Acad. Sci. U.S.***39,** 366–371.

Buchinger, A. (1929). Der Einfluss hoher Anfangstemperaturen auf die Keimung, dargestellt an *Trifolium pratense*. *Jahrb. Wiss. Bot.* **71,** 149–153.

Budnitskaya, E. V., and Borisova, I. G. (1961). Oxidative conversions of lipids in irradiated plant leaves. *Biokhimiya* **26,** 122–127.

Bünning, E., and Herdtle, H. (1946). Physiologische Untersuchungen an thermophilen Blaualgen. *Z. Naturforsch.* **1,** 93–99.

Bugaevsky, M. F. (1939a). Contribution to the study of causes of death in root crops subjected to low temperatures. *C. R. Acad. Sci. URSS* **25,** 527–530.

Buhlert, H. (1906). Untersuchungen über das Auswintern des Getreides. *Landwirt. Jahrb.* **35,** 837–887.

Bula, R. J., and Smith, D. (1954). Cold resistance and chemical composition in overwintering alfalfa, red clover, and sweet clover. *Agron. J.* **46,** 397–401.

Bull, H. B., and Breese, K. (1968). Protein hydration I. Binding sites. *Arch. Biochem. Biophys.* **128,** 488–496.

Burcik, E. (1950). Über die Beziehungen zwischen Hydratur und Wachstum bei Bakterien und Hefen. *Arch. Mikrobiol.* **15,** 203–235.

Burns, V. W. (1964). Reversible sonic inhibition of protein, purine, and pyrimidine biosynthesis in the living cell. *Science* **146,** 1056–1058.

Burrows, W. J., and Carr, D. J. (1969). Effects of flooding the root system of sunflower plants on the cytokinin content in the xylem sap. *Physiol. Plant.* **22,** 1105–1112.

Burström, H. (1929). Undersokningar over oorganiska amnens giftverkningar vid betning av *Tilletia tritici* sporer. *Med. N:r 356 Centralst. Forsoksvas. jordbruksomradet. Avdel. lantbruksbotanik N:r* **46,** 1–19.

Butera, M. K. (1970). A comparison of the hydrophobicity of protein extracted from hardy and tender plants. Master's Thesis, University of Missouri.

Byfield, J. E., and Scherbaum, O. H. (1967). Temperature effect on protein synthesis in a heat-synchronized protozoan treated with actinomycin D. *Science* **156,** 1504–1505.

Caldwell, J. S. (1913). The relation of environmental conditions to the phenomenon of permanent wilting in plants. *Physiol. Res.* **1,** 1–56.

Caldwell, M. M. (1968). Solar ultraviolet radiation as an ecological factor for alpine plants. *Ecol. Monogr.* **38,** 243–268.

Calvert, J. (1935). Drought resistance in wheat. The "bound" and "free" water of expressed sap from wheat leaves in relation to time and soil moisture. *Protoplasma* **24,** 505–524.

Campbell, C. A. (1968). Influence of soil moisture stress applied at various stages of growth on the yield components of Chinook wheat. *Can. J. Plant Sci.* **48,** 313–320.

Campbell, L. L., and Pace, B. (1968). Physiology of growth at high temperatures. *J. Appl. Bacteriol.* **31,** 24–35.

Candolle, C. de. (1895). Sur la vie latente des graines. *Bibl. Univ. Arch. Sci. Phys. Nat.* (3) **33,** 497–512.

Capeillare-Blandin, C. (1969). Interactions substrat-flavine-hemoproteine et stabilité thermique du cytochrome b_2 (L-lactate deshydrogenase de la levure). *FEBS Lett.* **4,** 311–315.

Carlier, G. (1961). Étude des variations de l'intensité respiratoire des feuilles de Pommier sur pied, provoquées par le dessechement et la réhumectation du sol. *C. R. Acad. Sci.* **253,** 898–900.

Carlier, G. (1968). Action of X-rays on the respiratory intensity of leaves of *Pelargonium zonale. Physiol. Veg.* **6,** 203–211.

Carmichael, J. W. (1962). Viability of mold cultures stored at −20°C. *Mycologia* **54,** 432–436.

Carroll, J. C. (1943). Effects of drought, temperature, and nitrogen on turf grasses. *Plant Physiol.* **18,** 19–36.

Casas, I. A., Redshaw, E. S., and Ibanez, M. L. (1965). Respiratory changes in the cacao seed cotyledon coincident with seed (cold) death. *Nature* (*London*) **208,** 1348–1349.

Caspary, R. (1854). Auffallende Eisbildung auf Pflanzen. *Bot. Zeitung* **12,** 665–674, 681–690, 697–706.

Caspary, R. (1857). Bewirkt die Sonne Risse in Rinde und Holze der Bäume ? *Bot. Zeitung* **15,** 153–156, 329–335, 345–350, 361–371.

Castenholz, R. W. (1967). Aggregation in a thermophilic Oscillatoria. *Nature* (*London*) **215,** 1285–1286.

Castenholz, R. W. (1969). The thermophilic cyanophytes of Iceland and the upper temperature limit. *J. Phycol.* **5,** 360–368.

Catsky, J. (1960). Determination of water deficit in disks cut out form leaf blades. *Biol. Plant.* **2,** 76–78.

Cervigni, T., Bassanelli, C., Caserta, G., and Laitano, R. F. (1969). Thermoluminescence in complex biochemicals: A possible correlation with radiosensitivity. *Radiat. Res.* **38,** 579–587.

Chambers, R., and Hale, H. P. (1932). The formation of ice in protoplasm. *Proc. Roy. Soc.* **B110,** 337–352.

Chandler, W. H. (1913). The killing of plant tissue by low temperature. *Mo. Agr. Expt. Sta. Res. Bull.* No. 8,171 pp.

Chaney, W. R., and Kozlowski, T. T. (1969). Diurnal expansion and contraction of leaves and fruits of English Morello cherry. *Ann. Bot.* **33,** 991–999.

Chao, L., and Walker, D. R. (1967). Effects of a magnetic field on the germination of apple, apricot, and peach seeds. *Hort. Sci.* **2,** 152–153.

Chapman, A., Sanner, T., and Pihl, A. (1969). X-ray inactivation of phosphofructokinase. The relative radiation sensitivity of the catalytic and the allosteric properties. *Biochim. Biophys. Acta* **178,** 74–82.

Charlier, S., Grosjean, H., Lurquin, P., Vanhunbeck, J., and Werenne, J. (1969). Comparative studies on the isoleucyl- and leucyl-tRNA synthetases from *Bacillus stearothermophilus* and *Escherichia coli:* Thermal stability of the aminoacyl-adenylate-enzyme complexes. *FEBS Lett.* **4,** 239–242.

Chatagner, F., Durieu-Trautmann, O., and Rain, M. C. (1968). Influence of pyridoxal phosphate and some other derivatives of pyridoxine on the stability *in vitro* of cystathionase and of cysteine sulfinic acid decarboxylase. *Bull. Soc. Chim. Biol.* **50,** 129–141.

Chatterton, N. J., McKell, C. M., Bingham, F. T., and Clawson, W. J. (1970a). Absorption of Na, Cl, and B by desert salt-bush in relation to composition of nutrient solution culture. *Agron. J.* **62,** 351–352.

Chatterton, N. J., McKell, C. M., and Strain, B. R. (1970b). Intraspecific differences in temperature-induced respiratory acclimation of desert saltbush. *Ecology* **51,** 545–547.

Chaudhuri, I. I., and Wiebe, H. H. (1968). Influence of calcium pretreatment on wheat germination on saline media. *Plant Soil* **18,** 208–216.

Cheban, A. I. (1968). Content and accumulation of nucleic acids in buds of wintering grape plant eyes of different stages on its seasonal development. *Fiziol. Rast.* **15,** 329–335.

Chen, A. W. (1964). Soil fungi with high salt tolerance. *Trans. Kans. Acad. Sci.* **67,** 36–40.

Chen, D., Sarid, S., and Katchalski, E. (1968). The role of water stress in the inactivation of messenger RNA of germinating wheat embryos. *Proc. Nat. Acad. Sci. U.S.* **61**, 1378–1383.

Chen, S. S. C., and Varner, J. E. (1970). Respiration and protein synthesis in dormant and after-ripened seeds of *Avena fatua. Plant Physiol.* **46**, 108–112.

Chen, T. M., and Ries, S. K. (1969). Effect of light and temperature on nitrate uptake and nitrate reductase actvity in rye and oat seedlings. *Can. J. Bot.* **47**, 341–343.

Chian, L. C., and Wu, S. H. (1965). Cytological studies on the cold resistance of plants. The changes of the intercellular materials of wheat in the overwintering period. *Acta Bot. Sinica* **13**, 196–207.

Chibnall, A. C. (1939). "Protein Metabolism in the Plant." Yale Univ. Press, New Haven, Connecticut.

Ching, Te. May, and Slabaugh, W. H. (1966). X-ray diffraction analysis of ice crystals in coniferous pollen. *Cryobiology* **2**, 321–327.

Chinoy, J. J. (1960). Physiology of drought resistance in wheat. I. Effect of wilting at different stages of growth on survival values of eight varieties of wheat belonging to seven species. *Phyton (Buenos Aires)* **14**, 147–157.

Chinoy, J. J., Jani, B. M., and John, D. (1965). Effect of water stress and pre-treatment on growth and yield of four varieties of barley. *In* "Growth and Development of Plants," (R. D. Asana and K. K. Nanda, eds.), pp. 169–186. Today and Tomorrow's Book Agency, New Delhi.

Chomel, M. (1710). Sur les arbres morts par la gelée de 1709. "Histoire de l'Academie Royal des Sciences" (Avec les memoires, etc.)., pp. 59–61.

Choudhuri, G. N. (1968). Effect of soil salinity on germination and survival of some steppe plants in Washington. *Ecology* **49**, 465–471.

Christ, H. (1911). Die Vegetation unter dem Einfluss des trockenen Sommers 1911 im nördlichen Jura. *Ber. Schweiz. bot. Ges.* **20**, 254–258.

Christiansen, M. N. (1968). Induction and prevention of chilling injury to radicle tips of imbibing cotton seed. *Plant Physiol.* **43**, 743–746.

Christophersen, B. O. (1969). Reduction of linolenic acid hydroperoxide by a glutathione peroxidase. *Biochim. Biophys. Acta* **176**, 463–470.

Christopherson, J. (1963). The effect of the adaptation temperature on resistance to heat and activity of the hexokinase in yeast cells. *Arch. Mikrobiol.* **45**, 58–64.

Christophersen, J., and Precht, H. (1952). Untersuchungen zum Problem der Hitzeresistenz. II. Untersuchungen an Hefezellen. *Biol. Zentr.* **71**, 585–601.

Chrominski, A., Belt, H., and Michniewicz, M. (1969). Effect of (2-chloroethyl)trimethylammonium chloride (CCC) on frost resistance, yield and seed quality indexes of winter rape. *Rocz. Nauk. Roln. Ser.* **A95**, 191–197.

Chu, Chien-Ren, (1936). Der Einfluss des Wassergehaltes der Blätter der Waldbäume auf ihre Saugkräfte und ihren Turgor. *Flora* **130**, 384–437.

Chuikov, A. G., and Skazkin, F. D. (1968). Effect of soil water deficiency at the critical period on the state of cell plasma of vegetative and generative organs in spring cereal crops. *Dokl. Akad. Nauk. SSSR* **178**, 1208–1211.

Chuvaev, P. P. (1967). Effect of seed orientation with respect to the point of the compass in their germination rate and the nature of seedling growth. *Fiziol. Rast.* **14**, 540–543.

Chuvashina, N. P. (1962). Using cytochemical methods in studies of the winter hardiness of apple trees. *Tr. Tsentr. Genet. Lab. Im. I.V. Michurina* **8**, 67–72.

Civinskij, V. (1934). Capacity of cotton to withstand cold. *C. R. Acad. Sci. URSS* **1**, 149–150.

Clark, J. A. (1956). An investigation of the drought hardening of the soybean plant. Ph.D. Thesis. Univ. of Missouri, Columbia, Missouri.

Clark, J. A., and Levitt, J. (1956). The basis of drought resistance in the soybean plant. *Physiol. Plant.* **9,** 598–606.

Clark, R. B. (1966). Effect of metal cations on aldolase from leaves of *Zea mays* L. seedlings. *Crop. Sci.* **6,** 593–596.

Clarkson, D. T. (1966). Aluminum tolerance within the genus *Agrostis. J. Ecol.* **54,** 167–178.

Clements, H. F. (1937a). Studies in the drought resistance of the sunflower and potato. *Bot. Dept. State Coll. Wash. Contr.* **59,** 81–98.

Clements, H. F. (1937b). Studies in the drought resistance of the soybean. *Res. Stud. State Coll. Wash.* **5,** 1–16.

Clements, H. F. (1938). Mechanisms of freezing resistance in the needles of *Pinus ponderosa* and *Pseudotsuga mucronata. Res. Stud. State Coll. Wash.* **6,** 3–45.

Cline, M. G., Conner, G. I., and Salisbury, F. B. (1969). Simultaneous reactivation of ultraviolet damage in Xanthium leaves. *Plant Physiol.* **44,** 1674–1678.

Clowes, F. A. L. (1970). The immediate response of the quiescent center to X-rays. *New Phytol.* **69,** 1–18.

Codaccioni, M. (1968). Réactions aux brusques abaisements thermiques des plantes du *Mentha viridis* cultivés in vitro. *C. R. Acad. Sci.* (*Paris*) **276,** 1499–1502.

Codaccioni, M., and Le Saint-Quervel, A. M. (1966). Endurcissement artificiel au gel chez le Chou de Milan. Role eventuel de l'amidon et sa localisation dans les parties aeriennes. *C. R. Acad. Sci.* (*Paris*) **263,** 1837–1840.

Coffman, F. A. (1955). Results from uniform winter hardiness nurseries of oats for the five years 1947–1951, inclusive. *Agron. J.* **47,** 54–57.

Coffman, F. A. (1957). Cold resistant oat varieties also resistant to heat. *Science* **125,** 1298–1299.

Cohn, F. and David, G. (1871). Wirkung der Kälte auf Pflanzenzellen. *Naturforscher* **39,** 316.

Coleman, E. A., Bula, R. J., and Davis, R. L. (1966). Electrophoretic and immunological comparisons of soluble root proteins of *Medicago sativa L.* genotypes in the cold hardened and nonhardened condition. *Plant Physiol.* **41,** 1681–1685.

Collander, R. (1924). Beobachtungen über die quantitativen Beziehungen zwischen Tötungsgeschwindigkeit und Temperatur beim Wärmetod pflanzlicher Zellen. *Soc. Sci. Fenn. Commentat. Biol.* **1,** 1–12.

Collis-George, N., and Williams, J. (1968). Comparison of the effects of soil matric potential and isotropic effective stress on the germination of *Lactuca sativa. Aust. J. Soil Res.* **6,** 179–192.

Collison, R. C., and Harlan, J. D. (1934). Winter injury of Baldwin apple trees and its relation to previous tree performances and nutritional treatment. *N. Y. Agr. Exp. Sta. Geneva Tech. Bull.* **647,** 1–13.

Conger, B. V., Nilan, R. A., and Konzak, C. F. (1969). The role of water content in the decay of radiation induced oxygen-sensitive sites in barley seeds during post-irradiation hydration (*Hordeum vulgare*). *Radiat. Res.* **39,** 45–56.

Connell, J. J. (1960a). Changes in the adenosinetriphosphatase activity and sulfhydryl groups of cod flesh during frozen storage. *J. Sci. Food. Agr.* **11,** 245–249.

Connell, J. J. (1960b). Changes in the actin of cod flesh during storage at $-14°$. *J. Sci. Food Agr.* **11,** 515–519.

Connor, D. J. (1969). Growth of grey mangrove (*Avicennia marina*) in nutrient culture. *Biotropica* **1,** 36–40.

Connor, D. J., and Turnstall, B. R. (1968). Tissue water relations for bigelow and mulga. *Aust. J. Bot.* **16,** 487–490.
Constantinescu, E. (1933). Weitere Beiträge zur Physiologie der Kälteresistenz bei Wintergetreide. *Planta* **21,** 304–323.
Conway, V. M. (1940). Aeration and plant growth in wet soils. *Bot. Rev.* **6,** 149–163.
Cooper, W. C. (1959). Cold hardiness in citrus as related to dormancy. *Proc. Fl. State Hort. Soc.* **72,** 61–66.
Cooper, W. C., Gorton, B. S., and Taylor, S. D. (1954). Freezing tests with small trees and detached leaves of grapefruit. *Proc. Amer. Soc. Hort. Sci.* **63,** 167–172.
Cooper, W. C., Peynado, A., and Otey, G. (1955). Effects of plant regulators on dormancy, coldhardiness and leaf form of grapefruit trees. *Proc. Amer. Soc. Hort. Sci.* **66,** 100–110.
Cooper, W. C., Rasmussen, G. K., and Waldon, E. S. (1969). Ethylene evolution stimulated by dulling in citrus and Persea sp. *Plant Physiol.* **44,** 1194–1196.
Cooperstein, S. J. (1963). The effect of disulfide bond reagents on cytochrome oxidase. *Biochim. Biophys. Acta* **73,** 343–346.
Cope, F. W. (1969). Nuclear magnetic resonance evidence using D_2O for structured water in muscle and brain. *Biophys. J.* **9,** 303–319.
Copeland, E. S., Sanner, T., and Pihl, A. (1968). Role of intermolecular reactions in the formation of secondary radicals in proteins irradiated in the dry state. *Radiat. Res.* **35,** 437–450.
Corns, G. and Schwerdtfeger, G. (1954). Improvement in low temperature resistance of sugar beet and garden beet seedlings with sodium TCA and Dalapon. *Can. J. Agr. Sci.* **34,** 639–641.
Cotrufo, C., and Berry, C. R. (1970). Some effects of a soluble NPK fertilizer on sensitivity of eastern white pine to injury from SO_2 air pollution. *Forest Sci.* **16,** 72–73.
Coutinho, L. M. (1969). Novas observações sobre a ocorrencia do "efeito de De Saussure" e suas relações com a suculencia a temperatura folhear e os movimentos estomaticos. *Bol. Fac. Fil. Cien. Letras Univ. Sao Paulo Bot.* **24,** 77–102.
Coville, F. V. (1920). The influence of cold in stimulating the growth of plants. *J. Agr. Res.* **20,** 151–160.
Cox, W., and Levitt, J. (1969). Direct relation between growth and frost hardening in cabbage leaves. *Plant Physiol.* **44,** 923–928.
Craig, D. L., Gass, D. A., and Fensom, D. S. (1970). Red raspberry growth related to electrical impedance studies. *Can. J. Plant Sci.* **50,** 59–66.
Craigie, J. S. (1969). Some salinity-induced changes in growth pigments, and cyclohexanetetrol content of *Monochrysis lutheri*. *J. Fish Res. Bd. Can.* **26,** 2959–2967.
Craker, L. E., Gusta, L. V., and Weiser, C. J. (1969). Soluble proteins and cold hardiness of two woody species. *Can. J. Plant. Sci.* **49,** 279–286.
Criddle, R. S. (1966). *In* "Biochemistry of Chloroplasts" (T. W. Goodwin, ed.), pp. 203–231. Academic Press, London.
Crosier, W., (1956). Longevity of seeds exposed to dry heat. *Proc. Ass. Off. Seed. Anal.* **46,** 72–74.
Cross, T. (1968). Thermophilic actinomycetes. *J. Appl. Bacteriol.* **31,** 36–53.
Currier, H. B., and Webster, D. H. (1964). Callose formation and subsequent disappearance: Studies in ultrasound stimulation. *Plant Physiol.* **39,** 843–847.
Curtis, B., Massey, V., and Zrnudka, M. (1968). Inactivation of snake venom L-amino acid oxidase by freezing. *J. Biol. Chem.* **243,** 2306–2314.
Curtis, O. F. (1936a). Comparative effects of altering leaf temperatures and air humidities on vapor pressure gradients. *Plant Physiol.* **11,** 595–603.

Curtis, O. F. (1936b). Leaf temperatures and the cooling of leaves by radiation. *Plant Physiol.* **11,** 343–364.

Curtis, O. F. (1938). Wallace and Clum, "Leaf Temperatures." A critical analysis with additional data. *Amer. J. Bot.* **25,** 761–771.

Curtis, O. F., and Clark, D. G. (1950). "An Introduction to Plant Physiology." McGraw-Hill, New York.

Czapek, F. (1901). Der Kohlenhydrat-Stoffwechsel der Laubblätter in Winter. *Ber. Deut. Bot. Ges.* **19,** 120–127.

Czerski, J. (1968). Gasometric method of water deficit measurement in leaves. *Biol. Plant.* **10,** 275–283.

Dalmer, M. (1895). Über Eisbildung in Pflanzen mit Rucksicht auf die anatomische Beschaffenheit derselben. *Flora (Jena)* **80,** 436–444.

Damjanovich, S., Sanner, T., and Pihl, A. (1967). The role of the allosteric sites in the X-ray inactivation of phosphorylase b. *Europ. J. Biochem.* **1,** 347–352.

Dangeard, P. (1951a). Observations sur la résistance des radicules de diverses plantes à des températures entre 40 et 60°. *C. R. Acad. Sci. (Paris)* **232,** 913–915.

Dangeard, P. (1951b). Observations sur la destruction du chondriome par la chaleur. *C. R. Acad. Sci. (Paris)* **232,** 1274–1276.

Dangeard, P. (1951c). Observations sur la résistance des radicules à des temperatures entre 40 et 60°. *Botaniste* **35,** 237–243.

Daniell, J. W., Chappell, W. E., and Couch, H. B. (1969). Effect of sublethal and lethal temperatures on plant cells. *Plant Physiol.* **44,** 1684–1689.

Dantuma, G., and Andrews, J. E. (1960). Differential response of certain barley and wheat varieties to hardening and freezing during sprouting. *Can. J. Bot.* **38,** 133–151.

Das, H. C., and Weaver, G. M. (1968). Modification of ozone damage to *Phaseolus vulgaris* by antioxidants, thiols, and sulfhydryl reagents. *Can. J. Plant Sci.* **48,** 569–574.

Das, N. K., and Alfert, M. (1961). Accelerated DNA synthesis in onion root meristem during X-irradiation. *Proc. Nat. Acad. Sci. U.S.* **47,** 1–6.

Dastane, N. G. (1957). Evaluation of root efficiency in moisture extraction. *J. Aust. Inst. Agr. Sci.* **23,** 223–226.

Davidson, B., and Fasman, G. (1967). The conformational transitions of unchanged poly-L-lysine: α-helix-random coil-B structure. *Biochemistry* **6,** 1616–1629.

Davis, D. L., and Gilbert, W. B. (1970). Winter hardiness and changes in soluble protein fractions of bermudagrass. *Crop Sci.* **10,** 7–8.

Davson, H., and Danielli, J. E. (1943). "The Permeability of Natural Membranes." Cambridge Univ. Press, New York.

Dawalkar, M. P., and Joshi, G. V. (1962). Effects of saline environment on ion uptake and distribution in *Ipomoea pes-caprae. J. Biol. Sci.* **5,** 26–30.

Day, W. R., and Peace, T. R. (1937). The influence of certain accessory factors on frost injury to forest trees. II. Temperature conditions before freezing. III. Time factors. *Forestry* **11,** 13–29.

De Long, W. A., Beaumont, J. H., and Willaman, J. J. (1930). Respiration of apple twigs in relation to winter hardiness. *Plant Physiol.* **5,** 509–534.

Deering, R. A. (1968). *Dictyostelium discoideum:* A gamma-ray resistant organism. *Science* **162,** 1289–1290.

De Jong, D. W., Olson, A. C., Hawker, K. M., and Jansen, E. F. (1968). Effect of cultivation temperature on peroxidase isozymes of tobacco plant cells grown in suspension. *Plant Physiol.* **43,** 841–844.

De Lis, B. R., Cavagnaro, J. B., Tizio, A. R., Urbieta, A., and Morales, H. (1969). Studies on water requirements of horticultural crops. V. Influence of drought at different growth stages of pepper (*Capsicum annuum* L.). *Rev. Invest. Agropecuar. Ser. 2* **6,** 145–153.

Del Rivero, J. M. (1966). Importance of the macro- and micro-elements in the citrus tolerance to and recovery from cold weather. *Bol. Patol. Veg. Entomol. Agr.* **29,** 405–411.

Den Dooren de Jong, L. E. (1965). Tolerance of *Chlorella vulgaris* for metallic and non-metallic ions. *Antonie Leeuwenhoek J. Microbiol. Serol.* **31,** 301–313.

Denna, D. W. (1970). Leaf wax and transpiration in *Brassica oleracea* L. *J. Amer. Soc. Hort. Sci.* **95,** 30–32.

Deo, R., and Kanwar, J. S. (1969). Effect of saline irrigation water on the growth and chemical composition of wheat. *J. Indian Soc. Soil Sci.* **16,** 365–370.

Devera, N. F., Marshall, D. R., and Balaam, L. M. (1969). Genetic variability in root development in relation to drought tolerance in spring wheats. *Exp. Agr.* **5,** 327–337.

Deutsch, J., and Carlier, G. (1965). Evolution des glucides au cours du fletrissement et de la rehydratation de tissus folaires du *Pelargonium zonale* L. *C. R. Acad. Sci. (Paris)* **261,** 2712–2715.

Dévay, M. (1965). The biochemical processes of vernalization. III. The changes of ascorbic acid oxidizing capacity in the course of vernalization. *Acta Agron. Acad. Sci. Hung.* **14,** 93–97.

De Visser Smits, D. (1926). Einfluss der Temperatur auf die Permeabilität des Protoplasmas bei *Beta vulgaris* L. *Rec. Trav. Bot. Neer.* **23,** 104–199.

De Vries, A. L. (1970). Cold resistance in fishes in relation to protective glycoproteins. *Cryobiology* **6,** 585.

De Vries, H. (1870). Matériaux pour la connaissance de l'influence de la température sur les plantes. *Arch. Neer. Sci. Exactes Natur.* **5,** 385–401.

De Vries, H. (1871). Sur la mort des cellules végétales par l'effet d'une température élevée. *Arch. Neer. Sci. Exactes Natur.* **6,** 245–295.

Devyatov, A. S. (1962). Changes with age in the salt tolerance of fruit crops. *Agrobiologiya* **3,** 383–387.

Dewey, D. L. (1965). Cysteine and the radiosensitivity of bacteria. *Progr. Biochem. Pharmacol.* **1,** 59–64.

Dewey, D. R. (1960). Salt tolerance of twenty-five strains of Agropyron. *Agron. J.* **52,** 631–634.

DeWit, M. C. J. (1969). Differential effect of low oxygen tension on protein and carbohydrate metabolism in barley roots. *Acta Bot. Neer.* **18,** 558–560.

Dexter, S. T. (1932). Studies of the hardiness of plants: a modification of the Newton pressure method for small samples. *Plant Physiol.* **7,** 721–726.

Dexter, S. T. (1933). Effect of several environmental factors on the hardening of plants. *Plant Physiol.* **8,** 123–139.

Dexter, S. T. (1935). Growth, organic nitrogen fractions and buffer capacity in relation to hardiness of plants. *Plant Physiol.* **10,** 149–158.

Dexter, S. T., Tottingham, W. E., and Graber, L. F. (1932). Investigations of hardiness of plants by measurement of electrical conductivity. *Plant Physiol.* **7,** 63–78.

de Zeeuw, D., and Leopold, A. C. (1957). The prevention of auxin responses by ultraviolet light. *Amer. J. Bot.* **44,** 225–228.

Diskus, A. (1958). Das Osmoseverhalten einiger Peridineen des Susswassers. *Protoplasma* **49,** 187–196.

Dobrenz, A. K., Wright, L., Humphrey, A. B., Massengale, M. A., and Kneebone, W. R. (1969). Stomate density and its relationship to water-use efficiency of blue panicgrass (*Panicum antidotale* Retz.). *Crop Sci.* **9,** 354–357.

Dobroserdova, I. V. (1968). The influence of drought on transpiration, water-retention capacity and growth processes of some tree species in the Volgograd Region. *Tr. Inst. Ekol. Rast. Zhivotn. Ural Fil. Akad. Nauk SSSR* **62,** 139–142.

Döring, H. (1932). Beiträge zur Frage der Hitzeresistenz pflanzlicher Zellen. *Planta* **18,** 405–434.

Dörr, M. (1941). Temperaturmessungen an Pflanzen des Frauensteins bei Mödling. *Beih. Bot. Centr.* **60,** 679–728.

Domien, F. (1949). Influence de la deshydratation sur las respiration des feuilles de végétaux aeriens. *Rev. Gen. Bot.* **56,** 285–317.

Dorsey, M. J. (1934). Ice formation in the fruit bud of the peach. *Proc. Amer. Soc. Hort. Sci.* **31,** 22–27.

Dorsey, M. J., and Strausbaugh, P. D. (1923). Winter injury to plum during dormancy. *Bot. Gaz.* (*Chicago*), **76,** 113–142.

Dounce, A. L. (1971). Nuclear gels and chromosomal structure. *Amer. Sci.* **59,** 74–83.

Dove, L. D. (1968). Nitrogen distribution in tomato plants during drought (*Lycopersicum esculentum* Marglobe). *Phyton* **25,** 49–52.

Dove, L. D. (1969). Phosphate absorption by air-stressed root systems. *Planta* **86,** 1–9.

Doyle, J., and Clinch, P. (1927). Seasonal changes in conifer leaves, with special reference to enzymes and starch formation. *Proc. Roy. Irish Acad. Sect.* **B37,** 373–414.

Drake, B. G., Raschke, K., and Salisbury, F. B. (1970). Temperatures and transpiration resistances of Xanthium leaves as affected by air temperature, humidity, and wind speed. *Plant Physiol.* **46,** 324–330.

Dubrov, A. P. (1968). "The Genetic and Physiological Effect of the Action of Ultraviolet Radiation on Higher Plants," 250 pp. Nauka, Moscow.

Duff, A. W. (ed.) (1937). "Physics," 715 pp. Blakiston, Philadelphia, Pennsylvania.

Dugger, W. M., Jr., Taylor, O. C., Cardiff, E., and Thompson, C. R. (1962). Stomatal action in plants as related to damage from photochemical oxidants. *Plant Physiol.* **37,** 487–491.

Du Hamel, H. L., and de Buffon, G. L. L. (1740). Observations des différents effets que produisent sur les Végétaux les grandes gelées d'Hiver et les petites gelées du Printemps. 1737. *Mem. Math. Phys. Acad. Roy. Sci.* (*Paris*), pp. 273–298.

Duisberg, P. C. (1952). Some relationships between xerophytism and the content of resin, nordihydroguaiaretic acid and protein of *Larrea divaricata* Cav. *Plant Physiol.* **27,** 769–777.

Duncan, C. J., and Bowler, K. (1969). Permeability and photoxidative damage of membrane enzymes. *J. Cell Physiol.* **74,** 259–272.

Durzan, D. J. (1968a). Nitrogen metabolism of *Picea glauca.* I. Seasonal changes of free amino acids in buds, shoot apices, and leaves and the metabolism of uniformly labelled ^{14}C-L-arginine by buds during the onset of dormancy. *Can. J. Bot.* **46,** 909–919.

Durzan, D. J. (1968b). Nitrogen metabolism of *Picea glauca.* II. Diurnal changes of free amino acids, amides, and guanidine compounds in roots, buds, and leaves during the onset of dormancy of white spruce saplings. *Can. J. Bot.* **46,** 921–928.

Durzan, D. J. (1969). Nitrogen metabolism of *Picea glauca.* IV. Metabolism of uniformly labelled ^{14}C-L-arginine (carbamyl-^{14}C)-L citrulline, and (1,2,3,4-^{14}C)-γ-guanidinobutyric acid during diurnal changes in the soluble and protein nitrogen associated with the onset of expansion of spruce buds. *Can. J. Biochem.* **47,** 771–783.

Dutrochet, M. (1839). Récherches sur la température propre des végétaux. *Ann. Sci. Nat.* **12,** 77–84.

Dutrochet, M. (1840). Récherches sur la chaleur propre des êtres vivants à basse température. *Ann. Sci. Nat.* **13,** 5–49, 65–85.

Eaks, I. L., and Morris, L. L., (1956). Respiration of cucumber fruits associated with physiological injury at chilling temperatures. *Plant Physiol.* **31,** 308–314.

Eaton, F. M., and Ergle, D. R. (1948). Carbohydrate accumulation in the cotton plants at low moisture levels. *Plant Physiol.* **23,** 169–187.

Ebert, M., and Barber, D. A. (1961). The effects of X-rays on the extension growth of barley roots. *Int. J. Radiat. Biol.* **3,** 587–593.

Echobichon, D. J., and Saschenbrecker, P. W. (1967). Dechlorination of DDT in frozen blood. *Science* **156,** 663–665.

Eckardt, F. (1952). Rapports entre la grandeur des feuilles et le comportement physiologique chez les xerophytes. *Physiol. Plant.* **5,** 52–69.

Eckardt, F. (1953). Transpiration et photosynthèse chez un xerophyte mesomorphe. *Physiol. Plant.* **6,** 253–261.

Eggert, R. (1944). Cambium temperatures of peach and apple trees in winter. *Proc. Amer. Soc. Hort. Sci.* **45,** 33–36.

Ehlig, C. F. (1964). Salt tolerance of raspberry, boysenberry, and blackberry. *Proc. Amer. Soc. Hort. Sci.* **85,** 318–324.

Ehlig, C. F., and Bernstein, L. (1958). Salt tolerance of strawberries. *Proc. Amer. Soc. Hort. Sci.* **72,** 198–206.

Ehrenberg, A., Ehrenberg, L., and Strom, G. (1969). Radiation-induced free radicals in embryo and endosperm of barley kernels. *Radiat. Bot.* **9,** 151–158.

Ehrler, W. (1960). Some effects of salinity on rice. *Bot. Gaz.* (*Chicago*) **122,** 102–104.

Eldjarn, L. (1965). Some biochemical effects of S-containing protective agents and the development of suitable SH/SS systems for the *in vitro* studies of such effects. *Progr. Biochem. Pharmacol.* **1,** 173–185.

El Gibaly, H., and Goumah, H. (1969). The effect of salinization on the growth and yield of sugar cane. *Beitr. Trop. Subtrop. Landwirt Tropenveterinaermed.* **7,** 27–39.

Ellis, R. J. (1969). Sulphate activation in higher plants. *Planta* **88,** 34–42.

El Nadi, A. H. (1969). Efficiency of water use by irrigated wheat in the Sudan. *J. Agr. Sci.* **73,** 261–266.

El Rahman, A. A. Abd., El Gamassy, A. M., and Mandour, M. S. (1968). Water economy of *Agave sisalana* under desert conditions. *Flora* (*Jena*) **B157,** 355–378.

Emmert, F. H., and Howlett, S. (1953). Electrolytic determination of the resistance of fifty-five apple varieties to low temperatures. *Proc. Amer. Soc. Hort. Sci.* **62,** 311–318.

Emmett, J. M., and Walker, D. A. (1969). Thermal uncoupling in chloroplasts. *Biochim. Biophys. Acta* **180,** 424–425.

Engelbrecht, L., and Mothes, K. (1960). Kinetin als Faktor der Hitzeresistenz. *Ber. Deut. Bot. Ges.* **73,** 246–257.

Engelbrecht, L., and Mothes, K. (1964). Weitere Untersuchungen zur experimentelle Beeinflussung der Hitzewirkung bei Blättern von *Nicotiana rustica*. *Flora* (*Jena*) **154,** 279–298.

Epel, B., and Butler, W. L. (1969). Cytochrome a: destruction by light. *Science* **166,** 621–622.

Eshel, Y., and Waisel, Y. (1965). The salt relations of *Prosopis farcata* (Banks et Sol.) Eig. *Isr. J. Bot.* **14,** 50.

Esterak, K. B. (1935). Resistenz-Gradienten in Elodea Blättern. *Protoplasma* **23,** 367–383.

Ewart, A. J. (1898). The action of cold and of sunlight upon aquatic plants. *Ann. Bot.* **12,** 363–397.

Ewart, M. H., Siminovitch, D., and Briggs, D. R. (1953). Studies on the chemistry of the living bark of the black locust tree in relation to frost hardiness. VI. Amylase and phosphorylase systems of the bark tissues. *Plant Physiol.* **28,** 629–644.

Farrant, J., and Woolgar, A. E. (1970). Possible relationships between the physical properties of solutions and cell damage during freezing. *In* "The Frozen Cell" (G. E. Wolstenholme and M. O'Connor, eds.), pp. 97–130. Churchill, London.

Fedorova, G. M. (1967). The effect of growth rate and degree of resistance to unfavorable conditions on the low-frequency electrical resistance of plants. *Uch. Zap. Mosk. Obl. Ped. Inst.* **169,** 166–173.

Fedoseeva, G. P. (1966). The adaptation of photosynthesis in cucumbers to high temperatures in protected sites. *Uch. Zap. Ural Gos. Univ.* **58,** 67–73.

Feinstein, R. N., Sacher, G. A., Howard, J. B., and Braun, J. T. (1967). Comparative heat stability of blood catalase. *Arch. Biochem. Biophys.* **122,** 338–343.

Feldman, N. L. (1962). The influence of sugars on the cell stability of some higher plants to heating and high hydrostatic pressure. *Tsitologya* **4,** 633–643.

Feldman, N. L. (1966). Increase in the thermostability of urease in the thermal hardening of leaves. *Dokl. Akad. Nauk. SSSR* **167,** 946–949.

Feldman, N. L. (1968). The effect of heat hardening on the heat resistance of some enzymes from plant leaves. *Planta* **78,** 213–225.

Feldman, N. L., and Lutova, M. I. (1963). Variations de la thermostabilité cellulaire des algues en fonction des changements de la température du millieu. *Cah. Biol. Mar.* **4,** 436–458.

Feldman, N. L., and Kamentseva, I. E. (1967). Urease thermostability in leaf extracts of two Leucojum species with different periods of vegetation. *Tsitologiya* **9,** 886–889.

Feldman, N. L., Kamentseva, I. E., and Yurashevskaya, K. N. (1966). Acid phosphatase thermostability in the extracts of cucumber and wheat seedling leaves after heat hardening. *Tsitologya* **8,** 755–759.

Feldman, A. L., Kobeleva, S. M., and Bespal'ko, E. N. (1968). Effect of tetramethylthiuramide disulfide on the properties of the potato. *Izv. Vyssh. Ucheb. Zaved. Pishch. Tekhnol.* **5,** 15–18.

Feldmann, A. 1969. Concerning x-ray stimulation. II. On the effect of the length of the day on the promotion of growth of *Lemna minor* L. which is induced by x-rays. *Radiat. Bot.* **9,** 459–471.

Fennema, O. (1966). An over-all view of low temperature food preservation. *Cryobiology* **3,** 197–213.

Ferguson, W. S. (1966). Salt-induced changes in the composition of lipid classes in barley roots. *Can. J. Plant Sci.* **46,** 639–646.

Ferri, M. G. (1953). Water balance of plants from the "Caatinga." *Rev. Brasil. Biol.* **13,** 237–244.

Ferri, M. G. (1955). Contribuicao ao Conhecimento da Ecologiea do Cerrado e da Caatings. *Bol. Fac. Fil. Cien. Letras. Univ. Sao Paulo Bot.* **12,** 170pp.

Ferri, M. G. (1960). Contribution to the knowledge of the ecology of the "Rio Negro Caatinga" (Amazon). *Bull. Res. Counc. Isr. Sect.***D8,** 195–208.

Feughelmann, M. (1966). Sulphydryl-disulfide interchange and the stability of keratin structure. *Nature (London)* **211,** 1259–1260.

Field, C. P. (1939). Low temperature injury to fruit blossom. I. On the damage caused to fruit blossom by varying degrees of cold. *Ann. Rep. (26th year) 1938. East Malling Res. Sta.,* pp. 127–138.

Filippovich, Yu, B., and Strashnova, M. L. (1969). Dynamics of soluble carbohydrates in the buds, leaves, and shoots of the mulberry tree in relation to its frost resistance. *Fiziol. Rast.* **16,** 49–54.

Fishbein, W. N., and Stowell, R. E. (1968). Studies on the mechanism of freezing damage to mouse liver using a mitochondrial enzyme assay. I. Temporal localization of the injury phase during slow freezing. *Cryobiology* **4,** 283–289.

Fleischmann, L. (1959). The action of refrigeration on some biochemical and enzymatic aspects of seeds in the course of germination. *Ric. Sci.* **29,** 287–294.

Flohe, L., and Brand, I. (1969). Kinetics of glutathione peroxidase. *Biochim. Biophys. Acta* **191,** 541–549.

Flowers, T. J., and Hanson, J. B. (1969). The effect of reduced water potential on soybean mitochondria. *Plant Physiol.* **44,** 939–945.

Foard, D. E., and Haber, A. H. (1961). Anatomic studies of gamma-irradiated wheat growing without cell division. *Amer. J. Bot.* **48,** 438–446.

Foard, D. E., and Haber, A. H. (1970). Physiologically normal senescence in seedlings grown without cell division after massive γ-irradiation of seeds. *Radiat. Res.* **42,** 372–380.

Fomenko, B. S., and Medvedev, A. I. (1969). Effect of gamma-irradiation of seeds of different plants on the metabolism of tyrosine, phenylalanine, and 3,4-dioxyphenylalanine in the shoots. *Radiobiologiya* **9,** 129–133.

Forssberg, A., and Nybom, N. (1953). Combined effects of cysteine and irradiation on growth and cytology of *Allium cepa* roots. *Physiol. Plant.* **6,** 78–95.

Forssberg, A., Brümmer, R., and Pehap, A. (1967). Gas exchange, growth, and biochemical reactions in *Phycomyces Blakesleeanus. Radiat. Bot.* **7,** 507–512.

Foy, C. D., Burns, G. R., Brown, J. C., and Fleming, A. L. (1965). Differential aluminum tolerance of 2 wheat varieties associated with plant-induced pH changes around their roots. *Soil Sci. Soc. Amer. Proc.* **29,** 64–67.

Foy, C. D., Armiger, W. H., Fleming, A. L., and Lewis, C. F. (1967). Differential tolerance of cotton varieties to an acid soil high in exchangeable aluminum. *Agron. J.* **59,** 415–418.

Foye, W. O., and Mickles, J. (1965). Metal binding as a mechanism of radioprotection by the mercaptan bases. *Progr. Biochem. Pharmacol.* **1,** 152–160.

Franco, C. M. (1958). "Influence of Temperature on Growth of Coffee Plant", 24 pp. IBEC Res. Inst. Rockefeller Plaza, New York.

Franco, C. M. (1961). Lesao do colo do cafeeiro, causada pelo calor. *Bragantia* **20,** 645–652.

Franco, C. M., and Bacchi, O. (1960). Estudos sobre a conservaçao des sementes. VII. Fumo. *Bragantia* **19,** 105–107.

Friedberg, F., Long, J. E., and Brecher, A. S. (1969). The protection of alpha-Chymotrypsin by hydrocinnamate from denaturation by ethanol and urea. *Proc. Soc. Exp. Biol. Med.* **130,** 1046–1047.

Friedman, H., Lu, P., and Rich, A. (1969). Ribosomal subunits produced by cold sensitive initiation of protein synthesis. *Nature (London)* **223,** 909–913.

Friedman, S. M. (1968). Protein-synthesizing machinery of thermophilic bacteria. *Bacteriol. Rev.* **32,** 27–38.

Frischenschlager, B. (1937). Versüche uber die Keimstimmung an einigen Gemüsearten. *Gartenbauwissenschaft* **11**, 159–166.

Fritz-Niggli, H., and Blattmann, H. (1969). Verschiedene mechanismen der Strahlen-Inaktivierung im Abhängigkeit van der Dosis und physiologischen Parameters. *Biophysik* **6**, 46–62.

Fritzsche, G. (1933). Untersuchungen über die Gewebetemperaturen von Strandpflanzen unter dem Einfluss der Insolation. *Bot. Centr. Beih.* **50**, 251–322.

Fuchs, W. H. (1935). Worauf beruht die Erhöhung der Kälteresistenz durch reichliche Kaliernährung. *Ernaehr. Pflanze* **31**, 233–234.

Füchtbauer, W. (1957a). Trockenresistenzsteigerung nach osmotischer Adaptation bei Saccharomyces und Chlorella. *Arch. Mikrobiol.* **26**, 209–230.

Füchtbauer, W. (1957b). Über den Zusammenhang der Trockenresistenz einiger Einzeller mit ihrem Retentionswasser und Mineralstoffgehalt. *Arch. Mikrobiol.* **26**, 231–253.

Füchtbauer, W., and Simonis, W. (1961). On the effect of β-rays on the photosynthetic phosphorylation and the Hill reaction of isolated chloroplasts. *Z. Naturforsch.* **B16**, 39–43.

Fujil, T., and Kondo, N. (1969). Changes in the levels of nicotinamide nucleotides and in activities of NADP-dependent dehydrogenases after a brief illumination of red light. *Develop. Growth Differ.* **11**, 40–45.

Furr, J. R., and Ream, C. L. (1968). Salinity effects on growth and salt uptake of seedlings of the date, *Phoenix dactylifera* L. *Proc. Amer. Soc. Hort. Sci.* **92**, 268–273.

Futrell, M. C., Lyles, W. E., and Pilgrim, A. J. (1962). Ascorbic acid and cold hardiness in oats. *Plant Physiol. Suppl.* **37**, 70.

Gadgil, R. L. (1969). Tolerance of heavy metals and the reclamation of industrial waste. *J. Appl. Ecol.* **6**, 247–259.

Gaff, D. F. (1966). The Sulphydryl-Disulphide hypothesis in relation to desiccation injury of cabbage leaves. *Aust. J. Biol. Sci.* **19**, 291–299.

Gale, J. (1961). Studies on plant antitranspirants. *Physiol. Plant.* **14**, 777–786.

Gale, J., Kohl, H. C., and Hagan, R. M. (1967). Changes in the water balance and photosynthesis of onion, bean, and cotton plants under saline conditions. *Physiol. Plant.* **20**, 408–420.

Galston, A. W., and Hand, M. E. (1949). Adenine as a growth factor for etiolated peas and its relation to the thermal inactivation of growth. *Arch. Biochem. Biophys.* **22**, 434–443.

Galston, A. W., and Kaur, R. (1959). An effect of auxins on the heat coagulability of the proteins of growing plant cells. *Proc. Nat. Acad. Sci. U.S.* **45**, 1587–1590.

Galston, A. W., Kaur, R., Maheshwari, N., and Maheshwari, S. C. (1963). Pectin-protein interaction as a basis for auxin-induced alteration of protein heat coagulability. *Amer. J. Bot.* **50**, 487–494.

Garibaldi, J. A., Donovan, J. W., Davis, J. G., and Cimino, S. L. (1968). Heat denaturation of the ovomucin-lysozyme electrostatic complex—a source of damage to the whipping properties of pasteurized egg whites. *J. Food Sci.* **33**, 514–524.

Gaskin, M. H. (1959). Effects of certain plant growth regulators upon cold hardiness of lychees. *Proc. Fl. State Hort. Soc.* **72**, 353–356.

Gassner, G., and Grimme, C. (1913). Beiträge zur Frage der Frosthärte der Getreidepflanzen. *Ber. Deut. Bot. Ges.* **31**, 507–516.

Gates, C. T., and Bonner, J. (1959). The response of the young tomato plant to a brief period of water shortage. IV. Effects of water stress on the ribonucleic acid metabolism of tomato leaves. *Plant Physiol.* **34**, 49–55.

Gates, C. T., and Muirhead, W. (1967). Studies of the tolerance of Atriplex species. I. Environmental characteristics and plant response of *A. vesicaria*, *A. nummularia* and *A. semibaccata*. *Aust. J. Exp. Agr. Anim. Husb.* **7,** 39–49.

Gates, C. T., Haydock, K. P., and Little, I. P. (1966a). Response to salinity in Glycine. I. *G. javanica*. *Aust. J. Exp. Agr. Anim. Husb.* **6,** 261–265.

Gates, C. T., Haydock, K. P., and Claringbold, P. J. (1966b). Response to salinity in Glycine. II. Differences in cultivars of *Glycine javanica* in dry weight, nitrogen, and water content. *Aust. J. Exp. Agr. Anim. Husb.* **6,** 374–379.

Gates, C. T., Haydock, K. P., and Robins, M. F. (1970). Response to salinity in Glycine. IV. Salt concentration and the content of phosphorus, potassium, sodium and chloride in cultivars of *Glycine wightii* (*G. javanica*). *Aust. J. Exp. Agr. Anim. Husb.* **10,** 99–110.

Gates, D. M. (1963). Leaf temperature and energy exchange. *Arch. Meteorol. Geophys. Bioklimatol. Ser.* **B12,** 321–336.

Gates, D. M. (1965). Heat transfer in plants. *Sci. Amer.* **213,** 76–84.

Gates, D. M., and Tantraporn, W. (1952). The reflectivity of deciduous trees and herbaceous plants in the infrared to 25 microns. *Science* **115,** 613–616.

Gäumann, E., and Jaag, O. (1936). Untersuchungen über die pflanzliche Transpiration. *Ber. Schweiz. Bot. Ges.* **45,** 411–518.

Gauch, H. G., and Eaton, F. M. (1942). Effect of saline substrate on hourly levels of carbohydrates and inorganic constituents of barley plants. *Plant Physiol.* **17,** 422–434.

Gauch, H. G., and Wadleigh, C. H. (1944). Effects of high salt concentrations on growth of bean plants. *Bot. Gaz.* (*Chicago*) **105,** 379–387.

Gauhl, E. (1970). Leaf factors affecting the rate of light saturated photosynthesis in ecotypes of *Solanum dulcamara*. *Carnegie Washington Year.* **68,** 633–636.

Gaur, Y. D. (1968). Preliminary studies on titrable acidity in xerophytic plants: *Salvadora persica* Linn. and *Prosopis juliflora* D.C. *Experientia* **24,** 239–240.

Gaur, B. K., Joshi, R. K., and Notani, N. K. (1969). Effect of heavy water on radiosensitivity of maize and barley seeds. *Radiat. Bot.* **9,** 61–67.

Gaur, B. K., Joshi, R. K., and Joshi, V. G. (1970). Potentiation of gamma-radiation effect by 2,4-dinitrophenol in barley seeds. *Radiat. Bot.* **10,** 29–34.

Gausman, H. W., and Cardenas, R. (1969). Effect of leaf pubescence of *Gynura aurantiaca* on light reflectance. *Bot. Gaz.* (*Chicago*) **130,** 158–162.

Gautreau, J. (1970). Comparative study of relative transpiration in two varieties of groundnut. *Oleagineux* **25,** 23–28.

Geiger, D. R. (1969). Chilling and translocation inhibition. *Ohio J. Sci.* **69,** 356–366.

Gej, B. (1962). Researches on the resistance of two varieties of spring wheat to periodical water deficit. *Acta Agrobot.* **11,** 31–46.

Genevès, L. (1955). Recherches sur les effets cytologiques du froid. *Rev. Cytol. Biol. Veg.* **16,** 1–207.

Genevès, L. (1957). Sur le role des ecailles dans la résistance au froid des bourgeons de Marronier: *Aesculus Hippocastanus*. *C.R. Acad. Sci.* (*Paris*) **244,** 2083–2085.

Genkel, P. A., and Oknina, E. Z. (1954). Diagnosis of plant frost resistance by the depth of the rest of the tissues and cells. *Akad. Nauk. SSSR Timiriazev Inst. Plant Physiol. Moscow,* pp. 1–34.

Genkel, P. A., and Pronina, N. D. (1968). Factors underlying the dehydration resistance of poikiloxerophytes. *Fiziol. Rast.* **15,** 84–92.

Genkel, P. A., and Tsvetkova, I. V. (1955). Increasing the heat resistance of plants. *Dokl. Akad. Nauk SSSR* **102,** 383–386.

Genkel, P. A., and Zhivukhnia, G. M. (1959). The process of protoplasm isolation as the second phase of hardening of winter wheat. *Dokl. Bot. Sci.* **127,** 216–219.

Genkel, P. A., Badanov, K. A., and Andreeva, I. I. (1967a). Significance of respiration for the water content of plant cells under drought conditions. *Fiziol. Rast.* **14,** 494–499.

Genkel, P. A., Satarova, N. A., and Tvorus, E. K. (1967b). Effect of drought on protein synthesis and the state of ribosomes in plants. *Fiziol. Rast.* **14,** 898–907.

George, L. Y. (1967). Accumulation of sodium and calcium by seedlings of some cereal crops under saline conditions. *Agron. J.* **59,** 297–299.

Gerloff, E. D., Richardson, T., and Stahmann, M. A. (1966). Changes in fatty acids of alfalfa roots during cold hardening. *Plant Physiol.* **41,** 1280–1284.

Gerloff, E. D., Stahmann, M. A., and Smith, D. (1967). Soluble proteins in alfalfa roots as related to cold hardiness. *Plant Physiol.* **42,** 895–899.

Geslin, H. (1939). La lutte contre les gelées et les seuils de résistance des principales cultures fruitières. *Ann. Epiphyt. Phytogenet.* **5,** 7–16.

Gessner, F. (1952). Der Druck in seiner Bedeutung fur das Wachstums submerser Wasserpflanzen. *Planta* **40,** 391–397.

Gessner, F. (1956a). Die Wasseraufnahme durch Blätter und Samen. Handb. *Pflanzen-physiol.* **3,** 215–246.

Gessner, F. (1956b). Wasserspeicherung und Wasserverscheibung. *Handb. Pflanzen-physiol.* **3,** 247–256.

Gessner, F., and Hammer, L. (1968). Exosmosis and "free space" in marine benthic algae. *Mar. Biol.* **2,** 89–91.

Getman, F. H., and Daniels, F. (1937). "Outlines of Theoretical Chemistry." John Wiley, New York.

Ghetie, V., and Buzila, L. (1964a). Cryoprecipitation in the cytoplasmic fluids of some seeds. *Rev. Roum. Biochim.* **1,** 41–49.

Ghetie, V., and Buzila, L. (1964b). Localization of cryoproteins in seeds. *Rev. Roum. Biochim.* **1,** 293–301.

Ghisleni, P. L. (1956). Further research concerning the action of ultrasounds, both of continuous and impulse type; Effect on asparagus seeds (*Asparagus officinalis L.*). *Allionia* **3,** 69–74.

Giles, C. H., and McKay, R. B. (1962). Studies in hydrogen bond formation. XI. Reactions between a variety of carbohydrates and proteins in aqueous solutions. *J. Biol. Chem.* **237,** 3288–3392.

Gilet, R., and Terrier, J. (1969). The effect of the process of restoration makes it possible to distinguish acute and subacute irradiations in Chlorella. *C.R. Acad. Sci.* (*Paris*) **D269,** 383–385.

Gilles, E. (1940). Effets des rayons ultra-violets sur les végétaux supérieurs. *C.R. Acad Agr.* (*Paris*). **26,** 38–42.

Gindel, I. (1969a). Stomatal number and size as related to soil moisture in tree xerophytes in Israel. *Ecology* **50,** 263–267.

Gindel, I. (1969b). Stomata constellation in the leaves of cotton, maize, and wheat plants as a function of soil moisture and environment. *Physiol. Plant* **22,** 1143–1151.

Gingrich, J. R., and Russell, M. B. (1956). Effect of soil moisture tension and oxygen concentration on the growth of corn roots. *Agron. J.* **48,** 517–520.

Gingrich, J. R., and Russell, M. B. (1957). A comparison of effects of soil moisture tension and osmotic stress on root growth. *Soil Sci.* **84,** 185–194.

Ginzburg, M. (1969). The unusual membrane permeability of two halophilic unicellular organisms. *Biochim. Biophys. Acta* **173,** 370–376.

Giorgi, M. C. Di., Fichera, P., and Tropea, M. (1967). New considerations on the use of saline water: III. The influence of the potassium-sodium relationship on a saline-sensitive culture (*Citrus aurantium*). *Agrochimica* **11,** 166–175.

Giosan, N., Biucan, D., and Lupas, V. (1962). Contribution to the study of the absorption of micronutrients by winter wheat at low temperatures. *Stud. Cercet. Agron.* **12,** 33–36.

Giraud, G. (1958). Sur la vitesse de croissance d'une Rhodophycee monocellulaire marine, le *Rhodosorus marinus* Geitler, cultivée en milieu synthétique. *C. Rend. Acad. Sci.* (*Paris*) **246,** 3501–3504.

Goas, M. (1966). On the nitrogen metabolism of halophytes: Study of amino acids and free amides. *Bull. Soc. Fr. Physiol. Veg.* **11,** 309–316.

Goas, M. (1968). Contribution to the study of nitrogenous metabolism of halophytes. Amino acids and free amides of *Aster tripolium L.* in aquaculture. *C.R. Acad. Sci.* (*Paris*) **D265,** 1049–1052.

Godman, R. M. (1959). Winter sunscald of yellow birch. *J. Forest.* **57,** 368–369.

Godnev, T. N., Khodasevich, E. V., and Arnautova, A. I. (1966). Biosynthesis of pigments at below freezing temperatures in lichens and hibernating plants. *Dokl. Akad. Nauk. SSSR* **167,** 451–453.

Göppert, H. R. (1830). "Über die Wärme-Entwickelung in den Pflanzen, deren Gefrieren und die Schützmittel gegen dasselbe." Max and Comp. Berlin.

Göppert, H. R. (1883). "Über das Gefrieren, Erfrieren der Pflanzen und Schutzmittel dagegen." Altes und Neues," pp. 1–87. Enke, Stuttgart.

Goetz, A., and Goetz, S. S. (1938). Vitrification and crystallization of protophyta at low temperatures. *Proc. Amer. Phil. Soc.* **97,** 361–388.

Goodale, G. L. (1885). "Gray's Botanical Textbook," Vol. 2: Physiological Botany. Ivison, Blakeman, Taylor, New York.

Goodin, R. (1969). On the cryoaggregation of bovine serum albumin. Thesis, Univ. of Missouri, Columbia, Missouri.

Goodin, R., and Levitt, J. (1970). Cryoaggregation of bovine serum albumin. Cryobiology **6,** 333–338.

Goodspeed, T. H. (1911). The temperature coefficient of the duration of life of barley grains. *Bot. Gaz.* (*Chicago*) **51,** 220–224.

Gorbanj, I. S. (1962). On the correlation between the growth and thermostability of plant cells. *Tsitologya* **4,** 182–192.

Gorbanj, I. S. (1968). Effect of beta-indoleacetic acid and maleic hydrazide on the elongation and thermostability of wheat coleoptile. *Tsitologiya* **10,** 76–87.

Gordon, L. Kh., and Bichurina, A. A. (1968). Can the P/O ratio serve as the main index of respiration efficiency of plants? *Dokl. Akad. Nauk. SSSR* **183,** 727–729.

Gordon, S. A. (1957). The effects of ionizing radiation on plants: Biochemical and physiological aspects. *Quart. Rev. Biol.* **32,** 3–14.

Gordon, S. A., and Weber, R. P. (1955). Studies on the mechanism of phytohormone damage by ionizing radiation. I. The radiosensitivity of indoleacetic acid. *Plant Physiol.* **30,** 200–210.

Gorin, G. (1965). Mercaptan-disulfide interchange and radioprotection. *Progr. Biochem. Pharmacol.* **1,** 142–151.

Goris, I. Ya. (1969). Respiratory metabolism of seeds germinating under salinization conditions. *Sel'skokhoz. Biol.* **4,** 246–251.

Gorke, H. (1906). Über chemische Vorgänge beim Erfrieren der Pflanzen. *Landwirtsch. Vers. Sta.* **65,** 149–160.

Gortner, R. A. (1938). "Outlines of Biochemistry." Wiley, New York.

Goryshina, T. K., and Kovaleva, J. A. (1967). Seasonal temperature adaptations of preverral, remoral ephemeroids. *Bot. Zh.* (*Leningrad*) **52,** 629–640.

Goujon, C., Maia, N., and Doussinault, G. (1968). Frost resistance in wheat. II. Reactions at the coleoptile stage studied under artificial conditions. *Ann. Amelior Plant.* **18,** 49–57.

Grahle, A. (1933). Vergleichende Untersuchungen über strukturelle und osmotische Eigenschaften der Nadeln verschiedener Pinus-Arten. *Jahrb. Wiss. Bot.* **78,** 203–294.

Grandfield, C. O. (1943). Food reserves and their translocation to the crown buds as related to cold and drought resistance in alfalfa. *J. Agr. Res.* **67,** 33–47.

Granhall, I. (1943). Genetical and physiological studies in interspecific wheat crosses. *Hereditas* **29,** 269–380.

Grant, D. W., Sinclair, N. A., and Nash, C. H. (1968). Temperature-sensitive glucose fermentation in the obligatory psychrophilic yeast *Candida gelida*. *Can J. Microbiol.* **14,** 1105–1110.

Grant, N. H. (1966). The biological role of ice. *Discovery* **27** (No. 8), 26–30.

Grant, N. H. (1969). Biochemical reactions in essentially nonaqueous systems. *Cryobiology* **6,** 182–187.

Grant, N. H., and Alburn, H. E. (1967). Reactions in frozen systems. VI. Ice as a possible model for biological structured-water systems. *Arch. Biochem. Biophys.* **118,** 292–296.

Greathouse, G. A., and Stuart, N. W. (1937). Enzyme activity in cold hardened and unhardened red clover. *Plant Physiol.* **12,** 685–702.

Greb, H. (1957). Der Einfluss tiefer Temperatur auf die Wasser und Stickstoffaufnahme der Pflanzen und ihre Bedeutung fur das "Xeromorphieproblem." *Planta* **48,** 523–563.

Greeley, A. W. (1901). On the analogy between the effect of loss of water and lowering of temperature. *Amer. J. Physiol.* **6,** 122–128.

Green, D. E., and MacLennan, D. H. (1969). Structure and function of the mitochondrial cristael membrane. *Bioscience* **19,** 213–222.

Green, D. G., Ferguson, W. S., and Warder, F. G. (1970). Effects of decenylsuccinic acid on ^{32}P uptake and translocation by barley and winter wheat. *Plant Physiol.* **45,** 1–3.

Greenham, C. G., and Daday, H. (1957). Electrical determination of cold hardiness in *Trifolium repens L.* and *Medicago sativa* L. *Nature* (*London*) **180,** 541–543.

Greenham, C. G., and Daday, H. (1960). Further studies on the determination of cold hardiness in *Trifolium repens* L. and *Medicago sativa* L. *Aust. J. Agr. Res.* **11,** 1–15.

Greenway, H. (1962). Plant response to saline substrates. I. Growth and ion uptake of several varieties of Hordeum during and after sodium chloride treatment. *Aust. J. Biol. Sci.* **5,** 16–38.

Greenway, H. (1963). Plant responses to saline substrates. III. Effect of nutrient concentration on the growth and ion uptake of *Hordeum vulgare* during a sodium chloride stress. *Aust. J. Biol. Sci.* **16,** 616–628.

Greenway, H. (1965a). Plant responses to saline substrates. IV. Chloride uptake by *Hordeum vulgare* as affected by inhibitors, transpiration, and nutrients in the medium. *Aust. J. Biol. Sci.* **18,** 249–268.

Greenway, H. (1965b). Plant response to saline substrates. VII. Growth and ion uptake throughout plant development in two varieties of *Hordeum vulgare*. *Aust. J. Biol. Sci.* **18,** 763–779.

Greenway, H. (1968). Growth stimulation by high chloride concentrations in halophytes. *Isr. J. Bot.* **17,** 169–177.

Greenway, H., and Rogers, A. (1963). Growth and ion uptake of *Agropyron elongatum* on saline substrates, as compared with a salt-tolerant variety of *Hordeum vulgare*. *Plant Soil* **18,** 21–30.

Greenway, H., and Thomas, D. A. (1965). Plant response to saline substrates. V. Chloride regulation in the individual organs of *Hordeum vulgare* during treatment with sodium chloride. *Aust. J. Biol. Sci.* **18,** 505–524.

Greenway, H., Gunn, A., Pitman, M. G., and Thomas, D. A. (1965). Plant response to saline substrates. VI. Chloride sodium and potassium uptake and distribution within the plant during ontogenesis of *Hordeum vulgare*. *Aust. J. Biol. Sci.* **18,** 525–540.

Greenway, H., Gunn, A., and Thomas, D. A. (1966). Plant responses to saline-substrates. VIII. Regulation of ion concentrations in salt-sensitive and halophytic species. *Aust. J. Biol. Sci.* **19,** 741–756.

Greenway, H., Klepper, B., and Hughes, P. G. (1968). Effects of low water potential on ion uptake and loss for excised roots. *Planta* **80,** 129–141.

Greenway, H., Hughes, P. G., and Klepper, B. (1969). Effects of water deficit on phosphorus nutrition of tomato plants. *Physiol. Plant.* **22,** 199–207.

Gregory, R. P. G., and Bradshaw, A. D. (1965). Heavy metal tolerance in populations of *Agrostis tenuis* Sibth. and other grasses. *New Phytol.* **64,** 131–143.

Grieve, B. J. (1953). The physiology of sclerophyll plants. *J. Roy. Soc. West Aust.* **39,** 31–45.

Grieve, B. J. (1956). Studies in the water relations of plants. I. Transpiration of western Australian (Swan Plain) sclerophylls. *J. Roy. Soc. West Aust.* **40,** 15–30.

Griggs, W. H., Harris, R. W., and Iwakiri, B. T. (1956). The effectiveness of growth-regulators in reducing fruit loss of Bartlett pears caused by freezing temperatures. *Proc. Amer. Soc. Hort. Sci.* **67,** 95–101.

Gross, R. E., Pugno, P., and Dugger, W. M. (1970). Observations on the mechanism of copper damage in Chlorella. *Plant Physiol.* **46,** 183–185.

Grossweiner, L. I., and Usui, Y. (1970). The role of the hydrated electron in photoreduction of cystine in the presence of indole. *Photochem. Photobiol.* **11,** 53–56.

Groves, J. F. (1917). Temperature and life duration of seeds. *Bot. Gaz.* (*Chicago*) **63,** 169–189.

Gudkov, I. N., and Grodzinskii, D. M. (1968). Oxygen aftereffects from the irradiation of barley seeds under low temperature conditions. *Radiobiologiya* **8,** 883–887.

Guern, N., and Gautheret, R. (1969). The hardening of the tissues of the Jerusalem artichoke at high temperatures. *C.R. Acad. Sci.* (*Paris*) **D269,** 332–334.

Gunar, I. I., and Sileva, M. N. (1954). Variation of the sugar content of winter wheats during the hardening-off process. *Fiziol. Rast.* **1,** 141–145.

Gunning, B. E. S., Steer, M. W., and Cochrane, M. P. (1968). Occurrence, molecular structure, and induced formation of the "stromacentre" in plastids. *J. Cell Sci.* **3,** 445–456.

Gupta, R. K., Narayan, R., and Gollukota, K. G. (1970). Differentiation between heat resistance and octyl alcohol resistance of the cells of *Bacillus cereus* T. *Biochem. Biophys. Res. Commun.* **38,** 23–30.

Gusev, N. A., Khokhlova, L. P., Gordon, L. Kh., and Sedykh, H. V. (1969). Problems regarding water metabolism in plants. *Bot. Zh.* (*Leningrad*) **54,** 53–66.

Haas, W. (1969). Amino acid composition of proteins in sun and shade leaves of the copper beach. (*Fagus sylvatica* L. cv. *Atropunicea*). *Planta* **87,** 95–101.

Haber, A. H., and Rothstein, B. E. (1969). Radiosensitivity and rate of cell division. Law of Bergonie and Tribondeau. *Science* **163,** 1338–1339.

Haber, A. H., Carrier, L., and Foard, D. E. (1961). Metabolic studies of gamma-irradiated wheat growing without cell division. *Amer. J. Bot.* **48,** 431–438.

Haberditzl, W. (1967). Enzyme activity in high magnetic fields. *Nature* (*London*) **213,** 72–73.

Haberlandt, F. (1875). "Wissenschaftlich-Praktische Untersuchungen auf dem Gebiete des Pflanzenbaues." Herausg, v. F. Haberlandt.

Hachisuka, Y., Tochikubo, K., Yokoi, Y., and Murachi, T. (1967). The action of dipicolinic acid and its chemical analogues on the heat stability of glucose dehydrogenase of *Bacillus subtilis* spores. *J. Biochem.* (*Tokyo*) **61,** 659–661.

Halevy, A. H. (1964). Effect of hardening and chemical treatment on drought resistance of gladiolus plants. *Proc. 16th Int. Hort. Congr.*, 1962 **4,** 252–258.

Halevy, A. H. (1967). Effect of growth retardants on drought resistance and longevity of various plants. *Proc. 17th Int. Hort. Congr.* **3,** 277–283.

Halevy, A. H., and Kessler, B. (1963). Increased tolerance of bean plants to soil drought by means of growth-retarding substances. *Nature* (*London*) **197,** 310–311.

Halevy, A. H., and Zieslin, N. (1968). The development and causes of petal blackening and malformation of Baccara rose flowers. *Floricultural Symp. Int. Soc. Hort. Sci. Vollebeck, Norway.*

Hall, D. O. (1963). Photosynthetic phosphorylation above and below 0°C. *Diss. Abstr.* **24,** 4962–4963.

Hall, E. J., and Cavanagh, J. (1969). The effect of hypoxia on recovery of sublethal radiation damage in Vicia seedlings. *Brit. J. Radiol.* **42,** 270–277.

Hall, T. C., McLeester, R. C., McCown, B. H., and Beck, G. E. (1969). Enzyme changes during acclimation. *Cryobiology* **6,** 263.

Halldal, P. (1961). Photoinactivation and their reversals in growth and mobility of the green alga Platymonas (Volvocales). *Physiol. Plant.* **14,** 558–575.

Hammer, L. (1968). Salinity and photosynthesis in marine plants. *Mar. Biol.* **1,** 185–190.

Hammonda, M., and Lange, O. L. (1962). Zur Hitzeresistenz der Blätter hoherer Pflanzen in Abhängigkeit von ihrem Wassergehalt. *Naturwissenschaften* **49,** 500–501.

Hanafusa, N. (1967). Denaturation of enzyme protein by freeze-thawing. p. 33. *In* "Cellular Injury and Resistance in Freezing Organisms" (E. Asahina, ed.). Inst. Low.Temp. Sci., Hokkaido, Japan.

Hanson, G. P., and Stewart, W. S. (1970). Photochemical oxidants: effect on starch hydrolysis in leaves. *Science* **168,** 1223–1224.

Harder, R. (1930). Beobachtungen über die Temperature der Assimilations-organe sommergrüner Pflanzen der algerischen Wüste. *Z. Bot.* **23,** 703–744.

Harder, R., Filzer, P., and Lorenz, A. (1932). Über versuche zur Bestimmung der Kohlensäureassimilation immergrüner Wüstenpflanzen während der Trockenzeit in Beni Unif (algerische Sahara). *Jahrb. Wiss. Bot.* **75,** 45–194.

Harder, W., and Veldkamp, H. (1968). Physiology of an obligately psychrophilic marine Pseudomonas sp. *J. Appl. Bacteriol.* **31,** 12–23.

Harper, J. E., and Paulsen, G. M. (1967). Changes in reduction and assimilation of nitrogen during the growth cycle of winter wheat. *Crop Sci.* **7,** 205–209.

Harrington, G. T., and Crocker, W. (1918). Resistance of seeds to desiccation. *J. Agr. Res.* **14,** 525–532.

Harris, P., and James, A. T. (1969). Effect of low temperature on fatty acid biosynthesis in seeds. *Biochim. Biophys. Acta* **187,** 13–18.

Harvey, R. B. (1918). Hardening process in plants and developments from frost injury. *J. Agr. Res.* **15,** 83–112.

Harvey, R. B. (1922). Varietal differences in the resistance of cabbage and lettuce to low temperatures. *Ecology* **3,** 134–139.

Harvey, R. B. (1923). Relation of the color of bark to the temperature of the cambium in winter. *Ecology* **4,** 391–394.

Harvey, R. B. (1930). Length of exposure to low temperature as a factor in the hardening process in tree seedlings. *J. Forest.* **28,** 50–53.

Hason-Porath, E., and Poljakoff-Mayber, A. (1969). The effect of salinity on the malic dehydrogenase of pea roots. *Plant Physiol.* **44,** 1031–1034.

Hason-Porath, E., and Poljakoff-Mayber, A. (1970). Effect of chloride and sulphate types of salinity on the nicotinamide-adenine-dinucleotides in pea root tips. *J. Exp. Bot.* **21,** 300–303.

Hatakeyama, I. (1957). Relation between the freezing point and the cold hardiness of plant tissue. *Physiol. Ecol.* (*Jap.*) **7,** 89–97.

Hatakeyama, I. (1960). The relation between growth and cold hardiness of leaves of *Camellia sinensis*. *Biol. J. Nara Women's Univ.* **10,** 65–69.

Hatakeyama, I. (1961). Studies on the freezing of living and dead tissues of plants, with special reference to the colloidally bound water in living state. *Mem. Coll. Sci. Kyoto Imp. Univ. Ser.* **B28,** 401–429.

Hatch, F. T., and Bruce, A. L. (1968). Amino-acid composition of soluble and membranous lipoproteins. *Nature* (*London*) **218,** 1166–1168.

Hatefi, Y., and Hanstein, W. G. (1970). Lipid oxidation in biological membranes. I. Lipid oxidation in submitochondrial particles and microsomes induced by chaotropic agents. *Arch. Biochem. Biophys.* **138,** 73–86.

Haurowitz, F. (1959). "Progress in Biochemistry since 1949" S. Karger, Basel; Interscience, New York.

Haxo, F. T., and Blinks, L. R. (1950). Photosynthetic action spectra of marine algae. *J. Gen. Physiol.* **33,** 389–422.

Hayashi, K., Kugimiya, M., and Fanatsu, M. (1968). Heat stability of lysozyme-substrate complex. *J. Biochem.* (*Tokyo*) **64,** 93–97.

Hayden, R. I., Moyse, C. A., Calder, F. W., Crawford, D. P., and Fensom, D. S. (1969). Electrical impedance studies on potato and alfalfa tissue. *J. Exp. Bot.* **20,** 177–200.

Hayes, H. K., and Aamodt, O. S. (1927). Inheritance of winter hardiness and growth habit in crosses of Marquis with Minhardi and Minturki wheats. *J. Agr. Res.* **35,** 223–236.

Heber, U. (1958a). Ursachen der Frostresistenz bei Winterweizen. I. Die Bedeutung der Zucker für die Frostresistenz. *Planta* **52,** 144–172.

Heber, U. (1958b). Ursachen der Frostresistenz bei Winterweizen. II. Die Bedeutung von Aminosäuren und Peptiden für die Frostresistenz. *Planta* **52,** 431–446.

Heber, U. (1959a). Beziehungen zwischen der Grosse von Chloroplasten und ihrem Gehalt an löslichen Eiweissen und Zuckern im Zusammenhang mit dem Frostresistenz Problem. *Protoplasma* **51,** 284–298.

Heber, U. (1959b). Ursachen der Frostresistenz bei Winterweizen. III. Die Bedeutung von Proteinen fur die Frostresistenz. *Planta* **54,** 34–67.

Heber, U. (1967). Freezing injury and uncoupling of phosphorylation from electron transport in chloroplasts. *Plant Physiol.* **42,** 1343–1350.

Heber, U. (1968). Freezing injury in relation to loss of enzyme activities and protection against freezing. *Cryobiol.* **5,** 188–201.

Heber, U. (1970). Proteins capable of protecting chloroplast membranes against freezing. *In* "The Frozen Cell" (G. E. W. Wolstenholme and M. O'Connor, eds.), pp. 175–188. J. and A. Churchill, London.

Heber, U., and Santarius, K. A. (1964). Loss of adenosine triphosphate synthesis caused by freezing and its relationship to frost hardiness problems. *Plant Physiol.* **39,** 712–719.

Hedlund, T. (1917). Über die Möglichkeit von der Ausbildung des Weizens im Herbst auf die Winterfestigkeit der verschiedenen Sorten zu schliessen. *Bot. Centr.* **135,** 222–224.

Heinricher, E. (1896). Über die Widerstandsfähigkeit der Adventivknospen von *Cystopderis bulbifera* (L.) Bernhardi gegen das Austrocknen. *Ber. Deut. Bot. Ges.* **14,** 234–244.

Heinrichs, D. H. (1959). Germination of alfalfa varieties in solutions of varying osmotic pressure and relationship to winter hardiness. *Can. J. Plant Sci.* **39,** 384–394.

Heitmann, P. (1968). A model for sulfhydryl groups in proteins. Hydrophobic interactions of the cysteine side chains in micelles. *Europ. J. Biochem.* **3,** 346–350.

Hellmuth, E. O. (1968). Eco-physiological studies on plants in arid and semi-arid regions in Western Australia: I. Autecology of *Rhagodia baccata* (Labill.) moq. *J. Ecol.* **56,** 319–344.

Henckel, P. A. (1964). Physiology of plants under drought. *Annu. Rev. Plant Physiol.* **15,** 363–386.

Henckel, P. A., and Margolin, K. P. (1948). Reasons of resistance of succulents to high temperatures. *Bot. Zh.* (*Leningrad*) **33,** 55–62.

Hendershott, C. H. (1962a). The influence of maleic hydrazide on citrus trees and fruits. *Proc. Amer. Soc. Hort. Sci.* **80,** 241–246.

Hendershott, C. H. (1962b). The responses of orange trees and fruits to freezing temperatures. *Proc. Amer. Soc. Hort. Sci.* **80,** 247–254.

Henn, S. W., and Ackers, G. K. (1969). Molecular sieve studies of interacting protein systems. V. Association of subunits of D-amino acid oxidase apoenzyme. *Biochemistry* **8,** 3829–3838.

Henrici, M. (1945). The effect of wilting on the direct assimilates of lucerne and other fodder plants. *S. Africa J. Sci.* **41,** 204–212.

Henrici, M. (1946). Effect of excessive water loss and wilting on the life of plants. *Union S. Afr. Dept. Agr. Forest Sci. Bull.* **256,** 1–22.

Henrici, M. (1955). Temperatures of karroo plants. *S. Afr. J. Sci.* **51,** 245–248.

Henriksen, T., and Sanner, T. (1967). Free radicals and their interactions in solid molecular mixtures. *Radiat. Res.* **32,** 164–175.

Hermbstädt, S. F. (1808). Über die Fähigkeit der lebenden Pflanzen im Winter Wärme zu erzeugen. *Ges. Naturforsch. Freunde Berlin zweiter Jahrgang,* pp. 316–319.

Heyne, E. G., and Laude, H. H. (1940). Resistance of corn seedlings to high temperatures in laboratory tests. *J. Amer. Soc. Agron.* **32,** 116–126.

Hickel, B. (1967). Contributions to the knowledge of a xerophilic water plant, *Chamaegigas intrepidus* Dtr. from southwest Africa. *Int. Rev. Gesamten Hydrobiol.* **53,** 361–400.

Highkin, H. R. (1959). Effect of vernalization on heat resistance in two varieties of peas. *Plant Physiol.* **34,** 643–644.

Highkin, H. R. 1967. Effect of temperature on formation of peroxidase enzymes. *Plant Physiol. Supp.* **42,** 5–16.

Higinbotham, N., and Mika, E. S. (1954). The effect of x-rays on uptake and loss of ions by potato tuber tissue. *Plant Physiol.* **29,** 174–177.

Hilbrig, H. (1900). Ueber den Einfluss supramaximaler Temperatur auf das Wachstum der Pflanzen. Inaug. Diss., Universität Leipzig, Leipzig.

Hildreth, A. C. (1926). Determination of hardiness in apple varieties and the relation of some factors to cold resistance. *Minn. Agr. Exp. Sta. Tech. Bull.*, p. 42.

Hilliard, J. H., and West, S. H. (1970). Starch accumulation associated with growth reduction at low temperatures in a tropical plant. *Science* **168,** 494–496.

Hindak, F., and Komarek, J. (1968). Cultivation of the cryosestonic alga *Koliella tatrae* (Kol) Hind. *Biol. Plant.* **10,** 95–97.

Hirsch, H. M. (1954). Temperature-dependent cellulose production by *Neurospora crassa* and its ecological implications. *Experientia* **10,** 180–182.

Hluchovsky, B., and Srb, V. (1963). Veränderungen der Zellpermeabilität bei *Allium cepa.* L. nach Rontgenbestrahlung. *Biol. Zentralbl.* **82,** 73–94.

Hock, B. (1969). Inhibition of isocitrate lyase in watermelon seedlings by white light. *Planta* **85,** 340–350.

Hodges, J. D., and Lorio, P. L., Jr. (1969). Carbohydrate and nitrogen fractions of the inner bark of loblolly pines under moisture stress. *Can. J. Bot.* **47,** 1651–1657.

Höfler, K. (1942a). Über die Austrocknungsfähigkeit des Protoplasmas. *Ber. Deut. Bot. Ges.* **60,** 94–107.

Höfler, K. (1942b). Über die Austrocknungsgrenzen des Protoplasmas. *Anz. Akad. Wiss. Wien. Math. Naturwiss. Kl.* **79,** 56–59.

Höfler, K. (1945). Über Trockenhartung und Hartungsgrenzen des Protoplasmas einiger Lebermoose. *Anz. Akad. Wiss. Wien. Math. Naturwiss. Kl.* **82,** 5–8.

Höfler, K. (1950). Über Trockenhärtung des Protoplasmas. *Ber. Deut. Bot. Ges.* **63,** 3–10.

Höfler, K., Migsche, H., and Rottenberg, W. (1941). Über die Austrocknungsresistenz landwirtschaftlicher Kulturpflanzen. Forschungsdienst **12,** 50–61.

Hoffman, H. (1857). "Witterung und Wachsthum, oder Grundzüge der Pflanzenklimatologie," pp. 312–334. Leipzig.

Hoffman, P., and Sachert, H. (1967). Effect of urea on the development of *Salicornia brachystachya.* A contribution to the halophyte problem. *Ber. Deut. Bot. Ges.* **80,** 437–446.

Hoffmann, P. (1968). Pigment content and gas exchange in desiccated and resaturated leaves of myrothamnus. *Photosynthetica* **2,** 245–252.

Hofstee, B. H. J., and Bobb, D. (1968). Heat denaturation of chymotrypsinogen A in the presence of polyanions. *Biochem. Biophys. Acta* **168,** 564–566.

Holle, H. (1915). Untersuchungen über Welken, Vertrocknen und Wiederstraffwerden. *Flora* **108,** 73–126.

Holm-Hansen, O. (1967). Factors affecting the viability of lyophilized algae. *Cryobiology* **4,** 17–23.

Holt, E. C., Faubion, J. L., Allen, W. W., and McBee, G. G. (1967). Arsenic translocation in nutsedge tuber systems and its effect on tuber viability. *Weeds* **15,** 13–15.

Holubowicz, T., and Boe, A. A. (1969). Development of cold hardiness in apple seedlings treated with gibberellic acid and abscisic acid. *J. Amer. Soc. Hort. Sci.* **94,** 661–664.

Holzl, J., and Bancher, E. (1968). Untersuchungen zur Frage des Uberlebens pflanzlicher Epidermen in flüssigen Stickstoff und Freon. *Z. Pflanzenphysiol.* **58,** 310–326.

Hopkins, E. F. (1924). Relation of low temperatures to respiration and carbohydrate changes in potato tubers. *Bot. Gaz.* (*Chicago*) **78,** 311–325.

Hopp, R. (1947). Internal temperatures of plants. *Proc. Amer. Soc. Hort. Sci.* **50,** 103–108.

Horn, T. (1923). Das gegenseitige Mengenverhältnis der Kohlenhydrate im Laubblatt in seiner Abhängigkeit vom Wassergehalt. *Bot. Arch.* **3,** 137–173.

Horovitz, C. T., and Waisel, Y. (1970). Cation activated ATPase and salt uptake by plants with different salt tolerance. *Proc. 18th Int. Hort. Congr. Tel Aviv, Isr.*, p. 173.

Hosokawa, T., and Kubota, H. (1957). On the osmotic pressure and resistance to desiccation of epiphytic mosses from a beech forest, southwest Japan. *J. Ecol.* **45,** 579–591.

Hostetter, H. P., and Hoshaw, R. W. (1970). Environmental factors affecting resistance to desiccation in the diatom *Stauroneis anceps. Amer. J. Bot.* **57,** 512–518.

Howard, A. (1968). The oxygen requirement for recovery in split-dose experiments with Oedogonium. *Int. J. Radiat. Biol.* **14,** 341–350.

Hsiao, T. C. (1970). Rapid changes in levels of polyribosomes in *Zea mays* in response to water stress. *Plant Physiol.* **46,** 281–285.

Hsiao, T. C., Acevedo, E., and Henderson, D. W. (1970). Maize leaf elongation: continuous measurements and close dependence on plant water status. *Science* **168,** 590–591.

Hubac, C. (1967). Increase in seedlings of resistance to desiccation by the preliminary effect of proline. *C.R. Acad. Sci.* (*Paris*) **D264,** 1286–1289.

Hubac, C., Guerrier, D., and Ferran, J. (1969). Résistance à la sécheresse du *Carex pachystylis* (J. Gay) plante du désert du Negev. *Oecol. Plant.* **4,** 325–346.

Huber, B. (1931). Die Trockenanpassungen in der Wipfelregion der Bäume und ihre Bedeutung für das Xerophytenproblem. *J. Ecol.* **19,** 283–291.

Huber, H. (1932). Einige Grundfragen des Wärmehaushalts der Pflanzen I. Die Ursache der hohen Sukkulenten-Temperaturen. *Ber. Deut. Bot. Ges.* **50,** 68–76.

Huber, H. (1935). Der Wärmehaushalt der Pflanzen. *Naturwiss. Landwirtsch.* **17,** 148.

Hudock, G. A., Mellin, D. B., and Fuller, R. C. (1965). Alternative forms of triosephosphate dehydrogenase in Chromatium. *Science* **150,** 776–777.

Humphries, E. C., Welbank, P. J., and Williams, E. D. (1967). Interaction of CCC and water deficit on wheat yield. *Nature* (*London*) **215,** 782.

Hunt, O. J. (1965). Salt tolerance in intermediate wheatgrass (*Agropyron intermedium*) *Crop Sci.* **5,** 407–409.

Hurd-Karrer, A. (1939). Antagonism of certain elements essential to plants toward chemically related toxic elements. *Plant Physiol.* **14,** 9–29.

Husain, I., May, L. H., and Aspinall, D. (1968). The effect of soil moisture stress on the growth of barley. IV. The response to pre-sowing treatment. *Aust. J. Agr. Res.* **19,** 213–220.

Huystee, R. B., Jachymuzyk, W., Tester, C. F., and Cherry, J. H. (1968). X-irradiation effects on protein synthesis and synthesis of messenger ribonucleic acid from peanut cotyledons. *J. Biol. Chem.* **243,** 2315–2320.

Hwang, Shuh-Wei, and Horneland, W. (1965). Survival of alga cultures after freezing by controlled and uncontrolled cooling. *Cryobiology* **1,** 305–311.

Hwang, Shuh-Wei, and Howells, A. (1968). Investigation of ultra-low temperatures for fungal cultures. II. Cryo-protection afforded by glycerol and dimethyl sulfoxide to 8 selected fungal cultures. *Mycologia* **60,** 622–626.

Hylmö, B. (1943). Disackaridbildning hos viktål vid kall väderlek. *Medd. N:R* **17** *Från Statens Trädgårdsförsök.*, pp. 1–37.

Ibanez, M. L. (1964). Role of cotyledon in sensitivity to cold of Cacao seed. *Nature (London)* **201,** 414–415.

Idso, S. B. (1968). Atmospheric and soil induced water stresses in plants and their effects on transpiration and photosynthesis. *J. Theoret. Biol.* **21,** 1–12.

Iljin, W. S. (1923a). Der Einfluss des Wassermangels auf die Kohlenstoff-assimilation durch die Pflanzen. *Flora (Jena)* **116,** 360–378.

Iljin, W. S. (1923b). Einfluss des Welkens auf die Atmung der Pflanzen. *Flora* **116,** 379–403.

Iljin, W. S. (1927). Uber die Austroknungsfähigkeit des lebenden Protoplasmas der vegetativen Pflanzenzellen. *Jahrb. Wiss. Bot.* **66,** 947–964.

Iljin, W. S. (1929a). Der Einfluss der Standortsfeuchtigkeit auf den osmotischen Wert bei Pflanzen. *Planta* **7,** 45–58.

Iljin, W. S. (1929b). Standortsfeuchtigkeit und der Zuckergehalt in den Pflanzen. *Planta* **7,** 59–71.

Iljin, W. S. (1930a). Die Ursache der Resistenz von Pflanzenzellen gegen Austrocknung. *Protoplasma* **10,** 379–414.

Iljin, W. S. (1930b). Der Einfluss des Welkens auf den Ab- und Aufbau der Starke in der Pflanze. *Planta* **10,** 170–184.

Iljin, W. S. (1931). Austrocknungsresistenz des Farnes *Notochlaena Marantae* R. Br. *Protoplasma* **13,** 322–330.

Iljin, W. S. (1932). Anpassung der Halophyten an konzentrierte Salzlösungen. *Planta* **16,** 352–366.

Iljin, W. S. (1933a). Über Absterben der Pflanzengewebe durch Austrocknung und über ihre Bewahrung vor dem Trockentode. *Protoplasma* **19,** 414–442.

Iljin, W. S. (1933b). Über den Kältetod der Pflanzen und seine Ursachen. *Protoplasma* **20,** 105–124.

Iljin, W. S. (1934). The point of death of plants at low temperatures. *Bull. Ass. Russe Rech. Sci. Prague Sect. Sci. Nat. Math.* **1** (6) (No. 4), 135–160.

Iljin, W. S. (1935a). The relation of cell sap concentration to cold resistance in plants. *Bull. Ass. Russe Rech. Sci. Prague, Sect. Sci. Nat. Math.* **13** 3(8), 33–55.

Iljin, W. S. (1935b). Lebensfähigkeit der Pflanzenzellen in trockenem Zustand. *Planta* **24,** 742–754.

Illert, H. (1924). Botanische Untersuchungen über Hitzetod und Stoffwechselgifte. *Bot. Arch.* **7,** 133–141.

Il'na, G. V., Kuznetsova, N. N., and Rydkii, S. G. 1968. Study of the radioprotectant properties of some antibiotics with regard to plants. *Vestn. Mosk. Univ. Ser. Biol. Pochvoved.* **23,** 56–63.

Imber, D., and Tal, M. (1970). Phenotypic reversion of Flacca, a wilty mutant of tomato, by abscisic acid. *Science* **169,** 592–593.

Imre, P. (1968). The salt tolerance of plants in the germination period and irrigation. *Agrokem. Talajtan* **17,** 61–76.

Imshenetskii, A. A., Komolova, K. S., Lysenko, S. V., and Gamulya, G. D. 1968. Effect of a high vacuum on the activity of some enzymes. *Dokl. Akad. Nauk. SSSR* **182,** 971–972.

Inglis, A. S., and Liu, T. Y. (1970). The stability of cysteine and cystine during acid hydrolysis of proteins and peptides. *J. Biol. Chem.* **245,** 112–116.

Ingraham, J. L. (1969). Factors which preclude growth of bacteria at low temperature. *Cryobiology* **6,** 188–193.

Irias, J. J., Olmsted, M. R., and Utter, M. F. (1969). Pyruvate carboxylase: Reversible inactivation by cold. *Biochemistry* **8,** 5136–5148.

Irmscher, E. (1912). Über die Resistenz der Laubmoose gegen Austrocknung und Kälte. *Jahrb. Wiss. Bot.* **50,** 387–449.

Irving, R. M. (1969a). Characterization and role of an endogenous inhibitor in the induction of cold hardiness in *Acer negundo. Plant Physiol.* **44,** 801–805.

Irving, R. M. (1969b). Influence of growth retardants on development and loss of hardiness of *Acer negundo. J. Amer. Soc. Hort. Sci.* **94,** 419–422.

Irving, R. M., and Lanphear, F. O. (1967a). Environmental control of cold hardiness in woody plants. *Plant Physiol.* **42,** 1191–1196.

Irving, R. M., and Lanphear, F. O. (1967b). Dehardening and the dormant condition in Acer and Viburnum. *Proc. Amer. Soc. Hort. Sci.* **91,** 699–705.

Irving, R. M., and Lanphear, F. O. (1968). Regulation of cold hardiness in *Acer negundo. Plant Physiol.* **43,** 9–13.

Isaac, W. E. (1933). Some observations and experiments on the drought resistance of *Pelvetia canaliculata. Ann. Bot.* **47,** 343–348.

Ishida, M., and Mizushima, S. (1969). The membrane ATPase of *Bacillus megaterium.* II. Purification of membrane ATPase and their recombination with membrane. *J. Biochem.* **66,** 133–138.

Itai, C., and Vaadia, Y. (1965). Kinetin-like activity in root exudate of water stressed sunflower. *Physiol. Plant.* **18,** 941–944.

Itai, C., and Vaadia, Y. (1968). The role of root cytokinins during water and salinity stress. *Isr. J. Bot.* **17,** 187–195.

Ivanov, S. M. (1939). Activity of growth process-principal factor in frost resistance of citrus plants. *C. R. Acad. Sci. URSS* **22,** 277–281.

Iwanoff, L. (1924). Über die Transpiration der Holzgewächse im Winter I. *Ber. Deut. Bot. Ges.* **42,** 44–49. II. pp. 210–218.

Jaarma, M. (1968). Studies on some chemical, enzymatical and physiological changes in plant material caused by irradiation. *Ark. Kemi.* **28,** 223–243.

Jacob, H. S., Brain, M. C., Dacil, J. V., Carrell, R. W., and Lehmann, H. (1968). Abnormal haem binding and globin SH group blockade in unstable haemoglobins. *Nature (London)* **218,** 1214–1217.

Jacobson, K. B. (1968). Alcohol dehydrogenase of Drosophila: interconversion of isoenzymes. *Science.* **159,** 324–325.

Jacoby, B. (1964). Function of bean roots and stems in sodium retention. *Plant Physiol.* **39,** 445–449.

Jacoby, B., and Dagan, J. (1969). Effects of age on sodium fluxes in primary bean leaves. *Physiol. Plant.* **22,** 29–36.

Jacoby, B., and Oppenheimer, H. R. (1962). Pre-sowing treatment of sorghum grains and its influence on drought resistance of the resulting plants. *Phyton* **19,** 109–113.

James, A. P., and Werner, M. M. (1969). β-mercaptoethanol-induced recovery of x-irradiated yeast cells. *Can. J. Genet. Cytol.* **11,** 848–856.

Janes, B. E. (1968). Effects of extended periods of osmotic stress on water relationships of pepper. *Physiol. Plant.* **21,** 334–345.

Jarvis, M. S. (1963). A comparison between the water relations of species with contrasting types of geographical distribution in the British Isles. *In* "The Water Relations of Plants" (A. J. Rutter, and F. W. Whitehead, eds.). Wiley, New York.

Jarvis, P. G., and Jarvis, M. S. (1963a). The water relations of tree seedlings. I. Growth and water use in relation to soil water potential. *Physiol. Plant.* **16,** 215–235.

Jarvis, P. G., and Jarvis, M. S. (1963b). The water relations of tree seedlings. IV. Some aspects of the tissue water relations and drought resistance. *Physiol. Plant.* **16,** 501–516.

Jefferies, R. L., and Willis, A. J. (1964). Studies on the calcicole-calcifuge habit. II. The influence of calcium on the growth and establishment of four species in soil and sand cultures. *J. Ecol.* **52,** 691–707.

Jenny, J. (1953). Trials of measuring winter temperatures of apple and pear buds. *Rev. Viticult. Arboricult.* **9,** 26–27.

Jensen, I. J. (1925). Winter wheat studies in Montana with special reference to winter killing. *J. Amer. Soc. Agron.* **17,** 630–631.

Jeremias, K. (1956). Zur Physiologie der Frosthärtung (Unter besonderer Berucksichtigung von Winterweizen). *Planta* **47,** 81–104.

Jockush, H. 1968. Stability and genetic variation of a structural protein. *Naturwissenschaften* **55,** 514–518.

Johansson, N.-O. (1970). Ice formation and frost hardiness in some agricultural plants. *Nat. Swed. Inst. Plant Protection Contrib.* **14** (132), 364–382.

Johansson, N.-O., and Krull, E. (1970). Ice formation, cell contraction, and frost killing of wheat plants. *Nat. Swed. Inst. Plant Protection Contrib.* **14**(131), 343–362.

Johansson, N.-O., and Torsell, B. (1956). Field trials with a portable refrigerator. *Acta Agr. Scand.* **6,** 81–99.

Johansson, N.-O., Albertson, C. E., and Månsson, T. (1955). Undersökningar över höstetets härdning och avhärdning. *Sver. Utsaedesfoeren. Tidsk.*, pp. 82–96.

Johnson, E. L. (1948). Response of *Kalanchoe tubiflora* to X-radiation. *Plant Physiol.* **23,** 544–556.

Johnson, M. K., Johnson, E. J., MacElroy, R. D., Speer, H. L., and Bruff, B. S. (1968). Effects of salts on the halophilic alga *Dunaliella viridis. J. Bacteriol.* **95,** 1461–1468.

Johnson, T. B., and Strobel, G. A. (1970). The active site on the phytotoxin of *Corynebacterium sepedonicum. Plant Physiol.* **45,** 761–764.

Johnson, W. H., and Henry, J. E. (1967). Response of germinating corn to temperature and pressure. *Trans. ASAE* (*Amer. Soc. Agr. Eng.*) **10,** 539–542.

Jollès, P. (1967). Rapports entre la structure et l'activité de quelques lysozymes. *Bull. Soc. Chim. Biol.* **49,** 1001–1012.

Jonard, R. (1968). Effect of 2,alkylating agents: methane methyl sulfonate and nitrogenous mustard on the Jerusalem artichoke tissues grown *in vitro*. Protection by cysteamine and restoration by kinetin. *Radiat. Bot.* **8,** 467–471.

Jorgensen, E. G. (1968). The adaptation of plankton Algae. II. Aspects of the temperature adaptation of *Skeletonema costatum. Physiol. Plant.* **21,** 423–427.

Julander, O. (1945). Drought resistance in range and pasture grasses. *Plant Physiol.* **20,** 573–599.

Jung, G. A. (1962). Effect of uracil, thiouracil, and guanine on cold resistance and nitrogen metabolism of alfalfa. *Plant Physiol.* **37,** 768–774.

Jung, G. A., and Smith, D. (1960). Influence of extended storage at constant low temperature on cold resistance and carbohydrate reserves of alfalfa and medium red clover. *Plant Physiol.* **35,** 123–125.

Jung, G. A., and Smith, D. (1962). Trends of cold resistance and chemical changes over winter in the roots and crowns of alfalfa and medium red clover. I. Changes in certain nitrogen and carbohydrate fractions. *Agron. J.* **53,** 359–364.

Jung, G. A., Shih, S. C., and Shelton, D. C. (1967a). Seasonal changes in soluble protein, nucleic acids, and tissue pH related to cold hardiness of alfalfa. *Cryobiology* **4,** 11–16.

Jung, G. A., Shih, S. C., and Shelton, D. C. (1967b). Influence of purines and pyrimidines on cold hardiness of plants. III. Associated changes in soluble protein and nucleic acid content and tissue pH (*Medicago sativa* varieties "Vernal" and "Caliverde") *Plant Physiol.* **42,** 1653–1657.

Just, L. (1877). Ueber die Einwirkung hoher Temperaturen auf die Erhaltung der Keimfähigeit der Samen. *Beitr. Biol. Pflanz.* **2,** 311–348.

Kacperska-Palacz, A., Blaziak, M., and Wcislinska, B. (1969). The effect of growth retardants CCC and B-9 on certain factors related to cold acclimation of plants. *Bot. Gaz.* (*Chicago*) **130,** 213–221.

Kaddah, M. T. (1962). Tolerance of berseem clover to salt. *Agron. J.* **54,** 421–425.

Kaddah, M. T., and Ghowail, S. I. (1964). Salinity effects on the growth of corn at different stages of development. *Agron. J.* **56,** 214–217.

Kadman, A. (1963). The uptake and accumulation of chloride in avocado leaves and the tolerance of avocado seedlings under saline conditions. *Proc. Amer. Soc. Hort. Sci.* **83,** 280–286.

Kärcher, H. (1931). Über die Kälteresistenz einiger Pilze und Algen. *Planta* **14,** 515–516.

Kahane, I., and Poljakoff-Mayber, A. (1968). Effect of substrate salinity on the ability for protein synthesis in pea roots. *Plant Physiol.* **43,** 1115–1119.

Kaho, H. (1921). Über die Beeinflussung der Hitzekoagulation des Pflanzenprotoplasmas durch Neutralsalze. I. *Biochem. Z.* **117,** 87–95.

Kaho, H. (1924). Über die Beeinflussung der Hitzekoagulation des Pflanzenplasmas durch die Salze der Erdalkalien. VI. *Biochem. Z.* **151,** 102–111.

Kaho, H. (1926). Über den Einfluss der Temperatur auf die koagulierende Wirkung einiger Alkalisalze auf das Pflanzenplasma. VIII. *Biochem. Z.* **167,** 182–194.

Kaku, S. (1963). Autocological studies on the hydrature accommodation in higher plants. *Sieboldia* **3,** 1–38.

Kaku, S., and Salt, R. W. (1968). Relation between freezing temperature and length of conifer needles. *Can J. Bot.* **46,** 1211–1213.

Kaloyereas, S. A. (1958). A new method of determining drought resistance. *Plant Physiol.* **33,** 232–233.

Kaltwasser, J. (1938). Assimilation und Atmung von Submersen als Ausdruck ihrer Entquellungsresistenz. *Protoplasma* **29,** 498–535.

Kamra, O. M. P., Kamra, K., Nilan, R. A., and Konzak, C. F. (1960). Radiation response of soaked barley seeds. I. Substances lost by leaching. *Hereditas* **46,** 152–172.

Kanwisher, J. (1957). Freezing and drying in intertidal algae. *Biol. Bull.* **113,** 275–285.

Kappen, L. (1966). Der Einfluss des Wassergehaltes auf die Widerstandsfähigkeit von Pflanzen gegegenüber hohen und tiefen Temperaturen, untersucht an Blättern einiger Farne und von *Ramonda myconi*. *Flora* **156,** 427–445.

Kappen, L. (1969a). Frost resistance of native halophytes in relation to their salt, sugar, and water content in summer and winter. *Flora Abt.* **B158,** 232–260.

Kappen, L. (1969b). Kälteverträglichkeit und Zuckergehalt von Salzpflanzen. *Ber. Deut. Bot. Ges.* **82,** 103–106.

Kappen, L., and Lange, O. L. (1968). Heat resistance of half-dried leaves of *Commelina africana*: Two research methods compared. *Protoplasma* **65,** 119–132.

Karmarkar, S. M., and Joshi, G. V. (1969). Effect of sand culture and sodium chloride on growth, physical structure and organic acid metabolism in *Bryophyllum pinnatum*. *Plant Soil* **30,** 41–48.

Kasarda, D. D., and Black, D. R. (1968). Thermal degradation of proteins studied by mass spectrometry. *Biopolymers* **6,** 1001–1004.

Katz, M., Ledingham, G. A., and Harris, A. E. (1939). "Effects of SO_2 on Vegetation." *Nat. Res. Counc. Can.*

Katz, S., and Reinhold, L. (1965). Changes in the electrical conductivity of Coleus tissue as a response to chilling temperatures. *Isr. J. Bot.* **13,** 105–114.

Kaufmann, M. R., and Ross, K. J. (1970). Water potential temperature and kinetin effects on seed germination in soil and solute systems. *Amer. J. Bot.* **57,** 413–419.

Kaul, B. L. (1970). Studies on radioprotective role of dimethyl sulfoxide in plants. *Radiat. Bot.* **10,** 69–78.

Kaul, R. (1966a). Relative growth rates of spring wheat, oats and barley under polyethylene glycol-induced water stress. *Can. J. Plant Sci.* **46,** 611–617.

Kaul, R. (1966b). Effect of water stress on respiration of wheat. *Can. J. Bot.* **44,** 623–632.

Kausch, W., and Ehrig, H. (1959). Beziehungen zwischen Transpiration und Wurzelwerk. *Planta* **53,** 434–448.

Keller, W., and Black, A. T. (1968). Preplanting treatment to hasten germination and emergence of grass seed. *J. Range Manage* **21,** 213–216.

Kemble, A. R., and MacPherson, H. T. (1954). Liberation of amino acids in perennial rye grass during wilting. *Biochem. J.* **58,** 46–49.

Kenefick, D. G., and Swanson, C. R. (1963). Mitochondrial activity in cold acclimated winter barley. *Crop Sci.* **3,** 202–205.

Kentzer, T. (1967). Further investigations concerning the influence of chlorocholine chloride (CCC) on the frost resistance of tomato plants. *Rocz. Nauk. Roln. Ser.* **A93,** 511–522.

Kerner von Marilaun, A. (1894). "The Natural History of Plants," Vol. 1, Part 2 (translated by F. W. Oliver), pp. 539–557. Holt, New York.

Kessler, B. (1961). Nucleic acids as factors in drought resistance of higher plants. *Recent Advan. Bot.* pp. 1153–1159.

Kessler, B., Engelberg, N., Chen, D., and Greenspan, H. (1964). Studies on physiological and biochemical problems of stress in higher plants. *Spec. Bull. Isr. Ministry Agr. Rehovot* **64,** 74 pp.

Kessler, W. (1935). Über die inneren Ursachen der Kälteresistenz der Pflanzen. *Planta* **24,** 312–352.

Kessler, W., and Ruhland, W. (1938). Weitere Untersuchungen über die inneren Ursachen der Kälteresistenz. *Planta* **28,** 159–204.

Kessler, W., and Ruhland, W. (1942). Über die inneren Ursachen der Kälteresistenz der Pflanzen. *Forschungsdienst* **16,** 345–351.

Ketellapper, H. J. (1969). Diurnal periodicity and plant growth. *Physiol. Plant* **22,** 899–907.

Ketellapper, H. J. and Bonner, J. (1961). The chemical basis of temperature responses in plants. *Plant Physiol. Suppl.* **36,** XXI.

Khan, A. W., Davidkova, E., and van den Berg, L. (1968). On cryodenaturation of chicken myofibrillar proteins. *Cryobiology* **7,** 184–188.

Khan, R. A., and Laude, H. M. (1969). Influence of heat stress during seed maturation on germinability of barley seed at harvest. *Crop Sci.* **9,** 55–58.

Khokhlova, L. P., Alekseeva, V. Ya., and Murav'eva, A. S. (1969). Effect of weak dehydration on the properties of amino acid-activating enzymes in wheat leaves. *Fiziol. Rast.* **16,** 66–70.

Kiefer, J. (1970). Effect of split-dose U.V. irradiation on diploid yeast. *Photochem. Photobiol.* **11,** 37–47.

Kiesselbach, T. A., and Ratcliff, J. A. (1918). Freezing injury of seed corn. *Nebr. Agr. Exp. Sta. Bull.* **163,** 1–16.

Kilen, T. C., and Andrew, R. H. (1969). Measurement of drought resistance in corn. *Agron. J.* **61,** 669–672.

Kilian, C., and Lemée, G. (1956). Les xerophytes: Leur economie d'eau. *Handb. Pflanzenphysiol.* **3,** 787–824.

Kim, C. M. (1958). Effect of saline and alkaline salts on the growth and internal components of selected vegetable plants. *Physiol. Plant.* **11,** 441–450.

Kim, S., and Paik, W. K. (1968). Effect of glutathione on ribonuclease. *Biochem. J.* **106,** 707–710.

Kinbacher, E. J. (1963). Relative high-temperature resistance of winter oats at different relative humidities. *Crop Sci.* **2,** 466–468.

Kinbacher, E. J. (1969). The physiology and genetics of heat tolerance. *In:* Physiological Limitations on Crop Production under Temperature and moisture Stress. (E. R. Lemon, *et al.*, eds.). Nat. Acad. Sci., Washington, D.C.

Kinbacher, E. J., Sullivan, C. Y., and Knull, H. R. (1967). Thermal stability of malic dehydrogenase from heat hardened *Phaseolus acutifolius.* Tepary. Buff. *Crop Sci.* **7,** 148–151.

Kinzel, H. (1963). Zellsaft-Analysen zum pflanzlichen Calcium-und Sauerestoffwechsel und zum Problem der Kalk-und Silikat-pflanzen. *Protoplasma* **57,** 522–555.

Kirkham, M. B., Gardner, W. R., and Gerloff, G. C. (1969). Leaf water potential of differentially salinized plants. *Plant Physiol.* **44,** 1378–1382.

Kislyuk, I. M. (1964). Influence of light on injury of *Cucumis sativus* leaves during cooling. *Dokl. Akad. Nauk. SSSR* **158,** 1434–1436.

Kisselew, N. N. (1928). Der Temperatureinfluss auf die Stärkehydrolyse in Mesophyll- und Schliesszellen. *Planta* **6,** 135–161.

Kisselew, N. N. (1935). Dürreresistenz und Saugkraft der Pflanzen. *Planta* **23,** 760–773.

Kitaura, K. (1967). Supercooling and ice formation in mulberry trees. *In* "Cellular Injury and Resistance in Freezing Organisms" (E. Asahina, ed.), pp. 143–156. Inst. Low Temp. Sci., Hokkaido.

Klages, K. H. (1926). Relation of soil moisture content to resistance of wheat seedlings to low temperatures. *J. Amer. Soc. Agron.* **18,** 184–193.

Klein, R. M., and Edsall, P. C. (1965). On the reported effects of sound (music) on the growth of plants (*Tagetes erecta*). *Bioscience* **15,** 125–126.

Klein, R. M., and Edsall, P. C. (1966). Substitution of redox chemicals for radiation in phytochrome-mediated photomorphogenesis. *Plant Physiol.* **41,** 949–952.

Klein, S., and Pollack, B. M. (1968). Cell fine structure of developing lima bean seeds related to seed desiccation. *Amer. J. Bot.* **55,** 658–672.

Kleinschmidt, M. G., and McMahon, V. A. (1970a). Effect of growth temperature on the lipid composition of *Cyanidium caldorium.* I. Class separation of lipids. *Plant Physiol.* **46,** 286–289.

Kleinschmidt, M. G., and McMahon, V. A. (1970b). Effect of growth temperature on the lipid composition of *Cyanidium caldorium.* II. Glycolipid and phospholipid components. *Plant Physiol.* **46,** 290–293.

Klingmuller, W. (1961). Radiation damage in *Vicia faba* seeds. *In* "Effects of Ionizing Radiation on Seeds, 1960. A symposium," pp. 67–74. Inter. Atomic Energy Agency, Vienna.

Klyshev, L. K., and Rakova, N. M. (1964). Effect of salinization of the substrate on the protein composition of the roots in pea shoots. *Tr. Bot. Inst. Akad. Nauk. Kaz. SSR* **20,** 156–165.

Kneen, E., and Blish, M. J. (1941). Carbohydrate metabolism and winter hardiness of wheat. *J. Agr. Res.* **62,** 1–26.

Knowlton, H. E., and Dorsey, M. J. (1927). A study of the hardiness of the fruit buds of the peach. *West Virginia Exp. Sta. Bull.* **211,** 1–28.

Koblitz, H., Grutzmann, K., and Hagen, I. (1967). Tissue cultures of alkaloid plants. I. *Papaver somniferum* L. *Z. Pflanzenphysiol.* **56,** 27–32.

Kochkar', N. T. (1961). The effect of ultrasonics on germination of pine and larch seeds. *Les. Khoz.* **6,** 38–39.

Koeppe, D. E., Rohrbaugh, L. M., and Wender, S. H. (1969). The effect of varying UV intensities on the concentration of scopolin and caffeoylquinic acids in tobacco and sunflower. *Phytochemisty* **8,** 889–896.

Koeppe, D. E., Rohrbaugh, L. M., Rice, E. L., and Wender, S. H. (1970). The effect of age and chilling temperatures on the concentration of scopolin and caffeoylquinic acids in tobacco. *Physiol. Plant.* **23,** 258–266.

Koffler, H., Mallett, G. E., and Adye, J. (1957). Molecular basis of biological stability to high temperatures. *Proc. Nat. Acad. Sci. U.S.* **43,** 464–477.

Kohn, H. (1959). Experimenteller Beitrag zur Kenntnis der Frostresistenz von Rinde, Winterknospen und Bluten verschiedener Apfelsorten. *Gartenbauwissenschaft* **24,** 314–329.

Kohn, H., and Levitt, J. (1965). Frost hardiness studies on cabbage grown under controlled conditions. *Plant Physiol.* **40,** 476–480.

Kohn, H., and Levitt, J. (1966). Interrelations between photoperiod, frost hardiness and sulfhydryl groups in cabbage. *Plant Physiol.* **41,** 792–796.

Kohn, H., Waisel, Y., and Levitt, J. (1963). Sulfhydryls—a new factor in frost resistance. V. Direct measurements on proteins and the nature of the change in SH during vernalization of wheat. *Protoplasma* **57,** 556–568.

Kol, E. (1969). Die Binnegewasser. Vol. XXIV. Kryobiologie. Biologie und Limnologie des Schnees und Eises. I. Kryovegetation. 216 p. Schweizerbat'she, Verlagsbuchhandlung (Nagele u. Obermiller), Stuttgart, Germany.

Kolomycev, G. G. (1936). Winter hardiness and earliness of wheats. *C. R. Acad. Sci. URSS* **12,** 351–356.

Kolosha, O. I., and Reshetnikova, T. P. (1967). The effect of the temperature on the content of macroenergetic phosphorus in the tillering nodes and on the frost resistance of winter wheat. *Rast. Ustoich. Rast.* **SB3,** 188–193.

Konev, S. V., Volotovskii, I. D., and Voskresenskaya, L. G. (1970). Dependence of photosensitivity of some proteins to ultraviolet light on molecule conformation. *Mol. Biol.* **4,** 395–400.

Kongsrud, K. L. (1969). Effects of soil moisture tension on growth and yield in black currants and apples. *Acta Agr. Scand.* **19,** 245–257.

Konis, E. (1949). The resistance of maquis plants to supramaximal temperatures. *Ecology* **30,** 425–429.

Konis, E. (1950). On the temperature of Opuntia joints. *Palestine J. Bot. Jerusalem Ser.* **5,** 46–55.

Konovalov, I. N. (1955). Experiments on increasing the frost resistance of stock and cabbage by the action of extracts of winter-resisting plants. *Dokl. Akad. Nauk. SSSR* **101,** 767–770.

Korogodin, V. I., Malinovskii, O. V., Poryadkova, N. A., and Izmozherov, N. A. (1959). On the reversibility of various forms of radiation damage in diploid yeast cells. *Tsitologiya* **1,** 306–315.

Korovin, A. I., and Frolov, I. N. (1968). The effects of potassium and calcium on the resistance of buckwheat to low soil temperatures and freezing during the initiation of vegetation. *Izv. Sib. Otd. Akad. Nauk. SSSR Ser. Biol. Med. Nauk.* **3,** 48–53.

Koshland, D. E., and Kirtley, M. E. (1966). Protein structure in relation to cell dynamics and differentiation. *In* "Major problems in Developmental Biology" (M. Locke, ed.), pp. 217–250. Academic Press, New York.

Koster, J. F., and Veeger, C. (1968). The relation between temperature-inducible allosteric effects and the activation energies of amino-acid oxidases. *Biochim. Biophys. Acta* **167,** 48–63.

Kotaka, S., and Krueger, A. P. (1968). Studies on the air-ion-induced growth increase in higher plants. *Advan. Front. Plant Sci.* **20,** 115–208.

Kotaka, S., Krueger, A. P., and Andriese, P. C. (1968). The effect of air ions on light-induced swelling and dark-induced shrinking of isolated (*Spinacia oleracea*) chloroplasts. *Int. J. Bioclimatol. Biometeorol.* **12,** 85–92.

Kotlyar, G. I., Lebedeva, G. Ya., and Zhuravskaya, T. G. (1969). Sulfhydryl groups and disulfide bonds in water-soluble proteins of heated germinated grain. *Priklad. Biokhim. Mikrobiol.* **5,** 366–368.

Kovalskaia, E. M. (1958). Change in salt resistance of plants during ontogenesis. *Fiziol. Rast.* **5,** 437–446.

Kovarik, U. (1963). Zur Permeabilität und Salzresistenz einiger Diatomeen des Salzlachengebietes am Neusiedler See. *Oesterr. Akad. Wiss. Math Naturwiss. Kl. Abt. 1. Sitzungsber.* **172,** 121–166.

Kramer, P. J. (1937). Photoperiodic stimulation of growth by artificial light as a cause of winter killing. *Plant Physiol.* **12,** 881–883.

Krasavtsev, O. A. (1961). Acclimatization of arboreal plants to extremely low temperatures. *Izv. Akad. Nauk. SSSR Ser. Biol.* **2,** 228–232.

Krasavtsev, O. A. (1962). Fluorescence of woody plant cells in the frozen state. *Sov. Plant Physiol.* **9,** 282–288.

Krasavtsev, O. A. (1968). Amount of nonfrozen water in woody plants at different temperatures. *Fiziol. Rast.* **15,** 225–235.

Krasavtsev, O. A. (1969). Effect of prolonged frost on trees. *Fiziol. Rast.* **16,** 228–236.

Kraus, M. P. (1969). Resistance of blue-green algae to ^{60}Co gamma radiation. *Radiat. Bot.* **9,** 481–489.

Kretschmer, G., and Beyer, B. (1970). Einfluss von CCC auf die Winterhärte von Weizen. *Albrecht Thaer Arch.* **14,** 93–104.

Kriedemann, P. E., (1968). Some photosynthetic characteristics of citrus leaves. *Aust. J. Biol. Sci.* **21,** 895–905.

Kriedemann, P. E., and Neales, T. F. (1963). Studies on the use of cetyl alcohol as a transpiration suppressor. *Aust. J. Biol. Sci.* **16,** 743–750.

Krishna, P. V. (1955). Observations on the ionic composition of blue-green algae growing in saline lagoons. *Proc. Nat. Inst. Sci. India* **B21,** 90–102.

Krull, E. (1961). A chamber for measuring drought resistance. M.A. Thesis, Univ. of Missouri, Columbia, Missouri.

Krull, E. (1966). Investigations of the frost hardiness of cabbage in relation to the sulfhydryl hypothesis. Ph.D. Thesis. Univ. of Missouri, Columbia, Missouri.

Kuijper, J. (1910). Ueber den Einfluss der Temperatur auf die Atmung höhere Pflanzen. *Rec. Trav. Bot. Neer.* **7,** 131–240.

Kuiper, P. J. C. (1964). Water transport across root cell membranes: Effect of alkenyl-succinic acids. *Science* **143,** 690–691.

Kuiper, P. J. C. (1967). Surface-active chemicals as regulators of plant growth, membrane permeability and resistance to freezing. *Mededeel. Landbouwhoogesh. Wageningen* **67,** 1–23.

Kuiper, P. J. C. (1969). Surface-active chemicals membrane permeability and resistance to freezing. *Proc.* 1*st Int. Citrus Symp.* **2,** 593–595.

Kuiper, P. J. C. (1970). Lipids in alfalfa leaves in relation to cold hardiness. *Plant Physiol.* **45,** 684–686.

Kuksa, I. N. (1939). The effect of mineral nutrition on winter hardiness and yield of winter wheat. *Himiz. Soc. Zemled.* **1,** 70–79. (*Herb. Abstr.* **9,** 635, 1939).

Kumar, H. D. (1964). Adaptation of a blue-green alga to sodium selenate and chloramphenicol. *Plant Cell Physiol.* **5,** 465–472.

Kunisch, H. (1880). Über die tödliche Einwirkung niederer Temperaturen auf die Pflanzen. Inaug. Diss. 55 pp., Breslau.

Kuntz, I. D., Jr., Brassfield, T. S., Law, G. D., and Purcell, G. V. (1969). Hydration of macromolecules. *Science* **163,** 1329–1331.

Kuokol, J., Dugger, W. M., Jr., and Palmer, R. L. (1967). Inhibitory effect of peroxyacetyl nitrate on cyclic photophosphorylation by chloroplasts from "Black Valentine" bean leaves. *Plant Physiol.* **42,** 1419–1422.

Kuraishi, S., Arai, N., Ushijima, T., and Tazaki, T. (1968). Oxidized and reduced nicotinamide adenine dinucleotide phosphate levels of plants hardened and unhardened against chilling injury. *Plant Physiol.* **43,** 238–242.

Kurkova, E. B. (1967). The effect of elevated temperatures on the oxidative phosphorylation of the mitochondria of corn roots. *Uch. Zap. Mosk. Abl. Pedagog. Inst.* **169,** 185–191.

Kurkova, E. B., and Andreeva, I. N. (1966). Changes in the morphology and biochemical activity of mitochondria due to temporary injury (in 3-day-old corn shoots). *Fiziol. Rast.* **13,** 1019–1023.

Kurtin, S. L., Mead, C. A., Mueller, W. A., Kurtin, B. C., and Wolf, E. D. (1970). "Polywater:" a hydrosol? *Science* **167,** 1720–1722.

Kurtz, E. B., Jr. (1958). Chemical basis for adaptation in plants. Understanding of heat tolerance in plants may permit improved yields in arid and semiarid regions. *Science* **128,** 1115–1117.

Kushmrenko, S. V., and Morozova, R. S. (1963). The effect of positive low temperatures on the structure of plastids in cucumbers hardened to cold. *Bot. Zh.* (*Leningrad*) **48,** 720–724.

Kustova, A. Kh. (1964). Some data on the effects of trace elements in increasing the salt tolerance of cotton. *Izv. Akad. Nauk. Turkm. SSR Ser. Biol.* Nauk. **6,** 3–8.

Kuzin, A. M., Chi, S., and Saenko, G. N. (1958). The functional radiosensitivity of chloroplasts. *Biofizika* **3,** 308–314.

Kuzin, A. M., and Ivanitskaya, E. A. (1967). Great radiosensitivity of the allosteric center of enzymes. *Radiobiologiya* **7,** 628–629.

Kuzin, A. M., Kriukova, L. M., Saenko, G. N., and Lazykova, V. A. (1959). The inhibiting effect of substances formed in irradiated plants on the cell division growth and development of unirradiated plants. *Biofizika* **4,** 104–108.

Kylin, H. (1917). Über die Kälteresistenz der Meeresalgen. *Ber Deut. Bot. Ges.* **35,** 370–384.

Kylin, A., and Gee, R. (1970). Adenosine triphosphatase activities in leaves of the mangrove *Avicennia nitrida* Jacq. Influence of sodium to potassium ratios and salt concentrations. *Plant Physiol.* **45,** 169–172.

Laetsch, W. M. (1968). Chloroplast specialization in dicotyledons possessing the C^4-dicarboxylic acid pathway of photosynthetic CO_2 fixation. *Amer. J. Bot.* **55,** 875–883.

Lagerwerff, J. V. (1969). Osmotic growth inhibition and electronic salt-tolerance evaluation of plants. *Plant Soil* **31,** 77–95.

Lagerwerff, J. V., and Eagle, H. E. (1961). Osmotic and specific effects of excess salts on beans. *Plant Physiol.* **36,** 472–477.

LaHaye, P. A., and Epstein, E. (1969). Salt toleration by plants. Enhancement with calcium. *Science* **166,** 395–396.

Lahiri, A. N., and Singh, S. (1968). Studies on plant-water relationships. IV. Impact of water deprivation on the nitrogen metabolism of *Pennisetum typhoides. Proc. Nat. Inst. Sci. India Part* **B34,** 313–322.

Lahiri, A. N., and Singh, S. (1969). Effect of hyperthermia on the nitrogen metabolism of *Pennisetum typhoides. Proc. Nat. Inst. Sci. India Part.* **B35,** 131–138.

Lake, J. V., Postlethwaite, J. D., and Slack, G. (1969). Transpiration of *Helxine solierolli* and the effect of drought. *J. Appl. Ecol.* **6,** 277–291.

Lamotte, C. E., Gochnauer, C., LaMotte, L. R., Mathur, J. R., and Davies, L. L. R. (1969). Pectin esterase in relation to leaf abscission in Coleus and Phaseolus. *Plant Physiol.* **44,** 21–26.

Lane, T. R., and Stiller, M. (1970). Glutamic acid decarboxylation in Chlorella. *Plant Physiol.* **45,** 558–562.

Lange, O. L. (1953). Hitze und Trockenresistenz der Flechten in Beziehung zu ihrer Verbreitung. *Flora (Jena)* **140,** 39–97.

Lange, O. L. (1955). Untersuchungen über die Hitzeresistenz der Moose in Beziehung zu ihrer Verbreitung. I. Die Resistenz stark ausgetrockneter Moose. *Flora (Jena)* **142,** 381–399.

Lange, O. L. (1958). Hitzeresistenz in Blattemperaturen mauretanisher Wüstenpflanzen. *Ber. Deut. Bot. Ges.* **70,** (31)–(32).

Lange, O. L. (1959). Untersuchungen über Warmehaushalt und Hitzeresistenz mauretanischer Wusten-und Savannenpflanzen. *Flora (Jena)* **147,** 595–651.

Lange, O. L. (1961). Die Hitzeresistenz einheimischer immer-und wintergrüner Pflanzen im Jahreslauf. *Planta* **56,** 666–683.

Lange, O. L. (1962a). Über die Beziehung zwischen Wasser-und Warmehaushalt von Wüsten pflanzen. *Veroeff. Geobot. Inst. Rueb. Zurich* **37,** 155–168.

Lange, O. L. (1962b). Die Photosynthese der Flechten bei tiefen Temperaturen und nach Frostperioden. *Ber. Deut. Bot. Ges.* **75,** 351–352.

Lange, O. L. (1962c). Versuche zur Hitzeresistenz—Adaptation bei hoheren Pflanzen. *Naturwiss.* **49,** 20–21.

Lange, O. L. (1964). Die Hitzeresistenz einheimischer immer-und wintergrüner Pflanzen im Jahreslauf. *Planta* **56,** 666–683.

Lange, O. L. (1965a). Der CO_2 Gaswechsel von Flechten bei tiefen Temperaturen. *Planta* **64,** 1–19.

Lange, O. L. (1965b). Leaf temperatures and methods of measurement. *In* "Methodology of Plant Ecophysiology" (F. E. Eckardt, ed.), pp. 203–209. Proc. Montpellier Symp. Unesco, Paris.

Lange, O. L. (1965c). The heat resistance of plants, its determination and variability. *In* "Methodology of Plant Eco-physiology" (F. Eckhardt, ed.), pp. 399–405. Proc. Montpellier Symp. Unesco, Paris.

Lange, O. L. (1967). Investigations on the variability of heat-resistance in plants. pp. 131–141. *In* "The Cell and Environmental Temperatures" (A. S. Troshin, ed.), pp. 131–141. Pergamon, New York.

Lange, O. L. (1969). Experimentell-ökologische Untersuchungen an Flechten der Negev Wüste. I. CO_2-Gaswechsel von *Ramalina maciformis* (Del.) Bory unter kontrollierten Bedingungen im Laboratorium. *Flora (Jena)* **158,** 324–359.

Lange, O. L., and Bertsch, A. (1965). Photosynthese der Wüstenflechte *Ramalina maciformis* nach Wasserdampfaufnahme aus dem Luftraum. *Naturwissenschaften* **9,** 1–2.

Lange, O. L., and Lange, R. (1963). Untersuchungen über Blattemperaturen, Transpiration and Hitzeresistenz an Pflanzen mediterraner Standorte (Costa brava, Spanien). *Flora (Jena)* **153,** 387–425.

Lange, O. L., and Metzner, H. (1965). Lichtabhängiger Kohlenstoff-Einbau in Flechten bei tiefen Temperaturen. *Naturwissenschaften* **52,** 191–192.

Lange, O. L., and Schwemmle, B. (1960). Untersuchungen zur Hitzeresistenz Vegetativer und Blühender Pflanzen von *Kalanchoe blossfeldiana. Planta* **55,** 208–225.

Lange, O. L., Schulze, E. D., and Koch, W. (1968). Photosynthese von Wüstenflechten am naturlichen standort nach Wasserdampfaufnahme ausden Luftraum. *Naturwissenschaften* **55,** 658–659.

Lange, O. L., Koch, W., and Schulze, E. D. (1969). CO_2-gas exchange and water relationships of plants in the Negev desert at the end of the dry period. *Ber. Deut. Bot. Ges.* **82,** 39–61.

Langlet, O. (1934). Om variationen hos tallen (P. silvestris L.) och dess samband med klimatet. *Sv. Skogsvardsf. Tidskr.* **32,** 87–110.

Langridge, J. (1963). Biochemical aspects of temperature response. *Ann. Rev. Plant Physiol.* **14,** 441–462.

Langridge, J., and Griffing, B. (1959). A study of high temperature lesions in *Arabidopsis thaliana. Aust. J. Biol. Sci.* **12,** 117–135.

Lapina, L. P. (1966). Effect of high iso-osmotic concentrations of dextran and sodium chloride on nitrogen and carbohydrate metabolism in corn. *Fiziol. Rast.* **13,** 1029–1040.

Lapina, L. P. (1968). Dextran as an osmotic agent during studies of salt hardiness of plants (bean). *Fiziol. Rast.* **15,** 182–187.

Lapins, K. (1962). Artificial freezing as a routine test of cold hardiness of young apple seedlings. *Proc. Amer. Soc. Hort. Sci.* **81,** 26–34.

Lapins, K. O., and Hough, L. F. (1970). Effects of gamma rays on apple and peach leaf buds at different stages of development. II. Injury to apical and axillary meristems and regeneration of shoot apices. *Radiat. Bot.* **10,** 59–68.

Larcher, W. (1954). Die Kälteresistenz mediterraner Immergrüner und ihre Beeinflussbarkeit. *Planta* **44,** 607–635.

Larcher, W. (1957). Frosttrocknis an der Waldgrenze und in der alpinen Zwergstrauchheide auf dem Patscherkofel bei Innsbruck. *Veroeff. Ferdinandeum Innsbruck* **37,** 49–81.

Larcher, W. (1959). Das Assimilationsvermögen von *Quercus ilex* und *Olea europea* im Winter. *Ber. Deut. Bot. Ges.* **72,** (18).

Larcher, W. (1960). Transpiration and photosynthesis of detached leaves and shoots of *Quercus pubescens and Q. ilex* during desiccation under standard conditions. *Bull. Res. Counc. Isr.* **8D,** 213–224.

Larcher, W. (1963). Zur Frage des Zusammenhanges zwischen Austrocknungsresistenz und Frosthärte bei Immergrünen. *Protoplasma* **57,** 569–587.

Larcher, W. (1968). Die Temperaturresistenz als Konstitutionsmerkmal der Pflanzen. *In* "Klimaresistenz Photosynthese und Stoffproduktion" (H. Polster, ed.), pp. 7–21. Deut. Akad. Landwirtsch., Berlin.

Larcher, W. (1969). Zunahme des Frostabhärtungsvermögen von *Quercus ilex* im Laufe der Individualentwicklung. *Planta* **88,** 130–135.

Larcher, W., and Eggarter, H. (1960). Anwendung des Triphenyltetrazoliumchlorids zur Beurteilung von Frostschäden in verschiedenen Achsengeweben bei *Pinus*-Arten, und Jahresgang der Resistenz. *Protoplasma* **41,** 595–619.

Larcher, W., and coworkers, (1969). Anwendung und Zuverlössigkeit der Tetrazoliummethode zur Feststellung von Schäden in pflanzlichen Gewebe. *Mikroskopie* **25,** 207–218.

Larikova, G. A., and Gal'tsova, R. D. (1968). Lipid biosynthesis by radioresistant and radiosensitive yeast organisms. *Mikrobiologiya* **37,** 448–453.

Larson, M. M., and Schubert, G. H. (1969). Effect of osmotic water stress on germination and initial development of ponderosa pine seedlings (*Pinus ponderosa*). *Forest Sci.* **15,** 30–36.

Latterell, R. L. (1966). Nitrogen- and helium-induced anoxia: Different lethal effects on rye seeds. *Science* **153,** 69–70.

Laude, H. M., and Chaugule, B. A. (1953). Effect of stage of seedling development upon heat tolerance in bromegrasses. *J. Range Manage.* **6,** 320–324.

Laue, E. (1938). Untersuchungen an Pflanzenzellen an Dampfraum. *Flora (Jena)* **132,** 193–224.

Lautenschlager-Fleury, D. (1955). On the transparency for ultra-violet light of leaf epidermis. *Ber. Schweiz. Bot. Ges.* **65,** 343–386.

Lavee, S., and Galston, A. W. (1968). Hormonal control of peroxidase activity in cultured Pelargonium pith. *Amer. J. Bot.* **55,** 890–893.

Lawlor, D. W. (1969). Plant growth in polyethylene glycol solutions in relation to the osmotic potential of the root medium and the leaf water balance. *J. Exp. Bot.* **20,** 895–911.

Layne, R. E. C. (1963). Effect of vacuum drying, freeze-drying and storage environment on the viability of pea pollen. *Crop. Sci.* **3,** 433–436.

Leary, J. V., Morris, A. J., and Ellingboe, A. H. (1969). Isolation of functional ribosomes and polysomes from lyophilized fungi. *Biochim. Biophys. Acta* **182,** 113–120.

Lebedincev, E. (1930). Untersuchungen über die wasserbindenden Kräfte der Pflanzen im Zusammenhang mit ihrer Dürre-und Kälteresistenz. *Protoplasma* **10,** 53–81.

Leclerc, J. C. (1969). Pigments and salinity in Porphyridium. *Physiol. Plant.* **22,** 1013–1024.

Lee, J. A., and Woolhouse, H. W. (1969). A comparative study of bicarbonate inhibition of root growth in calcicole and calcifuge grasses. *New Phytol.* **68,** 1–11.

Lee, T. T. (1967). Inhibition of oxidative phosphorylation and respiration by ozone in tobacco mitochondria. *Plant Physiol.* **42,** 691–696.

Leisle, V. F., and Nikulin, A. V. (1967). The effect of a low-tension magnetic field on the growth processes of corn, sunflower, and sugar beets. *Zap. Voronezh. Sel'skokhoz. Inst.* **34,** 113–115.

Leone, I. A., Brennan, E. G., Daines, R. H., and Robbins, W. R. (1948). Some effects of fluorine on peach, tomato, and buckwheat when absorbed through the roots. *Soil Sci.* **66,** 259–266.

Lepeschkin, W. W. (1912). Zur Kenntnis der Einwirkung supramaximaler Temperaturen auf die Pflanze. *Ber. Deut. Bot. Ges.* **30,** 703–714.

Lepeschkin, W. W. (1935). Zur Kenntnis des Hitzetodes des Protoplasmas. *Protoplasma* **23,** 349–366.

Lepeschkin, W. W. (1937). Zell-Nekrobiose und Protoplasma-Tod. *Protoplasma Monogr.* No. 12.

Le Saint (Quervel), A.-M. (1956). Quelques expériences sur la résistance au gel et la surfusion de jeunes plantes étiolées placées a −4°C. *Rev. Gen. Bot.* **63,** 514–523.

Le Saint, A.-M. (1957). Mise en évidence d'une chute de la perméabilité aux gaz des tissus de pomme, pendant le gel, a partir de la cessation de la surfusion. *Rev. Gen. Bot.* **64,** 334–338.

Le Saint, A.-M. (1958). Comparaison de la résistance au froid de jeunes plantes de pois étiolées ou chlorophylliennes. *Rev. Gen. Bot.* **65,** 471–477.

Le Saint (-Quervel), A.-M. (1960). Études des variations comparées des acides aminés libres et des glucides solubles, au cours de l'acquisition et de la perte de l'aptitude à résister au gel chez le chou de Milan. *C. R. Acad. Sci.* (*Paris*) **251,** 1403–1405.

Le Saint, A.-M. (1966). Observations physiologiques sur le gel et l'endurcissement au gel chez le chou de Milan. Thèses présenté à la faculté des sciences de l'Université de Paris, 93 pp.

Le Saint, A.-M. (1969a). Comparison of free protein and soluble carbohydrate levels in relation to the unequal sensitivity to freezing of the savoy cabbage cv. "Pontoise." *C.R. Acad. Sci.* (*Paris*) **D268,** 310–313.

Le Saint (-Quervel), A.-M. (1969b). Corrélations entre la résistance au gel, l'éclairement et la teneur en proline libre chez le chou de Milan Cult. Pontoise. *C. R. Acad. Sci.* (*Paris*) **269,** 1423–1426.

Le Saint, A.-M., and Catesson, A. M. (1966). Variations simultanées des teneurs en eau, en sucre solubles, en acide aminés et de la pression osmotique dans la phloeme et la cambium de Sycamore pendant les périodes de repos apparent et de reprise de la croissance. *C. R. Acad. Sci.* (*Paris*) **263,** 1463–1466.

Levitt, J. (1939). The relation of cabbage hardiness to bound water, unfrozen water, and cell contraction when frozen. *Plant Physiol.* **14,** 93–112.

Levitt, J. (1941). "Frost Killing and Hardiness of Plants," 211 pp. Burgess, Minneapolis, Minnesota.

Levitt, J. (1954). Investigations of the cytoplasmic particulates and proteins of potato tubers. III. Protein synthesis during the breaking of the rest period. *Physiol. Plant.* **7,** 597–601.

Levitt, J. (1956). "The Hardiness of Plants". 278 pp. Academic Press, New York.

Levitt, J. (1957a). The moment of frost injury. *Protoplasma* **48,** 289–302.

Levitt, J. (1957b). The role of cell sap concentration in frost hardiness. *Plant Physiol.* **32,** 237–239.

Levitt, J. (1958). Frost, drought, and heat resistance. *Protoplasmatologia* **6,** 87 pp.

Levitt, J. (1959a). Effects of artificial increases in sugar content on frost hardiness. *Plant Physiol.* **34,** 401–402.

Levitt, J. (1959b). Bound water and frost hardiness. *Plant Physiol.* **33,** 674–677.

Levitt, J. (1962). A sulfhydryl-disulfide hypothesis of frost injury and resistance in plants. *J. Theoret. Biol.* **3,** 355–391.

Levitt, J. (1963). Hardiness and the survival of extremes. A uniform system for measuring resistance and its two components. *In* "Environmental Control of Plant Growth" (L. T. Evans, ed.), pp. 351–366. Academic Press, New York.

Levitt, J. (1965). Thiogel—a model system for demonstrating intermolecular disulfide bond formation on freezing. *Cryobiology* **1,** 312–316.

Levitt, J. (1966a). Cryochemistry of plant tissue: protein interactions. *Cryobiology* **3,** 243–251.

Levitt, J. (1966b). Winter hardiness in plants. *In* "Cryobiology" (H.T. Meryman, ed.), pp. 495–563. Academic Press, New York.

Levitt, J. (1967a). Status of the sulfhydryl hypothesis of freezing injury and resistance. *In* "Molecular Mechanisms of Temperature Adaptation." (C. Ladd Prosser, ed.), pp. 41–51. Amer. Assoc. Adv. Sci., Washington, D.C.

Levitt, J. (1967b). The mechanism of hardening on the basis of the SH $\rightleftharpoons$ SS hypothesis of freezing injury. *In* "Cellular Injury and Resistance in Freezing Organisms." (E. Asahina, ed.), pp. 51–61. Inst. Low Temp. Sci. Hokkaido Univ., Sapporo, Japan.

Levitt, J. (1967c). The role of SH and SS in the resistance of cells to high and low temperatures. *In* "The Cell and Environmental Temperature" (A. S. Troshin, ed.), pp. 269–274. Pergamon, Oxford.

Levitt, J. (1969a). The effect of sulfhydryl reagents on freezing resistance of hardened and unhardened cabbage cells. *Cryobiology* **5,** 278–280.

Levitt, J. (1969b). Growth and survival of plants at extremes of temperature- a unified concept. *Proc. Soc. Exp. Biol.* **23,** 395–448.

Levitt, J., (1971). The effect of sulfhydryl reagents on cell permeability. In press.

Levitt, J., and Dear, J. (1970). The role of membrane proteins in freezing injury and resistance. *In* "The Frozen Cell." (G. E. W. Wolstenholme and M. O'Connor, eds.), pp. 148–174. Ciba Foundation Symp., J. & A. Churchill, London.

Levitt, J., and Hasman, M. (1964). Mechanism of protection by non-penetrating and non-toxic solutes against freezing injury to plant cells. *Plant Physiol.* **39,** 409–412.

Levitt, J., and Scarth, G. W. (1936). Frost-hardening studies with living cells. I. Osmotic and bound water changes in relation to frost resistance and the seasonal cycle. *Can. J. Res.* **C14,** 267–284.

Levitt, J., and Siminovitch, D. (1940). The relation between frost resistance and the physical state of protoplasm. I. The protoplasm as a whole. *Can. J. Res.* **C18,** 550–561.

Levitt, J., Scarth, G. W., and Gibbs, R. D. (1936). Water permeability of isolated protoplasts in relation to volume change. *Protoplasma* **26,** 237–248.

Levitt, J., Sullivan, C. Y., and Krull, E. (1960). Some problems in drought resistance. *Bull. Res. Counc. Isr.* **8D,** 173–179.

Levitt, J., Sullivan, C. Y., Johansson, N-O., and Pettit, R. M. (1961). Sulfhydryls—a new factor in frost resistance. Changes in SH content during frost hardening. *Plant Physiol.* **36,** 611–616.

Levitt, J., Sullivan, C. Y., and Johansson, N-O. (1962). Sulfhydryls—a new factor in frost resistance. III. Relation of SH increase during hardening to protein, glutathione, and glutathione oxidizing activity. *Plant Physiol.* **37,** 266–271.

Lewis, D. A. (1956). Protoplasmic streaming in plants sensitive and insensitive to chilling temperatures. *Science* **124,** 75–76.

Lewis, D. A., and Morris, L. L. (1956). Effects of chilling storage on respiration and deterioration of several sweet potato varieties. *Proc. Amer. Soc. Hort. Sci.* **68,** 421–428.

Lewis, F. J., and Tuttle, G. M. (1920). Osmotic properties of some plant cells at low temperatures. *Ann. Bot.* **34,** 405–416.

Lewis, F. J., and Tuttle, G. M. (1923). On the phenomena attending seasonal changes in the organisation in leaf cells of *Picea canadensis.* (Mill.) B.S.P. *New Phytol.* **22,** 225–232.

Lewis, J. C., and Powers, W. L. (1941). Antagonistic action of chlorides on the toxicity of iodides to corn. *Plant Physiol.* **16,** 393–397.

Lewis, T. L., and Workman, M. (1964). The effect of low temperature on phosphate esterification and cell membrane permeability in tomato fruit and cabbage leaf tissue. *Aust. J. Biol. Sci.* **17,** 147–152.

Levy, H. M., and Ryan, E. M. (1967). Heat inactivation of the relaxing site of actomyosin. *Science* **156,** 73.

Leyton, L., and Armitage, I. P. (1968). Cuticle structure and water relations of the needles of *Pinus radiata* (D.Don.). *New Phytol.* **67,** 31–38.

Li, P. H., and Weiser, C. J. (1967). Evaluation of extraction and assay methods for nucleic acids from red-osier dogwood and RNA, DNA, and protein changes during cold acclimation. *Proc. Amer. Soc. Hort. Sci.* **91,** 716–727.

Li, P. H., and Weiser, C. J. (1969a). Metabolism of nucleic acids in one-year-old apple twig during cold hardening and dehardening. *Plant Cell Physiol.* **10,** 21–30.

Li, P. H., and Weiser, C. J. (1969b). Influence of photoperiod and temperature on potato foliage protein and 4S RNA. *Plant Cell Physiol.* **10,** 929–934.

Liboff, R. L. (1965). A biomagnetic hypothesis. *Biophys. J.* **5,** 845–853.

Lidforss, B. (1896). Zur Physiologie und Biologie der wintergrünen Flora. *Bot. Centr.* **68,** 33–44.

Lidforss, B. (1907). Die wintergrüne Flora. *Lunds. Univ. Arsskr. Afd.* **2,** 1–76.

Lieberman, M., Craft, C. C., Audia, W. V., and Wilcox, M.S. (1958). Biochemical studies of chilling injury in sweet potatoes. *Plant Physiol.* **33,** 307–311.

Liebig, G. F., Jr., Vanselow, A. P., and Chapman, H. D. (1942). Effects of aluminum on copper toxicity as revealed by solution culture and spectrographic studies with citrus. *Soil Sci.* **53,** 341–351.

Limar', R. S. (1961a). The effect of ultrasound on germination in lentil. *Bot. Zh. (Leningrad)* **46,** 1166–1168.

Limar', R. S. (1961b). Effect of ultrasound on the seed of certain plants. *Referat. Zhur. Biol.* 1962. No. 17G177.

Lindley, J. (1842). Observations upon the effects produced on plants by the frost which occurred in England in the winter of 1837–38. (Read in 1838). *Trans. Hort. Soc. (London)* **3,** 225–315.

Lindner, J. (1915). Über den Einfluss günstiger Temperaturen auf gefrorene Schimmelpilze. (Zur Kenntnis der Kälteresistenz von *Aspergillus niger*) *Jahrb. Wiss. Bot.* **55,** 1–52.

Ling, G. N. (1968). The physical state of water in biological systems. *Food Technol.* **22,** 1254–1258.

Lipman, C. B. (1936a). The tolerance of liquid air temperatures by dry moss protonema. *Bull. Torrey Bot. Club* **63,** 515–518.

Lipman, C. B. (1936b). Normal viability of seeds and bacterial spores after exposure to temperatures near the absolute zero. *Plant Physiol.* **11,** 201–205.

Lipman, C. B. (1937). Tolerance of liquid air temperatures by spore-free and very young cultures of fungi and bacteria growing on agar media. *Bull. Torrey Bot. Club* **64,** 537–546.

Lipman, C. B., and Lewis, G. N. (1934). Tolerance of liquid-air temperatures by seeds of higher plants for sixty days. *Plant Physiol.* **9,** 392–394.

Lippincott, E. R., Stromberg, R. R., Grant, W. H., and Cessac, G. L. (1969). Polywater. *Science* **164,** 1482–1487.

Little, C. H. A., and Eidt, D. C. (1968). Effect of abscisic acid on budbreak and transpiration in woody species. *Nature* (*London*) **220,** 498–499.

Little, C., Sanner, T., and Pihl, A. (1969). X-ray modification of the catalytic and allosteric functions of fructose-1,6-diphosphatase. *Biochim. Biophys. Acta* **178,** 83–92.

Livingston, B. E. (1903). "The Role of Diffusion and Osmotic Pressure in Plants." Univ. Chicago Press, Chicago, Illinois.

Livingston, B. E. (1911). Light intensity and transpiration. *Bot. Gaz.* (*Chicago*) **52,** 417–438.

Livne, A., and Levin, N. (1967). Tissue respiration and mitochondrial oxidative phosphorylation of NaCl-treated pea seedlings. *Plant Physiol.* **42,** 407–414.

Livne, A. (1968). Membrane lipids as site for freezing injury. *Isr. J. Chem.* 152p.

Livne, A., and Vaadia, Y. (1965). Stimulation of transpiration rate in barley leaves by kinetin and gibberellic acid. *Physiol. Plant.* **18,** 658–664.

Ljunger, C. (1962). Introductory investigations of ions and thermal resistance. *Physiol. Plant.* **15,** 148–160.

Ljunger, C. (1970). On the nature of the heat resistance of thermophilic bacteria. *Physiol. Plant.* **23,** 351–364.

Lochmann, E.-R., and Stein, W. (1968). Mechanism of photodynamic inactivation of *Saccharomyces* cells. *Biophysik* **4,** 243–251.

Lockett, M. C., and Luyet, B. J. (1951). Survival of frozen seeds of various water contents. *Biodynamica* **7**(134), 67–76.

Lockhart, J. A. (1961a). Photoinhibition of stem elongation by full solar radiation. *Amer. J. Bot.* **48,** 387–392.

Lockhardt, J. A. (1961b). The hormonal mechanism of growth inhibition by visible radiation. pp. 543–556. *In* "Plant Growth Regulation." (R. M. Klein, ed.), pp. 543–556. Univ. Iowa Press, Ames, Iowa.

Loginova, L. G., and Tashpulatov, Zh. (1967). Multicomponent cellulolytic enzymes of thermotolerant and mesophilic fungi related to *Aspergillus fumigatus*. *Mikrobiologiya* **36,** 988–992.

Loginova, L. G., and Verkhovtseva, M. I. (1963). Amino acid requirements of thermotolerant yeasts. *Mikrobiologiya* **32,** 216–222.

Loginova, L. G., Gerasimova, N. F., and Seregina, L. M. (1962). Requirement of thermotolerant yeasts for supplementary growth factors. *Mikrobiologiya* **31,** 21–25.

Lomagin, A. G., and Antropova, T. A. (1966). Photodynamic injury to heated leaves. *Planta* **68,** 297–309.

Lomagin, A. G., and Antropova, T. A. (1968). A study of the capacity of *Physarum polycephalum* to adapt to different temperatures. *Tsitologiya* **10,** 1094–1104.

Lomagin, A. G., Antropova, T. A., and Semenichina, L. V. (1966). Phototaxis of chloroplasts as a criterion of viability of leaf parenchyma. *Planta* **71,** 119–124.

Lona, F. (1962). Resistenza al freddo in relazione all'effeto delle gibberelline, antigibberelline e delle radiazioni morfogenetische. *Ateneo Parmense* **33** (Suppl. 6), 209–213.

Lona, F., Squarza, A., Bocchi, A., and Cantoni, G. (1956). Le reazione al freddo di piante erbacee in rapporto all'azione di sostanze stimolanti ed inibenti l'attivita plastimatica. *Pubb. Chim. Biol. Med. Ist. Carlo Erba Ric. Ter.* **2,** 473–494.

Lorenz, R. W. (1939). High temperature tolerance of forest trees. *Univ. Minn. Agr. Sta. Tech. Bull.* **141,** 1–25.

Lovelock, J. E. (1953). The mechanism of the protective action of glycerol against haemolysis by freezing and thawing. *Biochim. Biophys. Acta* **11,** 28–36.

Lowenstein, A. (1903). Über die Temperaturgrenzen des Lebens bei der Thermalalge *Mastigocladus laminosus* Cohn. *Ber. Deut. Bot. Ges.* **21,** 317–323.

Lozeron, H., and Jadassohn, W. (1967). Radioprotective effect of hydrocortisone and prednisolone on the *Vicia faba equina. Experientia* **23,** 1045.

Lucas, J. W. (1954). Subcooling and ice nucleation in lemons. *Plant Physiol.* **29,** 245–251.

Lue, P. F., and Kaplan, J. G. (1970). Heat-induced disaggregation of a multifunctional enzyme complex catalyzing the first steps in pyrimidine biosynthesis in baker's yeast. *Can. J. Biochem.* **48,** 155–159.

Lukicheva, E. L. (1968). The changes in some oxidation-reduction enzymes of spring wheat in drought. *Tr. Inst. Bot. Akad. Nauk. Azerb. Kaz. SSR* **25,** 23–29.

Luknitskaya, A. F. (1967). Do chlamydomonads have a heat hardening capacity? *Tsitologiya* **9,** 800–803.

Lund, D. B., Fennema, O., and Powrie, W. D. (1969). Enzymic and acid hydrolysis of sucrose as influenced by freezing. *J. Food. Sci.* **34,** 378–382.

Lundegårdh, H. (1914). Einige Bedingungen der Bildung und Auflösung der Stärke. *Jahrb. Wiss. Bot.* **53,** 421–463.

Lundegårdh, H. (1949). "Klima und Boden," 3rd ed. Fischer, Jena.

Lunt, O. R., Oertli, J. J., and Kohl, H. C., Jr. (1960). Influence of certain environmental conditions on the salinity tolerance of *Chrysanthemum morifolium. Proc. Amer. Soc. Hort. Sci.* **75,** 676–687.

Lusena, C. V., and Cook, W. H. (1953). Ice propagation in systems of biological interest. I. Effect of membranes and solutes in a model cell system. *Arch. Biochem. Biophys.* **46,** 232–240.

Lutova, M. I., and Zavadskaya, I. G. (1966). Effects of the plant keeping duration at different temperatures on the cell heat resistance. *Tsitologiya* **8,** 484–493.

Luyet, B. J. (1937). The vitrification of organic colloids of protoplasm. *Biodynamica* **29,** 1–14.

Luyet, B. J. (1940). "Life and Death at Low Temperatures," Monogr. No. 1. Biodynamica, Normandy, Missouri.

Luyet, B. J. (1951). Survival of cells, tissues, and organisms after ultrarapid freezing. *In* "Freezing and Drying" (R. J. C. Harris, ed.), pp. 3–23. Institute of Biology, London.

Luyet, B. J., and Galcs, G. (1940). The effect of the rate of cooling on the freezing point of living tissues. *Biodynamica* **3**(65), 157–169.

Luyet, B. J., and Gehenio, P. M. (1937). The double freezing point of living tissues. *Biodynamica.* **1**(30), 1–23.

Luyet, B. J. and Gehenio, P. M. (1938). The lower limit of vital temperatures, a review. *Biodynamica* **33,** 1–92.

Luyet, B. J., and Grell, Sister M. (1936). A study with the ultracentrifuge of the mechanism of death in frozen cells. *Biodynamica* **23,** 1–16.

Luyet, B. J., and Hodapp, E. L. (1938). On the effect of mechanical shocks on the congelation of subcooled plant tissues. *Protoplasma* **30,** 254–257.

Luzikov, V. N., Rakhimov, M. M., Saks, V. A., and Berezin, I.V. (1967). Heat-inactivation of succinate oxidase and its fragments. *Biokhimiya* **32,** 1032–1035.

Lybeck, B. R. (1959). Winter freezing in relation to the rise of sap in tall trees. *Plant Physiol.* **34,** 482–486.

Lyons, J. M., and Asmundson, C. M. (1965). Solidification of unsaturated-saturated fatty acid mixtures and its relationship to chilling sensitivity in plants. *J. Amer. Oil. Chem. Soc.* **42,** 1056–1058.

Lyons, J. M., and Raison, J. K. (1970a). Oxidative activity of mitochondria isolated from plant tissues sensitive and resistant to chilling injury. *Plant Physiol.* **45,** 386–389.

Lyons, J. M., and Raison, J. K. (1970b). Changes in activation energy of mitochondrial oxidation induced by chilling temperatures in cold sensitive plants and homeothermic animals. *Cryobiology* **6,** 585.

Lyons, J. M., Wheaton, T. A., and Pratt, H. K. (1964). Relationship between the physical nature of mitochondrial membranes and chilling sensitivity in plants. *Plant Physiol.* **39,** 262–268.

Lyscov, V. N. (1969). Kinetic aspects of DNA cryolysis. *Biochim. Biophys. Acta* **190,** 111–115.

Lyscov, V. N., and Moshkovsky, Y. S. (1969). DNA cryolysis. *Biochim. Biophys. Acta* **190,** 101–110.

McBrien, D. C. H., and Hassall, K. A. (1967). The effect of toxic doses of copper upon respiration, photosynthesis and growth of *Chlorella vulgaris. Physiol. Plant.* **20,** 113–117.

McCown, B. H., Beck, G. E., and Hall, T. C. (1969a). The hardening of three clones of Dianthus and the corresponding complement of peroxidase isoenzymes. *J. Amer. Soc. Hort. Sci.* **94,** 691–693.

McCown, B. H., Hall, T. C., and Beck, G. E. (1969b). Plant leaf and stem proteins. II. Isozymes and environmental change. *Plant Physiol.* **44,** 210–216.

McCown, B. H., McLeester, R. C., Beck, G. E., and Hall, T. C. (1969c). Environment-induced changes in peroxidase zymograms in the stems of deciduous and evergreen plants. *Cryobiology* **5,** 410–412.

McCree, K. J., and Troughton, J. H. (1966). Prediction of growth rate at different light levels from measured photosynthesis and respiration rates. *Plant Physiol.* **41,** 559–566.

Macfayden, A. (1900). On the influence of the temperature of liquid air on bacteria. *Proc. Roy. Soc.* **66,** 180–182, 339–340.

McDonough, W. T. (1967). Arsenite-BAL as an inhibitor of germination. *Physiol. Plant.* **20,** 455–462.

McGee, J. M. (1916). The effect of position upon the temperature and dry weight of joints of Opuntia. *Carnegie Inst. Wash. Yearb.* **15,** 73–74.

McGinnis, W. J. (1960). Effects of moisture stress and temperature on germination of six range grasses. *Agron. J.* **52,** 159–162.

McGuire, J. J., and Flint, H. L. (1962). Effects of temperature and light on frost hardiness of conifers. *Proc. Amer. Soc. Hort. Sci.* **80,** 630–635.

Mack, A. R., and Ferguson, W. S. (1968). A moisture stress index for wheat by means of a modulated soil moisture budget. *Can. J. Plant Sci.* **48,** 535–544.

McLachlan, J. (1960). The culture of *Dunaliella tertiolecta* Butcher.—a euryhaline organism. *Can. J. Microbiol.* **6,** 367–379.

McLaren, A. D. (1969). Progress in the photochemistry of enzymes. *Enzymologia* **37,** 273–281.

McLean, R. J. (1967). Desiccation and heat resistance of the green alga *Spongichloris typica*. *Can. J. Bot.* **45,** 1933–1938.

McLeester, R. C., Weiser, C. J., and Hall, T. C. (1968). Seasonal variations in freezing curves of stem sections of *Cornus stolonifera*. Mischx. *Plant Cell Physiol.* **9,** 807–817.

McLeester, R. C., Weiser, C. J., and Hall, T. C. (1969). Multiple freezing points as a test for viability of plant stems in the determination of frost hardiness. *Plant. Physiol.* **44,** 37–44.

McMillan, C., and Moseley, F. N. (1967). Salinity tolerance of five marine spermatophytes of Redfish Bay, Texas. *Ecology* **48,** 503–506.

McNaughton, S. J. (1966). Thermal inactivation properties of enzymes from *Typha latifolia* L. ecotypes. *Plant Physiol.* **41,** 1736–1738.

McRostie, G. P. (1939). The thermal death point of corn from low temperatures. *Sci. Agr.* **19,** 687–699.

McWilliam, J. R., and Kramer, P. J. (1968). The nature of the perennial response in Mediterranean grasses. I. Water relations and summer survival in Phalaris. *Aust. J. Agr. Res.* **19,** 381–395.

Madhok, O. M. P., and Walker, R. B. (1969). Magnesium nutrition of two species of sunflower. *Plant Physiol.* **44,** 1016–1022.

Magistad, O. G., and Truog, E. (1925). Influence of fertilizers in protecting corn against freezing. *J. Amer. Soc. Agron.* **17,** 517–526.

Magness, J. R., Regeimbal, L. O., and Degman, E. S. (1932). Accumulation of carbohydrates in apple foliage, bark, and wood as influenced by moisture supply. *Proc. Amer. Soc. Hort. Sci.* **29,** 246–252.

Maianu, A., Aksenova, I., and Albescu, I. (1965). Tolerance to salinity of the main chloride plants on meadow soils with chloride salinization. *An. Inst. Cent. Cercet. Agr. Sect. Pedol.* **33,** 357–373.

Mair, B. (1968). A gradient in cold resistance of ash bud sequences. *Planta* **82,** 164–169.

Majernik, O., and Mansfield, T. A. (1970). Direct effect of SO_2 pollution on the degree of opening of stomata. *Nature* (*London*) **227,** 377–378.

Maki, T. E., Marshall, H., and Ostrom, C. E. (1946). Effects of naphthalene-acetic acid sprays on the development and drought resistance of pine seedlings. *Bot. Gaz.* (*Chicago*) **107,** 297–312.

Maksimova, E. V., and Matukhin, G. R. (1965). Effect of soil salinization on the respiration rate and terminal oxidase activity of millet leaves. *Fiziol. Rast.* **12,** 540–542.

Malcolm, N. L. (1968). A temperature-induced lesion in amino acid-transfer ribonucleic acid attachment in a psychrophile. *Biochim. Biophys. Acta* **157,** 493–503.

Malcolm, N. L. (1969). Molecular determinants of obligate psychrophily. *Nature* (*London*) **221,** 1031–1033.

Mallett, G. E., and Koffler, H. (1957). Hypothesis concerning the relative stability of flagella from thermophilic bacteria. *Arch. Biochem. Biophys.* **67,** 254–256.

Mandelli, E. F. (1969). The inhibitory effects of copper on marine phytoplankton. *Contrib. Mar. Sci.* **14,** 47–57.

Mani, R. S., and Zalik, S. (1970). Physicochemical studies of bean and wheat chloroplast structural protein. *Biochim. Biophys. Acta* **200,** 132–137.

Manis, R. E., and Knight, R. J., Jr. (1967). Avocado germ plasm evaluation: Technique used in screening for cold tolerance. *Proc. Fla. State Hort. Soc.* **80,** 387–391.

Manohar, M. S., Bhan, S., and Prasad, R. (1968). Germination in lower osmotic potential as an index of drought resistance in crop plants—a review. *Ann. Arid Zone* **7,** 80–92.

Mantai, K. E. (1970a). Some effects of hydrolytic enzymes on coupled and uncoupled electron flow in chloroplasts. *Plant Physiol.* **45,** 563–566.

Mantai, K. E. (1970b). Electron transport and degradation of chloroplasts by hydrolytic enzymes and ultraviolet irradiation. *Carnegie Inst. Yearb.* **68,** 598–603.

Mark, J. J. (1936). The relation of reserves to cold resistance in alfalfa. *Iowa Agr. Expt. Sta. Res. Bull.* **208,** 305–335.

Markert, C. L. (1965). Lactate dehydrogenase isozymes: Dissociation and recombination of subunits. *Science* **140,** 1329–1330.

Markowski, A., Myczkowski, J., and Lebek, J. (1962). Preliminary investigations on changes in nitrogen compounds of wheat embryos in the course of germination under various temperature conditions. *Bull. Acad. Pol. Sci. Cl. V* **10,** 145–150.

Marlangeon, R. C. (1969). Effects of gibberellic acid and other drugs on the permanence of green autumn foliage and the induction of resistance to frost in peach flowers. *Phyton* **26,** 107–111.

Marré, E., and Servettaz, O. (1956). Richerche sull' adattamento proteico in organismi termoresistent. I. Sul limite di resistenza all' inattivazione termica dei sistemi fotosintetico a respiratorio di alghe di acque termali. *Rend. Accad. Naz. Lincei.* **20,** (Ser. 8), 72–77.

Marré, E., and Servettaz, O. (1959). On the mechanism of adaptation to extreme osmotic conditions in *Dunaliella salina.* II. Composition of the cell at different salt concentrations in the medium. *Rend. Accad. Naz. Lincei* VIII. **26,** (Ser. 8), 272–278.

Marré, E., Albertario, M., and Vaccari, E. (1958a). Richerche sull' adattamento proteico in organismi termoresistenti. III. Relativa insensibilita di enzimi di Cianoficee a denaturanti che agiscomo rompendo i legami di idrogeno. *Rend. Accad. Naz. Lincei* **24**(Ser. 8), 351–353.

Marré, E., Servettaz, O., and Albergone, F. (1958b). On the mechanism of adaptation to extreme osmotic conditions in *Dunaliella salina.* I. Physiological responses to changes of external osmotic pressure. *Rend. Accad. Naz. Lincei* **25** (Ser. 8), 567–575.

Marschner, H., and Schafarczyk, W. (1967a). Comparative studies of the net uptake of sodium and potassium by corn and sugar beet plants. *Z. Pflanzenernähr. Bodenk.* **118,** 172–186.

Marschner, H., and Schafarczyk, W. (1967b). Comparative studies of influx and efflux of sodium and potassium by corn and sugar beet plants. *Z. Pflanzenernähr. Bodenk.* **118,** 187–201.

Marshall, D. C. (1961). The freezing of plant tissue. *Aust. J. Biol. Sci.* **14,** 368–390.

Martin, J. F. (1932). The cold resistance of Pacific Coast spring wheats at various stages of growth as determined by artificial refrigeration. *J. Amer. Soc. Agron.* **24,** 871–880.

Maslova, G. M., Maslov, S. P., and Shnol', S. E. (1965). Acceleration of the germination of pollen of *Tradescantia paludosa* under the influence of sound vibrations in the auditory range. *Biofizika* **10,** 538–539.

Massey, V., Hofmann, T., and Palmer, G. (1962). The relation of function and structure in lipoyl dehydrogenase. *J. Biol. Chem.* **237,** 3820–3828.

Matikhin, G. R., and Boyko, L. A. (1957). Chlorine content in plants adapting to salination of soil. *Uch. Zap. Rostovsk. na Dony Gos. Univ.* **28,** 79–84.

Matsubara, H. (1967). Some properties of thermolysin. *In* "Molecular Mechanisms of Temperature Adaptation" (C. Ladd Prosser, ed.), pp. 283–294. Amer. Assoc. Adv. Sci., Wash. D.C.

Matukhin, G. R., Zhukovskaya, N. V., and Gomzhina, S. I. (1968). Effect of salinization of different quality on accumulation of free nucleotides in barley. *Fiziol. Rast.* **15,** 197–202.

Maxie, E. C. (1957). Heat injury in prunes. *Proc. Amer. Soc. Hort. Sci.* **69,** 116–121.
Maximov, N. A. (1908). Zur Frag über das Erfrieren der Pflanzen. *J. Botan. Ed. Sect. Bot. Soc. Imp. Nat. St. Petersburg,* 32–46. (*Bot. Centr.* **110,** 597–598.)
Maximov, N. A. (1912). Chemische Schutzmittel der Pflanzen gegen Erfrieren. *Ber. Deut. Bot. Ges.* **30,** 52–65, 293–305, 504–516.
Maximov, N. A. (1914). Experimentelle und kritische Untersuchungen über das Gefrieren und Erfrieren der Pflanzen. *Jahrb. Wiss. Bot.* **53,** 327–420.
Maximov, N. A. (1929a). Internal factors of frost and drought resistance in plants. *Protoplasma* **7,** 259–291.
Maximov, N. A. (1929b). "The Plant in Relation to Water" (R. H. Yapp, ed.). Allen and Unwin, London.
Maynard, D. N., and Barker, A. V. (1969). Studies on the tolerance of plants to ammonium nutrition. *J. Amer. Soc. Hort. Sci.* **94,** 235–239.
Mazelis, M., and Fowden, L. (1969). Conversion of ornithine into proline by enzymes from germinating peanut cotyledons. *Phytochemistry* **8,** 801–809.
Mazur, P. (1960). Physical factors implicated in the death of microorganisms at subzero temperatures. *Ann. N.Y. Acad. Sci.* **85** (Art. 2), 610–629.
Mazur, P. (1961). Physical and temporal factors involved in the death of yeast at subzero temperatures. *Biophys. J.* **1,** 247–264.
Mazur, P. (1963). Kinetics of water ions from cells at subzero temperatures and the likelihood of intracellular freezing. *J. Gen. Physiol.* **47,** 347–369.
Mazur, P. (1970). Cryobiology: the freezing of biological systems. *Science* **168,** 939–949.
Mazur, P., and Schmidt, J. J. (1968). Interactions of cooling velocity, temperature, and warming velocity on the survival of frozen and thawed yeast. *Cryobiology* **5,** 1–17.
Mazur, P., Rhian, M. A., and Mahlandt, B. G. (1957). Survival of *Pasteurella Tulariensis* in sugar solutions after cooling and warming at subzero temperatures. *J. Bacteriol.* **73,** 394–397.
Measures, M., and Weinberger, P. (1970). The effect of four audible sound frequencies on the growth of Marquis spring wheat. *Can. J. Bot.* **48,** 659–662.
Mehrishi, J. N., and Grassetti, D. R. (1969). Sulfhydryl groups on the surface of intact Ehrlich ascites tumor cells, human blood platelets, and lymphocytes. Nature (*London*) **224,** 563–564.
Mehta, P. D., and Maisel, H. (1966). The effect of heat on bovine lens proteins. *Experientia* **22,** 818–820.
Meindl, T. (1934). Weitere Beiträge zur protoplasmatischen Anatomie des Helodea-Blattes. *Protoplasma* **21,** 362–393.
Meinzer, O. E. (1927). "Plants as Indicators of Ground Water," U.S. Geol. Survey. Water Supply Paper 577, 95 pp. U.S. Government Printing Office, Wash. D.C.
Meiri, A., and Poljakoff-Mayber, A. (1967). The effect of chlorine salinity on growth of bean leaves in thickness and in area. *Isr. J. Bot.* **16,** 115–123.
Meiri, A., and Poljakoff-Mayber, A. (1969). Effect of variations in substrate salinity on the water balance and ionic composition of bean leaves. *Isr. J. Bot.* **18,** 99–112.
Meiri, A. and Poljakoff-Mayber, A. (1970). Effect of various salinity regimes on growth, leaf expansion and transpiration rate of bean plants. *Soil Sci.* **109,** 26–34.
Melchior, D. L., Morowitz, H. J., Sturtevant, J. M., and Tsong, T. Y. (1970). Characterisation of the plasma membrane of *Mycoplasma laidlowii. Biochim. Biophys. Acta* **219,** 114–122.
Meletti, P., and D'Amato, F. (1960). The embryo transplantation technique in the study of embryo-endosperm relations in irradiated seeds. *In* "Effects of Ionizing Radia-

tions on Seeds, A Symposium," pp. 47–55. International Atomic Energy Agency, Vienna, Austria.
Mercado, B. T. (1970). On NaCl resistance of *Beta vulgaris*, and *Zea mays*. Protoplasma **69,** 151–170.
Meryman, H. T. (1966a). "Cryobiology," 775 pp. Academic Press, New York.
Meryman, H. T. (1966b). Review of biological freezing. *In* "Cryobiology" (H. T. Meryman, ed.), Academic Press, New York. pp. 3–114.
Meryman, H. T. (1967). The relationship between dehydration and freezing injury in the human erythrocyte. *In* "Cellular Injury and Resistance in Freezing Organisms" (E. Asahina, ed.), pp. 231–244. Inst. Low Temp. Sci. Sapporo, Hokkaido.
Meryman, H. T. (1968). Modified model for the mechanism of freezing injury in erythrocytes. *Nature* (*London*) **218,** 333–336.
Meryman, H. T. (1970). The exceeding of a minimum tolerable cell volume in hypertonic suspension as a cause of freezing injury. *In* "The Frozen Cell" (G. E. W. Wolstenholme and M. O'Connor, eds.), pp. 51–67. Ciba Found. Symp., J. and A. Churchill, London.
Mesdag, J., and Slootmaker, L. A. J. (1969). Classifying wheat varieties for tolerance to high soil acidity. *Euphytica* **18,** 36–42.
Mewissen, D. J., Damblon, J., and Bacq, Z. M. (1959). Comparative sensitivity to radiation of seeds from a wild plant grown on uraniferous and nonuraniferous soils. *Nature* (*London*) **183,** 1449.
Meyer, B. S. (1932). Further studies on cold resistance in evergreens, with special reference to the possible role of bound water. *Bot. Gaz.* (*Chicago*) **94,** 297–321.
Meyer, H. W., and Winkelmann, H. (1969). Die Gefrierätzung und die Struktur biologischer Membranen. *Protoplasma* **68,** 253–270.
Michaelis, P. (1934). Okologische Studien an der alpinen Baumgrenze. IV. Zur Kenntnis des winterlichen Wasserhaushaltes. *Jahrb. Wiss. Bot.* **80,** 169–247.
Michaelis, G. P. (1935). Okologische Studien an der alpinen Baumgrenze. *Bot. Centr. Beih.* **52b,** 333–377.
Michel-Durand, E. (1919). Variation des substances hydrocarbonées dans les feuilles. *Rev. Gen. Bot.* **31,** 145–156, 196–204.
Miehe, H. (1907). *Thermoidium sulfureum* n.g.n.sp., ein neuer Wärmepilz. *Ber. Deut. Bot. Ges.* **25,** 510–515.
Migahid, A. M. (1938). Binding of water in relation to drought resistance. *Fouad I. Univ. Bul. Fac. Sci.* **18,** 5–28.
Milborrow, B. V., and Noddle, R. C. (1970). Conversion of 5-(1,2-Epoxy-2,6,6-trimethylcyclohexyl)-3-methylpenta-cis-2-trans-4-dienoic acid into abscisic acid in plants. *Biochem. J.* **119,** 727–734.
Miller, A. A. (1969). Glass transition temperature of water. *Science* **163,** 1325–1326.
Miller, A. T. (1968). The postradiation effect in plants and modifying it. *Radiobiol. Inform. Byull.* **11,** 40–43.
Miller, E. C. (1939). A physiological study of the winter wheat plant in different stages of its development. *Kansas Agr. Expt. Sta. Tech. Bull.*, p. 47.
Miller, L. K. (1969). Freezing tolerance in an adult insect. *Science* **166,** 105–106.
Miller, M. W. (1968). The radiosensitivity of gemmae of *Lunularia cruciata* to acute gamma irradiation. *Bryologist* **71,** 73–81.
Millerd, A., and McWilliam, J. R. (1968). Studies on a maize mutant sensitive to low-temperature I. Influence of temperature and light on the production of chloroplast pigments. *Plant Physiol.* **43,** 1967–1972.

Millerd, A. D., Goodchild, J., and Spencer, D. (1969). Studies on a maize mutant sensitive to low temperature. II. Chloroplast structure, development, and physiology. *Plant Physiol.* **44,** 567–583.

Milner, H. W., and Hiesey, W. M. (1964). Photosynthesis in climatic races of Mimulus (Scrophulariaceae). I. Effect of light intensity and temperature on rate. *Plant Physiol.* **39,** 208–213.

Milthorpe, F. L. (1950). Changes in the drought resistance of wheat seedlings during germination. *Ann. Bot. (London)* **14,** 79–89.

Minamikawa, T., Akazawa, T., and Uritani, I. (1961). Mechanism of cold injury in sweet potatoes. II. Biochemical mechanism of cold injury with special reference to mitochondrial activities. *Plant Cell Physiol.* **2,** 301–309.

Mironyuk, V. L., and Einor, L. O. (1968). Oxygen exchange and the content of pigments in various forms of *Dunaliella salina* Teod, under conditions of increased NaCl content. *Gidrobiol. Zh.* **4,** 23–29.

Mishiro, Y., and Ochi, M. (1966). Effect of dipicolinate on the heat denaturation of proteins. *Nature (London)* **211,** 1190.

Mishustina, P. S. (1967). The effect of low temperatures on the respiration rate and carbohydrate metabolism in corn. *Rost. Ustoich. Rast.* **3,** 227–232.

Misra, D. (1956). Study of drought resistance in certain crop plants. *Ind. J. Agron.* **1,** 25–39.

Mitra, S. K. (1921). Seasonal changes and translocation of carbohydrate materials in fruit spurs and two-year-old seedlings of apple. *Ohio J. Sci.* **21,** 89–90.

Mittelheuser, C. J. and van Steveninck, R. F. M. (1969). Stomatal closure and inhibition of transpiration induced by (RS)-abscisic acid. *Nature (London)* **221,** 281–282.

Miyamoto, T. (1962). Antagonistic effect of urea and 2-chloroethanol on the resistance to high salt concentration in wheat seedlings. *Nature (London)* **196,** 491–492.

Miyamoto, T. (1963). The effect of seed treatment with the extracts of organisms and the solutions of some chemical substances on the resistance to salt concentration in wheat seedlings. *Physiol. Plant.* **16,** 333–336.

Mizrahi, J., Blumenfeld, A., and Richmond, A. E. (1970). Abscisic acid and transpiration in leaves in relation to osmotic root stress. *Plant Physiol.* **46,** 169–171.

Modlibowska, I. (1968). Effects of some growth regulators on frost damage. *Cryobiology* **5,** 175–187.

Modlibowska, I., and Rogers, W. S. (1955). Freezing of plant tissues under the microscope. *J. Exp. Bot.* **6,** 384–391.

Möbius, M. (1907). Die Erkältung der Pflanzen. *Ber. Deut Bot. Ges.* **25,** 67–70.

Mohr, W. P., and Stein, M. (1969). Effect of different freeze-thaw regimes on ice formation and ultrastructural changes in tomato fruit parenchyma tissue. *Cryobiology* **6,** 15–31.

Molisch, H. (1896). Das Erfrieren von Pflanzen bei Temperaturen über dem Eispunkt. *Sitzber. Kaiserlichen Akad. Wiss. Wien. Math. Naturwiss. Kl.* **105,** 1–14.

Molisch, H. (1897). "Untersuchungen über das Erfrieren der Pflanzen," pp. 1–73. Fischer, Jena.

Molisch, H. (1926). "Pflanzenbiologie in Japan." Fischer, Jena.

Monk, W., and Wiebe, H. H. (1961). Salt tolerance and protoplasmic salt hardiness of various woody and herbaceous ornamental plants. *Plant Physiol.* **36,** 478–482.

Montfort, C., and Hahn, H. (1950). Atmung und Assimilation als dynamische Kennzeichen abgestüfter Trockenresistenz bei Farnen und höheren Pflanzen. *Planta* **38,** 503–515.

Montfort, C., Reid, A., and Reid, I. (1957). Gradation of functional heat resistance in marine algae and its relation to the environment and hereditary advantage. *Biol. Zentralbl.* **76,** 257–289.

Mooney, H. A. (1969). Dark respiration of related evergreen and deciduous Mediterranean plants during induced drought. *Bull. Torrey Bot. Club* **96,** 550–555.

Moor, H. (1960). Reaktionsweisen der Pflanzen auf Kälteeinflüsse. *Z. Schweiz. Forstv.* **30** (Festschrift. Prof. Frey-Wyssling), 211–222.

Morchiladze, Z. N. (1969). Transformation of $3C^{14}$-series in the grapevine. *Soobshch. Soobshch. Akad. Nauk. Gruz. SSR* **54,** 705–708.

Moretti, A. (1953). Physiological effects of winter treatments of chemicals upon grapevine. *Riv. Frutticolt. Viticolt. Orticolt.* **15,** 2–25.

Morey, M., and Gonzalez, F. (1966). Aspects of water utilization by plants in the Canary Islands. *Bol. Real. Soc. Espan. Hist. Natur. Secc. Biol.* **64,** 369–383.

Morgan, P. W., Joham, H. E., and Amin, J. V. (1966). Effect of manganese toxicity on the indoleacetic acid oxidase system of cotton. *Plant Physiol.* **41,** 718–724.

Morosov, A. S. (1939). Effect of temperature on the reversible activity of invertase in forage grasses as dependent on their cold and heat resistance. *C. R. Acad. Sci. U.R.S.S.* **23,** 949–951.

Morozovski, V. V., and Kabanov, V. V., (1968). Effect of chloride salinization on the nucleotide content in pea leaves. *Fiziol. Rast.* **15,** 909–913.

Morré, D. J. (1970). Auxin effects on the aggregation and heat coagulability of cytoplasmic proteins and lipoproteins. *Physiol. Plant.* **23,** 38–50.

Morren, C. (1838). Observations anatomiques sur la congélation des organes des végétaux. *Bull. Acad. Roy. Sci. Belles-lett. Bruxelles* **5,** 65–66, 93–111.

Morris, C. J., Thompson, J. F., and Johnson, C. M. (1969). Metabolism of glutamic acid and *N*-acetylglutamic acid in leaf discs and cell-free extracts of higher plants. *Plant Physiol.* **44,** 1023–1026.

Morris, J. Y., and Tranquillini, W. (1969). Über den Einfluss des osmotischen Potentiales des Wurzelsubstrates auf die photosynthese von *Pinus contorta*-Samlingen im Wechsel der Jahreszeiten. *Flora Abt.* B**158,** 277–287.

Morton, W. (1969). Effects of freezing and hardening on the sulfhydryl groups of protein fractions from cabbage leaves. *Plant Physiol.* **44,** 168–172.

Moschkov, B. S. (1935). Photoperiodismus und Frosthärte ausdauernder Gewächse. *Planta* **23,** 774–803.

Moser, W. (1969). Die Photosyntheseleistung von Nivalpflanzen. *Ber. Deut. Bot. Ges.* **82,** 63–64.

Mothes, K. (1928). Die Wirkung des Wassermangels auf den Eiweissumsatz in hoheren Pflanzen. *Ber. Deut. Bot. Ges.* (Generalversam) **46,** 59–67.

Mothes, K. (1956). Der Einfluss des Wasserzustandes auf Fermentprozesse und Stoffumsatz. *Handb. Pflanzenphysiol.* **3,** 656–664.

Mouravieff, I. (1967). First observations on the resistance to dehydration of detached foliar epidermal cells in some species of the Mediterranean Midi in the course of the dry season. *Bull. Soc. Bot. Fr.* **114,** 353–359.

Mouravieff, I. (1969). On the protoplasmic characteristics of leaf epidermal cells subjected to the influence of progressive dehydration: Experiments with tetraceycline as a vital fluorochrome. *Physiol. Veg.* **7,** 191–200.

Moyse, A., and Guyon, D. (1963). Effet de la température sur l'efficacité de la phycocyanine et de la chlorophylle chez *Aphanocapsa* (*Cyanophyceae*). *In* "Studies on Microalgae and Photosynthetic Bacteria," (S. Miyachi, ed.), pp. 253–270. Univ. Tokyo Press, Tokyo.

Mozafar, A., and Goodin, J. R. (1970). Vesiculated hairs: a mechanism for salt tolerance in *Atriplex halimus* L. *Plant Physiol.* **45,** 62–65.

Mozafer, A., Goodin, J. R., and Oertli, J. J. (1970). Sodium and potassium interactions in increasing the salt tolerance of *Atriplex halimus* L.: II. Na^+ and K^+ uptake characteristics. *Agron. J.* **62,** 481–484.

Mudd, J. B., Leavitt, R., and Kersey, W. H. (1966). Reaction of peroxyacetyl nitrate with sulfhydryl groups of proteins. *J. Biol. Chem.* **241,** 4081–4085.

Mueller, I. M., and Weaver, J. E. (1942). Relative drought resistance of seedlings of dormant prairie grasses. *Ecology* **23,** 387–398.

Müller-Thurgau, H. (1880). Über das Gefrieren und Erfrieren der Pflanzen. *Landwirtsch. Jahrb.* **9,** 133–189.

Müller-Thurgau, H. (1882). Über Zuckeranhäufung in Pflanzentheilen in Folge niederer Temperatur. *Landwirtsch. Jahrb.* **11,** 751–828.

Müller-Thurgau, H. (1886). Über das Gefrieren und Erfrieren der Pflanzen. II. Theile. *Landwirtsch. Jahrb.* **15,** 453–610.

Münch, E. (1913). Hitzeschaden an Waldpflanzen. *Naturwiss. Z. Forst. Landwirtsch.* **11,** 557–562.

Münch, E. (1914). Nochmals Hitzeschaden an Waldpflanzen. *Naturw. Z. Forst. Landwirt.* **12,** 169–188.

Muller, B., and Ziegler, H. (1969). The light-induced activation of $NADP^+$-dependent glyceraldehyde-3-phosphate dehydrogenase. IX. The reaction in isolated chloroplasts. *Planta* **85,** 96–104.

Muller, H. P. (1969). Studies of indirect irradiation of roots of *Vicia faba*. *Radiat. Bot.* **9,** 49–59.

Mumma, R. O., Fergus, C. L., and Sekura, R. D. (1970). The lipids of thermophilic fungi: Lipid composition comparisons between thermophilic and mesophilic fungi. *Lipids* **5,** 100–103.

Munamikawa, T., Akazawa, T., and Uritani, J. (1961). Mechanism of cold injury in sweet potatoes. II. Biochemical mechanism of cold injury with special reference to mitochondrial activities. *Plant–Cell Physiol.* **2,** 301–309.

Munda, I. (1964). The effect of salinity on the respiration and photosynthesis of the brown alga *Ascophyllum nodosum*. *Biol. Vest.* **11,** 3–13.

Murata, T. (1969). Physiological and biochemical studies of chilling injury in bananas. *Physiol. Plant.* **22,** 401–411.

Murr, L. E. (1964). Mechanism of plant-cell damage in an electrostatic field. *Nature* (*London*) **201,** 1305–1306.

Murr, L. E. (1965a). Plant growth response following exposure to a short duration electrostatic field. *Proc. Penn. Acad. Sci.* **38,** 44–46.

Murr, L. E. (1965b). Biophysics of plant growth in an electrostatic field. *Nature* (*London*) **206,** 467–470.

Murr, L. E. (1966). Plant physiology in simulated geoelectric and geomagnetic field. *Advan. Front. Plant Sci.* **15,** 97–120.

Murty, K. S., and Srinivasulu, K. (1968). Studies on the physiological basis of drought resistance in rice: II. Effect of drought on the concentration of sugars in rice plant. *Oryza J. Ass. Rice Res. Workers* **5,** 45–48.

Musaeva, L. D. (1957). Effect of water deficiency on respiration of barley at various periods of development. *Fiziol. Rast.* **4,** 229–236.

Musich, V. N. (1968). The content and composition of sugars in winter wheat plants on hardening. *Ref. Zh. Biol.* No. 36111.

Myers, D. K., and Church, M. L. (1967). Inhibition of stromal enzymes by x-radiation. *Nature* (*London*) **213,** 636–637.

Myrback, K. (1965). Studies on yeast B-fructofuranosidase (invertase). XV. The binding of cupric ion by the enzyme. *Arkiv. Kemi* **24,** 471–477.

Nägeli, C. (1861a). Über die Wirkung des Frostes auf die Pflanzenzellen. *Sitzber. Math. Phys. Kl. Bayer. Akad. Wiss. Munchen* pp. 264–271.

Nagatsu, T., Kuzuya, H., and Hidaka, H. (1967). Inhibition of dopamine β-hydroxylase by sulfhydryl compounds and the nature of the natural inhibitors (cow). *Biochim. Biophys. Acta* **139,** 319–327.

Naidu, K. M., and Bhagyalakshmi, K. V. (1967). Stomatal movement in relation to drought resistance in sugarcane. *Curr. Sci.* **36,** 555–556.

Naidu, K. M., Appala, B., and Venkateswarlu, B. (1967). Influence of soil moisture on sugar and nitrogen contents of *Pennisetum typhoides*, S. and H. Andhra. *Agr. J.* **14,** 143–148.

Nakanish, M., Wilson, A. C., Nolan, R. A., Gorman, G. C., and Bailey, G. S. (1969). Phenoxyethanol: Protein preservative for taxonomists. *Science* **163,** 681–683.

Nakata, S., and Suehisa, R. (1969). Growth and development of *Litchi chinensis* as affected by soil-moisture stress. *Amer. J. Bot.* **56,** 1121–1126.

Nakayama, S. (1963). Properties of a substance in Japanese radish leaf (*Raphanus sativus* L. var. *acanthiformis* Makino) which protects sweet potato B-amylase. *Agr. Biol. Chem.* **27,** 326–331.

Nakayama, S., and Kono, Y. (1957). Studies on the denaturation of enzymes. I. Effect of concentration on the rate of heat-inactivation of enzymes. *J. Biochem.* (*Tokyo*) **44,** 25–31.

Nash, C. H., Grant, D. W., and Sinclair, N. A. (1969). Thermolability of protein synthesis in a cell-free system from the obligately psychrophilic yeast *Candida gelida*. *Can. J. Microbiol.* **15,** 339–343.

Nations, C. (1967). The stability of wheat embryo glutamate decarboxylase under conditions of water stress. *Can. J. Bot.* **45,** 1917–1925.

Neales, T. F. (1968). Effects of high temperature and genotype on the growth of excised roots of *Arabidopsis thaliana*. *Aust. J. Biol. Sci.* **21,** 217–223.

Neales, T. F., Patterson, A. A., and Hartney, V. J. (1968). Physiological adaptation to drought in the carbon assimilation and water loss of xerophytes. *Nature* (*London*) **219,** 469–472.

Neger, F. W. (1915). Die Stärke-ökonomie der grünen Pflanze. *Naturwiss. Z. Forst. Landwirtsch.* **13,** 1.

Nei, T., Araki, T., and Matsusaka, T. (1967). The mechanism of cellular injury by freezing in microorganisms. *In* "Cellular Injury and Resistance in Freezing Organisms" (E. Asahina, ed.), pp. 157–170. Inst. Low Temp. Sci., Hokkaido.

Newman, E. I., and Kramer, P. J. (1966). Effect of decenylsuccinic acid on the permeability and growth of bean roots. *Plant Physiol.* **41,** 606–609.

Newton, R., and Anderson, J. A. (1931). Respiration of winter wheat plants at low temperatures. *Can. J. Res.* **5,** 337–354.

Newton, R., and Brown, W. R. (1931). Catalase activity of wheat leaf juice in relation to frost resistance. *Can. J. Res.* **5,** 333–336.

Newton, R., and Martin, W. M. (1930). Physico-chemical studies on the nature of drought resistance in crop plants. *Can. J. Res.* **3,** 336–427.

Newton, R., Brown, W. R., and Anderson, J. A. (1931). Chemical changes in nitrogen fractions of plant juices on exposure to frost. *Can. J. Res.* **5,** 327–332.

Nezgovorov, L. A., and Solov'ev, A. K. (1965). Increase of field cold resistance of maize produced by treating the seeds with large doses of TMTD (tetramethylthiuram disulfide). *Fiziol. Rast.* **12,** 1093–1103.

Nielsen, E. S., Kamp-Nielsen, L., and Wium-Anderson, S. (1969). The effect of deleterious concentrations of copper on the photosynthesis of *Chlorella pyrenoidosa*. *Physiol. Plant.* **22,** 1121–1133.

Nieman, R. H. (1962). Some effects of sodium chloride on growth, photosynthesis, and respiration of twelve crop plants. *Bot. Gaz.* (*Chicago*) **123,** 279–285.

Nieman, R. H. (1965). Expansion of bean leaves and its suppression by salinity. *Plant Physiol.* **40,** 156–161.

Nieman, R. H., and Bernstein, L. (1959). Interactive effects of gibberellic acid and salinity on the growth of beans. *Amer. J. Bot.* **46,** 667–670.

Nifontova, M. G. (1967). After-effects of dehydration and high temperature on photosynthesis in lichens. *Izv. Sib. Otd. Akad. Nauk. SSSR Ser. Biol. Med. Nauk.* **2,** 57–62.

Nigam, S. N., and McConnell, W. B. (1969). Seleno amino compounds from *Astragalus bisulcatus*; Isolation and identification of gamma-L-glutamyl-Se-methylseleno-L-cysteine and Se-methylseleno-L-cysteine. *Biochim. Biophys. Acta* **192,** 185–190.

Nikitina, A. N., Lebedeva, G. P., Yakubova, M. M., and Nasyrov, Yu. S. (1969). Effect of high-altitude UV radiation on the photochemical activity of chloroplasts. *Dokl. Akad. Nauk. Tadzh. SSR* **12,** 57–60.

Nikulina, G. N. (1969). Relative quantitative evaluation of the energy efficiency of respiration at high temperatures. *Bot. Zh.* **54,** 1242–1253.

Nilan, R. A., Konak, C. F., Legault, R. R., and Harle, J. R. (1961). The oxygen effect in barley seeds. *In* "Effects of Ionizing Radiation on Seeds, A Symposium," pp. 139–154. Internat. Atomic Energy Agency, Vienna.

Nir, I., Klein, S., and Poljakoff-Mayber, P. (1969). Effect of moisture stress on submicroscopic structure of maize roots. *Aust. J. Biol. Sci.* **22,** 17–33.

Nir, I., Poljakoff-Mayber, A., and Klein, S. (1970). The effect of water stress on mitochondria of root cells: A biochemical and cytochemical study. *Plant Physiol.* **45,** 173–177.

Noack, K. (1920). Der Betriebstoffwechsel der thermophilen Pilze. *Jahrb. Wiss. Bot.* **59,** 413–466.

Noda, L. (1958). Adenosine triphosphate-adenosine monophosphate transphosphorylase. *J. Biol. Chem.* **232,** 237–256.

Nollendorf, V. F. (1969). Effect of various manganese doses on tomato growth with relation to copper level in the nutrient solution. *Izv. Akad. Nauk. Latv. SSR* **5,** 86–92.

Nor-Arevyan, N. G. (1968). Effect of an increase in oxygen pressure on radiation damage of wheat seeds. *Biol. Zh. Arm.* **21,** 93–95.

Norberg, P. and Hofsten, B. V. (1970). Chromatography of a halophilic enzyme on hydroxylapatite in 3.4 *M* sodium chloride. *Biochim. Biophys. Acta* **220,** 132–133.

Norkrans, B. (1968). Studies on marine occurring yeasts: Respiration, fermentation and salt tolerance. *Arch. Mikrobiol.* **62,** 358–372.

Norkrans, B., and Kylin, A. (1969). Regulation of the potassium to sodium ratio and of the osmotic potential in relation to salt tolerance in yeasts. *J. Bacteriol.* **100,** 836–845.

Northen, H. T. (1943). Relationship of dissociation of cellular proteins by incipient drought to physiological processes. *Bot. Gaz.* (*Chicago*) **104,** 480–485.

Norton, T. A., and South, G. R. (1969). Influence of reduced salinity on the distribution of two laminarian algae. *Oikos* **20,** 320–326.

Novak, J., and Valek, L. (1965). Attempt at demonstrating the effect of a weak magnetic field on *Taraxacum officinale*. *Biol. Plant. Acad. Sci. Bohemoslov* **7,** 469–471.

Novikov, V. A. (1928). Cold resistance of plants. II. *J. Exp. Landwirtsch. Südosten Eur-Russlands* **6,** 71–100.

O'Brien, R. T. (1961). Radiation sensitivity studies on related fermenting and respiring yeasts. *Radiat. Bot.* **1,** 61–68.

Ogato, E., and Matsui, T. (1965). Photosynthesis in several marine plants of Japan as affected by salinity, drying and pH, with attention to their growth habitats. *Bot. Mar.* **8,** 199–217.

Ogawa, M., and Uritani, I. (1970). Effect of gamma radiation on peroxidase development in sweet potato disks. *Radiat. Res.* **41,** 324–351.

Ogo, T. (1956). Studies on saline injury in crops. III. The relation between the air supply to the root zones and the development of symptoms of saline injury in wheat plants. *Proc. Crop Sci. Soc. Jap.* **24,** 314.

Ogo, T., and Nishikawa, S. (1959). Changes of leaf water content in the "saline" crops. *Proc. Crop Sci. Soc. Jap.* **28,** 211–212.

Ohta, Y., Ogura, Y., and Wada, A. (1966). Thermostable protease from thermophilic bacteria. I. Thermostability, physicochemical properties, and amino acid composition. *J. Biol. Chem.* **241,** 5919–5925.

Ojha, M. N., and Turian, G. (1968). Thermostimulation of conidiation and succinic oxidative metabolism of *Neurospora crassa*. *Arch. Mikrobiol.* **62,** 232–241.

Oknina, E. Z., and Markovich, A. A. (1951). Means of increasing resistance to cold in *Rosa gallica*. *Izv. Akad. Nauk. SSSR Ser. Biol.*, pp. 107–114.

O'Leary, J. W. (1970). The influence of ground water salinization on plant growth. *Arid Lands Res. Newsletter* **34,** 4–7.

Oleinikova, T. V. (1965). High temperature and light effects on the permeability of cells of spring cereal leaves. *Sci. Counc. Cytolo. Problems Acad. Nauk. SSSR:* 70–81.

Olien, C. R. (1961). A method of studying stresses occurring in plant tissue during freezing. *Crop Sci.* **1,** 26–28.

Olien, C. R. (1965). Interference of cereal polymers and related compounds with freezing. *Cryobiology* **2,** 47–54.

Olien, C. R., Marchetti, B. L., and Chomyn, E. V. (1968). Ice structure in hardened winter barley. *Mich. Agr. Exp. Sta. Quart. Bull.* **50,** 440–448.

Olsen, R. H., and Metcalf, E. S. (1968). Conversion of mesophilic to psychrophilic bacteria. *Science* **162,** 1288–1289.

Omran, A. O., Atkins, I. M., and Gilmore, E. C. Jr. (1968). Heritability of cold hardiness in flax. *Crop Sci.* **8,** 716–719.

Onishi, H. (1959). Studies on osmophilic yeasts. IV. Change in permeability of cell membranes of the osmophilic yeasts and maintenance of their viability in the saline medium. *Bull. Agr. Chem. Soc. Jap.* **23,** 332–339.

Onoda, N. (1937). "Mikroskopische Beobachtungen über das Gefrieren einiger Pflanzenzellen in flüssigem Paraffin." Botan. Inst. der Kaiserlichen Univ. zu Kyoto. *Bot. and Zool.* **5,** 1845–2188.

Oppenheim, A., and Castelfranco, P. A. (1967). An acetaldehyde dehydrogenase from germinating seeds. *Plant Physiol.* **42,** 125–132.

Oppenheimer, H. R. (1932). Zur Kenntnis der hochsommerlichen Wasserbilanz mediterraner Geholze. *Ber. Deut. Bot. Ges.* **50**A, 185–245.

Oppenheimer, H. R. (1947). Studies on the water balance of unirrigated woody plants. *Palestine J. Bot. Rehovot Ser.* **6,** 63–77.

Oppenheimer, H. R. (1949). The water turnover of the Valonea oak. *Palestine J. Bot. Rehovot Ser.* **7,** 177–179.
Oppenheimer, H. R. (1951). Summer drought and water balance of plants growing in the near east. *J. Ecol.* **39,** 357–362.
Oppenheimer, H. R. (1953). An experimental study on ecological relationships and water expenses of mediterranean forest vegetation. *Palestine J. Bot. Rehovot. Seri.* **8,** 103–124.
Oppenheimer, H. R. (1960). Adaptation to drought: xerophytism. Plant–water relationships in arid and semi-arid conditions. *Arid Zone Res.* (*Unesco, Paris*) **15,** 105–138.
Oppenheimer, H. R. (1963). Zur Kenntnis kritischer Wasser-Sättigungsdefizite in Blattern und ihrer Bestimmung. *Planta* **60,** 51–69.
Oppenheimer, H. R. (1967). Mechanism of drought resistance in conifers of the Mediterranean zone and the arid west of the U.S.A. I. Physiological and anatomical investigations. Final Report, Project No. AID–FS7, Grant No. FG–IS–119.
Oppenheimer, H. R. (1968). Drought resistance of Monterey pine needles. *Isr. J. Bot.* **17,** 163–168.
Oppenheimer, H. R., and Elze, D. L. (1941). Irrigation of citrus trees according to physiological indicators. *Agr. Res. Sta. Bull. Rehovot* **31,** 1–28.
Oppenheimer, H. R., and Halevy, A. H. (1962). Anabiosis of *Ceterach officinarum* Lam. et OC. *Bull. Res. Counc. Isr.* **11D,** 127–147.
Oppenheimer, H. R., and Jacoby, B. (1963). Does plasmolysis increase the drought tolerance of plant cells? *Protoplasma* **57,** 619–627.
Oppenheimer, H. R., and Shomer-Ilan, A. (1963). A contribution to the knowledge of drought resistance of Mediterranean pine trees. *Mit. Flor.-Soz. Arbeitsg. Stolz.* (*Weser*) *N.S.* **10,** 42–55.
Ordin, L., and Skoe, B. P. (1964). Ozone effects on cell wall metabolism of Avena coleoptile sections. *Plant Physiol.* **39,** 751–755.
Ordin, L., Garber, M. J., and Kindinger, J. I. (1970). Effect of 2,4-D on cell wall metabolism of peroxyacetyl nitrate pretreated Avena coleoptile sections. *Physiol. Plant.* **23,** 117–123.
Osborne, L. D., and Da Costa, E. W. B. (1970). Variation in copper tolerance among homokaryotic and dikaryotic mycelia of *Polyporus palustris*. Berk. et Curt. *Ann. Bot.* **34,** 941–950.
Osmond, C. B. (1968). Ion absorption in Atriplex leaf tissue. I. Absorption by mesophyll cells. *Aust. J. Biol. Sci.* **21,** 1119–1130.
Osmond, C. B., Luttge, U., West, K. R., Pallaghy, C. K., and Shacher-Hill, B. (1969). Ion absorption in Atriplex leaf tissue. II. Secretion of ions to epidermal bladders. *Aust. J. Biol. Sci.* **22,** 797–814.
Ostaplyuk, E. D. K. (1967). A physiological characterization of the frost resistance of winter barley. *Rost. Ustoich. Rast.* **3,** 203–209.
Ota, T. (1964). Increasing tolerance of kidney bean plants (*Phaseolus vulgaris*) to toxic levels of sodium chloride and ammonium sulfate through application of B995. *Plant Cell Physiol.* **5,** 255–258.
Ota, K., and Yasue, T. (1959). Studies on the salt injury to crops. XIV. Relation between the temperature and salt injury in paddy rice. *Proc. Crop Sci. Soc. Jap.* **28,** 33–34.
Ouellette, G. L., and Dessureaux, L. (1958). Chemical composition of alfalfa as related to degree of tolerance to manganese and aluminium. *Can. J. Plant Sci.* **38,** 206–214.

Overath, P., Schairer, H. V., and Stoffel, W. (1970). Correlation of *in vivo* and *in vitro* phase transitions of membrane lipids in *Escherichia coli*. *Proc. Nat. Acad. Sci. U.S.* **67,** 606–612.

Overton, E. (1899). Beobachtungen und Versuche über das Auftreten von rothem Zellsaft bei Pflanzen. *Jahrb. Wiss. Bot.* **33**: 171–177.

Ozerskii, M. I. (1969). Effect of x-irradiation on the penetrability of the alga *Nitella flexilis* by sodium-22 ions. *Radiobiologiya* **9,** 120–122.

Padwal-Desai, S. R., Ahmed, E. M., and Dennison, R. A. (1969). Some properties of mitochondria from irradiated tomato fruit. *J. Food Sci.* **34,** 332–335.

Palenzona, D. L. (1961). Effects of high doses of X-rays on seedling growth in wheats of different ploidy. *In* "Effects of Ionizing Radiations on Seeds, A Symposium." pp. 533–542. Intern. Atomic Energy Agency, Vienna.

Palfi, G. (1968a). Changes in the amino acid content of detached wilting leaves of *Solanum laciniatum* Ait. in the light and in the dark. *Acta Agron. Acad. Sci. Hung.* **17,** 381–388.

Palfi, G. (1968b). Effect of kinetin, 2,4-DNP and antimetabolites on the changes of amino acid content of withering leaves, (*Solanum laciniatum, Capsicum annuum, Nicotiana tabacum, Oryza sativa, Triticum vulgare*). *Planta* **78,** 196–199.

Palfi, G., and Juhasz, J. (1969). Relationships among water deficiency, saline or cold root medium and proline, pipecolic acid and total amino acid contents of plants. *Z. Pflanzenernaehr. Bodenk.* **124,** 36–42.

Palfi, G., and Juhasz, J. (1970). Increase of the free proline level in water deficient leaves as a reaction to saline or cold root media. *Acta Agron. Acad. Sci. Hung.* **19,** 79–88.

Pallas, J. E., Jr., Michel, B. E., and Harris, D. G. (1967). Photosynthesis, transpiration, leaf temperature, and stomatal activity of cotton plants under varying water potentials. *Plant Physiol.* **42,** 76–88.

Pang, Shih-Chu'uan, Chang, H. C., and Wu, C. F. (1964). A preliminary study of increasing the salt tolerance of soybean and Italian millet (*Setaria italica*). *Acta Bot. Sinica* **12,** 64–73.

Panoyan, R. E. (1969). Effect of the radioprotectorant S, β-aminoethylizothiuronium bromide hydrobromide on the growth and development of barley at high doses of gamma-irradiation from Cesium-137. *Radiobiologiya* **9,** 134–138.

Pantanelli, E. (1918). Sur la resistanza delle piante al freddo. *Atti reale Accad. Ital. Mem. Cl. Sci. Fis. Mat. Nat.* **27,** 126–130, 148–153. (*Biol. Abstr.* **2,** 1135, 1919).

Paricha, D. C., and Levitt, J. (1967). Enhancement of drought tolerance by applied thiols. *Physiol. Plant* **20,** 83–89.

Parija, P., and Mallik, P. (1941). Nature of the reserve food in seeds and their resistance to high temperature. *J. Indian Bot. Soc.* **19,** 223–230.

Parija, P., and Pillay, K. P. (1945). Effect of pre-sowing treatment on the drought resistance in rice. *Proc. Nat. Acad. Sci. India* **15B,** 6–14.

Parish, R. W. (1969). The effects of light on peroxidase synthesis and indoleacetic acid oxidase inhibitors in coleoptiles and first-leaves of wheat. *Z. Pflanzenphysiol.* **60,** 90–97.

Parker, J. (1951). Moisture retention in leaves of conifers of the Northern Rocky Mountains. *Bot. Gaz.* (*Chicago*) **113,** 210–216.

Parker, J. (1953). Some applications and limitations of tetrazolium chloride. *Science* **118,** 77–79.

Parker, J. (1956). Drought resistance in woody plants. *Bot. Rev.* **22,** 241–289.

Parker, J. (1958). Sol-gel transitions in the living cells of conifers and their relation to resistance to cold. *Nature* (*London*) **182,** 1815.

Parker, J. (1959a). Seasonal variations in sugars of conifers with some observations on cold resistance. *Forest Sci.* **5,** 56–63.

Parker, J. (1959b). Seasonal changes in white pine leaves; a comparison of cold resistance and free-sugar fluctuations. *Bot. Gaz.* (*Chicago*) **121,** 46–50.

Parker, J. (1960). Survival of woody plants at extremely low temperatures. *Nature* (*London*) **187,** 1133.

Parker, J. (1962). Relationships among cold hardiness, water-soluble protein, anthocyanins, and free sugars in *Hedera helix* L. *Plant Physiol.* **37,** 809–813.

Parker, J. (1968). Drought resistance of roots of white ash, sugar maple, and red oak. *U.S. Forest Serv. Res. Paper NE*–**95,** 9 pp.

Parkes, A. S. (1964). Cryobiology *Cryobiology* **1,** 3.

Parsons, R. F. (1969). Physiological and ecological tolerances of *Eucalyptus incrassata* and *E. socialis* to edaphic factors. *Ecology* **50,** 386–390.

Partanen, C. R. (1960). Amino acid suppression of radiation-induced tumorization of fern prothalli. *Proc. Nat. Acad. Sci. U.S.* **46,** 1206–1210.

Pasternak, D., and Wilson, G. L. (1969). Effects of heat waves on grain sorghum at the stage of head emergence. *Aust. J. Exp. Agr. Anim. Husb.* **9,** 636–638.

Patt, H. M., Tyree, E. B., Straube, R. L., and Smith, D. E. (1949). Cysteine protection against x-irradiation. *Science* **110,** 213–214.

Paugh, R. L., and Gray, W. D. (1969). Studies on the growth of the osmophilic fungus *Eremascus albus. Mycologia* **61,** 281–288.

Pauli, A. W., and Mitchell, H. L. (1960). Changes in certain nitrogenous constituents of winter wheat as related to cold hardiness. *Plant Physiol.* **35,** 539–542.

Paulsen, G. M. (1968). Effect of photoperiod and temperature on cold hardening in winter wheat. *Crop Sci.* **8,** 29–32.

Peacocke, A. R., and Walker, I. O. (1962). The thermal denaturation of sodium deoxyribonucleate. II. Kinetics. *J. Mol. Biol.* **5,** 560–563.

Pearson, G. A., and Bernstein, L. (1959). Salinity effects at several growth stages of rice. *Agron. J.* **51,** 654–657.

Peisker, M. (1967). On the increase of radio-sensitivity in resting seeds through oxygen. *Kulturpflanze* **15,** 97–104.

Pellett, N. E., and White, D. B. (1969). Relationship of seasonal tissue changes to cold acclimation of *Juniperus chinensis* Hetzi. *J. Amer. Soc. Hort. Sci.* **94,** 460–462.

Peltier, G. L., and Kiesselbach, T. A. (1934). The comparative cold resistance of spring small grains. *J. Amer. Soc. Agron.* **26,** 681–686.

Penskoy, I. V. (1956). The salt resistance of cotton and its effect on the seasonal dynamics of salts in the Jura-Araksinskaya Plain. *Pochvovedenie* **8,** 86–91.

Pentzer, W. T., and Heinze, P. H. (1954). Post-harvest physiology of fruits and vegetables. *Ann. Rev. Plant Physiol.* **5,** 205–224.

Perkins, H. J., and Andrews, J. E. (1960). The effects of uptake of certain sugars and amino acids on the cold hardiness of young wheat plants. *Naturwissenschaften* **24,** 608–609.

Perner, E., von Ealck, S., and Jacobi, G. (1961). The effect of ionizing rays on chloroplasts and the Hill reaction. *Z. Naturforsch.* **16B,** 74–75.

Peters, D. B. (1963). Use of octa-hexadecanol as a transpiration suppressant. *Agron. J.* **55,** 79.

Peters, R. A. (1963). "Biochemical Lesions and Lethal Synthesis," 321p. Macmillan (Pergamon), New York.

Petinov, N. S., and Molotkovsky, U. G. (1957). Protective reactions in heat-resistant plants induced by high temperatures. *Fiziol. Rast.* **4,** 221–228.

Petinov, N. S., and Molotkovsky, U. G. (1961). The protective processes of heat-resistant plants. pp. 275–283 *In* "Arid Zone Research, Plant-water Relationships in Arid and Semiarid Conditions, A Symposium." Vol. XVI pp. 275–283. UNESCO, Columbia Univ. Press, New York.

Petinov, N. S., and Molotkovsky, Yu, G. (1962). Heat stability of plants and ways of increasing it. *Vest. Akad. Nauk. SSSR* **8,** 62–64.

Petit-Thouars, A. du. (1817). "Le verger Français ou traité général de la culture des arbres fruitiers, etc," pp. 6–45. Paris.

Petrie, A. H. K., and Arthur, J. I. (1943). Physiological Ontogeny in the Tobacco Plant. The effect of varying water supply on the drifts in dry weight and leaf area and on various components of the leaves. *Austr. J. Exp. Biol. Med. Sci.* **21,** 191–200.

Petrova, O. V. (1967). Free amino acids and the form of nitrogen in the leaves during development of corn hybrids differing in resistance to cold. *Rost. Ustoich. Rast.* **SB3,** 233–240.

Pfeiffer, M. (1933). Frostuntersuchungen an Fichtentrieben. *Tharandter Forstl. Jahrb.* **84,** 664–695.

Phillips, G. O. (1968). "Energetics and Mechanisms in Radiation Biology." Academic Press, New York.

Pieniazek, J., and Wisniewska, J. (1954). The properties of the protoplasm in the tissues of one-year old fruit-tree shoots in the course of different phenophases. *Bull. Acad. Pol. Sci. Cl.* 2 **2,** 149–152.

Pihl, A., and Sanner, T. (1963a). Protection of sulfhydryl compounds against ionizing radiation. *Biochim. Biophys. Acta* **78,** 537–539.

Pihl, A., and Sanner, T. (1963b). X-ray inactivation of papain in solution. *Radiat. Res.* **19,** 27–41.

Pilet, P. E., and Dubois, J. (1968). Variations in content of acid-soluble sulfhydryl compounds in cultured tissues. *Physiol. Plant.* **21,** 445–454.

Pilet, P. E., and Braun, R. (1970). Ribonuclease activity and auxin effects in the Lens root. *Physiol. Plant.* **23,** 245–250.

Pincock, R. E., and Kiovsky, T. E. (1966). Reactions in frozen solutions. XI. The reaction of ethylene chlorohydrin with hydroxyl ion in ice. *J. Amer. Chem. Soc.* **88,** 4455–4459.

Pireyre, N. (1961). Contribution to the morphology, histology and physiology of cystoliths. *Rev. Cytol. Biol. Veg.* **23,** 93–320.

Pisek, A. (1950). Frosthärte und Zusammensetzung des Zellsaftes bei *Rhododendron ferrungineum, Pinus cembra* und *Picea excelsa. Protoplasma* **39,** 129–146.

Pisek, A. (1953). Wie schutzen sich die Alpenpflanzen gegen Frost? *Umschau* Heft, 21.

Pisek, A. (1958). Versuche zur Frostresistenzprüfung von Reinde, Winterknospen und Blüten einiger Arten von Obsthölzern. *Gartenbauwissenschaft* **23,** 54–74.

Pisek, A. (1962). Frostresistenz von Bäumen. *Meded. Inst. Vered. Tuenbougew, Wageningen* **182,** 74–79.

Pisek, A., and Kemnitzer, R. (1967). Der Einfluss von Frost auf die Photosynthese der Weisstanne (*Abies alba* Mill). *Flora* (*Jena*) **157,** 315–326.

Pisek, A., and Larcher, W. (1954). Zusammenhang zwischen Austrocknungsresistenz und Frosthärte bei Immergrünenpflanzen. *Protoplasma* **44,** 30–46.

Pisek, A., and Winkler, E. (1953). Die Schliessbewegung der Stomata bei ökologisch verschiedenen Pflanzentypen in Abhängigkeit vom Wassersättigungszustand der Blätter und vom Licht. *Planta* **42,** 253–278.

Pisek, A., Sohm, H., and Cartellieri, E. (1935). Untersuchungen über osmotischen Wert und Wassergehalt von Pflanzen und Planzengesellschaften der aplinen Stufe. *Beitr. Bot. Centr.* **52,** 634–675.

Pisek, A., Larcher, W., Pack, I., and Unterholzner, R. (1968). Kardinale Temperaturbereiche der Photosynthese und Grenztemperaturen des Lebens der Blätter verschiedener Spermatophyten. II. Temperaturmaximum der Netto-Photosynthese und Hitzeresistenz der Blätter. *Flora (Jena)* **158,** 110–128.

Plaut, Z., and Halevy, A. H. (1966). Regeneration after wilting, growth and yield of wheat plants, as affected by two growth retarding compounds. *Physiol. Plant.* **19,** 1064–1072.

Plaut, Z., Halevy, A. H., and Shmueli, E. (1964). The effect of growth retarding chemicals on growth and transpiration of bean plants grown under various irrigation regimes. *Isr. J. Agr. Res.* **14,** 153–158.

Podin, V. S. (1966). Comparative study of the xanthophyll transformation reaction of some plants as a factor of temperature in light and darkness. *Izv. Akad. Nauk. Latv. SSR* **11,** 82–86.

Pogosyan, K. S., and Sklyarova, I. A. (1968). Effect of low below-zero temperature on the respiration rate of grapevine shoots and on the activity of their oxidative enzymes. *Biol. Zh. Arm.* **21,** 72–78.

Pojarkova, H. A. (1924). Winterruhe, Reservestoffe, und Kälteresistenz bei Holzpflanzen. *Ber. Deut. Bot. Ges.* **42,** 420–429.

Pokrovskaia, E. I. (1958). Salt hardiness and various metabolic pathways in glycophytes. *Fiziol. Rast.* **5,** 260–266.

Polishchak, L. K., Dibrova, L. S., Zablotskaya, K. M., and Lapchik, V. F. (1968). The importance of oxidation-reduction processes in the frost resistance of plants. *Rost. Ustoich. Rast.* **SB4,** 122–129.

Pollock, B. M. (1969). Imbibition temperature sensitivity of lima bean seeds controlled by initial seed moisture. *Plant Physiol.* **44,** 907–911.

Pollock, G., and Waisel, Y. (1970). Salt secretion in *Aeluropus litoralis* (Willd.) Parl. *Ann. Bot.* **34,** 879–888.

Polster, H. (ed.) (1968). "Klimaresistenz Photosynthese und Stoffproduktion," 257 pp. Tagungsberichte Nr. 100. Deut. Akad. Landwirtsch., Berlin.

Pomerleau, R., and Ray, R. G. (1957). Occurrence and effects of summer frost in a conifer plantation. *Forest Res. Division Tech. Note (Ottawa, Can.)* **51,** 1–15.

Pontremoli, S., Traniello, S., Enser, M., Shapiro, S., and Horecker, B. L. (1967). Regulation of fructose diphosphatase activity by disulfide exchange. *Proc. Nat. Acad. Sci. U.S.* **58,** 286–293.

Porath, E., and Poljakoff-Mayber, A. (1965). Effect of salinity on metabolic pathways in pea root tips. *Isr. J. Bot.* **13,** 115–121.

Porath, E., and Poljakoff-Mayber, A. (1968). The effect of salinity in the growth medium on carbohydrate metabolism in pea root tips. *Plant Cell Physiol.* **9,** 195–203.

Porodko, T. M. (1926a). Über die Absterbegeschwindigkeit der erhitzten Samen. *Ber. Deut. Bot. Ges.* **44,** 71–80.

Porodko, T. M. (1926b). Einfluss der Temperatur auf die Absterbegeschwindigkeit der Samen. *Ber. Deut. Bot. Ges.* **44,** 80–84.

Porto, F., and Siegel, S. M. (1960). Effects of exposures of seeds to various physical agents. III. Kinetin-reversible heat damage in lettuce seed. *Bot. Gaz. (Chicago)* **122,** 70–71.

Powell, R. D., and Amin, J. V. (1969). Effect of chilling temperatures on the growth and development of cotton. *Cotton Growing Rev.* **46,** 21–28.

Prat, S., and Kubin, S. (1956). Photosynthesis and respiration of thermophilic blue-green algae. *Fiziol. Rast.* **3,** 508–515.
Precht, H. (1967). A survey of experiments on resistance adaptation. *In* "The Cell and Environmental Temperature" (A. S. Troshin, ed.), pp. 307–321. Pergamon Press, Oxford.
Precht, H., Christophersen, J., and Hensel, H. (1955). "Temperatur und Leben." Heidelberg.
Pressey, R. (1969). Potato sucrose synthetase: purification, properties, and changes in activity associated with maturation. *Plant Physiol.* **44,** 759–764.
Prianishnikov, D. N. (1951). "Nitrogen in the Life of Plants," 109 pp. (Transl. by S. A. Wilde). Kramer Business Service Inc. Madison, Wisconsin.
Prikhod'ko, L. S. (1968). Oxy- and keto-acid metabolism in pea and maize seedlings on saline substrates. *Fiziol. Rast.* **15,** 806–812.
Prikhod'ko, L. S., and Klyshev, L. K. (1963). The change in metabolism of pea seedlings caused by salinization of the substrate. *Tr. Inst. Bot. Akad. Nauk. Kaz. SSR* **16,** 81–96.
Prillieux, E. (1869). Sur la formation de glaçons a l'intérieur des plantes. *Ann. Sci. Nat. Paris* (Ser. 5) **12,** 125–134.
Prillieux, E. (1872). Coloration en bleu des fleurs de quelques orchidées sous l'influence de la gelée. *Bull. Soc. Bot. Fr.* **19,** 152–155.
Pringsheim, E. (1906). Wasserbewegung und Turgoregulation in welkenden Pflanzen. *Jahrb. Wiss. Bot.* **43,** 89–144.
Proctor, J. (1970). Magnesium as a toxic element. *Nature (London)* **227,** 742–743.
Proebsting, E. L., Jr. (1959). Cold hardiness of Elberta peach fruit buds during four winters. *Proc. Amer. Soc. Hort. Sci.* **74,** 144–153.
Prothero, J. W., Bulger, R., Jr., Chambers, J., and Mack, N. (1970). A kinetic model of membrane repair. *Can. J. Physiol. Pharmacol.* **48,** 422–432.
Protsenko, D. F., and Rubanyuk, E. A. (1967). Amino acid metabolism in winter rye and wheat during the overwintering period. *Rost. Ustoich. Rast.* Sb. **3,** 161–169.
Protsenko, D. F., Shmat'ko, I. G., and Rubanyuk, E. A. (1968). Drought hardiness of winter wheat varieties as related to their amino acid composition. *Fiziol. Rast.* **15,** 680–687.
Pruzsinszky, S. (1960). Über Trocken-und Feuchtluftresistenz des Pollens. *Sitzber. Oesterr Akad. Wiss. Math. Nat. Kl. Abt. I.* **169,** 43–100.
Pullman, M. E., and Monroy, G. C. (1963). A naturally occurring inhibitor of mitochondrial adenosine triphosphatase. *J. Biol. Chem.* **238,** 3762–3769.
Puscas, M., Stoiciu, T., Baia, V., Otarasanu, A., and Puscasu, A. (1966). Characteristics of some physiological processes in bean under the toxic action of some salts. *Inst. Agron. Timisoara Lucr. Stiint. Ser. Agron.* **9,** 295–309.
Pushpalata. (1967). Effect of cobalt on Biloxi soybean leaf. *Indian J. Exp. Biol.* **5,** 56–57.
Pyykko, M. (1966). The leaf anatomy of East Patagonian xeromorphic plants. *Ann. Bot. Fenn.* **3,** 453–622.
Quastler, H., and Baer, M. (1949). Inhibition of plant growth by irradiation. II. Sensitivity and development. *J. Cell. Comp. Physiol.* **33,** 349–363.
Quatrano, R. S. (1968). Freeze preservation of cultured flax cells utilizing dimethyl sulfoxide. *Plant Physiol.* **43,** 2057–2061.
Quisenberry, K. S., and Bayles, B. B. (1939). Growth habit of some winter wheat varieties and its relation to winter hardiness and earliness. *J. Amer. Soc. Agron.* **31,** 785–789.

Rabe, F. (1905). Über die Austrocknungsfähigkeit gekeimter Samen und Sporen. *Flora (Jena)* **95,** 253–324.

Rabinovitch, B., Maling, J. E., and Weissbluth, M. (1967). Enzyme–substrate reactions in very high magnetic fields. I. *Biophys. J.* **7,** 187–204.

Rachie, K. O., and Schmid, A. R. (1955). Winter hardiness of birdsfoot trefoil strains and varieties. *Agron. J.* **47,** 155–157.

Radchenko, S. I., Konovalov, I. N., and Pozdova, L. M. (1964). Frost resistant corn on the Korelian Isthmus. *Tr. Bot. Inst. Akad. Nauk. SSR. Ser.* **4,** 17: 53–72.

Radzievsky, G. B., and Shekhtman, Ya L. (1955). The application of roentgeno-structural analysis for the study of ice formation in plant grains. *Dokl. Akad. Nauk. SSSR* **101,** 1051–1053.

Raheja, P. C. (1951). Recent physiological investigations on drought resistance in crop plants. *Indian J. Agr. Sci.* **21,** 335–346.

Rains, D. W., and Epstein, E. (1967). Preferential absorption of potassium by leaf tissue of the mangrove, *Avicennia marina:* An aspect of halophytic competence in coping with salt. *J. Biol. Sci.* **20,** 847–857.

Raju, M. R., Amer, N. M., Gnanapurani, M., and Richman, C. (1970). The oxygen effect of π-Mesons in *Vicia faba*. *Radiat. Res.* **41,** 135–144.

Rakova, N. M., Klyshev, L. K., and Strogonov, B. P. (1969). Effect of sodium sulfate and sodium chloride on the protein composition of pea roots. *Fiziol. Rast.* **16,** 22–28.

Ramaiah, A., and Shanmugasundaram, E. R. B. (1962). Effect of molybdenum toxicity on sulphur metabolism in *Neurospora crassa*. *Biochim. Biophys. Acta* **60,** 373–385.

Ramakrishnan, P. S. (1969). Nutritional factors influencing the distribution of the calcareous and acidic populations in *Hypericum perforatum*. *Can. J. Bot.* **47,** 175–181.

Rao, D. N., and Leblanc, F. (1966). Effects of sulfur dioxide on the lichen alga, with special reference to chlorophyll. *Bryologist* **69,** 69–75.

Rao, K. S., Rao, M. B., and Rao, B. S. (1949). Drought resistance of plants in relation to hysteresis in sorption. I. Hydration and dehydration of the leaves of certain drought resistant and drought sensitive plants. *Proc. Nat. Inst. Sci. India* **15,** 41–49, 51–58.

Raschke, K. (1956). Über die physikalischen Beziehungen zwischen Wärmeübergangszahl, Strahlungsaustausch, Temperatur und Transpiration eines Blattes. *Planta* **48,** 200–238.

Raschke, K. (1960). Heat transfer between the plant and the environment. *Ann. Rev. Plant. Physiol.* **11,** 111–126.

Rauser, W. E., and Hanson, J. B. (1966). The metabolic status of ribonucleic acid in soybean roots exposed to saline media. *Can. J. Bot.* **44,** 759–776.

Rawlins, S. L., Gardner, W. R., and Dalton, F. N. (1968). *In situ* measurement of soil and plant leaf water potential. *Soil Sci. Soc. Amer. Proc.* **32,** 468–470.

Razmaev, I. I. (1965). After-effect of low temperatures above 0°C on nitrogen metabolism in wheat and corn. *Izv. Sib. Otol. Akad. Nauk. SSSR Ser. Biol. Med. Nauk.* **1,** 59–63.

Read, J. (1959). "Radiation Biology of *Vicia faba* in Relation to the General Problem." Thomas, Springfield, Illinois.

Redshaw, E. S., and Zalek, S. (1968). Changes in lipids of cereal seedlings during vernalization. *Can. J. Biochem.* **46,** 1093–1097.

Reid, D. A., Armiger, W. H., Foy, C. D., Koch, E. J., and Starling, T. M. (1969). Differential aluminum tolerance of winter barley varieties and selections in associated greenhouse and field experiments. *Agron. J.* **6,** 218–220.

Reilly, C. (1969). The uptake and accumulation of copper by *Becium homblei* (De Wild.) Duvig. et Plancke. *New Phytol.* **68,** 1081–1087.
Rein, R. (1908). Untersuchungen über den Kältetod der Pflanzen. *Z. Naturforsch.* **80,** 1–38.
Reinert, J. C., and Steim, J. M. (1970). Calorimetric detection of a membrane-lipid phase transition in living cells. *Science* **168,** 1580–1582.
Reisner, A. H., Rowe, J., and Sleigh, R. W. (1969). Concerning the tertiary structure of the soluble surface proteins of Paramecium. *Biochemistry* **8,** 4637–4644.
Repp, G. (1958). Die Salztoleranz der Pflanzen. I. Salzhaushalt und Salzresistenz von Marschpflanzen der Nordseeküstr Dänemarks in Beziehung zum Standort. *Oesterr. Bot. Z.* **104,** 454–490.
Repp, G. (1963). Copper resistance of the protoplasm of higher plants on copper soils. *Protoplasma* **57,** 643–659.
Repp, G., McAllister, D. R., and Wiebe, H. (1959). Salt resistance of protoplasm as a test for the salt tolerance of agricultural plants. *Agron. J.* **51,** 311–314.
Resnick, M. E. (1970). Effect of mannitol and polyethylene glycol on phosphorus uptake my maize plants. *Ann. Bot.* **34,** 497–504.
Reum, J. A. (1835). "Pflanzenphysiologie, oder das Leben Wachsen und Verhalten der Pflanzen," pp. 168–169. Arnoldische Buchhandlung. Dresden u. Leipzig.
Rey, L. (1961). Automatic regulation of the freeze-drying of complex systems. *Biodynamica* **8,** 241–260.
Richardson, B., Wagner, F. W., and Welch, B. E. (1969). Growth of *Chlorella sorokiniana* at hyperbaric oxygen pressures. *Appl. Microbiol.* **17,** 135–138.
Richter, H. (1968a). Die Gefrierresistenz glyzerinbehandelter Campanulazellen. *Protoplasma* **66,** 63–78.
Richter, H. (1968b). Die Reaktion hochpermeable Pflanzenzellen auf drei Gefrierschutzstoffe (Glyzerin, Aethylenglukal, Dimethylsulfoxid). *Protoplasma* **65,** 155–166.
Rickard, D. S., and Fitzgerald, P. D. (1969). The estimation and occurrence of agricultural drought. *J. Hydrol.* **8,** 11–16.
Ried, A. (1960a). Thallusbau und Assimilationshaushalt von Laub-und Krustenflechten. *Biol. Zentralbl.* **79,** 129–151.
Ried, A. (1960b). Stoffwechsel und Verbreitungsgrenzen von Flechten. II. Wasser und Assimilationshaushalt, Entquellungs-und Submersionsresistenz von Krustenflechten benachbarter Standorte. *Flora (Jena)* **149,** 345–385.
Rieth, A. (1966). Zur Kenntnis der Lebensbedingungen von *Porphyridium cruentum* (Ag.) Naeg. VI. Wachstum in einum Rhythmus von Frost und Wärmeperioden. *Biol. Zentralbl.* **85,** 569–578.
Rigby, B. J. (1967). Correlation between serine and thermal stability of collagen. *Nature (London)* **214,** 87–88.
Rimpau, R. H. (1958). Untersuchungen über die Wirkung von kritischer Photoperiode und Vernalisation auf die Kältresistenz von *Triticum aestivum* L. *Z. Pflanzenzuecht* **40,** 275–318.
Riov, J., and Goren, R. (1970). Effects of gamma radiation and ethylene on protein synthesis in peel of mature grapefruit. *Radiat. Bot.* **10,** 155–160.
Ritchie, D. (1957). Salinity optima for marine fungi affected by temperature. *Amer. J. Bot.* **44,** 870–874.
Rivera, V., and Corneli, E. (1931). Rassegna die casi fitopatologici osservati nel 1929 (danni da freddo e da crittogame). *Riv. Patol. Veg.* **21,** 65–100.
Robbins, W. J., and Petsch, K. F. (1932). Moisture content and high temperature in relation to the germination of corn and wheat grains. *Bot. Gaz. (Chicago)* **93,** 85–92.

Robelin, M. (1967). Effects and after-effects of drought on the growth and yield of sunflowers. *Ann. Agron.* **18,** 579–599.

Roberts, D. W. A. (1967). The temperature coefficient of invertase from the leaves of cold-hardened and cold-susceptible wheat plants. *Can. J. Bot.* **43,** 1347–1357.

Roberts, D. W. A. (1969a). A comparison of the peroxidase isozymes of wheat plants grown at 6°C and 20°C. *Can. J. Bot.* **47,** 263–265.

Roberts, D. W. A. (1969b). Some possible roles for isozymic substitutions during cold hardening in plants. *Int. Rev. Cytol.* **26,** 303–328.

Roberts, D. W. A., and Grant, M. N. (1968). Changes in cold hardiness accompanying development in winter wheat. *Can. J. Plant Sci.* **48,** 369–376.

Roberts, K., and Siegel, S. M. (1967). Experimental microbiology of saturated salt solutions and other harsh environments. III. Growth of salt-tolerant *Penicillium notatum* in boron-rich media. *Plant Physiol.* **42,** 1215–1218.

Roberts, R. H. (1922). The development and winter injury of cherry blossom buds. *Wisc. Agr. Expt. Sta. Res. Bull.*, No. 52.

Roberts, W. J. (1961). Reduction of transpiration. *J. Geophys. Res.* **66,** 3309–3312.

Robertson, A. H. (1927). Thermophile and thermoduric microorganisms, with special reference to species isolated from milk. *N. Y. Agr. Expt. Sta. Tech. Bull.*, No. 130.

Robinson, T. W. (1957). The phraeatophyte problem. *Symp. Phraeatophytes*, pp. 1–12. Rep. Southwest Reg. Meeting Amer. Geophys. Union, Sacramento, California.

Robson, M. G., and Jewiss, O. R. (1968). A comparison of British and North African varieties of tall fescue (*Festuca arundinacea.*) II. Growth during winter and survival at low temperatures. *J. Appl. Ecol.* **5,** 179–190.

Rogers, W. S. (1954). Some aspects of spring frost damage to fruit and its control. *J. Roy. Hort. Soc.* **79,** 29–36.

Rogers, W. S., Modlibowska, I., Ruxton, J. P., and Slater, C. H. W. (1954). Low temperature injury to fruit blossom. IV. Further experiments on water-sprinkling as an anti-frost measure. *J. Hort. Sci.* **29,** 126–141.

Romani, R. J., and Ku, L. L. (1970). Effects of ionizing radiation on the ribosomal system in relation to intracellular repair of radiation damage. *Radiat. Res.* **41,** 217–225.

Romani, R. J., Yu, I. K., Ku, L. L., Fisher, L. K., and Dehgan, N. (1968). Cellular senescence, radiation damage to mitochondria, and the compensatory response in ripening pear fruits. *Plant Physiol.* **43,** 1089–1096.

Romanova, L. N. (1967). The physiological bases for the resistance to cold of winter crops. *Ref. Zh. Biol.* 6G132.

Rosa, J. T. (1921). Investigation on the hardening process in vegetable plants. *Mo. Agr. Expt. Sta. Res. Bull.*, No. 48.

Rose, A. H. (1968). Physiology of micro-organisms at low temperatures. *J. Appl. Microbiol.* **31,** 1–11.

Rose, R. J., and Adamson, D. (1969). A sequential response to growth substances in coleoptiles from gamma-irradiated wheat. *Planta* **88,** 274–281.

Rosenberg, A., and Enberg, J. (1969). Studies of hydrogen exchange in protein: II. The reversible thermal unfolding of chymotrypsinogen A as studied by exchange kinetics. *J. Biol. Chem.* **244,** 6153–6159.

Rothstein, A., and Weed, R. I. (1963). The functional significance of sulfhydryl groups in the cell membrane. AEC Res. Div. Rep. UR–633, p. 35.

Rottenberg, W. (1968). Die Standardisierung von Frostresistenz Untersuchungen angewandt an Aussenepidermiszellen von *Allium cepa* L. *Protoplasma* **65,** 37–48.

Rouschal, E. (1938a). Eine physiologische Studie an *Ceterach officinarum*. *Flora (Jena)* **132,** 305–318.

Rouschal, E. (1938b). Zum Wärmehaushalt der Macchienpflanzen. *Oesterr. Bot. Z.* **87,** 42–50.

Rouschal, E. (1939). Zur Ökologie der Macchien. I. Der sommerliche Wasserhaushalt der Macchienpflanzen. *Jahrb. Wiss. Bot.* **87,** 436–523.

Rousseau, D. L., and Porto, S. P. S. (1970). Polywater: polymer or artifact? *Science* **167,** 1715–1719.

Roy, D. K. (1956). Heat stability of fungal alpha-amylase. *Ann. Biochem. Exp. Med.* **16,** 111–112.

Roy, R. M., and Clark, G. M. (1970). Carbon dioxide fixation and translocation of photoassimilates in *Vicia faba* following x-irradiation. *Radiat. Bot.* **10,** 101–111.

Rudorf, W. (1938). Keimstimmung und Photoperiode in ihrer Bedeutung fur die Kälteresistenz. *Zuechter* **10,** 238–246.

Rutherford, P. P., and Weston, E. W. (1968). Carbohydrate changes during cold storage of some inulin-containing roots and tubers. *Phytochemistry* **7,** 175–180.

Saakyan, R. G., and Petrosyan, G. P. (1964). Effect of soil salinity on the level of nucleic acids and nitrogenous substances in grape leaves. *Fiziol. Rast.* **11,** 681–688.

Sabnis, D. D., Gordon, M., and Galston, A. W. (1970). Localization of adenosine triphosphatase activity on the chloroplast envelope in tendrils of *Pisum sativum*. *Plant Physiol.* **45,** 25–52.

Sachs, J. (1860). Krystallbildungen bei dem Gefrieren und Veränderung der Zellhäute bei dem Aufthauen saftiger Pflanzentheile, mitgetheilt von W. Hofmeister. *Ber. Verhandl. Sächs. Akad. Wiss. Leipzig. Math. Phys. Kl.* **12,** 1–50.

Sachs, J. (1864). Ueber die obere Temperatur-Grenze der Vegetation. *Flora (Jena)* **47,** 5–12, 24–29, 33–39, 65–75.

Saito, N., and Werbin, H. (1969). Action spectrum for a DNA-photoreactivating enzyme isolated from higher plants. *Radiat. Bot.* **9,** 421–424.

Sakai, A. (1955a). The relationship between the process of development and the frost hardiness of the mulberry tree. *Low Temp. Sci. Ser.* **B13,** 21–31.

Sakai, A. (1955b). The seasonal changes of the hardiness and the physiological state of the cortical parenchyma cells of mulberry tree. *Low Temp. Sci. Ser.* **B13,** 33–41.

Sakai, A. (1956a). The effect of temperature on the maintenance of the frost hardiness. *Low Temp. Sci. Ser.* **B14,** 1–6.

Sakai, A. (1956b). The effect of temperature on the hardening of plants. *Low Temp. Sci. Ser.* **B14,** 7–15.

Sakai, A. (1957). The effect of maleic hydrazide upon the frost hardiness of twig of mulberry tree. *J. Sericult. Sci. Jap.* **26,** 13–20.

Sakai, A. (1958). Survival of plant tissue at super-low temperature. II. *Low Temp. Sci. Ser.* **B16,** 41–53.

Sakai, A. (1960a). The frost hardening process of woody plant. VII. Seasonal variations in sugars. *Low Temp. Sci. Ser.* **B18,** 1–14.

Sakai, A. (1960b). The frost hardening process of woody plant. VIII. Relation of polyhydric alcohol to frost hardiness. *Low Temp. Sci. Ser.* **B18,** 15–22.

Sakai, A. (1961). Effect of polyhydric alcohols to frost hardiness in plants. *Nature (London)* **189,** 416–417.

Sakai, A. (1962). Studies on the frost-hardiness of woody plants. I. The causal relation between sugar content and frost-hardiness. *Contr. Inst. Low Temp. Sci. Ser.* **B11,** 1–40.

Sakai, A. (1966). Temperature fluctuation in wintering trees. *Physiol. Plant.* **19,** 105–114.

Sakai, A. (1967). Mechanism of frost damage on basal stems in young trees. *Low Temp. Sci. Ser.* **B25,** 45–57.

Sakai, A. (1968). Survival of plant tissue at super-low temperatures. VII. Methods for maintaining viability of less hardy plant cells at super-low temperatures. *Low Temp. Sci. Ser.* **B26,** 1–11.

Sakai, A. (1970a). Freezing resistance in willows from different climates. *Ecology* **51,** 485–491.

Sakai, A. (1970b). Mechanism of desiccation damage of conifers wintering in soil-frozen areas. *Ecology* **51,** 657–664.

Sakai, A., and Yoshida, S. (1968a). The role of sugar and related compounds in variations of freezing resistance. *Cryobiology* **5,** 160–174.

Sakai, A., and Yoshida, S. (1968b). Protective actions of various compounds against freezing injury in plant cells. *Low Temp. Sci. Ser.* **B26,** 13–21.

Sakazaki, N., Ihara, Y., Tachibana, Y., Nagai, S., and Takada, H. (1954). Physiology of *Metasequoia glyptostroboides* and related species of conifers. II. Comparative studies of salt tolerance. *J. Inst. Polytech. Osaka City Univ. Ser. D. Biol.* **5,** 67–80.

Sakharova, A. S., and Yakupov, N. A. (1969). A morphophysiological method for predicting the winter-hardiness of woody plants. From *Ref. Zh. Biol.* No. 5G136.

Salcheva, G., and Samygin, G. (1963). A microscopic study of freezing of the tissues of winter wheat. *Fiziol. Rast.* **10,** 65–72.

Salim, M. H., and Todd, G. W. (1965). Transpiration patterns of wheat, barley, and oat seedlings under varying conditions of soil moisture. *Agron. J.* **57,** 593–596.

Salim, M. H., and Todd, G. W. (1968). Seed soaking as a pre-sowing, drought-hardening treatment in wheat and barley seeds. *Agron. J.* **60,** 179–182.

Salim, M. H., Todd, G. W., and Schlehuber, A. M. (1965). Root development of wheat, oats, and barley under conditions of soil moisture stress. *Agron. J.* **57,** 603–607.

Salim, M. H., Todd, G. W. and Stutte, C. A. (1969). Evaluation of techniques for measuring drought avoidance in cereal seedlings. *Agron. J.* **61,** 182–185.

Salt, R. W. (1950). Time as a factor in the freezing of undercooled insects. *Can. J. Res.* **D28,** 285–291.

Salt, R. W. (1955). Extent of ice formation in frozen tissues and a new method for its measurement. *Can. J. Zool.* **33,** 391–403.

Salt, R. W. (1957). Natural occurrence of glycerol in insects and its relation to their ability to survive freezing. *Can. Entomol.* **89,** 491–494.

Salt, R. W. (1958). Role of glycerol in producing abnormally low supercooling and freezing points in an insect, *Bracon cephi* (Gahan). *Nature (London)* **181,** 1281.

Salt, R. W. (1961). Principles of insect cold-hardiness. *Ann. Rev. Entomol.* **6,** 55–74.

Salt, R. W. (1962). Intracellular freezing in insects. *Nature (London)* **193,** 1207–1208.

Salt, R. W., and Kaku, S. (1967). Ice nucleation and propagation in spruce needles. *Can. J. Bot.* **45,** 1335–1346.

Saltykovskij, M. I., and Saprygina, E. S. (1935). The frost-resistance of winter cereals at different stages of development. *C. R. Acad. Sci. URSS* **4,** 99–103.

Sampson, M., Clarkson, D., and Davies, D. D. (1965). DNA synthesis in aluminum-treated roots of barley. *Science* **148,** 1476–1477.

Samygin, G. A., and Matveeva, N. M. (1962). Comparative resistance of cells to freezing, desiccation, and plasmolysis, and its variation during hardening. *Fiziol. Rast.* **8,** 381–386.

Samygin, G. A., and Matveeva, N. M. (1967). Protective effect of glycerine and other substances which easily penetrate protoplasts during the freezing of plant cells. *Fiziol. Rast.* **14,** 1048–1056.

Samygin, G. A., and Matveeva, N. M. (1968). Protective effect of solutions during cell drying. *Fiziol. Rast.* **15,** 1038–1044.

Samygin, G. A., and Matveeva, N. M. (1969). Protective action of salt solutions during freezing of plant cells. *Fiziol. Rast.* **15,** 552–560.

Sands, K., and Rutter, A. J. (1958). The relation of leaf water deficit to soil tension in *Pinus sylvestris* L. II. Variation in the relation caused by developmental and environmental factors. *New Phytol.* **58,** 387–399.

Sands, K., and Rutter, A. J. (1959). Studies in the growth of young plants of *Pinus sylvestris* L. II. The relation of growth to soil moisture tension. *Ann. Bot.* (*London*) **23,** 269–284.

Sanner, T. (1965). Reduced pyridine coenzymes as hydrogen donors in repair of radiation damage. *Radiat. Res.* **26,** 95–106.

Sanner, T., and Pihl, A. (1967). Identification of the water radicals in x-ray inactivation of enzymes in solution and determination of their rate of interaction with the enzyme. *Biochim. Biophys. Acta* **146,** 298–301.

Sanner, T., and Pihl, A. (1968). Sulfhydryl groups in radiation damage. *Scand. J. Clin. Lab. Invest.* **22** (Suppl. 106), 53–63.

Sanner, T., and Pihl, A. (1969). Significance and mechanism of the indirect effect in bacterial cells. The relative protective effect of added compounds in *Escherichia coli* B, irradiated in liquid and in frozen suspension. *Radiat. Res.* **37,** 216–227.

Sanner, T., Henricksen, T., and Pihl, A. (1967). Transfer of radiation energy between different solutes in frozen aqueous solutions. *Radiat. Res.* **32,** 463–474.

Santarius, K. A. (1967). Das Verhalten von CO_2-Assimilation, NADP und PGS-Reduktion und ATP-Synthese intakter Blattzellen in Abhängigkeit von Wasser-gehalt. *Planta* **73,** 228–242.

Santarius, K. A. (1969). Der Einfluss von Elektrolyten auf Chloroplasten beim Gefrieren und Trocknen. *Planta* **89,** 23–46.

Santarius, K. A., and Ernst, R. (1967). Das Verhalten von Hill-Reaktion und Photo-phosphorylierung isolierter Chloroplasten in Abhängigkeit vom Wassergehalt. I. Wasserentzug mittels konzentrierter Lösungen. *Planta* **73,** 91–108.

Santarius, K. A., and Heber, U. (1967). Das Verhalten von Hill-Reaktion und Photo-phosphorylierung isolierter Chloroplasten in Abhängigkeit von Wassergehalt. II. Wasserentzug über $CaCl_2$. *Planta* **73,** 109–137.

Sapper, I. (1935). Versuche zur Hitzresistenz der Pflanzen. *Planta* **23,** 518–556.

Saprygina, E. S. (1935). Frost resistance of spring wheats. (On the effect of length of the "light" stage on the hardiness of wheats). *C. R. Acad. Sci. URSS* **3,** 325–328.

Saric, M. (1961). The effects of irradiation in relation to the biological traits of the seed irradiated. p. 103–116 *In* "Effects of Ionizing Radiations on Seeds, A Symposium." pp. 103–116. Int. Atomic Energy Agency, Vienna.

Sarin, M. N. (1961a). Physiological studies on salt tolerance in crop plants. XII. Influence of sodium sulphate on early seedling growth of wheat and gram varieties. *Agra Univ. J. Res. Sci.* **10,** 41–60.

Sarin, M. N. (1961b). Physiological studies on salt tolerance in crop plants. XIII. Influence of IAA on the deleterious effect of sodium sulphate on root growth in wheat. *Proc. Nat. Acad. Sci. India Sect.* **B31,** 287–295.

Sarin, M. N. (1961c). Physiological studies on salt tolerance in crop plants. XIV. Further studies on the effect of sodium sulphate on respiration of wheat and gram seedlings. *Indian J. Plant Physiol.* **4,** 38–46.

Sarin, M. N. (1962). Physiological studies on salt tolerance of crop plants. V. Use of IAA to overcome depressing effect of sodium sulphate on growth and maturity of wheat. *Agra Univ. J. Res. Sci.* **11,** 187–196.

Sarin, M. N., and Narayanan, A. (1968). Effects of soil salinity and growth regulators on germination and seedling metabolism of wheat. *Physiol. Plant.* **21,** 1201–1209.

Sarin, M. N., and Rao, I. M. (1958). Physiological studies on salt tolerance in crop plants. III. Influence of sodium sulphate on seedling respiration in wheat and gram. *Indian J. Plant Physiol.* **1,** 30–38.

Sarin, M. N., and Rao, I. M. (1961). Physiological studies in salt tolerance of crop plants. VIII. Influence of IAA spraying on the deleterious effect of sodium sulphate on growth and maturity of wheat. *Agra Univ. J. Res. Sci.* **10,** 7–16.

Sastry, K., Krishna, S., and Appaiah, K. M. (1968). Effect of thiamine on growth of roots of *Dolichos biflorus* and *Eleusine coracana*. *Mysore J. Agr. Sci.* **2,** 106–110.

Satoo, T. (1956). Drought resistance of some conifers at the first summer after their emergence. *Bull. Tokyo Univ. Forests* **51,** 1–108.

Satyanarayana, K., and Rao, C. C. (1963). Physiological basis of alkali tolerance. I. Absorption of Na, K and Ca. *Soil Cult.* **29,** 410–411.

Saunier, R. E., Hull, H. M., and Ehrenreich, J. H. (1968). Aspects of the drought tolerance in creosotebush (*Larrea divaricata*). *Plant Physiol.* **43,** 401–404.

Saussure, T. de. (1827). De l'influence du dessèchement sur la germination de plusieurs grains alimentaires. *Ann. Sci. Nat.* **10,** 68–93.

Savin, V. N., and Stepanenko, O. G. (1967). Changes in the drought resistance of plants exposed to cobalt-60 gamma-rays. *Radiobiologiya* **7,** 619–622.

Savin, V. N., and Stepanenko, O. G. (1968). Effect of gamma-rays from cobalt-60 on water conditions and drought hardiness of peppermint plants (*Mentha piperita*). *Friziol. Rast.* **15,** 546–551.

Savitskaya, N. N. (1967). Problem of accumulation of free proline in barley plants under conditions of soil water deficiency. *Fiziol. Rast.* **14,** 737–739.

Sawano, M. (1965). Studies on the frost resistance of chestnut trees. II. Relation between hardiness and water and sugar contents of bark tissue. *Sci. Rep. Hyogo Univ. Agr. Ser. Agr. Hort.* **7,** 77–81.

Scarth, G. W., (1936). The yearly cycle in the physiology of trees. *Trans. Roy. Soc. Can. Sect.* 5 **30,** 1–10.

Scarth, G. W. (1941). Dehydration injury and resistance. *Plant Physiol.* **16,** 171–179.

Scarth, G. W., and Levitt, J. (1937). The frost-hardening mechanism of plant cells. *Plant Physiol.* **12,** 51–78.

Scarth, G. W., and Lloyd, F. E. (1930). "Elementary Course in General Physiology." Wiley, New York.

Schacht, H. (1857). "Lehrbuch der Anatomie und Physiologie der Gewächse," pp. 525–529. Berlin.

Schaffnit, E. (1910). Studien über der Einfluss neiderer Temperaturen auf die pflanzliche Zell. *Mitt. Kaiser-Wilhelm Inst. Landwirtsch. Bromberg* **3,** 93–144.

Schander, R., and Schaffnit, E. (1919). Untersuchungen über das Auswintern des Getreides. *Landwirtsch. Jahrb.* **52,** 1–66.

Schanderl, H. (1955). Studies on the internal temperature of submerged water plants. *Ber. Deut. Bot. Ges.* **68,** 28–34.

Scheffer, F., and Lorenz, H. (1968). Pool-Aminosäuren während des Wachstums und der Entwichlung einiger Weizensorten. I. Pool-Aminosaüren in keimenden Samen, in Blättern mehrerer Entwicklungsstadien sowie im wachsenden und reifenden Korn. *Phytochemistry* **7,** 1279–1288.

Scheibmair, G. (1937). Hitzresistenz-Studien an Moos-Zellen. *Protoplasma* **29,** 394–424.

Scheumann, W. (1968). Die Dynamik der Frostresistenz und ihre Bestimmung an Gehölzen im Massentest. *In* "Klimaresistenz Photosynthese und Stoffproduktion." (H. Polster, ed.), pp. 45–54. Deut. Akad. Landwirtsch. Berlin.

Scheumann, W., and Börtitz, S. (1965). Studien zur Physiologie der Frosthärtung bei Koniferen. *Biol. Zentralbl.* **84,** 489–500.

Schiebel, W., Chayka, T. G., DeVries, A., and Rusch, H. P. (1969). Decrease of protein synthesis and breakdown of polyribosomes by elevated temperature in *Physarum polycephalum. Biochem. Biophys. Res. Commun.* **35,** 338–345.

Schlösser, L. (1936). Frosthärte und Polyploidie. *Zuechter* **8,** 75–80.

Schmalz, H. (1957). Untersuchungen über den Einfluss von photoperiodischer Induktion und Vernalisation auf die Winterfestigkeit von Winterweizen. *Z. Pflanzenzucht.* **38,** 147–180.

Schmalz, H. (1958). Die generative Entwicklung von Winterweizensorten mit unterschiedlicher Winterfestigkeit bei Fruhjahrsaussaat nach Vernalisation mit Temperaturen unter-und oberhalb des Gefrierpunktes. *Zuechter* **28,** 193–203.

Schmid, G. H. (1969). The effect of blue light on glycolate oxidase of tobacco. *Hoppe-Seyler's Z. Physiol. Chem.* **350,** 1035–1046.

Schmidt, H. (1939). Plasmazustand und Wasserhaushalt bei *Lamium maculatum. Protoplasma* **33,** 25–43.

Schmidt, H., Duwald, K., and Stocker, O. (1940). Plasmatische Untersuchungen an dürreempfindlichen und dürreresistenten Sorten landwirtschaftlicher Kulturpflanzen. *Planta* **31,** 559–596.

Schmuetz, W. (1962). Weitere Untersuchungen über die Beziehung zwischen Sulfhydryl-Gehalt und Winterfestigkeit von 15 Weizensorten. *Z. Acker Pflanzenbau* **115,** 1–11.

Schmuetz, W. (1969). Zu Fragen der physiologischen Resistenz bei Getreide. *Vortr. Pflanzenzuechter* **12,** 105–122.

Schmuetz, W., Sullivan, C. Y., and Levitt, J. (1961). Sulfhydryls—A new factor in frost resistance. II. Relation between sulfhydryls and relative resistance of fifteen wheat varieties. *Plant Physiol.* **36,** 617–620.

Schneider, E. (1925). Über die Plasmolyse als Kennzeichen lebender Zellen. *Z. Wiss. Mikroskop.* **42,** 32–54.

Schneider, G. W., and Childers, N. F. (1941). Influence of soil moisture on photosynthesis, respiration, and transpiration of apple leaves. *Plant Physiol.* **16,** 565–583.

Schneider-Orelli, O. (1910). Versuche über die Widerstandsfähigket gewisser Medicago-Samen (Wollkletten) gegen hohe Temperaturen. *Flora* (*Jena*) **100,** 305–311.

Schölm, H. E. (1968). Untersuchungen zur Hitze-und Frostresistenz einheimscher Susswasseralgen. *Protoplasma* **65,** 97–118.

Schönbohm, E. (1969). Inhibition of light-induced movement of Mougeotia chloroplasts by *p*-chloromercuribenzoate: Investigations on the mechanics of chloroplast movement. *Z. Pflanzenphysiol.* **61,** 250–260.

Scholander, P. F. (1968). How mangroves desalinate seawater. *Physiol. Plant.* **21,** 251–261.

Scholander, P. F., Bradstreet, E. D., Hammel, H. T., and Hemmingsen, E. A. (1966). Sap concentrations in halophytes and some other plants. *Plant Physiol.* **41,** 529–532.

Scholander, P. F., Flagg, W., Hock, R. J., and Irving, L. (1953). Studies on the physiology of frozen plants and animals in the Arctic. *J. Cell. Comp. Physiol.* **42,** 1–56.

Scholander, P. F., Hammel, H. T., Hemmingsen, E., and Garey, W. (1962). Salt balance in mangroves. *Plant Physiol.* **37,** 722–729.

Schoolar, A. I., and Edelman, J. (1970). Production and secretion of sucrose by sugarcane leaf tissue. *J. Exp. Bot.* **21,** 49–57.

Schopmeyer, C. S. (1939). Transpiration and physico-chemical properties of leaves as related to drought resistance in loblolly pine and shortleaf pine. *Plant Physiol.* **14,** 447–462.

Schratz, E. (1931). Vergleichende Untersuchungen über den Wasserhaushalt von Pflanzen im Trockengebiete des südlichen Arizona. *Jahrb. Wiss. Bot.* **74,** 153–290.

Schroeder, C. A. (1963). Induced temperature tolerance of plant tissue in vitro. *Nature* (*London*) **200,** 1301–1302.

Schröder, G. (1886). Über die Austrocknungsfähigkeit der Pflanzen. Inaug. Dissertation, Tübingen, pp. 1–51.

Schübler, G. (1827). Beobachtungen über die Temperatur der Vegetabilien und einige damit verwandte Gegenstände. *Ann. Phys. Chem.* **10,** 581–592.

Schröder, D. (1909). Über den Verlauf des Welkens und die Lebenszähigkeit der Laubblätter. Inaug. Dissertation. Gottingen.

Schulz, E. D., Mooney, H. A., and Dunn, E. L. (1967). Wintertime photosynthesis of bristle-cone pine (*Pinus aristata*) in the White Mountains of California. *Ecology* **48,** 1044–1047.

Schumacher, E. (1875). II. Beiträge zur Morphologie und Biologie der Hefe. *Sitzber. Akad. Wiss. Wien Math. Naturwiss. Kl. Abt. I* **70,** 157–188.

Schwanitz, F., and Schwanitz, H. (1968). Enhanced germination and better growth of young plants that have been treated with substances that are also used to control damage done by irradiation. *Landwirt. Forsch.* **21,** 231–240.

Schwarz, W. (1968). Der Einfluss der Temperatur und Tageslänge auf die Frosthärte der Zirbe. *In* "Klimaresistenz Photosynthese und Stoffproduktion" (H. Polster, ed.), pp. 55–63. Deut. Akad. Landwirtsch. Berlin.

Schwemmle, B., and Lange, O. L. (1959a). Endogen-Tagesperiodische Schwankungen der Hitzeresistenz bei *Kalanchoe blossfeldiana*. *Planta* **53,** 134–144.

Schwemmle, B., and Lange, O. L. (1959b). Neue Beobachtungen über die Endogene Tagesrhythmik. *Nachr. Akad. Wiss. Gott.* II. *Math Phys. Kl. Nr. 3.*, pp. 31–35.

Scifres, C. J., and Brock, J. H. (1969). Moisture-temperature interrelations in germination and early seedling development of mesquite. *J. Range Manage.* **22,** 334–337.

Sebbah, M. (1967). Nature of xerophytism in *Corynephorus canescens* P.B. *Bull. Soc. Hist. Nat. Toulouse* **103,** 138–158.

Sedykh, N. V., and Khakhlova, L. P. (1967). Effect of drought on the structure of cytoplasmic proteins in plant leaves as studied by a dielectric technique at UHF. *Fiziol. Rast.* **14,** 507–511.

Sehgal, P. P., and Naylor, A. W. (1966). Purification and properties of urease derived from hydrated seeds of Jack Bean, *Canavalia ensiformis* (L) D.C. *Plant Physiol.* **41,** 567–572.

Seible, D. (1939). Ein Beitrag zur Frage der Kälteschäden an Pflanzen bei Temperaturen über dem Gefrierpunkt. *Beitr. Biol. Pflanz.* **26,** 289–330.

Seifriz, W. (1949). Toxicity and chemical properties of ions. *Science* **110,** 193–196.

Sellschop, J. P. F., and Salmon, S. C. (1928). The influence of chilling above the freezing point on certain crop plants. *J. Agr. Res.* **37,** 315–338.

Selwyn, M. J. (1966). Temperature and photosynthesis. II. A mechanism for the effects of temperature on carbon dioxide fixation. *Biochim. Biophys. Acta* **126,** 214–224.

Selwyn, M. J., and Chappell, J. B. (1962). Properties of mitochondrial adenosine triphosphatase before and after solubilization. *Biochem. J.* **84,** 63P.

Semenenko, V. E., Vladimirova, M. G., Orleanskaya, O. B., Raikov, N. I., and Kovanova, E. S. (1969). Physiological characteristics of Chlorella during disturbance of cell functions by high temperatures. *Fiziol. Rast.* **16,** 210–220.

Senebier, J. (1800). "Physiologie Végétale," Vol. 3, pp. 282–304. Paschoud, Genève.

Senn, G. (1922). Untersuchungen über die Physiologie der Alpenpflanzen. *Verhandl. Schweiz. Naturforsch. Ges.* **103,** 154–168.

Seon, B. K. (1967). Stepwise reduction of disulfide bonds in takaamylase A. *J. Biochem.* **61,** 606–614.

Sergeeva, K. A. (1968). The phosphorus metabolism of woody plants in relation to their winter hardiness. *Tr. Inst. Ekol. Rast. Zhivotn. Ural. Fil. Akad. Nauk. SSSR* **62,** 30–36.

Sergeev, L. I., Sergeeva, K. A., and Kandarova, I. V. (1959). The appearance of starch in the generative buds of woody plants in winter. *Byull. Glavnogo Bot. Sada Akad. Nauk. SSSR* **35,** 70–75.

Serrano, M. (1956). The elements Li, Na, K, Rb, and Cs in the germination and growth of wheat seedlings. *Farmacognosia* **14,** 27–51.

Sestakov, V. E. (1936). Frost resistance of winter crops during the light stage. *C. R. Acad. Sci. URSS* **3,** 395–398.

Sestakov, V. E., and Sergeev, L. I. (1937). Changes in frost resistance and in the properties of cell protoplasm in winter wheat during the photo-stage. *Bot. Zh. SSR* **22,** 351–363.

Sestakov, V. E., and Smirnova, A. D. (1936). Temperature hardening and the differentiation of the embryonic spike in winter wheats during the light stage of development. *C. R. Acad. Sci. URSS* **3,** 399–403.

Shah, B. H. (1967). Effect of sodium chloride and potassium chloride on growth, succulence and salt contents of *Suaeda fruticosa* Forsk. *Sci. Ind.* (Karachi) **5,** 379–390.

Shantz, H. L., and Piemeisel, N. (1927). The water requirement of plants at Akron, Colorado. *J. Agr. Res.* **34,** 1093–1190.

Shapiro, B., Kollmann, G., and Amen, J. (1966). Mechanism of the effect of ionizing radiation on sodium uptake by human erythrocytes. *Radiat. Res.* **27,** 139–158.

Shapiro, N. I., and Bocharova, E. M. (1961). Two kinds of radiational after effects in barley seeds. *Dokl. Akad. Nauk. SSSR* **133,** 512–514.

Shardakov, V. C. (1957). Principles for determining the watering periods of the cotton plant in relation to the magnitude of the suction force of the leaves. *In* "Physiological Problems of Cotton and Grass," (V. C. Shardakov, ed.), pp. 5–32. Institute Gen. and Physiology, Tashkent, USSR.

Shaw, R. H. (1954). Leaf and air temperatures under freezing conditions. *Plant Physiol.* **29,** 102–104.

Sherman, F. (1959). The effects of elevated temperatures on yeast. I. Nutrient requirements for growth at elevated temperatures. *J. Cell. Comp. Physiol.* **54,** 29–36.

Sherman, J. K. (1962). Questionable protection by intracellular glycerol during freezing and thawing. *J. Cell. Comp. Physiol.* **61,** 67–83.

Shevyakova, N. I. (1966). Effect of salts on the biosynthesis of certain amino acids of broad-bean leaves. *Dokl. Akad. Nauk. SSSR* **167,** 471–473.

Shevyakova, N. I., and Komizerko, E. I. (1969). Amino acid metabolism in a culture of callus tissue from cabbage leaves (*Brassica oleracea* var. *capitata*) under salinization conditions. *Dokl. Akad. Nauk. SSSR* **186,** 1441–1444.

Shevyakova, N. I., and Loshadkina, A. P. (1965). Changes in the sulfhydryl group content in plants under conditions of salinization. *Fiziol. Rast.* **12,** 332–339.

Shevyakova, N. I., and Popova, M. N. (1968). New formation of sulfate in plants in the presence of Na_2SO_4. *Dokl. Akad. Nauk. SSSR* **182,** 975–977.

Shibaoka, H. (1961). Studies on the mechanism of growth inhibiting effect of light. *Plant Cell Physiol.* **2,** 175–197.

Shields, M. (1950). Leaf xeromorphy as related to physiological and structural influences. *Bot. Rev.* **16,** 399–446.

Shih, S. C., Jung, G. A., and Shelton, D. C. (1967). Effects of temperature and photoperiod on metabolic changes in alfalfa in relation to cold hardiness. *Crop Sci.* **7,** 385–389.

Shikama, K. (1963). Denaturation of catalase and myosin by freezing and thawing. *Sci. Rep. Tohoku Univ. Ser.* 4 **29,** 91–106.

Shikama, K. (1965). Effect of freezing and thawing on the stability of double helix of DNS. *Nature (London)* **207,** 529–530.

Shimabukuro, R. H., Swanson, H. R., and Walsh, W. C. (1970). Glutathione conjugation. *Plant Physiol.* **46,** 103–107.

Shimshi, D. (1963). Effect of chemical closure of stomata on transpiration in varied soil and atmcspheric environments. *Plant Physiol.* **38,** 709–712.

Shinn, J. H., and Lemon, E. R. (1968). Photosynthesis under field conditions: XI. Soil-plant-water relations during drought stress in corn. *Agron. J.* **60,** 337–343.

Shiralipour, A., and Anthony, D. S. (1970). Chemical prevention of growth reduction caused by supraoptimal temperatures in *Arabidopsis thaliana*. *Phytochemistry* **9,** 463–469.

Shiroya, T., Lister, G. R., Slankis, V., Krotkov, G., and Nelson, C. D. (1966). Seasonal changes in respiration, photosynthesis and translocation of the ^{14}C labelled products of photosynthesis in young *Pinus strobus* L. plants. *Ann. Bot.* **30,** 81–90.

Shishenkova, L. K. (1967). The nature of the change in radiosensitivity during the swelling and sprouting of seeds. *Inform. Byull. Nauch. Sov. Probl. Radio. Biol. Akad. Nauk. SSSR* **10,** 39–41.

Shive, J. W. (1941). Significant roles of trace elements in the nutrition of plants. *Plant Physiol.* **16,** 435–445.

Shmat'ko, I. H., and Rubanyuk, O. O. (1969). Exosmosis of metabolites from leaves of winter wheat in varying degrees of moisture. *Ukr. Bot. Zh.* **26,** 66–71.

Shmatok, I. D. (1958). Seasonal dynamics of ascorbic acid in leaves of plants under polar conditions. *Fiziol. Rast.* **5,** 341–344.

Shmelev, I. K. (1935). Frost resistance of fruit trees. *Bull. Appl. Bot. Genet. Plant Breeding (Leningrad) Ser.* 3 No. 6, pp. 263–277.

Shmueli, E. (1953). Irrigation studies in the Jordan Valley. I. Physiological activity of the banana in relation to soil moisture. *Bull. Res. Counc. Isr.* **3,** 228–247.

Shmueli, E. (1960). Chilling and frost damage in banana leaves. *Bull. Res. Counc. Isr. Sect. Bot.* **8D,** 225–238.

Shnaider, T., and Vakher, Yu. (1969). Periodical changes of radiosensitivity of rape seeds during their swelling and germination. *Izv. Akad. Nauk. Est. SSR Ser. Biol.* **18,** 343–346.

Shvedskaya, I. M., and Kruzhilin, A. S. (1968). Changes in the nucleic acid content in biennial plants during vernalization or differentiation of the buds. *Fiziol. Rast.* **15,** 798–805.

Sidaway, G. H. (1966). Influence of electrostatic fields on seed germination (*Lactuca sativa*). *Nature (London)* **211,** 303.

Sidaway, G. H., and Asprey, G. F. (1968). Influence of electrostatic fields on plant

respiration (*Arum maculatum, Triticum vulgare, Vicia faba, Ulmus glabra*). *Int. J. Biometeorol.* **12,** 321–329.
Siegel, S. M. (1969a). Further studies on factors affecting the efflux of betacyanin from beet root: A note on the thermal effects. *Physiol. Plant.* **22,** 327–331.
Siegel, S. M. (1969b). Microbiology of saturated salt solutions and other harsh environments. V. Relation of inosine-5-phosphate and carbohydrate to growth of wildtype and mutant Penicillium in boric acid and potassium chloride selective media. *Physiol. Plant.* **22,** 1152–1157.
Siegel, S. M., and Daly, O. W. (1966). Experimental biology of ammonia-rich environments. Germination of allium seed, a novel capability among angiosperms. *Plant Physiol.* **41,** 1218–1221.
Siegel, S. M., Roberts, K., Lederman, M., and Daly, O. (1967). Microbiology of saturated salt solutions and other harsh environments. II. Ribonucleotide dependency in the growth of a salt-habituated *Penicillium notatum* in salt-free nutrient media. *Plant Physiol.* **42,** 201–204.
Siegel, S. M., Speitel, T., and Stoecker, R. (1969). Life in earth extreme environments: a study of cryobiotic potentialities. *Cryobiology* **6,** 160–181.
Siew, D., and Klein, S. (1968). The effect of sodium chloride on some metabolic and fine structural changes during the greening of etiolated (bean) leaves. *J. Cell. Biol.* **37,** 590–595.
Siminovitch, D. (1963). Evidence from increase in ribonucleic acid and protein synthesis in autumn for increase in protoplasm during the frost-hardening of black locust bark cells. *Can. J. Bot.* **41,** 1301–1308.
Siminovitch, D., and Briggs, D. R. (1949). The chemistry of the living bark of the black locust tree in relation to frost hardiness. I. Seasonal variations in protein content. *Arch. Biochem. Biophys.* **23,** 8–17.
Siminovitch, D., and Briggs, D. R. (1953a). Studies on the chemistry of the living bark of the black locust in relation to its frost hardiness. III. The validity of plasmolysis and desiccation tests for determining the frost hardiness of bark tissue. *Plant Physiol.* **28,** 15–34.
Siminovitch, D., and Briggs, D. R. (1953b). Studies on the chemistry of the living bark of the black locust tree in relation to frost hardiness. IV. Effects of ringing on translocation, protein synthesis and the development of hardiness. *Plant Physiol.* **28,** 177–200.
Siminovitch, D., and Briggs, D. R. (1954). Studies on the chemistry of the living bark of the black locust in relation to its frost hardiness. VII. A possible direct effect of starch on the susceptibility of plants to freezing injury. *Plant Physiol.* **29,** 331–337.
Siminovitch, D., and Levitt, J. (1941). The relation between frost resistance and the physical state of protoplasm. II. *Can. J. Res.* C**19,** 9–20.
Siminovitch, D., and Scarth, G. W. (1938). A study of the mechanism of frost injury to plants. *Can. J. Res.* C**16,** 467–481.
Siminovitch, D., Wilson, C. M., and Briggs, D. R. (1953). Studies on the chemistry of the living bark of the black locust in relation to its frost hardiness. V. Seasonal transformation and variations in the carbohydrates: starch-sucrose interconversions. *Plant Physiol.* **28,** 383–400.
Siminovitch, D., Therrien, H., Wilner, J., and Gfeller, F. (1962). The release of amino acids and other ninhydrin-reacting substances from plant cells after injury by freezing; a sensitive criterion for the estimation of frost injury in plant tissues. *Can. J. Bot.* **40,** 1267–1269.

Siminovitch, D., Rhéaume, B., and Sachar, R. (1967a). Seasonal increase in protoplasm and metabolic capacity in the cells during adaptation to freezing. 3–40. *In* "Molecular Mechanisms of Temperature Adaptation." (C. L. Prosser, ed.) pp. 3–40. Pub. No. 84. Amer. Assoc. Adv. Sci. Washington, D.C.

Siminovitch, D., Gfeller, F., and Rhéaume, B. (1967b). *In* "Cellular Injury and Resistance in Freezing Organisms," (E. Asahina, ed.), pp. 93–118. Inst. Low Temp. Sci. Hokkaido Univ., Sapporo, Japan.

Siminovitch, D., Rhéaume, B., Pomeroy, K., and Lepage, M. (1968). Phospholipid, protein, and nucleic acid increases in protoplasm and membrane structures associated with development of extreme freezing resistance in black locust tree cells. *Cryobiology* **5,** 202–225.

Simonis, W. (1947). CO_2-Assimilation und Stoffproduktion trocken gezogener Pflanzen. *Planta* **36,** 188–224.

Simonis, W. (1952). Untersuchungen zum Dürreeffekt. I. Morphologische Struktur, Wasserhaushalt, Atmung und Photosynthese feucht und trocken gezogener Pflanzen. *Planta* **40,** 313–332.

Simonis, W., and Werk, O. (1958). Untersuchungen zum Dürreeffekt. 3. Mitteilung. Über den Kalium und Calcium-anteil in verschiedenen Blattfraktionen bei feucht- und trockengezogenen Pflanzen von *Vicia faba*. *Flora (Jena)* **146,** 493–511.

Simonneau, P., and Aubert, G. (1963). Utilization of saline waters in the Sahara. *Ann. Agron. (Paris)* **14,** 859–872.

Simonova, O. N. (1968). A cytochemical study of the effect of drought on the RNA content during development of the pollen grain in barley. *Uch. Zap. Leningrad Gos. Pedagog. Inst. Im. A I. Gertsena* **333,** 206–212.

Singh, A., and Singh, P. N. (1966). Effect of boron on drought resistance, growth, and yield of *Hordeum vulgare* L. *Proc. Nat. Acad. Sci. India Biol. Sci.* B**36,** 537–543.

Sisakjan, N. M., and Rubin, B. A. (1939). The action of low temperature on the reversibility of enzymatic reaction in relation to winter hardiness of plants. *Biokhimiya* **4,** 149–153.

Sivtsev, M. V., and Kabuzenko, S. N. (1967). Some elements of water metabolism and the shape of plants as related to drought resistance. *Rost. Ustoich. Rast.* **3,** 256–262.

Skogqvist, I., and Fries, N. (1970). Induction of thermosensitivity and salt sensitivity in wheat roots and the effect of kinetin. *Experientia* **26,** 1160–1162.

Skoss, J. D. (1955). Structure and composition of plant cuticle in relation to environmental factors and permeability. *Bot. Gaz. (Chicago)* **117,** 55–72.

Skujins, J. J., and McLaren, A. D. (1967). Enzyme reaction rates at limited water activities. *Science* **158,** 1569–1570.

Slatyer, R. O. (1955). Studies of the water relations of crop plants grown under natural rainfall in northern Australia. *Austr. J. Agr. Res.* **6,** 365–377.

Slatyer, R. O. (1956). Absorption of water from atmospheres of different humidity and in transport through plants. *Aust. J. Biol. Sci.* **9,** 552–558.

Slatyer, R. O. (1957a). The significance of the permanent wilting percentage in studies of plant and soil water relations. *Bot. Rev.* **23,** 585–636.

Slatyer, R. O. (1957b). The influence of progressive increases in total soil moisture stress on transpiration, growth, and internal water relationships of plants. *Aust. J. Biol. Sci.* **10,** 320–336.

Slatyer, R. O. (1960). Aspects of the tissue water relationships of an important arid zone species (*Acacia aneura* F. Meull.) in comparison with two mesophytes. *Bull. Res. Counc. Isr.* **8D,** 159–167.

Slatyer, R. O. (1964). Efficiency of water utilization by arid zone vegetation. *Ann. Arid Zone* **3,** 1–12.

Slatyer, R. O., and Bierhuizen, J. F. (1964). The influence of several transpiration suppressants on transpiration, photosynthesis, and water-use efficiency of cotton leaves. *Aust. J. Biol. Sci.* **17,** 131–146.

Slonov, L. Kh. (1966). Effect of trace elements on the physico-chemical properties of plant cell protoplasm biocolloids under salinization conditions. *Fiziol. Rast.* **13,** 1024–1028.

Sluyterman, L. A. E., and De Graaf, M. J. M. (1969). The activity of papain in the crystalline state. *Biochim. Biophys. Acta* **171,** 277–287.

Smirnova, I. S. (1959). The problem of the winter resistance of the root systems of apple and pear trees in a nursery. *Trudy Plodoovshch. Inst. I.V. Michurina* **10,** 43–50.

Smith, A. M. (1915). The respiration of partly dried plant organs. *Brit. Ass. Advan. Sci. Rep.*, p. 725.

Smith, A. U., Polge, C., and Smiles, J. (1951). Microscopic observation of living cells during freezing and thawing. *J. Roy. Microsc. Soc.* **71,** 186–195.

Smith, D. (1959). Differential survival of ladino and common white clover encased in ice. *Agron. J.* **41,** 230–234.

Smith, D., (1952). The survival of winter-hardened legumes encased in ice. *Agron. J.* **44,** 469–473.

Smith, D. (1968a). Carbohydrates in grasses: IV. Influence of temperature on the sugar and fructosan composition of Timothy (*Phleum pratense*) plant parts at anthesis. *Crop Sci.* **8,** 331–334.

Smith, D. (1968b). Varietal chemical differences associated with freezing resistance in forage plants. *Cryobiology* **5,** 148–159.

Smith, F. H., and Silen, R. R. (1963). Anatomy of heat-damaged Douglas-fir seedlings. *Forest Sci.* **9,** 15–32.

Smith, R. A. H., and Bradshaw, A. D. (1970). Reclamation of toxic metalliforous wastes using tolerant populations of grass. *Nature* (*London*) **227,** 376–377.

Smith, T. A. (1965). *N*-carbamylputrescine amidohydrolase of higher plants and its relation to potassium nutrition. *Phytochemistry* **4,** 599–607.

Smith, T. J. (1942). Responses of biennial sweet clover to moisture, temperature, and length of day. *J. Amer. Soc. Agron.* **34,** 865–876.

Smith, W. H. (1954). Non-freezing injury in plant tissues with particular reference to the detached plum fruit. *8th Int. Congr. Bot.* **11,** 280–285.

Smol'skaya, E. M. (1964). Changes with age and season in the chlorophyll content of the shoots of woody plants. *In* "The effect of Soil Conditions on the Growth of Woody Plants," pp. 100–114. Nauka Tekhnika, Minsk. From *Ref. Zh. Biol.* 1964. 23G3 (Translation).

Snell, K. (1932). Die Beschleunigung der Keimung bei der Kartoffelknolle. *Ber. Deut. Bot. Ges.* **52**A, 146–161.

Snope, A. J., and Ellison, J. H. (1963). Freeze-drying improves preservation of pollen. *N. J. Agr.* **45,** 8–9.

Sokawa, Y., and Hase, E. (1968). Suppressive effect of light on the formation of DNA and on the increase of deoxythymidine monophosphate kinase activity in *Chlorella protothecoides. Plant Cell Physiol.* **9,** 461–466.

Solomonovskii, L. Y., and Pomazova, E. N. (1967). The effect of a physiologically active mixture on the water regime and cold resistance of thermophilic plants. *Izv. Sib. Otd. Akad. SSR Ser. Biol. Med. Nauk.* **1,** 72–78.

Solov'ev, V. A. (1967). Ways of regulating a surplus of absorbed ions in plant tissues (Na ions used as examples). *Fiziol. Rast.* **14,** 1093–1103.

Solov'ev, V. A. (1969a). Distribution of cations in plants depending on the degree of salinization of the substrate. *Fiziol. Rast.* **16,** 498–504.

Solov'ev, V. A. (1969b). Plant growth and water and mineral nutrient supply under NaCl salinization conditions. *Fiziol. Rast.* **16,** 870–876.

Solov'ev, A. K., and Nezgovorov, L. A. (1968). Differences in the response of shade plants to injuring and hardening temperatures as judged by extraction of cellular sap from leaves. *Fiziol. Rast.* **15,** 1045–1054.

Somero, G. N., and Hochachka, P. W. (1969). Isoenzymes and short-term temperature compensation in poikilotherms: Activation of lactate dehydrogenase isoenzymes by temperature decreases. *Nature* (*London*) **223,** 194–195.

Sorauer, P. (1924). "Handbuch der Pflanzenkrankheiten," Vol. 1. Parey, Berlin.

Sparrow, A. H., and Woodwell, G. M. (1962). Prediction of sensitivity of plants to chronic gamma irradiation. *Radiat. Bot.* **2,** 9–26.

Sparrow, A. H., Cauny, R. L., Miksche, J. P., and Schairer, L. A. (1961). Some factors affecting the responses of plants to acute and chronic radiation exposures. *Radiat. Bot.* **1,** 10–34.

Sparrow, A. H., Cauny, R. L., Miksche, J. P., and Schairer, L. A. (1960). Some factors affecting the responses of plants to acute and chronic radiation exposure. *In* "Effects of Ionizing Radiations on Seeds, A Symposium," pp. 289–320. Intern. Atomic Energy Agency, Vienna.

Sparrow, A. H., Furuya, M., and Schwemmer, S. S. (1968). Effects of x- and gamma-radiation on anthocyanin content in leaves of Rumex and other plant genera. *Radiat. Bot.* **8,** 7–16.

Sparrow, A. H., Schwemmer, S. S., Klug, E. E., and Puglielli, L. (1970). Woody plants: changes in survival in response to long-term (8 years) chronic gamma irradiation. *Science* **169,** 1082–1084.

Spencer, E. L. (1937). Frenching of tobacco and thallium toxicity. *Amer. J. Bot.* **24,** 16–24.

Spradlin, J., and Thoma, J. A. (1970). β-Amylase thiol group. Possible regulator sites. *J. Biol. Chem.* **245,** 117–127.

Spradlin, J. E., Thoma, J. A., and Filmer, D. (1969). Beta amylase thiol groups. Possible regulator sites *Arch. Biochem. Biophys.* **134,** 262–264.

Spragg, S. P., Lievesley, P. M., and Wilson, K. M. (1962). The relationship between glutathione and protein sulphydryl groups in germinating pea seeds. *Biochem. J.* **83,** 314–318.

Sprague, M. A. (1955). The influence rate of cooling and winter cover on the winter survival of ladino clover and alfalfa. *Plant Physiol.* **30**: 447–451.

Sprague, V. G., and Graber, L F. (1940). Physiological factors operative in ice sheet injury of alfalfa. *Plant Physiol.* **15,** 661–673.

Sprague, V. G., and Graber, L. F. (1943). Ice sheet injury to alfalfa. *J. Amer. Soc. Agron.* **35,** 881–894.

Srb, V., and Hluchovsky, B. (1963). Changes in cellular permeability of different coats of *Allium cepa* L. after x-ray irradiation. *Exp. Cell Res.* **29,** 261–267.

Srb, V., and Hluchovsky, B. (1967). Cell permeability and X-irradiation. *Sb. Ved. Kel. Fak. Karlovy Univ. Hradci. Kralove Suppl.* **10,** 491–507.

Stadelmann, E. J. (1968). Effect of ionizing radiation on a monolayer of cells. Final rep. Coo–1634–3, AEC Contract AT (11–1)-1634. Atomic Energy Commission, Wash., D.C.

Stadelmann, E. S., and Wattendorff, J. (1966). Plasmolysis and permeability of alpha-irradiated epidermal cells. of *Allium cepa*. L. *Protoplasma* **62,** 86–116.

Stark, N., and Love, L. D. (1969). Water relations of three warm desert species. *Isr. J. Bot.* **18,** 175–190.

Starr, P. R., and Parks, L. W. (1962). The effect of temperature on sterol metabolism in yeast. *J. Cell. Comp. Physiol.* **59,** 107–110.

Stasevs'ka, L. P. (1969). Free amino acids in the growing points of winter wheat under the effect of lower temperatures. *Ukr. Bot. Zh.* **26,** 123–125.

Steemann,-Nielsen, E., and Jorgensen, C. K. (1968). The adaptation of plankton algae. I. General part. *Physiol. Plant.* **21,** 401–413.

Steeman-Nielsen, E., and Kamp-Nielsen, L. (1970). Influence of deleterious concentrations of copper on the growth of *Chlorella pyrenoidosa. Physiol. Plant.* **23,** 828–840.

Steeman-Nielsen, E., Kamp-Nielsen, L., and Wuim-Andersen, S. (1969). The effect of deleterious concentrations of copper on the photosynthesis of *Chlorella pyrenoidosa. Physiol. Plant.* **22,** 1121–1133.

Steinbauer, G. (1926). Difference in resistance to low temperature shown by clover varieties. *Plant Physiol.* **1,** 281–286.

Steinbrinck, C. (1900). Zur Terminologie der Volumänderungen pflanzlicher Gewebe und organischer Substanzen bei wechselndem Flussigkeitsgehalt. *Ber. Deut. Bot. Ges.* **18,** 217–224.

Steinbrinck, C. (1903). Versuche über die Luftdurchlässigkeit der Zellwände von Farn-und Selaginella-Sporangien, sowie von Moosblättern. *Flora (Jena)* **92,** 102–131.

Steinbrinck, C. (1906). Über Schrumpfungs-und Kohäsionsmechanismen von Pflanzen. *Biol. Centr.* **26,** 657–677; 721–744.

Steiner, M. (1933). Zum Chemismus der osmotischen Jahresschwankungen einiger immergrüner Holzgewächse. *Jahrb. Wiss. Bot.* **78,** 564–622.

Steinhübel, G., and Halas, L. (1969). Seasonal trends in the rates of dry matter production in the evergreen and winter green broadleaf woody plants. *Photosynthetica* **3,** 244–254.

Stembridge, G. E., and Larue, J. H. (1969). The effect of potassium gibberellate on flower bud development in the Redskin peach. *J. Amer. Soc. Hort. Sci.* **94,** 492–495.

Steponkus, P. L. (1968). Cold acclimation of *Hedra helix*—a two-step process. *Cryobiology* **4,** 276–277.

Steponkus, P. L. (1969a). Protein-sugar interactions during cold acclimation. *Cryobiology* **6,** 285.

Steponkus, P. L. (1969b). Protein sugar interactions during cold-acclimation. *Abstr.* 11th *Int. Bot. Congr. Seattle, Wash.*, p. 209.

Steponkus, P. L., and Lanphear, F. O. (1966). The role of light in cold acclimation of woody plants. *Proc.* 17*th Int. Hort. Congr.* **1,** 93.

Steponkus, P. L., and Lanphear, F. O. (1967a). Light stimulation of cold acclimation: Production of a translocatable promoter. *Plant Physiol.* **42,** 1673–1679.

Steponkus, P. L., and Lanphear, F. O. (1967b). Factors influencing artificial cold acclimation and artificial freezing of *Hedera Helix* "Thorndale." *Proc. Amer. Soc. Hort. Sci.* **91,** 735–741.

Steponkus, P. L., and Lanphear, F. O. (1968a). The role of light in cold acclimation of *Hedera helix* var. "Thorndale." *Plant Physiol.* **43,** 151–156.

Steponkus, P. L., and Lanphear, F. O. (1968b). The relationship of carbohydrates to cold acclimation of *Hedera helix* L. v. Thorndale. *Physiol. Plant.* **2,** 777–791.

Stewart, C. R., Morris, C. J., and Thompson, J. F. (1966). Changes in amino acid content of excised leaves during incubation. *Plant Physiol.* **41,** 1585–1590.

Stewart, I., and Leonard, C. D. (1960). Increased winter hardiness in citrus from maleic hydrazide sprays. *Proc. Amer. Soc. Hort. Sci.* **75,** 253–256.

Stewart, J. McD., and Guinn, G. (1969). Chilling injury and changes in adenosine triphosphate of cotton seedlings. *Plant Physiol.* **44,** 605–608.

Stocker, O. (1929). Das Wasserdefizit von Gefässpflanzen in verschiedenen Klimazonen. *Planta* **7,** 382–387.

Stocker, O. (1956). Die Dürreresistenz. *Handb. Pflanzenphysiol.* **3,** 696–741.

Stocker, O. (1961). Contributions to the problem of drought resistance of plants. *Ind. J. Plant Physiol.* **4,** 87–102.

Stone, E. C. (1957). Dew as an ecological factor. I–II. The effect of artificial dew on the survival of *Pinus ponderosa* and associated species. *Ecology* **38,** 407–422.

Straib, W. (1946). Beträge zur Frosthärte des Weizens. *Zuechter* **17/18,** 1–12.

Strain, B. R. (1969). Seasonal adaptations in photosynthesis and respiration in four desert shrubs growing in situ. *Ecology* **50,** 511–513.

Strömer, M. (1749). Gedanken über die Ursache warum die Bäume bei starkem Winter erfrieren, wobei die Möglichkeit solchem vorzubeugen erwiesen wird. *Der Kgl. Schwed. Akad. Wiss. Abhandl. Nat. Haushalt. Mech. Jahre* 1739 und 1740 (Aus dem Schwed. übersetzt) **1,** 116–121.

Strogonov, B. P. (1964). "Physiological Basis of Salt Tolerance of Plants." Acad. Sci. USSR., Davey and Co. New York.

Strogonov, B. P., Ivanitskaya, E. F., and Kartashova, I. V. (1963). The water relations of cotton with different kinds of salinity. *In* "The Water Relations of Plants as Related to Metabolism and Productivity, pp. 192–199 Akad. Nauk. SSSR, Moscow.

Strogonov, B. P., and Lapina, L. P. (1964). A possible method for the separate study of toxic and osmotic action of salts on plants. *Fiziol. Rast.* **11,** 674–680.

Stuart, T. S. (1968). Revival of respiration and photosynthesis in dried leaves of *Polypodium polypodioides*. *Planta* **83,** 185–206.

Stuber, C. W., and Levings, C. S., III (1969). Auxin induction and repression of peroxidase isozymes in oats (*Avena sativa* L). *Crop Sci.* **9,** 415–416.

Stuckey, I. H., and Curtis, O. F. (1938). Ice formation and the death of plant cells by freezing. *Plant Physiol.* **13,** 815–833.

Sturani, E., Cocucci, S., and Marré, E. (1968). Hydration dependent polysome-monosome interconversion in the germinating castor bean endosperm. *Plant Cell Physiol.* **9,** 783–795.

Stutte, C. A., and Todd, G. W. (1967). Effects of water stress on soluble leaf proteins in *Triticum aestivum* L. *Phyton* **24,** 67–75.

Stutte, C. A., and Todd, G. W. (1968). Ribonucleotide compositional changes in wheat leaves caused by water stress. *Crop Sci.* **8,** 319–321.

Stutte, C. A., and Todd, G. W. (1969). Some enzyme and protein changes associated with water stress in wheat leaves. *Crop Sci.* **9,** 510–512.

Subbotina, N. V. (1959). Influence of wilting on redox conditions in plants. *Fiziol. Rast.* **6,** 39–43.

Sugiyama, T., and Akazawa, T. (1967). Structure and function of chloroplast proteins. I. Subunit structure of wheat Fraction I protein. *J. Biochem.* (*Tokyo*) **62,** 474–482.

Sugiyama, N., and Simura, T. (1966). Studies on varietal differentiation of the frost resistance in the tea plant. II. Artificial hardening and dehardening. *Jap. J. Breed.* **16,** 165–173.

Sugiyama, N., and Simura, T. (1967a). Studies on varietal differentiation of the frost resistance of the tea plant. III. With special emphasis on the relation between their frost resistance and chloroplast soluble protein. *Jap. J. Breed.* **17,** 37–42.

Sugiyama, N., and Simura, T. (1967b). Studies on the varietal differentiation of frost resistance of the tea plant. IV. The effects of sugar level combined with protein in chloroplasts on the frost resistance. *Jap. J. Breed.* **17,** 292–296.

Sugiyama, N., and Simura, T. (1968a). Studies on the varietal differentiation of frost resistance of the tea plant. V. The histochemical observations of the variation of glucose content in the stems of the tea plant during frost hardening. *Jap. J. Breed.* **18,** 37–41.

Sugiyama, N., and Simura, T. (1968b). Studies on the varietal differentiation of frost resistance of the tea plant. VI. The distribution of sugars in cells during frost hardening in the tea plant with special reference to histochemical observations by using ^{14}C sucrose. *Jap. J. Breed.* **18,** 27–32.

Sulakadze, T. S. (1961). Growth substances and frost resistance in plants. *Izv. Akad. Nauk. SSSR Ser. Biol.* **4,** 551–560.

Sullivan, C. Y., and Kinbacher, E. J. (1967). Thermal stability of Fraction I protein from heat-hardened *Phaseolus acutifolius* Gray, "Tepary Buff." *Crop Sci.* **7,** 241–244.

Sun, C. N. (1958). The survival of excised pea seedlings after drying and freezing in liquid nitrogen. *Bot. Gaz.* (*Chicago*) **119,** 234–236.

Sundaram, T. K., Cazzulo, J. J., and Kornberg, H. L. (1969). Anaplerotic CO_2 fixation in mesophilic and thermophilic bacilli. *Biochim. Biophys. Acta* **192,** 355–357.

Suneson, C. A., and Peltier, G. L. (1934). Cold resistance adjustments of field hardened winter wheats as determined by artificial freezing *J. Amer. Soc. Agron.* **26,** 50–58.

Suneson, C. A., and Peltier, G. L. (1938). Effect of weather variants on field hardening of winter wheat. *J. Amer. Soc. Agron.* **30,** 769–778.

Surrey, K. (1965). Modification of the relationship between growth and metabolism in seeds by x-irradiation. *Radiat. Res.* **25,** 470–479.

Sutherland, R. M., and Pihl, A. (1967). Repair of radiation damage to membrane sylfhydryl groups of human erythrocytes. *Biochim. Biophys. Acta* **133,** 568–570.

Sutherland, R. M., and Pihl, A. (1968). Repair of radiation damage to erythrocyte membranes: The reduction of radiation-induced disulfide groups. *Radiat. Res.* **34,** 300–314.

Sutherland, R. M., Rothstein, A., and Weed, R. I. (1967a). Erythrocyte membrane sulfhydryl groups and cation permeability. *J. Cell Physiol.* **69,** 185–198.

Sutherland, R. M., Stannard, J. N., and Weed, R. I. (1967b). Involvement of sulfhydryl groups in radiation damage to the human erythrocyte membrane. *Int. J. Radiat. Biol.* **12,** 551–564.

Swanson, C. R., and Adams, M. W. (1959). Some metabolic responses of alfalfa seedlings to freezing. *Plant Physiol.* **34,** 372–376.

Swartz, H. (1970). Effect of oxygen on freezing damage. I. Effect on survival of *Escherichia coli* B/r and *Escherichia coli* B_{s-1} *Cryobiology* **6,** 546–551.

Szalai, I. (1959). Quantitative distribution of free amino acids in rindite-forced new potato tubers in various phases of sprouting. *Acta Biol. Acad. Sci. Hung.* **9,** 253–264.

Szirmai, J. (1938). Die Dörrfleckenkrankheit (Hitzeschaden) des Paprikas. *Phytopathol. Z.* **11,** 1–13.

Tadros, T. M. (1936). The osmotic pressure of Egyptian desert plants in relation to water supply. *Egypt. Univ. Bull. Fac. Sci.* **7,** 1–35.

Tagawa, T., and Ishizaka, N. (1963). Physiological studies on the tolerance of rice plants to salinity. II. Effects of salinity on the absorption of water and chloride ion. *Proc. Crop Sci. Soc. Jap.* **31,** 337–341.

Takada, H. (1954). Ion accumulation and osmotic value of plants, with special reference to strand plants. *J. Inst. Polytech. Osaka City Univ. Ser. Biol.* **D5,** 81–96.

Takaoki T. (1962). Osmotic value, bound water, and respiration of the plant in relationship to soil moisture variations. *J. Sci. Hiroshima Univ. Ser. B Div.* **2** (*Bot.*) **9**, 209–238.

Takaoki, T. (1966). Relationship between drought tolerance and ageing in higher plants. I. Mineral content. *Bot. Mag.* (*Tokyo*) **79**, 414–421.

Takaoki, T. (1968). Relationship between drought tolerance and ageing in higher plants. II. Some enzyme activities. *Bot. Mag.* (*Tokyo*) **81**, 297–309.

Takehara, I., and Asahina, E. (1960a). Glycerol in the overwintering prepupa of slug moth, a preliminary note. *Low Temp. Sci. Ser.* B**18**, 51–56.

Takehara, I., and Asahina, E. (1960b). Frost resistance and glycerol content in overwintering insects. *Low Temp. Sci. Ser.* **B18**, 57–65.

Takehara, I., and Asahina, E. (1961). Glycerol in a slug caterpillar. I. Glycerol formation, diapause, and frost resistance in insect reared at various graded temperatures. *Low Temp. Sci. Ser.* **B19**, 29–36.

Taleinshik, E. D., and Usol'tseva, S. M. (1967). Some data on the water balance and heat resistance of Manchu, sour, ground and Sakhalin cherry trees. *Fiziol. Rast.* **14**, 1065–1070.

Talli, A. R., and Durgham, A. H. (1968). Dew as a biological factor in arid and semiarid regions. *Beitr. Trop. Subtrop. Landwirt. Tropenveterinaermed.* **6**, 57–62.

Tanada, T., and Hendricks, S. B. (1953). Photoreversal of ultraviolet effects in soybean leaves. *Amer. J. Bot.* **40**, 634–637.

Taniyama, T., and Arikado, H. (1968). Studies on the mechanism of injurious effects of toxic gases on crop plant. II. Relation between injurious effects of sulfur dioxide and illumination in barley and rape. *Proc. Crop Soc. Jap.* **37**, 608–613.

Tarakanova, G. A., Strekova, V. Yu, and Prudnikova, V. P. (1965). Some physiological and cytological changes in germinating seeds in a stationary magnetic field. *Fiziol. Rast.* **12**, 1029–1038.

Taube, O. (1970). Ultraviolet inactivation and photoreactivation in synchronized *Chlorella*. *Physiol. Plant.* **23**, 755–761.

Taylor, N. W., Orton, N. L., and Babcock, G. E. (1968). Sexual agglutination in yeast. VI. Role of disulfide bonds in 5-agglutinin. *Arch. Biochem. Biophys.* **123**, 265–270.

Tazaki, T. (1960a). Studies on the dehydration resistance of higher plants. I. Determination of the measures related to the dehydration resistance of mulberry plants. *Bot. Mag.* (*Tokyo*) **73**, 148–155.

Tazaki, T. (1960b). Studies on the dehydration resistance of higher plants. II. Theoretical considerations of dehydration resistance. *Bot. Mag.* (*Tokyo*) **73**, 205–211.

Tazaki, T. (1960c). Studies on the dehydration resistance of higher plants. III. Discussions on general analysis focussed on the dehydration resistance of pine yearlings. *Bot. Mag.* (*Tokyo*). **73**, 269–277.

Tazaki, T., and Ushijima, T. (1963). Studies on the dehydration resistance of higher plants. IV. Dehydration resistance of the leaves in young poplar plants with special reference to the water economy of other plant parts. *Bot. Mag.* (*Tokyo*) **76**, 237–245.

Tennant, J. R. (1954). Some preliminary observations on the water relations of a mesophytic moss, *Thamnium alopecurum* (Hedw.) B. and S. *Trans. Brit. Bryol. Soc.* **2**, 439–445.

Terumoto, I. (1957a). The relation between the activity of phosphorylase and the frost hardiness in table beet. *Low Temp. Sci. Ser.* **B15**, 31–38.

Terumoto, I. (1957b). The frost hardiness of onion. *Low Temp. Sci. Ser.* **B15**, 39–44.

Terumoto, J. (1959). Frost resistance in plant cells after immersion in a solution of various inorganic salts. *Low Temp. Sci. Ser.* **B17**, 9–19.

Terumoto, I. (1960a). Ice masses in root tissue of table beet. *Low Temp. Sci. Ser.* **B18,** 39–42.

Terumoto, I. (1960b). Effect of protective agents against freezing injury in lake ball. *Low Temp. Sci. Ser.* **B18,** 43–50.

Terumoto, I. (1961). Frost resistance in a marine alga *Enteromorpha intestinalis* (L.) Link. *Low Temp. Sci. Ser.* **B19,** 23–28.

Terumoto, I. (1962). Frost resistance in a fresh water alga, *Aegagropila sauteri* (Nees) Kütz. I. *Low Temp. Sci. Ser.* **B20,** 1–24.

Terumoto, I. (1967). Frost resistance in algae cells. pp. 191–210. *In* "Cellular Injury and Resistance in Freezing Organisms" (E. Asahina, ed.), pp. 191–210. Inst. Low Temp. Sci. University of Hokkaido, Sapporo, Japan.

Thiselton-Dyer, W. (1899). The influence of the temperature of liquid hydrogen on the germinative power of seeds. *Proc. Roy. Soc.* **65,** 361–368.

Thoday, D. (1921). On the behavior during drought of leaves of two Cape Species of Passerina, with some notes on their anatomy. *Ann. Bot.* (*London*) **35,** 585–601.

Thoday, D. (1922). On the organisation of growth and differentiation in the stem of the sunflower. *Ann. Bot.* (*London*) **36,** 489–510.

Thoday, D. (1931). The significance of reduction in the size of leaves. *J. Ecol.* **19,** 297–303.

Thomas, M. D., and Hill, G. R. (1949). Photosynthesis under field conditions. *In* "Photosynthesis in Plants" (J. Franck and W. E. Loomis, eds.), pp. 19–25. Iowa State College Press, Ames, Iowa.

Thomas, W. D., and Lazenby, A. (1968). Growth cabinet studies into cold-tolerance of *Festuca arundinacea* populations. II. Response to pretreatment conditioning and to number and deviation of low temperature periods. *J. Agr. Sci.* **30,** 347–353.

Thompson, K. F., and Taylor, J. P. (1968). Chemical composition and cold hardiness of the pith in marrow-stem kale. *J. Agr. Sci.* **30,** 375–381.

Thomson, W. W., Berry, W. L., and Liu, L. L. (1969). Localization and secretion of salt by the salt gland of *Tamarix aphylla. Proc. Nat. Acad. Sci. U.S.* **63,** 310–317.

Thomson, W. W., Dugger, Jr., W. M., and Palmer, R. L. (1966). Effects of ozone on the fine structure of the palisade parenchyma cells of bean leaves. *Can. J. Bot.* **44,** 1677–1682.

Thorup, R. M. (1969). Root development and phosphorus uptake by tomato plants under controlled soil moisture conditions. *Agron. J.* **61,** 808–811.

Thouin, A. (1806). Observations sur l'effet des gelées précoces qui ont eu lieu les 18, 19, 20 vendémiaire an XIV (11, 12, et 13 Octobre 1805). *Ann. Museum Hist. Nat. tome septieme,* pp. 85–114.

Thren, R. (1934). Jahreszeitliche Schwankungen des osmotischen Wertes verschiedener ökologischer Typen in der Umgebung von Heidelberg. Mit einen Beitrag zur Methodik der Pressaftuntersuchung. *Z. Bot.* **26,** 448–526.

Till, O. (1956). Über die Frosthärte von Pflanzen sommergrüner Laubwälder. *Flora* (*Jena*) **143,** 499–542.

Timofejeva, M. (1935). Frost resistance of winter cereals in connection with phasic development and hardening of plants. *C. R. Acad. Sci. URSS* **1,** 64–67.

Tinklin, R. (1968). Measurement of the water status of attached leaves (*Lycopersicum esculentum, Phaseolus vulgaris, Nicotiana glutinosa*). *Advan. Front. Plant Sci.* **21,** 157–161.

Tinklin, R., and Weatherley, P. E. (1968). The effect of transpiration rate on the leaf water potential of sand and soil rooted plants. *New Phytol.* **67,** 605–615.

Tjurina, M. M. (1968). Untersuchungen zur Frostresistenz von Holzgewächsen bei Künstlicher Frostung. *In* "Klimaresistenz Photosynthese und Stoffproduktion." (H. Polster, ed.), pp. 77–86. Deut. Akad. Landwirtsch, Berlin.

Toda, M. (1962). Studies on the chilling injury in wheat plants–I. Some researches on the mechanism of occurrence of the sterile phenomenon caused by the low temperature. *Proc. Crop. Sci. Japan* **30,** 241–249.

Todd, G. W., and Levitt, J. (1951). Bound water in *Aspergillus niger*. *Plant Physiol.* **26,** 331–336.

Todd, G. W., and Yoo, B. Y. (1964). Enzymatic changes in detached wheat leaves as affected by water stress. *Phyton* **21,** 61–68.

Tokarskaya, V. L., Kopylov, V. A., and Mel'nikova, S. K. (1966). The effect of radiotoxins on the synthesis of deoxyribonucleic acid in plants. *In* "Radiotoxins, Their Nature and Role in the Biological Action of High-Energy Radiation" pp. 71–75. Atomizdat, Moscow.

Toman, F. R. (1968). Effects of cycocel and B- nine on cold hardiness of wheat plants. *J. Agr. Food Chem.* **16,** 771–772.

Toman, F. R., and Mitchell, H. L. (1968). Soluble proteins of winter wheat crown tissues, and their relationship to cold hardiness. *Phytochemistry* **7,** 365–373.

Tomlinson, H., and Rich, S. (1968). The ozone resistance of leaves as related to their sulfhydryl and adenosine triphosphate content. *Phytopathology* **58,** 808–810.

Tong, Min-Min, and Pincock, R. E. (1969). Denaturation and reactivity of invertase in frozen solutions. *Biochemistry* **8,** 908–913.

Tonzig, S. (1941). "I Muco-proteidi e la vita della cellula vegetale." Libreria Universitaria di G. Randi, Padova.

Torssell, B. (1959). Hardiness and survival of winter rape and winter turnip rape. Växtodling **15,** 1–168.

Torssell, B., and Hellstrom, N. (1955). Investigations on oil turnips and oil rape. IV. Estimation of plant status. *Acta Agr. Scand.* **5,** 31–38.

Townsend, G. (1961). Photoreactivation by colored light of ultraviolet inactivated yeast cells. *Trans. Amer. Microsc. Soc.* **80,** 70–73.

Tranquillini, W. (1957). Standortsklima, Wasserbilanz und CO_2- Gaswechsel junger Zirben (*Pinus cembra* L.) an der alpinen Waldgrenze. *Planta* **49,** 612–661.

Tranquillini, W. (1958). Die Frosthärte der Zirbe unter besonderer Berücksichtigung autochthoner und aus Forstgärten stammender Jungpflanzen. *Forstwiss. Zentralbl.* **77,** 65–128.

Tranquillini, W. (1965). Das Klimahaus auf dem Patscherkofel in Rahmen der Forstlichen Forschung. Vortragsreihe des 2. *Sympo. Industriellen Pflanzenbau Wien. Band* **II,** 147–154.

Tranquillini, W. (1969). Photosynthese und Transpiration einiger Holzarten bei verschieden starkem Wind. *Centralbl. Ges. Forstwiss.* **86,** 35–48.

Tranquillini, W., and Unterholzner, R. (1968). Dürreresistenz und Anpflanzungserfolg von Junglarchen verschiedenen Entwicklungszustandes. *Centralbl. Ges. Forstwiss.* **85,** 97–110.

Treharne, K. J., and Cooper, J. P. (1969). Effect of temperature on the activity of carboxylases in tropical and temperate Gramineae. *J. Exp. Bot.* **20,** 170–175.

Treshow, M. (1970). "Environment and Plant Response" 422pp. McGraw-Hill, New York.

Troughton, J. H. (1969). Plant water status and carbon dioxide exchange of cotton leaves. *Aust. J. Biol. Sci.* **22,** 289–302.

Trukhin, N. V. (1967). Salt tolerance of *Chlorella pyrenoidosa*. *Bot. Zh.* **52,** 1325–1330.

Trunova, T. I. (1963). The significance of different forms of sugars in increasing the frost resistance of the coleoptiles of winter wheat. *Fiziol. Rast.* **10,** 495–499.

Trunova, T. I. (1968). Changes in the content of acid soluble phosphorus-containing substances during hardening of winter wheat. *Fiziol. Rast.* **15,** 103–109.

Trunova, T. I. (1969). Effect of 2,4-dinitrophenol on the frost resistance of Ulyanovka variety winter wheat. *Fiziol. Rast.* **16,** 237–240.

Tsuda, M. (1961). Studies on the halophilic characters of the strand dune plants and of the halophytes in Japan. *Jap. J. Bot.* **17,** 332–370.

Tsunoda, K., Fujimura, K., Nakahari, T., and Oyamado, Z. (1968). Studies on the testing method for cooling tolerance in rice plants. I. An improved method by means of short turn treatment with cool and deep water. *Jap. J. Breed.* **18,** 33–40.

Tumanov, I. I. (1927). Ungenügende Wasserversorgung und das Welken der Pflanzen als Mittel zur Erhöhung ihrer Dürreresistenz. *Planta* **3,** 391–480.

Tumanov, I. I. (1930). Welken und Dürreresistenz. *Arch. Pflanzenbau* **3,** 389–419.

Tumanov, I. I. (1931). Das Abhärten winterannueller Pflanzen gegen niedrige Temperaturen. *Phytopathol. Z.* **3,** 303–334.

Tumanov, I. I. (1967). Physiological mechanisms of frost resistance of plants. *Fiziol. Rast.* **14,** 520–539.

Tumanov, I. I. (1969). Physiology of plants not killed by frost. *Izv. Akad. Nauk. SSSR Ser. Biol.* **4,** 469–480.

Tumanov, I. I., and Borodin, I. N. (1930). Untersuchungen über die Kälteresistenz von Winterkulturen durch direktes Gefrieren und indirekt Methoden. *Phytopathol. Z.* **1,** 575–604.

Tumanov, I. I., and Khvalin, N. N. (1967). Causes of weak frost resistance of fruit tree roots. *Fiziol. Rast.* **14,** 908–918.

Tumanov, I. I., and Trunova, T. I. (1957). Hardening tissues of winter plants with sugar absorbed for the external solution. *Fiziol. Rast.* **4,** 379–388.

Tumanov, I. I., and Trunova, T. I. (1958). The effect of growth processes on the capacity for hardening. *Fiziol. Rast.* **5,** 108–117.

Tumanov, I. I., and Trunova, T. I. (1963). First phase of frost hardening of winter plants in the dark on sugar solutions. *Fiziol. Rast.* **10,** 140–149.

Tumanov, I. I., Krasavtsev, O. A., and Khvalin, N. N. (1959). Promotion of resistance of birch and black currant to −253° temperatures by frost hardening. *C.R. Acad. Sci. URSS* **127,** 1301–1303.

Tumanov, I. I., Krasavtsev, O. A., and Trunova, T. I. (1969). Study of ice formation in plants by measurement of heat evolution. *Fiziol. Rast.* **16,** 907–916.

Turner, R. G., and Gregory, R. P. G. (1967). The use of radioisotopes to investigate heavy metal tolerance in plants. *In* "Proceedings of the Symposium on Isotopes in Plant Nutrition and Physiology," pp. 493–509. Int. Atomic Energy Agency, Vienna.

Tyankova, L. A. (1967). Distribution of the free and bound proline and of the free hydroxyproline in the separate organs of wheat plants during drought. *C.R. Acad. Bulg. Sci.* **20,** 583–586.

Tysdal, H. M. (1933). Influence of light, temperature, and soil moisture on the hardening process in alfalfa. *J. Agr. Res.* **46,** 483–515.

Tysdal, H. M. (1934). Determination of hardiness in alfalfa varieties by their enzymatic responses. *J. Agr. Res.* **48,** 219–240.

Tysdal, H. M., and Pieters, A. J. (1934). Cold resistance of three species of lespedeza compared to that of alfalfa, red clover, and crown vetch. *J. Amer. Soc. Agron.* **26,** 923–928.

Uhlig, S. K. (1968). Studies on the Simazin resistance of plants. *Arch. Pflanzenschutz* **4,** 215–227.

Ullrich, H., and Heber, U. (1957). Über die Schutzwirkung der Zucker bei der Frostresistenz von Winterweizen. *Planta* **48,** 724–728.

Ullrich, H., and Heber, U. (1958). Über das Denaturieren pflanzlicher Eiweisse durch Ausfrieren und seine Verbinderung. Ein Beitrag zur Klarung der Frostresistenz bei Pflanzen. *Planta* **51,** 399–413.

Ullrich, H., and Heber, U. (1961). Frostresistenz bei Winterweizen IV. Das Verhalten von Fermenten und Fermentsystemen gegenüber tiefen Temperaturen. *Planta* **57,** 370–390.

Ulmer, W. (1937). Über den Jahresgang der Frosthärte einiger immergruner Arten der alpinen Stufe, sowie der Zirbe und Fichte. Unter Berucksichtigung von osmotischen Wert, Zuckerspiegel und Wassergehalt. *Jahrb. Wiss. Bot.* **84,** 553–592.

Ungar, I. A. (1962). The influence of salinity on seed germination in succulent halophytes. *Ecology* **43,** 763–764.

Ungar, I. A. (1967). Influence of salinity and temperature on seed germination. *Ohio J. Sci.* **67,** 120–123.

Ungar, I. A., Hogan, W., and McClelland, M. (1969). Plant communities of saline soils at Lincoln, Nebraska. *Amer. Midl. Natur.* **82,** 564–577.

Ungerson, J., and Scherdin, G. (1965). Untersuchungen über Photosynthese und Atmung unter naturlichen Bedingungen während des Winterhalbjahres bei *Pinus sylvestris.* L. *Picea excelsa* Link und *Juniperus communis.* L. *Planta* **67,** 136–137.

Ungerson, J., and Scherdin, G. (1967). Jahresgang von Photosynthese und Atmung unter naturlichen Bedingungen bei *Pinus silvestris.* L. an ihrer Nordgrenze in der Subarktis. *Flora* (*Jena*) **157,** 391–434.

Uribe, E. G., and Jagendorf, A. T. (1968). Membrane permeability and internal volume as factors in ATP synthesis by spinach chloroplasts. *Arch. Biochem. Biophys.* **128,** 351–359.

Url, W. (1957). Kenntnis der Todeszonen im konzentrationsgestuften Resistenzversuch. *Physiol. Plant.* **10,** 318–327.

Url, W. (1959). Comparative experiments concerning the resistance of plant cytoplasm to sodium carbonate. *Protoplasma* **51,** 338–370.

Url, W., and Fetzmann, E. (1959). Wärmeresistenz und chemische resistenz der Grünalge *Gloeococcus bavaricus,* Skuja. *Protoplasma* **50,** 471–482.

Ushakov, B. P. (1964). Thermostability of cells and proteins of poikilotherms and its significance in speciation. *Physiol. Rev.* **44,** 518–560.

Ushakov, B. P. (1966). The problem of associated changes in protein thermostability during the process of speciation. *Helgolaender Wiss. Meeresunters.* **14,** 466–481.

Ushakov, B. P., and Glushankova, M. A. (1961). On the lack of a deficit correlation between the iodine number of protoplasmic lipids and the cell thermostability. *Tsitolog'ya* **3,** 707–710.

Ushijima, T. (1961). A study on the drought resistance in successive leaves of mulberry seedlings. *J. Sericult. Sci. Jap.* **30,** 469–474.

Vaadia, Y., and Itai, C. (1969). Interrelationship of growth with reference to the distribution of growth substances. *In* "Root Growth" (W. J. Whittington, ed.), pp. 65–79. Butterworths, London.

Vaartaja, O. (1960). Effect of photoperiod on drought resistance of white spruce seedlings. *Can. J. Bot.* **38,** 597–599.

Van den Driessche, R. (1969a). Measurement of frost hardiness in 2-year-old Douglas Fir seedlings. *Can. J. Plant Sci.* **49,** 159–172.

Van den Driessche, R. (1969b). Influence of moisture supply, temperature, and light on frost hardiness changes in Douglas Fir seedlings. *Can. J. Bot.* **47,** 1765–1772.

Van der Veen, R., and Meijer, G. (1959). "Light and Plant Growth." Philips Tech. Lib., Eindhoven, Holland.

Van Fleet, D. S. (1954). The significance of the histochemical localization of quinones in the differentiation of plant tissues. *Phytomorphology* **4,** 300–310.

Van Halteren, P. (1950). Effets d'un choc thermique sur le métabolisme des levures. *Bull. Soc. Chim. Biol.* **32,** 458–463.

Van Huystee, R. B., Weiser, C. J., and Li, P. H. (1967). Cold acclimation in *Cornus stolonifera* under natural and controlled photoperiod and temperature. *Bot. Gaz. (Chicago)* **128,** 200–205.

Varga, M., and Szoldos, F. (1963). The effect of nitrogen supply on the indoleacetic acid oxidase activity of the roots of rice plants. *Acta Bot. Acad. Sci. Hung.* **9,** 171–176.

Vasile, P. D. (1963). On the influence of NaCl and Na_2SO_4 on the activities of some enzymes. *Lucrarile Gradinii Bot. Bucuresti Acta Bot. Horti Bucurestiensis,* **2,** 683–792.

Vassiliev, I. M. (1939). Winter wheats as lagging behind the spring varieties in growth intensity when subjected to low temperatures. *C. R. Acad. Sci. URSS* **24,** 85–87.

Vassiliev, I. M. (1962). Effect of ionizing radiations on plants. Acad. Sci. USSR. Moscow. AEC–tr–5836. Atomic Energy Commission.

Vassiliev, I. M., and Vassiliev, M. G. (1936). Changes in carbohydrate content of wheat plants during the process of hardening for drought resistance. *Plant Physiol.* **11,** 115–125.

Veihmeyer, F. J., (1956). Soil moisture. *Handb. Pflanzenphysiol.* **3,** 64–123.

Verduin, J., and Loomis, W. E. (1944). Absorption of carbon dioxide by maize. *Plant Physiol.* **19,** 278–293.

Vesell, E. S. (1968). Multiple molecular forms of enzymes. *Ann. N.Y. Acad. Sci.* **151** (Art. 1), 1–689.

Vetuhova, A. (1936). Winter hardiness of winter wheat during winter in relation to phasic development of plants. *Zbirnik Prac. Agrofiziol.* **2,** 83–102. (*Herb. Abstr.* **8,** 1861, 1938).

Vetuhova, A. (1938). On the internal factors of resistance to frost in winter plants. *Z. Inst. Bot. Akad. Nauk. URSS* No. 18–19 (26–27): 57–59. (*Herb. Abstr.* **9,** 633, 1939.)

Vetuhova, A. (1939). Colloidal changes in plants of winter wheat in relation to the dynamics of frost resistance. *Kolloid-Z.* **4,** 511–521. (*Herb. Abstr.* **10,** 173. 1940.)

Vieira-da-Silva, J. B. (1968a). The osmotic potential of the culture medium and the soluble and latent activity of acid phosphatase in *Gossypium thurberi*. *C.R. Acad. Sci. (Paris)* **D267,** 729–732.

Vieira-da-Silva, J. B. (1968b). Influence of osmotic potential of the nutrient solution on the soluble carbohydrate and starch content of three species of Gossypium. *C.R. Acad. Sci.(Paris)* **D267,** 1289–1292.

Vieira-da-Silva, J. B. (1969). Comparison among five Gossypium species as to acid phosphatase activity after an osmotic treatment: Study of the speed of solubilization and enzyme formation. *Z. Pflanzenphysiol.* **60,** 385–387.

Vijver, G. van de (1969). Studies on the metabolism of *Tetrahymena pyriformis* GL. IV. Irradiation effect on cell permeability. *Enzymologia* **36,** 371–374.

Vitvitskii, V. N. (1969). Effect of methyl orange on the tryptic hydrolysis and heat denaturation of serum albumin. *Mol. Biol.* **3,** 678–682.

Vlasyuk, P. A., Shmat'ko, I. G., and Rubanyuk, E. A. (1968). Role of the trace elements zinc and boron in amino acid metabolism and drought resistance of winter wheat. *Fiziol. Rast.* **15,** 281–287.

Von Wangenheim, K. H. (1969). Effects of x-rays on the development and ultrastructure of wheat seedling cells. *Radiat. Bot.* **9,** 179–193.

Von Wangenheim, K. H., and Walther, F. (1968). Radiation-biological investigations on cereals. II. Variable radiation sensitivity with the same nuclear volume and DNA content. *Radiat. Bot.* **8,** 251–258.

Vose, P. B., and Randall, P. J. (1962). Resistance to aluminum and manganese toxicities in plants related to variety and cation-exchange capacity. *Nature* (*London*) **196,** 85–86.

Vouk, V. (1923). Die Probleme der Biologie der Thermen. *Int. Rev. Hydrobiol.* **11,** 89–99.

Vrolik, G., and de Vriese, W. H. (1839). Nouvelles expériences sur l'élévation de température du spadice d'une Colocasia odora (*Caladium odorum*) faites au Jardin Botanique d'Amsterdam. *Ann. Sci. Nat.* **11,** 65–85.

Wagenbreth, D. (1965). Durch Hitzeschocks induzierte Vitalitätsänderungen bei Laubholzblättern. *Flora* (*Jena*) **156,** 63–75.

Waggoner, P. E., and Shaw, R. H. (1953). Temperature of potato and tomato leaves. *Plant Physiol.* **27,** 710–724.

Waisel, Y. (1958a). Dew absorption by plants of arid zones. *Bull. Res. Counc. Isr.* **6D,** 180–185.

Waisel, Y. (1958b). Germination behaviour of some halophytes. *Bull. Res. Counc. Isr.* **6D,** 187–189.

Waisel, Y. (1959). Endurance of a drought period beyond the wilting point. *Bull. Res. Counc. Isr.* **7D,** 44–46.

Waisel, Y. (1960). Ecological studies on *Tamarix aphylla* (L.) Karst. II. The water economy. *Phyton* **15**, 19–28.

Waisel, Y. (1962a). Ecotypic differentiation in the flora of Israel. IV. Seedling behavior of some ecotype pairs. *Phyton* **18,** 151–156.

Waisel, Y. (1962b). Presowing treatments and their relation to growth and to drought, frost and heat resistance. *Physiol. Plant.* **15,** 43–46.

Waisel, Y., and Pollak, G. (1969). Estimation of water stresses in the root zone by the double-root system technique. *Isr. J. Bot.* **18,** 123–128.

Waisel, Y., and Pollak, G. (1970). Estimation of water stresses in the active root zone of some native halophytes in Israel. *Isr. J. Bot.* **19,** 789–794.

Waisel, Y., Borger, G. A., and Kozlowski, T. T. (1969). Effects of phenylmercuric acetate on stomatal movement and transpirations of excised *Betula papyrifera* marsh leaves. *Plant Physiol.* **44,** 685–690.

Waisel, Y., Kohn, H., and Levitt, J. (1962). Sulfhydryls- a new factor in frost resistance. IV. Relation of GSH-oxidizing activity to flower induction and hardiness. *Plant Physiol.* **37,** 272–276.

Walter, H. (1929). Die osmotischen Werte und die Kälteschaden unserer wintergrünen Pflanzen während der Winterperiode 1929. *Ber. Deut. Bot. Ges.* **47,** 338–348.

Walter, H. (1931). "Hydratur der Pflanze und ihre physiologischökologische Bedeutung." Fischer, Jena.

Walter, H. (1949). Über die Assimilation und Atmung der Pflanzen im Winter bei tiefen Temperaturen. *Ber. Deut. Bot. Ges.* **62,** 47–50.

Walter, H. (1950). Einführung in die Phytologie. I. Die Grundlagen des Pflanzenlebens, pp. 72–78, 326–356. E. Ulmer, Stuttgart.

Walter, H. (1955). The water economy and the hydrature of plants. *Ann. Rev. Plant Physiol.* **6,** 239–252.

Walter, H. (1956). Die heutige ökologische Problemstellung und der Wettbewerb zwischen der mediterranen Hartlaubvegetation und den sommergrünen Laubwäldern. *Ber. Deut. Bot. Ges.* **69,** 263–273.

Walter, H. (1963a). Zur Klärung des spezifischen Wasserzustandes im Plasma und in der Zellwand der hoheren Pflanze und seine Bestimmung. *Ber. Deut. Bot. Ges.* **76,** 40–71.

Walter, H. (1963b). The water supply of desert plants. *In* "The Water Relations of Plants." (A. J. Rutter, and F. W. Whitehead, eds.), pp. 199–205. Wiley, New York.

Walter, H., and Schall, K. (1957). Das Verhalten des osmotischen Wertes beim Welken abgeschnittener Blätter. *Z. Bot.* **45,** 263–272.

Walter, H., and Stadelmann, E. (1968). The physiological prerequisites for the transition of autotrophic plants from water to terrestrial life. *Bioscience* **18,** 694–701.

Walter, H., and Weismann, O. (1935). Über die Gefrierpunkte und osmotischen Werte lebender und tote pflanzlicher Gewebe. *Jahrb. Wiss. Bot.* **82,** 273–310.

Ward, E. W. B. (1968). The low maximum temperature for growth of the psychrophile *Sclerotinia borealis.* Evidence for the uncoupling of growth from respiration. *Can. J. Bot.* **46,** 385–390.

Wardlaw, I. F. (1969). The effect of water stress on translocation in relation to photosynthesis and growth. II. Effect during leaf development in *Lolium temulentum* L. *Aust. J. Biol. Sci.* **22,** 1–16.

Warren, J. C., and Cheatum, S. G. (1966). Effect of neutral salts on enzyme activity and structure. *Biochemistry* **5,** 1702–1707.

Wasserman, A. R., and Fleischer, S. (1968). The stabilization of spinach chloroplast function. *Biochim. Biophys. Acta* **153,** 154–169.

Wassink, E. C., and Kuiper, P. J. C. (1959). Some remarks on factors determining the water requirement in tomato plants under controlled conditions. *Mededeel. Landbouwhcgeskh. Wageningen* **59,** 1–8.

Watanabe, M. (1953). Effect of heat application upon the pollen viability of Japanese black pine and Japanese red pine. *J. Jap. Forest. Soc.* **35,** 248–251.

Wattendorff, J. (1964). Changes in urea permeability of Allium epidermal cells induced by α-irradiation during deplasmolysis. *Naturwissenschaften* **51,** 247–248.

Wattendorff, J. (1965). Immediate alteration of urea permeability after alpha irradiation of Allium epidermis cells. *Protoplasma* **60,** 162–168.

Wattendorff, J. (1970). Permeabilität für Wasser, Malonsäurediamid und Harnstoff nach α-Bestrahlung von Convallaria-Zellen. *Ber. Deut. Bot. Ges.* **83,** 3–17.

Weatherley, P. E., and Slatyer, R. O. (1957). Relationship between relative turgidity and diffusion pressure deficit in leaves. *Nature (London)* **179,** 1085–1086.

Weaver, G. M., and Jackson, H. O. (1969). Assessment of winter hardiness in peach by a liquid nitrogen system. *Can. J. Plant Sci.* **49,** 459–463.

Weaver, G. M., Jackson, H. O., and Stroud, F. D. (1968). Assessment of winter hardiness in peach cultivars by electric impedance, scion diameter and artificial freezing studies. *Can. J. Plant Sci.* **48,** 37–47.

Webb, J. L. (1966). "Enzyme and Metabolic Inhibitors," Vol. III, p. 892. Academic Press, New York.

Webb, K. L., and Burley, J. W. A. (1965). Dark fixation of $C^{14}O_2$ by obligate and facultative salt marsh halophytes. *Can. J. Bot.* **43,** 281–285.

Weber, F. (1909). Untersuchungen über die Wandlungen des Starke-und Fettgehaltes der Pflanzen, insbesondere der Bäume. *Sitzber. Akad. Wiss. Wien Math. Naturwiss. Kl. Abt. I* **118,** 967–1031.

Weber, F. (1926a). Hitzeresistenz funktionierender Schliesszellen. *Planta* **1**, 553–557.

Weber, F. (1926b). Hitzeresistenz funktionierender Stomata-Nebenzellen. *Planta* **2**, 669–677.

Weber, S. J. (1965). "Bound Water in Biological Integrity." Charles C. Thomas, Springfield, Illinois.

Webster, C. C. (1967). "The Effects of Air Pollution on Plant and Soil," 53 pp. Agr. Res. Counc. London.

Weetall, H. H. (1969). Trypsin and papain covalently coupled to porous glass: preparation and characterization. *Science* **166**, 615–616.

Wehmer, C. (1904). Über die Lebensdauer eingetrockneter Pilzkulturen. *Ber. Deut. Bot. Ges.* **22**, 476–478.

Weijer, J. (1961). Protective action of calcium gluconate against after-effects of X-irradiation on conidia of *Neurospora crassa. Nature* (*London*) **189**, 760–761.

Weimer, J. L. (1929). Some factors involved in the winterkilling of alfalfa. *J. Agr. Res.* **39**, 263–283.

Weinberger, P., and Measures, M. (1968). The effect of two audible sound frequencies on the germination and growth of a spring and winter wheat. *Can. J. Bot.* **46**, 1151–1160.

Weise, G., and Polster, H. (1962). Investigations on the influence of cold hardiness on the physiological activity of forest growth. II. Metabolic-physiological ivestigation on the question of frost resistance of spruce and Douglas fir origin. *Biol. Zentralbl.* **81**, 129–243.

Weiser, C. J. (1970). Cold resistance and injury in woody plants. *Science* **169**, 1269–1278.

Weissenböck, G. (1969). The effect of the salt content of the soil upon the morphology and ion accumulation of halophytes. *Flora* (*Jena*) **B158**, 369–389.

Wendt, C. W., Haas, R. H., and Runkles, J. R. (1968). Influence of selected environmental variables on the transpiration rate of mesquite. *Agron. J.* **60**, 382–384.

Wessels, J. S. C. (1966). Mechanism of the reduction of organic nitro compounds by chloroplasts. *Biochim. Biophys. Acta* **109**, 357–371.

West, F. L., and Edlefsen, N. E. (1917). The freezing of fruit buds. *Utah Agr. Expt. Sta. Bull.* **151**, 1–24.

West, F. L., and Edlefsen, N. E. (1921). Freezing of fruit buds. *J. Agr. Res.* **20**, 655–662.

Wheaton, T. A., and Morris, L. L. (1967). Modification of chilling sensitivity by temperature condition. *Proc. Amer. Soc. Hort. Sci.* **91**, 529–533.

Whiteman, P. C., and Koller, D. (1964). Saturation deficit of the mesophyll evaporating surfaces in a desert halophyte. *Science* **146**, 1320–1321.

Whiteman, T. M. (1957). Freezing points of fruits, vegetables, and florist stocks. *U.S.Dept. Agr. Marketing Res. Rep.* No. 196.

Whiteside, A. G. O. (1941). Effect of soil drought on wheat plants. *Sci. Agr.* **21**, 320–334.

Whitman, W. C. (1941). Seasonal changes in bound water content of some prairie grasses. *Bot. Gaz.* (*Chicago*) **103**, 38–63.

Wiegand, K. M. (1906a). Some studies regarding the biology of buds and twigs in winter. *Bot. Gaz.* (*Chicago*) **41**, 373–424.

Wiegand, K. M. (1906b). The occurrence of ice in plant tissue. *Plant World* **9**, 25–39.

Wild, H. (1970). Geobotanical anomalies in Rhodesia: III. The vegetation of nickel bearing soils. *Kirkia Suppl.* **7**, 1–62.

Wilde, S. A., Voigt, G. K., and Persidsky, D. J. (1957). Transmitted effect of allyl alcohol on growth of Monterey pine seedlings. *Forest Sci.* **2**, 58–59.

Wilding, M. D., Stahmann, M. A., and Smith, D. (1960a). Free amino acids in alfalfa as related to cold hardiness. *Plant Physiol.* **35**, 726–732.

Wilding, M. D., Stahmann, M. A., and Smith, D. (1960b). Free amino acids in red clover as related to flowering and winter survival. *Plant Physiol.* **35,** 733–735.

Wilhelm, A. F. (1935). Untersuchungen über das Verhalten sogenannter nicht eisbeständiger Kulturpflanzen bei niederen-Temperaturen, unter besonderer Berücksichtigung des Einflusses verschiedener Mineralsalzernährung und des N-Stoffwechsels. *Phytopathol. Z.* **8,** 337–362.

Willard, C. J. (1922). Root reserves of alfalfa with special reference to time of cutting on yield. *J. Amer. Soc. Agron.* **22,** 595–602.

Willey, C. R. (1970). Effects of short periods of anaerobic and near-anaerobic conditions on water uptake by tobacco roots. *Agron. J.* **62,** 224–229.

Williams, R. J. P. (1967). Heavy metals in biological systems. *Endeavour* **26,** 96–100.

Williams, R. J., and Meryman, H. T. (1970). Freezing injury and resistance in spinach chloroplast grana. *Plant Physiol.* **45,** 752–755.

Williamson, R. E. (1963). The effect of a transpiration-suppressant on tobacco leaf temperature. *Soil Sci. Soc. Amer. Proc.* **27,** 106.

Wilner, J. (1952). A study of desiccation in relation to winter injury. *Sci. Agr.* **32,** 651–658.

Wilner, J. (1955). Results of laboratory tests for winter hardiness of woody plants by electrolytic methods. *Proc. Amer. Soc. Hort. Sci.* **66,** 93–99.

Wilner, J. (1960). Relative and absolute electrolytic conductance tests for frost hardiness of apple varieties. *Can. J. Plant Sci.* **40,** 630–637.

Wilson, A. M. (1970). Incorporation of ^{32}P in seeds at low water potentials. *Plant Physiol.* **45,** 524–526.

Wilson, A. M., and Harris, G. A. (1968). Phosphorylation in crested wheatgrass seeds at low water potentials. *Plant Physiol.* **43,** 61–65.

Wilson, J. H. (1968). Water relations of maize. I. Effects of severe soil moisture stress imposed at different stages of growth on grain yields of maize. *Rhodesian J. Agr. Res.* **6,** 103–105.

Wilson, J. R., Haydock, K. P., and Robins, M. F. (1970). Response to salinity in glycine. V. Changes in the chemical composition of three Australian species and *G. wightii* (*G. javanica*) over a range of salinity stresses. *Aust. J. Exp. Agr. Anim. Husb.* **10,** 156–165.

Wiman, L. G. (1964). "Protein Bound Sulfhydryl Groups in Pulmonary Cytodiagnosis and Their Significance in Papanicolaou Technique and Acridine-Orange Fluorescence Microscopy." Almquist and Wiksell, Stockholm.

Winkler, A. (1913). Über den Einfluss der Aussenbedingungen auf die Kälteresistenz ausdauernder Gewächse. *Jahrb. Wiss. Bot.* **52,** 467–506.

Wirth, H., Daniel, E. E., and Carroll, P. M. (1970). The effect of adenosine deaminase inhibitor and 2-mercaptoethanol on frozen-thawed uterine horns. *Cryobiology* **6,** 395–400.

Wolf, F. T. (1968). Enzymatic activities in the leaves of light-grown and dark-grown wheat seedlings. *Z. Pflanzenphysiol.* **59,** 39–44.

Wolff, S., and Sicard, A. M. (1961). Post-irradiation storage and the growth of barley seedlings. *In* "Effects of Ionizing Radiations on Seeds, A Symposium," pp. 171–179. Intern. Atomic Energy Agency, Vienna.

Wolpert, A. (1962). Heat transfer analysis of factors affecting plant leaf temperature. Significance of leaf hair. *Plant Physiol.* **37,** 113–120.

Wood, H., and Rosenberg, A. M. (1957). Freezing in yeast cells. *Biochim. Biophys. Acta* **25,** 78–87.

Woodroof, J. G. (1938). Microscopic studies of frozen fruits and vegetables. *Ga. Agr. Exp. Sta.* No. 201.

Woodwell, G. M. (1965). Radiation and the patterns of nature. *Brookhaven Lect. Ser.* **45,** 1–15.

Worzella, W. W. (1932). Root development in hardy and non-hardy winter wheat varieties. *J. Amer. Soc. Agron.* **24,** 626–637.

Wright, R. C. (1932). Some physiological studies of potatoes in storage. *J. Agr. Res.* **45,** 543–555.

Wright, L. N., and Dobrenz, A. K. (1970). Efficiency of water use and seedling drouth tolerance of boer lovegrass, *Eragrosits curvula* Nees. *Crop Sci.* **10,** 1–2.

Wright, R. C., and Taylor, G. F. (1923). The freezing temperatures of some fruits, vegetables, and cut flowers. *U.S. Dept. Agr. Bull. No.* 1133.

Wright, S. T. C. (1969). An increase in the "inhibitor-B" content of detached wheat leaves following a period of wilting. *Planta* **86,** 10–20.

Wright, S. T. C., and Hiron, R. W. P. (1969). (+)-Abscisic acid, the growth inhibitor induced in detached wheat leaves by a period of wilting. *Nature (London)* **224,** 719–720.

Wyen, N. V., Udvardy, J., Solymosy, F., Marré, E., and Farkas, G. L. (1969). Purification and properties of a ribonuclease from Avena leaf tissues. *Biochim. Biophys. Acta* **191,** 588–597.

Yabe, A., Kuse, G., Murata, T., and Takada, H. (1965). Physiology of halophytes I. The osmotic value of leaves and the osmotic role of each ion in cell sap. *Physiol. Ecol.* **13,** 25–33.

Yannas, I. V., and Tobolsky, A. V. (1967). Cross linking of gelatin by dehydration. *Nature (London)* **215,** 509–510.

Yariv, J., Kalb, A. J. Katchalski, E., Goldman, R., and Thomas, E. W. (1969). Two locations of the lac permease sulphydryl in the membrane of *E. coli. FEBS Lett.* **5,** 173–176.

Yaron, B., Zieslin, N., and Halevy, A. H. (1969). Response of Baccara roses to saline irrigation. *J. Amer. Soc. Hort. Sci.* **94,** 481–484.

Yarwood, C. E. (1961a). Acquired tolerance of leaves to heat. *Science* **134,** 941–942.

Yarwood, C. E. (1961b). Translocated heat injury in plants. *Nature (London)* **192,** 887.

Yarwood, C. E. (1962). Acquired sensitivity of (*Phaseolus vulgaris* var. Pinto) leaves to heat. *Plant Physiol. Suppl.* **37,** 70.

Yarwood, C. E. (1963). Sensitization of leaves to heat. *Adv. Front. Plant Sci.* **7,** 195–203.

Yarwood, C. E. (1967). Adaptations of plants and plant pathogens to heat. *In* "Molecular Mechanisms of Temperature Adaptation" (C. L. Prosser, ed.), pp. 75–92. Publ. No. 84. Amer. Ass. Adv. Sci., Washington, D.C.

Yasmykova, O. O., and Tolmachov, I. M. (1967). The conversion of stored substances in fruit and nut trees during the fall and winter. *Ref. Zh. Biol.* No. 9G139.

Yoshida, S. (1965). Chemical aspects of the role of silicon in physiology of the rice plant. *Bull. Nat. Inst. Agr. Sci. Ser. (Jap.)* **B15,** 1–58.

Yoshida, S. (1969a). Studies on the freezing resistance in plants. I. Seasonal changes of glycolipids in the bark tissue of black locust tree. *Low Temp. Sci. Ser.* **B27,** 109–117.

Yoshida, S. (1969b). Studies on the freezing resistance in plants. II. Seasonal changes in phospholipids in the bark tissues of the black locust tree. *Low Temp. Sci. Ser.* **B27,** 119–124.

Yoshida, S., and Sakai, A. (1967). The frost hardening process of woody plants. XII. Relation between frost resistance and various substances in stem bark of black locust trees. *Low Temp. Sci. Ser. Biol. Sci.* **B25,** 29–44.

Yoshida, S., and Sakai, A. (1968). The effect of thawing rate on freezing injury in plants. II. The change in the amount of ice in leaves as produced by the change in temperature. *Low Temp. Sci. Ser. Biol. Sci.* **B26**, 23–31.

Young, R. (1969a). Cold hardening in "Redblush" grapefruit as related to sugars and water soluble proteins. *J. Amer. Soc. Hort. Sci.* **94**, 252–254.

Young, R. (1969b). Cold hardening in citrus seedlings as related to artificial hardening conditions. *J. Amer. Soc. Hort. Sci.* **94**, 612–614.

Younis, M. E. (1969). Studies in the respiratory and carbohydrate metabolism of plant tissues. XXVII. The effect of iodoacetate on the utilization of (^{14}C) glucose in strawberry leaves and on permeability barriers. *New Phytol.* **68**, 1059–1067.

Yu, S. A., Sussman, A. S., and Wooley, S. (1967). Mechanisms of protection of trehalase against heat inactivation in Neurospora. *Bacteriology* **94**, 1306–1312.

Zacharowa, T. M. (1926). Über den Einfluss neidriger Temperaturen auf die Pflanzen. *Jahrb. Wiss. Bot.* **65**, 61–87.

Zavadskaya, I. G., and Den'ko, E. I. (1968). Effect of an insufficient water supply on the stability of leaf cells of some plants of the Pamirs. *Bot. Zh.* **53**, 795–805.

Zavitkovski, J., and Ferrell, W. K. (1968). Effect of drought upon rates of photosynthesis, respiration, and transpiration of seedlings of two ecotypes of Douglas-fir (*Pseudotsuga menziesii*). *Bot. Gaz.* (*Chicago*) **129**, 346–350.

Zehavi, U. (1969). Modification by low temperature of the transfer function of lysozyme. *Biochim. Biophys. Acta* **194**, 526–531.

Zehavi-Willner, T., Wax, R., and Kosower, E. M. (1970). The inhibition of ribonucleic acid synthesis by the thiol-oxidizing agent diamide, in *Escherichia coli*. *FEBS Lett.* **9**, 100–102.

Zelitch, I. (1964). Reduction of transpiration of leaves through stomatal closure induced by alkenylsuccinic acids. *Science* **143**, 692–693.

Zeller, O. (1951). Über Assimilation und Atmung der Pflanzen im Winter bei tiefen Temperaturen. *Planta* **39**, 500–526.

Zholkevich, V. N., and Rogacheva, A. Ya. (1968). Ratio of phosphorylation to oxidation in mitochondria from wilting plant tissues. *Fiziol. Rast.* **15**, 537–545.

Zholkevich, V. N., Prusakova, L. D., and Lizandr, A. A. (1958). Translocation of assimilants and respiration of conducting pathways in relation to soil moisture. *Fiziol. Rast.* **5**, 333–340.

Zhukovskaya, N. V. (1962). The phosphorus metabolism of tomatoes on saline soil. (Notes on the Fourth Sci. Conf. of Grad. Students) Rostov. Univ. Rostov-on-Don 239–241. *Ref. Zh. Biol.* 1963. No. 17G27.

Zhuravlev, Y. N., and Popova, L. I. (1968). Early spring photosynthesis in cryophilic plants. *Ref. Zh. Biol.* No. 5G136.

Ziganigirov, A. M. (1968). The morphological periodicity and winter hardiness of the dog rose. *Tr. Inst. Ekol. Rast. Zhivotn. Ural. Fil. Akad. Nauk. SSSR* **62**, 83–88.

Zimmer, G. (1970). Isolation and characterization of *in vitro* radioactively labelled SH-proteins from rat liver mitochondrial membranes. *FEBS Lett.* **9**, 108–112.

Zirkle, R. E. (1932). Some effects of alpha radiation on plant cells. *J. Cell Comp. Physiol.* **2**, 251–274.

Zobl, K. (1950). Ueber die Beziehungen zwischen chemischer Zusammensetzung von Pilzsporen und ihrem Verhalten gegen Erhitzen. *Sydowia* **4**, 175–184.

Zotochkina, T. V. (1962). A study of the redox processes in pears in relation to their winter-hardiness. *Tr. Tsentr. Genet. Lab. Im. I.V. Michurina* **8**, 160–170. *Ref. Zh. Biol.*, 1963, No. 15G64.

SUBJECT INDEX

B

C

E

F

G

M

Q

R

T